Prir

First published in 2006, this book has bec in its second edition, the text has been thoro opments and research directions. While the ov book remain unchanged, all existing chapters hav been added.

Adopting a broad perspective, the authors provide a ical and experimental concepts that are needed to unders across subfields ranging from quantum optics to biophysic include optical antennas, new imaging techniques, Fano interfe reciprocity, metamaterials, and cavity optomechanics.

With numerous end-of-chapter problem sets and illustrative materia discussed in the main text, this is an ideal textbook for graduate students It is also a valuable reference for researchers and course teachers.

Lukas Novotny is Professor of Optics and Physics at the University of Rochester, N where he heads the Nano-Optics Research Group at the Institute of Optics. He re his Ph.D. from the Swiss Federal Institute of Technology (ETH Zürich) in Switzerland later joined the Pacific Northwest National Laboratory (Washington, WA) as a researc fellow. In 1999, he joined the faculty of the Institute of Optics at the University of Rochester and developed a course on nano-optics that has been taught several times at the graduate level and forms the basis of this textbook. In 2012, he joined the ETH Zürich.

Bert Hecht is Professor of Experimental Physics at the University of Würzburg. After studying physics at the University of Konstanz, he joined the IBM Zurich Research Laboratory in Rüschlikon and worked in near-field optical microscopy and plasmonics. In 1996, he received his Ph.D. from the University of Basel and then joined the Physical Chemistry Laboratory of the Swiss Federal Institute of Technology, where he worked on the combination of single-molecule spectroscopy with scanning probe techniques. In 2001, he was awarded a Swiss National Science Foundation research professorship at the University of Basel. His research interests concern the enhancement of light–matter interaction on the nanometer scale.

Principles of Nano-Optics

Second Edition

LUKAS NOVOTNY
University of Rochester, New York
ETH Zürich, Switzerland

BERT HECHT
Universität Würzburg, Germany

CAMBRIDGE
UNIVERSITY PRESS

CAMBRIDGE
UNIVERSITY PRESS

Shaftesbury Road, Cambridge CB2 8EA, United Kingdom

One Liberty Plaza, 20th Floor, New York, NY 10006, USA

477 Williamstown Road, Port Melbourne, VIC 3207, Australia

314–321, 3rd Floor, Plot 3, Splendor Forum, Jasola District Centre, New Delhi – 110025, India

103 Penang Road, #05–06/07, Visioncrest Commercial, Singapore 238467

Cambridge University Press is part of Cambridge University Press & Assessment, a department of the University of Cambridge.

We share the University's mission to contribute to society through the pursuit of education, learning and research at the highest international levels of excellence.

www.cambridge.org
Information on this title: www.cambridge.org/9781107005464

First published 2006
Second edition 2012 (version 12, June 2023)

Printed in the United Kingdom by TJ Books Limited, Padstow Cornwall

A catalog record for this publication is available from the British Library

Library of Congress Catalog in Publication data
Novotny, Lukas.
Principles of nano-optics / Lukas Novotny, University of Rochester, New York, Bert Hecht, Universität Würzburg, Germany. – Second Edition.
p. cm.
ISBN 978-1-107-00546-4 (hardback)
1. Nanostructured materials. 2. Near-field microscopy. 3. Quantum optics. 4. Photonics. 5. Nanophotonics. I. Hecht, Bert, 1968– II. Title.
TA418.9.N35N68 2012
535′.15–dc23
2012005672

ISBN 978-1-107-00546-4 Hardback

To our families
(Jessica, Leonore, Jakob, David, Rahel, Rebecca, Nadja, Jan)

and our parents
(Annemarie, Werner, Miloslav, Vera)

. . . it *was* worth the climb
(B. B. Goldberg)

Contents

Preface to the first edition

Why should we care about nano-optics? For the same reason we care about optics! The foundations of many fields of the contemporary sciences have been established using optical experiments. To give an example, think of quantum mechanics. Blackbody radiation, hydrogen lines, or the photoelectric effect were key experiments that nurtured the quantum idea. Today, optical spectroscopy is a powerful means to identify the atomic and chemical structure of different materials. The power of optics is based on the simple fact that the energy of light quanta lies in the energy range of electronic and vibrational transitions in matter. This fact is at the core of our abilities for visual perception and is the reason why experiments with light are very close to our intuition. Optics, and in particular optical imaging, helps us to consciously and logically connect complicated concepts. Therefore, pushing optical interactions to the nanometer scale opens up new perspectives, properties and phenomena in the emerging century of the nanoworld.

Nano-optics aims at the understanding of optical phenomena on the nanometer scale, i.e. near or beyond the diffraction limit of light. It is an emerging new field of study, motivated by the rapid advance of nanoscience and nanotechnology and by their need for adequate tools and strategies for fabrication, manipulation and characterization at the nanometer scale. Interestingly, nano-optics predates the trend of nanotechnology by more than a decade. An optical counterpart to the scanning tunneling microscope (STM) was demonstrated in 1984 and optical resolutions had been achieved that were significantly beyond the diffraction limit of light. These early experiments sparked a field initially called *near-field optics*, since it was realized quickly that the inclusion of near-fields in the problem of optical imaging and associated spectroscopies holds promise for achieving arbitrary spatial resolutions, thus providing access for optical experiments on the nanometer scale.

The first conference on near-field optics was held in 1992. About seventy participants discussed theoretical aspects and experimental challenges associated with near-field optics and near-field optical microscopy. The subsequent years are characterized by a constant refinement of experimental techniques, as well as the introduction of new concepts and applications. Applications of near-field optics soon covered a large span ranging from fundamental physics and materials science to biology and medicine. Following a logical development, the strong interest in near-field optics gave birth to the fields of *single-molecule spectroscopy* and *plasmonics*, and inspired new theoretical work associated with the nature of optical near-fields. In parallel, relying on the momentum of the flowering nanosciences, researchers started to tailor nanomaterials with novel optical properties. Photonic crystals, single-photon sources and optical microcavities are products of this effort. Today, elements of nano-optics are scattered across the disciplines. Various review articles and books capture the state-of-the-art in the different subfields but there appears to be no dedicated textbook that introduces the reader to the general theme of nano-optics.

This textbook is intended to teach students at the graduate level or advanced undergraduate level about the elements of nano-optics encountered in different subfields. The book evolved from lecture notes that have been the basis for courses on nano-optics taught at the Institute of Optics of the University of Rochester, and at the University of Basel. We were happy to see that students from many different departments found interest in this course, which shows that nano-optics is important to many fields of study. Not all students were interested in the same topics and, depending on their field of study, some students needed additional help with mathematical concepts. The courses were supplemented with laboratory projects that were carried out in groups of two or three students. Each team picked the project that had most affinity with their interest. Among the projects were: surface enhanced Raman scattering, photon scanning tunneling microscopy, nanosphere lithography, spectroscopy of single quantum dots, optical tweezers, and others. Towards the end of the course, students gave a presentation on their projects and handed in a written report. Most of the problems at the end of individual chapters have been solved by students as homework problems or take-home exams. We wish to acknowledge the very helpful input and inspiration that we received from many students. Their interest and engagement in this course is a significant contribution to this textbook.

Nano-optics is an active and evolving field. Every time the course was taught new topics were added. Also, nano-optics is a field that easily overlaps with other fields such as physical optics or quantum optics, and thus the boundaries cannot be clearly defined. This first edition is an initial attempt to put a frame around the field of nano-optics. We would be grateful to receive input from our readers related to corrections and extensions of existing chapters and for suggestions of new topics.

Acknowledgments

We wish to express our thanks for the input we received from various colleagues and students. We are grateful to Dieter Pohl who inspired our interest in nano-optics. This book is a result of his strong support and encouragement. We received very helpful input from Andreas Lieb, Scott Carney, Jean-Jacques Greffet, Stefan Hell, Carsten Henkel, Mark Stockman, Gert Zumofen, Jer-Shing Huang, Paolo Bragioni, and Jorge Zurita-Sanchez. It was also a great pleasure to discuss various topics with Miguel Alonso, Joe Eberly, Robert Knox, and Emil Wolf at the University of Rochester.

Preface to the second edition

We are very pleased that this textbook found wide use and reasonably high demand. Since the first printing of the first edition in 2006, the field of nano-optics has gained considerable momentum and new research directions have been established. Among the new topics are metamaterials, optical antennas, and cavity optomechanics, to name but a few. The high field localization associated with metals at optical frequencies has given rise to the demonstration of truly nanoscale lasers and the high nonlinearity of metals is being used for frequency conversion in subwavelength volumes. These new trends define a clear motivation for a second edition of *Principles of Nano-Optics*.

The overall structure of the book has been left unchanged with the exception of a new chapter on optical antennas (Chapter 13). Chapter 2 (Theoretical foundations) has been adjusted to include topics such as reciprocity and energy density in lossy media, and Chapter 4 (Resolution and localization) has been extended by including new microscopy techniques, such as structured illumination and localization microscopy. Chapter 5 received a major polish: optical microscopy is now classified in terms of interaction orders between probe and sample. On the other hand, Chapter 6 has been condensed since some near-field techniques are no longer of general interest. Several new topics have been included in Chapter 8, which covers the theory of localized light–matter interactions. Among the new sections is a discussion of Fano interference, strong coupling between modes, and level crossing. Chapters 9 and 10 received only minor revisions, while Chapter 11 has been extended by a section on metamaterials and cavity optomechanics. Chapter 12 (Surface plasmons) has also been restructured: metals are discussed from a perspective of plasma physics leading to screening and to ponderomotive forces, which give rise to a wide range of optical nonlinearities. The chapter on optical forces (Chapter 14) has been adjusted to provide a more self-consistent perspective on dipole forces. Finally, various typos have been fixed. We thank our critical readers for pointing out several errors and for suggesting valuable changes.

Despite the changes and additions it is not possible to account for all the new results and directions in the field. However, the purpose of this book is not to provide a comprehensive review, but to present the necessary foundations and concepts to understand what's going on. In this sense, the book has remained a textbook and a reference for those seeking a conceptual understanding of the working principles of nano-optics.

1 Introduction

In the history of science, the first applications of optical microscopes and telescopes to investigate nature mark the beginnings of new eras. Galileo Galilei used a telescope to see for the first time craters and mountains on a celestial body, the Moon, and also discovered the four largest satellites of Jupiter. With this he opened the field of optical astronomy. Robert Hooke and Antony van Leeuwenhoek used early optical microscopes to observe certain features of plant tissue that were called "cells," and to observe microscopic organisms, such as bacteria and protozoans, thus marking the beginning of optical biology. The newly developed instrumentation enabled the observation of fascinating phenomena not directly accessible to human senses. Naturally, the question of whether the observed structures not detectable within the range of normal vision should be accepted as reality at all was raised. Today, we have accepted that, in modern physics, scientific proofs are verified by indirect measurements, and the underlying laws have often been established on the basis of indirect observations. It seems that as modern science progresses it withholds more and more findings from our natural senses. In this context, the use of optical instrumentation excels among ways to study nature. This is due to the fact that because of our ability to perceive electromagnetic waves at optical frequencies our brain is used to the interpretation of phenomena associated with light, even if the structures that are observed are magnified a thousandfold. This intuitive understanding is among the most important features that make light and optical processes so attractive as a means to reveal physical laws and relationships. The fact that the energy of light lies within the energy range of electronic and vibrational transitions in matter allows us to use light for gaining unique information about the structural and dynamical properties of matter and also to perform subtle manipulations of the quantum state of matter. These unique spectroscopic capabilities associated with optical techniques are of great importance for the study of biological and solid-state nanostructures.

Today we encounter a strong trend towards nanoscience and nanotechnology. This trend was originally driven by the benefits of miniaturization and integration of electronic circuits for the computer industry. More recently we have observed a paradigm shift that manifests itself in the notion that nanoscience and technology are more and more driven by the fact that, as we move to smaller and smaller scales, new physical effects that may be exploited in future technological applications become prominent. The advances in nanoscience and technology are due in large part to our newly acquired ability to measure, fabricate, and manipulate individual structures on the nanometer scale using scanning probe techniques, optical tweezers, high-resolution electron microscopes and lithography tools, focused ion-beam milling systems etc.

The increasing trend towards nanoscience and nanotechnology makes it inevitable that we will need to study optical phenomena on the nanometer scale. Since the diffraction limit does not allow us to focus light to dimensions smaller than roughly one half of the wavelength (200 nm), traditionally it was not possible to optically interact selectively with nanoscale features. However, in recent years, several new approaches have been put forth to "shrink" the diffraction limit or even overcome it. A central goal of nano-optics is to extend the use of optical techniques to length scales beyond the diffraction limit. The most obvious potential technological applications that arise from breaking the diffraction barrier are super-resolution microscopy and ultra-high-density data storage. But the field of nano-optics is by no means limited to technological applications and instrument design. Nano-optics also opens new doors to basic research on nanometer-sized structures.

Nature has developed various nanoscale structures to bring out unique optical effects. A prominent example is photosynthetic membranes, which use light-harvesting proteins to absorb sunlight and then channel the excitation energy to other neighboring proteins. The energy is guided to a so-called reaction center where it initiates charge transfer across the cell membrane. Other examples are sophisticated diffractive structures used by insects (butterflies) and other animals (peacocks) to produce attractive colors and effects. Also, nanoscale structures are used as antireflection coatings in the retinas of various insects, and naturally occurring photonic bandgaps are encountered in gemstones (opals). In recent years, we have succeeded in creating different artificial nanophotonic structures [1]. A few examples are depicted in Fig. 1.1. Single molecules are being used as local probes for electromagnetic fields and for biophysical processes, resonant metal nanostructures are

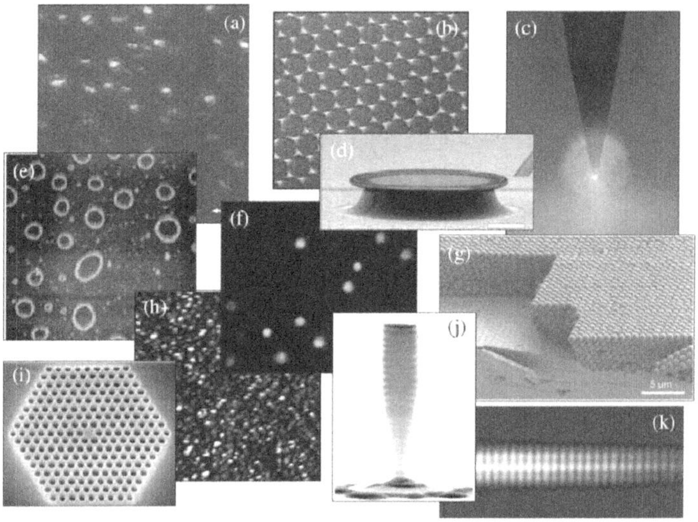

Fig. 1.1 A potpourri of man-made nanophotonic structures. (a) Strongly fluorescent molecules, (b) metal nanostructures fabricated by nanosphere lithography, (c) localized photon sources, (d) microdisk resonators (from [2]), (e) semiconductor nanostructures, (f) particle plasmons (from [3]), (g) photonic bandgap crystals (from [4]), (h) nanocomposite materials, (i) laser microcavities (from [5]), (j) single-photon sources (from [6]), (k) surface plasmon waveguides (from [7]).

being exploited as sensor devices, localized photon sources are being developed for high-resolution optical microscopy, extremely high Q-factors are being generated with optical microdisk resonators, nanocomposite materials are being explored for generating increased nonlinearities and collective responses, microcavities are being built for single-photon sources, surface plasmon waveguides are being implemented for planar optical networks, and photonic bandgap materials are being developed to suppress light propagation in specific frequency windows. All of these nanophotonic structures are being created to provide unique optical properties and phenomena, and the aim of this book is to establish a basis for their understanding.

1.1 Nano-optics in a nutshell

Let us try to get a quick glimpse of the very basics of nano-optics just to show that optics at the scale of a few nanometers makes perfect sense and is not forbidden by any fundamental law. In free space, the propagation of light is determined by the dispersion relation $\hbar\omega = c \cdot \hbar k$, which connects the wavevector $k = \sqrt{k_x^2 + k_y^2 + k_z^2}$ of a photon with its angular frequency ω via the speed of propagation c. Heisenberg's uncertainty relation states that the product of the uncertainty in the spatial position of a microscopic particle in a certain direction and the uncertainty in the component of its momentum in the same direction cannot become smaller than $\hbar/2$. For photons this leads to the relation

$$\Delta\left(\hbar k_x\right) \cdot \Delta x \geq \hbar/2, \tag{1.1}$$

which can be rewritten as

$$\Delta x \geq \frac{1}{2\,\Delta k_x}. \tag{1.2}$$

The interpretation of this result is as follows. The spatial confinement that can be achieved for photons is inversely proportional to the spread in the magnitude of wavevector components in the respective spatial direction, here x. Such a spread in wavevector components occurs for instance in a light field that converges towards a focus, e.g. behind a lens. Such a field may be represented by a superposition of plane waves traveling at different angles (see Section 2.12). The maximum possible spread in the wavevector component k_x is the total length of the free-space wavevector $k = 2\pi/\lambda$.[1] This leads to

$$\Delta x \geq \frac{\lambda}{4\pi}, \tag{1.3}$$

which is very similar to the well-known expression for the Rayleigh diffraction limit. Note that the spatial confinement that can be achieved is limited only by the spread of wavevector components in a given direction. In order to increase the spread of wavevector components we can play a mathematical trick. If we choose two arbitrary perpendicular directions in space, e.g. x and z, we can increase one wavevector component to values beyond the total

[1] For a real lens this must be corrected by the numerical aperture.

wavevector while at the same time requiring the wavevector in the perpendicular direction to become purely imaginary. If this is the case, then we can still fulfill the requirement for the total length of the wavevector $k = \sqrt{k_x^2 + k_y^2 + k_z^2}$ to be $2\pi/\lambda$. If we choose to increase the wavevector in the x-direction then the possible range of wavevectors in this direction is also increased and the confinement of light is no longer limited by Eq. (1.3). However, the possibility of increased confinement has to be paid for and the currency is confinement also in the z-direction, resulting from the purely imaginary wavevector component in this direction that is necessary to compensate for the large wavevector component in the x-direction. On introducing the purely imaginary wavevector component into the expression for a plane wave we obtain $\exp(ik_z z) = \exp(-|k_z|z)$. In one direction this leads to an exponentially decaying field, an evanescent wave, while in the opposite direction the field is exponentially increasing. Since exponentially increasing fields have no physical meaning we may safely discard the strategy just outlined to obtain a solution, and state that in free space Eq. (1.3) is always valid. However, this argument holds only for infinite free space! If we divide our infinite free space into at least two half-spaces with different refractive indices, then the exponentially decaying field in one half-space can exist without needing the exponentially increasing counterpart in the other half-space. In the other half-space a different solution that satisfies the boundary conditions for the fields at the interface may be valid.

These simple arguments show that in the presence of an inhomogeneity in space the Rayleigh limit for the confinement of light is no longer strictly valid, but in principle infinite confinement of light becomes, at least theoretically, possible. This insight is the basis of nano-optics. One of the key questions in nano-optics is how material structures have to be shaped to actually realize the theoretically possible field confinement. Another key issue is the nature of the physical consequences of the presence of exponentially decaying and strongly confined fields, which we will discuss in some detail in the following chapters.

1.2 Historical survey

In order to put this text on nano-optics into the right perspective and context we deem it appropriate to start out with a very short introduction to the historical development of optics in general and the advent of nano-optics in particular.

Nano-optics builds on the achievements of classical optics, the origin of which goes back to antiquity. At that time, burning glasses and the reflection law were already known and Greek philosophers (Empedocles, Euclid) speculated about the nature of light. They were the first to do systematic studies on optics. In the thirteenth century the first magnifying glasses were used. There are documents reporting the existence of eye glasses in China several centuries earlier. However, the first optical instrumentation for scientific purposes was not built until the beginning of the seventeenth century, when modern human curiosity started to awaken. It is often stated that the earliest telescope was the one constructed by

Galileo Galilei in 1609, since there is definite confirmation of its existence. Likewise, the first prototype of an optical microscope (1610) is also attributed to Galilei [8]. However, it is known that Galilei knew of a microscope built in Holland (probably by Zacharias Janssen) and that his instrument was built according to existing plans. In the sixteenth century craftsmen were already using glass spheres filled with water for the magnification of small details. As in the case of the telescope, the development of the microscope extends over a considerable period and cannot be attributed to any single inventor. A pioneer who advanced the development of the microscope, as has already been mentioned, was Antony van Leeuwenhoek. It is remarkable that the resolution of his microscope, built in 1671, was not exceeded for more than a century. At the time, his observation of red blood cells and bacteria was revolutionary. In the eighteenth and ninteenth centuries the development of the theory of light (polarization, diffraction, dispersion) helped to significantly advance optical technology and instrumentation. It was soon realized that optical resolution cannot be improved arbitrarily and that a lower bound is set by the diffraction limit. The theory of resolution was formulated by Abbe in 1873 [9] and Rayleigh in 1879 [10]. It is interesting to note, as we saw above, that there is a close relation to Heisenberg's uncertainty principle. Different techniques such as confocal microscopy [11] were invented over the years in order to stretch the diffraction limit beyond Abbe's limit. Today, confocal fluorescence microscopy is a key technology in biomedical research [12]. Highly fluorescent molecules that can be specifically attached to biological entities such as lipids, muscle fibers, and various cell organelles have been synthesized. This chemically specific labeling and the associated discrimination of different dyes in terms of their fluorescence emission allows scientists to visualize the interior of cells and study biochemical reactions in living environments. The invention of pulsed laser radiation propelled the field of nonlinear optics and enabled the invention of multiphoton microscopy [13]. However, multiphoton excitation is not the only nonlinear interaction that is exploited in optical microscopy. Second-harmonic, third-harmonic, and coherent anti-Stokes Raman scattering (CARS) microscopy [14] are other examples of extremely important inventions for visualizing processes with high spatial resolution. Besides nonlinear interactions, it has also been demonstrated that saturation effects can, in principle, be applied to achieve arbitrary spatial resolutions, provided that one knows what molecules are being imaged [15].

A different approach for boosting spatial resolution in optical imaging is provided by near-field optical microscopy. In principle, this technique does not rely on prior information. While it is restricted to imaging of features near the surface of a sample it provides complementary information about the surface topology similar to atomic force microscopy. A challenge in near-field optical microscopy is posed by the coupling of source (or detector) and the sample to be imaged. This challenge is absent in standard light microscopy where the light source (e.g. the laser) is not affected by the properties of the sample. Near-field optical microscopy was originally proposed in 1928 by Synge (Fig. 1.2). In a prophetic article he proposed an apparatus that comes very close to present implementations in scanning near-field optical microscopy [16, 17]. A minute aperture in an opaque plate illuminated from one side is placed in close proximity to a sample surface, thereby creating an illumination spot not limited by diffraction. The transmitted light is then collected with a microscope, and its intensity is measured with a photoelectric cell. In order to establish an image of the sample, the aperture is moved in small increments over the surface. The resolution of such an image should be limited by the size of the aperture and not by the wavelength of the illuminating light, as Synge correctly stated. It is known that Synge was in contact with Einstein about his ideas and Einstein encouraged Synge to publish his ideas. It is also known that later in his life Synge was no longer convinced about his idea and proposed alternative but, as we know today, incorrect ideas. Owing to the obvious experimental limitations at that time, Synge's idea was not realized and was soon forgotten. Later, in 1956, O'Keefe proposed a similar set-up without knowing of Synge's visionary idea [18]. The first experimental realization in the microwave region was performed in 1972 by Ash and Nichols, again without knowledge of Synge's paper [19]. Using a 1.5 mm aperture, illuminated with 10 cm waves, Ash and Nichols demonstrated subwavelength imaging with a resolution of $\lambda/60$.

The invention of scanning probe microscopy [20] at the beginning of the 1980s enabled distance regulation between probe and sample with high precision, and hence set the ground for a realization of Synge's idea at optical frequencies. In 1984 Massey proposed the use of piezoelectric position control for the accurate positioning of a minute aperture illuminated at optical frequencies [21]. Shortly after, Pohl, Denk and Lanz at the IBM Rüschlikon Research Laboratory managed to solve the remaining experimental difficulties of producing a subwavelength-sized aperture: a metal-coated pointed quartz tip was "pounded" against the sample surface until some light leakage through the foremost end could be detected. In 1984 the IBM group published the first subwavelength images at optical frequencies [22] and an independent development was undertaken by Lewis *et al.* [23] and Fischer *et al.* [24]. Subsequently, the technique was systematically advanced and extended to various applications mainly by Betzig *et al.* [25, 26], who demonstrated

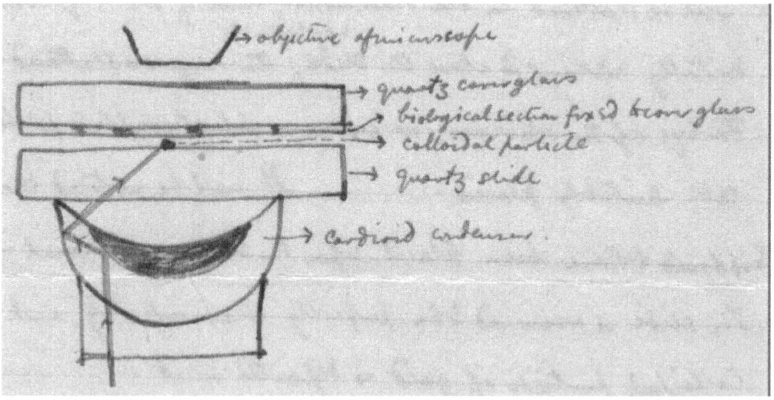

Fig. 1.2 In an April 1928 sketch sent to Albert Einstein, Edward Hutchinson Synge proposed a new microscopy method: using a tiny gold particle between two quartz slides to scatter incident light from below onto a sample. Light that didn't strike the particle would be totally internally reflected, and an objective lens of a microscope could be positioned to accept some of the gold-scattered light. That arrangement, Synge wrote, could be used to image a biological specimen fixed to the top cover slip at a resolution below the diffraction limit. (Courtesy of the Albert Einstein Archives, Hebrew University of Jerusalem, Israel.)

subwavelength magnetic data storage and detection of single fluorescent molecules. Over the years, various related techniques were proposed, such as the photon scanning tunneling microscope, the near-field reflection microscope, microscopes using luminescent centers as light-emitting sources, microscopes based on local plasmon interaction, microscopes based on local light scattering, and microscopes relying on the field enhancement effect near sharply pointed metal tips. All these techniques provide a confined photon flux between probe and sample. However, the confined light flux is not the only limiting factor for the achievable resolution. In order to be detectable, the photon flux needs to have a minimum intensity. These two requirements are to some extent contradictory and a compromise between light confinement and light throughput has to be found. The interested reader is referred to Ref. [17] for a more detailed account on the history of near-field optics.

1.3 Scope of the book

Traditionally, the field of optics is part of both the basic sciences (e.g. quantum optics) and applied sciences (e.g. optical communication and computing). Therefore, nano-optics can be defined as the broad spectrum of optics on the nanometer scale, ranging from nanotechnology applications to fundamental nanoscience.

On the nanotechnology side, we find topics like nanolithography, high-resolution optical microscopy, and high-density optical data storage. On the basic science end, we might mention atom–photon interactions in the optical near-field and their potential applications for atom trapping and manipulation experiments. Compared with free propagating light the optical near-field is enriched by so-called virtual photons that correspond to the exponentially decaying fields introduced before. The virtual-photon picture can be used to describe local, non-propagating fields in general. These virtual photons are the same sort of particles as is also responsible for molecular binding (van der Waals and Casimir forces) and therefore have potential for selective probing of molecular-scale structures. The consideration of virtual photons in the field of quantum optics will enlarge the range of fundamental experiments and will result in new applications. The present book provides an introduction to nano-optics that reflects the full breadth of the field between applied and basic science.

We start out by providing in Chapter 2 an overview of the theoretical foundations of nano-optics. Maxwell's equations, being scale-invariant, provide a secure basis for nano-optics. Since optical near-fields are always associated with matter, we review constitutive relations and complex dielectric constants. The systems that are investigated in the context of nano-optics, as we saw, must separate into several spatial domains that are separated by boundaries. Representations of Maxwell's equations valid in piecewise homogeneous media and the related boundary conditions for the fields are therefore derived. We then proceed with the discussion of fundamental theoretical concepts, such as the Green function and the angular spectrum representation, that are particularly useful for the discussion of nano-optical phenomena. The treatment of the angular spectrum representation leads to a thorough discussion of evanescent waves, which correspond to the new virtual-photon modes just mentioned.

Light confinement is a key issue in nano-optics. To set the basis for further discussions in Chapter 3, we analyze what is the smallest possible confinement of light that can be achieved by classical means, i.e. microscope objectives and other high-numerical-aperture focusing optics. Starting out with the treatment of focused fields in the paraxial approximation, which yields the well-known Gaussian beams, we proceed by discussing focused fields beyond the paraxial approximation as they occur, for example, in modern confocal microscopes.

Speaking of microscopy, spatial resolution is a key issue. There exist several definitions of the spatial resolution of an optical microscope that are related to the diffraction limit. An analysis of their physical foundations in Chapter 4 leads to the discussion of methods that can be used to enhance the spatial resolution of optical microscopy. Saturation effects and the difference between spatial position accuracy and resolution are discussed.

The following three chapters then deal with more practical aspects of nano-optics related to applications in the context of near-field optical microscopy. In Chapter 5 we discuss the basic technical realizations of high-resolution microscopes, starting with confocal microscopy and proceeding with various near-field techniques that have been developed over time. Chapter 6 then deals with the central technical question of how light can be squeezed into subwavelength regions. This is the domain of the so-called optical probes, material structures that typically have the shape of pointed tips and exhibit a confined and enhanced optical field at their apex. Finally, to complete the technical section, we show how such delicate optical probes can be approached and scanned in close proximity to a sample surface of interest. A method relying on the measurement of interaction (shear) forces between probe and sample is introduced and discussed. Taken together, these three chapters provide the technical basics for understanding the current methods used in scanning near-field optical microscopy.

We then proceed with a discussion of more fundamental aspects of nano-optics, i.e. light emission and optical interactions in nanoscale environments. As a starting point, we show that the light emission of a small particle (atom, molecule) with an electronic transition can be treated in the dipole approximation. We discuss the resulting fields of a radiating dipole and its interactions with the electromagnetic field in some detail. We proceed with the discussion of spontaneous decay in complex environments, which in the ultimate limit leads to the discussion of dipole–dipole interactions, energy transfer and excitonic coupling.

Having discussed dipolar emitters without mentioning a real-world realization, we discuss in Chapter 9 some experimental aspects of the detection of single-quantum emitters such as single fluorescent molecules and semiconductor quantum dots. Saturation count rates and the solutions of rate equation systems are discussed as well as fascinating issues such as the non-classical photon statistics of fields emitted by quantum emitters and coherent control of wave functions. Finally we discuss how single emitters can be used to map spatially confined fields in great detail.

In Chapter 10 we return to the issue of dipole emission in a nanoscale environment. Here, we treat in some detail the very important and illustrative case of dipole emission near a planar interface. We calculate radiation patterns and decay rates of dipolar emitters and also discuss the image-dipole approximation that can be used to obtain approximate and qualitative results.

If we consider multiple interfaces, instead of only one, that are arranged in a regular pattern, we obtain a so-called photonic crystal. The properties of such structures can be described in analogy to solid-state physics by introducing an optical band structure that may contain bandgaps in certain directions where propagating light cannot exist. Defects in photonic crystals lead to localized states, much like their solid-state counterparts, which are of particular interest in nano-optics since they can be considered as microscopic cavities with very high quality factors. In the same chapter we discuss optical resonators and their interaction with mechanical oscillators. This interaction makes it possible either to amplify the motion of a mechanical system or to slow it down.

Chapter 12 then takes up the topic of surface plasmons. Resonant collective oscillations of the free surface charge density in metal structures of various geometries can couple efficiently to optical fields and, due to the occurrence of resonances, are associated with strongly enhanced and confined optical near-fields. We give a basic introduction to the topic, covering the optical properties of noble metals, thin-film plasmons, and particle plasmons. In the following chapter we discuss optical antennas, devices designed to convert free-propagating radiation to localized energy, and vice versa.

The next chapter concentrates on optical forces. We formulate a theory based on Maxwell's stress tensor that allows us to calculate forces of particles of arbitrary shape once the field distribution is known. We then specialize the discussion and introduce the dipole approximation valid for small particles. Practical applications discussed include the optical-tweezers principle. Finally, the transfer of angular momentum using optical fields is discussed, as well as forces exerted by optical near-fields.

Another type of forces is discussed in the subsequent chapter, i.e. forces that are related to fluctuating electromagnetic fields which include the Casimir–Polder force and electromagnetic friction. On the way we also discuss the emission of radiation by fluctuating sources.

The present textbook is concluded by a summary of theoretical methods used in the field of nano-optics. Hardly any predictions can be made in the field of nano-optics without using adequate numerical methods. A selection of the most powerful theoretical tools is presented and their advantages and drawbacks are discussed.

References

[1] A. J. Haes and R. P. Van Duyne, "A nanoscale optical biosensor: sensitivity and selectivity of an approach based on the localized surface plasmon resonance spectroscopy of triangular silver nanoparticles," *J. Am. Chem. Soc.* **124**, 10596 (2002).

[2] D. K. Armani, T. J. Kippenberg, S. M. Spillane, and K. J. Vahala, "Ultra-high-*Q* toroid microcavity on a chip," *Nature* **421**, 925–928 (2003).

[3] J. J. Mock, M. Barbic, D. R. Smith, D. A. Schultz, and S. Schultz, "Shape effects in plasmon resonance of individual colloidal silver nanoparticles," *J. Chem. Phys.* **116**, 6755–6759 (2002).
[4] Y. A. Vlasov, X. Z. Bo, J. C. Sturm, and D. J. Norris, "On-chip natural assembly of silicon photonic bandgap crystals," *Nature* **414**, 289–293 (2001).
[5] O. J. Painter, A. Husain, A. Scherer, *et al.*, "Two-dimensional photonic crystal defect laser," *J. Lightwave Technol.* **17**, 2082–2089 (1999).
[6] J. M. Gérard, B. Sermage, B. Gayral, *et al.*, "Enhanced spontaneous emission by quantum boxes in a monolithic optical microcavity," *Phys. Rev. Lett.* **81**, 1110–1114 (1998).
[7] W. L. Barnes, A. Dereux, and T. W. Ebbesen, "Surface plasmon subwavelength optics," *Nature* **424**, 824–830 (2003).
[8] M. Born and E. Wolf, *Principles of Optics*, 6th edn. Oxford: Pergamon (1970).
[9] E. Abbe, "Beiträge zur Theorie des Mikroskops und der mikroskopischen Wahrnehmung," *Arch. Mikroskop. Anat.* **9**, 413–420 (1873).
[10] L. Rayleigh, "Investigations in optics, with special reference to the spectroscope," *Phil. Mag.* **8**, 261–274, 403–411, and 477–486 (1879).
[11] M. Minsky, "Memoir on inventing the confocal scanning microscope," *Scanning* **10**, 128–138 (1988).
[12] J. B. Pawley (ed.) *Handbook of Biological Confocal Microscopy*. New York: Plenum Press (1995).
[13] W. Denk, J. H. Strickler, and W. W. Webb, "2-Photon laser scanning fluorescence microscopy," *Science* **248**, 73–76 (1990).
[14] A. Zumbusch, G. R. Holtom, and X. S. Xie, "Three-dimensional vibrational imaging by coherent anti-Stokes Raman scattering," *Phys. Rev. Lett.* **82**, 4142–4145 (1999).
[15] T. A. Klar, S. Jakobs, M. Dyba, A. Egner, and S. W. Hell, "Fluorescence microscopy with diffraction resolution barrier broken by stimulated emission," *Proc. Nat. Acad. Sci.* **97**, 8206–8210 (2000).
[16] E. H. Synge, "A suggested model for extending microscopic resolution into the ultra-microscopic region," *Phil. Mag.* **6**, 356–362 (1928).
[17] L. Novotny, "The history of near-field optics," in *Progress in Optics*, vol. 50, ed. E. Wolf. Amsterdam: Elsevier, pp. 137–180 (2007).
[18] J. A. O'Keefe, "Resolving power of visible light," *J. Opt. Soc. Am.* **46**, 359–360 (1956).
[19] E. A. Ash and G. Nicholls, "Super-resolution aperture scanning microscope," *Nature* **237**, 510–513 (1972).
[20] G. Binnig, H. Rohrer, C. Gerber, and E. Weibel, "Tunneling through a controllable vacuum gap," *Appl. Phys. Lett.* **40**, 178–180 (1982).
[21] G. A. Massey, "Microscopy and pattern generation with scanned evanescent waves," *Appl. Opt.* **23**, 658–660 (1984).
[22] D. W. Pohl, W. Denk, and M. Lanz, "Optical stethoscopy: image recording with resolution $\lambda/20$," *Appl. Phys. Lett.* **44**, 651–653 (1984).
[23] A. Lewis, M. Isaacson, A. Harootunian, and A. Muray, "Development of a 500 Å spatial resolution light microscope," *Ultramicroscopy* **13**, 227–231 (1984).

[24] U. Ch. Fischer, "Optical characteristics of 0.1 μm circular apertures in a metal film as light sources for scanning ultramicroscopy," *J. Vac. Sci. Technol.* **B3**, 386–390 (1985).

[25] E. Betzig, M. Isaacson, and A. Lewis, "Collection mode nearfield scanning optical microscopy," *Appl. Phys. Lett.* **61**, 2088–2090 (1987).

[26] E. Betzig and R. J. Chichester, "Single molecules observed by near-field scanning optical microscopy," *Science* **262**, 1422–1425 (1993).

2 Theoretical foundations

Light embraces the most fascinating spectrum of electromagnetic radiation. This is mainly due to the fact that the energy of light quanta (photons) lies within the energy range of electronic transitions in matter. This gives us the beauty of color and is the reason why our eyes adapted to sense the optical spectrum.

Light is also fascinating because it manifests itself in the forms of waves and particles. In no other range of the electromagnetic spectrum are we more confronted with the wave–particle duality than in the optical regime. While long wavelength radiation (radiofrequencies, microwaves) is well described by wave theory, short wavelength radiation (X-rays) exhibits mostly particle properties. The two worlds meet in the optical regime.

To describe optical radiation in nano-optics it is mostly sufficient to adopt the wave picture. This allows us to use classical field theory based on Maxwell's equations. Of course, in nano-optics the systems with which the light fields interact are small (single molecules, quantum dots), which necessitates a quantum description of the material properties. Thus, in most cases we can use the framework of semiclassical theory, which combines the classical picture of fields and the quantum picture of matter. However, occasionally, we have to go beyond the semiclassical description. For example the photons emitted by a quantum system can obey non-classical photon statistics in the form of photon-antibunching (no two photons arriving simultaneously).

This section summarizes the fundamentals of electromagnetic theory forming the necessary basis for this book. Only the basic properties are discussed and for more detailed treatments the reader is referred to standard textbooks on electromagnetism such as the books by Jackson [1] and Stratton [2]. The starting point is Maxwell's equations, which were established by James Clerk Maxwell in 1873.

2.1 Macroscopic electrodynamics

In macroscopic electrodynamics the singular character of charges and their associated currents is avoided by considering charge densities ρ and current densities $\mathbf{j}$. In differential form and in SI units the *macroscopic* Maxwell's equations have the form

$$\nabla \times \mathbf{E}(\mathbf{r}, t) = -\frac{\partial \mathbf{B}(\mathbf{r}, t)}{\partial t}, \tag{2.1}$$

$$\nabla \times \mathbf{H}(\mathbf{r}, t) = \frac{\partial \mathbf{D}(\mathbf{r}, t)}{\partial t} + \mathbf{j}(\mathbf{r}, t), \tag{2.2}$$

$$\nabla \cdot \mathbf{D}(\mathbf{r}, t) = \rho(\mathbf{r}, t), \tag{2.3}$$

$$\nabla \cdot \mathbf{B}(\mathbf{r}, t) = 0. \tag{2.4}$$

where **E** denotes the electric field, **D** the electric displacement, **H** the magnetic field, **B** the magnetic induction, **j** the current density, and ρ the charge density. The components of these vector and scalar fields constitute a set of 16 unknowns. Depending on the medium considered, the number of unknowns can be reduced considerably. For example, in linear, isotropic, homogeneous and source-free media the electromagnetic field is entirely defined by two scalar fields. Maxwell's equations combine and complete the laws formerly established by Faraday, Ampère, Gauss, Poisson, and others. Since Maxwell's equations are differential equations they do not account for any fields that are constant in space and time. Any such field can therefore be added to the fields. It has to be emphasized that the concept of fields was introduced to explain the transmission of forces from a source to a receiver. The physical observables are therefore forces, whereas the fields are definitions introduced to explain the troublesome phenomenon of "action at a distance." Notice that the macroscopic Maxwell's equations deal with fields that are local spatial averages over microscopic fields associated with discrete charges. Hence, the microscopic nature of matter is not included in the macroscopic fields. Charge and current densities are considered as continuous functions of space. In order to describe the fields on an atomic scale it is necessary to use the microscopic Maxwell's equations which consider all matter to be made of charged and uncharged particles.

The conservation of charge is implicitly contained in Maxwell's equations. Taking the divergence of Eq. (2.2), noting that $\nabla \cdot \nabla \times \mathbf{H}$ is identically zero, and substituting Eq. (2.3) for $\nabla \cdot \mathbf{D}$ one obtains the continuity equation

$$\nabla \cdot \mathbf{j}(\mathbf{r}, t) + \frac{\partial \rho(\mathbf{r}, t)}{\partial t} = 0. \tag{2.5}$$

The electromagnetic properties of the medium are most commonly discussed in terms of the macroscopic polarization **P** and magnetization **M** according to

$$\mathbf{D}(\mathbf{r}, t) = \varepsilon_0 \mathbf{E}(\mathbf{r}, t) + \mathbf{P}(\mathbf{r}, t), \tag{2.6}$$

$$\mathbf{H}(\mathbf{r}, t) = \mu_0^{-1} \mathbf{B}(\mathbf{r}, t) - \mathbf{M}(\mathbf{r}, t), \tag{2.7}$$

where ε_0 and μ_0 are the permittivity and the permeability of vacuum, respectively. These equations do not impose any conditions on the medium and are therefore always valid.

2.2 Wave equations

After substituting the fields **D** and **B** in Maxwell's *curl* equations by the expressions (2.6) and (2.7) and combining the two resulting equations we obtain the inhomogeneous wave equations

$$\nabla \times \nabla \times \mathbf{E} + \frac{1}{c^2}\frac{\partial^2 \mathbf{E}}{\partial t^2} = -\mu_0 \frac{\partial}{\partial t}\left(\mathbf{j} + \frac{\partial \mathbf{P}}{\partial t} + \nabla \times \mathbf{M}\right), \tag{2.8}$$

$$\nabla \times \nabla \times \mathbf{H} + \frac{1}{c^2}\frac{\partial^2 \mathbf{H}}{\partial t^2} = \nabla \times \mathbf{j} + \nabla \times \frac{\partial \mathbf{P}}{\partial t} - \frac{1}{c^2}\frac{\partial^2 \mathbf{M}}{\partial t^2}. \tag{2.9}$$

The constant c was introduced for $(\varepsilon_0\mu_0)^{-1/2}$ and is known as the vacuum speed of light. The expression in the brackets of Eq. (2.8) can be associated with the *total current density*

$$\mathbf{j}_{\mathrm{t}} = \mathbf{j}_{\mathrm{s}} + \mathbf{j}_{\mathrm{c}} + \frac{\partial \mathbf{P}}{\partial t} + \nabla \times \mathbf{M}, \tag{2.10}$$

where $\mathbf{j}$ has been split into a *source current density* $\mathbf{j}_{\mathrm{s}}$ and an induced *conduction current density* $\mathbf{j}_{\mathrm{c}}$. The terms $\partial\mathbf{P}/\partial t$ and $\nabla \times \mathbf{M}$ are recognized as the *polarization current density* and the *magnetization current density*, respectively. The wave equations as stated in Eqs. (2.8) and (2.9) do not impose any conditions on the media considered and hence are generally valid.

2.3 Constitutive relations

Maxwell's equations define the fields that are generated by currents and charges in matter. However, they do not describe how these currents and charges are generated. Thus, to find a self-consistent solution for the electromagnetic field, Maxwell's equations must be supplemented by relations that describe the behavior of matter under the influence of the fields. These material equations are known as constitutive relations. In a non-dispersive linear and isotropic medium they have the form

$$\mathbf{D} = \varepsilon_0\varepsilon\,\mathbf{E} \qquad (\mathbf{P} = \varepsilon_0\,\chi_{\mathrm{e}}\,\mathbf{E}), \tag{2.11}$$

$$\mathbf{B} = \mu_0\mu\,\mathbf{H} \qquad (\mathbf{M} = \chi_{\mathrm{m}}\,\mathbf{H}), \tag{2.12}$$

$$\mathbf{j}_{\mathrm{c}} = \sigma\,\mathbf{E}. \tag{2.13}$$

with χ_{e} and χ_{m} denoting the electric and magnetic susceptibility, respectively. For *nonlinear* media, the right hand sides can be supplemented by terms of higher power. *Anisotropic* media can be considered using tensorial forms for ε and μ. In order to account for general *bianisotropic* media, additional terms relating **D** and **E** to both **B** and **H** have to be introduced. For such complex media, solutions to the wave equations can be found for very special situations only. The constituent relations given above account for *inhomogeneous* media if the material parameters ε, μ, and σ are functions of space. The medium is called *temporally dispersive* if the material parameters are functions of frequency, and *spatially dispersive* if the constitutive relations are convolutions over space. An electromagnetic field in a linear medium can be written as a superposition of monochromatic fields of the form

$$\mathbf{E}(\mathbf{r},t) = \mathbf{E}(\mathbf{k},\omega)\cos(\mathbf{k}\cdot\mathbf{r}-\omega t), \tag{2.14}$$

where $\mathbf{k}$ and ω are the wavevector and the angular frequency, respectively. In its most general form, the amplitude of the induced displacement $\mathbf{D}(\mathbf{r},t)$ can be written as[1]

$$\mathbf{D}(\mathbf{k},\omega) = \varepsilon_0\,\varepsilon(\mathbf{k},\omega)\,\mathbf{E}(\mathbf{k},\omega). \tag{2.15}$$

Since $\mathbf{E}(\mathbf{k},\omega)$ is equivalent to the Fourier transform $\hat{\mathbf{E}}$ of an arbitrary time-dependent field $\mathbf{E}(\mathbf{r},t)$, we can apply the inverse Fourier transform to Eq. (2.15) and obtain

$$\mathbf{D}(\mathbf{r},t) = \varepsilon_0 \iint \tilde{\varepsilon}(\mathbf{r}-\mathbf{r}',t-t')\mathbf{E}(\mathbf{r}',t')\,\mathrm{d}\mathbf{r}'\,\mathrm{d}t'. \tag{2.16}$$

Here, $\tilde{\varepsilon}$ denotes the response function in space and time. The displacement $\mathbf{D}$ at time t depends on the electric field at all times t' previous to t (temporal dispersion). Additionally, the displacement at a point $\mathbf{r}$ also depends on the values of the electric field at neighboring points $\mathbf{r}'$ (spatial dispersion). A spatially dispersive medium is therefore also called a *non-local* medium. Non-local effects can be observed at interfaces between different media or in metallic objects with sizes comparable to the mean-free path of electrons. In general, it is very difficult to account for spatial dispersion in field calculations. In most cases of interest the effect is very weak and we can safely ignore it. Temporal dispersion, on the other hand, is a widely encountered phenomenon and it is important to take it accurately into account.

2.4 Spectral representation of time-dependent fields

The spectrum $\hat{\mathbf{E}}(\mathbf{r},\omega)$ of an arbitrary time-dependent field $\mathbf{E}(\mathbf{r},t)$ is defined by the Fourier transform

$$\hat{\mathbf{E}}(\mathbf{r},\omega) = \frac{1}{2\pi}\int_{-\infty}^{\infty} \mathbf{E}(\mathbf{r},t)\,\mathrm{e}^{\mathrm{i}\omega t}\,\mathrm{d}t. \tag{2.17}$$

In order that $\mathbf{E}(\mathbf{r},t)$ is a real-valued field we have to require that

$$\hat{\mathbf{E}}(\mathbf{r},-\omega) = \hat{\mathbf{E}}^{*}(\mathbf{r},\omega). \tag{2.18}$$

Applying the Fourier transform to the time-dependent Maxwell's equations (2.1)–(2.4) gives

$$\nabla\times\hat{\mathbf{E}}(\mathbf{r},\omega) = \mathrm{i}\omega\hat{\mathbf{B}}(\mathbf{r},\omega), \tag{2.19}$$

$$\nabla\times\hat{\mathbf{H}}(\mathbf{r},\omega) = -\mathrm{i}\omega\hat{\mathbf{D}}(\mathbf{r},\omega) + \hat{\mathbf{j}}(\mathbf{r},\omega), \tag{2.20}$$

[1] In an anisotropic medium the dielectric constant $\varepsilon = \overleftrightarrow{\varepsilon}$ is a second-rank tensor.

$$\nabla \cdot \hat{\mathbf{D}}(\mathbf{r}, \omega) = \hat{\rho}(\mathbf{r}, \omega), \tag{2.21}$$

$$\nabla \cdot \hat{\mathbf{B}}(\mathbf{r}, \omega) = 0. \tag{2.22}$$

Once the solution for $\hat{\mathbf{E}}(\mathbf{r}, \omega)$ has been determined, the time-dependent field is calculated by the inverse transform as

$$\mathbf{E}(\mathbf{r}, t) = \int_{-\infty}^{\infty} \hat{\mathbf{E}}(\mathbf{r}, \omega)\, \mathrm{e}^{-\mathrm{i}\omega t}\, \mathrm{d}\omega. \tag{2.23}$$

Thus, the time dependence of a non-harmonic electromagnetic field can be Fourier transformed and every spectral component can be treated separately as a monochromatic field. The general time dependence is obtained from the inverse transform.

2.5 Fields as complex analytic signals

The relationship (2.18) indicates that the positive-frequency region contains all the information of the negative-frequency region. If we restrict the integration in Eq. (2.23) to positive frequencies, we obtain what is called a *complex analytic signal* [3]

$$\mathbf{E}^{+}(\mathbf{r}, t) = \int_{0}^{\infty} \hat{\mathbf{E}}(\mathbf{r}, \omega)\, \mathrm{e}^{-\mathrm{i}\omega t}\, \mathrm{d}\omega, \tag{2.24}$$

with the superscript "+" denoting that only positive frequencies are included. Similarly, we can define a complex analytic signal $\mathbf{E}^{-}$ that accounts only for negative frequencies. The truncation of the integration range causes $\mathbf{E}^{+}$ and $\mathbf{E}^{-}$ to become complex functions of time. Because $\mathbf{E}$ is real, we have $[\mathbf{E}^{+}]^{*} = \mathbf{E}^{-}$. By taking the Fourier transform of $\mathbf{E}^{+}(\mathbf{r}, t)$ and $\mathbf{E}^{-}(\mathbf{r}, t)$ we obtain $\hat{\mathbf{E}}^{+}(\mathbf{r}, \omega)$ and $\hat{\mathbf{E}}^{-}(\mathbf{r}, \omega)$, respectively. It turns out that $\hat{\mathbf{E}}^{+}$ is identical to $\hat{\mathbf{E}}$ for $\omega > 0$ and it is zero for negative frequencies. Similarly, $\hat{\mathbf{E}}^{-}$ is identical to $\hat{\mathbf{E}}$ for $\omega < 0$ and it is zero for positive frequencies. Consequently, $\hat{\mathbf{E}} = \hat{\mathbf{E}}^{+} + \hat{\mathbf{E}}^{-}$. In quantum mechanics, $\hat{\mathbf{E}}^{-}$ is associated with the creation operator $\hat{a}^{\dagger}$ and $\hat{\mathbf{E}}^{+}$ with the annihilation operator $\hat{a}$.

2.6 Time-harmonic fields

The time dependence in the wave equations can be easily separated to obtain a harmonic differential equation. A monochromatic field can then be written as[2]

$$\mathbf{E}(\mathbf{r}, t) = \mathrm{Re}\{\mathbf{E}(\mathbf{r})\, \mathrm{e}^{-\mathrm{i}\omega t}\} = \frac{1}{2}\left[\mathbf{E}(\mathbf{r})\, \mathrm{e}^{-\mathrm{i}\omega t} + \mathbf{E}^{*}(\mathbf{r})\, \mathrm{e}^{\mathrm{i}\omega t}\right], \tag{2.25}$$

[2] This can also be written as $\mathbf{E}(\mathbf{r}, t) = \mathrm{Re}\{\mathbf{E}(\mathbf{r})\}\cos(\omega t) + \mathrm{Im}\{\mathbf{E}(\mathbf{r})\}\sin(\omega t)$.

with similar expressions for the other fields. Notice that $\mathbf{E}(\mathbf{r}, t)$ is real, whereas the spatial part $\mathbf{E}(\mathbf{r})$ is complex. The symbol $\mathbf{E}$ will be used both for the real, time-dependent field and for the complex spatial part of the field. The introduction of a new symbol is avoided in order to keep the notation simple. It is convenient to represent the fields of a time-harmonic field by their complex amplitudes. Maxwell's equations can then be written as

$$\nabla \times \mathbf{E}(\mathbf{r}) = \mathrm{i}\omega \mathbf{B}(\mathbf{r}), \tag{2.26}$$

$$\nabla \times \mathbf{H}(\mathbf{r}) = -\mathrm{i}\omega \mathbf{D}(\mathbf{r}) + \mathbf{j}(\mathbf{r}), \tag{2.27}$$

$$\nabla \cdot \mathbf{D}(\mathbf{r}) = \rho(\mathbf{r}), \tag{2.28}$$

$$\nabla \cdot \mathbf{B}(\mathbf{r}) = 0, \tag{2.29}$$

which is equivalent to Maxwell's equations (2.19)–(2.22) for the spectra of arbitrary time-dependent fields. Thus, the solution for $\mathbf{E}(\mathbf{r})$ is equivalent to the spectrum $\hat{\mathbf{E}}(\mathbf{r}, \omega)$ of an arbitrary time-dependent field. It is obvious that the complex field amplitudes depend on the angular frequency ω, i.e. $\mathbf{E}(\mathbf{r}) = \mathbf{E}(\mathbf{r}, \omega)$. However, ω is usually not included in the argument. Also the material parameters ε, μ, and σ are functions of space and frequency, i.e. $\varepsilon = \varepsilon(\mathbf{r}, \omega)$, $\sigma = \sigma(\mathbf{r}, \omega)$, and $\mu = \mu(\mathbf{r}, \omega)$. For simpler notation, we will often drop the argument in the fields and material parameters. It is the context of the problem that determines which of the fields $\mathbf{E}(\mathbf{r}, t)$, $\mathbf{E}(\mathbf{r})$, or $\hat{\mathbf{E}}(\mathbf{r}, \omega)$ is being considered.

2.7 Longitudinal and transverse fields

For some problems it is favorable to represent a field vector $\mathbf{E}$ in terms of a transverse field $\mathbf{E}_{\perp}$ and a longitudinal field $\mathbf{E}_{\parallel}$, that is

$$\mathbf{E}(\mathbf{r}) = \mathbf{E}_{\perp}(\mathbf{r}) + \mathbf{E}_{\parallel}(\mathbf{r}), \tag{2.30}$$

with $\nabla \times \mathbf{E}_{\parallel} = 0$ and $\nabla \cdot \mathbf{E}_{\perp} = 0$. The meaning of 'transverse' and 'longitudinal' is best seen in reciprocal space, where $\mathbf{E}(\mathbf{r}) = \int_{\mathbf{k}} \hat{\mathbf{E}}(\mathbf{k}) \exp(\mathrm{i}\mathbf{k} \cdot \mathbf{r})\, \mathrm{d}\mathbf{k}$. We then obtain $\mathrm{i}\mathbf{k} \times \hat{\mathbf{E}}_{\parallel} = 0$ and $\mathrm{i}\mathbf{k} \cdot \hat{\mathbf{E}}_{\perp} = 0$ (c.f. Section 2.15), that is $\hat{\mathbf{E}}_{\parallel}$ points in the direction of the $\mathbf{k}$ vector and $\hat{\mathbf{E}}_{\perp}$ perpendicular to it. $\mathbf{E}_{\parallel}$ is also called solenoidal and $\mathbf{E}_{\perp}$ irrotational. The statement in Eq. (2.30) directly follows from Helmholtz's theorem, which states that any vector field can be written as $\mathbf{E} = -\nabla\phi + \nabla \times \mathbf{A}$. Here, $\nabla\phi$ is associated with a longitudinal field because $\nabla \times (\nabla\phi) = 0$. Similarly, $\nabla \times \mathbf{A}$ is transverse because $\nabla \cdot (\nabla \times \mathbf{A}) = 0$. Evidently, because $\nabla \cdot \mathbf{B} = 0$ the magnetic field is purely transverse. On the other hand, since $\nabla \cdot \mathbf{E} = -\rho/\varepsilon$, it follows that the electric field generated by charges is longitudinal. Note, however, that the current density $\mathbf{j} = \mathbf{j}_{\perp} + \mathbf{j}_{\parallel}$ gives rise to both transverse and longitudinal electric fields. It has to be emphasized that $\mathbf{E}_{\perp}$ and $\mathbf{E}_{\parallel}$ are a mathematical construct and that they have no physical meaning. Only when they are added together do we obtain a fully retarded and causal field.

2.8 Complex dielectric constant

With the help of the linear constitutive relations we can express Maxwell's curl equations (2.26) and (2.27) in terms of $\mathbf{E}(\mathbf{r})$ and $\mathbf{H}(\mathbf{r})$. We then multiply both sides of the first equation by μ^{-1} and then apply the curl operator to both sides. After the expression $\nabla \times \mathbf{H}$ has been substituted by the second equation we obtain

$$\nabla \times \mu^{-1} \nabla \times \mathbf{E} - \frac{\omega^2}{c^2} \left[\varepsilon + \mathrm{i}\sigma/(\omega\varepsilon_0)\right] \mathbf{E} = \mathrm{i}\omega\mu_0 \mathbf{j}_{\mathrm{s}}. \tag{2.31}$$

It is common practice to replace the expression in the brackets on the left-hand side by a complex dielectric constant, i.e.

$$\left[\varepsilon + \mathrm{i}\sigma/(\omega\varepsilon_0)\right] \rightarrow \varepsilon. \tag{2.32}$$

In this notation one does not distinguish between conduction currents and polarization currents. Energy dissipation is associated with the imaginary part of the dielectric constant. With the new definition of ε, the wave equations for the complex fields $\mathbf{E}(\mathbf{r})$ and $\mathbf{H}(\mathbf{r})$ in linear, isotropic, but *inhomogeneous* media are

$$\nabla \times \mu^{-1} \nabla \times \mathbf{E} - k_0^2 \varepsilon\, \mathbf{E} = \mathrm{i}\omega\mu_0\, \mathbf{j}_{\mathrm{s}}, \tag{2.33}$$

$$\nabla \times \varepsilon^{-1} \nabla \times \mathbf{H} - k_0^2 \mu\, \mathbf{H} = \nabla \times \varepsilon^{-1}\, \mathbf{j}_{\mathrm{s}}, \tag{2.34}$$

where $k_0 = \omega/c$ denotes the vacuum wavenumber. These equations are also valid for anisotropic media if the substitutions $\varepsilon \rightarrow \overleftrightarrow{\varepsilon}$ and $\mu \rightarrow \overleftrightarrow{\mu}$ are performed. The complex dielectric constant will be used throughout this book.

2.9 Piecewise homogeneous media

In many physical situations the medium is piecewise homogeneous. In this case the entire space is divided into subdomains in which the material parameters are independent of position $\mathbf{r}$. In principle, a piecewise homogeneous medium is inhomogeneous and the solution can be derived from Eqs. (2.33) and (2.34). However, the inhomogeneities are entirely confined to the boundaries and it is convenient to formulate the solution for each subdomain separately. These solutions must be connected with each other via the interfaces to form the solution for all space. Let the interface between two homogeneous domains D_i and D_j be denoted as ∂D_{ij}. If ε_i and μ_i designate the constant material parameters in subdomain D_i, the wave equations in that domain read as

$$(\nabla^2 + k_i^2)\mathbf{E}_i = -\mathrm{i}\omega\mu_0\mu_i\, \mathbf{j}_i + \frac{\nabla\rho_i}{\varepsilon_0\,\varepsilon_i}, \tag{2.35}$$

$$(\nabla^2 + k_i^2)\mathbf{H}_i = -\nabla \times \mathbf{j}_i, \tag{2.36}$$

where $k_i = (\omega/c)\sqrt{\mu_i \varepsilon_i}$ is the wavenumber and $\mathbf{j}_i$ and ρ_i are the sources in domain D_i. To obtain these equations, the identity $\nabla \times \nabla \times = -\nabla^2 + \nabla\nabla\cdot$ was used and Maxwell's equation (2.3) was applied. Equations (2.35) and (2.36) are also denoted as the inhomogeneous vector Helmholtz equations. In most practical applications, such as scattering problems, there are no source currents or charges present and the Helmholtz equations are homogeneous.

2.10 Boundary conditions

Since the material properties are discontinuous on the boundaries, Eqs. (2.35) and (2.36) are valid only in the interior of the subdomains. However, Maxwell's equations must also hold for the boundaries. Owing to the discontinuity it turns out to be difficult to apply the differential forms of Maxwell's equations, but there is no problem with the corresponding integral forms. The latter can be derived by applying the theorems of Gauss and Stokes to the differential forms (2.1)–(2.4), which yields

$$\int_{\partial S} \mathbf{E}(\mathbf{r},t) \cdot \mathrm{d}\mathbf{s} = -\int_S \frac{\partial}{\partial t}\mathbf{B}(\mathbf{r},t) \cdot \mathbf{n}_s \,\mathrm{d}a, \tag{2.37}$$

$$\int_{\partial S} \mathbf{H}(\mathbf{r},t) \cdot \mathrm{d}\mathbf{s} = \int_S \left[\mathbf{j}(\mathbf{r},t) + \frac{\partial}{\partial t}\mathbf{D}(\mathbf{r},t)\right] \cdot \mathbf{n}_s \,\mathrm{d}a, \tag{2.38}$$

$$\int_{\partial V} \mathbf{D}(\mathbf{r},t) \cdot \mathbf{n}_s \,\mathrm{d}a = \int_V \rho(\mathbf{r},t)\,\mathrm{d}V, \tag{2.39}$$

$$\int_{\partial V} \mathbf{B}(\mathbf{r},t) \cdot \mathbf{n}_s \,\mathrm{d}a = 0. \tag{2.40}$$

In these equations, $\mathrm{d}a$ denotes a surface element, $\mathbf{n}_s$ the normal unit vector to the surface, $\mathrm{d}\mathbf{s}$ a line element, ∂V the surface of the volume V, and ∂S the border of the surface S. The integral forms of Maxwell's equations lead to the desired boundary conditions if they are applied to a sufficiently small part of the considered boundary. In this case the boundary looks flat and the fields are homogeneous on both sides (Fig. 2.1). Consider a small rectangular path ∂S along the boundary as shown in Fig. 2.1(a). As the area S (enclosed by the path ∂S) is arbitrarily reduced, the electric and magnetic fluxes through S become zero. This does not necessarily apply for the source current, since a surface current density $\mathbf{K}$ might be present. The first two of Maxwell's equations then lead to the boundary conditions for the tangential field components[3]

$$\mathbf{n} \times (\mathbf{E}_i - \mathbf{E}_j) = \mathbf{0} \quad \text{on } \partial D_{ij}, \tag{2.41}$$

$$\mathbf{n} \times (\mathbf{H}_i - \mathbf{H}_j) = \mathbf{K} \quad \text{on } \partial D_{ij}, \tag{2.42}$$

[3] Notice that $\mathbf{n}$ and $\mathbf{n}_s$ are different unit vectors: $\mathbf{n}_s$ is perpendicular to the surfaces S and ∂V, whereas $\mathbf{n}$ is perpendicular to the boundary ∂D_{ij}.

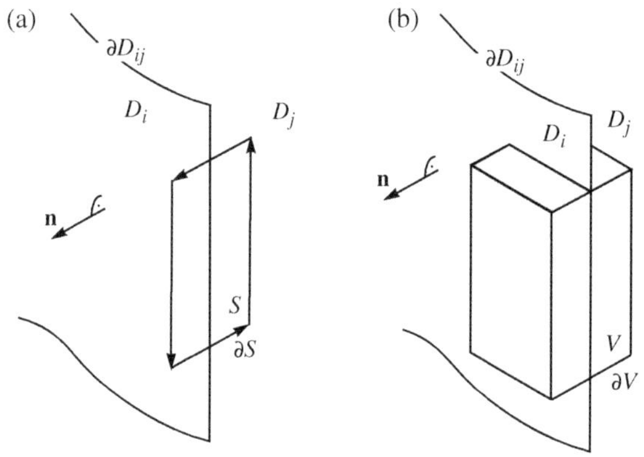

Fig. 2.1 Integration paths for the derivation of the boundary conditions on the interface ∂D_{ij} between two adjacent domains D_i and D_j.

where $\mathbf{n}$ is the unit normal vector on the boundary. A relation for the normal field components can be obtained by considering an infinitesimal rectangular box with volume V and surface ∂V according to Fig. 2.1(b). If the fields are considered to be homogeneous on both sides and if a surface charge density σ is assumed, Maxwell's third and fourth equations lead to the boundary conditions for the normal field components:

$$\mathbf{n} \cdot (\mathbf{D}_i - \mathbf{D}_j) = \sigma \quad \text{on } \partial D_{ij}, \tag{2.43}$$

$$\mathbf{n} \cdot (\mathbf{B}_i - \mathbf{B}_j) = 0 \quad \text{on } \partial D_{ij}. \tag{2.44}$$

In most practical situations there are no sources in the individual domains, and $\mathbf{K}$ and σ consequently vanish. The four boundary conditions (2.41)–(2.44) are not independent of each other since the fields on both sides of ∂D_{ij} are linked by Maxwell's equations. It can be easily shown, for example, that the conditions for the normal components are automatically satisfied if the boundary conditions for the tangential components hold everywhere on the boundary and Maxwell's equations are fulfilled in both domains.

2.10.1 Fresnel reflection and transmission coefficients

Applying the boundary conditions to a simple plane wave incident on a single planar interface leads to the familiar Fresnel reflection and transmission coefficients. A detailed derivation can be found in many textbooks, e.g. [4], pages 36ff. We only briefly mention the results.

An arbitrarily polarized plane wave $\mathbf{E}_1 \exp(\mathrm{i}\mathbf{k}_1 \cdot \mathbf{r} - \mathrm{i}\omega t)$ can always be written as the superposition of two orthogonally polarized plane waves. It is convenient to choose these polarizations parallel or perpendicular to the *plane of incidence* defined by the $\mathbf{k}$-vector of the plane wave and the surface normal $\mathbf{n}$ of the plane interface

$$\mathbf{E}_1 = \mathbf{E}_1^{(s)} + \mathbf{E}_1^{(p)}. \tag{2.45}$$

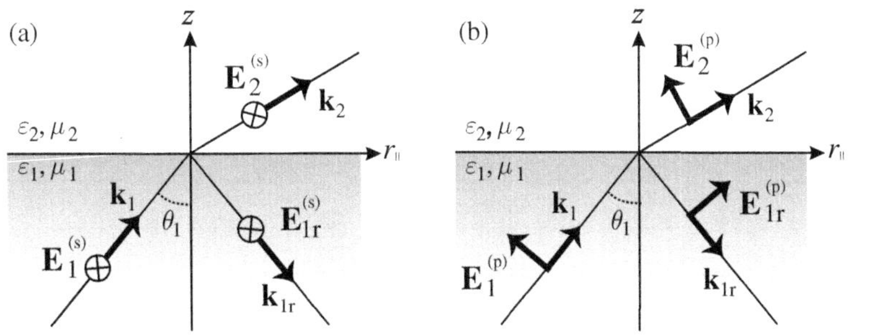

Fig. 2.2 Reflection and refraction of a plane wave at a plane interface: (a) s-polarization and (b) p-polarization.

$\mathbf{E}_1^{(s)}$ is parallel to the interface while $\mathbf{E}_1^{(p)}$ is perpendicular to the wavevector $\mathbf{k}$ and $\mathbf{E}_1^{(s)}$. The indices (s) and (p) stand for the German words *senkrecht* (perpendicular) and *parallel* (parallel), respectively, and refer to the plane of incidence. Upon reflection or transmission at the interface, the polarizations (s) and (p) are conserved.

As shown in Fig. 2.2, we denote the dielectric constants of the medium of incidence and the medium of transmittance as ε_1 and ε_2, respectively. The same designation applies to the magnetic permeability μ. Similarly, we distinguish between incident and reflected as well as transmitted wavevectors $\mathbf{k}_1, \mathbf{k}_{1r}$, and $\mathbf{k}_2$. Using the coordinate system shown in Fig. 2.2, it follows from the boundary conditions that

$$\mathbf{k}_1 = (k_x, k_y, k_{z_1}), \qquad |\mathbf{k}_1| = k_1 = \frac{\omega}{c}\sqrt{\varepsilon_1\mu_1}, \tag{2.46}$$

$$\mathbf{k}_2 = (k_x, k_y, k_{z_2}), \qquad |\mathbf{k}_2| = k_2 = \frac{\omega}{c}\sqrt{\varepsilon_2\mu_2}. \tag{2.47}$$

Thus, the transverse components of the wavevector (k_x, k_y) are conserved and the magnitudes of the longitudinal wavenumbers are given by

$$k_{z_1} = \sqrt{k_1^2 - (k_x^2 + k_y^2)}, \qquad k_{z_2} = \sqrt{k_2^2 - (k_x^2 + k_y^2)}. \tag{2.48}$$

The transverse wavenumber $k_\parallel = \sqrt{k_x^2 + k_y^2}$ can be expressed conveniently in terms of the angle of incidence θ_1 as

$$k_\parallel = \sqrt{k_x^2 + k_y^2} = k_1 \sin\theta_1, \tag{2.49}$$

which, according to Eqs. (2.48), also allows us to express k_{z_1} and k_{z_2} in terms of θ_1.

It follows from the boundary conditions that the amplitudes of the reflected and transmitted waves can be represented as

$$\begin{aligned} E_{1r}^{(s)} &= E_1^{(s)} r^s(k_x, k_y), \qquad & E_{1r}^{(p)} &= E_1^{(p)} r^p(k_x, k_y), \\ E_2^{(s)} &= E_1^{(s)} t^s(k_x, k_y), \qquad & E_2^{(p)} &= E_1^{(p)} t^p(k_x, k_y), \end{aligned} \tag{2.50}$$

where the Fresnel reflection and transmission coefficients are defined as[4]

$$r^{s}(k_x,k_y)=\frac{\mu_2 k_{z_1}-\mu_1 k_{z_2}}{\mu_2 k_{z_1}+\mu_1 k_{z_2}},\qquad r^{p}(k_x,k_y)=\frac{\varepsilon_2 k_{z_1}-\varepsilon_1 k_{z_2}}{\varepsilon_2 k_{z_1}+\varepsilon_1 k_{z_2}},\tag{2.51}$$

$$t^{s}(k_x,k_y)=\frac{2\mu_2 k_{z_1}}{\mu_2 k_{z_1}+\mu_1 k_{z_2}},\qquad t^{p}(k_x,k_y)=\frac{2\varepsilon_2 k_{z_1}}{\varepsilon_2 k_{z_1}+\varepsilon_1 k_{z_2}}\sqrt{\frac{\mu_2\varepsilon_1}{\mu_1\varepsilon_2}}.\tag{2.52}$$

As indicated by the superscripts, these coefficients depend on the polarization of the incident plane wave. The coefficients are functions of k_{z_1} and k_{z_2}, which can be expressed in terms of k_x and k_y and thus in terms of the angle of incidence θ_1. The sign of the Fresnel coefficients depends on the definition of the electric field vectors shown in Fig. 2.2. For a plane wave at normal incidence ($\theta_1 = 0$), r^s and r^p differ by a factor of -1. Notice that the transmitted waves can be either plane waves or evanescent waves. This aspect will be discussed in Section 2.14.

2.11 Conservation of energy

The equations established so far describe the behavior of electric and magnetic fields. They are a direct consequence of Maxwell's equations and the properties of matter. Although the electric and magnetic fields were initially postulated to explain the forces in Coulomb's and Ampère's laws, Maxwell's equations do not provide any information about the energy or forces in a system. The basic Lorentz law describes the forces acting on moving charges only. As the Abraham–Minkowski controversy shows, the forces acting on an arbitrary object cannot be extracted from a given electrodynamic field in a consistent way. It is also interesting, that Coulomb's and Ampère's laws were sufficient to establish the Lorentz force law. Although later the field equations were completed by adding the Maxwell displacement current, the Lorentz law remained unchanged. There is less controversy regarding the energy. Although also not a direct consequence of Maxwell's equations, Poynting's theorem provides a plausible relationship between the electromagnetic field and its energy content. For later reference, Poynting's theorem will be outlined below.

If the scalar product of the field $\mathbf{E}$ and Eq. (2.2) is subtracted from the scalar product of the field $\mathbf{H}$ and Eq. (2.1) the following equation is obtained:

$$\mathbf{H}\cdot(\nabla\times\mathbf{E})-\mathbf{E}\cdot(\nabla\times\mathbf{H})=-\mathbf{H}\cdot\frac{\partial\mathbf{B}}{\partial t}-\mathbf{E}\cdot\frac{\partial\mathbf{D}}{\partial t}-\mathbf{j}\cdot\mathbf{E}.\tag{2.53}$$

On noting that the expression on the left is identical to $\nabla\cdot(\mathbf{E}\times\mathbf{H})$, integrating both sides over space and applying Gauss's theorem, the above equation becomes

$$\int_{\partial V}(\mathbf{E}\times\mathbf{H})\cdot\mathbf{n}\,\mathrm{d}a=-\int_{V}\left[\mathbf{H}\cdot\frac{\partial\mathbf{B}}{\partial t}+\mathbf{E}\cdot\frac{\partial\mathbf{D}}{\partial t}+\mathbf{j}\cdot\mathbf{E}\right]\mathrm{d}V.\tag{2.54}$$

[4] For symmetry reasons, some authors omit the square-root term in the coefficient t^p. In this case, t^p refers to the ratio of transmitted and incident *magnetic field*. We adopt the definition from Born & Wolf [4].

Although this equation already forms the basis of Poynting's theorem, more insight is provided when **B** and **D** are substituted by the generally valid equations (2.6) and (2.7). Equation (2.54) then reads

$$\int_{\partial V} (\mathbf{E} \times \mathbf{H}) \cdot \mathbf{n} \, da + \frac{1}{2} \frac{\partial}{\partial t} \int_V \Big[\mathbf{D} \cdot \mathbf{E} + \mathbf{B} \cdot \mathbf{H} \Big] \mathrm{d}V$$

$$= -\int_V \mathbf{j} \cdot \mathbf{E} \, \mathrm{d}V - \frac{1}{2} \int_V \left[\mathbf{E} \cdot \frac{\partial \mathbf{P}}{\partial t} - \mathbf{P} \cdot \frac{\partial \mathbf{E}}{\partial t} \right] \mathrm{d}V - \frac{\mu_0}{2} \int_V \left[\mathbf{H} \cdot \frac{\partial \mathbf{M}}{\partial t} - \mathbf{M} \cdot \frac{\partial \mathbf{H}}{\partial t} \right] \mathrm{d}V. \tag{2.55}$$

This equation is a direct conclusion of Maxwell's equations and has therefore the same validity. Poynting's theorem is more or less an interpretation of the equation above. It states that the first term is equal to the net energy flow in or out of the volume V, the second term is equal to the time rate of change of electromagnetic energy inside V and the remaining terms on the right-hand side are equal to the rate of energy dissipation inside V. According to this interpretation

$$\mathbf{S} = (\mathbf{E} \times \mathbf{H}) \tag{2.56}$$

represents the energy flux density and

$$W = \frac{1}{2} \Big[\mathbf{D} \cdot \mathbf{E} + \mathbf{B} \cdot \mathbf{H} \Big] \tag{2.57}$$

is the density of electromagnetic energy. If the medium within V is linear and non-dispersive, the two last terms in Eq. (2.55) equal zero and the only term accounting for energy dissipation is $\mathbf{j} \cdot \mathbf{E}$. The vector **S** is called the Poynting vector. In principle, the curl of any vector field can be added to **S** without changing the conservation law (2.55), but it is convenient to make the choice as stated in (2.56).

Of special interest is the mean time value of **S**. This quantity describes the net power flux density and is needed for the evaluation of radiation patterns. Assuming that the fields are harmonic in time, linear, and non-dispersive, the time average of Eq. (2.55) becomes

$$\int_{\partial V} \langle \mathbf{S} \rangle \cdot \mathbf{n} \, da = -\frac{1}{2} \int_V \mathrm{Re}\big\{ \mathbf{j}^* \cdot \mathbf{E} \big\} \mathrm{d}V, \tag{2.58}$$

where we have used complex notation. The term on the right defines the mean energy dissipation within the volume V. $\langle \mathbf{S} \rangle$ represents the time average of the Poynting vector,

$$\langle \mathbf{S} \rangle = \frac{1}{2} \, \mathrm{Re}\big\{ \mathbf{E} \times \mathbf{H}^* \big\}. \tag{2.59}$$

In the far-field, the electromagnetic field is purely transverse. Furthermore, the electric and magnetic fields are in phase and the ratio of their amplitudes is constant. In this case $\langle \mathbf{S} \rangle$ can be expressed in terms of the electric field alone as

$$\langle \mathbf{S} \rangle = \frac{1}{2} \sqrt{\frac{\varepsilon_0 \varepsilon}{\mu_0 \mu}} \, |\mathbf{E}|^2 \mathbf{n}_\mathrm{r}, \tag{2.60}$$

where $\mathbf{n}_\mathrm{r}$ represents the unit vector in the radial direction and the inverse of the square root denotes the wave impedance.

Energy density in dispersive and lossy media

The two last terms in Eq. (2.55) strictly vanish only in a linear medium with no dispersion and no losses. The only medium fulfilling these conditions is vacuum. For all other media, the last two terms only vanish approximately. In this section we consider a linear medium with a frequency-dependent and complex ε and μ.

Let us return to the Poynting theorem stated in Eq. (2.54). While the left-hand side denotes the power flowing in or out of the volume V, the right-hand side denotes the power dissipated or generated in the volume V. The three terms on the right-hand side are of similar form and so we start by considering first the electric energy term $\mathbf{E}\cdot(\partial\mathbf{D}/\partial t)$. The electric energy density w_{E} at the time t is

$$w_{\mathrm{E}}(\mathbf{r},t) \;=\; \int_{-\infty}^{t} \mathbf{E}(\mathbf{r},t')\cdot\frac{\partial\mathbf{D}(\mathbf{r},t')}{\partial t'}\,dt'. \tag{2.61}$$

We now express the fields $\mathbf{E}$ and $\mathbf{D}$ in terms of their Fourier transforms as $\mathbf{E}(t') = \int \hat{\mathbf{E}}(\omega)\exp[-\mathrm{i}\omega t']\mathrm{d}\omega$ and $\mathbf{D}(t') = \int \hat{\mathbf{D}}(\omega)\exp[-\mathrm{i}\omega t']\mathrm{d}\omega$, respectively. In the last expression we substitute $\omega = -\omega'$ and obtain $\mathbf{D}(t') = \int \hat{\mathbf{D}}^{*}(\omega')\exp[\mathrm{i}\omega' t']\mathrm{d}\omega'$, where we used $\hat{\mathbf{D}}(-\omega') = \hat{\mathbf{D}}^{*}(\omega')$ since $\mathbf{D}(t)$ is real (c.f. Eq. (2.18)). Using the linear relation $\hat{\mathbf{D}} = \varepsilon_0\varepsilon\,\hat{\mathbf{E}}$ and inserting the Fourier transforms in Eq. (2.61) yields

$$w_{\mathrm{E}}(\mathbf{r},t) \;=\; \varepsilon_0 \int_{-\infty}^{\infty}\int_{-\infty}^{\infty} \frac{\omega'\,\varepsilon^{*}(\omega')}{\omega'-\omega}\,\hat{\mathbf{E}}(\omega)\cdot\hat{\mathbf{E}}^{*}(\omega')\,\mathrm{e}^{i(\omega'-\omega)t}d\omega'\,d\omega, \tag{2.62}$$

where we have carried out the differentiation and integration over time and assumed that the fields were zero at $t\rightarrow -\infty$. For later purposes it is advantageous to represent the above result in different form. Using the substitutions $u' = -\omega$ and $u = -\omega'$ and making use of $\hat{\mathbf{E}}(-u) = \hat{\mathbf{E}}^{*}(u)$ and $\varepsilon(-u) = \varepsilon^{*}(u)$ gives an expression similar to Eq. (2.62) but in terms of u and u'. Finally, we add this expression to Eq. (2.62) and take one half of the resulting sum, which yields [5]

$$w_{\mathrm{E}}(\mathbf{r},t) \;=\; \frac{\varepsilon_0}{2}\int_{-\infty}^{\infty}\int_{-\infty}^{\infty}\left[\frac{\omega'\varepsilon^{*}(\omega')-\omega\varepsilon(\omega)}{\omega'-\omega}\right]\hat{\mathbf{E}}(\omega)\cdot\hat{\mathbf{E}}^{*}(\omega')\,\mathrm{e}^{i(\omega'-\omega)t}\,d\omega'\,d\omega. \tag{2.63}$$

Similar expressions are obtained for the magnetic term $\mathbf{H}\cdot(\partial\mathbf{B}/\partial t)$ and the dissipative term $\mathbf{j}\cdot\mathbf{E}$ in Eq. (2.54).

If $\varepsilon(\omega)$ is a complex function then w_{E} accounts not only for the energy density built up in the medium but also for the energy transferred to the medium, such as heat dissipation. This contribution becomes indistinguishable from the term $\mathbf{j}\cdot\mathbf{E}$ in Eq. (2.54) as has already been discussed in Section 2.8. Thus, the imaginary part of ε can be included in the conductivity σ (c.f. Eq. (2.32)) and accounted for in the term $\mathbf{j}\cdot\mathbf{E}$ through the linear relationship $\hat{\mathbf{j}}=\sigma\hat{\mathbf{E}}$. Therefore, to discuss the energy density it suffices to consider only the real part of ε, which we're going to denote as ε'.

Let us now consider a monochromatic field represented by $\hat{\mathbf{E}}(\mathbf{r},\omega) = \mathbf{E}_0(\mathbf{r})\,[\delta(\omega-\omega_0)+\delta(\omega+\omega_0)]/2$. Inserting into Eq. (2.63) yields four terms: two that are constant in time and two that oscillate in time. Upon averaging over an oscillation period $2\pi/\omega_0$ the oscillatory terms vanish and only the constant terms survive. For these terms we must view the expression in brackets in Eq. (2.63) as a limit; that is,

$$\lim_{\omega'\rightarrow\omega}\left[\frac{\omega'\,\varepsilon'(\omega')-\omega\varepsilon'(\omega)}{\omega'-\omega}\right] \;=\; \left.\frac{d\,[\omega\,\varepsilon'(\omega)]}{d\omega}\right|_{\omega=\omega 0}. \tag{2.64}$$

Thus, the cycle average of Eq. (2.63) yields

$$\bar{w_E}(\mathbf{r}) = \left.\frac{\varepsilon_0 d\,[\omega\,\varepsilon'(\omega)]}{4d\omega}\right|_{\omega=\omega 0} |\mathbf{E}_0(\mathbf{r})|^2. \tag{2.65}$$

A similar result can be derived for the magnetic term $\mathbf{H}\cdot(\partial\mathbf{B}/\partial t)$.

It can be shown that Eq. (2.65) also holds for quasi-monochromatic fields that have frequency components ω only in a narrow range about a center frequency ω_0. Such fields can be represented as

$$\mathbf{E}(\mathbf{r},t) = \mathrm{Re}\{\tilde{\mathbf{E}}(\mathbf{r},t)\} = \mathrm{Re}\{\mathbf{E}_0(\mathbf{r},t)\,\mathrm{e}^{-i\omega_0 t}\}, \tag{2.66}$$

which is known as the *slowly varying amplitude approximation*. Here, $\mathbf{E}_0(\mathbf{r},t)$ is the slowly varying (complex) amplitude and ω_0 is the "carrier" frequency. The envelope $\mathbf{E}_0$ spans over many oscillations of frequency ω_0.

Expressing the field amplitudes in terms of time-averages, that is $|\mathbf{E}_0|^2 = 2\,\langle\mathbf{E}(t)\cdot\mathbf{E}(t)\rangle$, we can express the total cycle-averaged energy density $\bar{W}$ as

$$\bar{W} = \left[\varepsilon_0\frac{d\,[\omega\,\varepsilon'(\omega)]}{d\omega}\Big\langle\mathbf{E}\cdot\mathbf{E}\Big\rangle + \mu_0\frac{d\,[\omega\,\mu'(\omega)]}{d\omega}\Big\langle\mathbf{H}\cdot\mathbf{H}\Big\rangle\right], \tag{2.67}$$

where $\mathbf{E} = \mathbf{E}(\mathbf{r},t)$ and $\mathbf{H} = \mathbf{H}(\mathbf{r},t)$ are the time-dependent fields. Notice that ω is the center frequency of the spectra of $\mathbf{E}$ and $\mathbf{H}$. For a medium with negligible dispersion this expression reduces to the familiar $\bar{W} = (1/2)\left[\varepsilon_0\varepsilon'|\mathbf{E}_0|^2 + \mu_0\mu'\,|\mathbf{H}_0|^2\right]$, which follows from Eq. (2.57) using the dispersion-free constitutive relations. Because of $d(\omega\varepsilon')/d\omega > 0$ and $d(\omega\mu')/d\omega > 0$ the energy density is always positive, even for metals with $\varepsilon' < 0$. A detailed discussion on energy density in dispersive and lossy materials can be found in Refs. [5, 6].

2.12 Dyadic Green functions

An important concept in field theory is the Green function: the fields due to a point source. In electromagnetic theory, the dyadic Green function $\overleftrightarrow{\mathbf{G}}$ is essentially defined by the electric field $\mathbf{E}$ at the field point $\mathbf{r}$ generated by a radiating electric dipole $\mathbf{p}$ located at the source point $\mathbf{r}'$. In mathematical terms this reads as

$$\mathbf{E}(\mathbf{r}) = \omega^2\mu_0\mu\,\overleftrightarrow{\mathbf{G}}(\mathbf{r},\mathbf{r}')\mathbf{p}. \tag{2.68}$$

To understand the basic idea of Green functions we will first consider a general mathematical point of view.

2.12.1 Mathematical basis of Green functions

Consider the following general, inhomogeneous equation:

$$\mathcal{L}\mathbf{A}(\mathbf{r}) = \mathbf{B}(\mathbf{r}). \tag{2.69}$$

$\mathcal{L}$ is a linear operator acting on the vector field **A** representing the unknown response of the system. The vector field **B** is a known source function and makes the differential equation inhomogeneous. A well-known theorem for linear differential equations states that the general solution is equal to the sum of the complete homogeneous solution ($\mathbf{B}=0$) and a particular inhomogeneous solution. Here, we assume that the homogeneous solution ($\mathbf{A}_0$) is known. We thus need to solve for an arbitrary particular solution.

Usually it is difficult to find a solution of Eq. (2.69) and it is easier to consider the special inhomogeneity $\delta(\mathbf{r}-\mathbf{r}')$, which is zero everywhere, except in the point $\mathbf{r}=\mathbf{r}'$. Then, the linear equation reads as

$$\mathcal{L}\mathbf{G}_i(\mathbf{r},\mathbf{r}') = \mathbf{n}_i\delta(\mathbf{r}-\mathbf{r}') \qquad (i=x,y,z), \tag{2.70}$$

where $\mathbf{n}_i$ denotes an arbitrary constant unit vector. Here $\mathbf{G}_i$ is the solution of $\mathcal{L}$ for the source $\mathbf{n}_i\delta(\mathbf{r}-\mathbf{r}')$, while **A** is the solution of $\mathcal{L}$ for the source **B**. In general, the vector field $\mathbf{G}_i$ depends on the location $\mathbf{r}'$ of the inhomogeneity $\delta(\mathbf{r}-\mathbf{r}')$. Therefore, the vector $\mathbf{r}'$ has been included in the argument of $\mathbf{G}_i$. The three equations given by Eq. (2.70) can be written in closed form as

$$\mathcal{L}\overleftrightarrow{\mathbf{G}}(\mathbf{r},\mathbf{r}') = \overleftrightarrow{\mathbf{I}}\,\delta(\mathbf{r}-\mathbf{r}'), \tag{2.71}$$

where the operator $\mathcal{L}$ acts on each column of $\overleftrightarrow{\mathbf{G}}$ separately and $\overleftrightarrow{\mathbf{I}}$ is the unit dyad. The function $\overleftrightarrow{\mathbf{G}}$ fulfilling Eq. (2.71) is known as the dyadic Green function.

Next, assume that Eq. (2.71) has been solved and that $\overleftrightarrow{\mathbf{G}}$ is known. Postmultiplying Eq. (2.71) by $\mathbf{B}(\mathbf{r}')$ on both sides and integrating over the volume V in which $\mathbf{B}\neq 0$ gives

$$\int_V \mathcal{L}\overleftrightarrow{\mathbf{G}}(\mathbf{r},\mathbf{r}')\mathbf{B}(\mathbf{r}')\mathrm{d}V' = \int_V \mathbf{B}(\mathbf{r}')\delta(\mathbf{r}-\mathbf{r}')\mathrm{d}V'. \tag{2.72}$$

The right-hand side simply reduces to $\mathbf{B}(\mathbf{r})$ and with Eq. (2.69) it follows that

$$\mathcal{L}\mathbf{A}(\mathbf{r}) = \int_V \mathcal{L}\overleftrightarrow{\mathbf{G}}(\mathbf{r},\mathbf{r}')\mathbf{B}(\mathbf{r}')\mathrm{d}V'. \tag{2.73}$$

If on the right-hand side the operator $\mathcal{L}$ is taken out of the integral, the solution of Eq. (2.69) can be expressed as

$$\mathbf{A}(\mathbf{r}) = \int_V \overleftrightarrow{\mathbf{G}}(\mathbf{r},\mathbf{r}')\mathbf{B}(\mathbf{r}')\mathrm{d}V'. \tag{2.74}$$

Thus, the solution of the original equation can be found by integrating the product of the dyadic Green function and the inhomogeneity **B** over the source volume V.

The assumption that the operators $\mathcal{L}$ and $\int \mathrm{d}V'$ can be interchanged is not strictly valid and special care must be applied if the integrand is not well-behaved. Most often $\overleftrightarrow{\mathbf{G}}$ is singular at $\mathbf{r}=\mathbf{r}'$ and an infinitesimal exclusion volume surrounding $\mathbf{r}=\mathbf{r}'$ has to be introduced [7, 8]. Depolarization of the principal volume must be treated separately, resulting in a term ($\overleftrightarrow{\mathbf{L}}$) depending on the geometrical shape of the volume. Furthermore, in numerical schemes the principal volume has a finite size, giving rise to a second correction term, which is commonly designated by $\overleftrightarrow{\mathbf{M}}$. As long as we consider field points outside of the source volume V, i.e. $\mathbf{r}\notin V$, we do not need to consider these tricky issues. However, the topic of the principal volume will be taken up in Chapter 16.

2.12.2 Derivation of the Green function for the electric field

The derivation of the Green function for the electric field is most conveniently accomplished by considering the time-harmonic vector potential $\mathbf{A}$ and the scalar potential ϕ in an infinite and homogeneous space characterized by the constants ε and μ. In this case, $\mathbf{A}$ and ϕ are defined by the relationships

$$\mathbf{E}(\mathbf{r}) = \mathrm{i}\omega\mathbf{A}(\mathbf{r}) - \nabla\phi(\mathbf{r}), \tag{2.75}$$

$$\mathbf{H}(\mathbf{r}) = \frac{1}{\mu_0\mu}\,\nabla\times\mathbf{A}(\mathbf{r}). \tag{2.76}$$

We can insert these equations into Maxwell's second equation (2.27) and obtain

$$\nabla\times\nabla\times\mathbf{A}(\mathbf{r}) = \mu_0\mu\mathbf{j}(\mathbf{r}) - \mathrm{i}\omega\mu_0\mu\varepsilon_0\varepsilon[\mathrm{i}\omega\mathbf{A}(\mathbf{r}) - \nabla\phi(\mathbf{r})], \tag{2.77}$$

where we used $\mathbf{D} = \varepsilon_0\varepsilon\mathbf{E}$. The potentials $\mathbf{A}$ and ϕ are not uniquely defined by Eqs. (2.75) and (2.76). We are still free to define the value of $\nabla\cdot\mathbf{A}$, which we choose as

$$\nabla\cdot\mathbf{A}(\mathbf{r}) = \mathrm{i}\omega\mu_0\mu\varepsilon_0\varepsilon\phi(\mathbf{r}). \tag{2.78}$$

A condition that fixes the redundancy of Eqs. (2.75) and (2.76) is called a gauge condition. The gauge chosen through Eq. (2.78) is the so-called Lorenz gauge. Using the mathematical identity $\nabla\times\nabla\times = -\nabla^2 + \nabla\nabla\cdot$ together with the Lorenz gauge we can rewrite Eq. (2.77) as

$$\left[\nabla^2 + k^2\right]\mathbf{A}(\mathbf{r}) = -\mu_0\mu\mathbf{j}(\mathbf{r}), \tag{2.79}$$

which is the inhomogeneous Helmholtz equation. It holds independently for each component A_i of $\mathbf{A}$. A similar equation can be derived for the scalar potential ϕ,

$$\left[\nabla^2 + k^2\right]\phi(\mathbf{r}) = -\rho(\mathbf{r})/(\varepsilon_0\varepsilon). \tag{2.80}$$

Thus, we obtain four scalar Helmholtz equations of the form

$$\left[\nabla^2 + k^2\right]f(\mathbf{r}) = -g(\mathbf{r}). \tag{2.81}$$

To derive the *scalar* Green function $G_0(\mathbf{r},\mathbf{r}')$ for the Helmholtz operator we replace the source term $g(\mathbf{r})$ by a single point source $\delta(\mathbf{r}-\mathbf{r}')$ and obtain

$$\left[\nabla^2 + k^2\right]G_0(\mathbf{r},\mathbf{r}') = -\delta(\mathbf{r}-\mathbf{r}'). \tag{2.82}$$

The coordinate $\mathbf{r}$ denotes the location of the field point, i.e. the point at which the fields are to be evaluated, whereas the coordinate $\mathbf{r}'$ designates the location of the point source. Once we have determined G_0 we can state the particular solution for the vector potential in Eq. (2.79) as

$$\mathbf{A}(\mathbf{r}) = \mu_0\mu\int_V \mathbf{j}(\mathbf{r}')\,G_0(\mathbf{r},\mathbf{r}')\,\mathrm{d}V'. \tag{2.83}$$

A similar equation holds for the scalar potential. Both solutions require the knowledge of the Green function defined through Eq. (2.82). In free space, the only physical solution of this equation is [1]

$$G_0(\mathbf{r},\mathbf{r}') = \frac{\mathrm{e}^{\pm ik|\mathbf{r}-\mathbf{r}'|}}{4\pi\,|\mathbf{r}-\mathbf{r}'|}. \tag{2.84}$$

The solution with the plus sign denotes a spherical wave that propagates out of the origin, whereas the solution with the minus sign is a wave that converges towards the origin. In the following we retain only the outwards propagating wave. The scalar Green function can be introduced into Eq. (2.83) and the vector potential can be calculated by integrating over the source volume V. Thus, we are in a position to calculate the vector potential and scalar potential for any given current distribution $\mathbf{j}$ and charge distribution ρ. Notice that the Green function in Eq. (2.84) applies only to a homogeneous three-dimensional space. The Green function of a two-dimensional space or a half-space will have a different form.

So far we have reduced the treatment of Green functions to the potentials $\mathbf{A}$ and ϕ because this allows us to work with scalar equations. The formalism becomes more involved when we consider the electric and magnetic fields. The reason for this is that a source current in the x-direction leads to an electric and magnetic field with x-, y-, and z-components. This is different for the vector potential: a source current in x gives rise to a vector potential with just an x-component. Thus, in the case of the electric and magnetic fields we need a Green function that relates all components of the source to all components of the fields, or, in other words, the Green function must be a tensor. This type of Green function is called a *dyadic Green function* and has been introduced in the previous section. To determine the dyadic Green function we start with the wave equation for the electric field Eq. (2.33). In a homogeneous space it reads as

$$\nabla\times\nabla\times\mathbf{E}(\mathbf{r}) - k^2\mathbf{E}(\mathbf{r}) = \mathrm{i}\omega\mu_0\mu\,\mathbf{j}(\mathbf{r}). \tag{2.85}$$

We can define for each component of $\mathbf{j}$ a corresponding Green function. For example, for j_x we have

$$\nabla\times\nabla\times\mathbf{G}_x(\mathbf{r},\mathbf{r}') - k^2\mathbf{G}_x(\mathbf{r},\mathbf{r}') = \delta(\mathbf{r}-\mathbf{r}')\mathbf{n}_x, \tag{2.86}$$

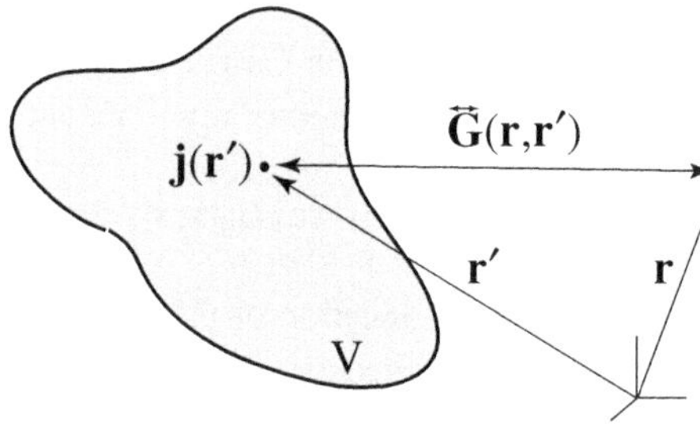

Fig. 2.3 Illustration of the dyadic Green function $\overleftrightarrow{\mathbf{G}}(\mathbf{r},\mathbf{r}')$. The Green function renders the electric field at the field point $\mathbf{r}$ due to a single point source $\mathbf{j}$ at the source point $\mathbf{r}'$. Since the field at $\mathbf{r}$ depends on the orientation of $\mathbf{j}$ the Green function must account for all possible orientations in the form of a tensor.

where $\mathbf{n}_x$ is the unit vector in the x-direction. A similar equation can be formulated for a point source in the y- and z-directions. In order to account for all orientations we write as the general definition of the dyadic Green function for the electric field [9]

$$\nabla \times \nabla \times \overleftrightarrow{\mathbf{G}}(\mathbf{r}, \mathbf{r}') - k^2 \overleftrightarrow{\mathbf{G}}(\mathbf{r}, \mathbf{r}') = \overleftrightarrow{\mathbf{I}} \delta(\mathbf{r} - \mathbf{r}'), \tag{2.87}$$

$\overleftrightarrow{\mathbf{I}}$ being the unit dyad (unit tensor). The first column of the tensor $\overleftrightarrow{\mathbf{G}}$ corresponds to the field due to a point source in the x-direction, the second column to the field due to a point source in the y-direction, and the third column is the field due to a point source in the z-direction. Thus a dyadic Green function is just a compact notation for three vectorial Green functions.

As before, we can view the source current in Eq. (2.85) as a superposition of point currents. Thus, if we know the Green function $\overleftrightarrow{\mathbf{G}}$ we can state a particular solution of Eq. (2.85) as

$$\mathbf{E}(\mathbf{r}) = \mathrm{i}\omega\mu\mu_0 \int_V \overleftrightarrow{\mathbf{G}}(\mathbf{r}, \mathbf{r}')\mathbf{j}(\mathbf{r}')\mathrm{d}V'. \tag{2.88}$$

However, this is a particular solution and we need to add any homogeneous solutions $\mathbf{E}_0$. Thus, the general solution turns out to be

$$\mathbf{E}(\mathbf{r}) = \mathbf{E}_0(\mathbf{r}) + \mathrm{i}\omega\mu_0\mu \int_V \overleftrightarrow{\mathbf{G}}(\mathbf{r}, \mathbf{r}')\,\mathbf{j}(\mathbf{r}')\mathrm{d}V' \qquad \mathbf{r} \notin V. \tag{2.89}$$

The corresponding magnetic field reads as

$$\mathbf{H}(\mathbf{r}) = \mathbf{H}_0(\mathbf{r}) + \int_V \left[\nabla \times \overleftrightarrow{\mathbf{G}}(\mathbf{r}, \mathbf{r}')\right] \mathbf{j}(\mathbf{r}')\mathrm{d}V' \qquad \mathbf{r} \notin V. \tag{2.90}$$

These equations are called *volume integral equations*. They are very important since they form the basis for various formalisms such as the method of moments, the Lippmann–Schwinger equation, and the coupled-dipole method. We have limited the validity of the volume integral equations to the space outside the source volume V in order to avoid the apparent singularity of $\overleftrightarrow{\mathbf{G}}$ at $\mathbf{r} = \mathbf{r}'$. This limitation will be relaxed in Chapter 16.

In order to solve Eqs. (2.89) and (2.90) for a given distribution of currents, we still need to determine the explicit form of $\overleftrightarrow{\mathbf{G}}$. Introducing the Lorenz gauge Eq. (2.78) into Eq. (2.75) leads to

$$\mathbf{E}(\mathbf{r}) = \mathrm{i}\omega \left[1 + \frac{1}{k^2} \nabla\nabla\cdot\right] \mathbf{A}(\mathbf{r}). \tag{2.91}$$

The first column vector of $\overleftrightarrow{\mathbf{G}}$, i.e. $\mathbf{G}_x$, defined in Eq. (2.86) is simply the electric field due to a point source current $\mathbf{j} = (\mathrm{i}\omega\mu_0)^{-1}\delta(\mathbf{r} - \mathbf{r}')\mathbf{n}_x$. The vector potential originating from this source current is, according to Eq. (2.83),

$$\mathbf{A}(\mathbf{r}) \;=\; (\mathrm{i}\omega)^{-1} G_0(\mathbf{r}, \mathbf{r}')\mathbf{n}_x. \tag{2.92}$$

Upon inserting this vector potential into Eq. (2.91) we find

$$\mathbf{G}_x(\mathbf{r}, \mathbf{r}') = \left[1 + \frac{1}{k^2}\nabla\nabla\cdot\right] G_0(\mathbf{r}, \mathbf{r}')\mathbf{n}_x, \tag{2.93}$$

with similar expressions for $\mathbf{G}_y$ and $\mathbf{G}_z$. The only thing remaining to be done is to tie the three solutions together to form a dyad. With the definition $\nabla \cdot [G_0 \overleftrightarrow{\mathbf{I}}] = \nabla G_0$ the dyadic Green function $\overleftrightarrow{\mathbf{G}}$ can be calculated from the scalar Green function G_0 in Eq. (2.84) as

$$\overleftrightarrow{\mathbf{G}}(\mathbf{r}, \mathbf{r}') = \left[\overleftrightarrow{\mathbf{I}} + \frac{1}{k^2} \nabla\nabla \right] G_0(\mathbf{r}, \mathbf{r}'). \tag{2.94}$$

2.12.3 Time-dependent Green functions

The time dependence in the wave equations can be separated and the resulting harmonic differential equation for the time behavior is easily solved. A monochromatic field can be represented in the form of Eq. (2.25) and any other time-dependent field can be generated by a Fourier transform (sum of monochromatic fields). However, for the study of ultrafast phenomena it is of advantage to retain the explicit time behavior. In this case we have to generalize the definition of $\mathbf{A}$ and ϕ as[5]

$$\mathbf{E}(\mathbf{r}, t) = -\frac{\partial}{\partial t}\mathbf{A}(\mathbf{r}, t) - \nabla\phi(\mathbf{r}, t), \tag{2.95}$$

$$\mathbf{H}(\mathbf{r}, t) = \frac{1}{\mu_0 \mu} \nabla \times \mathbf{A}(\mathbf{r}, t), \tag{2.96}$$

from which we find the time-dependent Helmholtz equation in the Lorenz gauge (cf. Eq. (2.79))

$$\left[\nabla^2 - \frac{n^2}{c^2}\frac{\partial^2}{\partial t^2} \right] \mathbf{A}(\mathbf{r}, t) = -\mu_0 \mu\, \mathbf{j}(\mathbf{r}, t). \tag{2.97}$$

A similar equation holds for the scalar potential ϕ. The definition of the *scalar* Green function is now generalized to

$$\left[\nabla^2 - \frac{n^2}{c^2}\frac{\partial^2}{\partial t^2} \right] G_0(\mathbf{r}, \mathbf{r}';\, t, t') = -\delta(\mathbf{r} - \mathbf{r}')\delta(t - t'). \tag{2.98}$$

The point source is now defined with respect to space and time. The solution for G_0 is [1]

$$G_0(\mathbf{r}, \mathbf{r}';\, t, t') = \frac{\delta\Big(t' - \big[t \mp (n/c)|\mathbf{r} - \mathbf{r}'|\big]\Big)}{4\pi\, |\mathbf{r} - \mathbf{r}'|}, \tag{2.99}$$

where the minus sign is associated with the response at a time t later than t'. Using G_0 we can construct the time-dependent dyadic Green function $\overleftrightarrow{\mathbf{G}}(\mathbf{r}, \mathbf{r}';\, t, t')$ as in the previous case. Since we shall mostly work with time-independent Green functions we avoid further details and refer the interested reader to specialized books on electrodynamics. Working with time-dependent Green functions accounts for arbitrary-time behavior but it is very difficult to incorporate dispersion. Time-dependent processes in dispersive media are more conveniently solved using Fourier transforms of monochromatic fields.

[5] We assume a non-dispersive medium, i.e. $\varepsilon(\omega) = \varepsilon$ and $\mu(\omega) = \mu$.

2.13 Reciprocity

The reciprocity theorem generally states that the source and detector of electromagnetic fields can be interchanged without affecting the physical situation. The derivation of the reciprocity theorem is formally the same as the derivation of Poynting's theorem in Section 2.11. For simplicity we restrict the discussion to purely monochromatic fields represented by complex amplitudes. Let us consider two spatially separate volumes V_1 and V_2 with the current densities $\mathbf{j}_1$ and $\mathbf{j}_2$, respectively. $\mathbf{j}_1$ creates the fields $\mathbf{E}_1$ and $\mathbf{H}_1$, and $\mathbf{j}_2$ gives rise to fields $\mathbf{E}_2$ and $\mathbf{H}_2$. We write Maxwell's curl equations separately for the two fields as

$$\begin{aligned}\nabla \times \mathbf{E}_1 &= \mathrm{i}\omega \mathbf{B}_1,\\ \nabla \times \mathbf{H}_1 &= -\mathrm{i}\omega \mathbf{D}_1 + \mathbf{j}_1,\\ \nabla \times \mathbf{E}_2 &= \mathrm{i}\omega \mathbf{B}_2,\\ \nabla \times \mathbf{H}_2 &= -\mathrm{i}\omega \mathbf{D}_2 + \mathbf{j}_2.\end{aligned}$$

We now multiply the first equation by $\mathbf{H}_2$, the second by $\mathbf{E}_2$, the third by $\mathbf{H}_1$, and the fourth by $\mathbf{E}_1$, and then subtract the sum of the latter two equations from the sum of the first two equations, which yields

$$\begin{aligned}&(\mathbf{H}_2 \cdot \nabla \times \mathbf{E}_1 - \mathbf{E}_1 \cdot \nabla \times \mathbf{H}_2) + (\mathbf{E}_2 \cdot \nabla \times \mathbf{H}_1 - \mathbf{H}_1 \cdot \nabla \times \mathbf{E}_2)\\ &\quad = \mathrm{i}\omega(\mathbf{H}_2 \cdot \mathbf{B}_1 - \mathbf{H}_1 \cdot \mathbf{B}_2) - \mathrm{i}\omega(\mathbf{E}_2 \cdot \mathbf{D}_1 - \mathbf{E}_1 \cdot \mathbf{D}_2) + (\mathbf{j}_1 \cdot \mathbf{E}_2 - \mathbf{j}_2 \cdot \mathbf{E}_1).\end{aligned} \tag{2.100}$$

The left-hand side is identical to $\nabla \cdot (\mathbf{E}_1 \times \mathbf{H}_2 - \mathbf{E}_2 \times \mathbf{H}_1)$. Furthermore, assuming linear constitutive relations the first two terms on the right-hand side cancel out and we arrive at

$$\nabla \cdot (\mathbf{E}_1 \times \mathbf{H}_2 - \mathbf{E}_2 \times \mathbf{H}_1) = \mathbf{j}_1 \cdot \mathbf{E}_2 - \mathbf{j}_2 \cdot \mathbf{E}_1, \tag{2.101}$$

which is the Lorentz reciprocity theorem with sources [10, 11].

We now integrate Eq. (2.101) over a spherical volume with large radius and assume that all sources and objects, such as scatterers, are finite in size, Then, after making use of Gauss's theorem and the fact that far-fields are transverse to the surface normal of the spherical volume, the term on the left-hand side in Eq. (2.101) vanishes and we obtain

$$\int_{V_1} \mathbf{j}_1 \cdot \mathbf{E}_2 \, \mathrm{d}V = \int_{V_2} \mathbf{j}_2 \cdot \mathbf{E}_1 \, \mathrm{d}V, \tag{2.102}$$

where we have reduced the integration volume to regions where the currents are non-zero. Equation (2.102) is of central importance and is widely used in antenna theory. For lossless media, the reciprocity theorem is equivalent to time reversibility. However, in dissipative media time-reversibility is lost whereas the reciprocity theorem remains valid [11]. At first sight, the expressions in Eq. (2.102) look similar to the right-hand side of Eq. (2.58). However, there are no complex conjugates in Eq. (2.102) and reciprocity is *not* just another statement of energy conservation.

Let us now express the fields $\mathbf{E}_1$ and $\mathbf{E}_2$ in Eq. (2.102) in terms of their source currents. This can be done using Eq. (2.88) by means of the Green dyadic $\overleftrightarrow{\mathbf{G}}$. The equality in Eq. (2.102) then leads to

$$\overleftrightarrow{\mathbf{G}}(\mathbf{r}_1, \mathbf{r}_2) = \overleftrightarrow{\mathbf{G}}(\mathbf{r}_2, \mathbf{r}_1). \tag{2.103}$$

Thus, reciprocity implies that the Green dyadic is symmetric and that it isn't affected by interchanging the source and the detector.

2.14 Evanescent fields

Evanescent fields play a central role in nano-optics. The word *evanescent* derives from the Latin word *evanescere* and has meanings like *vanishing from notice* or *imperceptible*. Evanescent fields can be described by plane waves of the form $\mathbf{E}\mathrm{e}^{\mathrm{i}(\mathbf{kr}-\omega t)}$. They are characterized by the fact that at least one component of the wavevector $\mathbf{k}$ describing the direction of propagation is imaginary. In the spatial direction defined by the imaginary component of $\mathbf{k}$ the wave does not propagate but rather decays exponentially. Evanescent fields are of major importance for the understanding of optical fields that are confined to subwavelength dimensions. This section discusses the basic properties of evanescent waves and introduces simple experimental arrangements for their creation and measurement.

Evanescent waves never occur in a homogeneous medium but are inevitably connected to the interaction of light with inhomogeneities [12]. The simplest case of an inhomogeneity is a plane interface. Let us consider a plane wave impinging on such a flat interface between two media characterized by optical constants ε_1, μ_1 and ε_2, μ_2. As discussed in Section 2.10.1, the presence of the interface will lead to a reflected wave and a refracted wave whose amplitudes and directions are described by Fresnel coefficients and by Snell's law, respectively.

To derive the evanescent wave generated by total internal reflection at the surface of a dielectric medium, we refer to the configuration shown in Fig. 2.2. We choose the x-axis to be in the plane of incidence. Using the symbols defined in Section 2.10.1, the complex transmitted field vector can be expressed as

$$\mathbf{E}_2 = \begin{bmatrix} -\mathrm{E}_1^{(\mathrm{p})} t^{\mathrm{p}}(k_x)\, k_{z_2}/k_2 \\ \mathrm{E}_1^{(\mathrm{s})} t^{\mathrm{s}}\,(k_x) \\ \mathrm{E}_1^{(\mathrm{p})} t^{\mathrm{p}}(k_x)\, k_x/k_2 \end{bmatrix} \mathrm{e}^{\mathrm{i}k_x x + \mathrm{i}k_{z_2} z}, \tag{2.104}$$

which can be expressed entirely in terms of the angle of incidence θ_1 using $k_x = k_1 \sin\theta_1$. Note that we suppressed the harmonic time factor $\exp(-\mathrm{i}\omega t)$. With this substitution the longitudinal wavenumbers can be written as (cf. Eq. (2.48))

$$k_{z_1} = k_1 \sqrt{1 - \sin^2\theta_1}, \qquad k_{z_2} = k_2 \sqrt{1 - \tilde{n}^2 \sin^2\theta_1}, \tag{2.105}$$

where we introduced the relative index of refraction

$$\tilde{n} = \frac{\sqrt{\varepsilon_1 \mu_1}}{\sqrt{\varepsilon_2 \mu_2}}. \tag{2.106}$$

For $\tilde{n} > 1$, with increasing θ_1 the argument of the square root in the expression for k_{z_2} gets smaller and smaller and eventually becomes negative. The critical angle θ_c can be defined by the condition

$$1 - \tilde{n}^2 \sin^2\theta_1 = 0, \tag{2.107}$$

which describes a refracted plane wave with zero wavevector component in the z-direction ($k_{z_2} = 0$). Consequently, the refracted plane wave travels parallel to the interface. Solving for θ_1 yields

$$\theta_c = \arcsin(1/\tilde{n}). \tag{2.108}$$

For a glass/air interface at optical frequencies, we have $\varepsilon_2 = 1$, $\varepsilon_1 = 2.25$, and $\mu_1 = \mu_2 = 1$, yielding a critical angle $\theta_c = 41.8°$.

For $\theta_1 > \theta_c$, k_{z_2} becomes imaginary. Expressing the transmitted field as a function of the angle of incidence θ_1 results in

$$\mathbf{E}_2 = \begin{bmatrix} -\mathrm{i}E_1^{(p)} t^p(\theta_1) \sqrt{\tilde{n}^2 \sin^2\theta_1 - 1} \\ E_1^{(s)} t^s(\theta_1) \\ E_1^{(p)} t^p(\theta_1)\, \tilde{n} \sin\theta_1 \end{bmatrix} e^{\mathrm{i}\sin(\theta_1)\, k_1 x} e^{-\gamma z}, \tag{2.109}$$

where the decay constant γ is defined by

$$\gamma = k_2 \sqrt{\tilde{n}^2 \sin^2\theta_1 - 1}\,. \tag{2.110}$$

This equation describes a field that propagates along the surface but decays exponentially into the medium of transmittance. Thus, a plane wave incident at an angle $\theta_1 > \theta_c$ creates an evanescent wave. Excitation of an evanescent wave with a plane wave at supercritical incidence ($\theta_1 > \theta_c$) is referred to as *total internal reflection* (TIR). For the glass/air interface considered above and an angle of incidence of $\theta_1 = 45°$, the decay constant is $\gamma = 2.22/\lambda$. This means that already at a distance of $\approx\lambda/2$ from the interface the time-averaged field is a factor of e smaller than it is at the interface. At a distance of $\approx 2\lambda$ the field becomes negligible. The larger the angle of incidence θ_1 the faster the decay will be. Note that the Fresnel coefficients depend on θ_1. For $\theta_1 > \theta_c$ they become complex numbers and, consequently, the phase of the reflected and transmitted wave is shifted relative to the incident wave. This phase shift is the origin of the so-called Goos–Hänchen shift. Furthermore, for p-polarized excitation, it results in elliptic polarization of the evanescent wave with the field vector rotating in the plane of incidence (see e.g. [13] and Problem 2.5).

Evanescent fields as described by Eq. (2.109) can be produced by directing a beam of light into a glass prism as sketched in Fig. 2.4(b). Experimental verification for the existence of this rapidly decaying field in the optical regime relies on approaching a transparent body to within $\lambda/2$ of the interface that supports the evanescent field. As shown in Fig. 2.5, this can be accomplished, for example, by using a sharp transparent fiber that converts the

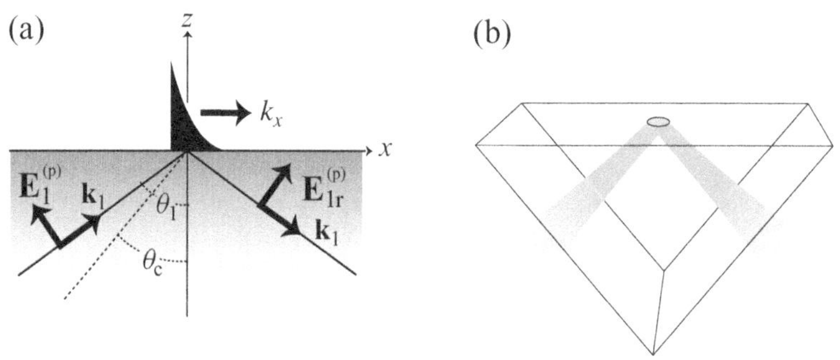

Fig. 2.4 Excitation of an evanescent wave by total internal reflection. (a) An evanescent wave is created in a medium if the plane wave is incident at an angle $\theta_1 > \theta_c$. (b) Actual experimental realization using a prism and a weakly focused Gaussian beam.

evanescent field at its tip into a guided mode propagating along the fiber [14]. This measurement technique is called *photon scanning tunneling microscopy* and will be discussed later in Chapter 5.

For p- and s-polarized evanescent waves, the intensity of the evanescent wave can be larger than that of the input beam. To see this we set $z = 0$ in Eq. (2.109) and we write for an s- and p-polarized plane wave separately the intensity ratio $|\mathbf{E}_2(z = 0)|^2/|\mathbf{E}_1(z = 0)|^2$. This ratio is equal to the absolute square of the Fresnel transmission coefficient $t^{p,s}$. These transmission coefficients are plotted in Fig. 2.6 for the example of a glass/air interface. For p- (s-)polarized light the transmitted evanescent intensity is up to a factor of 9 (4) larger than the incoming intensity. The maximum enhancement is found at the critical angle of TIR. The physical reason for this enhancement is a surface polarization that is induced by the incoming plane wave which is also represented by the boundary condition (2.43). A similar enhancement effect, but a much stronger one, can be obtained when the glass/air interface is covered by a thin layer of a noble metal. Here, so-called surface plasmon polaritons can be excited. We will discuss this and similar effects in more detail in Chapter 12.

2.14.1 Energy transport by evanescent waves

For non-absorbing media and for supercritical incidence, all the power of the incident wave is reflected. This effect is known as *total internal reflection* (*TIR*). One can predict that because no losses occur upon reflection at the interface there is no net energy transport into the medium of transmittance. In order to prove this fact we have to investigate the time-averaged energy flux across a plane parallel to the interface. This can be done by considering the z-component of the Poynting vector (cf. Eq. (2.59))

$$\langle S \rangle_z = \frac{1}{2} \,\mathrm{Re}\Big(E_x H_y^* - E_y H_x^*\Big), \qquad (2.111)$$

where all fields are evaluated in the upper medium, i.e. the medium of transmittance. Applying Maxwell's equation (2.26) to the special case of a plane or evanescent wave allows us to express the magnetic field in terms of the electric field as

$$\mathbf{H} = \sqrt{\frac{\varepsilon_0\varepsilon}{\mu_0\mu}}\left[\left(\frac{\mathbf{k}}{k}\right) \times \mathbf{E}\right]. \tag{2.112}$$

On introducing the expressions for the transmitted field components of **E** and **H** into Eq. (2.111), it is straightforward to prove that $\langle \mathbf{S} \rangle_z$ vanishes (Problem 2.4) and that there is no net energy transport in the direction *normal* to the interface.

On the other hand, when considering the energy transport along the interface ($\langle \mathbf{S} \rangle_x$), a non-zero result is found:

$$\langle \mathbf{S} \rangle_x = \frac{1}{2}\sqrt{\frac{\varepsilon_2\mu_2}{\varepsilon_1\mu_1}}\sin\theta_1\left(\left|t^{\mathrm{s}}\right|^2\left|\mathbf{E}_1^{(\mathrm{s})}\right|^2 + \left|t^{\mathrm{p}}\right|^2\left|\mathbf{E}_1^{(\mathrm{p})}\right|^2\right)\mathrm{e}^{-2\gamma z}. \tag{2.113}$$

Thus, an evanescent wave transports energy along the surface, in the direction of the transverse wavevector.

The absence of a net energy flow normal to the surface does not mean that there is no energy contained in an evanescent wave. For example, the local field distribution can be mapped out by using the fluorescence of a single molecule as a local probe.[6] The rate R at which the fluorophore emits photons when excited by the optical electric field is given by

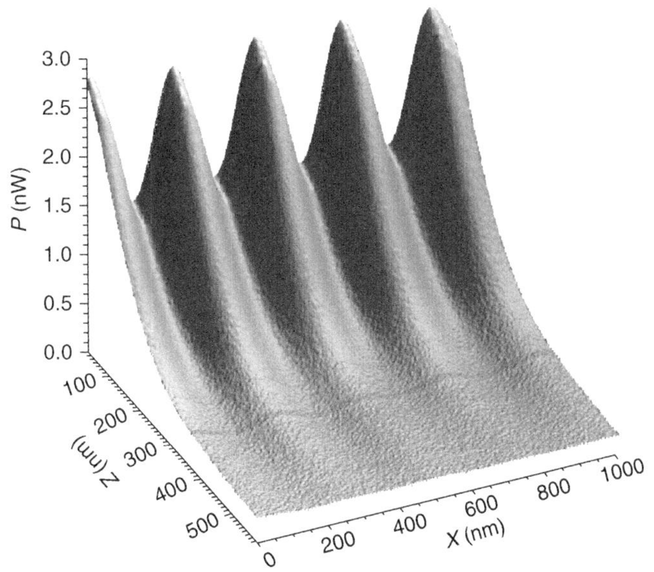

Fig. 2.5 Spatial modulation of the standing evanescent wave along the propagation direction of two interfering waves (x-axis) and the decay of the intensity in the z-direction. The ordinate represents the measured optical power. From [14].

[6] Excitation of fluorescence using evanescent waves is quite popular in biological imaging. Since only a thin slice of the sample is illuminated, background is drastically reduced. The technique is known as total internal reflection fluorescence (TIRF) microscopy [15].

$$R \sim |\mathbf{p} \cdot \mathbf{E}|^2 , \qquad (2.114)$$

where $\mathbf{p}$ is the absorption dipole moment of the molecule. As an example, for s-polarized fields the fluorescence rate of a molecule with a non-zero dipole component along the y-axis at a distance z above the interface will be

$$R(z) \sim \left| t^s \mathbf{E}_1^{(s)} \right|^2 e^{-2\gamma z}, \qquad (2.115)$$

decaying twice as fast as the electric field itself. Notice that a molecule can be excited even though the average Poynting vector vanishes.

2.14.2 Frustrated total internal reflection

Evanescent fields can be converted into propagating radiation if they interact with matter [12]. This phenomenon is among the most important effects in near-field optical microscopy since it explains how information about subwavelength structures is transported into the far-field. We shall discuss the physics behind this conversion by considering a very simple model. A plane interface will be used in order to create an evanescent wave by TIR as before. A second parallel plane interface is then advanced towards the first interface until the gap d is within the range of the typical decay length of the evanescent wave. A possible way to realize this experimentally is to close together two prisms with very flat or slightly curved surfaces as indicated in Fig. 2.7(b). The evanescent wave then interacts with the second interface and can be partly converted into propagating radiation. This situation is analogous to quantum mechanical tunneling through a potential barrier. The geometry of the problem is sketched in Fig. 2.7(a).

The fields are most conveniently expressed in terms of partial fields that are restricted to a single medium. The partial fields in media 1 and 2 are written as a superposition of incident and reflected waves, whereas for medium 3 there is only a transmitted wave. The propagation character of these waves, i.e. whether they are evanescent or propagating in either of the three media, can be determined from the magnitude of the longitudinal

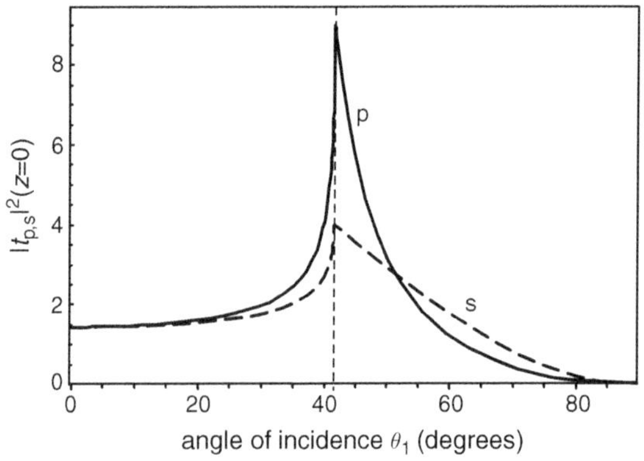

Fig. 2.6 Intensity enhancement on top of a glass surface irradiated by a plane wave with variable angle of incidence θ_1. For p- and s-polarized waves, the enhancement peaks at the critical angle $\theta_c = 41.8°$ marked by the dotted line.

wavenumber in each medium in analogy to Eq. (2.105). The longitudinal wavenumber in medium j reads

$$k_{j_z} = \sqrt{k_j^2 - k_\parallel^2} = k_j\sqrt{1 - (k_1/k_j)^2 \sin^2\theta_1}, \quad j \in \{1,2,3\}, \tag{2.116}$$

where $k_j = n_j k_0 = n_j(\omega/c)$ and $n_j = \sqrt{\varepsilon_j \mu_j}$. In the following a layered system with $n_2 < n_3 < n_1$ will be discussed, which includes the system sketched in Fig. 2.7. This leads to three regimes for the angle of incidence in which the transmitted intensity as a function of the gap width d shows different behavior.

1. For $\theta_1 < \arcsin(n_2/n_1)$ or $k_\parallel < n_2 k_0$, the field is entirely described by propagating plane waves. The intensity transmitted to a detector far away from the second interface (in the far-field) will not vary substantially with gapwidth, but will only show rather weak interference undulations.
2. For $\arcsin(n_2/n_1) < \theta_1 < \arcsin(n_3/n_1)$ or $n_2 k_0 < k_\parallel < n_3 k_0$ the partial field in medium 2 is evanescent, but in medium 3 it is propagating. At the second interface evanescent waves are converted into propagating waves. The intensity transmitted to a remote detector will decrease strongly with increasing gapwidth. This situation is referred to as *frustrated total internal reflection* (FTIR).
3. For $\theta_1 > \arcsin(n_3/n_1)$ or $k_\parallel > n_3 k_0$ the waves in layer 2 *and* in layer 3 are evanescent and no intensity will be transmitted to a remote detector in medium 3.

If we chose θ_1 such that case 2 is realized (FTIR), the transmitted intensity $I(d)$ will reflect the steep distance dependence of the evanescent wave(s) in medium 2. However, as shown in Fig. 2.8, $I(d)$ deviates from a purely exponential behavior because the field in medium 2 is a superposition of two evanescent waves of the form

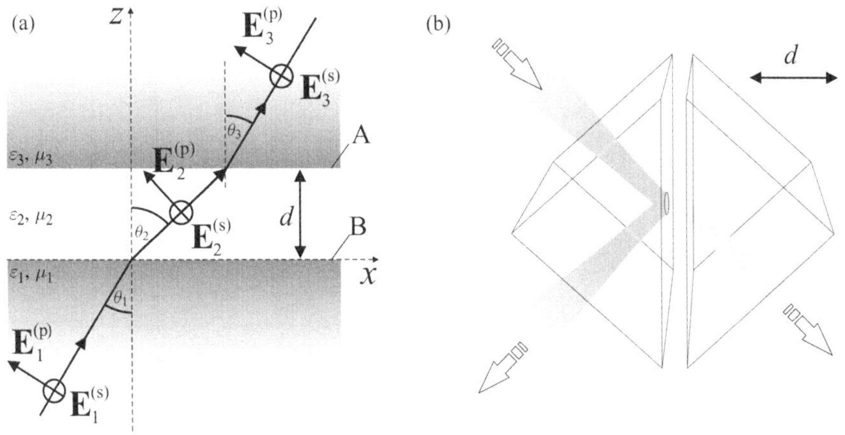

Fig. 2.7 Transmission of a plane wave through a system of two parallel interfaces. In frustrated total internal reflection (FTIR), the evanescent wave created at interface B is partly converted into a propagating wave by the interface A of a second medium. (a) Configuration and definition of parameters. A and B are interfaces between media 2 and 3 and 1 and 2, respectively. The reflected waves are omitted for clarity. (b) The experimental set-up used to observe frustrated total internal reflection.

$$c_1 e^{-\gamma z} + c_2 e^{+\gamma z}. \tag{2.117}$$

The second term originates from the reflection of the primary evanescent wave (first term) at the second interface and its magnitude (c_2) depends on the material properties. Figure 2.8 shows typical transmission curves for two different angles of incidence. This figure also shows that the decay measured in FTIR deviates from a simple exponential decay. In the next section, the importance of evanescent waves for the rigorous theoretical description of arbitrary optical fields near sources or material boundaries will be discussed.

2.15 Angular spectrum representation of optical fields

The angular spectrum representation is a mathematical technique to describe optical fields in homogeneous media. Optical fields are described as a superposition of plane waves and evanescent waves, both of which are physically intuitive solutions of Maxwell's equations. The angular spectrum representation has been found to be a very powerful method for the description of laser-beam propagation and light focusing. Furthermore, in the paraxial limit, the angular spectrum representation becomes identical with the framework of Fourier optics, which extends its importance even further. We will use the angular spectrum representation extensively in Chapters 3 and 4 to discuss strongly focused laser beams and limits of spatial resolution.

By the angular spectrum representation we understand the series expansion of an arbitrary field in terms of plane (and evanescent) waves with variable amplitudes and propagation directions. Assume we know the electric field $\mathbf{E}(\mathbf{r})$ at any point $\mathbf{r} = (x, y, z)$ in space.

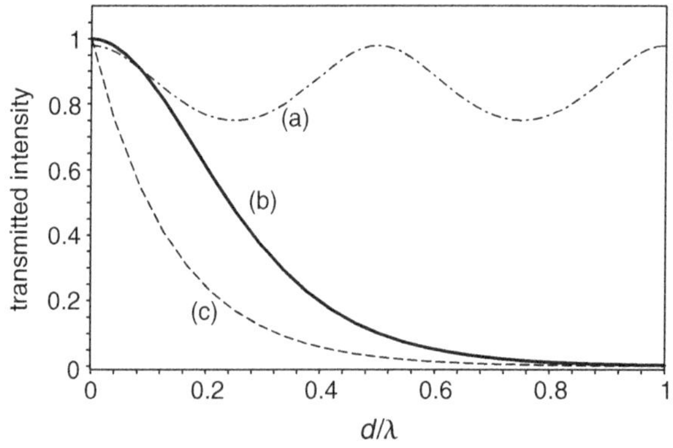

Fig. 2.8 Transmission in a system of three media with parallel interfaces as a function of the gap d between the two interfaces. A p-polarized plane wave excites the system. The material constants are $n_1 = 2, n_2 = 1$, and $n_3 = 1.51$. This leads to critical angles θ_c of 30° and 49.25°. For angles of incidence θ_1 between 0° and 30° the gap dependence shows interference-like behavior (here $\theta_1 = 0°$, dash–dotted line, curve (a)). For angles between 30° and 49.25° the transmission (monotonically) decreases with increasing gap width (here $\theta_1 = 35°$, full line, curve (b)). Curve (c) shows the intensity of the evanescent wave in the absence of the third medium.

For example, $\mathbf{E}(\mathbf{r})$ can be the solution of an optical scattering problem, as shown in Fig. 2.9, for which $\mathbf{E} = \mathbf{E}_{\text{inc}} + \mathbf{E}_{\text{scatt}}$. In the angular spectrum picture, we draw an arbitrary axis z and consider the field $\mathbf{E}$ in a plane z = constant transverse to the chosen axis. In this plane we can evaluate the two-dimensional Fourier transform of the field $\mathbf{E}$ as

$$\hat{\mathbf{E}}(k_x, k_y; z) = \frac{1}{4\pi^2} \iint_{-\infty}^{\infty} \mathbf{E}(x, y, z)\, \mathrm{e}^{-\mathrm{i}[k_x x + k_y y]}\, \mathrm{d}x\, \mathrm{d}y, \tag{2.118}$$

where x, y are the Cartesian transverse coordinates and k_x, k_y the corresponding spatial frequencies or reciprocal coordinates. Similarly, the inverse Fourier transform reads as

$$\mathbf{E}(x, y, z) = \iint_{-\infty}^{\infty} \hat{\mathbf{E}}(k_x, k_y; z)\, \mathrm{e}^{\mathrm{i}[k_x x + k_y y]}\, \mathrm{d}k_x\, \mathrm{d}k_y. \tag{2.119}$$

Notice that in the notation of Eqs. (2.118) and (2.119) the field $\mathbf{E} = (E_x, E_y, E_z)$ and its Fourier transform $\hat{\mathbf{E}} = (\hat{E}_x, \hat{E}_y, \hat{E}_z)$ represent vectors. Thus, the Fourier integrals hold separately for each vector component.

So far we have imposed no requirements on the field $\mathbf{E}$, but we will assume that in the transverse plane the medium is homogeneous, isotropic, linear and source-free. Then, a time-harmonic, optical field with angular frequency ω has to satisfy the vector Helmholtz equation

$$(\nabla^2 + k^2)\mathbf{E}(\mathbf{r}) = 0, \tag{2.120}$$

where k is determined by $k = (\omega/c)n$ and $n = \sqrt{\mu\varepsilon}$ is the index of refraction. In order to get the time-dependent field $\mathbf{E}(\mathbf{r}, t)$ we use the convention

$$\mathbf{E}(\mathbf{r}, t) = \text{Re}\{\mathbf{E}(\mathbf{r})\mathrm{e}^{-\mathrm{i}\omega t}\}. \tag{2.121}$$

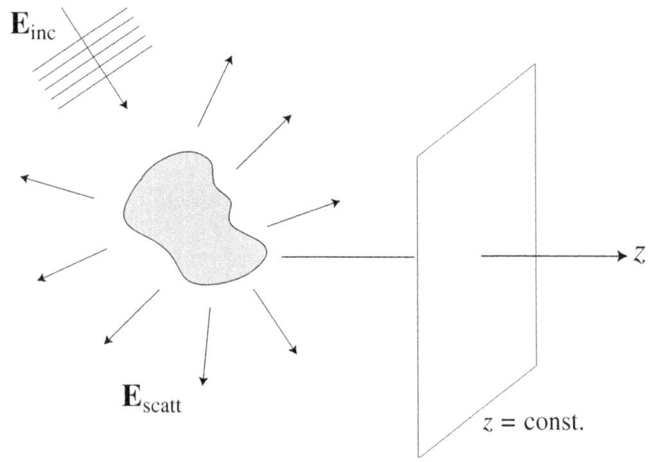

Fig. 2.9 In the angular spectrum representation the fields are evaluated in planes (z = constant) perpendicular to an arbitrarily chosen axis z.

On inserting the Fourier representation of $\mathbf{E}(\mathbf{r})$ (Eq. (2.119)) into the Helmholtz equation and defining

$$k_z \equiv \sqrt{(k^2 - k_x^2 - k_y^2)} \qquad \text{with} \quad \mathrm{Im}\{k_z\} \geq 0 \tag{2.122}$$

we find that the Fourier spectrum $\hat{\mathbf{E}}$ evolves along the z-axis as

$$\hat{\mathbf{E}}(k_x, k_y; z) = \hat{\mathbf{E}}(k_x, k_y; 0)\, \mathrm{e}^{\pm \mathrm{i} k_z z}. \tag{2.123}$$

The $\pm$ sign specifies that we have two solutions that need to be superimposed: the $+$ sign refers to a wave propagating into the half-space $z > 0$ whereas the $-$ sign denotes a wave propagating into $z < 0$. Equation (2.123) tells us that the Fourier spectrum of $\mathbf{E}$ in an arbitrary *image plane* located at $z =$ constant can be calculated by multiplying the spectrum in the *object plane* at $z = 0$ by the factor $\exp(\pm \mathrm{i}\, k_z z)$. This factor is called the *propagator* in reciprocal space. In Eq. (2.122) we defined that the square root leading to k_z renders a result with positive imaginary part. This ensures that the solutions remain finite for $z \to \pm\infty$. On inserting the result of Eq. (2.123) into Eq. (2.119) we finally find for arbitrary z

$$\mathbf{E}(x, y, z) = \iint_{-\infty}^{\infty} \hat{\mathbf{E}}(k_x, k_y; 0)\, \mathrm{e}^{\mathrm{i}\,[k_x x + k_y y \pm k_z z]}\, \mathrm{d}k_x\, \mathrm{d}k_y, \tag{2.124}$$

which is known as the *angular spectrum representation*. In a similar way, we can also represent the magnetic field $\mathbf{H}$ by an angular spectrum as

$$\mathbf{H}(x, y, z) = \iint_{-\infty}^{\infty} \hat{\mathbf{H}}(k_x, k_y; 0)\, \mathrm{e}^{\mathrm{i}\,[k_x x + k_y y \pm k_z z]}\, \mathrm{d}k_x\, \mathrm{d}k_y. \tag{2.125}$$

By using Maxwell's equation $\mathbf{H} = (\mathrm{i}\omega\mu\mu_0)^{-1}(\nabla \times \mathbf{E})$ we find the following relationship between the Fourier spectra $\hat{\mathbf{E}}$ and $\hat{\mathbf{H}}$:

$$\begin{aligned} \hat{H}_x &= Z_{\mu\varepsilon}^{-1}[(k_y/k)\hat{E}_z - (k_z/k)\hat{E}_y], \\ \hat{H}_y &= Z_{\mu\varepsilon}^{-1}[(k_z/k)\hat{E}_x - (k_x/k)\hat{E}_z], \\ \hat{H}_z &= Z_{\mu\varepsilon}^{-1}[(k_x/k)\hat{E}_y - (k_y/k)\hat{E}_x], \end{aligned} \tag{2.126}$$

where $Z_{\mu\varepsilon} = \sqrt{(\mu_0\mu)/(\varepsilon_0\varepsilon)}$ is the wave impedance of the medium. Although the angular spectra of $\mathbf{E}$ and $\mathbf{H}$ satisfy the Helmholtz equation they are not yet rigorous solutions of Maxwell's equations. We still have to require that the fields are divergence-free, i.e. $\nabla \cdot \mathbf{E} = 0$ and $\nabla \cdot \mathbf{H} = 0$. These conditions restrict the k-vector to directions perpendicular to the spectral amplitudes ($\mathbf{k} \cdot \hat{\mathbf{E}} = \mathbf{k} \cdot \hat{\mathbf{H}} = 0$).

For the case of a purely dielectric medium with no losses the index of refraction n is a real and positive quantity. The wavenumber k_z is then either real or imaginary and turns the factor $\exp(\pm \mathrm{i}\, k_z z)$ into an oscillatory or exponentially decaying function. For a certain (k_x, k_y) pair we then find two different characteristic solutions:

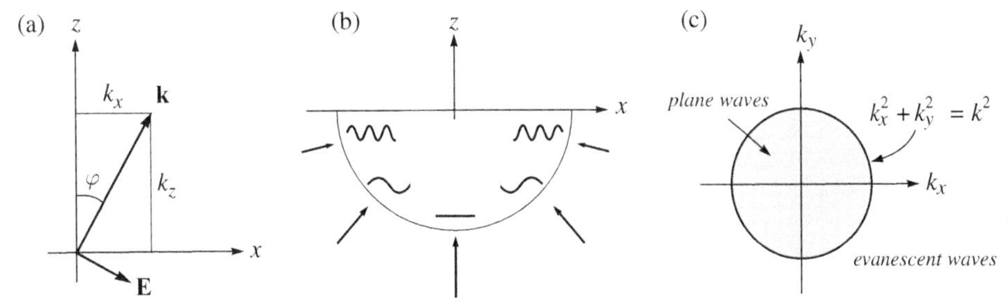

Fig. 2.10 (a) Representation of a plane wave propagating at an angle φ to the z axis. (b) Illustration of the transverse spatial frequencies of plane waves incident from different angles. The transverse wavenumber $(k_x^2 + k_y^2)^{1/2}$ depends on the angle of incidence and is limited to the interval $[0 \ldots k]$. (c) The transverse wavenumbers k_x and k_y of plane waves are restricted to a circular area with radius k. Evanescent waves fill the space outside the circle.

$$
\begin{aligned}
\text{Plane waves:} \quad & e^{i[k_x x + k_y y]} e^{\pm i|k_z|z}, \quad & k_x^2 + k_y^2 \leq k^2, \\
\text{Evanescent waves:} \quad & e^{i[k_x x + k_y y]} e^{-|k_z||z|}, \quad & k_x^2 + k_y^2 > k^2.
\end{aligned}
\tag{2.127}
$$

Hence, we find that the angular spectrum is indeed a superposition of *plane waves* and *evanescent waves*. Plane waves are oscillating functions in z and are restricted by the condition $k_x^2 + k_y^2 \leq k^2$. On the other hand, for $k_x^2 + k_y^2 > k^2$ we encounter evanescent waves with an exponential decay along the z-axis. Figure 2.10 shows that the larger the angle between the **k**-vector and the z-axis is, the larger the oscillation frequency in the transverse plane will be. A plane wave propagating in the direction of z has no oscillation frequency in the transverse plane $(k_x^2 + k_y^2 = 0)$, whereas, in the other limit, a plane wave propagating at right angles to z exhibits the highest spatial oscillation frequency in the transverse plane $(k_x^2 + k_y^2 = k^2)$. Even higher spatial frequencies are covered by evanescent waves. In principle, an infinite bandwidth of spatial frequencies can be achieved. However, the higher the spatial frequencies of an evanescent wave are, the faster the field decay along the z-axis will be. Therefore, practical limitations make the bandwidth finite.

2.15.1 Angular spectrum representation of the dipole field

Strongly localized sources such as dipoles are most conveniently described in a spherical coordinate system. The corresponding solutions of the wave equation are called multipoles. In order to couple these solutions with the angular spectrum picture we need to express the localized sources in terms of plane waves and evanescent waves. Let us start with the vector potential **A** of an oscillating dipole with its axis aligned along an arbitrary z-axis. The vector potential can be expressed as a one-component vector field as (cf. Eq. (2.92))

$$
\mathbf{A}(x, y, z) = A(x, y, z)\mathbf{n}_z = \frac{-ikZ_{\mu\varepsilon}}{4\pi} \frac{e^{ik\sqrt{x^2 + y^2 + z^2}}}{\sqrt{x^2 + y^2 + z^2}} \mathbf{n}_z. \tag{2.128}
$$

Besides a constant factor, the expression on the right-hand side corresponds to the scalar Green function (2.84). According to Eqs. (2.76) and (2.91) the electric and magnetic fields are obtained from **A** as

$$\mathbf{E}(x,y,z) = \mathrm{i}\omega\left(1+\frac{1}{k^2}\nabla\nabla\cdot\right)\mathbf{A}(x,y,z), \tag{2.129}$$

$$\mathbf{H}(x,y,z) = \frac{1}{\mu_0\mu}\nabla\times\mathbf{A}(x,y,z). \tag{2.130}$$

Thus, the electromagnetic field of the dipole can be constructed from the function $\exp(\mathrm{i}kr)/r$, where $r = (x^2+y^2+z^2)^{1/2}$ is the radial distance from the dipole's origin. To find an angular spectrum representation of the dipole's electric and magnetic field we need first to find the angular spectrum of the function $\exp(\mathrm{i}kr)/r$. This is not a trivial task because the function $\exp(\mathrm{i}kr)/r$ is singular at $r = 0$ and therefore not divergence-free at its origin. The homogeneous Helmholtz equation is therefore not valid in the present case. Nevertheless, using complex contour integration it is possible to derive an angular spectrum representation of the function $\exp(ikr)/r$. Since the derivation can be found in other textbooks [3] we state here only the result, which is

$$\frac{\mathrm{e}^{\mathrm{i}k\sqrt{x^2+y^2+z^2}}}{\sqrt{x^2+y^2+z^2}} = \frac{\mathrm{i}}{2\pi}\iint\limits_{-\infty}^{\infty}\frac{\mathrm{e}^{\mathrm{i}k_x x+\mathrm{i}k_y y+\mathrm{i}k_z|z|}}{k_z}\,\mathrm{d}k_x\,\mathrm{d}k_y. \tag{2.131}$$

We have to require that the real and imaginary parts of k_z stay positive for all values of k_x and k_y in the integration. The result (2.131) is known as the Weyl identity [16]. In Chapter 10 we shall use the Weyl identity to calculate dipole emission near planar interfaces.

Problems

2.1 Derive the dyadic Green function $\overleftrightarrow{\mathbf{G}}$ by substituting the scalar Green function G_0 into Eq. (2.94). Discuss the distance dependence $|\mathbf{r}-\mathbf{r}'|$.

2.2 Consider an interface between two media 1 and 2 with dielectric constants $\varepsilon_1 = 2.25$ and $\varepsilon_2 = 1$, respectively. The magnetic permeabilities are equal to unity. A p-polarized plane wave with wavelength $\lambda = 532$ nm is incident from medium 1 at an angle of incidence of θ_1. Express the Fresnel reflection coefficient in terms of amplitude A and phase Φ. Plot A and Φ as functions of θ_1. What are the consequences for the reflected wave?

2.3 Consider the refraction of a plane wave at a plane interface and derive Snell's law by using the invariance of the transverse wavevector $\mathbf{k}_{\parallel}$.

2.4 Show that the z-component of the time-averaged Poynting vector $\langle\mathbf{S}\rangle_z$ vanishes for an evanescent field propagating in the x-direction.

2.5 Analyze the polarization state of an evanescent field propagating in the x-direction created by total internal reflection of a p-polarized plane wave. Calculate the time-dependent electric field $\mathbf{E}_2(x,t) = (E_{2,x}(x,t), 0, E_{2,z}(x,t))$ just on top of the interface

($z = 0$). For a fixed position x, the electric field vector $\mathbf{E}_2$ defines a curve in the (x, z) plane as the time runs from 0 to λ/c. Determine and plot the shapes of these curves as a function of the position x. For numerical values choose $\theta_1 = 60°$ and $\tilde{n} = 1.5$.

2.6 Calculate the transmitted intensity for a system of two glass half-spaces ($n = 1.5$) separated by an air gap (d) and as a function of the angle of incidence θ_1. Determine the transmission function for s-polarized excitation. Normalize the transmission function with the value obtained for $\theta_1 = 0°$. Repeat for p-polarized excitation.

2.7 Derive Eq. (2.123) by inserting the inverse Fourier transform in Eq. (2.119) into the Helmholtz equation (2.120). Assume that the Fourier spectrum is known in the plane $z=0$.

2.8 Using the Weyl identity (2.131), derive the spatial spectrum $\hat{E}(k_x, k_y; z)$ of an electric dipole at $\mathbf{r_0} = (0, 0, z_0)$ with dipole moment $\mathbf{p} = (p, 0, 0)$. Consider the asymptotic limit $z \to \infty$ and solve for the electric field $\mathbf{E}$.

2.9 Apply Eq. (2.67) to a small metallic particle described by a free-electron gas. For which frequency is the energy density highest? How do losses scale with frequency? When is the ratio of energy density to energy loss smallest?

References

[1] J. D. Jackson, *Classical Electrodynamics*, 2nd edn. New York: Wiley (1975).

[2] J. A. Stratton, *Electromagnetic Theory*. New York: McGraw-Hill (1941).

[3] L. Mandel and E. Wolf, *Optical Coherence and Quantum Optics*. New York: Cambridge University Press (1995).

[4] M. Born and E. Wolf, *Principles of Optics*, 7th edn. New York: Cambridge University Press (1999).

[5] F. S. S. Rosa, D. A. R. Dalvit, and P. W. Milonni, "Electromagnetic energy, absorption, and Casimir forces: Uniform dielectric media in thermal equilibrium," *Phys. Rev. A* **81**, 033812 (2010); "Electromagnetic energy, absorption, and Casimir forces. II. Inhomogeneous dielectric media," *Phys. Rev. A* **84**, 053813 (2011).

[6] L. D. Landau, E. M. Lifshitz, and L. P. Pitaevskii, *Electrodynamics of Continuous Media*, 2nd edn. Amsterdam: Elsevier (1984).

[7] A. D. Yaghjian, "Electric dyadic Green's functions in the source region," *Proc. IEEE* **68**, 248–263 (1980).

[8] J. V. Bladel, "Some remarks on Green's dyadic for infinite space," *IRE Trans. Antennas Propag.* **9**, 563–566 (1961).

[9] C. T. Tai, *Dyadic Green's Functions in Electromagnetic Theory*, 2nd edn. New York: IEEE Press (1993).

[10] H. A. Lorentz, *Versl. Gewone Vergad. Afd. Natuurkd. Koninkl. Ned. Akad. Wetenschap* **4**, 176–188 (1896); H. A. Lorentz, "The theorem of Poynting concerning the energy in the electromagnetic field and two general propositions concerning the propagation of light," in *Collected Papers*, vol. III. Den Haag: Martinus Nijhoff, pp. 1–11 (1936).

[11] R. Carminati, M. Nieto-Vesperinas, and J.-J. Greffet, "Reciprocity of evanescent electromagnetic waves," *J. Opt. Soc. Am. A* **15**, 706–712 (1998).

[12] E. Wolf and M. Nieto-Vesperinas, "Analyticity of the angular spectrum amplitude of scattered fields and some of its consequences," *J. Opt. Soc. Am. A* **2**, 886–889 (1985).

[13] S. Sund, J. Swanson, and D. Axelrod, "Cell membrane orientation visualized by polarized total internal reflection fluorescence," *Biophys. J.* **77**, 2266–2283 (1999).

[14] A. Meixner, M. Bopp, and G. Tarrach, "Direct measurement of standing evanescent waves with a photon scanning tunneling microscope," *Appl. Opt.* **33**, 7995–8000 (1994).

[15] D. Axelrod, N. Thompson, and T. Burghardt, "Total internal reflection fluorescent microscopy," *J. Microsc.* **129**, 19–28 (1983).

[16] H. Weyl, "Ausbreitung elektromagnetischer Wellen über einem ebenen Leiter," *Ann. Phys.* **60**, 481–500 (1919).

3 Propagation and focusing of optical fields

In this chapter we use the angular spectrum representation outlined in Section 2.15 to discuss field distributions in strongly focused laser beams. The same formalism is applied to understand how the fields in a given reference plane are mapped to the far-field. The theory is relevant for the understanding of confocal and multiphoton microscopy, single-emitter experiments, and the understanding of resolution limits. It also defines the framework for different topics to be discussed in later chapters.

3.1 Field propagators

In Section 2.15 we have established that, in a homogeneous space, the spatial spectrum $\hat{\mathbf{E}}$ of an optical field $\mathbf{E}$ in a plane z = constant (the image plane) is uniquely defined by the spatial spectrum in a different plane $z = 0$ (the object plane) according to the linear relationship

$$\hat{\mathbf{E}}(k_x, k_y; z) = \hat{H}(k_x, k_y; z)\, \hat{\mathbf{E}}(k_x, k_y; 0), \tag{3.1}$$

where $\hat{H}$ is the so-called *propagator in reciprocal space*

$$\hat{H}(k_x, k_y; z) = \mathrm{e}^{\pm \mathrm{i} k_z z}, \tag{3.2}$$

which is also referred to as the *optical transfer function* (OTF) in free space. Remember that the longitudinal wavenumber is a function of the transverse wavenumber, i.e. $k_z = [k^2 - (k_x^2 + k_y^2)]^{1/2}$, where $k = nk_0 = n\omega/c = n2\pi/\lambda$. The $\pm$ sign indicates that the field can propagate in the positive- and/or negative-z direction. Equation (3.1) can be interpreted in terms of linear response theory: $\hat{\mathbf{E}}(k_x, k_y; 0)$ is the input, $\hat{H}$ is a filter function, and $\hat{\mathbf{E}}(k_x, k_y; z)$ is the output. The filter function describes the propagation of an arbitrary spectrum through space. $\hat{H}$ can also be regarded as the response function because it describes the field at z due to a point source at $z = 0$. In this sense, it is directly related to the Green function $\overleftrightarrow{\mathbf{G}}$.

The filter $\hat{H}$ is an oscillating function for $(k_x^2 + k_y^2) < k^2$ and an exponentially decreasing function for $(k_x^2 + k_y^2) > k^2$. Thus, if the image plane is sufficiently separated from the object plane, the contribution of the decaying parts (evanescent waves) is zero and the integration can be reduced to the circular area $(k_x^2 + k_y^2) \leq k^2$. In other words, the image at z is a *low-pass-filtered* representation of the original field at $z = 0$. The spatial frequencies $(k_x^2 + k_y^2) > k^2$ of the original field are filtered out during propagation and the information on high spatial variations gets lost. Hence, there is always a loss of information on propagating from near- to far-field and only structures with lateral dimensions larger than

$$\Delta x \approx \frac{1}{k} = \frac{\lambda}{2\pi n} \tag{3.3}$$

can be imaged with sufficient accuracy. Here, n is the index of refraction. This equation is qualitative and we will provide a more detailed discussion in Chapter 4. In general, higher resolution can be obtained by using a higher index of refraction of the embodying system (substrate, lenses, etc.) or shorter wavelengths. Theoretically, resolutions down to a few nanometers can be achieved by using far-ultraviolet radiation or X-rays. The central idea of *near-field optics* is to increase the bandwidth of spatial frequencies by retaining the evanescent components of the source fields.

Let us now determine how the fields themselves evolve. For this purpose we denote the transverse coordinates in the object plane at $z = 0$ as (x', y') and those in the image plane at $z =$ constant as (x, y). The fields in the image plane are described by the angular spectrum (2.124). We just have to express the Fourier spectrum $\hat{\mathbf{E}}(k_x, k_y; 0)$ in terms of the fields in the object plane. Similarly to Eq. (2.118), this Fourier spectrum can be represented as

$$\hat{\mathbf{E}}(k_x, k_y; 0) = \frac{1}{4\pi^2} \iint_{-\infty}^{\infty} \mathbf{E}(x', y', 0)\, \mathrm{e}^{-\mathrm{i}[k_x x' + k_y y']}\, \mathrm{d}x'\, \mathrm{d}y'. \tag{3.4}$$

After insertion into Eq. (2.124) we find the following expression for the field $\mathbf{E}$ in the image plane $z =$ constant:

$$\begin{aligned} \mathbf{E}(x, y, z) &= \frac{1}{4\pi^2} \iint_{-\infty}^{\infty} \mathbf{E}(x', y'; 0) \iint_{-\infty}^{\infty} \mathrm{e}^{\mathrm{i}[k_x(x-x') + k_y(y-y') \pm k_z z]}\, \mathrm{d}x'\, \mathrm{d}y'\, \mathrm{d}k_x\, \mathrm{d}k_y \\ &= \mathbf{E}(x, y; 0) * H(x, y; z). \end{aligned} \tag{3.5}$$

This equation describes an invariant filter with the following impulse response (*propagator in direct space*)

$$H(x, y; z) = \iint_{-\infty}^{\infty} \mathrm{e}^{\mathrm{i}[k_x x + k_y y \pm k_z z]}\, \mathrm{d}k_x\, \mathrm{d}k_y. \tag{3.6}$$

H is simply the inverse Fourier transform of the propagator in reciprocal space $\hat{H}$ (3.2). The field at $z =$ constant is represented by the convolution of H with the field at $z = 0$.

3.2 Paraxial approximation of optical fields

In many optical problems the light fields propagate along a certain direction z and spread out only slowly in the transverse direction. Examples include laser-beam propagation and optical waveguide applications. In these examples the wavevectors $\mathbf{k} = (k_x, k_y, k_z)$ in the angular spectrum representation are almost parallel to the z-axis and the transverse wavenumbers (k_x, k_y) are small compared with k. We can then expand the square root of Eq. (2.122) in a series as

$$k_z = k\sqrt{1 - (k_x^2 + k_y^2)/k^2} \approx k - \frac{k_x^2 + k_y^2}{2k}. \tag{3.7}$$

This approximation is called the paraxial approximation and it considerably simplifies the analytical integration of the Fourier integrals. In the following we shall apply the paraxial approximation to find a description for weakly focused laser beams.

3.2.1 Gaussian laser beams

We consider a fundamental laser beam with a linearly polarized, Gaussian field distribution in the beam waist

$$\mathbf{E}(x', y', 0) = \mathbf{E}_0 \mathrm{e}^{-\frac{x'^2 + y'^2}{w_0^2}}, \tag{3.8}$$

where $\mathbf{E}_0$ is a constant-field vector in the transverse (x, y) plane. We have chosen $z = 0$ at the beam waist. The parameter w_0 denotes the beam-waist radius. We can calculate the spatial Fourier spectrum at $z = 0$ as[1]

$$\begin{aligned}\hat{\mathbf{E}}(k_x, k_y; 0) &= \frac{1}{4\pi^2}\iint_{-\infty}^{\infty} \mathbf{E}_0 \mathrm{e}^{-\frac{x'^2 + y'^2}{w_0^2}} \mathrm{e}^{-\mathrm{i}[k_x x' + k_y y']}\,\mathrm{d}x'\,\mathrm{d}y' \\ &= \mathbf{E}_0 \frac{w_0^2}{4\pi} \mathrm{e}^{-(k_x^2 + k_y^2)\frac{w_0^2}{4}},\end{aligned} \tag{3.9}$$

which is again a Gaussian function. We now insert this spectrum into the angular spectrum representation Eq. (2.124) and replace k_z by its paraxial expression in Eq. (3.7)

$$\mathbf{E}(x, y, z) = \mathbf{E}_0 \frac{w_0^2}{4\pi} \mathrm{e}^{\mathrm{i}kz} \iint_{-\infty}^{\infty} \mathrm{e}^{-(k_x^2 + k_y^2)\left(\frac{w_0^2}{4} + \frac{\mathrm{i}z}{2k}\right)} \mathrm{e}^{\mathrm{i}[k_x x + k_y y]}\,\mathrm{d}k_x\,\mathrm{d}k_y. \tag{3.10}$$

This equation can be integrated and gives as a result the familiar paraxial representation of a Gaussian beam

$$\mathbf{E}(x, y, z) = \frac{\mathbf{E}_0\,\mathrm{e}^{\mathrm{i}kz}}{1 + 2\mathrm{i}z/(kw_0^2)} \mathrm{e}^{-\frac{(x^2 + y^2)}{w_0^2}\frac{1}{1 + 2\mathrm{i}z/(kw_0^2)}}. \tag{3.11}$$

[1] We have $\int_{-\infty}^{\infty} \exp(-ax^2 + \mathrm{i}bx)\mathrm{d}x = \sqrt{\pi/a}\,\exp[-b^2/(4a)]$ and $\int_{-\infty}^{\infty} x\exp(-ax^2 + \mathrm{i}bx)\mathrm{d}x = \mathrm{i}b\sqrt{\pi}\,\exp[-b^2/(4a)]/(2a^{3/2})$.

To get a better feeling for a paraxial Gaussian beam we set $\rho^2 = x^2 + y^2$, define a new parameter z_0 as

$$z_0 = \frac{k w_0^2}{2}, \tag{3.12}$$

and rewrite Eq. (3.11) as

$$\mathbf{E}(\rho, z) = \mathbf{E}_0 \frac{w_0}{w(z)}\, \mathrm{e}^{-\frac{\rho^2}{w^2(z)}}\, \mathrm{e}^{\mathrm{i}\,[kz - \eta(z) + k\rho^2/(2R(z))]} \tag{3.13}$$

with the following symbols:

$$\begin{aligned} w(z) &= w_0 (1 + z^2/z_0^2)^{1/2} && \text{beam radius,} \\ R(z) &= z(1 + z_0^2/z^2) && \text{wavefront radius,} \\ \eta(z) &= \arctan(z/z_0) && \text{phase correction.} \end{aligned} \tag{3.14}$$

The transverse size of the beam is usually defined by the value of $\rho = \sqrt{x^2 + y^2}$ for which the electric field amplitude has decreased to a value of $1/\mathrm{e}$ of its center value:

$$|\mathbf{E}(x, y, z)| / |\mathbf{E}(0, 0, z)| = 1/\mathrm{e}. \tag{3.15}$$

It can be shown that the surface defined by this equation is a hyperboloid whose asymptotes enclose an angle

$$\theta = \frac{2}{k w_0} \tag{3.16}$$

with the z-axis. From this equation we can directly find the correspondence between the numerical aperture ($\mathrm{NA} = n \sin\theta$) and the beam angle as $\mathrm{NA} \approx 2n/(kw_0)$. Here we used the fact that, in the paraxial approximation, θ is restricted to small beam angles. Another property of the paraxial Gaussian beam is that, close to the focus, the beam stays roughly collimated over a distance $2z_0$. z_0 is called the *Rayleigh range* and denotes the distance from the beam waist to where the beam radius has increased by a factor of $\sqrt{2}$. It is important to notice that along the z-axis ($\rho = 0$) the phases of the beam deviate from those of a plane wave. If at $z \to -\infty$ the beam was in phase with a reference plane wave, then at $z \to +\infty$ the beam will be exactly out of phase with the reference wave. This phase

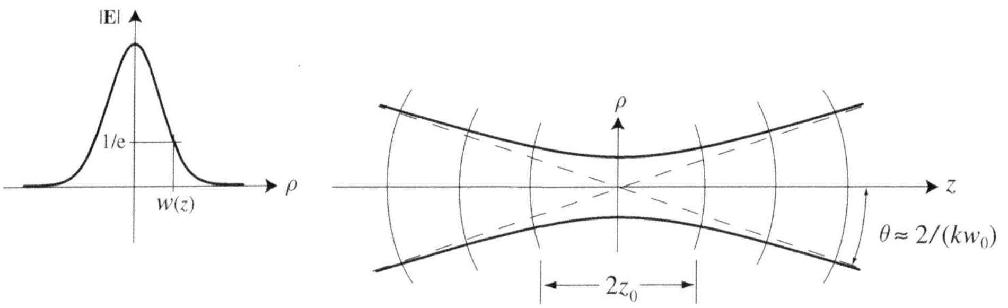

Fig. 3.1 Illustration and main characteristics of a paraxial Gaussian beam. The beam has a Gaussian field distribution in the transverse plane. The surfaces of constant field strength form a hyperboloid along the z-axis.

shift is called the *Gouy phase shift* and has practical implications in nonlinear confocal microscopy [1]. The 180° phase change happens gradually as the beam propagates through its focus. The phase variation is described by the factor $\eta(z)$ in Eq. (3.14). The tighter the focus the faster the phase variation will be.

A qualitative picture of a paraxial Gaussian beam and some of its characteristics are shown in Fig. 3.1 and more detailed descriptions can be found in other textbooks [2, 3]. It is important to notice that, once the paraxial approximation is introduced, the field **E** no longer satisfies Maxwell's equations. The error becomes larger the smaller the beam-waist radius w_0 is. When w_0 becomes comparable to the reduced wavelength λ/n we have to include higher-order terms in the expansion of k_z in Eq. (3.7). However, the series expansion converges very poorly for strongly focused beams and one needs to find a more accurate description. We shall return to this topic at a later stage.

Another important aspect of Gaussian beams is that they do not exist, no matter how rigorous the theory that describes them! The reason is that a Gaussian beam profile demands a Gaussian Fourier spectrum. However, the Gaussian Fourier spectrum is infinite and contains evanescent components that are not available in a realistic situation. Thus, a Gaussian beam must always be regarded as an approximation. The tighter the focus, the broader the Gaussian spectrum and the more contradictory the Gaussian beam profile will be. Hence, it actually does not make much sense to include higher-order corrections to the paraxial approximation.

3.2.2 Higher-order laser modes

A laser beam can exist in different transverse modes. It is the laser cavity that determines which type of transverse mode is emitted. The most commonly encountered higher beam modes are Hermite–Gaussian and Laguerre–Gaussian beams. The former are generated in cavities with rectangular end mirrors whereas the latter are observed in cavities with circular end mirrors. In the transverse plane, the fields of these modes extend over larger distances and have sign variations in the phase.

Since the fundamental Gaussian mode is a solution of a linear homogeneous partial differential equation, namely the Helmholtz equation, any combinations of spatial derivatives of the fundamental mode are also solutions to the same differential equation. Zauderer [4] pointed out that Hermite–Gaussian modes $\mathbf{E}^{\mathrm{H}}_{nm}$ can be generated from the fundamental mode **E** according to

$$\mathbf{E}^{\mathrm{H}}_{nm}(x, y, z) = w_0^{n+m}\,\frac{\partial^n}{\partial x^n}\,\frac{\partial^m}{\partial y^m}\,\mathbf{E}\,(x, y, z), \tag{3.17}$$

where n and m denote the order and degree of the beam, respectively. Laguerre–Gaussian modes $\mathbf{E}^{\mathrm{L}}_{n,m}$ are derived in a similar way as

$$\mathbf{E}^{\mathrm{L}}_{nm}(x, y, z) = k^n w_0^{2n+m} \mathrm{e}^{\mathrm{i}kz}\,\frac{\partial^n}{\partial z^n}\left(\frac{\partial}{\partial x} + \mathrm{i}\frac{\partial}{\partial y}\right)^m \left\{\mathbf{E}\,(x, y, z)\,\mathrm{e}^{-\mathrm{i}kz}\right\}. \tag{3.18}$$

Thus, any higher-order modes can be generated by simply applying Eqs. (3.17) and (3.18). It can be shown that Laguerre–Gaussian modes can be generated as a

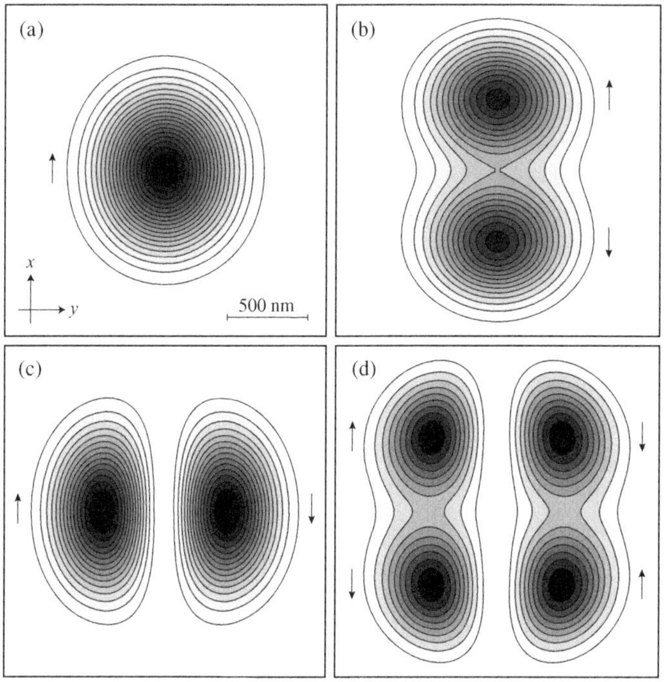

Fig. 3.2 Intensity ($|\boldsymbol{E}|^2$) in the focal plane ($z = 0$) of the first four Hermite–Gaussian modes: (a) (00) mode (Gaussian mode), (b) (10) mode, (c) (01) mode, and (d) (11) mode. The wavelength and beam angle are $\lambda = 800$ nm and $\theta = 28.65°$, respectively. The arrows indicate the polarization directions of the individual lobes.

superposition of a finite number of Hermite–Gaussian modes and vice versa. The two sets of modes are therefore not independent. Note that the parameter w_0 represents the beam waist only for the Gaussian beam and that for higher-order modes the amplitude E_0 does not correspond to the field at the focal point. Figure 3.2 shows the fields in the focal plane ($z = 0$) for the first four Hermite–Gaussian modes. As indicated by the arrows, the polarizations of the individual maxima are either in phase or 180° out of phase with each other.

The commonly encountered doughnut modes with a circular intensity profile can be described by a superposition of Hermite–Gaussian or Laguerre–Gaussian modes. Linearly polarized doughnuts are simply defined by the fields $\mathbf{E}_{01}^{\mathrm{L}}$ or $\mathbf{E}_{11}^{\mathrm{L}}$. An azimuthally polarized doughnut mode is a superposition of two perpendicularly polarized $\mathbf{E}_{01}^{\mathrm{H}}$ fields and a radially polarized doughnut mode is a superposition of two perpendicularly polarized $\mathbf{E}_{10}^{\mathrm{H}}$ fields.

3.2.3 Longitudinal fields in the focal region

The paraxial Gaussian beam is a transverse electromagnetic (TEM) beam, i.e. it is assumed that the electric and magnetic fields are always transverse to the propagation direction.

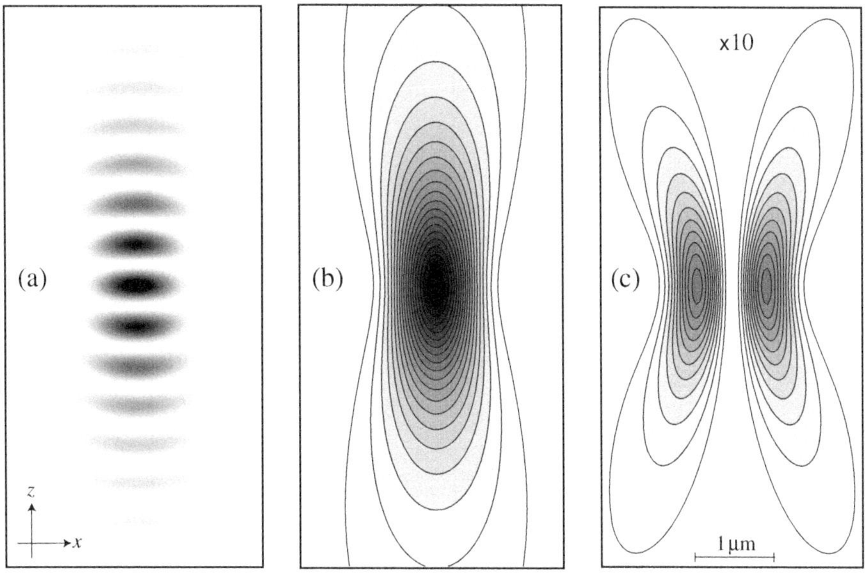

Fig. 3.3 Fields of the Gaussian beam depicted in the polarization plane (x, z). The wavelength and beam angle are $\lambda = 800$ nm and $\theta = 28.65°$, respectively. (a) Time-dependent power density; (b) total electric field intensity ($|\mathbf{E}|^2$); (c) longitudinal electric field intensity ($|E_z|^2$).

However, in free space the only true TEM solutions are infinitely extended fields such as plane waves. Therefore, even a Gaussian beam must possess field components polarized in the direction of propagation. In order to estimate these longitudinal fields we apply the divergence condition $\nabla \cdot \mathbf{E} = 0$ to the x-polarized Gaussian beam, i.e.

$$E_z = -\int \left[\frac{\partial}{\partial x} E_x\right] \mathrm{d}z. \tag{3.19}$$

E_z can be derived using the angular spectrum representation of the paraxial Gaussian beam Eq. (3.10). In the focal plane $z = 0$ we obtain

$$E_z(x, y, 0) = -\mathrm{i}\frac{2x}{kw_0^2}\, E_x(x, y, 0), \tag{3.20}$$

where E_x corresponds to the Gaussian beam profile defined in Eq. (3.8). The prefactor shows that the longitudinal field is 90° out of phase with respect to the transverse field and that it is zero on the optical axis. Its magnitude depends on the tightness of the focus. Figures 3.3 and 3.4 show the calculated total and transverse electric field distributions for the Gaussian beam and the Hermite–Gaussian (10) beam, respectively. While the longitudinal electric field of the fundamental Gaussian beam is always zero on the optical axis, it exhibits two lobes to the sides of the optical axis. Displayed on a cross-section through the beam waist, the two lobes are aligned along the polarization direction. The longitudinal electric field of the Hermite–Gaussian (10) mode, on the other hand, has its maximum

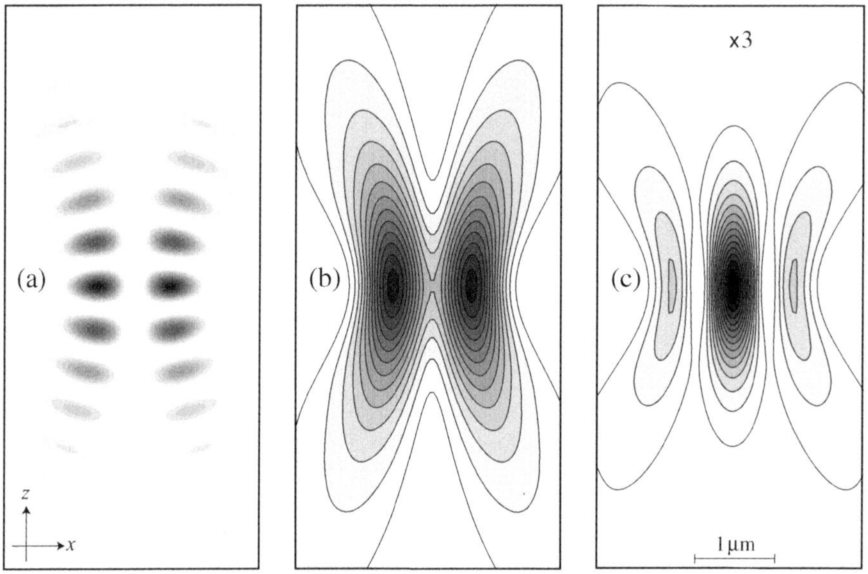

Fig. 3.4 Fields of the Hermite–Gaussian (10) mode. Same scaling and definitions as in Fig. 3.3.

at the beam focus with a much larger field strength. This longitudinal field qualitatively follows from the 180° phase difference and the polarization of the two corresponding field maxima in Fig. 3.2, since the superposition of two similarly polarized plane waves propagating at angles $\pm\varphi$ to the z-axis with 180° phase difference also leads to a longitudinal field component. It has been proposed that one could use the longitudinal fields of the Hermite–Gaussian (10) mode to accelerate charged particles along the beam axis in linear particle accelerators [5]. The longitudinal (10) field has also been applied to image the spatial orientation of molecular transition dipoles [6, 7]. In general, the (10) mode is important for all experiments that require the availability of a longitudinal field component. We shall see in Section 3.6 that the longitudinal field strength of a strongly focused higher-order laser beam can even exceed the transverse field strength.

3.3 Polarized electric and polarized magnetic fields

If we send an optical beam through a polarizer, we eliminate one of the two transverse field components. The transmitted field is then called *polarized electric*.

In fact, any propagating optical field can be split into a polarized electric (PE) and a polarized magnetic (PM) field:

$$\mathbf{E} = \mathbf{E}^{\mathrm{PE}} + \mathbf{E}^{\mathrm{PM}}. \tag{3.21}$$

For a PE field, the electric field is linearly polarized when projected onto the transverse plane. Similarly, for a PM field the magnetic field is linearly polarized when projected onto the transverse plane. Let us first consider a PE field for which we can choose $\mathbf{E}^{\mathrm{PE}} = (E_x, 0, E_z)$. On requiring that the field is divergence free ($\nabla \cdot \mathbf{E}^{\mathrm{PE}} = 0$) we find that

$$\hat{E}_z(k_x, k_y; 0) = -\frac{k_x}{k_z}\hat{E}_x(k_x, k_y; 0), \tag{3.22}$$

which allows us to express the fields $\mathbf{E}^{\mathrm{PE}}$ and $\mathbf{H}^{\mathrm{PE}}$ in the form

$$\mathbf{E}^{\mathrm{PE}}(x, y, z) = \iint_{-\infty}^{\infty} \hat{E}_x(k_x, k_y; 0)\, \frac{1}{k_z}\, [k_z \mathbf{n}_x - k_x \mathbf{n}_z]\, \mathrm{e}^{\mathrm{i}[k_x x + k_y y \pm k_z z]}\, \mathrm{d}k_x\, \mathrm{d}k_y, \tag{3.23}$$

$$\begin{aligned}\mathbf{H}^{\mathrm{PE}}(x, y, z) = Z_{\mu\varepsilon}^{-1} \iint_{-\infty}^{\infty} \hat{E}_x(k_x, k_y; 0) \frac{1}{kk_z} [-k_x k_y \mathbf{n}_x + (k_x^2 + k_z^2)\mathbf{n}_y \\ - k_y k_z \mathbf{n}_z]\, \mathrm{e}^{\mathrm{i}[k_x x + k_y y \pm k_z z]}\, \mathrm{d}k_x\, \mathrm{d}k_y,\end{aligned} \tag{3.24}$$

where $\mathbf{n}_x$, $\mathbf{n}_y$, $\mathbf{n}_z$ are unit vectors along the x, y, z axes. To derive $\mathbf{H}^{\mathrm{PE}}$ we used the relations in Eq. (2.126).

To derive the corresponding PM fields we require that $\mathbf{H}^{\mathrm{PM}} = (0, H_y, H_z)$. After following the same procedure as before one finds that in the PM solution the expressions for the electric and magnetic fields are simply interchanged:

$$\begin{aligned}\mathbf{E}^{\mathrm{PM}}(x, y, z) = Z_{\mu\varepsilon} \iint_{-\infty}^{\infty} \hat{H}_y(k_x, k_y; 0) \frac{1}{kk_z} [(k_y^2 + k_z^2)\mathbf{n}_x - k_x k_y \mathbf{n}_y \\ + k_x k_z \mathbf{n}_z]\, \mathrm{e}^{\mathrm{i}[k_x x + k_y y \pm k_z z]}\, \mathrm{d}k_x\, \mathrm{d}k_y,\end{aligned} \tag{3.25}$$

$$\mathbf{H}^{\mathrm{PM}}(x, y, z) = \iint_{-\infty}^{\infty} \hat{H}_y(k_x, k_y; 0)\, \frac{1}{k_z}\, [k_z \mathbf{n}_y - k_y \mathbf{n}_z]\, \mathrm{e}^{\mathrm{i}[k_x x + k_y y \pm k_z z]}\, \mathrm{d}k_x\, \mathrm{d}k_y. \tag{3.26}$$

It is straightforward to demonstrate that in the paraxial limit the PE and PM solutions are identical. In this case they become identical with a TEM solution.

The decomposition of an arbitrary optical field into a PE field and a PM field has been achieved by setting one transverse field component to zero. The procedure is similar to the commonly encountered decomposition into transverse electric (TE) and transverse magnetic (TM) fields for which one longitudinal field component is set to zero (see Problem 3.2).

3.4 Far-fields in the angular spectrum representation

In this section we will derive the important result that Fourier optics and geometrical optics naturally emerge from the angular spectrum representation.

Consider a particular (localized) field distribution in the plane $z = 0$. The angular spectrum representation tells us how this field propagates and how it is mapped onto other planes $z = z_0$. Here, we ask what the field will be in a very remote plane. Vice versa, we can ask what field will result when we focus a particular far-field onto an image plane. Let us start with the familiar angular spectrum representation of an optical field

$$\mathbf{E}(x, y, z) = \iint_{-\infty}^{\infty} \hat{\mathbf{E}}(k_x, k_y; 0)\, \mathrm{e}^{\mathrm{i}\,[k_x x + k_y y \pm k_z z]}\, \mathrm{d}k_x\, \mathrm{d}k_y. \tag{3.27}$$

We are interested in the asymptotic far-zone approximation of this field, i.e. in the evaluation of the field at a point $\mathbf{r} = \mathbf{r}_\infty$ at an infinite distance from the object plane. The dimensionless unit vector $\mathbf{s}$ in the direction of $\mathbf{r}_\infty$ is given by

$$\mathbf{s} = (s_x, s_y, s_z) = \left(\frac{x}{r}, \frac{y}{r}, \frac{z}{r}\right), \tag{3.28}$$

where $r = (x^2 + y^2 + z^2)^{1/2}$ is the distance of $\mathbf{r}_\infty$ from the origin. To calculate the far-field $\mathbf{E}_\infty$ we require that $r \to \infty$ and rewrite Eq. (3.27) as

$$\mathbf{E}_\infty(s_x, s_y) = \lim_{kr\to\infty} \iint_{(k_x^2+k_y^2)\le k^2} \hat{\mathbf{E}}(k_x, k_y; 0)\, \mathrm{e}^{\mathrm{i}kr\left[\frac{k_x}{k}s_x + \frac{k_y}{k}s_y \pm \frac{k_z}{k}s_z\right]}\, \mathrm{d}k_x\, \mathrm{d}k_y, \tag{3.29}$$

where $s_z = \sqrt{1 - (s_x^2 + s_y^2)}$. Because of their exponential decay, evanescent waves do not contribute to the fields at infinity. We therefore reject their contribution and reduce the integration range to $(k_x^2 + k_y^2) \le k^2$. The asymptotic behavior of the double integral as $kr \to \infty$ can be evaluated by the method of *stationary phase*. For a clear outline of this method we refer the interested reader to Section 3.3 of Ref. [3]. Without going into details,

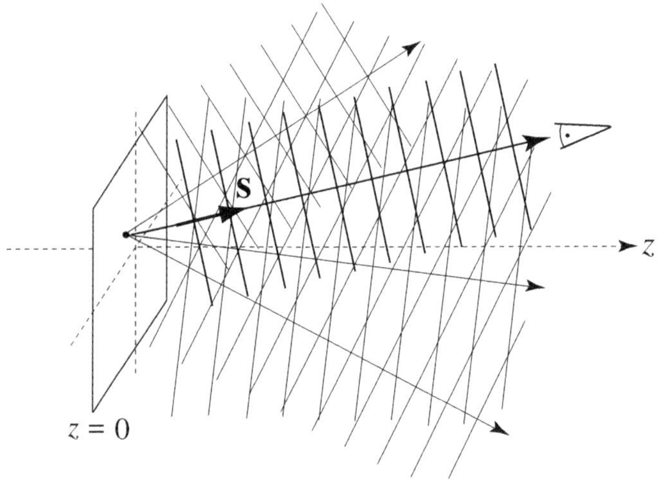

Fig. 3.5 Illustration of the far-field approximation. According to the angular spectrum representation, a point in the source plane $z = 0$ emits plane waves in all possible directions. However, a distant detector ($kr \gg 1$) measures only the plane wave that propagates towards it (in the direction of the unit vector $\mathbf{s}$). The fields of all other plane waves are cancelled out by destructive interference.

the result of Eq. (3.29) can be expressed as

$$\mathbf{E}_\infty(s_x, s_y) = -2\pi \,\mathrm{i} k s_z \,\hat{\mathbf{E}}(ks_x, ks_y; 0)\, \frac{\mathrm{e}^{\mathrm{i}kr}}{r}. \tag{3.30}$$

This equation tells us that the far-fields are entirely defined by the Fourier spectrum of the fields $\hat{\mathbf{E}}(k_x, k_y; 0)$ in the object plane if we make the replacements $k_x \to ks_x$ and $k_y \to ks_y$. This simply means that the unit vector **s** satisfies

$$\mathbf{s} = (s_x, s_y, s_z) = \left(\frac{k_x}{k}, \frac{k_y}{k}, \frac{k_z}{k}\right), \tag{3.31}$$

which implies that only one plane wave with the wavevector $\mathbf{k} = (k_x, k_y, k_z)$ of the angular spectrum at $z = 0$ contributes to the far-field at a point located in the direction of the unit vector **s** (see Fig. 3.5). The effect of all other plane waves is cancelled out by destructive interference. This beautiful result allows us to treat the field in the far-zone as a collection of rays with each ray being characterized by a particular plane wave of the original angular spectrum representation (geometrical optics). On combining Eqs. (3.30) and (3.31) we can express the Fourier spectrum $\hat{\mathbf{E}}$ in terms of the far-field as

$$\hat{\mathbf{E}}(k_x, k_y; 0) = \frac{\mathrm{i} r \mathrm{e}^{-\mathrm{i}kr}}{2\pi k_z} \mathbf{E}_\infty\left(\frac{k_x}{k}, \frac{k_y}{k}\right), \tag{3.32}$$

keeping in mind that the vector **s** is entirely defined by k_x and k_y. This expression can be substituted into the angular spectrum representation (Eq. 3.27) as

$$\mathbf{E}(x, y, z) = \frac{\mathrm{i} r \mathrm{e}^{-\mathrm{i}kr}}{2\pi} \iint\limits_{(k_x^2 + k_y^2) \le k^2} \mathbf{E}_\infty\left(\frac{k_x}{k}, \frac{k_y}{k}\right) \mathrm{e}^{\mathrm{i}[k_x x + k_y y \pm k_z z]} \frac{1}{k_z} \,\mathrm{d}k_x \,\mathrm{d}k_y. \tag{3.33}$$

Thus, as long as evanescent fields are not part of our system, the field **E** and its far-field $\mathbf{E}_\infty$ form essentially a Fourier-transform pair at $z = 0$. The only deviation is given by the k_z terms. In the approximation $k_z \approx k$, the two fields form a perfect Fourier-transform pair. This is the limit of *Fourier optics*.

As an example consider the diffraction at a rectangular aperture with sides $2L_x$ and $2L_y$ in an infinitely thin conducting screen, which we choose to be our object plane ($z = 0$). A plane wave illuminates the aperture at normal incidence from the back. For simplicity we assume that the field in the object plane has a constant field amplitude $\mathbf{E}_0$, whereas the screen blocks all of the field outside of the aperture. The Fourier spectrum at $z = 0$ is then

$$\begin{aligned}\hat{\mathbf{E}}(k_x, k_y; 0) &= \frac{\mathbf{E}_0}{4\pi^2} \int_{-L_y}^{+L_y} \int_{-L_x}^{+L_x} \mathrm{e}^{-\mathrm{i}[k_x x' + k_y y']} \,\mathrm{d}x' \,\mathrm{d}y' \\ &= \mathbf{E}_0 \frac{L_x L_y}{\pi^2} \frac{\sin(k_x L_x)}{k_x L_x} \frac{\sin(k_y L_y)}{k_y L_y},\end{aligned} \tag{3.34}$$

With Eq. (3.30) we now determine the far-field as

$$\mathbf{E}_\infty(s_x, s_y) = -\mathrm{i} k s_z \mathbf{E}_0 \frac{2L_x L_y}{\pi} \frac{\sin(ks_x L_x)}{ks_x L_x} \frac{\sin(ks_y L_y)}{ks_y L_y} \frac{\mathrm{e}^{\mathrm{i}kr}}{r}, \tag{3.35}$$

which, in the paraxial limit $k_z \approx k$, agrees with Fraunhofer diffraction.

Equation (3.30) is an important result. It links the near-fields of an optical problem with the corresponding far-fields. While in the near-field a rigorous description of fields is necessary, the far-fields are well approximated by the laws of geometrical optics.

3.5 Focusing of fields

The limit of classical light confinement is achieved with highly focused laser beams. Such beams are used in fluorescence spectroscopy to investigate molecular interactions in solutions and the kinetics of single molecules on interfaces [6]. Highly focused laser beams also play a key role in confocal microscopy and optical data storage, where resolutions on the order of $\lambda/4$ are achieved. In optical tweezers, focused laser beams are used to trap particles and to move and position them with high precision [8]. All these fields require a theoretical understanding of strongly focused light.

The fields of a focused laser beam are determined by the boundary conditions of the focusing optical element and the incident optical field. In this section we will study the focusing of a paraxial optical field by an aplanatic optical lens as shown in Fig. 3.6. In our theoretical treatment we will follow the theory established by Richards and Wolf [9, 10]. The fields near the optical lens can be formulated by the rules of geometrical optics. In this approximation the finiteness of the optical wavelength is neglected ($k \rightarrow \infty$) and the energy is transported along light rays. The average energy density is propagated with the velocity $v = c/n$ in the direction perpendicular to the geometrical wavefronts. To describe an aplanatic lens we need two rules: (1) the sine condition and (2) the intensity law. These rules are illustrated in Fig. 3.7. The *sine condition* states that each optical ray that emerges from or converges to the focus F of an aplanatic optical system intersects its conjugate ray on a sphere of radius f (the Gaussian reference sphere), where f is the focal length of the lens. By "conjugate ray," one understands the refracted or incident ray that propagates parallel to the optical axis. The distance h between the optical axis and the conjugate ray is given by

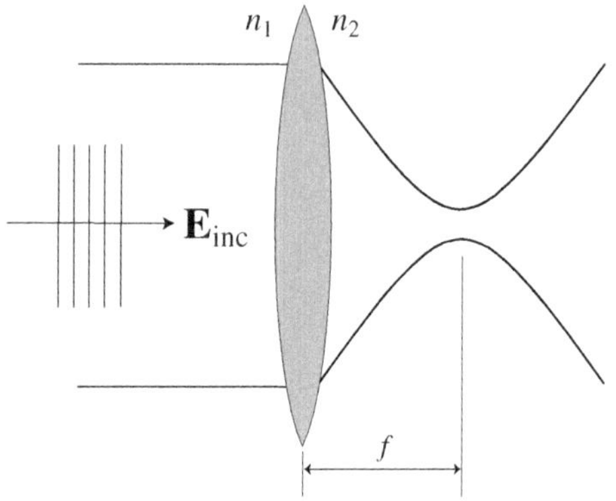

Fig. 3.6 Focusing of a laser beam by an aplanatic lens.

$$h = f \sin\theta, \tag{3.36}$$

θ being the divergence angle of the conjugate ray. Thus, the sine condition is a prescription for the refraction of optical rays at the aplanatic optical element. The *intensity law* is nothing other than a statement of energy conservation: the energy flux along each ray must remain constant. As a consequence, the electric field strength of a spherical wave has to scale as $1/r$, r being the distance from the origin. The intensity law ensures that the energy incident on the aplanatic lens equals the energy that leaves the lens. We know that the power transported by a ray is $dP = (1/2)Z_{\mu\varepsilon}^{-1}|\mathbf{E}|^2 dA$, where $Z_{\mu\varepsilon}$ is the wave impedance and dA is an infinitesimal cross-section perpendicular to the ray propagation. Thus, as indicated in Fig. 3.7(b), the fields before and after refraction must satisfy

$$|\mathbf{E}_2| = |\mathbf{E}_1| \sqrt{\frac{n_1}{n_2}} \sqrt{\frac{\mu_2}{\mu_1}}\, (\cos\theta)^{1/2}. \tag{3.37}$$

Since in practically all media the magnetic permeability at optical frequencies is equal to one ($\mu = 1$), we will drop the term $\sqrt{\mu_2/\mu_1}$ for the sake of having more convenient notation.

Using the sine condition, our optical system can be represented as shown in Fig. 3.8. The incident light rays are refracted by the reference sphere of radius f. We denote an arbitrary point on the surface of the reference sphere by $(x_\infty, y_\infty, z_\infty)$ and an arbitrary field point near the focus by (x, y, z). The two points are also represented by the spherical coordinates (f, θ, ϕ) and (r, ϑ, φ), respectively.

To describe refraction of the incident rays at the reference sphere we introduce the unit vectors $\mathbf{n}_\rho$, $\mathbf{n}_\phi$, and $\mathbf{n}_\theta$, as shown in Fig. 3.8. $\mathbf{n}_\rho$ and $\mathbf{n}_\phi$ are the unit vectors of a cylindrical coordinate system, whereas $\mathbf{n}_\theta$ and $\mathbf{n}_\phi$ are the unit vectors of a spherical coordinate system. We recognize that the reference sphere transforms a cylindrical coordinate system (incoming beam) into a spherical coordinate system (focused beam). Refraction at the reference sphere is most conveniently calculated by splitting the incident vector $\mathbf{E}_{\text{inc}}$ into two components denoted as $\mathbf{E}_{\text{inc}}^{(s)}$ and $\mathbf{E}_{\text{inc}}^{(p)}$. The indices (s) and (p) stand for s-polarization and p-polarization, respectively. In terms of the unit vectors we can express the two fields as

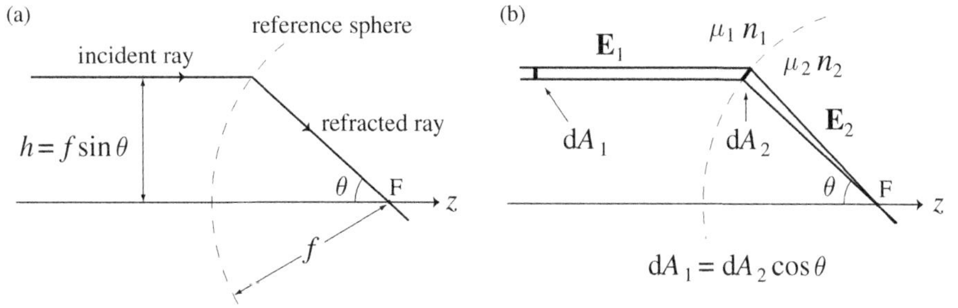

Fig. 3.7 (a) The sine condition of geometrical optics. The refraction of light rays at an aplanatic lens is determined by a spherical surface with radius f. (b) The intensity law of geometrical optics. The energy carried along a ray must stay constant.

$$\mathbf{E}_{\text{inc}}^{(\text{s})} = \left[\mathbf{E}_{\text{inc}} \cdot \mathbf{n}_{\phi}\right] \mathbf{n}_{\phi}, \qquad \mathbf{E}_{\text{inc}}^{(\text{p})} = \left[\mathbf{E}_{\text{inc}} \cdot \mathbf{n}_{\rho}\right] \mathbf{n}_{\rho}. \tag{3.38}$$

As shown in Fig. 3.8 these two fields refract differently at the spherical surface. While the unit vector $\mathbf{n}_{\phi}$ remains unaffected, the unit vector $\mathbf{n}_{\rho}$ is mapped into $\mathbf{n}_{\theta}$. Thus, the total refracted electric field, denoted by $\mathbf{E}_{\infty}$, can be expressed as

$$\mathbf{E}_{\infty} = \left[t^{\text{s}}\left[\mathbf{E}_{\text{inc}} \cdot \mathbf{n}_{\phi}\right] \mathbf{n}_{\phi} + t^{\text{p}}\left[\mathbf{E}_{\text{inc}} \cdot \mathbf{n}_{\rho}\right] \mathbf{n}_{\theta}\right] \sqrt{\frac{n_1}{n_2}}\,(\cos\theta)^{1/2}. \tag{3.39}$$

For each ray we have included the corresponding transmission coefficients t^{s} and t^{p} as defined in Eqs. (2.52). The factor outside the brackets is a consequence of the intensity law to ensure energy conservation. The subscript ∞ was added to indicate that the field is evaluated at a large distance from the focus $(x, y, z) = (0, 0, 0)$.

The unit vectors $\mathbf{n}_{\rho}$, $\mathbf{n}_{\phi}$, $\mathbf{n}_{\theta}$ can be expressed in terms of the Cartesian unit vectors $\mathbf{n}_x$, $\mathbf{n}_y$, $\mathbf{n}_z$ using the spherical coordinates θ and ϕ defined in Fig. 3.8:

$$\mathbf{n}_{\rho} = \cos\phi\; \mathbf{n}_x + \sin\phi\; \mathbf{n}_y, \tag{3.40}$$

$$\mathbf{n}_{\phi} = -\sin\phi\; \mathbf{n}_x + \cos\phi\; \mathbf{n}_y, \tag{3.41}$$

$$\mathbf{n}_{\theta} = \cos\theta\; \cos\phi\; \mathbf{n}_x + \cos\theta\; \sin\phi\; \mathbf{n}_y - \sin\theta\; \mathbf{n}_z. \tag{3.42}$$

On inserting these vectors into Eq. (3.39) we obtain

$$\begin{aligned} \mathbf{E}_{\infty}(\theta,\phi) = {} & t^{\text{s}}(\theta)\left[\mathbf{E}_{\text{inc}}(\theta,\phi) \cdot \begin{pmatrix} -\sin\phi \\ \cos\phi \\ 0 \end{pmatrix}\right] \begin{pmatrix} -\sin\phi \\ \cos\phi \\ 0 \end{pmatrix} \sqrt{\frac{n_1}{n_2}}\,(\cos\theta)^{1/2} \\ & + t^{\text{p}}(\theta)\left[\mathbf{E}_{\text{inc}}(\theta,\phi) \cdot \begin{pmatrix} \cos\phi \\ \sin\phi \\ 0 \end{pmatrix}\right] \begin{pmatrix} \cos\phi\; \cos\theta \\ \sin\phi\; \cos\theta \\ -\sin\theta \end{pmatrix} \sqrt{\frac{n_1}{n_2}}\,(\cos\theta)^{1/2}, \end{aligned} \tag{3.43}$$

which is the field in Cartesian vector components just to the right of the reference sphere of the focusing lens. We can also express $\mathbf{E}_{\infty}$ in terms of the spatial frequencies k_x and k_y by using the substitutions

$$k_x = k\sin\theta\; \cos\phi, \qquad k_y = k\sin\theta\; \sin\phi, \qquad k_z = k\cos\theta. \tag{3.44}$$

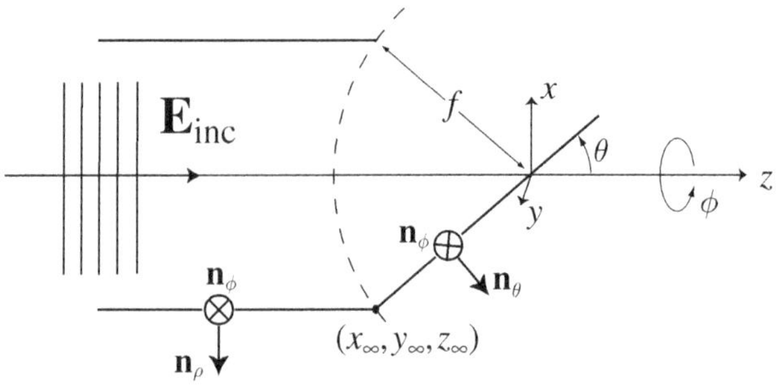

Fig. 3.8 Geometrical representation of the aplanatic system and definition of coordinates.

The resulting far-field on the reference sphere is then of the form $\mathbf{E}_\infty(k_x/k,\, k_y/k)$ and can be inserted into Eq. (3.33) to rigorously calculate the focal fields. Thus, the field $\mathbf{E}$ near the focus of our lens is entirely determined by the far-field $\mathbf{E}_\infty$ on the reference sphere. All rays propagate from the reference sphere towards the focus $(x, y, z) = (0, 0, 0)$ and there are no evanescent waves involved.

Owing to the symmetry of our problem it is convenient to express the angular spectrum representation Eq. (3.33) in terms of the angles θ and ϕ instead of k_x and k_y. This is easily accomplished by using the substitutions in Eq. (3.44) and expressing the transverse coordinates (x, y) of the field point as

$$x = \rho \cos\varphi, \qquad y = \rho \sin\varphi. \tag{3.45}$$

In order to replace the planar integration over k_x, k_y by a spherical integration over θ, ϕ we must transform the differentials as

$$\frac{1}{k_z}\, \mathrm{d}k_x\, \mathrm{d}k_y = k \,\sin\theta\, \mathrm{d}\theta\, \mathrm{d}\phi, \tag{3.46}$$

which is illustrated in Fig. 3.9. We can now express the angular spectrum representation of the focal field (Eq. 3.33) as

$$\mathbf{E}(\rho, \varphi, z) = -\frac{\mathrm{i}kf\mathrm{e}^{-\mathrm{i}kf}}{2\pi} \int_0^{\theta_{\max}} \int_0^{2\pi} \mathbf{E}_\infty(\theta, \phi)\, \mathrm{e}^{\mathrm{i}kz\cos\theta}\, \mathrm{e}^{\mathrm{i}k\rho\sin\theta\cos(\phi-\varphi)} \sin\theta\, \mathrm{d}\phi\, \mathrm{d}\theta. \tag{3.47}$$

We have replaced the distance r_∞ between the focal point and the surface of the reference sphere by the focal length f of the lens.[2] We have also limited the integration over θ to the finite range $[0 \ldots \theta_{\max}]$ because any lens will have a finite size. Furthermore, since all fields propagate in the positive-z direction we retained only the $+$ sign in the exponent of Eq. (3.33). Equation (3.47) is the central result of this section. Together with Eq. (3.43), it allows us to calculate the focusing of an arbitrary optical field $\mathbf{E}_{\mathrm{inc}}$ by an aplanatic lens with focal length f and numerical aperture

$$\mathrm{NA} = n \,\sin\theta_{\max} \qquad (0 < \theta_{\max} < \pi/2), \tag{3.48}$$

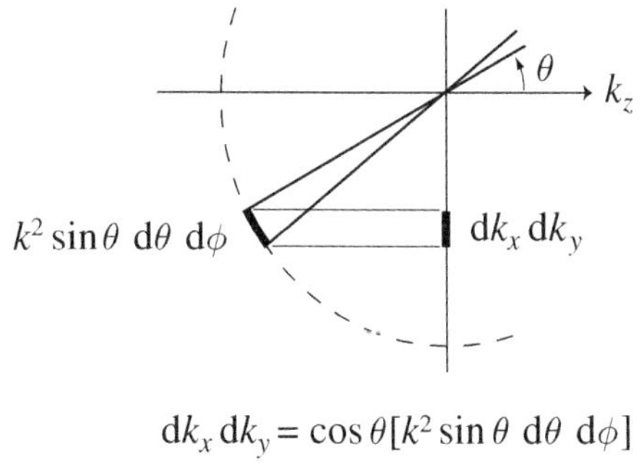

Fig. 3.9 Illustration of the substitution $(1/k_z)\mathrm{d}k_x\, \mathrm{d}k_y = k \,\sin\theta\, \mathrm{d}\theta\, \mathrm{d}\phi$. The factor $1/k_z = 1/(k\cos\theta)$ ensures that the differential areas on the plane and on the sphere stay equal.

[2] The '$-$' sign originates from taking the farfield at $z \to -\infty$. In Eq. (3.33) the farfield is evaluated at $z \to +\infty$.

where $n = n_2$ is the index of refraction of the surrounding medium. The field distribution in the focal region is entirely determined by the far-field $\mathbf{E}_\infty$. As we shall see in the next section, the properties of the laser focus can be engineered by adjusting the amplitude and phase profile of $\mathbf{E}_\infty$.

3.6 Focal fields

Typically, the back-aperture of a microscope objective is a couple of millimeters in diameter. In order to make use of the full NA of the objective, the incident field $\mathbf{E}_{\rm inc}$ has to fill or overfill the back-aperture. Thus, because of the large diameter of the incident beam, it is reasonable to treat it in the paraxial approximation. Let us assume that $\mathbf{E}_{\rm inc}$ is entirely polarized along the x-axis, i.e.

$$\mathbf{E}_{\rm inc} = E_{\rm inc}\mathbf{n}_x. \tag{3.49}$$

Furthermore, we assume that the waist of the incoming beam coincides with the lens so that it hits the lens with a planar phase front. For simplicity we also assume that we have a lens with good antireflection coating so that we can neglect the Fresnel transmission coefficients,

$$t_\theta^{\rm s} = t_\theta^{\rm p} = 1. \tag{3.50}$$

With these assumptions the far-field $\mathbf{E}_\infty$ in Eq. (3.43) can be expressed as

$$\begin{aligned}\mathbf{E}_\infty(\theta,\phi) &= E_{\rm inc}(\theta,\phi)\left[\cos\phi\,\mathbf{n}_\theta - \sin\phi\,\mathbf{n}_\phi\right]\sqrt{n_1/n_2}\,(\cos\theta)^{1/2}\\ &= E_{\rm inc}(\theta,\phi)\,\frac{1}{2}\begin{bmatrix}(1+\cos\theta)-(1-\cos\theta)\cos(2\phi)\\ -(1-\cos\theta)\sin(2\phi)\\ -2\cos\phi\,\sin\theta\end{bmatrix}\sqrt{\frac{n_1}{n_2}}\,(\cos\theta)^{1/2},\end{aligned} \tag{3.51}$$

where the last expression is represented in Cartesian vector components. To proceed we need to specify the amplitude profile of the incoming beam $E_{\rm inc}$. We will concentrate on the three lowest Hermite–Gaussian modes displayed in Fig. 3.2. The first of these modes corresponds to the fundamental Gaussian beam and the other two can be generated according to Eq. (3.17) of Section 3.2.2. On expressing the coordinates $(x_\infty, y_\infty, z_\infty)$ in Fig. 3.8 in terms of the spherical coordinates (f,θ,ϕ) we find

(0,0) mode:

$$E_{\rm inc} = E_0\,{\rm e}^{-(x_\infty^2+y_\infty^2)/w_0^2} = E_0\,{\rm e}^{-f^2\sin^2\theta/w_0^2}, \tag{3.52}$$

(1,0) mode:

$$E_{\rm inc} = E_0(2x_\infty/w_0){\rm e}^{-(x_\infty^2+y_\infty^2)/w_0^2} = (2E_0 f/w_0)\,\sin\theta\cos\phi\;{\rm e}^{-f^2\sin^2\theta/w_0^2}, \tag{3.53}$$

(0,1) mode:

$$E_{\rm inc} = E_0(2y_\infty/w_0)\,{\rm e}^{-(x_\infty^2+y_\infty^2)/w_0^2} = (2E_0 f/w_0)\sin\theta\,\sin\phi\;{\rm e}^{-f^2\sin^2\theta/w_0^2}. \tag{3.54}$$

The factor $f_w(\theta) = \exp(-f^2 \sin^2\theta / w_0^2)$ is common to all modes. The focal field **E** will depend on how much the incoming beam is expanded relative to the size of the lens. Since the *aperture radius* of our lens is equal to $f \sin\theta_{max}$ we define the *filling factor* f_0 as

$$f_0 = \frac{w_0}{f \sin\theta_{max}}, \tag{3.55}$$

which allows us to write the exponential function in Eqs. (3.52)–(3.54) in the form

$$f_w(\theta) = \mathrm{e}^{-\frac{1}{f_0^2}\frac{\sin^2\theta}{\sin^2\theta_{max}}}. \tag{3.56}$$

This function is called the *apodization function* and can be viewed as a pupil filter. We now have all the necessary ingredients to compute the field **E** near the focus. With the mathematical relations

$$\int_0^{2\pi} \cos(n\phi)\mathrm{e}^{\mathrm{i}x\cos(\phi-\varphi)}\,\mathrm{d}\phi = 2\pi\,(\mathrm{i}^n)J_n(x)\cos(n\varphi), \tag{3.57}$$

$$\int_0^{2\pi} \sin(n\phi)\mathrm{e}^{\mathrm{i}x\cos(\phi-\varphi)}\,\mathrm{d}\phi = 2\pi\,(\mathrm{i}^n)J_n(x)\sin(n\varphi)$$

we can carry out the integration over ϕ analytically. Here, J_n is the nth-order Bessel function. The final expressions for the focal field now contain a single integration over the variable θ. It is convenient to use the following abbreviations for the integrals:

$$I_{00} = \int_0^{\theta_{max}} f_w(\theta)(\cos\theta)^{1/2}\sin\theta(1+\cos\theta)J_0(k\rho\sin\theta)\,\mathrm{e}^{\mathrm{i}kz\cos\theta}\,\mathrm{d}\theta, \tag{3.58}$$

$$I_{01} = \int_0^{\theta_{max}} f_w(\theta)(\cos\theta)^{1/2}\sin^2\theta\,J_1(k\rho\sin\theta)\,\mathrm{e}^{\mathrm{i}kz\cos\theta}\,\mathrm{d}\theta, \tag{3.59}$$

$$I_{02} = \int_0^{\theta_{max}} f_w(\theta)(\cos\theta)^{1/2}\sin\theta(1-\cos\theta)J_2(k\rho\sin\theta)\,\mathrm{e}^{\mathrm{i}kz\cos\theta}\,\mathrm{d}\theta, \tag{3.60}$$

$$I_{10} = \int_0^{\theta_{max}} f_w(\theta)(\cos\theta)^{1/2}\sin^3\theta\,J_0(k\rho\sin\theta)\,\mathrm{e}^{\mathrm{i}kz\cos\theta}\,\mathrm{d}\theta, \tag{3.61}$$

$$I_{11} = \int_0^{\theta_{max}} f_w(\theta)(\cos\theta)^{1/2}\sin^2\theta(1+3\cos\theta)J_1(k\rho\sin\theta)\,\mathrm{e}^{\mathrm{i}kz\cos\theta}\,\mathrm{d}\theta, \tag{3.62}$$

$$I_{12} = \int_0^{\theta_{max}} f_w(\theta)(\cos\theta)^{1/2} \sin^2\theta (1-\cos\theta) J_1(k\rho \sin\theta)\, e^{ikz\cos\theta}\, d\theta, \tag{3.63}$$

$$I_{13} = \int_0^{\theta_{max}} f_w(\theta)(\cos\theta)^{1/2} \sin^3\theta\, J_2(k\rho \sin\theta)\, e^{ikz\cos\theta}\, d\theta, \tag{3.64}$$

$$I_{14} = \int_0^{\theta_{max}} f_w(\theta)(\cos\theta)^{1/2} \sin^2\theta (1-\cos\theta) J_3(k\rho \sin\theta)\, e^{ikz\cos\theta}\, d\theta, \tag{3.65}$$

where the function $f_w(\theta)$ is given by Eq. (3.56). Notice, that these integrals are functions of the coordinates (ρ, z), i.e. $I_{ij} = I_{ij}(\rho, z)$. Thus, for each field point we have to evaluate these integrals numerically. Using these abbreviations we can now express the focal fields of the various modes as

(0,0) mode:

$$\begin{aligned} \mathbf{E}(\rho,\varphi,z) &= -\frac{ikf}{2}\sqrt{\frac{n_1}{n_2}}\, E_0 e^{-ikf} \begin{bmatrix} I_{00} + I_{02}\cos(2\varphi) \\ I_{02}\sin(2\varphi) \\ -2iI_{01}\cos\varphi \end{bmatrix}, \\ \mathbf{H}(\rho,\varphi,z) &= -\frac{ikf}{2Z_{\mu\varepsilon}}\sqrt{\frac{n_1}{n_2}}\, E_0 e^{-ikf} \begin{bmatrix} I_{02}\sin(2\varphi) \\ I_{00} - I_{02}\cos(2\varphi) \\ -2iI_{01}\sin\varphi \end{bmatrix}, \end{aligned} \tag{3.66}$$

(1,0) mode:

$$\begin{aligned} \mathbf{E}(\rho,\varphi,z) &= -\frac{ikf^2}{2w_0}\sqrt{\frac{n_1}{n_2}}\, E_0 e^{-ikf} \begin{bmatrix} iI_{11}\cos\varphi + iI_{14}\cos(3\varphi) \\ -iI_{12}\sin\varphi + iI_{14}\sin(3\varphi) \\ -2I_{10} + 2I_{13}\cos(2\varphi) \end{bmatrix}, \\ \mathbf{H}(\rho,\varphi,z) &= -\frac{ikf^2}{2w_0 Z_{\mu\varepsilon}}\sqrt{\frac{n_1}{n_2}}\, E_0 e^{-ikf} \begin{bmatrix} -iI_{12}\sin\varphi + iI_{14}\sin(3\varphi) \\ i(I_{11}+2I_{12})\cos\varphi - iI_{14}\cos(3\varphi) \\ 2I_{13}\sin(2\varphi) \end{bmatrix}, \end{aligned} \tag{3.67}$$

(0,1) mode:

$$\begin{aligned} \mathbf{E}(\rho,\varphi,z) &= -\frac{ikf^2}{2w_0}\sqrt{\frac{n_1}{n_2}}\, E_0 e^{-ikf} \begin{bmatrix} i(I_{11}+2I_{12})\sin\varphi + iI_{14}\sin(3\varphi) \\ -iI_{12}\cos\varphi - iI_{14}\cos(3\varphi) \\ 2I_{13}\sin(2\varphi) \end{bmatrix}, \\ \mathbf{H}(\rho,\varphi,z) &= -\frac{ikf^2}{2w_0 Z_{\mu\varepsilon}}\sqrt{\frac{n_1}{n_2}}\, E_0 e^{-ikf} \begin{bmatrix} -iI_{12}\cos\varphi - iI_{14}\cos(3\varphi) \\ iI_{11}\sin\varphi - iI_{14}\sin(3\varphi) \\ -2I_{10} - 2I_{13}\cos(2\varphi) \end{bmatrix}. \end{aligned} \tag{3.68}$$

For completeness, we have also listed the magnetic fields for the three modes. They can be derived in the same way by using the corresponding paraxial input fields H_∞ with the magnetic field axis along the y-axis. Notice that only the zeroth-order Bessel function

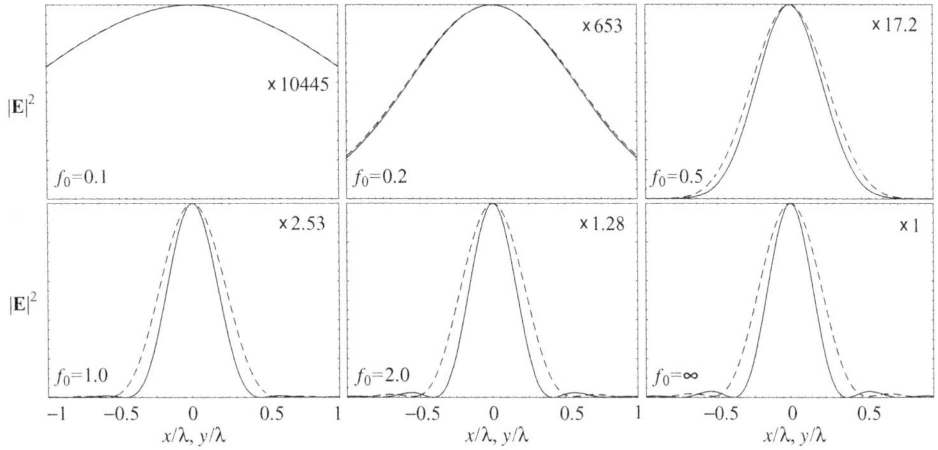

Fig. 3.10 Influence of the filling factor f_0 of the back-aperture on the sharpness of the focus. A lens with NA = 1.4 is assumed and the index of refraction is 1.518. The figure shows the magnitude of the electric field intensity $|\mathbf{E}|^2$ in the focal plane $z = 0$. The dashed curves have been evaluated along the x-direction (plane of polarization) and the solid curves along the y-direction. All curves have been scaled to an equal amplitude. The scaling factor is indicated in the figures. The larger the filling factor, the bigger the deviation between the solid and dashed curves, indicating the importance of polarization effects.

possesses a non-vanishing value at its origin. As a consequence, only the $(1,0)$ mode has a longitudinal electric field (E_z) at its focus.

In the limit $f_w = 1$ the fields for the $(0,0)$ mode are identical with the solutions of Richards and Wolf [10]. According to Eq. (3.56), this limit is reached for $f_0 \to \infty$, which corresponds to an infinitely overfilled back-aperture of the focusing lens. This situation is identical with that of a plane wave incident on the lens. Figure 3.10 demonstrates the effect of the filling factor f_0 on the confinement of the focal fields. In these examples we used an objective with a numerical aperture of 1.4 and an index of refraction of 1.518, which corresponds to a maximum collection angle of 68.96°. It is obvious that the filling factor is important for the quality of the focal spot and thus for the resolution in optical microscopy. It is important to notice that with increasing field confinement at the focus the focal spot becomes more and more elliptical. Whereas in the paraxial limit the spot is perfectly circular, a strongly focused beam has a spot that is elongated in the direction of polarization. This observation has important consequences: as we aim towards higher resolutions by using spatially confined light we need to take the vector nature of the fields into account. Scalar theories become insufficient. Figure 3.11 shows field plots for the electric field for a filling factor of $f_0 = 1$ and an NA = 1.4 objective lens. This figure depicts the total electric field intensity $\mathbf{E}^2$ in the plane of incident polarization (x, z) and perpendicular to it (y, z). The three images to the side show the intensities of the different field components in the focal plane $z = 0$. The maximum relative values are $\mathrm{Max}[E_y^2]/\mathrm{Max}[E_x^2] = 0.003$ and $\mathrm{Max}[E_z^2]/\mathrm{Max}[E_x^2] = 0.12$. Thus, an appreciable amount of the electric field energy is in the longitudinal field.

How can we experimentally verify the calculated focal fields? An elegant method is to use a single dipolar emitter, such as a single molecule, to probe the field (Fig. 3.12).

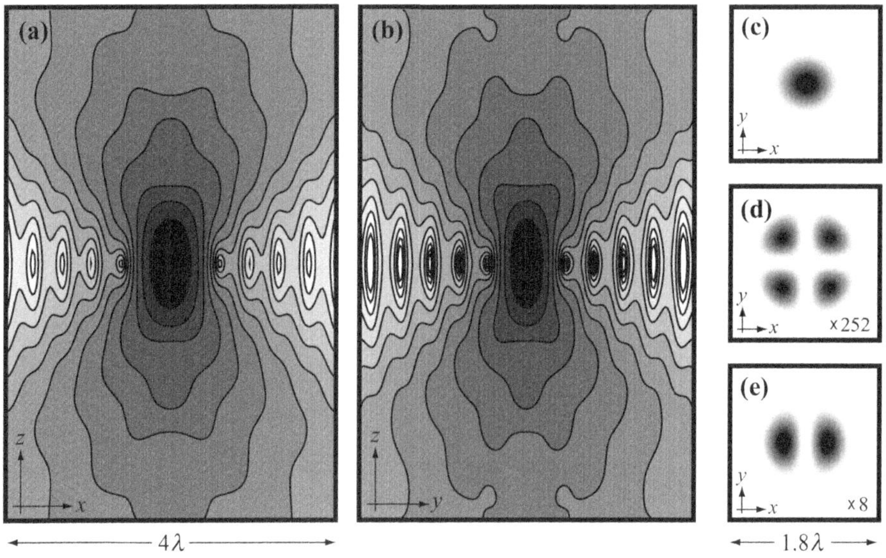

Fig. 3.11 Contour plots of constant $|\mathbf{E}|^2$ in the focal region of a focused Gaussian beam (NA $= 1.4$, $n = 1.518$, $f_0 = 1$): (a) in the plane of incident polarization (x, z); (b) in the plane perpendicular to the plane of incident polarization (y, z). A logarithmic scaling is used, with a factor of 2 difference between adjacent contour lines. Images (c), (d), and (e) show the magnitudes of the individual field components $|\mathbf{E}_x|^2$, $|\mathbf{E}_y|^2$, and $|\mathbf{E}_z|^2$, respectively, in the focal plane ($z = 0$).

The molecule can be embedded into the surrounding medium with index n and moved with accurate translators to any position $\mathbf{r} = (x, y, z) = (\rho, \varphi, z)$ near the laser focus. The excitation rate of the molecule depends on the product $\mathbf{E} \cdot \mathbf{p}$, with $\mathbf{p}$ being the transition dipole moment of the molecule. The excited molecule then relaxes with a certain rate and probability by emitting a fluorescence photon. We can use the same aplanatic lens to collect the emitted photons and direct them onto a photodetector. The fluorescence intensity (photon counts per second) will be proportional to $|\mathbf{E} \cdot \mathbf{p}|^2$. Thus, if we know the dipole orientation of the molecule, we can determine the field strength of the exciting field at the molecule's position. For example, a molecular dipole aligned with the x-axis will render the x-component of the focal field. We can then translate the molecule to a new position and determine the field at this new position. Thus, point by point we can establish a map of the magnitude of the electric field component that points along the molecular dipole axis. With the x-aligned molecule we should be able to reproduce the pattern shown in Fig. 3.11(c) if we scan the molecule point by point in the plane $z = 0$. This has been demonstrated in various experiments and will be discussed in Chapter 9.

3.7 Focusing of higher-order laser modes

So far, we have discussed focusing of the fundamental Gaussian beam. What about the (10) and (01) modes? We have calculated those in order to synthesize *doughnut modes* with arbitrary polarization. Depending on how we superimpose those modes, we obtain

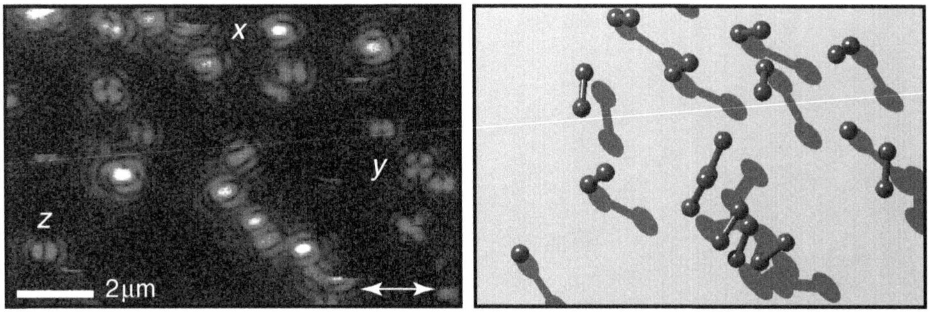

Fig. 3.12 Single-molecule excitation patterns. A sample with isolated single molecules is raster scanned in the focal plane of a strongly focused laser beam. For each pixel, the fluorescence intensity is recorded and encoded in the color scale. The excitation rate in each pixel is determined by the relative orientation of the local electric field vector and the molecular absorption dipole moment. Using the known field distribution in the laser focus allows the dipole moments to be reconstructed from the recorded patterns. Compare the patterns marked *x*, *y*, and *z* with those in Fig. 3.11.

Linearly polarized doughnut mode:

$$\mathrm{LP} = \mathrm{HG}_{10}\,\mathbf{n}_x + \mathrm{i}\,\mathrm{HG}_{01}\,\mathbf{n}_x \tag{3.69}$$

Radially polarized doughnut mode:

$$\mathrm{RP} = \mathrm{HG}_{10}\,\mathbf{n}_x + \mathrm{HG}_{10}\,\mathbf{n}_y \tag{3.70}$$

Azimuthally polarized doughnut mode:

$$\mathrm{AP} = -\mathrm{HG}_{01}\,\mathbf{n}_x + \mathrm{HG}_{01}\,\mathbf{n}_y. \tag{3.71}$$

Here, $\mathrm{HG}_{ij}\,\mathbf{n}_l$ denotes a Hermite–Gaussian (ij) mode polarized along the unit vector $\mathbf{n}_l$. The linearly polarized doughnut mode is identical with the Laguerre–Gaussian (01) mode defined in Eq. (3.18) and it is easily calculated by adding the fields of Eqs. (3.67) and (3.68) with a 90° phase delay. To determine the focal fields of the other two doughnut modes we need to derive the focal fields for the y-polarized modes. This is easily accomplished by rotating the existing fields in Eqs. (3.67) and (3.68) by 90° around the z-axis. The resulting focal fields turn out to be

Radially polarized doughnut mode:

$$\mathbf{E}(\rho,\varphi,z) = \frac{\mathrm{i}kf^2}{2w_0}\sqrt{\frac{n_1}{n_2}}\,E_0\mathrm{e}^{-\mathrm{i}kf}\begin{bmatrix}\mathrm{i}(I_{11}-I_{12})\cos\varphi\\ \mathrm{i}(I_{11}-I_{12})\sin\varphi\\ -4I_{10}\end{bmatrix},$$
$$\mathbf{H}(\rho,\varphi,z) = -\frac{\mathrm{i}kf^2}{2w_0Z_{\mu\varepsilon}}\sqrt{\frac{n_1}{n_2}}\,E_0\mathrm{e}^{-\mathrm{i}kf}\begin{bmatrix}-\mathrm{i}(I_{11}+3I_{12})\sin\varphi\\ \mathrm{i}(I_{11}+3I_{12})\cos\varphi\\ 0\end{bmatrix}, \tag{3.72}$$

Azimuthally polarized doughnut mode:

$$\mathbf{E}(\rho,\varphi,z) = -\frac{\mathrm{i}kf^2}{2w_0}\sqrt{\frac{n_1}{n_2}}\,E_0\mathrm{e}^{-\mathrm{i}kf}\begin{bmatrix}\mathrm{i}(I_{11}+3I_{12})\sin\varphi\\ -\mathrm{i}(I_{11}+3I_{12})\cos\varphi\\ 0\end{bmatrix},$$

$$\mathbf{H}(\rho,\varphi,z) = -\frac{\mathrm{i}kf^2}{2w_0 Z_{\mu\varepsilon}}\sqrt{\frac{n_1}{n_2}}\,E_0\mathrm{e}^{-\mathrm{i}kf}\begin{bmatrix}\mathrm{i}(I_{11}-I_{12})\cos\varphi\\ \mathrm{i}(I_{11}-I_{12})\sin\varphi\\ -4I_{10}\end{bmatrix}. \tag{3.73}$$

With the definition of the integrals

$$I_{\mathrm{rad}} = I_{11} - I_{12} = \int_0^{\theta_{\max}} f_w(\theta)(\cos\theta)^{3/2}\sin^2\theta\, J_1(k\rho\sin\theta)\,\mathrm{e}^{\mathrm{i}kz\cos\theta}\,\mathrm{d}\theta, \tag{3.74}$$

$$I_{\mathrm{azm}} = I_{11} + 3I_{12} = \int_0^{\theta_{\max}} f_w(\theta)(\cos\theta)^{1/2}\sin^2\theta\, J_1(k\rho\sin\theta)\,\mathrm{e}^{\mathrm{i}kz\cos\theta}\,\mathrm{d}\theta \tag{3.75}$$

we see that to describe the focusing of radially polarized and azimuthally polarized doughnut modes we need to evaluate totally two integrals. The radial and azimuthal symmetries are easily seen by transforming the Cartesian field vectors into cylindrical field vectors as

$$\begin{aligned} E_\rho &= \cos\varphi\, E_x + \sin\varphi\, E_y,\\ E_\phi &= -\sin\varphi\, E_x + \cos\varphi\, E_y, \end{aligned} \tag{3.76}$$

and similarly for the magnetic field. While the radially polarized focused mode has a rotationally symmetric longitudinal electric field E_z, the azimuthally polarized focused mode has a rotationally symmetric longitudinal magnetic field H_z. As shown in Fig. 3.13 the longitudinal field strength $|E_z|^2$ increases with increasing numerical aperture. At a numerical aperture of NA ≈ 1 the magnitude of $|E_z|^2$ becomes larger than the magnitude of the radial field $|E_\rho|^2$. This is important for applications that require strong longitudinal fields. Figure 3.14 shows field plots for the focused radially polarized beam using the same parameters and settings as in Fig. 3.11. More detailed discussions of the focusing of radially and azimuthally polarized beams are presented in Refs. [11–13]. The field distribution in the beam focus has been measured using single molecules as probes [7] and by the knife-edge method [13].

Although laser beams can be adjusted to a higher mode by manipulating the laser resonator, it is desirable to convert a fundamental Gaussian beam into a higher-order mode *externally* without perturbing the laser characteristics. Such a conversion can be realized by inserting phase plates into different regions in the beam cross-section [14]. As shown in Fig. 3.15, the conversion to a Hermite–Gaussian (10) mode is favored by bisecting the fundamental Gaussian beam with the edge of a thin phase plate which shifts the phase of one half of the beam by 180°. The incident beam has to be polarized perpendicular to the edge of the phase plate and subsequent spatial filtering has to be performed to reject higher-order modes. A related approach makes use of half-coated mirrors to delay one half of the laser

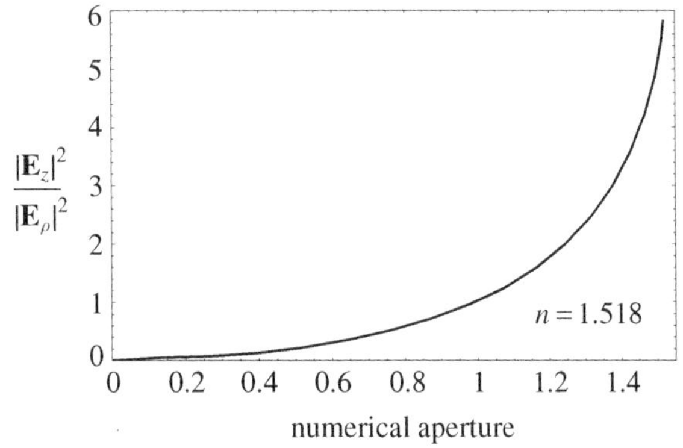

Fig. 3.13 Ratio of the longitudinal and transverse electric field intensities $|\mathbf{E}_z|^2/|\mathbf{E}_\rho|^2$ of a radially polarized doughnut mode as a function of the numerical aperture ($f_0 = 1, n = 1.518$). $|\mathbf{E}_\rho|^2$ has its maximum on a ring in the plane $z = 0$, whereas the maximum of $|\mathbf{E}_z|^2$ is at the origin $(x, y, z) = (0, 0, 0)$. According to the figure, the maximum longitudinal electric energy density can be more than five times larger than the maximum transverse electric energy density.

beam. In this case, the beam passes twice through the bisected part and hence the thickness of the coated part must be $\lambda/4$. Other mode-conversion schemes make use of external four-mirror ring cavities or interferometers [15, 16]. The approach shown in Fig. 3.16(a) was developed by Youngworth and Brown to generate azimuthally and radially polarized beams [11, 12]. It is based on a Twyman–Green interferometer with half-coated mirrors. The polarization of the incoming Gaussian beam is adjusted to 45°. A polarizing beamsplitter divides the power of the beam into two orthogonally polarized beams. Each of the beams passes a $\lambda/4$ phase plate which makes the beams circularly polarized. Each beam then reflects from an end mirror. One half of each mirror has a $\lambda/4$ coating which, after reflection, delays one half of the beam by 180° with respect to the other half. Each of the two reflected beams passes through the $\lambda/4$ plate again and becomes converted into equal amounts of orthogonally polarized Hermite–Gaussian (10) and (01) modes. Subsequently, one of these modes will be rejected by the polarizing beamsplitter whereas the other will be combined with the corresponding mode from the other interferometer arm. Whether a radially polarized mode or an azimuthally polarized mode is generated depends on the positioning of the half-coated end mirrors. To produce the other mode one needs to simply rotate the end mirrors by 90°. The two modes from the different interferometer arms need to be in phase, which requires adjustability of the path length. The correct polarization can always be verified by sending the output beam through a polarizer and by selectively blocking the beam in one of the two interferometer arms. Since the mode conversion is not 100% efficient one needs to spatially filter the output beam to reject any undesired modes. This is accomplished by focusing the output beam onto a pinhole with adjusted diameter.

To obviate the need for noise- and drift-sensitive interferometers, Dorn *et al.* implemented a single-path mode-conversion scheme for radially and azimuthally polarized beams [13]. As shown in Fig. 3.16(b), a laser beam is sent through a $\lambda/2$ waveplate

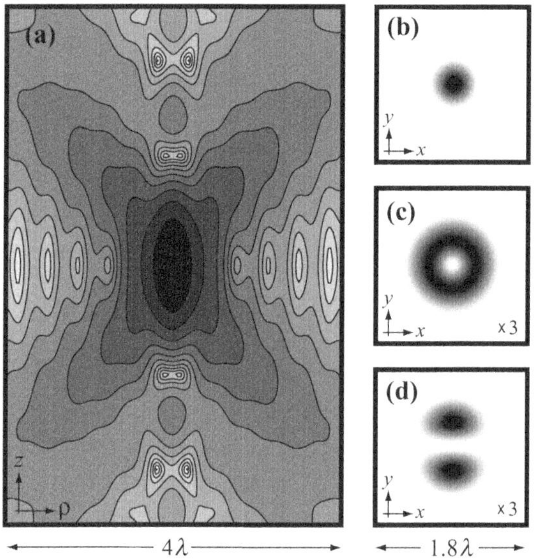

Fig. 3.14 (a) Contour plots of constant $|\mathbf{E}|^2$ in the focal region of a focused radially polarized doughnut mode (NA = 1.4, $n = 1.518$, $f_0 = 1$) in the (ρ, z) plane. The intensity is rotationally symmetric with respect to the z-axis. A logarithmic scaling is used with a factor of 2 difference between adjacent contour lines. Images (b), (c), and (d) show the magnitudes of the individual field components $|\mathbf{E}_z|^2$, $|\mathbf{E}_\rho|^2$, and $|\mathbf{E}_y|^2$, respectively, in the focal plane ($z = 0$). A linear scale is used.

consisting of four segments. The optical axis of each segment is oriented such that the field is rotated to point in the radial direction. Subsequent spatial filtering extracts the desired mode with very high purity. A phase plate as shown in Fig. 3.16(b) can be fabricated by cutting two $\lambda/2$ plates into four quadrants each, and then assembling the pieces into two new phase plates. This mode-conversion principle can be generalized to waveplates with many elements such as liquid-crystal spatial light modulators.

3.8 The limit of weak focusing

Before we proceed to the next section we need to verify that our formulas for the focused fields render the familiar paraxial expressions for the limit of small $\theta_{\max}$. In this limit we may make the approximations $\cos\theta \approx 1$ and $\sin\theta \approx \theta$. However, for the phase factor in the exponent of the integrals $I_{00} \ldots I_{14}$ we need to retain the second-order term, i.e. $\cos\theta \approx 1 - \theta^2/2$, because the first-order term alone would cancel out the θ dependence. For small arguments x, the Bessel functions behave like $J_n(x) \approx x^n$. Using these approximations, a comparison of the integrals $I_{00} \ldots I_{14}$ shows that the integral I_{00} is of lowest order in θ, followed by I_{11} and I_{12}. Whereas I_{00} defines the paraxial Gaussian mode, the other two remaining integrals determine the paraxial Hermite–Gaussian (1, 0) and (0, 1) modes. In

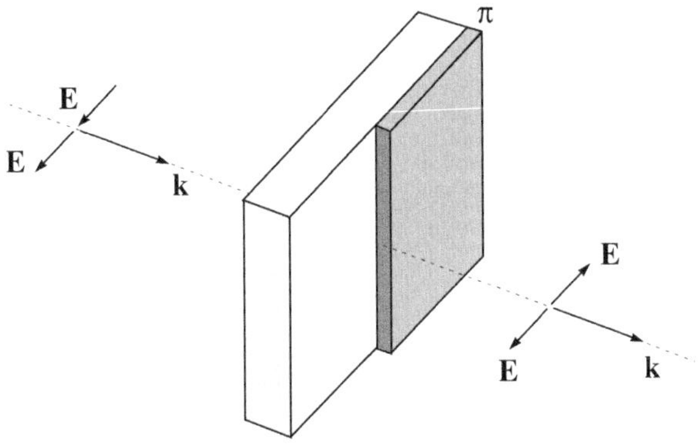

Fig. 3.15 Generation of a Hermite–Gaussian (10) beam. A fundamental Gaussian beam is bisected at the edge of a 180° phase plate. The polarization of the incident beam is perpendicular to the edge of the phase plate. The arrangement delays one half of the beam by 180° and therefore favors the conversion to the Hermite–Gaussian (10) mode. A subsequent spatial filter rejects any modes of higher order than the (10) mode.

principle, the integration of I_{00}, I_{10} and I_{11} can now be carried out analytically. However, since the results lead to inconvenient Lommel functions we reduce our discussion to the focal plane $z=0$. Furthermore, we assume an overfilled back-aperture of the lens ($f_0 \gg 1$) so that the apodization function $f_w(\theta)$ can be considered constant. Using the substitution $x = k\rho\theta$ we find

$$I_{00} \approx \frac{2}{k\rho} \int_0^{k\rho\theta_{\max}} x\, J_0(x)\, \mathrm{d}x = 2\theta_{\max}^2 \frac{J_1(k\rho\theta_{\max})}{k\rho\theta_{\max}}. \tag{3.77}$$

The paraxial field of the focused Gaussian beam in the focal plane turns out to be

$$\mathbf{E} \approx -\mathrm{i}kf\,\theta_{\max}^2 E_0\, \mathrm{e}^{-\mathrm{i}kf} \frac{J_1(k\rho\theta_{\max})}{k\rho\theta_{\max}} \mathbf{n}_x. \tag{3.78}$$

This is the familiar expression for the point-spread function in the paraxial limit. Abbe's and Rayleigh's definitions of the resolution limit are closely related to the expression above as we shall see in Section 4.1. The focal fields of the $(1,0)$ and $(0,1)$ modes in the paraxial limit can be derived in a similar way as

(1,0) mode:

$$\mathbf{E} \propto \theta_{\max}^3 [J_2(k\rho\theta_{\max})/(k\rho\theta_{\max})] \cos\varphi\; \mathbf{n}_x, \tag{3.79}$$

(0,1) mode:

$$\mathbf{E} \propto \theta_{\max}^3 [J_2(k\rho\theta_{\max})/(k\rho\theta_{\max})] \sin\varphi\; \mathbf{n}_x. \tag{3.80}$$

In all cases, the radial dependence of the paraxial focal fields is described by Bessel functions, *not* by the original Gaussian envelope. After passing through the lens the beam shape in the focal plane becomes oscillatory. These spatial oscillations can be viewed as diffraction lobes and are a consequence of the boundary conditions imposed by the aplanatic

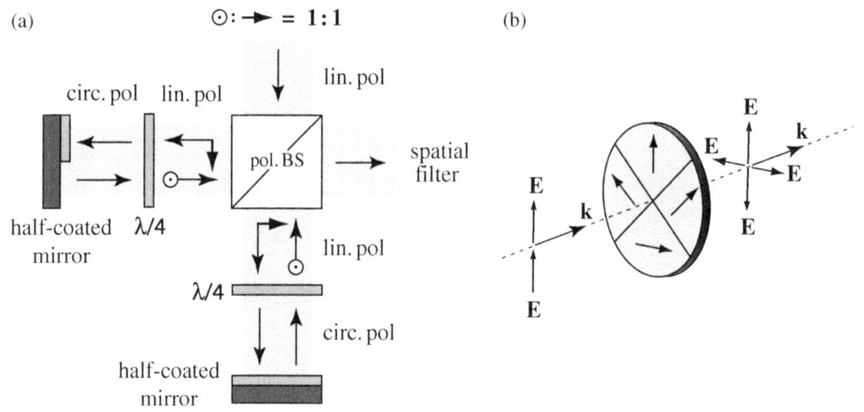

Fig. 3.16 Two different mode-conversion schemes for the generation of radially and azimuthally polarized modes. (a) Using a Twyman–Green interferometer. The incident beam is polarized at 45° and is split by a polarizing beamsplitter into two orthogonally polarized beams of equal power. Each beam is then turned circularly polarized and reflected off a half-coated end mirror. (b) Using a "composite waveplate" consisting of four quadrants with different optical axes. Each segment is oriented such that the field is rotated to point in the radial direction. In both schemes, the outgoing beam needs to be spatially filtered to reject unwanted higher-order modes. (Abbreviations: circ. pol, circular polarization; lin. pol, linear polarization). See the text for details.

lens. We have assumed $f_0 \rightarrow \infty$ and we can reduce the oscillatory behavior by reducing f_0 (see Fig. 3.10). However, this is at the expense of the spot size. The fact that the spot shape is described by an Airy function and not by a Gaussian function is very important. In fact, there are no freely propagating Gaussian beams! The reason is, as outlined in Section 3.2.1, that a Gaussian profile has a Gaussian Fourier spectrum, which is never zero and only asymptotically approaches zero as $k_x, k_y \rightarrow \infty$. Thus, for a Gaussian profile we need to include evanescent components, even if their contribution is small. The oscillations in the Airy profile arise from the hard cut-off at high spatial frequencies. The smoother this cut-off the less oscillatory the beam profile will be.

3.9 Focusing near planar interfaces

Many applications in optics involve laser beams that are strongly focused near planar surfaces. Examples are confocal microscopy, for which objective lenses with NA > 1 are used, optical microscopy or data storage based on solid immersion lenses, and optical tweezers, whereby laser light is focused into a liquid to trap tiny particles. The angular spectrum representation is well suited to solve for the fields since the planar interface is a constant coordinate surface. For simplicity we assume that we have a single interface between two dielectric media with indices n_1 and n_2 (see Fig. 3.17). The interface is located at $z = z_0$ and the focused field $\mathbf{E}_f$ illuminates the interface from the left ($z < z_0$). While the spatial frequencies k_x and k_y are the same on each side of the interface, k_z is not. Therefore, we

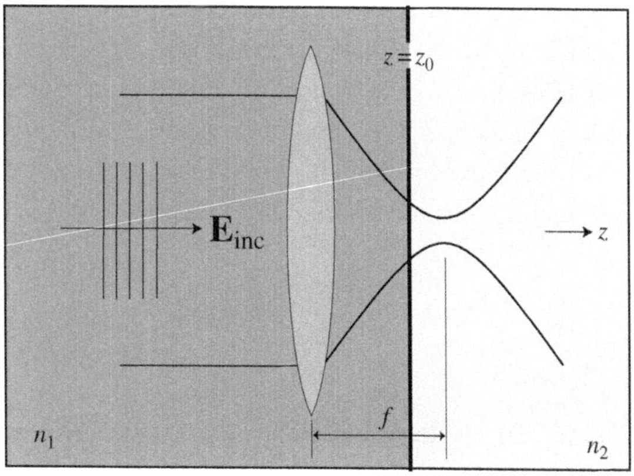

Fig. 3.17 Focusing of a laser beam near an interface at $z = z_0$ between two dielectric media with refractive indices n_1 and n_2.

specify k_z in the domain $z < z_0$ by k_{z_1} defined by $k_{z_1} = (k_1^2 - k_x^2 - k_y^2)^{1/2}$. Similarly we define $k_{z_2} = (k_2^2 - k_x^2 - k_y^2)^{1/2}$ for the domain $z > z_0$. The wavenumbers are determined by $k_1 = (\omega/c)n_1$ and $k_2 = (\omega/c)n_2$, respectively.

The interface leads to reflection and transmission. Therefore, the total field can be represented as

$$\mathbf{E} = \begin{cases} \mathbf{E}_\mathrm{f} + \mathbf{E}_\mathrm{r}, & z < z_0, \\ \mathbf{E}_\mathrm{t}, & z > z_0, \end{cases} \tag{3.81}$$

where $\mathbf{E}_\mathrm{r}$ and $\mathbf{E}_\mathrm{t}$ represent the reflected and transmitted fields, respectively. The refraction of plane waves at planar interfaces is described by Fresnel reflection coefficients (r^s, r^p) and transmission coefficients (t^s, t^p), which were defined in Chapter 2 (Eqs. (2.51) and (2.52)). As indicated by the superscripts, these coefficients depend on the polarization of the field. We therefore need to split each plane wave component in the angular spectrum representation of the field $\mathbf{E}$ into an s-polarized part and a p-polarized part,

$$\mathbf{E} = \mathbf{E}^{(\mathrm{s})} + \mathbf{E}^{(\mathrm{p})}. \tag{3.82}$$

$\mathbf{E}^{(\mathrm{s})}$ is parallel to the interface while $\mathbf{E}^{(\mathrm{p})}$ is perpendicular to the wavevector $\mathbf{k}$ and $\mathbf{E}^{(\mathrm{s})}$. The decomposition of the incoming focused field $\mathbf{E}_\mathrm{f}$ into s- and p-polarized fields has already been done in Section 3.5. According to Eq. (3.39) we obtain the s- and p-polarized fields by projecting $\mathbf{E}_\mathrm{f}$ along the unit vectors $\mathbf{n}_\theta$ and $\mathbf{n}_\phi$, respectively. Equation (3.43) represents the refracted far-field as a sum of s- and p-polarized fields expressed in terms of θ and ϕ. Using the substitutions of Eq. (3.44) we are able to express the far-field in terms of the spatial frequencies k_x and k_y.

In the case in which $\mathbf{E}_\mathrm{f}$ originates from a paraxial beam polarized in the x-direction we can express the far-field as (cf. Eq. (3.51))

$$\mathbf{E}_\infty = E_\mathrm{inc}\left(\frac{k_x}{k}, \frac{k_y}{k}\right)\begin{bmatrix} k_y^2 + k_x^2 k_{z_1}/k_1 \\ -k_x k_y + k_x k_y k_{z_1}/k_1 \\ 0 - (k_x^2 + k_y^2)k_x/k_1 \end{bmatrix}\frac{\sqrt{k_{z_1}/k_1}}{k_x^2 + k_y^2}, \tag{3.83}$$

where the first terms in the bracket specify the s-polarized field and the second ones the p-polarized field. Notice that according to Fig. 3.16 we consider a lens with the same medium on both sides, i.e. $n_1 = n = n'$. $\mathbf{E}_\infty$ is the asymptotic far-field in the direction of the unit vector $\mathbf{s} = (k_x/k, k_y/k, k_{z_1}/k)$ and corresponds to the field on the surface of the reference sphere of the focusing lens. In terms of $\mathbf{E}_\infty$, the angular spectrum representation of the incident focused beam is given by (c.f. Eq. (3.33))

$$\mathbf{E}_\mathrm{f}(x,y,z) = -\frac{\mathrm{i} f \mathrm{e}^{-\mathrm{i}k_1 f}}{2\pi} \iint\limits_{k_x,k_y} \mathbf{E}_\infty\left(\frac{k_x}{k},\frac{k_y}{k}\right) \frac{1}{k_{z_1}} \mathrm{e}^{\mathrm{i}[k_x x + k_y y + k_{z_1} z]}\, \mathrm{d}k_x\, \mathrm{d}k_y. \tag{3.84}$$

To determine the reflected and transmitted fields ($\mathbf{E}_\mathrm{r}$, $\mathbf{E}_\mathrm{t}$) we define the following angular spectrum representations:

$$\mathbf{E}_\mathrm{r}(x,y,z) = -\frac{\mathrm{i} f \mathrm{e}^{-\mathrm{i}k_1 f}}{2\pi} \iint\limits_{k_x,k_y} \mathbf{E}_\mathrm{r}^\infty\left(\frac{k_x}{k},\frac{k_y}{k}\right) \frac{1}{k_{z_1}} \mathrm{e}^{\mathrm{i}[k_x x + k_y y - k_{z_1} z]}\, \mathrm{d}k_x\, \mathrm{d}k_y, \tag{3.85}$$

$$\mathbf{E}_\mathrm{t}(x,y,z) = -\frac{\mathrm{i} f \mathrm{e}^{-\mathrm{i}k_1 f}}{2\pi} \iint\limits_{k_x,k_y} \mathbf{E}_\mathrm{t}^\infty\left(\frac{k_x}{k},\frac{k_y}{k}\right) \frac{1}{k_{z_2}} \mathrm{e}^{\mathrm{i}[k_x x + k_y y + k_{z_2} z]}\, \mathrm{d}k_x\, \mathrm{d}k_y. \tag{3.86}$$

Notice that in order to ensure that the reflected field propagates in the backward direction we had to change the sign of k_{z_1} in the exponent. We also made sure that the transmitted wave propagates with the longitudinal wavenumber k_{z_2}.

In the next step we invoke the boundary conditions at $z = z_0$, which leads to explicit expressions for the as-yet-undefined far-fields $\mathbf{E}_\mathrm{r}^\infty$ and $\mathbf{E}_\mathrm{t}^\infty$. Using the Fresnel reflection or transmission coefficients we obtain

$$\mathbf{E}_\mathrm{r}^\infty = -E_\mathrm{inc}\left(\frac{k_x}{k},\frac{k_y}{k}\right) \mathrm{e}^{2\mathrm{i}k_{z_1}z_0} \begin{bmatrix} -r^\mathrm{s} k_y^2 + r^\mathrm{p} k_x^2 k_{z_1}/k_1 \\ r^\mathrm{s} k_x k_y + r^\mathrm{p} k_x k_y k_{z_1}/k_1 \\ 0 + r^\mathrm{p}(k_x^2 + k_y^2) k_x/k_1 \end{bmatrix} \frac{\sqrt{k_{z_1}/k_1}}{k_x^2 + k_y^2}, \tag{3.87}$$

$$\mathbf{E}_\mathrm{t}^\infty = E_\mathrm{inc}\left(\frac{k_x}{k},\frac{k_y}{k}\right) \mathrm{e}^{\mathrm{i}(k_{z_1} - k_{z_2})z_0} \begin{bmatrix} t^\mathrm{s} k_y^2 + t^\mathrm{p} k_x^2 k_{z_2}/k_2 \\ -t^\mathrm{s} k_x k_y + t^\mathrm{p} k_x k_y k_{z_2}/k_2 \\ 0 - t^\mathrm{p}(k_x^2 + k_y^2) k_x/k_2 \end{bmatrix} \frac{k_{z_2}}{k_{z_1}} \frac{\sqrt{k_{z_1}/k_1}}{k_x^2 + k_y^2}. \tag{3.88}$$

These equations together with Eqs. (3.83)–(3.86) define the solution of our problem. They hold for an interface between two materials characterized by constant ε_i and μ_i. This is straightforward to verify by evaluating the boundary conditions at $z = z_0$ (Problem 3.7). We are now able to evaluate the field distribution near a plane interface illuminated by a strongly focused laser beam. The field depends on the amplitude profile E_inc of the incident paraxial beam (cf. Eqs. (3.52)–(3.54)) and on the defocus z_0. The defocus essentially introduces a phase factor into the expressions for $\mathbf{E}_\mathrm{r}^\infty$ and $\mathbf{E}_\mathrm{t}^\infty$. Although we concentrated on a single interface, the results are easily adapted to a multiply layered interface by introducing generalized Fresnel reflection/transmission coefficients that account for the total structure (cf. Ref. [17]).

In the next step, we can use the relations (3.44) to perform a transformation to spherical coordinates. As before, we are able to reduce the double integrals to single integrals by involving Bessel functions. We avoid going into further details and instead discuss some important aspects that result from this theory.

In the example of Fig. 3.18 a Gaussian beam is focused by an aplanatic objective lens of NA $=1.4$ on a glass/air interface at $z_0=0$. The most characteristic features in the field plots are the standing-wave patterns in the denser medium. These standing-wave patterns occur at angles θ beyond the critical angle of total internal reflection θ_c. To understand this let us have a look at a single plane wave in the angular spectrum representation of the incident

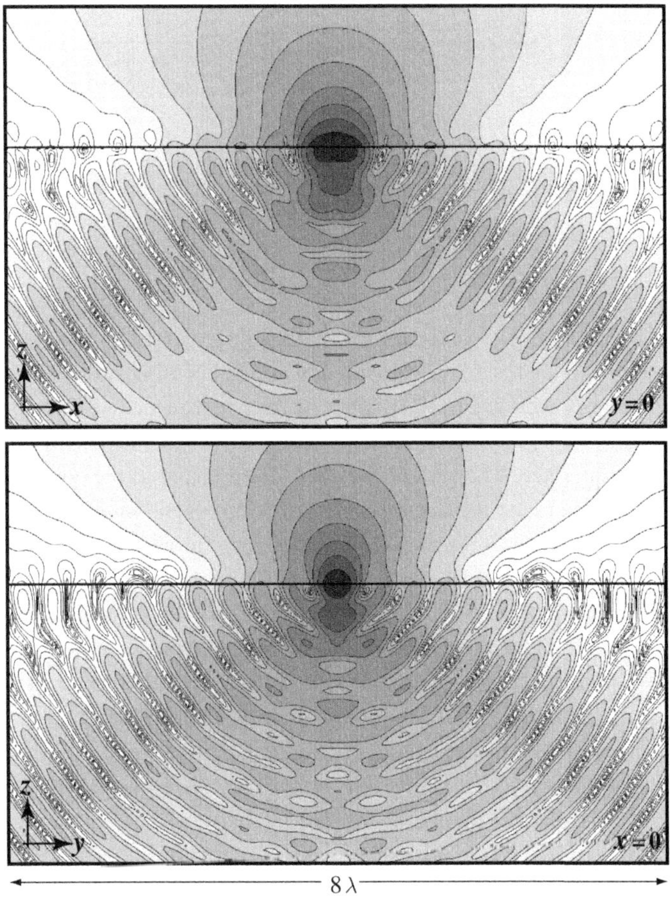

Fig. 3.18 Contour plots of constant $|\mathbf{E}|^2$ in the focal region of a Gaussian beam (NA $=1.4$, $n=1.518$, $f_0=2$) focused on a glass/air interface ($n_1=1.518$, $n_2=1$). A logarithmic scaling is used, with a factor of 2 difference between adjacent contour lines. The critical angle for total internal reflection is $\theta_c=41.2°$. All plane-wave components incident from angles larger than θ_c are totally reflected at the interface and interfere with the incoming waves.

focused field $\mathbf{E}_\mathrm{f}$. This plane wave is characterized by the two transverse wavenumbers k_x and k_y, its polarization, and the complex amplitude given by the Fourier spectrum $\hat{\mathbf{E}}_\mathrm{f}$. The transverse wavenumbers are the same on each side of the interface, but the longitudinal wavenumbers k_z are not, since they are defined as

$$k_{z_1} = \sqrt{k_1^2 - (k_x^2 + k_y^2)}, \qquad k_{z_2} = \sqrt{k_2^2 - (k_x^2 + k_y^2)}. \tag{3.89}$$

On eliminating k_x and k_y we obtain

$$k_{z_2} = \sqrt{k_{z_1}^2 + (k_2^2 - k_1^2)}. \tag{3.90}$$

Let θ denote the angle of incidence of the plane wave so that

$$k_{z_1} = k_1 \cos\theta. \tag{3.91}$$

Equation (3.90) can then be written as

$$k_{z_2} = k_2 \sqrt{1 - \frac{k_1^2}{k_2^2}\sin^2\theta}. \tag{3.92}$$

It follows that k_{z_2} can be either real or imaginary, depending on the sign of the expression under the square root. This in turn depends on the angle θ. We find that for angles larger than

$$\theta_\mathrm{c} = \arcsin\left(\frac{n_2}{n_1}\right) \tag{3.93}$$

k_{z_2} is imaginary. Thus, for $\theta > \theta_\mathrm{c}$ the plane wave considered is totally reflected at the interface, giving rise to an evanescent wave on the other side of the interface. The standing-wave patterns seen in Fig. 3.18 are a direct consequence of this phenomenon: all the supercritical ($\theta > \theta_\mathrm{c}$) plane wave components of the incident focused field are totally reflected at the interface. The standing-wave pattern is due to the equal superposition of incident and reflected plane-wave components. Owing to total internal reflection an appreciable amount of laser power is reflected at the interface. The ratio of reflected to transmitted power can be further increased by using a larger filling factor or a higher numerical aperture. For example, in applications based on solid immersion lenses with numerical apertures of 1.8–2 over 90% of the beam power is reflected at the interface.

An inspection of the focal spot reveals that the interface further increases the ellipticity of the spot shape. Along the polarization direction (x) the spot is almost twice as big as in the direction perpendicular to it (y). Furthermore, the interface enhances the strength of the longitudinal field component E_z. At the interface, just outside the focusing medium ($z > -z_0$), the maximum relative intensity values for the different field components are $\mathrm{Max}[E_y^2]/\mathrm{Max}[E_x^2] = 0.03$ and $\mathrm{Max}[E_z^2]/\mathrm{Max}[E_x^2] = 0.43$. Thus, compared with the situation in which no interface is present (cf. Fig. 3.11), the longitudinal field intensity is roughly four times stronger. How can we understand this phenomenon? According to the boundary conditions at the interface, the transverse field components E_x and E_y have to be continuous across the interface. However, the longitudinal field scales as

$$E_{z_1}\varepsilon_1 = E_{z_2}\varepsilon_2. \tag{3.94}$$

With $\varepsilon_2 = 2.304$ we find that E_z^2 changes by a factor of 5.3 from one side to the other side of the interface. This qualitative explanation is in reasonable agreement with the calculated values. In the focal plane, the longitudinal field has its two maxima just to the side of the optical axis. These two maxima are aligned along the polarization direction and give rise to the elongated spot size. The relative magnitude of $\mathrm{Max}[E_y^2]$ is still small, but it is increased by a factor of 10 by the presence of the interface.

In order to map the dipole orientation of arbitrarily oriented single molecules it is desirable that all three excitation field components (E_x, E_y, E_z) in the focus are of comparable magnitude. It has been demonstrated that this can be achieved by annular illumination for which the center part of the focused laser beam is suppressed [18]. This can be achieved by placing a central obstruction such as a circular disk in the excitation beam. In this situation, the integration of plane-wave components runs over the angular range $[\theta_{\mathrm{min}} \ldots \theta_{\mathrm{max}}]$ instead of, as before, over the full range $[0 \ldots \theta_{\mathrm{max}}]$. By using annular illumination we reject the plane-wave components with propagation directions close to the optical axis, thereby suppressing the transverse electric field components. As a consequence, the longitudinal field components in the focus will be enhanced compared with the transverse components. Furthermore, the local polarization of the interface due to the longitudinal fields gives rise to a strong enhancement of the E_y fields. Hence, strong longitudinal fields are a prerequisite for generating strong E_y fields close to interfaces. It is possible to prepare the annular beam such that the three patterns in Fig. 3.11(c)–(e) are of comparable magnitude [18].

3.10 The reflected image of a strongly focused spot

It is interesting to further investigate the properties of the reflected field $\mathbf{E}_{\mathrm{r}}$ given by Eqs. (3.85) and (3.87). The image of the reflected spot can be experimentally recorded as shown in Fig. 3.19. A 45° beamsplitter reflects part of the incoming beam upwards where it is focused by a high NA objective lens near a planar interface. The distance between the focus ($z = 0$) and the interface is designated by z_0. The reflected field is collected by the same lens, transmitted through the beamsplitter and then focused by a second lens onto the image plane. There are four different media involved and we specify them with the refractive indices defined in Fig. 3.19. We are interested in calculating the resulting field distribution in the image plane. It will be shown that, for the case in which the beam is incident from the optically denser medium, the image generated by the reflected light is strongly aberrated.

The reflected far-field $\mathbf{E}_{\mathrm{r}}^{\infty}$ before it is refracted by the first lens has been calculated in Eq. (3.87). It is straightforward to refract this field at the two lenses and refocus it onto the image plane. The two lenses perform transformations between spherical and cylindrical systems. In Section 3.5 it has been shown that the lens refracts the unit vector $\mathbf{n}_\rho$ into the unit vector $\mathbf{n}_\theta$, or vice versa, whereas the unit vector $\mathbf{n}_\phi$ remains unaffected. In order to oversee the entire imaging process we follow the light path from the beginning. The incoming field $\mathbf{E}_{\mathrm{inc}}$ is an x-polarized, paraxial beam defined as (Eq. (3.49))

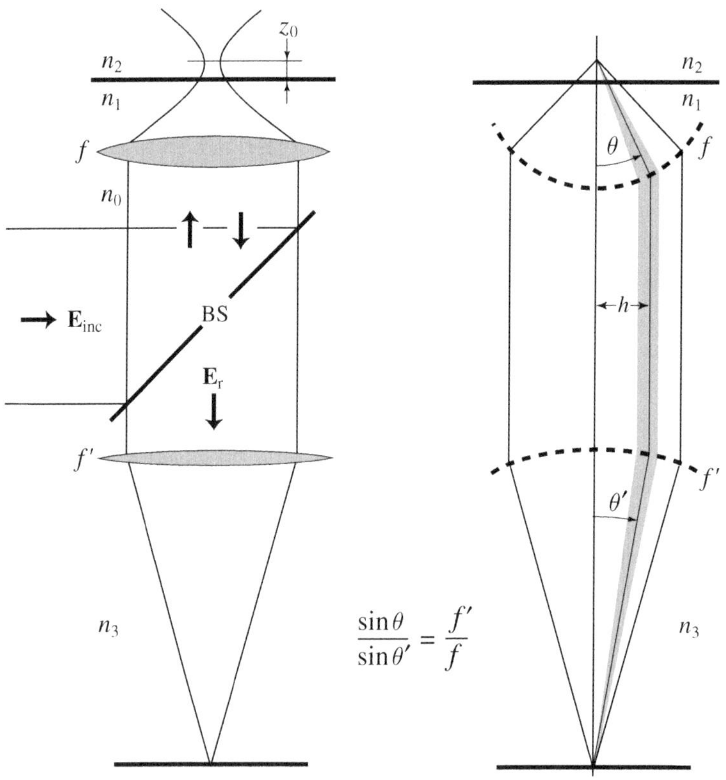

Fig. 3.19 Experimental set-up for the investigation of the reflected image of a diffraction-limited focused spot. A linearly polarized beam is reflected by a beamsplitter (BS) and focused by a high-NA objective lens with focal radius f onto an interface between two dielectric media with refractive indices n_1 and n_2. The reflected field is collected by the same lens, transmitted through the beamsplitter and refocused by a second lens with focal radius f'.

$$\mathbf{E}_{\text{inc}} = E_{\text{inc}}\mathbf{n}_x, \tag{3.95}$$

where E_{inc} is an arbitrary beam profile. Expressed in cylindrical coordinates the field has the form

$$\mathbf{E}_{\text{inc}} = E_{\text{inc}}\big[\cos\phi\,\mathbf{n}_\rho - \sin\phi\,\mathbf{n}_\phi\big]. \tag{3.96}$$

After refraction at the first lens f it turns into

$$\mathbf{E} = E_{\text{inc}}\big[\cos\phi\,\mathbf{n}_\theta - \sin\phi\,\mathbf{n}_\phi\big]\sqrt{\frac{n_0}{n_1}}\,(\cos\theta)^{1/2}. \tag{3.97}$$

The field is now reflected at the interface. The Fresnel reflection coefficient r^{p} accounts for the reflection of $\mathbf{n}_\theta$-polarized fields whereas r^{s} accounts for the reflection of $\mathbf{n}_\phi$-polarized fields. We obtain for the reflected field

$$\mathbf{E} = E_{\text{inc}}e^{2ik_{z_1}z_0}\big[-\cos\phi\,r^{\text{p}}\,\mathbf{n}_\theta - \sin\phi\,r^{\text{s}}\mathbf{n}_\phi\big]\sqrt{\frac{n_0}{n_1}}\,(\cos\theta)^{1/2}, \tag{3.98}$$

where z_0 denotes the defocus (cf. Eq. (3.87)). Next, the field is refracted by the same lens f as

$$\mathbf{E} = E_{\text{inc}} e^{2ik_{z_1} z_0} \left[-\cos\phi \, r^{\text{p}} \mathbf{n}_\rho - \sin\phi \, r^{\text{s}} \mathbf{n}_\phi \right], \tag{3.99}$$

and propagates as a collimated beam in the negative-z direction. Expressed in Cartesian field components the field reads as

$$\mathbf{E}_{\text{r}}^{\infty} = -E_{\text{inc}} e^{2ik_{z_1} z_0} \Big[[\cos^2\!\phi \, r^{\text{p}} - \sin^2\!\phi \, r^{\text{s}}] \mathbf{n}_x + \sin\phi \cos\phi [r^{\text{p}} + r^{\text{s}}] \mathbf{n}_y \Big]. \tag{3.100}$$

This is the field immediately after refraction at the reference sphere f. For an incident field focused on a perfectly reflecting interface located at $z_0 = 0$ the reflection coefficients are $r^{\text{p}} = 1$ and $r^{\text{s}} = -1$.[3] In this case we simply obtain $\mathbf{E}_{\text{ref}}^{\infty} = -E_{\text{inc}} \, \mathbf{n}_x$, which is, apart from the minus sign, identical with the assumed input field of Eq. (3.49). The difference in sign indicates that the reflected field is "upside down."

In order to calculate the reflected collimated beam anywhere along the optical axis we have to make the substitutions $\sin\theta = \rho/f$ and $\cos\theta = [1 - (\rho/f)^2]^{1/2}$, where ρ denotes the radial distance from the optical axis (see Problem 3.8). This allows us to plot the field distribution in a cross-sectional plane through the collimated reflected beam. We find that the Fresnel reflection coefficients modify the polarization and amplitude profile of the beam, and, more importantly, also its phase profile. For the case of no defocus ($z_0 = 0$) phase variations arise only at radial distances $\rho > \rho_{\text{c}}$ for which the Fresnel reflection coefficients become complex numbers. The critical distance corresponds to $\rho_{\text{c}} = f n_2/n_1$ and is the radial distance associated with the critical angle of total internal reflection ($\theta_{\text{c}} = \arcsin(n_2/n_1)$). Since $\rho_{\text{c}} < f$ there are no aberrations if $n_2 > n_1$.

We now proceed to the refraction at the second lens f'. Immediately after refraction the reflected field reads as

$$\mathbf{E} = E_{\text{inc}} e^{2ik_{z_1} z_0} \left[-\cos\phi \, r^{\text{p}} \mathbf{n}_{\theta'} - \sin\phi \, r^{\text{s}} \mathbf{n}_\phi \right] \sqrt{\frac{n_0}{n_3}} \, (\cos\theta')^{1/2}, \tag{3.101}$$

where we introduced the new azimuth angle θ' as defined in Fig. 3.19. The field now corresponds to the far-field $\mathbf{E}_{\text{r}}^{\infty}$ that we need in Eq. (3.33) to calculate the field distribution in the image space. We express this field in Cartesian field components using the relations in Eqs. (3.41)–(3.42) for $\mathbf{n}_{\theta'}$ and $\mathbf{n}_\phi$ and obtain

$$\mathbf{E}_{\text{r}}^{\infty} = -E_{\text{inc}} e^{2ik_{z_1} z_0} \begin{bmatrix} r^{\text{p}} \cos\theta' \cos^2\!\phi & - & r^{\text{s}} \sin^2\!\phi \\ r^{\text{p}} \cos\theta' \sin\phi \cos\phi & + & r^{\text{s}} \sin\phi \cos\phi \\ r^{\text{p}} \sin\theta' \cos\phi & + & 0 \end{bmatrix} \sqrt{\frac{n_0}{n_3}} \, (\cos\theta')^{1/2}. \tag{3.102}$$

This far-field can now be introduced into Eq. (3.47), which, after being adapted to the current situation, reads as

$$\mathbf{E}(\rho, \varphi, z) = -\frac{ik_3 f' e^{-ik_3 f'}}{2\pi} \int_0^{\theta'_{\max}} \int_0^{2\pi} \mathbf{E}_{\text{r}}^{\infty}(\theta', \phi) e^{-ik_3 z \cos\theta'} e^{ik_3 \rho \sin\theta' \cos(\phi - \varphi)} \sin\theta' \, d\phi \, d\theta'. \tag{3.103}$$

[3] Notice that the reflection coefficients r^{s} and r^{p} for a plane wave at normal incidence differ by a factor of -1, i.e. $r^{\text{s}}(\theta = 0) = -r^{\text{p}}(\theta = 0)$.

Notice that we had to change the sign in one of the exponents in order to ensure that the field propagates in the negative-z direction. To proceed, we could express the longitudinal wavenumbers k_{z_1} and k_{z_2} in terms of the angle θ'. This would also make the reflection and transmission coefficients functions of θ'. However, it is more convenient to work with θ and transform the integral in Eq. (3.103) correspondingly.

As indicated in Fig. 3.19 the angles θ and θ' are related by

$$\frac{\sin\theta}{\sin\theta'} = \frac{f'}{f}, \tag{3.104}$$

which allows us to express the new longitudinal wavenumber k_{z_3} in terms of θ as

$$k_{z_3} = k_3\sqrt{1-(f/f')^2\,\sin^2\theta}. \tag{3.105}$$

With these relationships we can perform a substitution in Eq. (3.105) and represent the integration variables by θ and ϕ. The Fresnel reflection coefficients $r_{\mathrm{s}}(\theta)$ and $r_{\mathrm{p}}(\theta)$ are given by Eqs. (2.51) together with the expressions for the longitudinal wavenumbers k_{z_1} and k_{z_2} in Eqs. (3.91) and (3.92). For the lowest three Hermite–Gaussian beams, explicit expressions for $E_{\mathrm{inc}}(\theta,\phi)$ have been stated in Eqs. (3.52)–(3.54) and the angular dependence in ϕ can be integrated analytically by using Eq. (3.57). Thus, we are now able to calculate the field near the image focus.

In practically all optical systems the second focusing lens has a much larger focal length than the first one, i.e. $f/f' \ll 1$. We can therefore reduce the complexity of the expressions considerably by making the approximation

$$\left[1 \pm (f/f')^2\sin^2\theta\right]^{1/n} \approx 1 \pm \frac{1}{n}\left(\frac{f}{f'}\right)^2 \sin^2\theta. \tag{3.106}$$

If we retain only the lowest orders in f/f', the image field can be represented by

$$\begin{aligned}\mathbf{E}(\rho,\varphi,z) = -\frac{\mathrm{i}k_3 f' \mathrm{e}^{-\mathrm{i}k_3(z+f')}}{2\pi}\frac{f^2}{f'^2}\int_0^{\theta_{\max}}\int_0^{2\pi} \mathbf{E}_{\mathrm{r}}^{\infty}(\theta,\phi)\mathrm{e}^{(\mathrm{i}/2)k_3 z(f/f')^2\sin^2\theta} \\ \times\, \mathrm{e}^{\mathrm{i}k_3\rho(f/f')\sin\theta\,\cos(\phi-\varphi)}\sin\theta\cos\theta\,\mathrm{d}\phi\,\mathrm{d}\theta,\end{aligned} \tag{3.107}$$

where $\mathbf{E}_{\mathrm{r}}^{\infty}$ reads as

$$\mathbf{E}_{\mathrm{r}}^{\infty}(\theta,\phi) = -E_{\mathrm{inc}}(\theta,\phi)\mathrm{e}^{2\mathrm{i}k_1 z_0\cos\theta}\begin{bmatrix} r^{\mathrm{p}}\cos^2\phi - r^{\mathrm{s}}\sin^2\phi \\ \sin\phi\cos\phi(r^{\mathrm{p}}+r^{\mathrm{s}}) \\ 0 \end{bmatrix}\sqrt{\frac{n_0}{n_3}}. \tag{3.108}$$

In order to keep the discussion within bounds we will assume that the incident field E_{inc} is a fundamental Gaussian beam as defined in Eq. (3.52). Using the relations in Eq. (3.57) we can integrate the ϕ dependence and finally obtain

$$\mathbf{E}(\rho,\varphi,z) = E_0\,\frac{k_3 f^2}{2f'i}\mathrm{e}^{-\mathrm{i}k_3(z+f')}\sqrt{\frac{n_0}{n_3}}\left[(I_{0\mathrm{r}} - I_{2\mathrm{r}}\cos(2\varphi))\mathbf{n}_x - I_{2\mathrm{r}}\sin(2\varphi)\mathbf{n}_y\right], \tag{3.109}$$

with

$$I_{0\mathrm{r}}(\rho, z) = \int_0^{\theta_{\max}} f_w(\theta)\cos\theta \ \sin\theta\big[r_\mathrm{p}(\theta) - r_\mathrm{s}(\theta)\big]J_0(k_3\rho \sin\theta f/f') \times \exp\Big[(\mathrm{i}/2)k_3 z(f/f')^2 \sin^2\theta + 2\mathrm{i}k_1 z_0 \cos\theta\Big] \,\mathrm{d}\theta, \tag{3.110}$$

$$I_{2\mathrm{r}}(\rho, z) = \int_0^{\theta_{\max}} f_w(\theta)\cos\theta \ \sin\theta\big[r_\mathrm{p}(\theta) + r_\mathrm{s}(\theta)\big]J_2(k_3\rho \sin\theta f/f') \times \exp\Big[(\mathrm{i}/2)k_3 z(f/f')^2 \sin^2\theta + 2\mathrm{i}k_1 z_0 \cos\theta\Big] \,\mathrm{d}\theta, \tag{3.111}$$

where f_w is the apodization function defined in Eq. (3.56). We find that the spot depends on the Fresnel reflection coefficients and the defocus defined by z_0. The latter simply adds an additional phase delay for each plane-wave component. If the upper medium n_2 is a perfect conductor we have $r_\mathrm{p} = -r_\mathrm{s} = 1$ and the integral $I_{2\mathrm{r}}$ vanishes. In this case the reflected spot is linearly polarized and rotationally symmetric.

In order to discuss the field distributions in the image plane we choose $n_1 = 1.518$ for the object space, $n_3 = 1$ for the image space, and a numerical aperture of 1.4 ($\theta_{\max} = 67.26°$) for the objective lens. For the ideally reflecting interface, the images in the lower row of Fig. 3.20 depict the electric field intensity $|\mathbf{E}_\mathrm{r}|^2$ as a function of slight defocus. It is evident that the spot shape and size are not significantly affected by the defocus. However, as shown in the upper row in Fig. 3.20 the situation is very different if the medium beyond the interface has a lower index than the focusing medium, i.e. if $n_2 < n_1$. In this case, the reflected spot changes strongly as a function of defocus. The spot shape deviates considerably from a Gaussian spot and resembles the spot of an optical system with axial astigmatism. The overall size of the spot is increased and the polarization is not preserved since $I_{0\mathrm{r}}$ and $I_{2\mathrm{r}}$ are of comparable magnitude. The patterns displayed in Fig. 3.20 can be verified in the laboratory. However, some care has to be applied when using dichroic beamsplitters since they have slightly different characteristics for s- and p-polarized light. In fact, the patterns in Fig. 3.20 depend sensitively on the relative magnitudes of the two superposed polarizations. Using a polarizer in the reflected beam path allows us to examine the two polarizations separately, as shown in Fig. 3.21. Notice that the focus does not coincide with the interface when the intensity of the reflected pattern is maximized. The focus coincides with the interface when the *center* of the reflected pattern ($I_0(\rho, z)$) has maximum intensity. The images in Figs. 3.20 and 3.21 display the electric energy density, which is the quantity that is detected by optical detectors such as a CCD. On the other hand, the *total* energy density, and the magnitude of the time-averaged Poynting vector, render rotationally symmetric patterns.

How can we understand the appearance of the highly aberrated spot in the case of a glass/air interface? The essence lies in the nature of total internal reflection. All plane-wave components with angles of incidence in the range $[0 \ldots \theta_\mathrm{c}]$, θ_c being the critical angle of total internal reflection ($\approx 41.2°$ for a glass/air interface), are partly transmitted and partly reflected at the interface. Both reflection coefficients r_s and r_p are real numbers

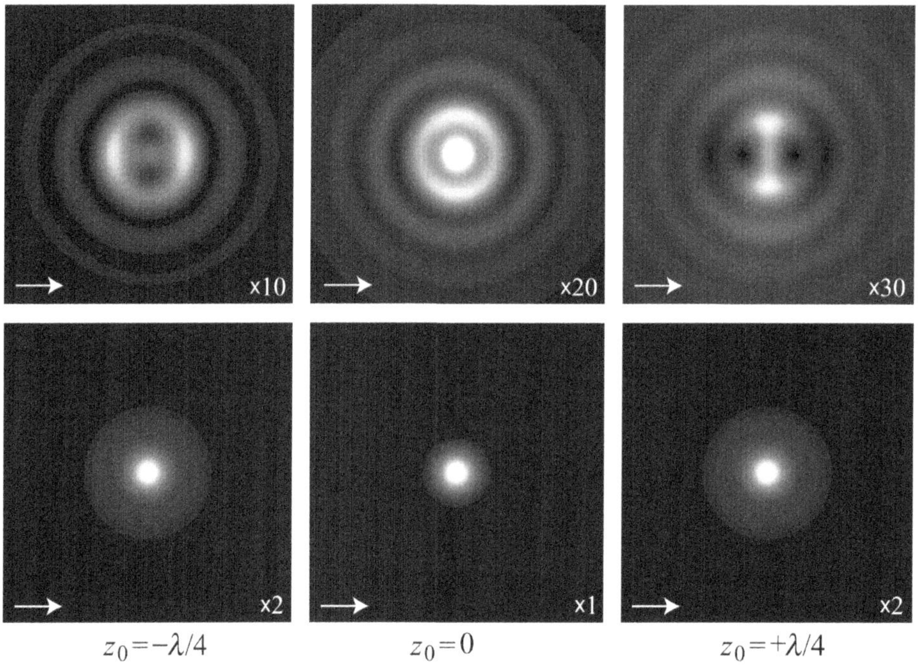

Fig. 3.20 Reflected images of a diffraction-limited focused spot. The spot is moved in steps of $\lambda/4$ across the interface. z_0 is positive (negative) when the focus is below (above) the interface. The primary focusing objective lens has a numerical aperture of 1.4. The index of refraction is $n_1 = 1.518$ and the filling factor $f_0 = 2$. The upper row shows the situation for a glass/air interface ($n_2 = 1$) and the lower row is for a glass/metal interface ($\varepsilon_2 \rightarrow -\infty$). Large aberrations are observed in the case of the glass/air interface because the totally internally reflected plane-wave components generate a second virtual focus above the interface. The arrow indicates the direction of polarization of the primary incoming beam, and the numbers indicate the factors by which the images have been multiplied to boost the contrast of the images.

and there are no phase shifts between incident and reflected waves. On the other hand, the plane-wave components in the range $[\theta_c \ldots \theta_{max}]$ are totally reflected at the interface. In this case the reflection coefficients become complex-valued functions imposing a phase shift between incident and reflected waves. This can be viewed as an additional path difference between incident and reflected waves similar to the Goos–Hänchen shift [19]. It displaces the apparent reflection point beyond the interface thereby creating a second, virtual focus [21]. In order to visualize this effect we plot in Fig. 3.22 only the scattered field (transmitted and reflected) of Fig. 3.18. If we detected this radiation on the surface of an enclosing sphere with large radius, the direction of radiation would appear as indicated by the two lines which obviously intersect above the interface. Although all reflected radiation originates at the interface, there is an apparent origin above the interface. If we follow the radiation maxima from the far-field towards the interface we see that close to the interface the radiation bends towards the focus to ensure that the origin of the radiation does indeed come from the focal spot.

We thus find the important result that the reflected light associated with the angular range $[0 \ldots \theta_c]$ originates from the real focal point on the interface, whereas the light associated

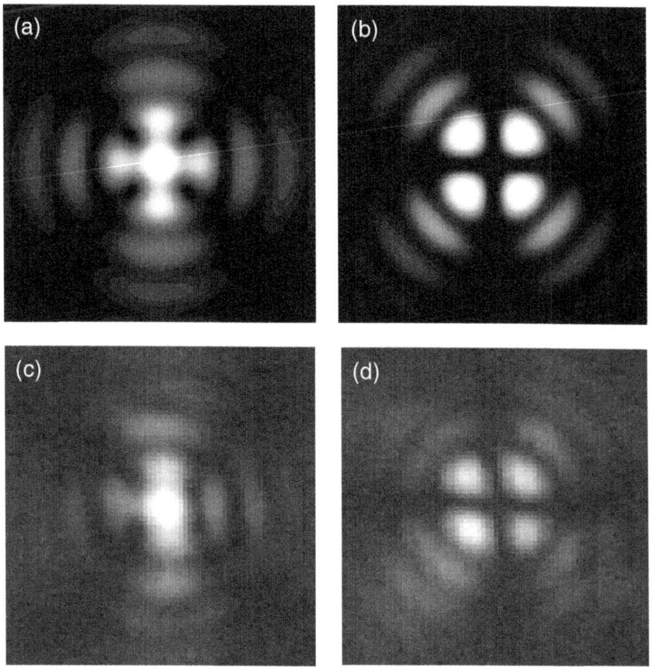

Fig. 3.21 Decomposition of the in-focus reflected image (center image of Fig. 3.20) into two orthogonal polarizations. Images (a) and (c) show polarization in the direction of incident polarization ($\boldsymbol{n}_x$); (b) and (d) show polarization perpendicular to incident polarization ($\boldsymbol{n}_y$). Images (a) and (b) are calculated patterns and (c) and (d) are experimental patterns. From [20].

with $[\theta_c \ldots \theta_{max}]$ originates from a virtual point located above the interface. To be correct, the "virtual" point above the interface is not really a geometrical point. Instead, it is made of many points distributed along the vertical axis. The waves that emanate from these points have different relative phases and give rise to a conically shaped wavefront similar to the Mach cone in fluid dynamics. The resulting toroidal aberration was first investigated by Maeker and Lehman [22].

The observation of the aberrations in the focal point's reflected image has important consequences for reflection-type confocal microscopy and data sampling. In these techniques the reflected beam is focused onto a pinhole in the image plane. Because of the aberrations of the reflected spot, most of the reflected light is blocked by the pinhole destroying the sensitivity and resolution. However, it has been pointed out that this effect can dramatically increase the contrast between metallic and dielectric sample features [21] because the reflected spot from a metal interface appears to be aberration-free. Finally, it has to be emphasized that the real focal spot on the interface remains greatly unaffected by the interface; the aberrations are associated with the reflected image alone. The understanding of the patterns in Figs. 3.20 and 3.21 proves to be very valuable for the alignment of an optical system, for example to ensure that the focal plane of a laser coincides with the glass/air interface (object plane).

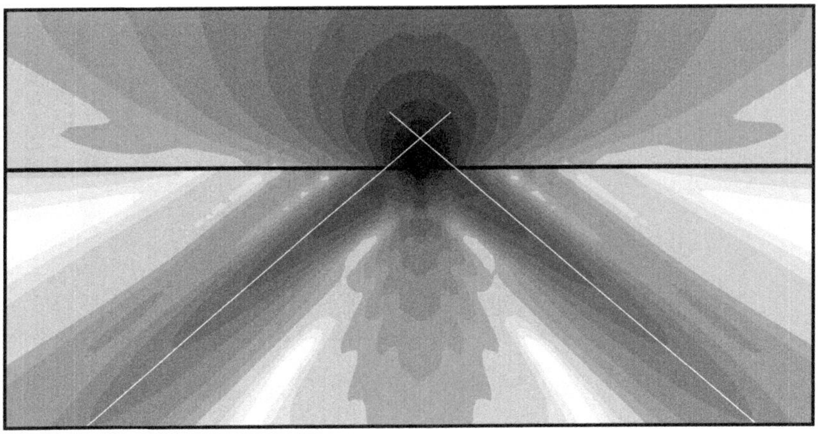

Fig. 3.22 Scattered radiation (reflected and transmitted) of a laser focused on a glass/air interface. Same parameters as in Fig. 3.18. The lines indicate the apparent direction of radiation as seen by an observer in the far-field. The lines intersect in a *virtual focus* located above the interface. While all plane-wave components in the angular range $[0 \ldots \theta_c]$ originate from the focal point on the interface, the supercritical plane-wave components emerge from an apparent spot above the interface, giving rise to the aberrations in Fig. 3.20. Image size: $16\lambda \times 31\lambda$. Logarithmic scale.

Problems

3.1 The paraxial Gaussian beam is not a rigorous solution of Maxwell's equations. Its field is therefore not divergence-free ($\nabla \cdot \mathbf{E} \neq 0$). By requiring $\nabla \cdot \mathbf{E} = 0$ one can derive an expression for the longitudinal field E_z. Assume that $E_y = 0$ everywhere and derive E_z to the lowest order for which the solution is non-zero. Sketch the distribution of $|E_z|^2$ in the focal plane.

3.2 Determine the decomposition of an arbitrary optical field into transverse electric (TE) and transverse magnetic (TM) fields. The longitudinal field E_z vanishes for the TE field, whereas H_z vanishes for the TM field.

3.3 Consider the fields emerging from a truncated hollow metal waveguide with a square cross-section and with ideally conducting walls. The side length a_0 is chosen in such a way that only the lowest-order TE_{10} mode polarized in the x-direction is supported. Assume that the fields are not influenced by the edges of the truncated side walls.

(a) Calculate the spatial Fourier spectrum of the electric field in the exit plane ($z=0$).

(b) Calculate and plot the corresponding far-field ($\mathbf{E} \cdot \mathbf{E}^*$).

3.4 Verify that energy is conserved for a strongly focused Gaussian beam as described in Section 3.6. To do this, compare the energy flux through transverse planes on both sides of the optical lens. It is advantageous to choose one plane at the origin of the focus ($z = 0$). The energy flux is calculated most conveniently by evaluating the z-component of the time-averaged Poynting vector $\langle S_z \rangle$ and integrating it over the area of the transverse plane. Hint: you will need the Bessel-function closure relation

$$\int_0^{\infty} J_n(a_1 bx) J_n(a_2 bx) x \,\mathrm{d}x = \frac{1}{a_1 b^2}\delta(a_1 - a_2). \tag{3.112}$$

Check the units!

3.5 Consider a small circular aperture with radius a_0 in an infinitely thin and ideally conducting screen that is illuminated by a plane wave at normal incidence and polarized along the x-axis. In the long-wavelength limit ($\lambda \gg a_0$) the electric field in the aperture ($z = 0$, $x^2 + y^2 \leq a_0^2$) has been derived by Bouwkamp [23] as

$$\begin{aligned} E_x(x,y) &= -\frac{4ikE_0}{3\pi}\frac{2a_0^2 - x^2 - 2y^2}{\sqrt{a_0^2 - x^2 - y^2}}, \\ E_y(x,y) &= -\frac{4ikE_0}{3\pi}\frac{xy}{\sqrt{a_0^2 - x^2 - y^2}}, \end{aligned} \tag{3.113}$$

where E_0 is the incident field amplitude. The corresponding spatial Fourier spectrum has been calculated by Van Labeke *et al.* [24] as

$$\hat{E}_x(k_x,k_y) = \frac{2ika_0^3E_0}{3\pi^2}\left[\frac{3k_y^2\cos(a_0k_\rho)}{a_0^2k_\rho^4} - \frac{(a_0^2k_x^4 + 3k_y^2 + a_0^2k_x^2k_y^2)\sin(a_0k_\rho)}{a_0^3k_\rho^5}\right], \tag{3.114}$$

$$\hat{E}_y(k_x,k_y) = -\frac{2ika_0^3E_0}{3\pi^2}\left[\frac{3k_xk_y\cos(a_0k_\rho)}{a_0^2k_\rho^4} - \frac{k_xk_y(3 - a_0^2k_\rho^2)\sin(a_0k_\rho)}{a_0^3k_\rho^5}\right], \tag{3.115}$$

with $k_\rho = (k_x^2 + k_y^2)^{1/2}$ being the transverse wavenumber.

(a) Derive the Fourier spectrum of the longitudinal field component E_z.

(b) Find expressions for the field $\mathbf{E} = (E_x, E_y, E_z)$ at an arbitrary field point (x, y, z).

(c) Calculate the far-field and express it in spherical coordinates (r, ϑ, φ) and spherical vector components $\mathbf{E} = (E_r, E_\vartheta, E_\varphi)$. Expand in powers of ka_0 and retain only the lowest orders. What does this field look like?

3.6 The reflected image of a laser beam focused on a dielectric interface is given by Eqs. (3.109)–(3.111). Derive these equations starting from Eq. (3.100) which is the collimated reflected field. Notice that the fields propagate in the negative z-direction.

3.7 Show that the field $\mathbf{E}$ defined through $\mathbf{E}_\mathrm{f}$, $\mathbf{E}_\mathrm{r}$, and $\mathbf{E}_\mathrm{t}$ in Section 3.9 satisfies the boundary conditions at the interface $z = z_0$. Furthermore, show that the Helmholtz equation and the divergence condition are satisfied in each of the two half-spaces.

3.8 In order to correct for the aberrations introduced by the reflection of a strongly focused beam from an interface we design a pair of phase plates. By using a polarizing beamsplitter, the collimated reflected beam (cf. Fig. 3.19 and Eq. (3.100)) is split into two purely polarized light paths. The phase distortion in each light path is corrected by a phase plate. After correction, the two light paths are recombined and refocused on the image plane. Calculate and plot the phase distribution of each phase plate if the incident field is a Gaussian beam ($f_0 \to \infty$) focused by an NA = 1.4

objective on a glass/air interface ($z_0 = 0$) and incident from the optically denser medium with $n_1 = 1.518$. What happens if the focus is displaced from the interface ($z_0 \neq 0$)?

References

[1] M. Muller, J. Squier, K. R. Wilson, and G. J. Brakenhoff, "3D microscopy of transparent objects using third-harmonic generation," *J. Microsc.* **191**, 266–274 (1998).

[2] A. E. Siegman, *Lasers*. Mill Valley, CA: University Science Books (1986).

[3] L. Mandel and E. Wolf, *Optical Coherence and Quantum Optics*. New York: Cambridge University Press (1995).

[4] E. Zauderer, "Complex argument Hermite–Gaussian and Laguerre–Gaussian beams," *J. Opt. Soc. Am. A* **3**, 465–469 (1986).

[5] E. J. Bochove, G. T. Moore, and M. O. Scully, "Acceleration of particles by an asymmetric Hermite–Gaussian laser beam," *Phys. Rev. A* **46**, 6640–6653 (1992).

[6] X. S. Xie and J. K. Trautman, "Optical studies of single molecules at room temperature," *Annu. Rev. Phys. Chem.* **49**, 441–480 (1998).

[7] L. Novotny, M. R. Beversluis, K. S. Youngworth, and T. G. Brown, "Longitudinal field modes probed by single molecules," *Phys. Rev. Lett.* **86**, 5251–5254 (2001).

[8] A. Ashkin, J. M. Dziedzic, J. E. Bjorkholm, and S. Chu, "Observation of a single-beam gradient force optical trap for dielectric particles," *Opt. Lett.* **11**, 288–290 (1986).

[9] E. Wolf, "Electromagnetic diffraction in optical systems. I. An integral representation of the image field," *Proc. Roy. Soc. A* **253**, 349–357 (1959).

[10] B. Richards and E. Wolf, "Electromagnetic diffraction in optical systems. II. Structure of the image field in an aplanatic system," *Proc. Roy. Soc. A* **253**, 358–379 (1959).

[11] K. S. Youngworth and T. G. Brown, "Focusing of high numerical aperture cylindrical-vector beams," *Opt. Express* **7**, 77–87 (2000).

[12] K. S. Youngworth and T. G. Brown, "Inhomogeneous polarization in scanning optical microscopy," *Proc. SPIE* **3919**, 75–85 (2000).

[13] R. Dorn, S. Quabis, and G. Leuchs, "Sharper focus for a radially polarized light beam," *Phys. Rev. Lett.* **91**, 233901 (2003).

[14] L. Novotny, E. J. Sanchez, and X. S. Xie, "Near-field optical imaging using metal tips illuminated by higher-order Hermite–Gaussian beams," *Ultramicroscopy* **71**, 21–29 (1998).

[15] M. J. Snadden, A. S. Bell, R. B. M. Clarke, E. Riis, and D. H. McIntyre, "Doughnut mode magneto-optical trap," *J. Opt. Soc. Am. B* **14**, 544–552 (1997).

[16] S. C. Tidwell, D. H. Ford, and D. Kimura, "Generating radially polarized beams interferometrically," *Appl. Opt.* **29**, 2234–2239 (1990).
[17] W. C. Chew, *Waves and Fields in Inhomogeneous Media*. New York: Van Nostrand Reinhold (1990).
[18] B. Sick, B. Hecht, and L. Novotny, "Orientational imaging of single molecules by annular illumination," *Phys. Rev. Lett.* **85**, 4482–4485 (2000).
[19] J. D. Jackson, *Classical Electrodynamics*, 3rd edn. New York: John Wiley & Sons (1998).
[20] L. Novotny, R. D. Grober, and K. Karrai, "Reflected image of a strongly focused spot," *Opt. Lett.* **26**, 789–791 (2001).
[21] K. Karrai, X. Lorenz, and L. Novotny, "Enhanced reflectivity contrast in confocal solid immersion lens microscopy," *Appl. Phys. Lett.* **77**, 3459–3461 (2000).
[22] H. Maecker and G. Lehmann, "Die Grenze der Totalreflexion. I–III," *Ann. Phys.* **10**, 115–128, 153–160, and 161–166 (1952).
[23] C. J. Bouwkamp, "On Bethe's theory of diffraction by small holes," *Philips Res. Rep.* **5**, 321–332 (1950).
[24] D. Van Labeke, D. Barchiesi, and F. Baida, "Optical characterization of nanosources used in scanning near-field optical microscopy," *J. Opt. Soc. Am. A* **12**, 695–703 (1995).

4 Resolution and localization

Localization refers to the precision with which the position of an object can be defined. Spatial resolution, on the other hand, is a measure of the ability to distinguish two separated point-like objects from a single object. The diffraction limit implies that optical resolution is ultimately limited by the wavelength of light. Before the advent of near-field optics it was believed that the diffraction limit imposes a hard boundary and that physical laws strictly prohibit resolution significantly better than $\lambda/2$. It was then found that this limit is not as strict as assumed and that access to evanescent modes of the spatial spectrum offers a direct route to overcome the diffraction limit. However, further critical analysis of the diffraction limit revealed that "super-resolution" can also be obtained by pure far-field imaging under certain constraints. In this chapter we analyze the diffraction limit and discuss the principles of different imaging modes with resolutions near or beyond the diffraction limit.

4.1 The point-spread function

The point-spread function is a measure of the resolving power of an optical system. The narrower the point-spread function the better the resolution will be. As the name implies, the point-spread function defines the spread of a point source. If we have a radiating point source then the image of that source will appear to have a finite size. This broadening is a direct consequence of spatial filtering. A point in space is characterized by a delta function that has an infinite spectrum of spatial frequencies k_x and k_y. On propagation from the source to the image, high-frequency components are filtered out. Usually the entire spectrum $(k_x^2 + k_y^2) > k^2$ associated with the evanescent waves is lost. Furthermore, not all plane-wave components can be collected, which leads to a further reduction in bandwidth. The reduced spectrum is not able to accurately reconstruct the original point source and the image of the point will have a finite size. The standard derivation of the point-spread function is based on scalar theory and the paraxial approximation. This theory is insufficient for many high-resolution optical systems. With the "angular spectrum" framework established thus far we are in a position to rigorously investigate image formation in an optical system.

Consider the situation in Fig. 4.1, which has been analyzed by Sheppard and Wilson [1] and by Enderlein [2]. An ideal electromagnetic point source is located at the focus of a high-NA aplanatic objective lens with focal length f. This lens collimates the rays

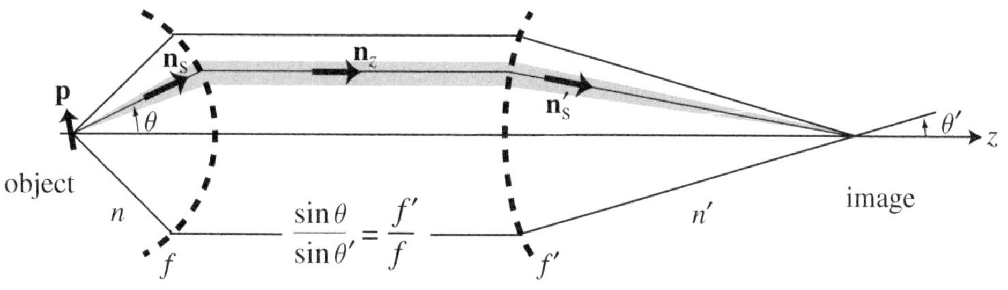

Fig. 4.1 Configuration used for the calculation of the point-spread function. The source is an arbitrarily oriented electric dipole with moment **p**. The dipole radiation is collected with a high-NA aplanatic objective lens and focused by a second lens on the image plane at $z = 0$.

emanating from the point source and a second lens with focal length f' focuses the fields on the image plane at $z = 0$. The situation is similar to the problem in Fig. 3.19. The only difference is that the source is a point source instead of the reflected field at an interface.

The smallest radiating electromagnetic unit is a dipole. In the optical regime most subwavelength-sized particles scatter as *electric* dipoles. On the other hand, small apertures radiate as *magnetic* dipoles. In the microwave regime, paramagnetic materials exhibit magnetic transitions, and in the infrared, small metal particles show magnetic dipole absorption caused by eddy currents of free carriers produced by the magnetic field. Nevertheless, we can restrict our analysis to an electric dipole since the field of a magnetic dipole is identical to the field of an electric dipole if we interchange the electric and magnetic fields, i.e. $\mathbf{E} \rightarrow \mathbf{H}$ and $\mathbf{H} \rightarrow -\mathbf{E}$.

In its most general form, the electric field at a point $\mathbf{r}$ of an arbitrarily oriented electric dipole located at $\mathbf{r}_0$ with dipole moment $\mathbf{p}$ is defined by the dyadic Green function $\overleftrightarrow{\mathbf{G}}(\mathbf{r}, \mathbf{r}_0)$ as (cf. Chapter 1)

$$\mathbf{E}(\mathbf{r}) = \frac{\omega^2}{\varepsilon_0 c^2} \overleftrightarrow{\mathbf{G}}(\mathbf{r}, \mathbf{r}_0)\mathbf{p}. \tag{4.1}$$

We assume that the distance between the dipole and the objective lens is much larger than the wavelength of the emitted light. In this case, we do not need to consider the evanescent components of the dipole field. Furthermore, we choose the dipole to be located at the origin $\mathbf{r}_0 = 0$ and surrounded by a homogeneous medium with index n. In this case, we can use the free-space far-field form of $\overleftrightarrow{\mathbf{G}}$, which, expressed in spherical coordinates (r, θ, ϕ), reads as (see Appendix D)

$$\overleftrightarrow{\mathbf{G}}_\infty(\mathbf{r}, 0) = \frac{\exp(ikr)}{4\pi r} \times \begin{bmatrix} 1 - \cos^2\phi \sin^2\theta & -\sin\phi\cos\phi\sin^2\theta & -\cos\phi\sin\theta\cos\theta \\ -\sin\phi\cos\phi\sin^2\theta & 1 - \sin^2\phi\sin^2\theta & -\sin\phi\sin\theta\cos\theta \\ -\cos\phi\sin\theta\cos\theta & -\sin\phi\sin\theta\cos\theta & \sin^2\theta \end{bmatrix}. \tag{4.2}$$

This is simply a 3 × 3 matrix that has to be multiplied by the dipole moment vector $\mathbf{p} = (p_x, p_y, p_z)$ to obtain the electric field.[1] To describe refraction at the reference sphere f we have to project the electric field vector along the vectors $\mathbf{n}_\theta$ and $\mathbf{n}_\phi$ as we did in Section 3.5. After being refracted, the field propagates as a collimated beam to the second lens f', where it is refracted once again. For a dipole aligned with the x-axis ($\mathbf{p} = p_x \mathbf{n}_x$) the field just after the second lens becomes

$$\mathbf{E}_\infty^{(x)}(\theta,\phi) = \frac{\omega^2 p_x}{\varepsilon_0 c^2} \frac{\exp(\mathrm{i}kf)}{8\pi f} \times \begin{bmatrix} 1 + \cos\theta\cos\theta' - (1-\cos\theta\cos\theta')\cos(2\phi) \\ -(1-\cos\theta\cos\theta')\sin(2\phi) \\ -2\cos\theta\sin\theta'\cos\phi \end{bmatrix} \sqrt{\frac{n\cos\theta'}{n'\cos\theta}}, \tag{4.3}$$

where

$$\sin\theta' = \frac{f}{f'}\sin\theta, \qquad \cos\theta' = g(\theta) = \sqrt{1 - (f/f')^2\sin^2\theta}\,. \tag{4.4}$$

The term $(\cos\theta'/\cos\theta)^{1/2}$ is a consequence of energy conservation as discussed in Section 3.5. In the limit $f \ll f'$ the contribution of $\cos\theta'$ can be ignored, but $\cos\theta$ cannot since we are dealing with a high-NA objective lens. The fields for a dipole p_y and a dipole p_z can be derived in a similar way. For an arbitrarily oriented dipole $\mathbf{p} = (p_x, p_y, p_z)$ the field is simply obtained by the superposition

$$\mathbf{E}_\infty(\theta,\phi) = \mathbf{E}_\infty^{(x)} + \mathbf{E}_\infty^{(y)} + \mathbf{E}_\infty^{(z)}. \tag{4.5}$$

To obtain the fields $\mathbf{E}$ near the focus of the second lens we insert the field $\mathbf{E}_\infty$ into Eq. (3.47). We assume that $f \ll f'$, which allows us to use the approximations in Eq. (3.106). The integration with respect to ϕ can be carried out analytically and the result can be written as

$$\mathbf{E}(\rho,\varphi,z) = \frac{\omega^2}{\varepsilon_0 c^2}\overleftrightarrow{\mathbf{G}}_{\mathrm{PSF}}(\rho,\varphi,z)\,\mathbf{p}, \tag{4.6}$$

where the *dyadic point-spread function* is given by

$$\overleftrightarrow{\mathbf{G}}_{\mathrm{PSF}} = \frac{\mathrm{i}k'}{8\pi}\frac{f}{f'}\mathrm{e}^{\mathrm{i}(kf - k'f')} \begin{bmatrix} \tilde{I}_{00} + \tilde{I}_{02}\cos(2\varphi) & \tilde{I}_{02}\sin(2\varphi) & 2\mathrm{i}\tilde{I}_{01}\cos\varphi \\ \tilde{I}_{02}\sin(2\varphi) & \tilde{I}_{00} - \tilde{I}_{02}\cos(2\varphi) & 2\mathrm{i}\tilde{I}_{01}\sin\varphi \\ 0 & 0 & 0 \end{bmatrix} \sqrt{\frac{n}{n'}}, \tag{4.7}$$

and the integrals $\tilde{I}_{00}$–$\tilde{I}_{02}$ are defined as

$$\tilde{I}_{00}(\rho,z) = \int_0^{\theta_{\max}} (\cos\theta)^{1/2}\sin\theta(1+\cos\theta)J_0(k'\rho\sin\theta f/f') \times \exp\left\{\mathrm{i}k'z[1 - (1/2)(f/f')^2\sin^2\theta]\right\}\,\mathrm{d}\theta, \tag{4.8}$$

[1] The far-field at $\mathbf{r}$ of a dipole located at $\mathbf{r}_0 = 0$ can also be expressed as $\mathbf{E} = -\omega^2\mu_0[\mathbf{r}\times\mathbf{r}\times\mathbf{p}]\,\exp(\mathrm{i}kr)/(4\pi r^3)$.

$$\tilde{I}_{01}(\rho, z) = \int_0^{\theta_{\max}} (\cos\theta)^{1/2} \sin^2\theta \, J_1(k'\rho \sin\theta f/f') \times \exp\left\{ik'z[1 - (1/2)(f/f')^2 \sin^2\theta]\right\} \, d\theta, \tag{4.9}$$

$$\tilde{I}_{02}(\rho, z) = \int_0^{\theta_{\max}} (\cos\theta)^{1/2} \sin\theta(1 - \cos\theta) J_2(k'\rho \sin\theta f/f') \times \exp\left\{ik'z[1 - (1/2)(f/f')^2 \sin^2\theta]\right\} \, d\theta. \tag{4.10}$$

The first column of $\overleftrightarrow{\mathbf{G}}_{\mathrm{PSF}}$ denotes the field of a dipole p_x, the second column the field of a dipole p_y, and the third column the field of a dipole p_z. The integrals $\tilde{I}_{00}$–$\tilde{I}_{02}$ are similar to the integrals I_{00}–I_{02} encountered in conjunction with the focusing of a Gaussian beam (cf. Eqs. (3.58–3.60)). The main differences are the arguments of the Bessel functions and the exponential functions. Furthermore, the longitudinal field E_z is zero in the present case because we required $f \ll f'$.

Equations (4.6)–(4.10) describe the mapping of an arbitrarily oriented electric dipole from its source to its image. The result depends on the numerical aperture NA of the primary objective lens

$$\mathrm{NA} = n \sin\theta_{\max} \tag{4.11}$$

and the (transverse) magnification M of the optical system defined as

$$M = \frac{n}{n'} \frac{f'}{f}. \tag{4.12}$$

In the following, we will use the quantity $|\mathbf{E}|^2$ to denote the point-spread function, since it is the quantity relevant to optical detectors. We first consider the situation of a dipole with its axis perpendicular to the optical axis. Without loss of generality, we can define the x-axis to be parallel with the dipole axis, i.e. $\mathbf{p} = p_x \mathbf{n}_x$. For a low-NA objective lens, $\theta_{\max}$ is sufficiently small to allow us to make the approximations $\cos\theta \approx 1$ and $\sin\theta \approx \theta$. Furthermore, in the image plane ($z=0$, $\vartheta = \pi/2$) the exponential terms in the integrals are equal to one and the second-order Bessel function J_2 goes to zero for small θ, making the integral $\tilde{I}_{02}$ disappear. We are then left with $\tilde{I}_{00}$, which can be integrated analytically using

$$\int x J_0(x) \mathrm{d}x = x J_1(x). \tag{4.13}$$

The *paraxial point-spread function* in the image plane for a dipole oriented along the x-axis turns out to be

$$\lim_{\theta_{\max} \ll \pi/2} \left|\mathbf{E}(x, y, z=0)\right|^2 = \frac{\pi^4}{\varepsilon_0^2 n n'} \frac{p_x^2}{\lambda^6} \frac{\mathrm{NA}^4}{M^2} \left[2 \frac{J_1(2\pi\tilde{\rho})}{(2\pi\tilde{\rho})}\right]^2, \quad \tilde{\rho} = \frac{\mathrm{NA}\rho}{M\lambda}. \tag{4.14}$$

The functional form is given by the term in brackets which is known as the *Airy function*. It is depicted in Fig. 4.2(a) as the solid curve. The dashed and the dotted curves

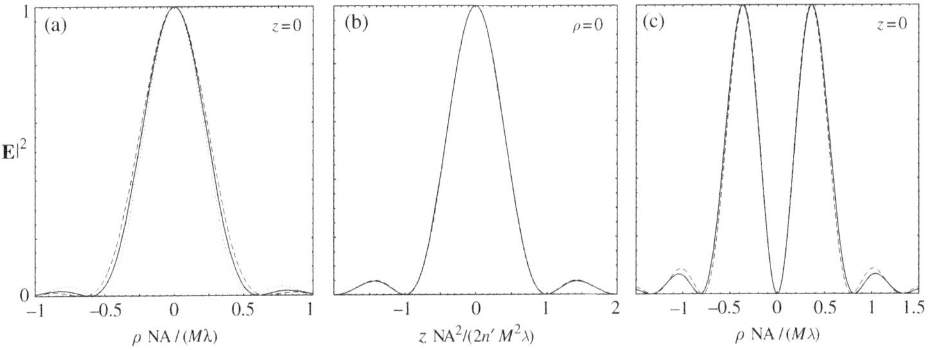

Fig. 4.2 (a) The point-spread function depicted in the image plane ($z = 0$) of a dipole with moment $\mathbf{p} = p_x \mathbf{n}_x$. The solid curve is the paraxial approximation whereas the dashed and dotted curves are the results of exact calculations for an NA $= 1.4$ ($n = 1.518$) objective lens. The dashed curve has been evaluated along the x-axis and the dotted curve along the y-axis. (b) The point-spread function evaluated along the optical axis z. The solid curve is the paraxial approximation and the dashed curve is the exact result for NA $= 1.4$. (c) The point-spread function depicted in the image plane of a dipole with moment $\mathbf{p} = p_z \mathbf{n}_z$. The solid curve is the paraxial approximation and the dashed curve is the exact result for NA $= 1.4$. The figures demonstrate that the paraxial point-spread function is a good approximation even for high-NA objective lenses!

show the exact calculation of the point-spread function for an NA $= 1.4$ objective lens according to Eqs. (4.7)–(4.10). The dashed curve is depicted along the x-axis (the direction of the dipole axis) and the dotted curve along the y-axis. The field is purely polarized ($\cos(2\varphi) = \pm 1$, $\sin(2\varphi) = 0$) along both axes, but the width along the x-axis is larger. This is caused by the term $\tilde{I}_{02}$, which is in one case subtracted from $\tilde{I}_{00}$ and in the other case added to $\tilde{I}_{00}$. The result is an elliptically shaped spot. The ellipticity increases with increasing NA. Nevertheless, it is surprising that the paraxial point-spread function is a very good approximation even for high-NA objective lenses! If the average between the curves along the x-axis and the y-axis is taken, the paraxial point-spread function turns out to be nearly a perfect fit. The point-spread function can be measured by using a single quantum emitter, such as a single molecule or a quantum dot, as a point emitter. Figure 4.3 shows such a measurement together with a fit according to Eq. (4.14). The point-spread function has been recorded by using an NA $= 1.3$ lens to collect the fluorescence photons from a single DiI molecule with a center wavelength of $\lambda \approx 580$ nm.

The width of the point-spread function Δx is usually defined as the radial distance for which the value of the paraxial point-spread function becomes zero, or

$$\Delta x = 0.6098 \, \frac{M\lambda}{\mathrm{NA}}. \tag{4.15}$$

This width is also denoted as the *Airy disk radius*. It depends in a simple manner on the numerical aperture, the wavelength, and the magnification of the system.

We defined the point-spread function as proportional to the electric energy density, the quantity to which optical detectors are sensitive. Since the magnetic field $\mathbf{H}$ is simply proportional to the electric field rotated by 90° around the z-axis, we find that the point-spread function for the magnetic field is also rotated by 90° compared with the point-spread

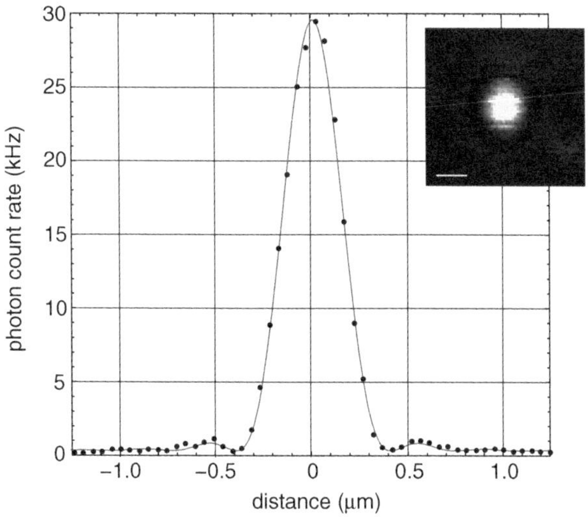

Fig. 4.3 Point-spread function measured with a single-molecule point source. Fluorescence photons emitted by a DiI molecule are collected with an NA = 1.3 objective lens. The center wavelength is $\lambda \approx 580$ nm. The data points correspond to a horizontal line cut through the center of the fluorescence rate image shown in the inset. The solid curve corresponds to the Airy function.

function for the electric field. The total energy density and the time-averaged Poynting vector are therefore rotationally symmetric with respect to the z-axis.

Let us now discuss the field strength along the optical axis z, denoted as the axial point-spread function. The only non-vanishing integral is $\tilde{I}_{00}$, implying that anywhere on the z-axis the field stays polarized along the direction of the dipole axis x. In the paraxial limit we can integrate $\tilde{I}_{00}$ and obtain the result

$$\lim_{\theta_{\max} \ll \pi/2} \left| \mathbf{E}(x=0, y=0, z) \right|^2 = \frac{\pi^4}{\varepsilon_0^2 n n'} \frac{p_x^2}{\lambda^6} \frac{\mathrm{NA}^4}{M^2} \left[\frac{\sin(\pi \tilde{z})}{\pi \tilde{z}} \right]^2, \quad \tilde{z} = \frac{\mathrm{NA}^2 z}{2n' M^2 \lambda}. \tag{4.16}$$

This result is compared with the exact calculation in Fig. 4.2(b) for NA = 1.4. The curves overlap perfectly, indicating that the paraxial result is an excellent fit even for large NA. The distance Δz for which the axial point-spread function becomes zero is

$$\Delta z = 2n' \frac{M^2 \lambda}{\mathrm{NA}^2}, \tag{4.17}$$

and is denoted as the *depth of field*. In contrast to the Airy disk, Δz depends on the index of refraction of the image space. Furthermore, it depends on the squares of M and NA. Therefore, the depth of field is usually much larger than the Airy disk radius. For a typical microscope objective with $M = 60\times$ and NA = 1.4 and for a wavelength of 500 nm we obtain $\Delta x \approx 13$ μm and $\Delta z \approx 1.8$ mm.

So far, we have considered a dipole with its axis perpendicular to the optical axis. The situation is very different for a dipole with its axis parallel to the optical axis, i.e. $\mathbf{p} = p_z \mathbf{n}_z$.

The focal fields turn out to be rotationally symmetric, radially polarized, and zero on the optical axis. In the paraxial limit we find

$$\lim_{\theta_{\max}\ll\pi/2} |\mathbf{E}(x,y,z=0)|^2 = \frac{\pi^4}{\varepsilon_0^2 n^3 n'} \frac{p_z^2\, \mathrm{NA}^6}{\lambda^6\, M^2} \left[2\,\frac{J_2(2\pi\tilde{\rho})}{2\pi\tilde{\rho}}\right]^2, \quad \tilde{\rho} = \frac{\mathrm{NA}\,\rho}{M\lambda}, \tag{4.18}$$

which is shown in Fig. 4.2(c). The comparison with the exact calculation using $\mathrm{NA} = 1.4$ demonstrates again that the paraxial expression is a good approximation. Because of the vanishing field amplitude on the optical axis it is difficult to define a characteristic width for the point-spread function of a dipole with its axis along the optical axis. However, the comparison between Fig. 4.2(a) and Fig. 4.2(c) shows that the image of a dipole p_z is wider than the image of a dipole p_x.

In many experimental situations it is desirable to determine the dipole orientation and dipole strength of an emitter. This is an inverse problem which can be solved in our configuration by detecting the field distribution in the image plane by using, for example, a CCD [3, 4]. With Eqs. (4.6)–(4.10) we can then calculate back and determine the parameters of the emitter. This analysis can be made more efficient by splitting the collected radiation into two orthogonal polarization states and focusing it onto two separate detectors. The detection and analysis of single molecules on the basis of their emission and absorption patterns will be further discussed in Chapter 9.

As a conclusion of this section we mention that the point-spread function depends strongly on the orientation of the dipole moment of the emitting point source. For dipoles aligned perpendicular to the optical axis we find excellent agreement with the familiar paraxial point-spread function, even for high NA.

4.2 The resolution limit(s)

Now that we have determined how a single point emitter is mapped from its source to its image, we ask ourselves how well are we able to distinguish two point emitters separated by a distance $\Delta r_{||} = (\Delta x^2 + \Delta y^2)^{1/2}$ in the object plane. Each point source will be identified on the basis of its point-spread function having some characteristic width. If we move the two emitters in the object plane closer and closer together, their point-spread functions in the image plane will start to overlap and then reach a point where they become indistinguishable. We might state that the two point-spread functions can be distinguished only if their maxima are separated by more than the characteristic width of one individual point-spread function (Fig. 4.4). Thus, the narrower the point-spread function is the better the resolution will be.

We have mentioned already in Section 3.1 that the resolving power of an optical system depends on the bandwidth of spatial frequencies $\Delta k_{||} = (\Delta k_x^2 + \Delta k_y^2)^{1/2}$ that are collected by the optical system. Simple Fourier mathematics leads to

$$\Delta k_{||}\, \Delta r_{||} \geq 1, \tag{4.19}$$

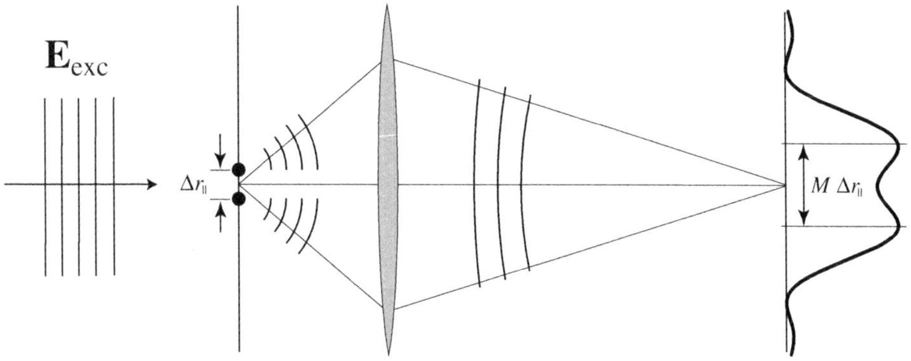

Fig. 4.4 Illustration of the resolution limit. Two simultaneously radiating point sources separated by $\Delta r_{\parallel}$ in the object plane generate a combined point-spread function in the image plane. The two point sources are optically resolved if they can be distinguished by examination of their image patterns.

similar to the Heisenberg uncertainty principle in quantum mechanics. The product of $\Delta r_{\parallel}$ and $\Delta k_{\parallel}$ is minimized for a Gaussian distribution of spatial frequencies. This Gaussian distribution is the analog of the minimum-uncertainty wavefunction in quantum mechanics.

In far-field optics, the upper bound for $\Delta k_{\parallel}$ is given by twice the wavenumber $k = (\omega/c)n = (2\pi/\lambda)\,n$ of the object medium because we discard spatial frequencies associated with evanescent-wave components. In this case the resolution cannot be better than[2]

$$\mathrm{Min}\big[\Delta r_{\parallel}\big] = \frac{\lambda}{4\pi n}. \qquad (4.20)$$

However, in practice we are not able to sample the entire spectrum of $\Delta k_{\parallel} = [-k \ldots k]$ and the upper limit will be defined by the numerical aperture of the system, i.e.

$$\mathrm{Min}\big[\Delta r_{\parallel}\big] = \frac{\lambda}{4\pi\,\mathrm{NA}}. \qquad (4.21)$$

This figure is the best case and, in fact, Abbe's and Rayleigh's formulations of the resolution limit are less optimistic.

Abbe's formulation considers the paraxial point-spread function of two dipoles with axes perpendicular to the optical axis (cf. Eq. (4.14)). The distance $\Delta r_{\parallel}$ between the two dipoles in the object plane is mapped onto a distance $M\,\Delta r_{\parallel}$ in the image plane. Abbe states that the minimum distance $\mathrm{Min}\big[M\,\Delta r_{\parallel}\big]$ corresponds to the distance between the two point-spread functions for which the maximum of one point-spread function coincides with the first minimum of the second point-spread function. This distance is given by the Airy disk radius defined in Eq. (4.15). We find according to Abbe [5]

$$\text{Abbe (1873):} \qquad \mathrm{Min}\big[\Delta r_{\parallel}\big] = 0.6098\,\frac{\lambda}{\mathrm{NA}}. \qquad (4.22)$$

2 We have to account for both positive and negative spatial frequencies.

This limit is a factor of ≈ 7.7 worse than the one defined in Eq. (4.21). It is based on the paraxial approximation and applies to the special case of two parallel dipoles oriented perpendicular to the optical axis. Things look quite different for two dipoles aligned parallel to the optical axis. We see that there is some arbitrariness in the definition of a resolution limit. This applies also to Rayleigh's criterion [6], which is based on the overlap of two point-spread functions in a two-dimensional geometry. Rayleigh's criterion was formulated in connection with a grating spectrometer rather than with an optical microscope. However, it is often adopted in conjunction with optical microscopy.

In Abbe's resolution limit the distance between the two point sources does not become distorted for dipoles with unequal strengths. This is because the maximum of one point-spread function overlaps with a minimum (zero) of the other point-spread function. Of course, we can overlap the two point-spread functions further and still be able to distinguish the two sources. In fact, in a noise-free system we will always be able to deconvolve the combined response into two separate point-spread functions even if we are not able to observe two separate maxima in the combined point-spread function. However, even if the two sources, the optical instrument, and the detector are all noise-free there is always shot-noise associated with the quantized nature of light, which puts a limit on this idealized view of resolution.

According to Eq. (4.19) there is no limit to optical resolution if the bandwidth $\Delta k_{\parallel}$ is arbitrarily large. However, going beyond the limit of Eq. (4.20) requires the involvement of evanescent field components. This is the subject of near-field optical microscopy and will be discussed in subsequent chapters.

Many tricks can also be applied to stretch the resolution limit if prior information on the properties of the point sources is available. For example, in Abbe's formulation, *prior knowledge* about the dipole orientation is necessary. If, in addition to Abbe's assumption, the two dipoles are perpendicular to each other, i.e. p_x and p_y, a polarizer in the detection path can increase the resolution further. Other prior knowledge might be available in regard to coherence properties of the two emitters, i.e. $|\mathbf{E}_1|^2 + |\mathbf{E}_2|^2$ versus $|\mathbf{E}_1 + \mathbf{E}_2|^2$. In all cases, prior knowledge about the properties of a sample reduces the set of possible configurations and thereby improves the resolution. Object reconstruction with prior knowledge about the properties of the object is one of the central topics of inverse scattering. In fluorescence microscopy prior knowledge is associated with the type of molecules used to label specific parts of a biological specimen. Knowledge of the absorption and emission properties of these molecules makes it possible to substantially increase resolution (see Section 5.2.3). A general theory of optical resolution must include a quantitative measure of prior information. Since, however, information can exist in a variety of different forms, it is certainly difficult to propose a generally valid concept.

4.2.1 Increasing resolution through selective excitation

In discussing the resolution limit we assumed that there were two radiating point sources separated by a distance $\Delta r_{\parallel}$ in the object plane. However, the sources do not radiate without any external excitation. If, for example, we can make only one dipole radiate at a certain time, then we are in a position to assign the detected field in the image plane to this particular dipole. We then scan the excitation to the other dipole and record its image in a similar way. Thus, we are perfectly able to distinguish the two point sources no matter how close they are. Therefore, the resolution criteria require some correction.

In practice, the point sources are excited by an excitation source $\mathbf{E}_{\mathrm{exc}}$ with finite spatial extent (Fig. 4.5). It is this extent which determines whether for a given dipole separation $\Delta r_{||}$ we are able to excite only one dipole at a time or not. The resolution criteria formulated before assume a broad illumination of the sample surface making all point sources radiate simultaneously. Hence, we need to incorporate the effect of the excitation profile. This can be done in a general way by considering the interaction between the excitation field $\mathbf{E}_{\mathrm{exc}}$ and a sample dipole

$$\mathbf{p}_{\mathrm{n}} = f\left[\text{material properties}, \mathbf{E}_{\mathrm{exc}}(\mathbf{r}_{\mathrm{s}} - \mathbf{r}_{\mathrm{n}})\right], \tag{4.23}$$

where $\mathbf{r}_{\mathrm{n}}$ is the (fixed) position vector of the dipole $\mathbf{p}_{\mathrm{n}}$ and $\mathbf{r}_{\mathrm{s}}$ is the (variable) position vector of the excitation field origin. The latter coordinate vector can be scanned in the object space to selectively excite individual dipoles. With the relationship of Eq. (4.23), the point-spread function becomes dependent on the excitation field and the specific light–matter interaction. The resolution of the optical system will therefore depend on the type of interaction. This increases the number of parameters in our analysis considerably. The problem becomes even more complicated if we have to consider interactions between the individual dipoles. To keep our feet on the ground, we need to restrict our analysis somewhat.

Let us assume that the interaction between the dipole and the excitation field is given by a general nonlinear relationship

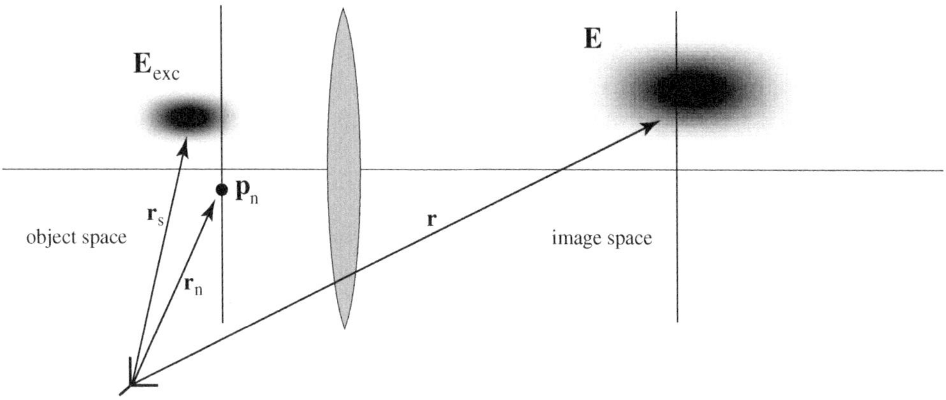

Fig. 4.5 Schematic representation of a general set-up using a confined excitation source for sample excitation. The dipole strength $\mathbf{p}_n$ of the point source depends on the excitation field $\mathbf{E}_{\mathrm{exc}}$. The point-spread function defined by the field $\mathbf{E}$ in the image space depends on the nature of the interaction between $\mathbf{p}_n$ and $\mathbf{E}_{\mathrm{exc}}$, and on the relative coordinates $\mathbf{r}_{\mathrm{n}} - \mathbf{r}_{\mathrm{s}}$.

$$\begin{aligned}\mathbf{p}_{\mathrm{n}}(\omega, 2\omega, \ldots; \mathbf{r}_{\mathrm{s}}, \mathbf{r}_{\mathrm{n}}) = {} & \alpha(\omega)\mathbf{E}_{\mathrm{exc}}(\omega, \mathbf{r}_{\mathrm{s}} - \mathbf{r}_{\mathrm{n}}) \\ & + \beta(2\omega)\mathbf{E}_{\mathrm{exc}}(\omega, \mathbf{r}_{\mathrm{s}} - \mathbf{r}_{\mathrm{n}})\mathbf{E}_{\mathrm{exc}}(\omega, \mathbf{r}_{\mathrm{s}} - \mathbf{r}_{\mathrm{n}}) \\ & + \gamma(3\omega)\mathbf{E}_{\mathrm{exc}}(\omega, \mathbf{r}_{\mathrm{s}} - \mathbf{r}_{\mathrm{n}})\mathbf{E}_{\mathrm{exc}}(\omega, \mathbf{r}_{\mathrm{s}} - \mathbf{r}_{\mathrm{n}})\mathbf{E}_{\mathrm{exc}}(\omega, \mathbf{r}_{\mathrm{s}} - \mathbf{r}_{\mathrm{n}}) \\ & + \cdots, \end{aligned} \tag{4.24}$$

where the multiplications between field vectors denote outer products. In its most general form, the polarizability α is a tensor of rank two, and the hyperpolarizabilities β and γ are tensors of rank three and four, respectively. It is convenient to consider the different nonlinearities separately by writing

$$\mathbf{p}_{\mathrm{n}}(\omega, 2\omega, \ldots; \mathbf{r}_{\mathrm{s}}, \mathbf{r}_{\mathrm{n}}) = \mathbf{p}_{\mathrm{n}}(\omega, \mathbf{r}_{\mathrm{s}}, \mathbf{r}_{\mathrm{n}}) + \mathbf{p}_{\mathrm{n}}(2\omega, \mathbf{r}_{\mathrm{s}}, \mathbf{r}_{\mathrm{n}}) + \mathbf{p}_{\mathrm{n}}(3\omega, \mathbf{r}_{\mathrm{s}}, \mathbf{r}_{\mathrm{n}}) + \cdots . \tag{4.25}$$

With the help of the dyadic point-spread function for a dipole in the object space at $\mathbf{r}_{\mathrm{n}}$, the focal field at $\mathbf{r}$ as a function of the position $\mathbf{r}_{\mathrm{s}}$ of the excitation beam becomes

$$\mathbf{E}(\mathbf{r}, \mathbf{r}_{\mathrm{s}}, \mathbf{r}_{\mathrm{n}}; n\omega) = \frac{(n\omega)^2}{\varepsilon_0 c^2} \overleftrightarrow{\mathbf{G}}_{\mathrm{PSF}}(\mathbf{r}, \mathbf{r}_{\mathrm{n}}; n\omega) \cdot \mathbf{p}_{\mathrm{n}}(n\omega, \mathbf{r}_{\mathrm{s}}, \mathbf{r}_{\mathrm{n}}). \tag{4.26}$$

For multiple dipoles we have to sum over n.

Equation (4.26) demonstrates in a quite general way how the point-spread function can be influenced by the excitation source. This tailoring of the point-spread function was named *point-spread function engineering* and plays an essential role in high-resolution confocal microscopy. The field in Eq. (4.26) depends on the coordinates of the excitation source, the coordinates of the dipole in the object space, and the coordinates of the field point in the image space. It is convenient to keep the coordinates of the excitation beam fixed and to collect, after some spatial filtering, the total intensity in the image plane (integration over $\mathbf{r}$). In this way, the detector signal will depend only on the coordinates $\mathbf{r}_{\mathrm{n}}$ of the dipole. Similarly, the field in the image plane can be evaluated at a single point such as on the optical axis. This is essentially what is done in confocal microscopy, which will be discussed in the next section. Notice that the field $\mathbf{E}$ depends not only on the spatial coordinates of the system but also on the material properties, represented by the polarizabilities α, β, and γ. Any optical image of the sample will therefore be a mixture of spectroscopic information and spatial information.

4.2.2 Axial resolution

To characterize the position of the dipole emitter, confocal microscopy uses the relative coordinate $\mathbf{r}_{\mathrm{n}} - \mathbf{r}_{\mathrm{s}}$ between the excitation beam and the dipole position. An image is generated by assigning to every coordinate $\mathbf{r}_{\mathrm{n}} - \mathbf{r}_{\mathrm{s}}$ some property of the emitter measured in the image plane.

To demonstrate the basic idea of axial resolution in confocal microscopy we discuss two special situations. First we assume that the properties of a dipole located on the optical axis are represented by the total integrated field intensity in the image plane. Using the Bessel-function closure relations (see Problem 3.4) we find

$$s_1(z) \equiv \int_0^{2\pi}\int_0^{\infty} \mathbf{E}(\rho,\varphi,z)\,\mathbf{E}^*(\rho,\varphi,z)\rho\,\mathrm{d}\rho\,\mathrm{d}\varphi$$
$$= \frac{\pi^4 n}{24\varepsilon_0^2\lambda^4 n'}\Big[(p_x^2+p_y^2)(28-12\cos\theta_{\max}-12\cos^2\theta_{\max}-4\cos^3\theta\,\mathrm{max}) + p_z^2(8-9\cos\theta_{\max}+\cos^3\theta_{\max})\Big]. \qquad (4.27)$$

The signal has units of V^2 and depends on the NA of the system through $\theta_{\max}$. The important point is that the signal does *not* depend on the axial coordinate z! Thus, if the position of the dipole is displaced from the object plane in the direction of the optical axis it will render the same signal s_1. There is no axial resolution associated with this type of detection.

In order to achieve axial resolution we need to spatially filter the fields in the image plane before they are sent to the detector. Usually, this is achieved by placing a pinhole with a radius on the order of the Airy disk radius (Eq. (4.15)) into the image plane. In this way, only the center part of the point-spread function reaches the detector. There are different strategies for the choice of the pinhole size [7] but to illustrate the effect we can assume that only the field on the optical axis passes through the pinhole. The resulting signal has been calculated in Eq. (4.16) and reads as

$$s_2(z) \equiv \mathbf{E}(\rho=0,z)\,\mathbf{E}^*(\rho=0,z)\,\delta A$$
$$= \frac{\pi^4}{\varepsilon_0^2 n n'}\frac{p_x^2+p_y^2}{\lambda^6}\frac{\mathrm{NA}^4}{M^2}\left[\frac{\sin(\pi\tilde{z})}{\pi\tilde{z}}\right]^2\delta A, \quad \tilde{z} = \frac{\mathrm{NA}^2\,z}{2n'M^2\lambda}. \qquad (4.28)$$

Here, dA denotes the infinitesimal area of the pinhole. We see that a dipole located on the optical axis with a dipole moment parallel to the optical axis is not detected in this scheme because its field is zero on the optical axis. In order to enable its detection we have to increase the pinhole size or displace the dipole from the optical axis. However, the important information in Eq. (4.28) is the dependence of the signal s_2 on the axial coordinate z which gives us axial resolution! To illustrate this axial resolution, let us consider two dipoles on the optical axis near the object plane. While we keep one of the dipoles in the image plane we move the other by a distance $\Delta r_\perp$ out of the image plane as shown in Fig. 4.6. The lens maps a longitudinal distance $\Delta r_\perp$ in the object space into a longitudinal distance $M_\mathrm{L}\,\Delta r_\perp$ in the image space, where M_L is the longitudinal magnification defined as

$$M_\mathrm{L} = \frac{n'}{n}M^2. \qquad (4.29)$$

It depends on the transverse magnification M defined in Eq. (4.12) and the refractive indices n and n' of object and image space, respectively. We place the detector into the image plane ($z=0$). According to Eq. (4.28), the signal of the in-plane dipole is maximized whereas the signal of the out-of-plane dipole gives[3]

[3] We assume that the two dipoles radiate incoherently, i.e. $|\mathbf{E}|^2 = |\mathbf{E}_1|^2 + |\mathbf{E}_2|^2$. The situation is essentially the same for coherently radiating dipoles, i.e. $|\mathbf{E}|^2 = |\mathbf{E}_1+\mathbf{E}_2|^2$.

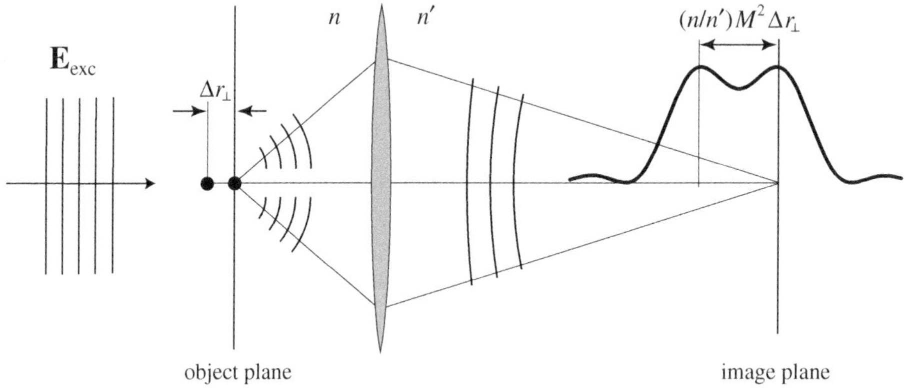

Fig. 4.6 Illustration of axial resolution in confocal microscopy. A pinhole on the optical axis in the image plane spatially filters the image before it is directed onto a detector. The pinhole passes only the fields near the optical axis, thereby generating axial resolution.

$$s_2(z) \propto \frac{\sin^2[\pi \mathrm{NA}^2\, \Delta r_\perp/(2n\lambda)]}{[\pi \mathrm{NA}^2\, \Delta r_\perp/(2n\lambda)]^2}. \tag{4.30}$$

To ensure that the entire signal can be assigned to the in-plane dipole we have to require that the contribution of the out-of-plane dipole cancels out. This is achieved for a separation $\Delta r_\perp$ between the dipoles of

$$\mathrm{Min}[\Delta r_\perp] = 2\frac{n\lambda}{\mathrm{NA}^2}. \tag{4.31}$$

This distance defines the axial resolution of the confocal system. Only dipoles within a distance of $\mathrm{Min}[\Delta r_\perp]$ from the image plane will lead to a significant signal at the detector. Therefore, $\mathrm{Min}[\Delta r_\perp]$ is called the *focal depth*. Besides providing lateral resolution on the order of $\mathrm{Min}[\Delta r_{||}]$, confocal microscopy also provides axial resolution on the order of $\mathrm{Min}[\Delta r_\perp]$. Hence, a sample can be imaged in three dimensions. While the lateral resolution scales linearly with NA, the axial resolution scales quadratically with NA. As an example, Fig. 4.7 shows a multiphoton confocal microscopy image of a spiky pollen grain [8]. The three-dimensional image was reconstructed from multiple sectional images that are displaced in the z-direction by roughly $2n\lambda/\mathrm{NA}^2$. More detailed experimental issues related to axial resolution will be discussed in Chapter 5.

4.2.3 Resolution enhancement through saturation

We have discussed how the point-spread function can be squeezed by using nonlinear optical interactions, i.e. the width of $E^{2n}(r_{||})$ is narrower than the width of $E^2(r_{||})$. A similar advantage can be achieved through saturation, as demonstrated in the pioneering work by Hell and coworkers [9]. The necessary ingredients are (1) an intensity zero located at the region of interest and (2) a target material with a reversible saturable linear transition.

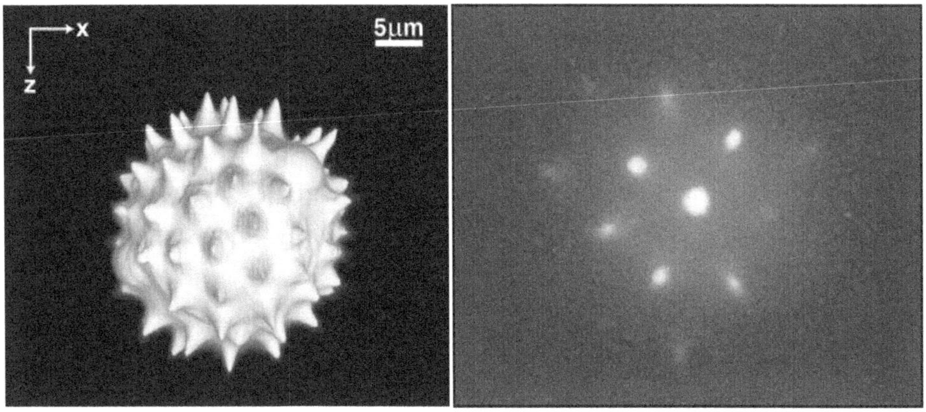

Fig. 4.7 Multiphoton confocal image of a spiky pollen grain of diameter 25 μm. Three-dimensional reconstruction based on multiple sectional images (left), and a single sectional image (right). From [8].

To illustrate how saturation can be used to increase resolution in fluorescence microscopy let us consider a dense sample made of randomly oriented molecules that are well approximated by two-level systems as shown in Fig. 4.8(a). Each two-level system interacts with two laser fields: (1) an excitation field $\mathbf{E}_{\mathrm{e}}$, which populates the excited state $|1\rangle$, and (2) a field $\mathbf{E}_{\mathrm{d}}$ used to deplete the excited state by stimulated emission. For sufficiently high intensities the depletion field saturates the ground state $|0\rangle$. Figure 4.8(b) shows typical intensity profiles of excitation and depletion fields. Far from saturation of the excited state $|1\rangle$, the excitation rate of the system is given by

$$\gamma_{\mathrm{e}}(\mathbf{r}) = \sigma\, I_{\mathrm{e}}(\mathbf{r})/(\hbar\omega_0), \tag{4.32}$$

where σ is the one-photon absorption cross-section and I_{e} is the intensity associated with the excitation field $\mathbf{E}_{\mathrm{e}}$. Once the system is in its excited state the probability of a spontaneous transition to the ground state $|0\rangle$ (emission of a fluorescence photon) is given by

$$\frac{\gamma_{\mathrm{r}}}{\gamma_{\mathrm{r}} + \gamma_{\mathrm{d}}}. \tag{4.33}$$

Here, γ_{r} is the spontaneous decay rate and γ_{d} the stimulated transition rate. The latter can be written as

$$\gamma_{\mathrm{d}}(\mathbf{r}) = \sigma I_{\mathrm{d}}(\mathbf{r})/(\hbar\omega_0), \tag{4.34}$$

with I_{d} being the intensity of the depletion field. By combining Eqs. (4.32) and (4.33) we can express the fluorescence rate of the system as

$$\gamma_{\mathrm{fl}}(\mathbf{r}) = \gamma_{\mathrm{e}}(\mathbf{r})\frac{\gamma_{\mathrm{r}}}{\gamma_{\mathrm{r}} + \gamma_{\mathrm{d}}(\mathbf{r})} = \frac{\sigma}{\hbar\omega_0}\frac{I_{\mathrm{e}}(\mathbf{r})}{1 + d_{\mathrm{p}}(\mathbf{r})}, \tag{4.35}$$

where we introduced the *depletion parameter*

$$d_{\mathrm{p}}(\mathbf{r}) \equiv \frac{\sigma}{\hbar\omega_0\gamma_{\mathrm{r}}}\, I_{\mathrm{d}}(\mathbf{r}), \tag{4.36}$$

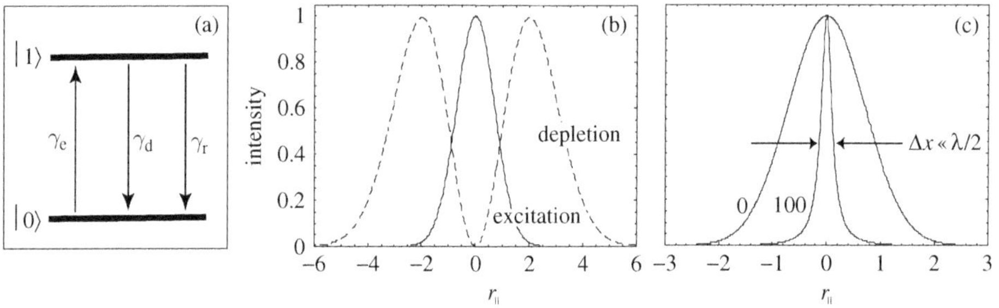

Fig. 4.8 Illustration of resolution enhancement through saturation. (a) Energy level diagram of a two-state molecule with excitation rate γ_e, radiative decay rate γ_r, and stimulated depletion rate γ_d. (b) Transverse intensity profiles of the excitation field and the depletion field. The zero of the depletion field is placed at the maximum of the excitation field. (c) Transverse fluorescence profiles (γ_r) for two different depletion parameters $d_p = 0$ and $d_p = 100$. The higher d_p the narrower the fluorescence peak will be.

which corresponds to the ratio of the rates of stimulated and spontaneous emission. For a weak depletion field the stimulated emission is weak ($d_p \rightarrow 0$) and the fluorescence rate reduces to the familiar expression given by Eq. (4.32).

Let us now discuss the relationship between this simple theory and the issue of resolution in optical microscopy. Obviously, for $d_p = 0$ the resolution in the fluorescence image will be determined by the width of the excitation field shown in Fig. 4.8(b). However, if we use a depletion field with a zero at the maximum of the excitation field then the width can be narrowed significantly, depending on the magnitude of d_p. This behavior is illustrated in Fig. 4.8(c) for $d_p = 100$. In principle, there is no limit for the narrowing of the fluorescent region and, in principle, arbitrary resolution can be achieved. We can introduce the depletion parameter into Abbe's resolution criterion and obtain approximately

$$\text{Min}\left[\Delta r_{||}\right] \approx 0.6098 \, \frac{\lambda}{\text{NA}\sqrt{1+d_p}}. \tag{4.37}$$

Thus, any $d_p > 0$ improves the spatial resolution. It should be noted that resolution enhancement arising from saturation is not limited to imaging. The same idea can be employed for lithography or for data storage, provided that a material with the desired saturation/depletion properties can be found. Finally, we have to realize that resolution enhancement through saturation makes use of very specific material properties as provided, for example, by a fluorophore. In this sense, the electronic structure of the target material has to be known in advance and hence there is no spectroscopic information to be gained. Nevertheless, information on biological samples is normally provided through chemically specific labeling with fluorophores.

4.3 Principles of confocal microscopy

Today, confocal microscopy is a technique that is applied in many scientific disciplines, ranging from solid state physics to biology. The central idea is to irradiate the sample with focused light originating from a point source (or a single-mode laser beam) and direct the response from the sample onto a pinhole as discussed in Section 4.2.2. The basic idea was put forward in a patent application by Minsky in 1955 [10]. Over the years, different variations of confocal microscopy have been developed. They differ mostly in the specific type of laser–matter interaction, such as scattering, fluorescence, multiphoton excited fluorescence, stimulated emission depletion, third-harmonic generation, or Raman scattering. In this section we will outline the general ideas behind confocal microscopy using the theoretical framework established so far. Experimental aspects will be covered later in Chapter 5. More detailed treatments can be found in dedicated books on confocal microscopy such as Refs. [11–13].

To understand image formation in confocal microscopy we will focus on the configuration shown in Fig. 4.9. This is a special case of the general situation shown in Fig. 4.5. In the present situation, excitation and detection are accomplished by the same objective lens using an *inverted* light path. A beamsplitter is used to split the excitation path and the detection path into two separate arms. In fluorescence microscopy, the beamsplitter is usually replaced by a dichroic mirror that transmits or reflects only specific spectral ranges thereby increasing the efficiency. To keep things as simple as possible we assume that a sample with one single dipolar particle is translated in all three dimensions relative to the fixed optical system. Thus, we can set $\mathbf{r}_s = 0$ and use the vector $\mathbf{r}_{\mathrm{n}} = (x_{\mathrm{n}}, y_{\mathrm{n}}, z_{\mathrm{n}})$ to denote the coordinates of the particle.

To generate an image we assign to each position $\mathbf{r}_{\mathrm{n}}$ a scalar quantity measured in the image space. In confocal microscopy, this quantity corresponds to the signal s_2 discussed previously. Similarly, for non-confocal microscopy we use the signal s_1. The process of image formation embraces the following three steps.

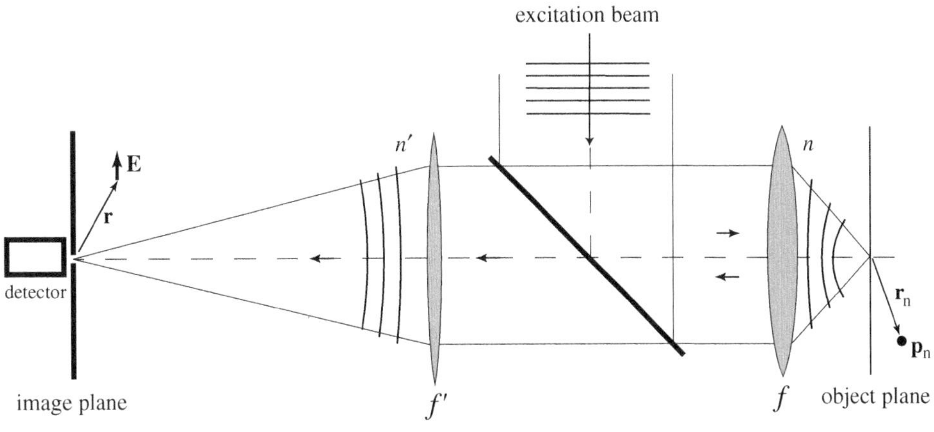

Fig. 4.9 Schematic representation of an inverted confocal microscope. In this set-up, the light path is held fixed whereas the sample is scanned in three dimensions. A beamsplitter divides the excitation path and detection path into two separate arms. A laser beam is focused into the sample by a high-NA objective lens to provide a spatially confined excitation source. The response of the sample is collected by the same objective lens and focused onto a pinhole in front of a detector.

1. Calculation of the excitation field in the object space (Sections 3.5 and 3.6) → **excitation point-spread function.**
2. Calculation of interaction.
3. Calculation of the response in the image space (Section 4.1) → **detection point-spread function.**

The first step provides the excitation field $\mathbf{E}_{\text{exc}}$. It depends not only on the parameters of the confocal system but also on the incident laser mode. For the interaction between the excitation field $\mathbf{E}_{\text{exc}}$ and the dipolar particle we first assume a linear relationship, which we write as

$$\mathbf{p}_{\text{n}}(\omega) = \overset{\leftrightarrow}{\alpha}\,\mathbf{E}_{\text{exc}}(\mathbf{r}_{\text{n}}, \omega). \tag{4.38}$$

Finally, the response of the dipole in the image space is determined by (cf. Eq. (4.6))

$$\mathbf{E}(\mathbf{r}) = \frac{\omega^2}{\varepsilon_0 c^2} \overset{\leftrightarrow}{\mathbf{G}}_{\text{PSF}} \cdot \mathbf{p}_{\text{n}}. \tag{4.39}$$

The combination of these equations allows us to eliminate $\mathbf{p}_{\text{n}}$ and thus to calculate the image field as a function of the excitation field, the particle polarizability and the system parameters.

To evaluate the equations above it is necessary to understand the mapping from object space to image space. A field point in the image space is defined by the vector $\mathbf{r}$. We have learned before that a dipole $\mathbf{p}_{\text{n}}$ at the origin ($\mathbf{r}_{\text{n}} = 0$) generates a field $\mathbf{E}(x, y, z)$ in the image space according to Eqs. (4.6)–(4.10). If we translate the dipole from its origin to an arbitrary position $\mathbf{r}_{\text{n}}$ in the object space, the field in the image space will transform as

$$\mathbf{E}(x, y, z) \rightarrow \mathbf{E}(x - x_{\text{n}}M,\ y - y_{\text{n}}M,\ z - z_{\text{n}}M^2 n'/n), \tag{4.40}$$

where M is the transverse magnification defined in Eq. (4.12). The pinhole filters this field and the detector behind it performs an integration over x and y. To keep things simple, we can assume that the pinhole is sufficiently small, allowing us to replace the detected signal by the field intensity at $\mathbf{r}=0$ multiplied by an infinitesimal detector area $\mathrm{d}A$ (cf. Eq. (4.28)). The detector signal is then dependent only on the coordinates of the dipole

$$s_2(x_{\text{n}}, y_{\text{n}}, z_{\text{n}}) = \left|\mathbf{E}(x_{\text{n}}M,\ y_{\text{n}}M,\ z_{\text{n}}M^2 n'/n)\right|^2 \delta A. \tag{4.41}$$

The field $\mathbf{E}\,(x_{\text{n}}M,\ y_{\text{n}}M,\ z_{\text{n}}M^2\ n'/n)$ is obtained from Eqs. (4.6)–(4.10) by using the substitutions $\rho \rightarrow \rho_{\text{n}}M,\ z \rightarrow z_{\text{n}}M^2 n'/n,$ and $\varphi \rightarrow \varphi_{\text{n}}$. Then, the detector signal becomes

$$s_2(x_{\text{n}}, y_{\text{n}}, z_{\text{n}}) = \frac{\omega^4}{\varepsilon_0^2 c^4} \left|\overset{\leftrightarrow}{\mathbf{G}}_{\text{PSF}}(\rho_{\text{n}}, \varphi_{\text{n}}, z_{\text{n}}) \cdot \mathbf{p}_{\text{n}}\right|^2 \delta A, \tag{4.42}$$

with

$$\overset{\leftrightarrow}{\mathbf{G}}_{\text{PSF}}(\rho_{\text{n}}, \varphi_{\text{n}}, z_{\text{n}}) \propto \frac{k}{8\pi}\frac{1}{M} \begin{bmatrix} \tilde{I}_{00} + \tilde{I}_{02}\cos(2\varphi_{\text{n}}) & \tilde{I}_{02}\sin(2\varphi_{\text{n}}) & -2\mathrm{i}\tilde{I}_{01}\cos\varphi_{\text{n}} \\ \tilde{I}_{02}\sin(2\varphi_{\text{n}}) & \tilde{I}_{00} - \tilde{I}_{02}\cos(2\varphi_{\text{n}}) & -2\mathrm{i}\tilde{I}_{01}\sin\varphi_{\text{n}} \\ 0 & 0 & 0 \end{bmatrix} \tag{4.43}$$

and the integrals $\tilde{I}_{00}$–$\tilde{I}_{02}$ are

$$\tilde{I}_{00}(\rho_n, z_n) = \int_0^{\theta_{max}} (\cos\theta)^{1/2} \sin\theta(1+\cos\theta) J_0(k\rho_n \sin\theta) e^{-\frac{i}{2}kz_n \sin^2\theta}\, d\theta,$$

$$\tilde{I}_{01}(\rho_n, z_n) = \int_0^{\theta_{max}} (\cos\theta)^{1/2} \sin^2\theta\, J_1(k\rho_n \sin\theta) e^{-\frac{i}{2}kz_n \sin^2\theta}\, d\theta, \tag{4.44}$$

$$\tilde{I}_{02}(\rho_n, z_n) = \int_0^{\theta_{max}} (\cos\theta)^{1/2} \sin\theta(1-\cos\theta) J_2(k\rho_n \sin\theta) e^{-\frac{i}{2}kz_n \sin^2\theta}\, d\theta.$$

The field depends on the magnitude and orientation of the dipole $\mathbf{p}_n$, which, in turn, depends on the nature of the interaction between the excitation field $\mathbf{E}_{exc}$ and the dipolar particle. The excitation field can be an arbitrary focused laser mode as discussed in Section 3.6. Let us choose a fundamental Gaussian beam as this is used in most confocal set-ups. We assume that the beam is focused on the object plane and that its propagation direction coincides with the optical axis. According to Eqs. (3.66) and (4.38) the dipole moment can be written as

$$\mathbf{p}_n(\omega) = ikfE_0 e^{-ikf} \frac{1}{2} \begin{bmatrix} \alpha_{xx}(I_{00} + I_{02}\cos(2\varphi_n)) \\ \alpha_{yy}(I_{02}\sin(2\varphi_n)) \\ \alpha_{zz}(-2iI_{01}\cos\varphi_n)\sqrt{n'/n} \end{bmatrix}, \tag{4.45}$$

where α_{ii} denote the diagonal elements of the polarizability and E_0 is the field amplitude of the incident paraxial Gaussian beam. The integrals I_{00}–I_{02} are defined in Eqs. (3.58)–(3.60) and read as

$$I_{00}(\rho_n, z_n) = \int_0^{\theta_{max}} f_w(\theta)(\cos\theta)^{1/2} \sin\theta(1+\cos\theta) J_0(k\rho_n \sin\theta) e^{ikz_n \cos\theta}\, d\theta,$$

$$I_{01}(\rho_n, z_n) = \int_0^{\theta_{max}} f_w(\theta)(\cos\theta)^{1/2} \sin^2\theta J_1(k\rho_n \sin\theta) e^{ikz_n \cos\theta}\, d\theta, \tag{4.46}$$

$$I_{02}(\rho_n, z_n) = \int_0^{\theta_{max}} f_w(\theta)(\cos\theta)^{1/2} \sin\theta(1-\cos\theta) J_2(k\rho_n \sin\theta) e^{ikz_n \cos\theta}\, d\theta,$$

where the function f_w defines the expansion of the incident beam relative to the back-aperture of the objective lens.

The integrals $\tilde{I}_{nm}$ and the integrals I_{nm} differ only by the term $f_w(\theta)$ and in the exponential terms, which become identical in the small-angle limit ($\cos\theta \approx 1 - \frac{1}{2}\theta^2, \sin^2\theta \approx \theta^2$). Using Eq. (4.42), we are now in a position to exactly calculate the confocal signal in the image plane. However, in order to see the essence of confocal microscopy we need to reduce the complexity somewhat. We assume that the incident beam is sufficiently expanded, i.e. $f_w(\theta) = 1$, and that the slight difference in the exponential terms is marginal so that the two sets of integrals become identical. Furthermore, we neglect the contribution of I_{02} relative to I_{00} and assume that the dipole is rigidly aligned along the polarization direction, i.e. $\alpha_{yy} = \alpha_{zz} = 0$. The resulting detector signal is then identical to the signal that would result from a purely scalar calculation and reads as

$$\text{confocal:} \qquad s_2(x_\text{n}, y_\text{n}, z_\text{n}; \omega) \propto \left|\alpha_{xx} I_{00}^2\right|^2 \delta A. \tag{4.47}$$

The important outcome is the fact that the integral appears squared. This means that the point-spread function in confocal microscopy is essentially the square of the point-spread function in ordinary microscopy! Thus, in addition to the axial resolution, confocal microscopy has increased transverse resolution – and this is simply the result from placing a pinhole in front of the detector. If the pinhole is removed and all radiation in the image plane is directed on the detector, the signal turns out to be

$$\text{non-confocal:} \qquad s_1(x_\text{n}, y_\text{n}, z_\text{n}; \omega) \propto \left|\alpha_{xx} I_{00}\right|^2 \delta A. \tag{4.48}$$

This seems somewhat surprising since in the previous section we concluded that ordinary far-field microscopy has no axial resolution. However, we assumed before that we have a uniform illumination of the object space. The axial resolution in the present case is achieved by the spatially confined excitation source provided by the focused laser beam and by having only a single dipolar emitter in the sample volume. If we had a dense sample of dipoles (see Problem 4.3) we would lose any axial resolution in non-confocal microscopy. Nevertheless, we clearly see that the pinhole in confocal microscopy increases both transverse and longitudinal resolution.

The total point-spread function of the system can be regarded as the product of an excitation point-spread function and a detection point-spread function

$$\text{TOTAL PSF} \approx \text{EXCITATION PSF} \times \text{DETECTION PSF}, \tag{4.49}$$

where the former is determined by the field distribution of the focused excitation beam and the latter by the spatial filtering properties of the pinhole in the image plane. However, we have to keep in mind that the increase in transverse resolution achieved by confocal microscopy is marginal, often only a small percentage. While the zeros of the point-spread function remain unchanged, the width of the central lobe becomes slightly narrower. The benefit of confocal microscopy lies much more in the axial sectioning capabilities in dense samples (see Problem 4.3). It has to be emphasized that it is a rough approximation to reduce the two sets of integrals in Eqs. (4.44) and (4.46) to a single set. This can only be done for a Gaussian excitation beam because the symmetries of detection and excitation turn out to be the same. The analysis becomes more complicated if we use a higher-order beam mode as an excitation source.

Figure 4.10 shows an experimentally measured point-spread function. It has been recorded by raster scanning a gold particle through the focal region of a focused excitation beam and recording, for each image pixel, the scattered light intensity. Because of

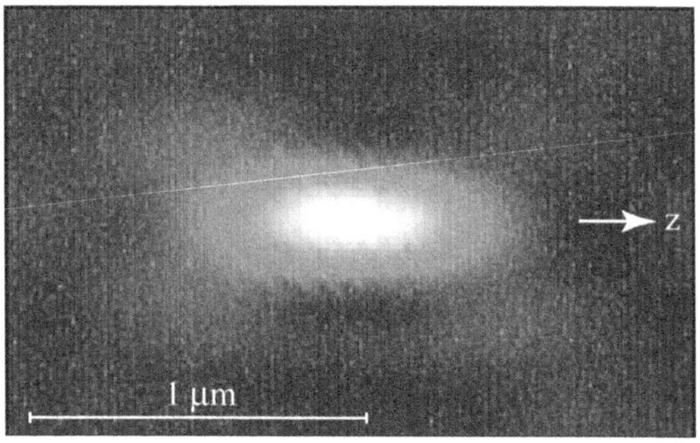

Fig. 4.10 The total point-spread function (PSF) measured by scanning a gold particle through the laser focus and detecting the scattered intensity at each position. From [9].

its spherical symmetry, the particle has no preferred dipole axis and hence $\alpha_{xx} = \alpha_{yy} = \alpha_{zz}$. Experimental aspects of confocal microscopy will be discussed in more detail in Section 5.2.1.

It is straightforward to extend the analysis to account for nonlinear interactions between the particle and the excitation beam. For example, with the same assumptions and approximations as before we find for a second-order nonlinear process

$$\text{confocal:} \quad s_2(x_n, y_n, z_n; 2\omega) \propto \left| \beta_{xxx} I_{00}(2\omega) I_{00}^2(\omega) \right|^2 \delta A, \tag{4.50}$$

$$\text{non-confocal:} \quad s_1(x_n, y_n, z_n; 2\omega) \propto \left| \beta_{xxx} I_{00}^2(\omega) \right|^2 \delta A. \tag{4.51}$$

Here, we had to consider that excitation occurs at a frequency ω, whereas detection occurs at a frequency of 2ω. It is often claimed that nonlinear excitation increases resolution. However, this is not true. Although a nonlinear process squeezes the point-spread function it requires longer excitation wavelengths. While the Airy disk radius scales proportionally with the wavelength it is not so strongly influenced by being multiplied by itself. Therefore, the wavelength scaling dominates.

4.4 Axial resolution in multiphoton microscopy

We have determined that the benefit of confocal microscopy is not necessarily an increase of the transverse resolution but rather an increase of the longitudinal resolution. This longitudinal resolution provides sectioning capability for true three-dimensional imaging. The same benefits are achieved in multiphoton microscopy even without using confocal arrangements. In multiphoton fluorescence microscopy the signal generated at a position $\mathbf{r}$ is qualitatively given by

$$s(\mathbf{r}) \propto \sigma_n \left[\mathbf{E}(\mathbf{r}) \cdot \mathbf{E}^*(\mathbf{r}) \right]^n, \tag{4.52}$$

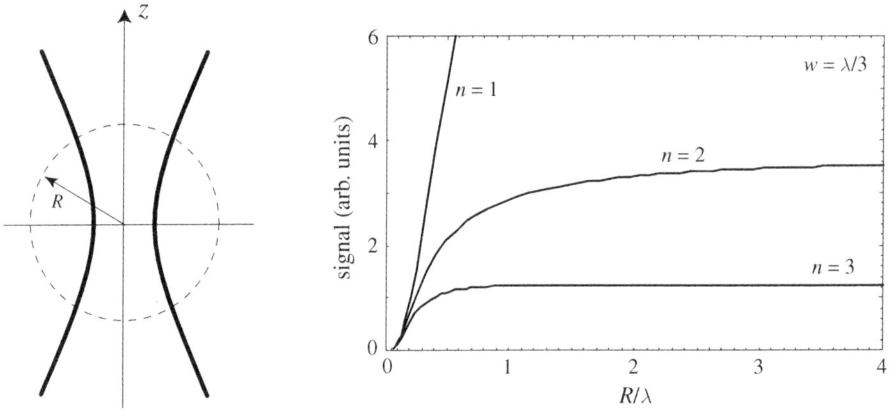

Fig. 4.11 Localization of the excitation volume in multiphoton microscopy. The figure depicts the signal that is generated in a dense sample within a sphere of radius R when excited by a focused Gaussian beam by n-photon excitation. In contrast to multiphoton excitation ($n > 1$), one-photon excitation ($n = 1$) cannot restrict the excitation volume without the use of a confocal pinhole.

where σ_n is the n-photon absorption cross-section and $\mathbf{E}$ is the excitation field. In a dense sample of fluorophores the total signal generated in a spherical volume of radius R is calculated as

$$s_{\text{total}} \propto \sigma_n \int_0^{2\pi}\int_0^{\pi}\int_0^{R} |\mathbf{E}(r,\theta,\phi)|^{2n} r^2 \sin\theta \, \mathrm{d}r \, \mathrm{d}\theta \, \mathrm{d}\phi. \tag{4.53}$$

For large distances from the exciting laser focus, the excitation fields decay as r^{-1} and consequently the integral does not converge for $n = 1$. Thus, without the use of a confocal pinhole, it is not possible to axially localize the signal in one-photon excitation. However, for $n > 1$ the situation is different. The signal is generated only in the vicinity of the laser focus. This is illustrated in Fig. 4.11 where we evaluated Eq. (4.53) for a Gaussian beam with beam waist radius $w_0 = \lambda/3$. Although we used the paraxial approximation and ignored the fact that longer wavelengths are used in multiphoton microscopy, it is a general finding that localization of the excitation volume requires a process with $n > 1$. It is this property that makes multiphoton microscopy such an attractive technique. Multiphoton microscopy will be discussed in more detail in Chapter 5.

4.5 Localization and position accuracy

We have seen that a dipole emitter, such as a fluorescent molecule, gives rise to a characteristic point-spread function in image space. Vice versa, the point-spread function can be recorded to reconstruct the position of the emitter [14–17]. The accuracy of determining the position of a point-like emitter is much better than the spatial extent of the point-spread

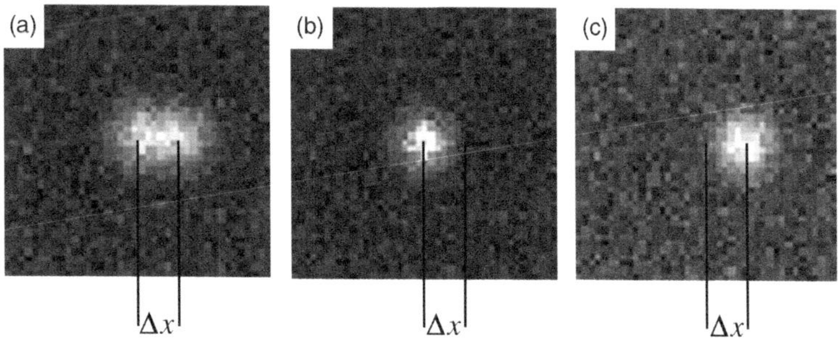

Fig. 4.12 Simulated CCD camera image patterns of two partially overlapping emission spots with background. The emitters are simulated to exhibit Gaussian patterns with Poissonian noise that is uncorrelated to the Poissonian background noise. (a) Two emission spots without discrimination of photons. Images (b) and (c) show Individual emission spots resolved by spectral or temporal discrimination of photons. The centroids of the individual patterns are displaced by finite distance Δx.

function and, as will be discussed in the following, is limited only by the "quality" of the data, that is the amount of noise present in the data. For example, a fluorescent molecule can be localized with a precision of a few nanometers even if the emitters are single fluorescent molecules. Furthermore, if the photons that arrive at the detector can be distinguished by any observable, e.g. energy, polarization, or arrival time, as discussed before, they may be attributed to separate objects even if two objects are very close and their image patterns overlap. This idea is illustrated in Fig. 4.12. In Fig. 4.12(a) a composite pattern consisting of two partially overlapping spots is shown. If the photons that contribute to these spots can be distinguished by any observable, e.g. color or arrival time, Figs. 4.12(b) and (c), then the individual positions and therefore also the distance between the two emitters can be determined with nearly molecular-scale precision, limited only by background and counting noise. This way of attaining subwavelength position accuracy even for very weak emitters, such as single fluorescent molecules, has important applications in astronomy [14], single-molecule localization and tracking [17], analytical chemistry [18], and localization microscopy [19–21].

4.5.1 Theoretical background

In principle, there are numerous ways to find the position of an isolated emitter. For example, one could calculate the "center of mass" or centroid of a given pattern from the intensities of the pixels or use appropriate correlation filtering techniques. In order to quantify the precision with which a position is found, a statement about the uncertainty in the position measurement is required. It is therefore common to approximate the point-spread function by a suitable model and to fit this model to the obtained data by minimizing χ^2, the sum of the squares of the deviation between data and model at each data point. Because χ^2 reflects the likelihood that a certain set of parameters is correct, it can be used

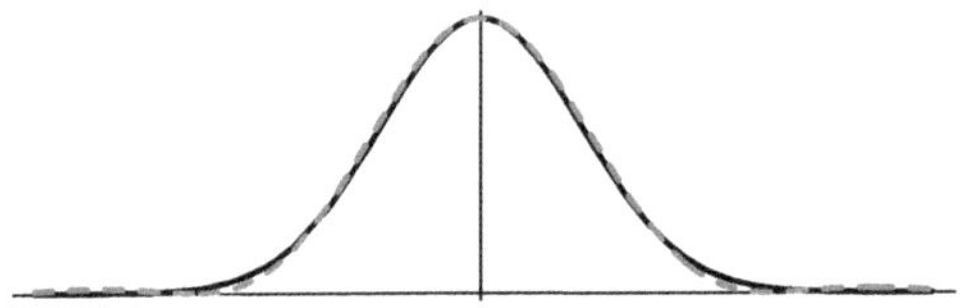

Fig. 4.13 Similarity of the Airy function and a fitted Gaussian. The deviations are negligible for noisy data.

to establish well-defined error limits to each fit parameter. Thus, by exploiting the χ^2 statistics it is possible to obtain not only the set of best fit parameters for a given model but also the standard deviations associated with this set based on the measured data. The analysis given here follows the work of Bobroff [14], which relies on a maximum likelihood criterion for data with a Gaussian error distribution. More general approaches are discussed in the literature [16]. We limit ourselves here to the specific case of least-squares fitting of two-dimensional Gaussian distributions. A two-dimensional Gaussian fits very well to the intensity patterns of subwavelength emitters obtained in optical microscopy. Although fitting an Airy pattern would be the more realistic choice, usually the signal quality is not good enough to result in significant systematic deviations (Fig. 4.13) [16]. In special cases, however, the use of more complex models might be necessary or advantageous depending on the problem. For example, the complex patterns obtained by annular illumination confocal microscopy and illumination with higher-order modes certainly have to be fitted by more complex models. The present analysis can be adapted to such cases.

For a two-dimensional Gaussian intensity distribution

$$G(x,y) = B + A\,\exp\left[-\frac{(x-x_0)^2+(y-y_0)^2}{2\gamma^2}\right] \tag{4.54}$$

there are five parameters that have to be determined, i.e. the spatial coordinates of the maximum x_0 and y_0 (i.e. the spot position), the amplitude A, the width γ, and the background B. Sometimes the width γ of the point-spread function is assumed to be known from independent measurements. This reduces the number of fit parameters and increases the accuracy of the remaining parameters by roughly 10% as shown below. Typically, experimental data are recorded at a finite number of points (x_i, y_j), e.g. corresponding to the pixels of a CCD chip or of a scan image. Each data point (x_i, y_j) is associated with a signal $D_{i,j}$ and a corresponding uncertainty $\sigma_{i,j}$, e.g. due to Poissonian counting statistics. The sum of the squares of the deviation between data and model, χ^2, over all data points (i,j) then reads as

$$\chi^2 = \sum_{i,j=1}^{N} \frac{1}{\sigma_{i,j}^2}\left[G_{i,j} - D_{i,j}\right]^2, \tag{4.55}$$

where N is the number of pixels in the x- and y-directions. Here, $G_{i,j}$ are the values of the model at the point (x_i, y_j), and $1/\sigma_{i,j}^2$ is used as a weighting factor to ensure that data points with small uncertainties are more important. The set of parameters that minimizes χ^2, which then becomes equal to χ^2_{min}, is denoted as $\left[x_{0,\text{min}}, y_{0,\text{min}}, \gamma_{\text{min}}, A_{\text{min}}, B_{\text{min}}\right]$. For this set of data $G_{i,j}$ is called $G_{i,j,\min}$. It is obvious that the uncertainty in each of the parameters depends on the shape of χ^2 around its minimum $\chi^2_{\min}$. To a good approximation, for small variations of a single parameter about the minimum, χ^2 has the shape of a parabola. Depending on whether the parabola has a small or large opening factor, the statistical error associated with the respective parameter is smaller or larger. In order to find these opening factors and thus quantify the uncertainties we write the Taylor expansion of χ^2 around its minimum $\chi^2_{\min}$:

$$\chi^2 \simeq \sum_{i,j=1}^{N} \frac{1}{\sigma_{i,j}^2} \Bigg[\left(G_{i,j,\min} - D_{i,j}\right) + \left(\frac{\partial G_{i,j}}{\partial x_0}\right)_{x_{0,\min}} (x_0 - x_{0,\min}) + \left(\frac{\partial G_{i,j}}{\partial y_0}\right)_{y_{0,\min}} \left(y_0 - y_{0,\min}\right) + \left(\frac{\partial G_{i,j}}{\partial \gamma}\right)_{\gamma_{\min}} (\gamma - \gamma_{\min}) + \left(\frac{\partial G_{i,j}}{\partial A}\right)_{A_{\min}} (A - A_{\min}) + \left(\frac{\partial G_{i,j}}{\partial B}\right)_{B_{\min}} (B - B_{\min}) \Bigg]^2, \quad (4.56)$$

where the first term describes $\chi^2_{\min}$. The deviation Δ of χ^2 from this minimum can then be expressed as

$$\begin{aligned}\Delta &= \chi^2 - \chi^2_{\min} \\ &\simeq \sum_{i,j=1}^{N} \frac{1}{\sigma_{i,j}^2} \Bigg[\left(\frac{\partial G_{i,j}}{\partial x_0}\right)^2_{x_{0,\min}} (x_0 - x_{0,\min})^2 \\ &\quad + \left(\frac{\partial G_{i,j}}{\partial y_0}\right)^2_{y_{0,\min}} \left(y_0 - y_{0,\min}\right)^2 + \left(\frac{\partial G_{i,j}}{\partial \gamma}\right)^2_{\gamma_{\min}} (\gamma - \gamma_{\min})^2 \\ &\quad + \left(\frac{\partial G_{i,j}}{\partial A}\right)^2_{A_{\min}} (A - A_{\min})^2 + \left(\frac{\partial G_{i,j}}{\partial B}\right)^2_{B_{\min}} (B - B_{\min})^2 \\ &\quad + \text{cross terms} \Bigg]. \end{aligned} \quad (4.57)$$

The cross terms can be shown to vanish [14]. Some contain the partial derivatives of χ^2 that vanish because χ^2 has a minimum at $\left[x_{0,\min}, y_{0,\min}, \gamma_{\min}, A_{\min}, B_{\min}\right]$. The other cross terms are negligible because they are sums over products of symmetric and antisymmetric functions. This leads to an important intermediate result, i.e. the approximative behavior of Δ for small deviations from the minimum

$$\begin{aligned}\Delta \simeq \sum_{i,j\simeq 1}^{N} \frac{1}{\sigma_{i,j}^2} \Bigg[&\left(\frac{\partial G_{i,j}}{\partial x_0}\right)^2_{x_{0,\min}} (x_0 - x_{0,\min})^2 + \left(\frac{\partial G_{i,j}}{\partial y_0}\right)^2_{y_{0,\min}} \left(y_0 - y_{0,\min}\right)^2 \\ &+ \left(\frac{\partial G_{i,j}}{\partial \gamma}\right)^2_{\gamma_{\min}} (\gamma - \gamma_{\min})^2 + \left(\frac{\partial G_{i,j}}{\partial A}\right)^2_{A_{\min}} (A - A_{\min})^2 \\ &+ \left(\frac{\partial G_{i,j}}{\partial B}\right)^2_{B_{\min}} (B - B_{\min})^2 \Bigg]. \end{aligned} \quad (4.58)$$

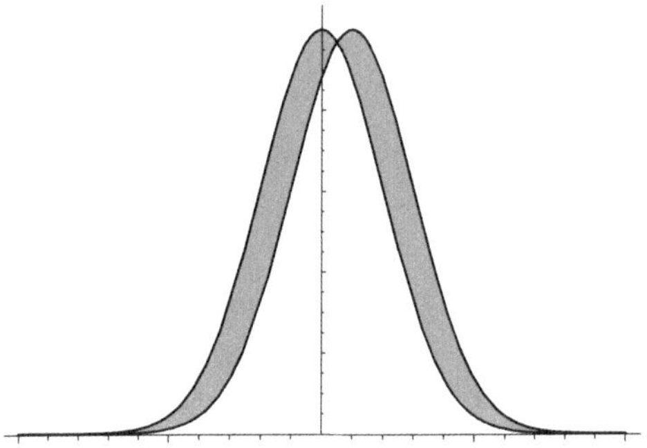

Fig. 4.14 Two Gaussians displaced by a small amount. It is obvious that the main contribution to the differences between the two curves (shaded in gray) occurs where the slope is large. This is expressed in Eq. (4.58).

This result describes by how much χ^2 is increased with respect to its minimal value by a variation of the parameters around their optimal values. The surfaces of constant Δ are "ellipses" in the parameter space. According to Eq. (4.58) the strongest contributions to χ^2 come from the regions where G has steep slopes. For the position parameters (x_0, y_0) this can be easily verified by displacing a Gaussian fit curve from the best fit parameters $(x_{0,\min}, y_{0,\min})$ illustrated in Fig. 4.14.

4.5.2 Estimating the uncertainties of fit parameters

As Δ increases, the statistical likelihood of the parameter set being the correct one decreases. It is possible to establish a connection between the magnitude of Δ and the statistical likelihood associated with the fit parameters [22, 23]. Once the value of Δ for a given level of confidence of the fit is substituted, Eq. (4.58) can be used to estimate the uncertainty in the parameters. The normalized probability distribution function for Δ with ν fitting parameters[4] is given by (see e.g. [22] Appendix C-4)

$$P(\Delta, \nu) = \frac{\Delta^{\frac{\nu-2}{2}} e^{-\frac{\Delta}{2}}}{2^{\nu/2}\Gamma(\nu/2)}. \tag{4.59}$$

If we integrate $P(\Delta, \nu)$ up to infinity starting from the value of Δ_a that leads to a value of the integral of 0.317,

$$\int_{\Delta_a}^{\infty} P(\Delta, \nu) \mathrm{d}\Delta = 0.317, \tag{4.60}$$

then with a probability of $1 - 0.317 = 0.683$ the correct parameters lie within the region of parameter space for which Δ is smaller than Δ_a, corresponding to a 1σ confidence level. The value of Δ_a increases with the number of free fit parameters ν since usually correlations between the different parameters exist. Table 4.1 provides the respective values

[4] Also called "degrees of freedom."

Table 4.1 Values of Δ_a obtained from Eq. (4.60) for up to seven fit parameters

ν	1	2	3	4	5	6	7
Δ_a	1	2.3	3.5	4.7	5.9	7.05	8.2

of Δ_a for up to seven fit parameters for a 68.3% confidence level. Other values can be calculated using Eqs. (4.59) and (4.60).

For example, in order to estimate the uncertainty of the position x_0 we assume that all parameters apart from x_0 have their optimum values. In Eq. (4.58), in this case all terms except the one containing x_0 vanish. From Eq. (4.58) we then obtain

$$\sigma_x \equiv \left(x_0 - x_{0,\min}\right) = \Delta_a^{1/2} \left(\sum_{i,j=1}^{N} \frac{1}{\sigma_{i,j}^2} \left(\frac{\partial G_{i,j}}{\partial x_0} \right)^2_{x_{0,\min}} \right)^{-\frac{1}{2}} . \tag{4.61}$$

The sum over i and j can either be calculated directly numerically from the result of the fit or be approximated by an integral to yield an analytical expression for the uncertainty σ_x. The latter approach has the advantage that it allows us to discuss the dependence of the positional uncertainty on various experimental parameters. To obtain an analytical expression we exploit the fact that

$$\frac{1}{N^2} \sum_{i,j=1}^{N} \frac{1}{\sigma_{i,j}^2} \left(\frac{\partial G_{i,j}}{\partial x_0} \right)^2_{x_{0,\min}} \approx \frac{1}{L^2} \iint_{-L/2}^{L/2} \delta x\, \delta y \, \frac{1}{\sigma^2(x,y)} \left(\frac{\partial G_{i,j}}{\partial x_0} \right)^2_{x_{0,\min}} , \tag{4.62}$$

where $L = N\,\delta x = N\,\delta y$ is the side length of the quadratic fitting area with δx and δy being the dimensions of individual square pixels,[5] and N is the number of pixels within the length L. To evaluate the integral on the right-hand side of Eq. (4.62) we have to make some assumptions about the noise of the data $\sigma^2(x, y)$. We assume uncorrelated Poissonian (or Gaussian) noise of the background and the signal. Thus we have $\sigma^2(x, y) = \sigma_B^2 + \sigma_A^2$, where, according to Eq. (4.54), $\sigma_B^2 = B$ and $\sigma_A^2 = A \exp\left[-((x - x_0)^2 + (y - y_0)^2)/(2\gamma^2)\right]$. On introducing this expression into Eq. (4.62) it is difficult to arrive at an analytical result. We therefore apply the following approximations. (i) We assume that the signal dominates the background around the maximum of the Gaussian peak up to a distance of $\kappa\gamma$. This means that only the Poissonian noise of the signal σ_A is assumed to contribute in this region. (ii) For distances larger than $\kappa\gamma$ we assume that the Poissonian noise of the background σ_B dominates. The parameter κ allows us to adjust the transition point depending on the relative magnitude of signal and background that may occur in specific experiments. The sum of Eq. (4.62) can now be approximated by a sum of three integrals as follows

[5] This assumption is not mandatory but simplifies the analysis.

$$\sum_{i,j=1}^{N} \frac{1}{\sigma_{i,j}^2}\left(\frac{\partial G_{i,j}}{\partial x_0}\right)^2_{x_{0,\min}} \approx \frac{N^2}{L^2}\iint\limits_{-\kappa\gamma}^{\kappa\gamma} \delta x\,\delta y\,\frac{1}{\sigma_A^2(x,y)}\left(\frac{\partial G_{i,j}}{\partial x_0}\right)^2_{x_{0,\min}}$$
$$+\frac{N^2}{L^2}\iint\limits_{-L/2}^{-\kappa\gamma} \delta x\,\delta y\,\frac{1}{\sigma_B^2}\left(\frac{\partial G_{i,j}}{\partial x_0}\right)^2_{x_{0,\min}}$$
$$+\frac{N^2}{L^2}\iint\limits_{\kappa\gamma}^{L/2} \delta x\,\delta y\,\frac{1}{\sigma_B^2}\left(\frac{\partial G_{i,j}}{\partial x_0}\right)^2_{x_{0,\min}}, \tag{4.63}$$

where the last two terms yield identical results due to the symmetry of the problem. With this approximative description using Eq. (4.61), we can write for the normalized uncertainty in the position in the x-direction

$$\frac{\sigma_x}{\gamma} = \frac{2t}{N}\sqrt{\frac{\Delta_a}{\left[c(\kappa)A + (A^2/B)F(t,\kappa)\right]}} = \frac{\delta x}{\gamma}\sqrt{\frac{\Delta_a}{\left[c(\kappa)A + (A^2/B)F(t,\kappa)\right]}}. \tag{4.64}$$

Here we have introduced the dimensionless parameter $t = L/(2\gamma)$ which describes the width of the fitting area in units of the width of the peak. The function $F(t,\kappa)$ and the constant $c(\kappa)$ in Eq. (4.64) are defined as

$$F(t,\kappa) = \frac{\sqrt{\pi}}{2}[\mathrm{Erf}(\kappa) - \mathrm{Erf}(t)]\left[\frac{\sqrt{\pi}}{2}[\mathrm{Erf}(\kappa) - \mathrm{Erf}(t)] + t\mathrm{e}^{-t^2} - \kappa\mathrm{e}^{-\kappa^2}\right],$$
$$c(\kappa) = 2\,\mathrm{Erf}\left(\frac{\kappa}{\sqrt{2}}\right)\left[\pi\,\mathrm{Erf}\left(\frac{\kappa}{\sqrt{2}}\right) - \sqrt{2\pi}\kappa\mathrm{e}^{-\frac{\kappa^2}{2}}\right], \tag{4.65}$$

with

$$\mathrm{Erf}(z) = \frac{2}{\sqrt{\pi}}\int_0^z \mathrm{e}^{-u^2}\,\mathrm{d}u \tag{4.66}$$

being the so-called error function. From our definitions it follows that $0 \leq \kappa \leq t$, where $\kappa = t$ and $\kappa = 0$ correspond to the limiting cases of totally negligible background noise and dominating background noise, respectively. In the first case we find that the uncertainty in the position scales as $1/\sqrt{A}$ whereas in the latter case it scales as $\sqrt{B}/A$. The volume of the Gaussian in Eq. (4.54), which describes the total number of signal counts n, is proportional to the amplitude A according to $n = 2\pi A\gamma^2$. Thus, decreasing the background becomes very important at low light levels.

We are now in a position to provide hard numbers for the uncertainty in the peak position σ_x/γ for a given experimental situation (see Problem 4.6). It is obvious that a similar analysis can be used to obtain uncertainties in other parameters such as the width of the spot (see Problem 4.7). To visualize the dependence of the normalized uncertainty in position σ_x/γ on the various parameters we plot σ_x as a function of the number of pixels, the signal amplitude, and the background level for a spot size of 250 nm (FWHM) as achieved by a high-NA oil immersion objective. We observe by inspection of Figs. 4.15(a)–(c), that a position accuracy down to a few nanometers can be achieved by increasing the number of pixels, increasing the signal and lowering the background level. On the other hand, increasing the width of the fitted area decreases the position accuracy linearly for $t \geq 2.5$, which is where $F(t, 1.6)$ saturates (see Fig. 4.15(d)) unless the number of pixels N is also increased. Also, the number of free parameters has an influence on the uncertainty. Roughly, increasing the number of parameters by one decreases the accuracy of all parameters by 10% which would be relevant e.g. for elliptical instead of circular spot shapes.

The conditions to achieve nanometer-scale position accuracy in single-molecule fluorescence detection according to Figs. 4.15(a)–(c) are a rather large signal amplitude A, e.g. about 1000 counts, and a background B that is small enough, e.g. around 100 counts. Furthermore, the number of pixels N that is used to display and fit a single Gaussian peak has to be rather large, e.g. around 16 with $t \approx 5$. All these conditions were met in [17]. Examples of measured spots are shown in Fig. 4.16(a). This plot shows the high quality of the data obtained from single fluorescent molecules during an integration time of 0.5 s. Using the parameter set just mentioned, we obtain a position accuracy of better than 3 nm using Eq. (4.64) with $\kappa = 1.6$. In [17] the step size of the molecular motor myosin V has been investigated. To this end the motor protein was labeled using single fluorescent molecules and the positions of the individual marker molecules were observed over time

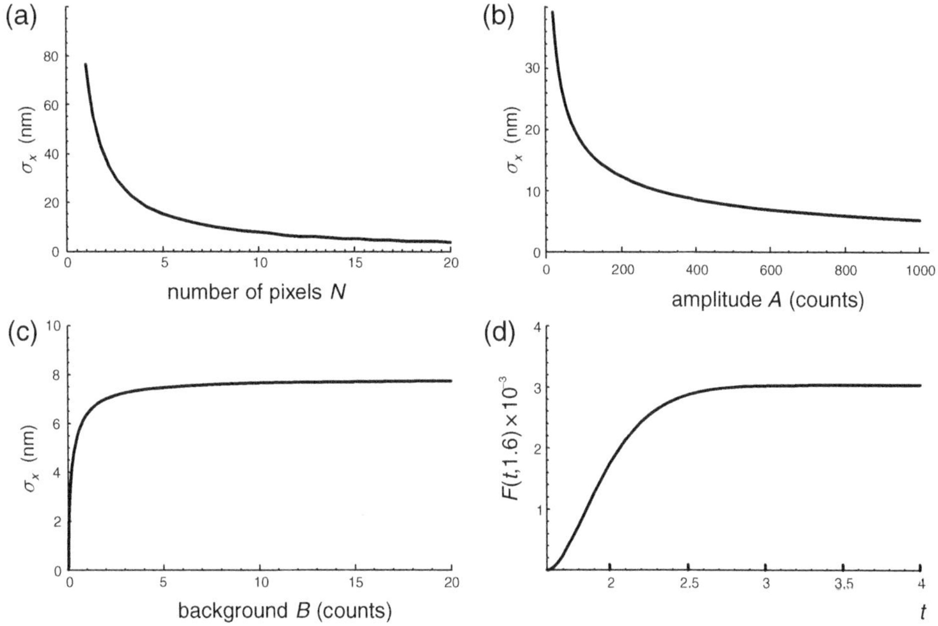

Fig. 4.15 Position uncertainly σ_x as a function of number of pixels (N), signal amplitude (A), and background (B). $\gamma = 125$ nm. (a) Plot of σ_x vs. number of pixels N. Other parameters: $A = 500, B = 10, t = 5, \Delta_a = 5.9, \kappa = 1.6$. (b) Plot of σ_x vs. the amplitude of the signal A. Other parameters: $B = 10, t = 5, \Delta_a = 5.9, N = 10, \kappa = 1.6$. (c) Plot of σ_x vs. the background level B. Other parameters: $A = 500, t = 5, \Delta_a = 5.9, N = 10, \kappa = 1.6$. (d) Plot of $F(t, \kappa)$ vs. t for $\kappa = 1.6$.

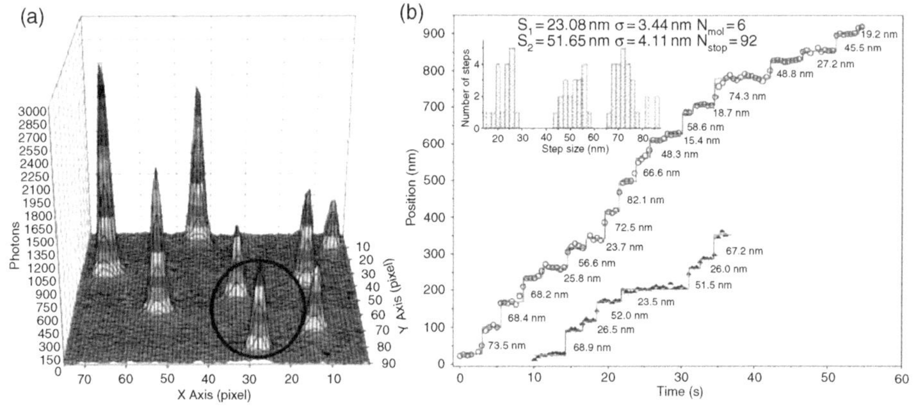

Fig. 4.16 Nanometer position accuracy with single dye labels. (a) A three-dimensional representation of an image of single Cy3-dyes recorded at an integration time of 0.5 s. Note the high amplitudes of up to 3000 counts and the low background. The variation in amplitude is due to non-uniform illumination. (b) Displacement of individual markers linked to a myosin V motor protein vs. time. The stepwise motion of the marker is clearly resolved. Adapted from [17].

while the motor was stepping ahead. Individual steps of down to ~25 nm could be easily discerned as shown in Fig. 4.16(b) [17]. The traces in Fig. 4.16(b) nicely show that the position accuracy is within the estimated range.

Apart from applications in tracing the motion of individual molecules, the high position accuracy can also be used to address questions such as whether two molecules that are distinguishable in a certain observable are co-localized or not. This question is of major importance e.g. in the evaluation of binding assays at the level of individual or few molecules [18].

We have shown that it is possible to achieve nanometer precision in position measurements using optical imaging. The precision depends on the noise level of the data and can be as high as a few nanometers even when detecting individual fluorescent molecules. It should be emphasized again that this type of precision is not to be confused with high resolution, although it can be used to determine distances between closely spaced individual emitters. The latter distance determination is possible only if *prior* information about the molecules exists, i.e. if the photons that are emitted can be assigned to one or other emitter by means of differences in a certain observable, such as the energy of the photon. Thus, this type of "resolution enhancement" falls into the categories of tricks discussed in Section 4.2. The application of these principles for the implementation of localization microscopy is discussed in Section 5.2.3.

4.6 Principles of near-field optical microscopy

So far we assumed that the spatial frequencies (k_x, k_y) associated with evanescent waves are lost upon propagation from source to detector. The loss of these spatial frequencies leads to the diffraction limit and hence to different criteria, which impose a limit on the spatial resolution, i.e. the ability to distinguish two separate point-like objects. The central idea of near-field optical microscopy is to *retain* the spatial frequencies associated with evanescent waves thereby increasing the bandwidth of spatial frequencies. In principle, arbitrary resolution can be achieved, provided that the bandwidth is infinite. However, this is at the expense of strong coupling between the source and the sample, a feature *not* present in standard microscopy, where the properties of the light source (e.g. laser) are negligibly affected by the light–matter interaction with the sample. In this section we will ignore this coupling mechanism and simply extend the concepts of confocal microscopy to include the optical near-field.

A near-field optical microscope is essentially a generalization of the confocal set-up shown in Fig. 4.9, where the same objective lens was used for excitation and collection. If we use two separate lenses we end up with the situation shown in Fig. 4.17(a). In general, for high optical resolution we require high spatial confinement of the light flux through the

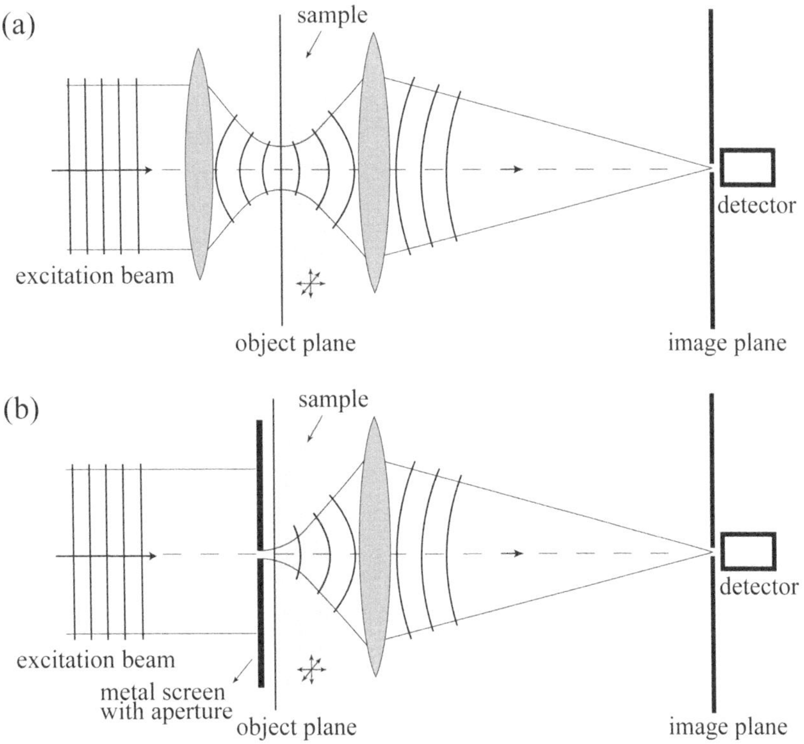

Fig. 4.17 Near-field optical microscopy viewed as a generalization of confocal microscopy. (a) In a far-field microscope the propagating field components are focused onto the object plane in the sample. The bandwidth of spatial frequencies is limited to $\Delta k_{\parallel} < 2k$, where $k = n2\pi/\lambda$, which sets a limit for the maximum achievable resolution. (b) In a near-field optical microscope the focusing lens is replaced by an object (aperture), which extends the bandwidth of spatial frequencies beyond k. Because the field components with spatial frequencies beyond k do not propagate, the object has to be placed close to the sample.

object plane. This spatial confinement can be viewed as the product of excitation confinement and detection confinement as stated in Eq. (4.49). To achieve a highly confined light flux we need to include a broad spectrum of spatial frequencies (k_x, k_y), which requires the use of high-NA objective lenses. However, in far-field optics we encounter a strict cut-off of the spatial spectrum: only the free propagating plane-wave components with $k_\| < k$ $(k = n2\pi/\lambda, k_\| = k_\rho = \sqrt{k_x^2 + k_y^2})$ can be included.

In order to extend the spectrum of spatial frequencies we need to include evanescent waves with $k_\| \geq k$. Unfortunately, these do not propagate and thus cannot be guided towards the sample by using standard optical elements. Evanescent waves are bound to the surfaces of material structures, which necessitates that we bring an "evanescent-wave-carrying object" close to the sample in order to extend the spectrum of spatial frequencies. Such an object can be a favorably illuminated metal tip or a tiny illuminated aperture in a metal screen as shown in Fig. 4.17(b). The price that we have to pay for the inclusion of evanescent waves is high! The object that is brought close to the sample becomes part of the system and the interactions between object and sample complicate data analysis considerably. Furthermore, the extended spatial spectrum is available only close to the object; since in most cases we cannot move with the object into the sample, near-field optical imaging is limited to sample surfaces.

Beyond the source plane the confined fields spread out very rapidly. Indeed, this is a general observation: the more we confine a field laterally the faster it will diverge. This is a consequence of diffraction and it can be nicely explained in terms of the angular spectrum representation. Let us consider a confined field in the plane $z = 0$ (the source plane). We assume that the x-component of this field has a Gaussian amplitude distribution according to Eq. (3.8). In Section 3.2.1 we have determined that the Fourier spectrum of E_x is also a Gaussian function, i.e.

$$E_x(x, y, 0) = E_0 \mathrm{e}^{-\frac{x^2+y^2}{w_0^2}} \rightarrow \hat{E}_x(k_x, k_y; 0) = E_0 \frac{w_0^2}{4\pi} \mathrm{e}^{-(k_x^2+k_y^2)\frac{w_0^2}{4}}. \tag{4.67}$$

Figures 4.18(a) and (b) demonstrate that for a field confinement better than $\lambda/(2n)$ we require the inclusion of evanescent field components with $k_\| \geq k$. The shaded area in Fig. 4.18(b) denotes the spectrum of spatial frequencies associated with evanescent waves. The better the confinement of the optical field is the broader the spectrum will be. Notice that we have displayed only the field component E_x and that in order to describe the distribution of the total field $|\mathbf{E}|$ we need to include the other field components as well (see Problem 4.4). Beyond the plane $z = 0$ the field spreads out as defined by the angular spectrum representation Eq. (3.23). Using cylindrical coordinates, the field component E_x evolves as

$$E_x(x, y, z) = E_0 \frac{w_0^2}{2} \int_0^\infty \mathrm{e}^{-k_\|^2 w_0^2/4} k_\| J_0(k_\| \rho) \mathrm{e}^{\mathrm{i}k_z z} \, \mathrm{d}k_\| . \tag{4.68}$$

This field distribution is plotted along the z-axis in Fig. 4.18(c). It can be observed that a highly confined field in the source plane decays very fast along the optical axis. The reason for this decay is the fact that the spectrum of a strongly confined field contains mainly

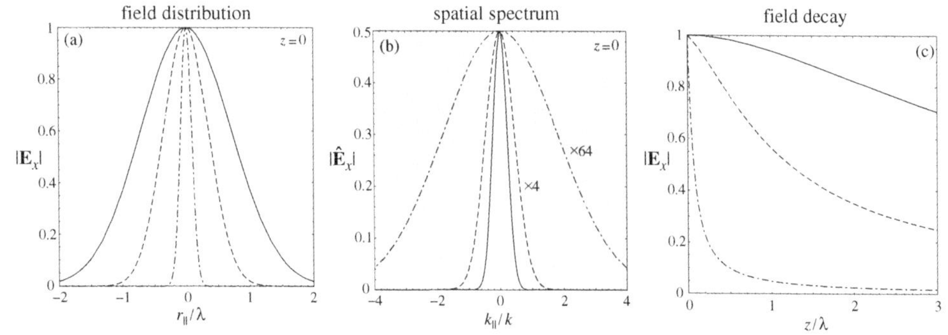

Fig. 4.18 (a) Gaussian field distributions with different confinements in the source plane $z = 0$: $w_0 = \lambda$ (solid curve), $w_0 = \lambda/2$ (dashed curve) and $w_0 = \lambda/8$ (dash–dotted curve). (b) Spectrum of spatial frequencies corresponding to the field distributions in (a). The shaded area denotes the range of spatial frequencies associated with evanescent fields. The better the confinement of the optical field is, the broader the spectrum of spatial frequencies will be. (c) Field decay along the optical axis z corresponding to the field distributions in (a). The better the confinement in the source plane is, the faster the field decay will be.

evanescent field components that do not propagate but exponentially decay along the z-axis. However, this is not the only reason. Another contribution to the fast decay stems from the fast divergence of a highly confined field. As shown in Fig. 4.19, the more we squeeze the fields at $z = 0$ the faster they spread out (like a bunch of half-cooked spaghetti). Thus, to achieve high resolution with a strongly confined light field we need to bring the source (aperture) very close to the sample surface. It has to be emphasized that E_x does not represent the total field strength. In fact, the inclusion of the other field components leads to even stronger field divergence than displayed in Fig. 4.19.

Notice that the conclusions of this section are consistent with the findings of Section 3.2, where we discussed the behavior of a Gaussian field distribution in the paraxial approximation. In particular we found that the Rayleigh range r_0 and the beam divergence angle θ are related to the beam confinement w_0 as

$$z_0 = \frac{k w_0^2}{2}, \qquad \theta = \frac{2}{k w_0}. \tag{4.69}$$

Hence, the stronger the field confinement is, the faster the decay along the optical axis will be and the faster the fields will spread out.

Each near-field source (tip, aperture, particle, ...) has its own unique field distribution. The electromagnetic properties of these sources will be discussed in Chapter 6. The unavoidable interaction between sample and source is also different for each source. To investigate these issues it is necessary to perform elaborate field computations. In general, the configurations need to be strongly simplified in order to achieve analytical solutions. On the other hand, the intuitive insight of such calculations is very valuable and provides helpful guidelines for experiments. Examples of analytical models are the fields near a small aperture as derived by Bethe and Bouwkamp [24, 25], and models for dielectric and metal tips as formulated by Barchiesi and Van Labeke [26, 27].

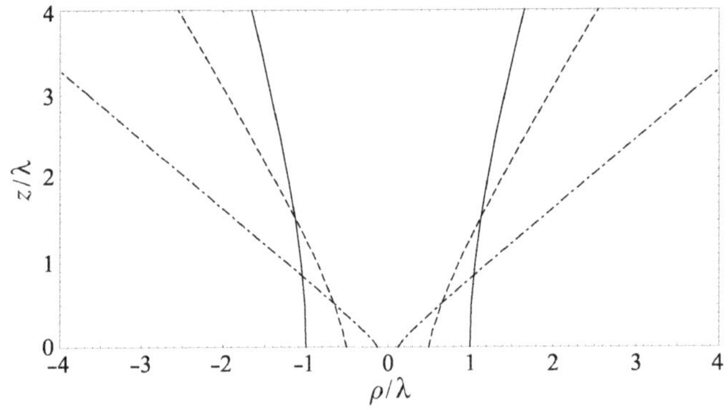

Fig. 4.19 Divergence of optical fields with different confinements in the source plane. The same parameters are used as in Fig. 4.18. A point on a line denotes the radial distance for which the field strength of E_x decays to $1/e$ of its on-axis value. The better the confinement in the source plane at $z = 0$, the faster the fields will diverge.

4.6.1 Information transfer from near-field to far-field

In near-field optics, the electromagnetic field of a source interacts with a sample surface in close proximity and then propagates to the far-field, where it is detected and analyzed. But how does information about subwavelength-sized structures get encoded in the radiation? How is it possible at all to retrieve near-field information in the far-field where evanescent waves do not contribute? We shall discuss the problem in a rather general way specifying neither the illumination field distribution, which may be due to a near-field probe or a focused spot, nor the specific properties of the sample. We also neglect interaction of probe and sample. A more detailed discussion can be found in Refs. [28, 29].

Let us consider three different planes as shown in Fig. 4.20: (1) the source plane at $z = -z_0$, (2) the sample plane at $z = 0$, and (3) the detection plane at $z = z_\infty$. The source plane corresponds to the end face of an optical probe used in near-field optical microscopy but it could also be the focal plane of a laser beam employed in confocal microscopy. The sample plane $z = 0$ forms the boundary between two different media characterized by indices n_1 and n_2, respectively. Using the framework of the angular spectrum representation (cf. Section 2.15), we express the source field in terms of its spatial spectrum as

$$\mathbf{E}_{\text{source}}(x, y; -z_0) = \iint_{-\infty}^{\infty} \hat{\mathbf{E}}_{\text{source}}(k_x, k_y; -z_0) e^{i[k_x x + k_y y]} \, dk_x \, dk_y. \tag{4.70}$$

Using the propagator (3.2), the field that arrives at the sample is given by

$$\mathbf{E}_{\text{source}}(x, y; 0) = \iint_{-\infty}^{\infty} \hat{\mathbf{E}}_{\text{source}}(k_x, k_y; -z_0) e^{i[k_x x + k_y y + k_{z_1} z_0]} \, dk_x \, dk_y, \tag{4.71}$$

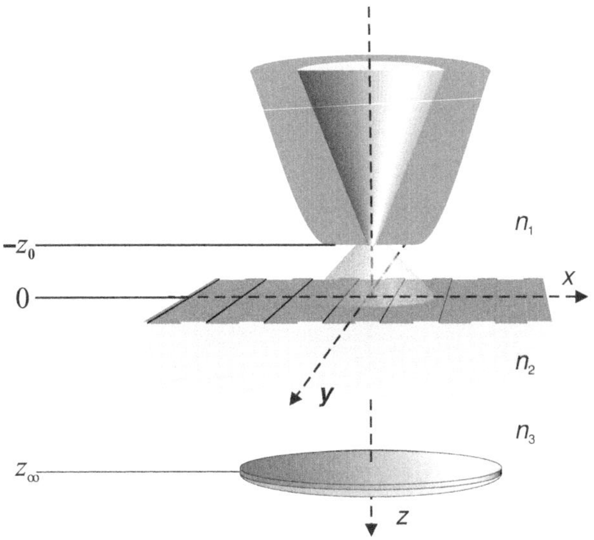

Fig. 4.20 Mapping of information from a sample plane ($z = 0$) to a detector plane ($z = z_\infty \gg \lambda$) using a confined source field at $z = -z_0$. The high spatial frequencies of the sample can be detected by using a strongly confined source field in close proximity to the sample ($z_0 \ll \lambda$).

where $\mathbf{E}_{\text{source}}(x, y; 0)$ is the field at the sample surface before any interaction takes place. Because of the proximity of the sample to the source ($z_0 \ll \lambda$), $\mathbf{E}_{\text{source}}$ is a superposition of plane and evanescent waves. However, as qualitatively shown in Fig. 4.21, the magnitude of the evanescent waves is attenuated as their transverse wavenumber increases. Since we know $\mathbf{E}_{\text{source}}$ at the surface of the sample, we can determine the interaction separately for each plane or evanescent wave and then obtain the total response by integrating over all incident waves, i.e. over the entire (k_x, k_y) plane.

To keep the discussion focused, we assume the sample to be an infinitely thin object characterized by a transmission function $T(x, y)$, which we aim to determine in this thought experiment. This choice allows us to ignore topography-induced effects [30]. Very thin samples can be produced, for example by microcontact printing [31]. Immediately after being transmitted, the field is calculated as

$$\mathbf{E}_{\text{sample}}(x, y; 0) = T(x, y) \cdot \mathbf{E}_{\text{source}}(x, y; 0) . \tag{4.72}$$

We have to keep in mind that this treatment is a rough approximation since e.g. the influence of the sample on the probe field is neglected. A more rigorous description could be accomplished e.g. by adopting the concept of the *equivalent surface profile* [28]. The multiplication of T and $\mathbf{E}_{\text{source}}$ in direct space becomes a convolution in Fourier space. Therefore, the Fourier spectrum of $\mathbf{E}_{\text{sample}}$ can be written as

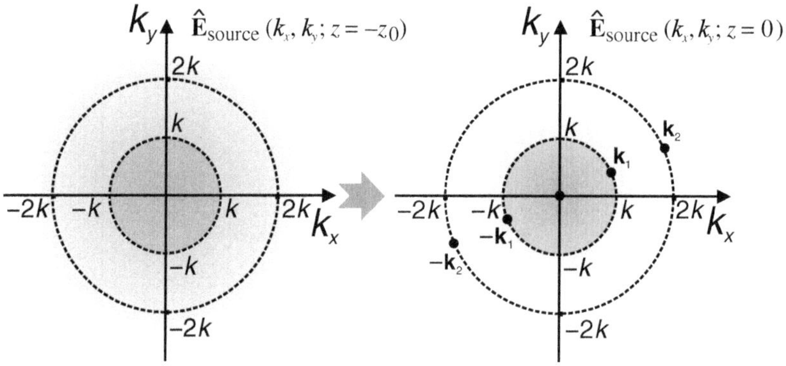

Fig. 4.21 Attenuation of the bandwidth of spatial frequencies upon propagation from the source ($z = -z_0$) to the sample ($z = 0$). Evanescent components ($|\boldsymbol{k}_\parallel| = |(k_x, k_y)| > k$) are exponentially attenuated. The attenuation is stronger the larger $|\boldsymbol{k}_\parallel|$ is. The spatial spectrum arriving at the sample can be written as a sum over discrete spatial frequencies represented by delta functions (see Eq. (4.77)). Five representative spatial frequencies are depicted as black dots for illustration: $\delta(\boldsymbol{k}_\parallel)$, $\delta(\boldsymbol{k}_\parallel \pm \boldsymbol{k}_1)$, and $\delta(\boldsymbol{k}_\parallel \pm \boldsymbol{k}_2)$, where $|\boldsymbol{k}_1| = k$ and $|\boldsymbol{k}_2| = 2k$.

$$\begin{aligned}\hat{\mathbf{E}}_{\text{sample}}(\kappa_x, \kappa_y; 0) &= \iint_{-\infty}^{\infty} \hat{T}(\kappa_x - k_x, \kappa_y - k_y)\hat{\mathbf{E}}_{\text{source}}(k_x, k_y; 0)\mathrm{d}k_x\,\mathrm{d}k_y,\\ &= \iint_{-\infty}^{\infty} \hat{T}(\kappa_x - k_x, \kappa_y - k_y)\hat{\mathbf{E}}_{\text{source}}(k_x, k_y; -z_0)\mathrm{e}^{\mathrm{i}k_{z_1}z_0}\,\mathrm{d}k_x\,\mathrm{d}k_y,\end{aligned} \tag{4.73}$$

with $\hat{T}(k'_x, k'_y)$ being the Fourier transform of T and $k'_i = \kappa_i - k_i$, $i \in \{x, y\}$.

We now propagate the sample field $\mathbf{E}_{\text{sample}}$ to the detector in the far-field at $z = z_\infty$. We have seen in Section 3.4 that the far-field simply corresponds to the spatial spectrum in the source plane. However, here we are interested in the spatial spectrum in the detector plane and therefore propagate $\hat{\mathbf{E}}_{\text{sample}}$ as

$$\mathbf{E}_{\text{detector}}(x, y; z_\infty) = \iint_{-\infty}^{\infty} \hat{\mathbf{E}}_{\text{sample}}(\kappa_x, \kappa_y; 0)\,\mathrm{e}^{\mathrm{i}[\kappa_x x + \kappa_y y]}\,\mathrm{e}^{\mathrm{i}\kappa_z z_\infty}\,\mathrm{d}\kappa_x\,\mathrm{d}\kappa_y. \tag{4.74}$$

Because of the propagator $\exp[\mathrm{i}\kappa_z z_\infty]$ only plane-wave components will reach the detector. These plane waves fulfill

$$|\kappa_\parallel| \le k_3 = \frac{\omega}{c} n_3, \tag{4.75}$$

where the transverse wavenumber $\kappa_\parallel$ is defined as $\kappa_\parallel = [\kappa_x^2 + \kappa_y^2]^{1/2}$. If the finite collection angle of a lens with numerical aperture NA is taken into account we obtain

$$|\kappa_\parallel| \le k_3 \mathrm{NA}. \tag{4.76}$$

Now, this appears just to be a restatement of the diffraction limit. What can we learn from this?

To simplify the interpretation, let us rewrite the spectrum of the source field as

$$\hat{\mathbf{E}}_{\text{source}}(k_x, k_y; 0) = \iint_{-\infty}^{\infty} \hat{\mathbf{E}}_{\text{source}}(\tilde{k}_x, \tilde{k}_y; 0)\, \delta(\tilde{k}_x - k_x)\, \delta(\tilde{k}_y - k_y)\, \mathrm{d}\tilde{k}_x\, \mathrm{d}\tilde{k}_y, \tag{4.77}$$

which, as illustrated in Fig. 4.21, is simply a sum over discrete spatial frequencies. Thus, we can imagine the source field as consisting of an infinite number of partial source fields with discrete spatial frequencies. For each pair of partial fields having spatial frequencies $\pm(\tilde{k}_x, \tilde{k}_y)$ we calculate separately the interaction with the sample and the resulting far-field at the detector. Pairs are considered here since they correspond to the equal amplitudes of plane or evanescent waves that interfere in the sample plane, leading to a stationary standing-wave pattern. In the end, we may sum over all the individual pair responses.

Recall that we performed a convolution of $\hat{T}(k'_x, k'_y)$ and $\hat{\mathbf{E}}_{\text{source}}(k_x, k_y; 0)$. A source field consisting of a single pair of spatial frequencies $\mathbf{k}_\parallel = \pm(k_x, k_y)$ will simply *shift* the transverse wavevectors of the sample $\mathbf{k}'_\parallel$ as

$$\boldsymbol{\kappa}_\parallel = \mathbf{k}_\parallel \pm \mathbf{k}'_\parallel, \tag{4.78}$$

i.e. it translates the spectrum $\hat{T}$ by $\pm\mathbf{k}_\parallel$. Figure 4.22 illustrates the shifting of the sample spectrum $\hat{T}$ for three pairs of transverse wavevectors of the source field: $\delta(k_\parallel)$, $\delta(k_\parallel \pm k)$, and $\delta(k_\parallel \pm 2k)$ already highlighted in Fig. 4.21. A plane wave at normal incidence is represented by $\delta(k_\parallel)$ and does not shift the original spectrum. The plane waves with the largest transverse wavevector travel parallel to the surface and are represented by $\delta(k_\parallel \pm k)$. This wavenumber shifts the original spectrum by $\pm k$, thereby bringing a range of previously inaccessible spatial frequencies of $T(kx', ky')$ into the circular region of the **k**-space corresponding to propagating waves $|\kappa_\parallel| < k$. Finally, $\delta(k_\parallel \pm 2k)$ represents a pair of evanescent waves. It shifts $\hat{T}$ by $\pm 2k$ and brings spatial frequencies up to $k'_\parallel = 3k$ into the range supported by the OTF. Hence, the large spatial frequencies of the sample are combined with the large spatial frequencies of the probe field, such that the difference wavevector corresponds to a propagating wave in the angular spectrum that travels towards the detector. The effect that occurs here is similar to the creation of the long-wavelength Moiré patterns that occur when the transmissions of two high-frequency gratings are multiplied by each other. We conclude that using a confined source field with a large bandwidth of spatial frequencies makes high spatial frequencies of the sample become accessible in the far-field! The better the confinement of the source field is, the better the resolution of the sample will be.

Let us estimate the highest spatial frequencies that can be sampled using a specific probe field. According to Eqs. (4.76) and (4.78)

$$\left|\mathbf{k}'_{\parallel,\max} \pm \mathbf{k}_{\parallel,\max}\right| = \frac{2\pi\,\mathrm{NA}}{\lambda}. \tag{4.79}$$

For a confined source field with a characteristic lateral dimension L (aperture diameter, tip diameter, ...) the highest spatial frequencies are on the order of $k_{\|,\max} \approx \pi/L$ and thus

$$k'_{\|,\max} \approx \left| \frac{\pi}{L} \mp \frac{2\pi \mathrm{NA}}{\lambda} \right| . \tag{4.80}$$

For $L \ll \lambda$ we can neglect the last term and find that the source confinement entirely defines the highest detectable spatial frequencies of the sample. However, one has to keep in mind that the detection bandwidth is restricted to a circle with radius k_3 and that the high spatial frequencies are always intermixed with low spatial frequencies, which makes image reconstruction a challenging task!

4.7 Structured-illumination microscopy

Structured-illumination microscopy (SIM) is a super-resolving far-field optical imaging technique based on the concepts we have just developed. It consists of illuminating the sample with sinusoidal standing wave patterns created by discrete pairs of plane waves that interfere in the sample plane. The fact that the source field spectrum now consists of three delta peaks can be exploited to recover – by means of simple mathematical operations in Fourier space – a larger portion of such parts of the sample's Fourier spectrum that usually would not contribute to the far field. Overall the procedure to be described exploits the information content of multiple images recorded under different illumination conditions.

The fact that structured illumination can lead to improved spatial resolution was first noted in 1963 [32] by Lukosz and Marchand, [32] who used a periodic intensity variation along the optical axis. Laterally structured illumination was suggested and experimentally demonstrated by Gustafsson [33, 34], Heintzmann & Cremer [35], and Frohn [36] as well as by Neil *et al.*, who demonstrated optical sectioning capability in addition [37].

To discuss the principle of the technique we assume an imaging system as sketched in Fig. 4.5 for which we set the magnification $M = 1$ for convenience. We assume that the sample contains a distribution of fluorescent molecules $S(x, y)$ that needs to be determined. For example, the fluorescent molecules could be specific biological labels that mark certain cellular structures. We further assume for now that the fluorescent molecules are excited in the linear regime (for details see Chapter 9). For illumination we use a one-dimensional sinusoidal intensity grating whose intensity distribution in the sample plane can be described by $I(x, y) = I_o\,[1 + \cos(ux + \Delta)]$ where $2\pi/u$ is the spatial wavelength of the modulation and Δ is an adjustable phase shift. The spatial distribution of fluorescence in the sample plane $F(x, y)$ is then given by $F(x, y) = S(x, y) \cdot I(x, y) = S(x, y)\,[1 + \cos(ux + \Delta)]$, where we suppress all proportionality constants. This multiplicative relation is in complete analogy to Eq. (4.72) and an equivalent analysis can be performed, except that now the particular form of the illumination field $I(x, y)$ allows us to obtain analytical expressions when performing the Fourier analysis. As discussed in the

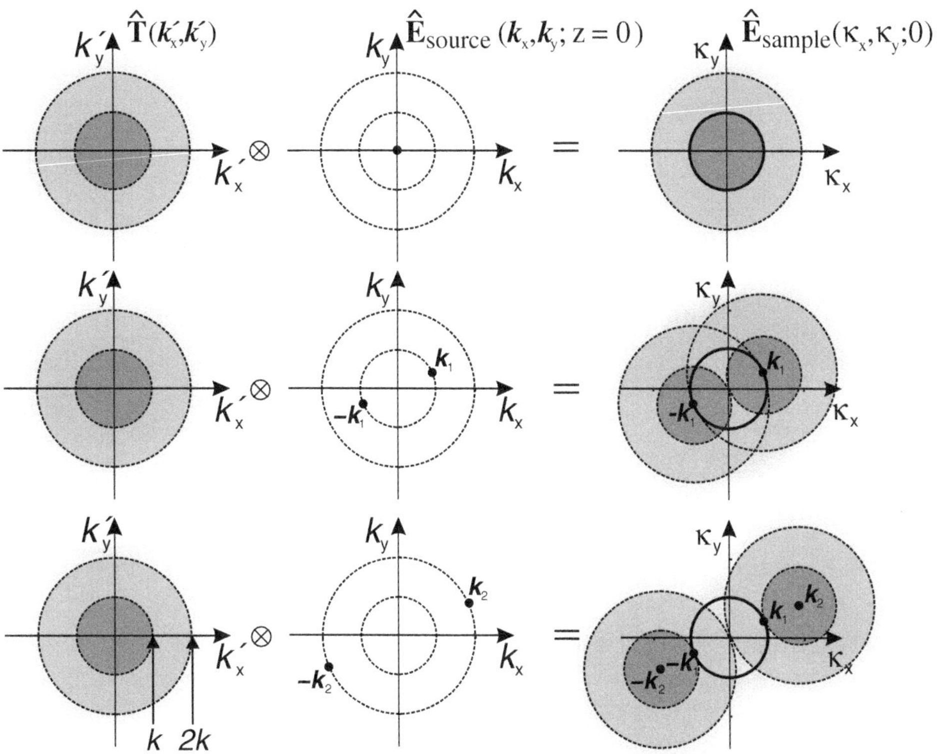

Fig. 4.22 Convolution of the spatial spectra of sample transmission ($\hat{T}$) and source field ($\hat{\boldsymbol{E}}_{source}$). Dashed circles have a radius of k or $2k$. Three pairs of spatial frequencies of $\hat{\boldsymbol{E}}_{source}$ are shown. Convolution with $\delta(k_{\|} \pm mk)$ shifts the spatial spectrum of $\hat{T}$ by $\pm mk$; $m = 0$ corresponds to a plane wave at normal incidence, $m = \pm 1$ to a pair of counterpropagating plane waves at parallel incidence, and $m = \pm 2$ to a pair of counterpropagating evanescent waves. In the far-field, the resulting spectrum of $\hat{\boldsymbol{E}}_{sample}$ can be detected only in the range $|k_{\|}| < k$. The figure illustrates that high spatial frequencies associated with the sample can be shifted into the propagating region supported by the optical transfer function, Eq. (3.2), which is marked by the thick black circle.

previous section, we need to calculate the Fourier spectrum $\hat{F}(\kappa_x, \kappa_y; 0)$ of $F(x, y)$ according to Eq. (4.73) to find the Fourier components of $S(x, y)$ that can be recovered in the far field. The resulting convolution integral can be evaluated to yield the discrete sum

$$\hat{F}(\kappa_x, \kappa_y; 0) = \sqrt{2\pi}\,\hat{S}(\kappa_x, \kappa_y) + \sqrt{\frac{\pi}{2}}e^{-i\Delta}\hat{S}(\kappa_x - u, \kappa_y) + \sqrt{\frac{\pi}{2}}e^{i\Delta}\hat{S}(\kappa_x + u, \kappa_y). \quad (4.81)$$

This sum consists of a term containing the unshifted Fourier spectrum $\hat{S}$ (representing the conventional far-field image) as well as two more terms containing copies of $\hat{S}$ that are shifted by $\pm u$ as sketched in Fig. 4.23(a). Owing to this shift, new parts of $\hat{S}$ fall within the range of spatial frequencies that can be converted to propagating waves depicted by the thick circle in Fig. 4.23(a) (representing far-field images containing information about higher spatial frequencies). The additional ranges of higher spatial frequencies that become available to far-field observation are highlighted by the hatched areas. If the three shifted

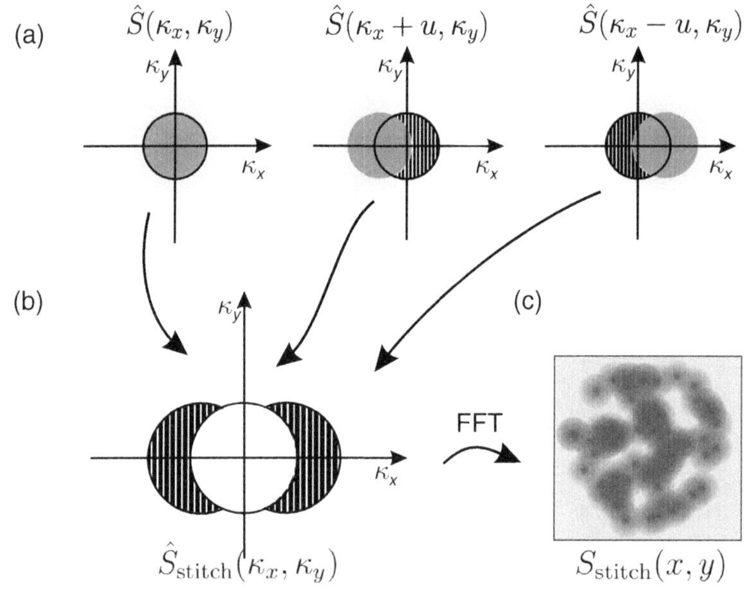

Fig. 4.23 Reconstruction of super-resolved images in structured-illumination microscopy. (a) Once the individual shifted and unshifted sample spectra $\hat{S}$ within the circle with radius k_0 have been determined by solving a system of linear equations they can (b) be used to reconstruct the extended $\hat{F}_{\mathrm{detector}}(\kappa_x, \kappa_y; z_\infty)$. The hatched areas in (a) and (b) highlight the additional portions of the Fourier plane that become accessible due to the shifting of $\hat{S}$ after correct stitching of the Fourier spectra. (c) A Fourier transformation finally yields the resulting image $F_{\mathrm{detector}}(x, y)$ with enhanced resolution mostly along the x-direction in this example.

and unshifted Fourier spectra $\hat{S}$ inside the thick circles were known, we could easily combine those parts of the Fourier spectra that are not redundant into a combined and extended Fourier spectrum as sketched in Fig. 4.23(b). Upon inverse Fourier transformation of this stitched Fourier spectrum a better-resolved image of the unknown distribution of fluorescent molecules $S(x, y)$ would be obtained (Fig. 4.23(c)). Unfortunately, the far-field fluorescence image that results from propagating the expression Eq. 4.81 to the far-field contains a superposition of far-field images corresponding to a complicated Moiré-pattern.

The strategy employed to separately determine the unknown shifted Fourier spectra $\hat{S}(\kappa_x \pm u, \kappa_y)$ is as follows. Experimentally $\hat{F}(\kappa_x, \kappa_y; 0)$ for the range of propagating waves can be determined from a recorded fluorescence image by applying a numerical Fourier transformation. Now Eq. (4.81) represents one equation for the three unknowns, $\hat{S}(\kappa_x, \kappa_y)$, $\hat{S}(\kappa_x - u, \kappa_y)$, and $\hat{S}(\kappa_x + u, \kappa_y)$. Two further independent equations are needed in order to obtain a system of equations that can be solved for the set of unknowns. This is achieved by recording two further images, each with a different phase shift Δ of the sinusoidal illumination pattern, however. This yields two more equations analogous to Eq. (4.81). It is convenient to set the phase shift of the first image to zero and the phase shifts for the two additional images e.g. to $\Delta = 2\pi/3$ and $\Delta = 4\pi/3$ to obtain a total of three linearly independent equations. Experimentally, phase shifts are adjusted by physically shifting the illumination pattern in the sample plane. Once the three unknowns have been determined, the extended Fourier spectrum $\hat{S}_{\text{stitch}}(\kappa_x, \kappa_y)$ can be reconstructed in the computer (see Fig. 4.23 (b)). This is then Fourier transformed to yield $S_{\text{stitch}}(x, y)$ with enhanced resolution along the x-direction (see Fig. 4.23 (c)). It is obvious that the method can be extended to using multiple orientations of the sinusoidal illumination gratings. By doing so nearly isotropic resolution enhancement can be achieved [34, 36].

To what degree can the spatial resolution be enhanced by structured illumination? The highest value of a modulation frequency in a standing-wave pattern that can theoretically be created by means of far-field optics is $u = 4\pi n/\lambda$. This could be achieved e.g. by the interference of two counter-propagating plane waves at grazing angle of incidence with respect to the sample surface. From this we conclude that compared with the standard optical microscope whose resolution limit is given by Eq. (4.21), in structured-illumination microscopy the optical resolution can be increased by up to a factor of 2. Let us also remark here that it was pointed out that a focused laser beam can be interpreted as a special case of structured illumination. Using a two-dimensional detector instead of a point detector and appropriate image reconstruction, the resolution in scanning confocal microscopy can reach that of structured-illumination microscopy [38].

In section 4.2.3 we saw that optical resolution can be enhanced if nonlinear optical effects are involved. Also in the case of structured-illumination microscopy the resolution can be further enhanced by making use of nonlinear effects, e.g. fluorescence saturation [39]. Of course, other nonlinear optical effects can in principle serve the same purpose. While increasing the modulation frequency in a standing wave beyond the limit of $u = 4\pi n/\lambda$ is impossible using linear far-field optics, it can be increased by taking advantage of the nonlinear relation between excitation intensity and fluorescence emission due to saturation (see chapter 9 for details). Such a nonlinear relation leads to a distorted effective excitation pattern, which therefore needs to be described with the help of higher harmonics of the modulation frequency u. Figure 4.24 illustrates this effect. The presence of higher

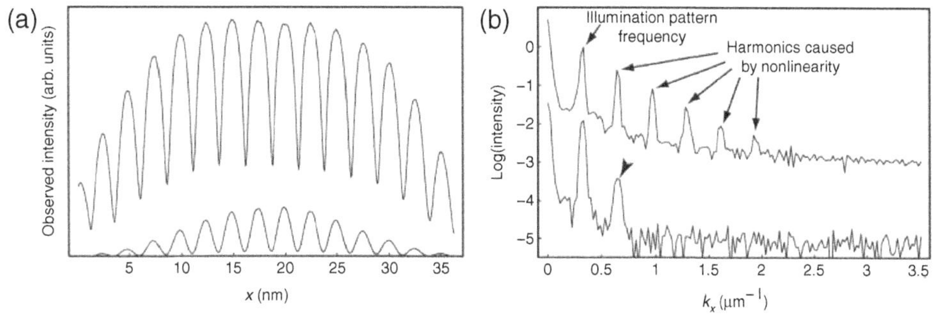

Fig. 4.24 Higher harmonics introduced into structured-illumination microscopy by saturated fluorescence. (a) Detected fluorescence from a thin fluorescent layer illuminated by a Gaussian beam modulated by sinusoidal stripes with a period of 2.5 μm for a higher and a lower excitation intensity. (b) Corresponding Fourier transformations showing five higher harmonics in addition to the fundamental frequency for the higher excitation intensity but only one for the lower intensity. From [39].

harmonics can be exploited to further increase spatial resolution in full analogy to what we have discussed for linear structured-illumination microscopy by adding the respective higher harmonic terms to Eq. (4.81). Since the number of unknowns increases it is necessary to record a correspondingly larger number of images. This is done by shifting the phase of the excitation pattern in even finer steps in order to be able to construct the stitched Fourier transform of the image by solving the resulting larger set of linear equations. Also, to guarantee isotropic enhancement of resolution, the illumination pattern has to be turned in finer angular steps. In general, the highest spatial frequency that becomes accessible by means of nonlinear structured illumination is

$$k_{\|,\max} = \frac{(2+l)2\pi n}{\lambda}, \tag{4.82}$$

where l is the number of higher harmonics present in the signal. According to Eq. (4.21) the smallest separation of two point-like particles that can be resolved is then given by

$$\mathrm{Min}\left[\Delta r_{\|}\right] = \frac{\lambda}{2\pi(2+l)\mathrm{NA}}, \tag{4.83}$$

being limited only by the number of harmonics that contribute to the image reconstruction. In experimental nonlinear structured-illumination microscopy a resolution of < 50 nm has been demonstrated using three harmonic orders [39].

Problems

4.1 A continuously fluorescing molecule is located at the focus of a high-NA objective lens. The fluorescence is imaged onto the image plane as described in Section 4.1. Although the molecule's position is fixed (there is no translational diffusion), it is rotating in all three dimensions (rotational diffusion) with high speed. Calculate and plot the averaged field distribution in the image plane using the paraxial approximation.

4.2 Consider the set-up of Fig. 4.1. Replace the single dipole emitter by a pair of incoherently radiating dipole emitters separated by a distance $\Delta x = \lambda/2$ along the x-axis. The two dipoles radiate at $\lambda = 500$ nm and they have the same dipole strength. One of the dipoles is oriented transverse to the optical axis whereas the other dipole is parallel to the optical axis. The two dipoles are scanned in the object plane and for each position of their center coordinate a signal is recorded in the image plane using an NA $= 1.4$ ($n = 1.518$), $M = 100\times$ objective lens.
   (a) Determine the total integrated field intensity (s_1) in the image plane.
   (b) Calculate and plot the recorded image (s_2) if a confocal detector is used. Use the paraxial approximation.
   (c) Discuss how s_1 and s_2 change if the dipoles are scanned at a constant height $\Delta z = \lambda/4$ above the image plane.

4.3 Consider a sample with a uniform layer of dipolar particles with fixed dipole orientations along the x-axis. The layer is transverse to the optical axis and each element of the layer has a constant polarizability α_{xx}. The sample is illuminated by a focused Gaussian beam and is translated along the optical axis z. We use both non-confocal (s_1) and confocal (s_2) detection. The two signals are well approximated by Eqs. (4.47) and (4.48), respectively.

(a) Calculate the non-confocal signal as a function of z.
(b) Calculate the confocal signal as a function of z.
(c) What is the conclusion?

Hint: use the Bessel-function closure relations of Eq. (3.112).

4.4 Calculate the longitudinal fields corresponding to the Gaussian field distribution in Eq. (4.67). Assume that $E_y = 0$ everywhere in space. Show how the longitudinal field evolves in transverse planes z = constant. State the result in cylindrical coordinates as in Eq. (4.68). Plot the longitudinal field strength in the planes $z = 0$ and $z = \lambda$.

4.5 Consider a plane z = constant transverse to the optical axis of a paraxial Gaussian beam $\mathbf{E}$ with focus at $z = 0$, beam waist $w_0 = \lambda$, and wavelength $\lambda = 500$ nm. Assume that the plane is covered with a layer of incoherently radiating fluorescent molecules. Calculate the power of the generated fluorescence P as a function of z by assuming that the fluorescence intensity generated at a point (x, y, z) is given by

(a) $I_\omega(x, y, z) = A\,|\mathbf{E}(x, y, z)|^2$ (one-photon excitation),
(b) $I_{2\omega}(x, y, z) = B\,|\mathbf{E}(x, y, z)|^4$ (two-photon excitation).

Plot P for the two cases. Normalize such that $P(z = 0) = 1$.

4.6 In order to verify the validity of Eq. (4.64) perform a Monte Carlo simulation of the fitting process. To this end, simulate a large number (~1000) of point images by creating Gaussian peaks with uncorrelated Poissonian noise superimposed on the background and on the amplitude. In terms of Eq. (4.54), in the absence of the background B, this means that for each data point a random number drawn from a Poissonian distribution with maximum at $G(x, y)$ and width $\sqrt{G(x, y)}$ is added to the originally calculated $G(x, y)$. Now perform a nonlinear least-squares fit on each of the peaks using a suitable software package (the use of a Levenberg–Marquard algorithm is recommended). Plot the distribution of positions $x_{0,\min}$ and $y_{0,\min}$ that results from the fits. Compare the width of this distribution with the value for σ obtained from Eq. (4.64).

4.7 Determine analytical expressions for the uncertainties of the other parameters in Eq. (4.54) using the same analysis as that which led to Eq. (4.64).

4.8 Structured illumination. Assume that you would like to obtain a far-field image of a fluorescent pattern in which the density of fluorescent labels varies sinusoidally with a spatial frequency of $3k/2$, which is larger than k and is therefore not resolvable in conventional far field microscopy. Apply an appropriate structured-illumination pattern with sinusoidal intensity distribution and plot the resulting Moiré pattern. Show by direct calculation that three exposures of the fluorescent pattern with appropriately (phase-)shifted illumination patterns will suffice to recover the Fourier spectrum of the fluorescent pattern. Assume all refractive indices to be equal to unity.

References

[1] C. J. R. Sheppard and T. Wilson, "The image of a single point in microscopes of large numerical aperture," *Proc. Roy. Soc. A* **379**, 145–158 (1982).

[2] J. Enderlein, "Theoretical study of detection of a dipole emitter through an objective with high numerical aperture," *Opt. Lett.* **25**, 634–636 (2000).

[3] R. M. Dickson, D. J. Norris, and W. E. Moerner, "Simultaneous imaging of individual molecules aligned both parallel and perpendicular to the optic axis," *Phys. Rev. Lett.* **81**, 5322–5325 (1998).

[4] M. A. Lieb, J. M. Zavislan, and L. Novotny, "Single molecule orientations determined by direct emission pattern imaging," *J. Opt. Soc. B* **21**, 1210–1215 (2004).

[5] E. Abbe, "Beiträge zur Theorie des Mikroskops und des mikroskopischen Wahrnehmung," *Arch. Mikrosk. Anat.* **9**, 413–468 (1873).

[6] Lord Rayleigh, "On the theory of optical images with special reference to the microscope," *Phil. Mag.* **5**, 167–195 (1896).

[7] R. H. Webb, "Confocal optical microscopy," *Rep. Prog. Phys.* **59**, 427–471 (1996).

[8] V. Andresen, A. Egner, and S. W. Hell, "Time-multiplexed multifocal multiphoton microscope," *Opt. Lett.* **26**, 75–77 (2001).

[9] T. A. Klar, S. Jakobs, M. Dyba, A. Egner, and S. W. Hell, "Fluorescence microscopy with diffraction resolution barrier broken by stimulated emission," *Proc. Nat. Acad. Sci.* **97**, 8206–8210 (2000).

[10] M. Minsky, "Memoir on inventing the confocal scanning microscope," *Scanning* **10**, 128–138 (1988).

[11] C. J. R. Sheppard, D. M. Hotton, and D. Shotton, *Confocal Laser Scanning Microscopy*. New York: BIOS Scientific Publishers (1997).

[12] G. Kino and T. Corle, *Confocal Scanning Optical Microscopy and Related Imaging Systems*. New York: Academic Press (1997).

[13] T. Wilson, *Confocal Microscopy*. New York: Academic Press (1990).

[14] N. Bobroff, "Position measurement with a resolution and noise-limited instrument," *Rev. Sci. Instrum.* **57**, 1152–1157 (1986).

[15] R. E. Thompson, D. R. Larson, and W. W. Webb, "Precise nanometer localization analysis for individual fluorescent probes," *Biophys. J.* **82**, 2775–2783 (2002).

[16] R. J. Ober, S. Ram, and E. S. Ward, "Localization accuracy in single-molecule microscopy," *Biophys. J.* **86**, 1185–1200 (2004).

[17] A. Yildiz, J. N. Forkey, S. A. McKinney, *et al.*, "Myosin V walks hand-over-hand: single fluorophore imaging with 1.5-nm localization," *Science* **300**, 2061–2065 (2003). Reprinted with permission from AAAS.

[18] W. Trabesinger, B. Hecht, U. P. Wild, *et al.*, "Statistical analysis of single-molecule colocalization assays," *Anal. Chem.* **73**, 1100–1105 (2001).

[19] E. Betzig, G. H. Patterson, R. Sougrat, *et al.*, "Imaging intracellular fluorescent proteins at nanometer resolution," *Science* **313**, 1642–1645 (2006).

[20] M. J. Rust, M. Bates, and X. Zhuang, "Sub-diffraction-limit imaging by stochastic optical reconstruction microscopy (STORM)," *Nature Methods* **3**, 793–795 (2006).

[21] S. T. Hess, T. P. K. Girirajan, and M. D. Mason, "Ultra-high resolution imaging by fluorescence photoactivation localization microscopy," *Biophys. J.* **91**, 4258–4272 (2006).

[22] P. R. Bevington and D. K. Robinson, *Data Reduction and Error Analysis for the Physical Sciences*. New York: McGraw-Hill, p. 212 (1994).

[23] M. Lampton, B. Margon, and S. Bowyer, "Parameter estimation in X-ray astronomy," *Astrophys. J.* **208**, 177–190 (1976).

[24] H. A. Bethe, "Theory of diffraction by small holes," *Phys. Rev.* **66**, 163–182 (1944).

[25] C. J. Bouwkamp, "On Bethe's theory of diffraction by small holes," *Philips Res. Rep.* **5**, 321–332 (1950).

[26] D. Van Labeke, D. Barchiesi, and F. Baida, "Optical characterization of nanosources used in scanning near-field optical microscopy," *J. Opt. Soc. Am. A* **12**, 695–703 (1995).

[27] D. Barchiesi and D. Van Labeke, "Scanning tunneling optical microscopy: theoretical study of polarization effects with two models of tip," in *Near-field Optics*, ed. D. W. Pohl and D. Courjon. Dordrecht: Kluwer, pp. 179–188 (1993).

[28] J.-J. Greffet and R. Carminati, "Image formation in near-field optics," *Prog. Surf. Sci.* **56**, 133–237 (1997).

[29] B. Hecht, H. Bielefeld, D. W. Pohl, L. Novotny, and H. Heinzelmann, "Influence of detection conditions on near-field optical imaging," *J. Appl. Phys.* **84**, 5873–5882 (1998).

[30] B. Hecht, H. Bielefeldt, L. Novotny, Y. Inouye, and D. W. Pohl, "Facts and artifacts in near-field optical microscopy," *J. Appl. Phys.* **81**, 2492–2498 (1997).

[31] Y. Xia and G. M. Whitesides, "Soft lithography," *Angew. Chem. Int. Edn. Engl.* **37**, 551–575 (1998).

[32] W. Lukosz and M. Marchand, "Optische Abbildung unter Überschreitung der beugungsbedingten Auflösungsgrenze," *J. Mod. Opt.* **10**, 241–255 (1963).

[33] M. G. L. Gustafsson, D. A. Agard, and J. W. Sedat, "Method and apparatus for three-dimensional microscopy with enhanced depth resolution," US patent 5671085, cols. 23–25 (1997).

[34] M. G. L. Gustafsson, "Surpassing the lateral resolution limit by a factor of two using structured illumination microscopy," *J. Microsc.* **198**, 82–87 (2000).

[35] R. Heintzmann and C. Cremer, "Laterally modulated excitation microscopy: improvement of resolution by using a diffraction grating," *Proc. SPIE* **3568**, 185–196 (1999).

[36] J. T. Frohn, H. F. Knapp, and A. Stemmer, "True optical resolution beyond the Rayleigh limit achieved by standing wave illumination," *Proc. Nat. Acad. Sci.* **97**, 7232–7236 (2000).

[37] M. A. A. Neil, R. Juskaitis, and T. Wilson, "Method of obtaining optical sectioning by using structured light in a conventional microscope," *Opt. Lett.* **22**, 1905–1907 (1997).

[38] C. B. Müller and J. Enderlein, "Image scanning microscopy," *Phys. Rev. Lett.* **104**, 198101 (2010).

[39] M. G. L. Gustafsson, "Nonlinear structured-illumination microscopy: wide-field fluorescence imaging with theoretically unlimited resolution," *Proc. Nat. Acad. Sci.* **102**, 13081–13086 (2005). Copyright 2005 National Academy of Sciences, U.S.A.

5 Nanoscale optical microscopy

Optical measurement techniques, and near-field optical microscopy in particular, exist in a broad variety of configurations. In the following we will derive an interaction series to understand and categorize different experimental configurations. The interaction series describes multiple scattering events between an optical probe and a sample and is similar to the Born series in light scattering. We will start out by discussing far-field microscopy first and then proceed with selected configurations encountered in near-field optical microscopy.

5.1 The interaction series

The interaction of light with matter can be discussed in terms of light scattering events [1, 2]. Figure 5.1 is a sketch of a generic geometry considered in the following. The sample and – in the case of near-field optical microscopy – also an optical probe, which is positioned in close proximity, are assumed to be described by dielectric susceptibilities $\eta(\mathbf{r})$ and $\chi(\mathbf{r})$, respectively. An incident light field $\mathbf{E}^{\mathrm{i}}$ is illuminating the probe–sample region. $\mathbf{E}^{\mathrm{i}}$ is assumed to be a solution of the homogeneous Helmholtz equation (2.35). The incoming field causes a scattered wave $\mathbf{E}^{\mathrm{s}}$, which is detectable in the far-field. The total field is then given by $\mathbf{E} = \mathbf{E}^{\mathrm{i}} + \mathbf{E}^{\mathrm{s}}$. In a qualitative picture, there are several processes that can convert an incoming photon into a scattered photon. For example, the incoming photon may be scattered only at the probe or only at the sample before traveling into the far-field. Alternatively the first scattering event may occur at the sample, directly followed by a second scattering event at the probe and only then by propagation into the far-field. More complicated multiple-scattering processes may of course also occur, e.g. probe–sample–probe scattering. One may assume that the overall scattering is well described by a sum of the contributions of different scattering processes and that the sum converges after a few orders since multiple scattering is expected to become negligible with increasing order. In the following we outline the derivation of a series representation that is based on the Born series of multiple-scattering processes which supports this intuitive picture [1].

Formally, the Helmholtz equation for the total field, $\left(\nabla^2 + k^2\right)\mathbf{E} = 0$, can be written as $\left(\nabla^2 + k_0^2\right)\mathbf{E} = -k_0[\varepsilon(\mathbf{r}) - 1]\mathbf{E} = -k_0^2[\eta(\mathbf{r}) + \chi(\mathbf{r})]\mathbf{E}$, where k_0 is the magnitude of the wavevector in vacuum. If we make use of the property of the incoming field $(\nabla^2 + k_0^2)\mathbf{E}^{\mathrm{i}} = 0$, we obtain

$$(\nabla^2 + k_0^2)\mathbf{E}^{\mathrm{s}} = -k_0^2[\eta(\mathbf{r}) + \chi(\mathbf{r})]\mathbf{E}, \tag{5.1}$$

which is an inhomogeneous Helmholtz equation for the scattered field $\mathbf{E}^{\mathrm{s}}$ in which the probe and sample susceptibilities appear in the source terms. Equation (5.1) can be formally solved by using $\mathbf{G}(\mathbf{r}, \mathbf{r}')$, the dyadic Green function of the system consisting of two semi-infinite half-spaces of background dielectric constants, which we assume to be $n_1 = 1$ and n_2, respectively. An explicit expression for this particular Green dyadic is given in Section 10.4. However, for the present purpose we need only assume that $\mathbf{G}(\mathbf{r}, \mathbf{r}')$ is known in order to write a formal solution for Eq. (5.1).

$$\mathbf{E}^{\mathrm{s}}(\mathbf{r}, \omega) = k_0^2 \int \mathrm{d}V' \, \mathbf{G}(\mathbf{r}, \mathbf{r}')[\eta(\mathbf{r}) + \chi(\mathbf{r})]\mathbf{E}. \tag{5.2}$$

In order to simplify the notation, we introduce some abbreviations.[1] We define $\mathbf{S} \cdot \mathbf{E} = \int \mathrm{d}V' \, \mathbf{S}(\mathbf{r}, \mathbf{r}')\mathbf{E}$ and $\mathbf{T}_0 \cdot \mathbf{E} = \int \mathrm{d}V' \, \mathbf{T}_0(\mathbf{r}, \mathbf{r}')\mathbf{E}$ with $\mathbf{S}(\mathbf{r}, \mathbf{r}') = k_0^2\mathbf{G}(\mathbf{r}, \mathbf{r}')\eta(\mathbf{r}')$ and $\mathbf{T}_0(\mathbf{r}, \mathbf{r}') = k_0^2\mathbf{G}(\mathbf{r}, \mathbf{r}')\chi(\mathbf{r}')$, which allows us to rewrite Eq. (5.2) as

$$\mathbf{E}^{\mathrm{s}}(\mathbf{r}, \omega) = (\mathbf{S} + \mathbf{T}_0) \cdot \mathbf{E}. \tag{5.3}$$

Equation (5.3) is a recursive integral equation for the scattered field $\mathbf{E}^{\mathrm{s}}$, which can be solved by iteration. The idea of such an iterative solution is that the lowest-order solution can be obtained by assuming that the total field $\mathbf{E}$ is well approximated by $\mathbf{E}^{\mathrm{i}}$. This then

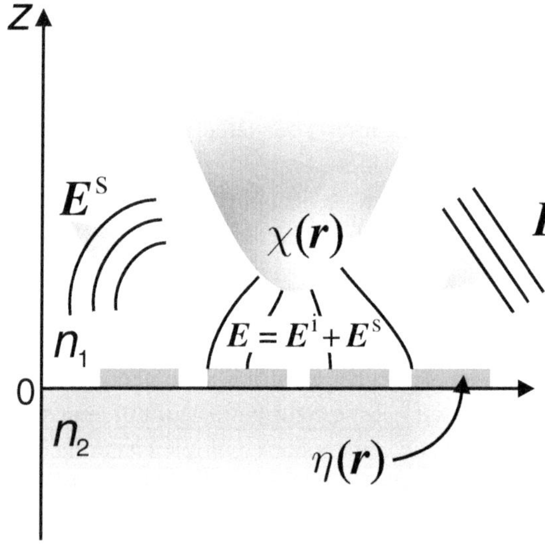

Fig. 5.1 Geometry and definitions for the discussion of light scattering from the probe–sample region.

[1] We use the symbol **T** for scattering processes at the probe ("tip") in order to be consistent with the literature.

allows us to calculate a first approximation to $\mathbf{E}^{\mathrm{s}}$, and can then be used to obtain a refined approximation for $\mathbf{E}$, and so on. By following this scheme we obtain

$$\mathbf{E}^{\mathrm{s}}(\mathbf{r}, \omega) = \sum_{n=1}^{\infty} (\mathbf{S} + \mathbf{T}_0)^n \cdot \mathbf{E}^{\mathrm{i}}, \tag{5.4}$$

which is the so-called Born series [2]. The expression for $\mathbf{E}^{\mathrm{s}}$ provided in Eq. (5.4) is still valid without approximation. We now consider the case in which the sample is absent (but the plane interface remains). This will provide us with an expression for the field scattered by the probe alone, which is assumed to be known. In this particular case, the scattered field according to Eq. (5.4) can be expressed as

$$\mathbf{E}^{\mathrm{S}}_{\mathrm{probe}}(\mathbf{r}, \omega) = \sum_{n=1}^{\infty} (\mathbf{T}_0)^n \cdot \mathbf{E}^{\mathrm{i}} = \mathbf{T} \cdot \mathbf{E}^{\mathrm{i}}, \tag{5.5a}$$

where

$$\mathbf{T} = \sum_{j=1}^{\infty} (\mathbf{T}_0)^j, \tag{5.5b}$$

the effective probe operator, is introduced.

We are now in a position to obtain a perturbative solution for the scattered field by assuming that the sample is scattering only weakly. As a consequence, in the series expansion of Eq. (5.4) only terms up to linear order in $\eta(\mathbf{r})$ (and therefore in $\mathbf{S}$) are retained. After proper rearrangement, and by making use of Eq. (5.5a), we obtain

$$\begin{aligned}\mathbf{E}^{\mathrm{s}}(\mathbf{r}, \omega) &= [\mathbf{T} + (\mathbf{I} + \mathbf{T}) \cdot \mathbf{S} \cdot (\mathbf{I} + \mathbf{T}) + \cdots] \cdot \mathbf{E}^{\mathrm{i}} \\ &= [\mathbf{T} + \mathbf{S} + \mathbf{TS} + \mathbf{ST} + \mathbf{TST} + \cdots] \cdot \mathbf{E}^{\mathrm{i}},\end{aligned} \tag{5.6}$$

where $\mathbf{I}$ is the identity operator. Equation (5.6) shows that the intuitive picture of multiple-scattering events occurring between the probe and the sample is fully supported within the Born approximation. We are now in a position to categorize different types of optical microscopy techniques according to which term of Eq. (5.6) is the leading term in the interaction series. The respective graphical representations of the dominant scattering processes are displayed in Fig. 5.2. For example, conventional far-field microscopy and related high-resolution derivatives such as stimulated emission depletion and localization microscopy techniques are characterized by the $\mathbf{S}$ term only; see Fig. 5.2(a), whereas illumination-mode near-field optical microscopy relies on the $\mathbf{ST}$ term of Eq. (5.6) and is sketched in Fig. 5.2(b). Collection-mode near-field optical microscopy, Fig. 5.2(c), and scattering-type near-field optical microscopy using antenna probes, Fig. 5.2 (d), are represented by the $\mathbf{TS}$ and $\mathbf{TST}$ terms, respectively. It is important to keep in mind that the operators $\mathbf{T}$ and $\mathbf{S}$ are tensorial in nature, which means that polarization effects play an important role and can even be used to suppress certain higher-order interactions in the Born series. Furthermore, one should keep in mind that the individual interaction steps can involve different frequencies, which is relevant in the context of spectroscopy.

5.2 Far-field optical microscopy techniques

5.2.1 Confocal microscopy

Confocal microscopy employs focused light to achieve a diffraction-limited illumination spot in combination with a point-like detector. The relevant theory has been discussed in Section 4.3. Despite the limited bandwidth of spatial frequencies imposed by far-field illumination and detection, confocal microscopy is successfully employed for high-precision localization measurements as explained in Section 4.5 and for high-resolution imaging by exploiting nonlinear or saturation effects discussed in Section 4.2.3. Let us start out here by considering experimental aspects of conventional confocal optical microscopy.

Experimental set-up

Figure 5.3 shows the set-up of the simplest type of scanning confocal microscope. Its beam path is fixed and the sample is raster scanned to record an image. In such an instrument, light from a laser source is typically spatially filtered, e.g. by sending it through a single-mode optical fiber or a pinhole. The purpose of the spatial filtering is to arrive at a beam with a perfect Gaussian beam profile. After propagating through the fiber or the pinhole, the light is collimated by a lens. The focal length of the collimation lens should be chosen such that the beam diameter is large enough to overfill the back-aperture of the microscope objective used to focus the light onto the sample. It is advantageous if the microscope objective is designed to work with collimated beams. Such objectives are called "infinity corrected". The spotsize Δx that is achieved at the sample depends on the numerical

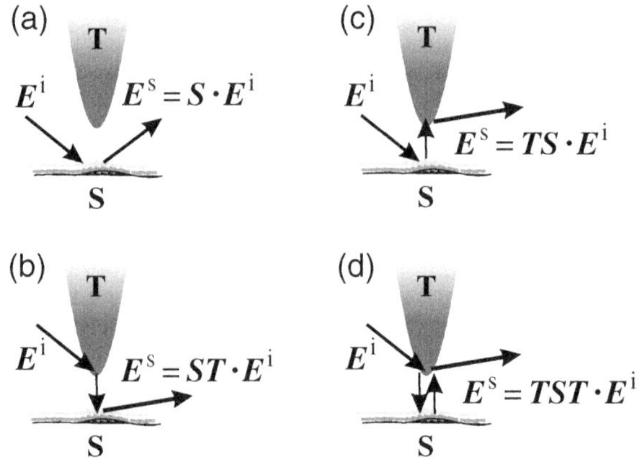

Fig. 5.2 Different types of nanoscale microscopy techniques categorized according to the leading term in the Born series.

aperture NA of the objective and the wavelength used for illumination (see Section 4.2). It is usually limited by diffraction of the laser light at the entrance aperture of the objective to (cf. Section 4.2)

$$\Delta x = 0.61 \frac{\lambda}{\mathrm{NA}}, \tag{5.7}$$

where λ is the light wavelength. For NA = 1.4 the lateral spotsize for green light (λ = 500 nm) is about 220 nm, slightly better than $\lambda/2$.

The laser light interacts with the sample and produces reflected and scattered light at the excitation wavelength and possibly also at wavelengths shifted with respect to the excitation. The microscope objective that is used for illumination can also be used to collect light emanating from the sample. It is possible to collect the light with a second objective facing the first one, however this is experimentally more demanding because it requires the alignment of two objectives with respect to each other with a precision much better than $\lambda/2$. On the other hand, the dual-objective configuration opens up new possibilities for excitation, e.g. by overlapping the foci of two counterpropagating beams [3]. We come back to these issues later on in this chapter.

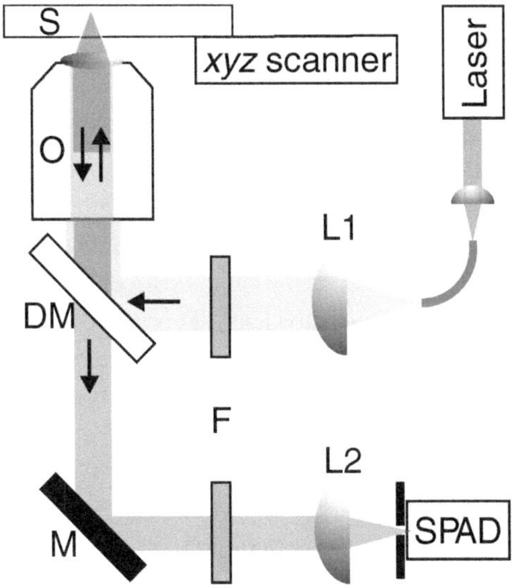

Fig. 5.3 Set-up of a simple scanning epi-illumination confocal optical microscope. A laser light source is spatially filtered, e.g. by sending the light through a single-mode optical fiber or a pinhole, and collimated by a lens. A (dichroic) beamsplitter reflects the light into a high-numerical-aperture microscope objective. The back-aperture of the objective should be overfilled to achieve the optimal spotsize (see Chapter 4). The optical signal (e.g. fluorescence) and scattered light created at the focus are collected by the same objective and converted into a collimated beam. The dichroic beamsplitter transmits light in a restricted spectral range, which is then filtered further and finally focused onto another pinhole in front of a detector. Images are obtained pixel by pixel by scanning the sample relative to the focus.

When using a single objective, once the incoming beam of light is collimated, the beam of collected light is also collimated for a chromatically corrected microscope objective. Working with collimated beams makes it possible to introduce filters and other optical elements anywhere into the beam path without introducing offsets in the light path. Ultrafast laser sources with few-femtosecond pulse durations deliver excitation light with a very broad spectrum (>400 nm). In order to work with such sources the use of Cassegrain objectives may be favorable.

The collected light has to be separated from the incoming light. This can be done by exploiting the difference in wavelength using a dichroic mirror, by exploiting changes in the polarization using a polarizing beamsplitter, by time gating if pulsed excitation is used, or simply by exploiting different directions of propagation using a non-polarizing beamsplitter. Figure 5.3 depicts the case in which a dichroic mirror that transmits e.g. red-shifted fluorescence is used. The filtered beam of collected light is now focused by a second lens onto a pinhole in front of a detector. Certain detectors such as the widely used single-photon counting avalanche photodiodes have rather small active areas. They can be used without an additional pinhole. The size of the detection pinhole must be correctly matched to the diameter of the focal spot (Airy disk) produced by the second lens in order to efficiently reject out-of-focus signals. A larger pinhole diameter impairs the rejection of out-of-focal-plane signals but can help to optimize the effective transmission of light through the pinhole. It is found that a spotsize two times smaller than the pinhole diameter still yields good results in terms of both lateral resolution and out-of-focal-plane rejection.

Another point of view one may take when designing the detection path is the following. The lateral spotsize from which to a good approximation light is efficiently and uniformly collected corresponds to the size of the demagnified image of the detection aperture in the focal plane of the microscope objective. Using geometrical optics, the demagnification factor is given by the ratio of the two focal distances of the objective lens and the lens focusing to the pinhole (tube lens). This point of view becomes very important when implementing e.g. a scanning probe near-field microscope, for which one has to make sure that the full scan range of the probe remains well within the detectable area.

At this point we note that the beam profile at the output of a single-mode optical fiber is a fundamental Gaussian mode. As discussed in Section 3.7, other beam modes can be created and some of them can lead to particular properties of the fields in the focal region including e.g. reduced spotsize or longitudinal polarization. If higher-order modes are required, a mode-conversion unit (see Section 3.7) can be introduced to the excitation beam path *before* the beamsplitter in order to keep the detection beam path unperturbed.

The confocal principle

Confocal detection is based on the fact that light not originating from the focal area will not be able to pass through the detection pinhole and hence cannot reach the detector. Laterally displaced beams will be blocked by the detector aperture and beams originating from points displaced along the optical axis will not be focused in the detection plane

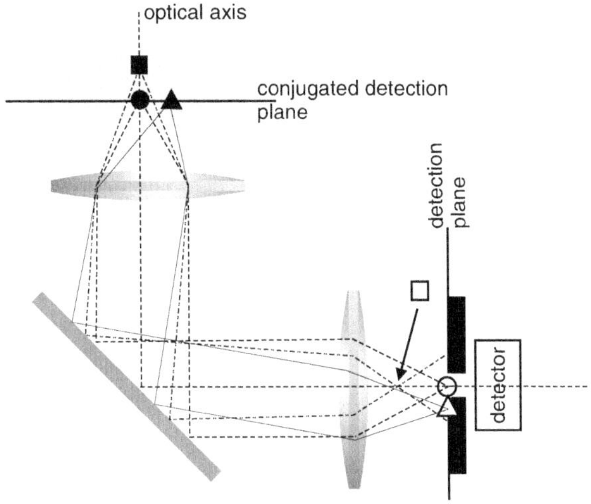

Fig. 5.4 The confocal principle. The detection path of a scanning confocal optical microscope is shown. Three objects in the sample are depicted. Only the object (circle) on the optical axis lying in the conjugated detection plane in the object space is imaged onto the pinhole and can be detected. The other objects (triangle and square) are either focused to the side of the pinhole (triangle) or arrive at the pinhole unfocused such that their signals are suppressed.

and therefore will be strongly attenuated by the detection pinhole. This effect has been discussed theoretically in Section 4.2.2 and is illustrated qualitatively in Fig. 5.4. The imaging properties of a confocal microscope are best discussed in terms of the *total* point-spread function introduced in Section 4.3, which is represented by the product of an excitation and a detection point-spread function. One may think of the total point-spread function as the volume out of which the probability of exciting *and* detecting a photon is larger than a chosen threshold value. It was discussed previously that the point-spread function of a confocal microscope has the shape of an ellipsoid that is elongated along the optical axis and whose center coincides with the geometrical focus of the objective lens. For an NA $= 1.4$ objective and visible light, its extent is 220 nm in the lateral direction and 750 nm along the optical axis, providing the possibility of optical sectioning. The lateral resolution of a confocal microscope is not significantly increased by the multiplication of illumination and detection point-spread function as compared with a wide-field illumination microscope due to the fact that the zero-field points in the total point-spread function remain unchanged. Squaring the Airy pattern merely reduces the fullwidth at half-maximum by a factor of 1.3. However, side lobes are suppressed significantly, leading to a significant increase in the dynamic range of images, meaning that weak signals may be detected in the proximity of strong ones. For a detailed discussion of these issues see e.g. Ref. [4].

Images can be recorded in a number of different ways in a confocal microscope by raster scanning either the sample or the excitation beam. At each pixel either the number of counts per integration time or the output voltage of a photomultiplier tube is sampled. The brightness (or color) of a pixel is defined by the sampled detector value. The information from all the pixels can then be represented in the form of a digital image. In particular, due to the finite extent of the confocal point-spread function, it is possible to perform optical slicing of thick samples. In this way, three-dimensional reconstructions of samples can be obtained. A more detailed description of instrumentation and reconstruction techniques can be found in Refs. [4, 5].

The spatial resolution in confocal microscopy can be optimized by "point-spread function engineering." The underlying idea is that the total point-spread function is the product of the illumination and detection point-spread functions. If either or both are modified, e.g. by means of nonlinear optical interactions, or by being displaced or tilted with respect to each other, their spatial extent and/or spatial overlap decreases. This can lead to an effective point-spread function with a smaller volume. In addition, interference effects between coherent counterpropagating beams can be exploited. These principles form the basis of confocal microscopy techniques known as *4π* [6], *theta* [7], and *4π–theta* confocal microscopy [8]. The respective configurations of detection and illumination point-spread functions are illustrated in Fig. 5.5.

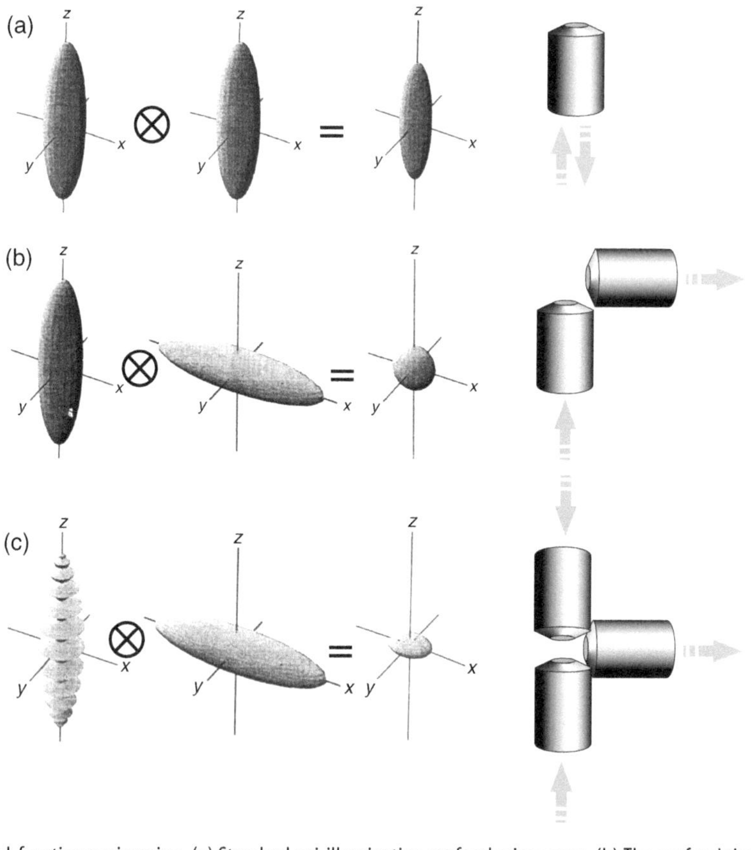

Fig. 5.5 Point-spread-function engineering. (a) Standard epi-illumination confocal microscopy. (b) The confocal theta configuration. (c) The 4π–theta confocal configuration. Adapted from [5].

Nonlinear excitation and saturation

The possibility that a transition in a quantum system could be achieved by the simultaneous absorption of two or more photons was first investigated theoretically by Maria Göppert-Mayer in 1929 [9]. The phenomenon could only be demonstrated experimentally in 1961 [10] after the invention of the laser, which provided the necessary high photon densities. Today, with the availability of femtosecond-pulsed lasers, two-photon and multiphoton excitation is a standard tool in high-resolution confocal microscopy [11]. Chromophores with transitions in the blue and green can be excited by using infrared light. At the same time, multiphoton microscopy leads to improved and simplified optical sectioning capabilities since excitation takes place only in the regions of highest intensity (Section 4.4), i.e. in a tight focus, which makes the technique an indispensable tool, not only in biology, for studying the three-dimensional morphology of samples.

Figure 5.6 summarizes the basics of two-photon excitation. Two low-energy photons are absorbed simultaneously and excite a molecule from the ground state to a vibronic level of the first excited electronic state. Much the same as for one-photon fluorescence, the excited molecule relaxes to the lowest vibrational level of the excited state and then, after a few nanoseconds, decays to the ground state either non-radiatively or by emitting a photon. While for one-photon excitation the fluorescence rate scales linearly with the excitation intensity (see Chapter 9), for two-photon excitation it scales as the excitation intensity squared. The low cross-section for two-photon excitation, which is on the order of 10^{-50} cm^4 s per photon,[2] requires the use of pulsed lasers with pulse duration $\sim$100 fs at high repetition rates. The pulses have to be short in order to limit the total irradiation dose of a sample and still provide the peak intensities required to make up for the low cross-section of two-photon excitation. The repetition rate has to be high since per pulse a maximum of one fluorescence photon is produced per molecule. Typically, 100-fs-pulsed

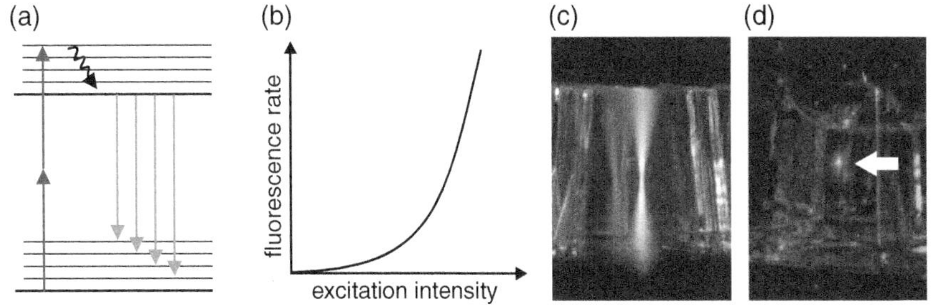

Fig. 5.6 Two-photon excitation of a fluorescent molecule. (a) The energy-level scheme. A fluorophore with a one-photon absorption in the blue is excited by simultaneous absorption of two near-infrared photons. The emission of the molecule occurs in the green. (b) The fluorescence rate increases as the square of the excitation intensity. This leads to the fact that, while for one-photon excitation the whole beam path in a fluorescent medium lights up (c), for two-photon excitation. (d) Notable fluorescence is excited only in regions of the highest field strength, e.g. in the focus of a laser beam (indicated by the arrow). Images (c) and (d) have been adapted from [12].

[2] Also denoted as 1 GM (Göppert-Mayer).

Ti:sapphire lasers operating at around 850 nm at repetition rates of 80 MHz are used to excite two-photon excited fluorescence of suitable dyes.

A method that exploits focal engineering is the so-called stimulated emission depletion (STED) technique already discussed in Section 4.2.3. The basic principle of STED is the use of stimulated emission to selectively reduce the excited-state population of suitable fluorescent dyes in certain spatial regions in the focal area, while in other regions it remains largely unchanged. In principle, this requires subwavelength control over the spatial field distribution which induces the stimulated emission. Such control is indeed possible by exploiting the pronounced saturation behavior of the degree of stimulated emission depletion as a function of the depletion beam power. Saturation allows the creation of extremely sharp transitions between regions with and without depletion of the excited state. In particular, if there exists in the focus a region where the intensity of the depletion beam is zero, a tiny volume of undiminished fluorescence is created around it.

The principle of STED microscopy is summarized in Fig. 5.7. The set-up includes two pulsed laser beams. One is used to induce one-photon excitation of dye molecules present in the focal volume. The second, more powerful laser beam is redshifted in order to produce stimulated emission from the excited to the ground state.[3] The delay between the pulses is chosen such that the vibrational relaxation in the first excited electronic state, which takes a few picoseconds, has time to complete. This ensures that the excited electron is

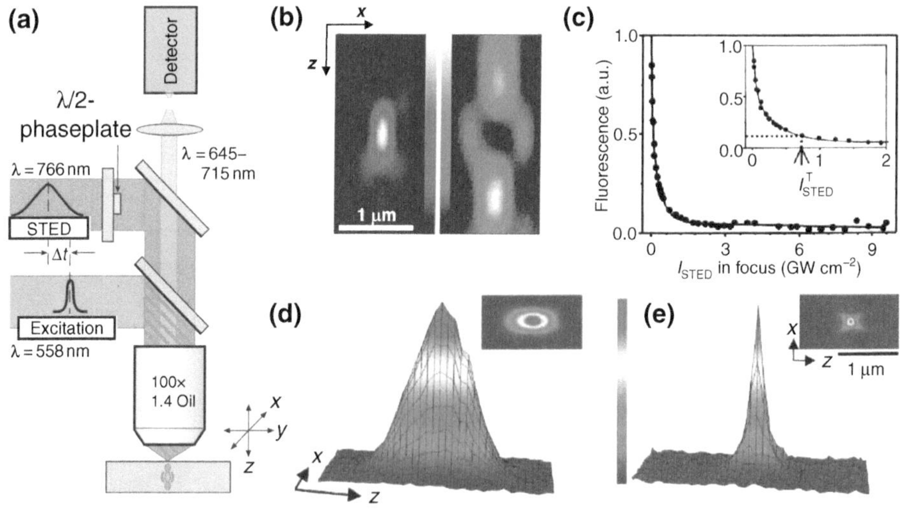

Fig. 5.7 The principle of stimulated emission depletion confocal microscopy. (a) Set-up of the STED confocal microscope. A short excitation pulse and a long depletion pulse are coupled into a microscope objective. The depletion beam is engineered so that it exhibits zero intensity at the geometrical focus (right panel of (b)) while the excitation beam shows the usual focus (left panel of (b)). (c) Fluorescence from the confocal volume as a function of the depletion beam intensity. Note the strongly nonlinear behavior. The point-spread function (d) without and (e) with the depletion beam (right panel of (b)) switched on. Adapted from [13].

[3] For a detailed discussion of molecular fluorescence see Chapter 9.

in a relatively long-lived state where stimulated emission can become effective. This is important since the probability of stimulated emission increases with time. This is also the reason why the STED pulse has to be substantially longer than the excitation pulse as indicated in Fig. 5.7(a). The wavelength of the depletion pulse has to be chosen such that it cannot excite fluorescence. This can be ensured by introducing a relatively large redshift. The large redshift has the additional advantage that it opens up a spectral window between the excitation and depletion wavelengths in which fluorescence can be recorded. Time gating of the fluorescence can be used to increase the signal-to-background ratio. In STED, the foci of the excitation and depletion beams are made to overlap, but the field distribution in the focal region of the STED beam is engineered such that the intensity is zero at the geometrical focus. This guarantees that the STED beam depopulates the excited states everywhere but in a small region centered around the zero-intensity point. Because of saturation, this region can be made smaller than a diffraction-limited spot. Thus, the spatial extent of the fluorescent region can be narrowed down substantially. This effect is illustrated in Figs. 5.7(d) and (e) (see Problem 5.3).

Confocal fluorescence microscopy methods, such as STED microscopy or multiphoton-excitation microscopy, rely on the presence of fluorescent markers in a sample, e.g. in a living cell. However, it is not always possible or even desirable to attach a dye marker to an entity of interest. This is especially true e.g. for small bio-molecules that would be significantly altered by the labeling. If chemical contrast via optical microscopy is the goal, an obvious way to go is to exploit the energy transfer between photons and molecular vibrations. Since the energies of molecular vibrations cover the far infrared spectral region, it is difficult to achieve high spatial resolution since the diffraction-limited spots are technically difficult to achieve and anyway quite large. A work-around for this problem is to use Raman spectroscopy. Here, photons interacting with the sample can either lose or accept quanta of vibrational energy (see Figs. 5.8(a)–(c)). In essence, Raman scattering is the analog of the amplitude modulation used in broadcasting: the frequency of the carrier (laser) is mixed with the frequencies of the signal (molecular vibrations). As a result, the frequencies of Raman scattered light correspond to sums and differences of the frequencies of laser and vibrations. Because a Raman scattering spectrum contains information about the characteristic molecular vibrations it constitutes a highly specific fingerprint for the chemical composition of the sample under investigation. The likelihood that a photon interacting with a molecule undergoes Raman scattering is very small. Typical Raman-scattering cross-sections are up to 14 orders of magnitude smaller than the cross-sections for fluorescence. These low cross-sections usually make the use of Raman scattering for microscopy very difficult. Long integration times, which require very stable and static samples, are necessary. However, the cross-section can be strongly increased near metal surfaces with nanoscale roughness or near metal nanoparticles. This effect, called surface-enhanced Raman scattering (SERS), is limited to regions near the very surface of a sample as discussed later on (see Section 12.4.3), and cannot be employed for long-range subsurface imaging and three-dimensional sectioning. Nevertheless, for bulk imaging the cross-section of Raman scattering can be enhanced by applying a coherent (resonant) pumping scheme. Coherent pumping gives rise to an in-phase oscillation of the molecular vibrations in the illuminated sample volume, leading to constructive interference in certain directions. The so-called coherent anti-Stokes Raman-scattering (CARS) process [14, 15] is a four-wave mixing process that uses two (pulsed) tunable lasers with a wavelength difference that can be adjusted to coincide with the energy of a molecular vibration, which then leads to an increased efficiency of the Raman-scattered signal. The CARS energy diagram and phase-matching condition are shown in Figs. 5.8(d) and (e), respectively. Owing to the fact that CARS is proportional to the intensity squared of the pump beam at ω_p and the intensity of the Stokes beam at ω_s a sizable signal is generated only in regions of high pump intensities. Therefore, the optical sectioning capabilities of CARS microscopy are similar to those of two-photon microscopy. Furthermore, a combination with point-spread-function engineering techniques as they are used in 4π and theta microscopy is conceivable to improve spatial resolution.

Another label-free spectroscopic imaging technique, which allows three-dimensional optical sectioning, is stimulated Raman scattering (SRS) microscopy [16]. In SRS the energy difference $\Omega = \omega_p - \omega_s$ between two picosecond-pulsed focused laser beams is adjusted such that the energy difference Ω coincides with the energy of a vibronic transition. The first beam is the so-called pump beam, which excites a molecule from the ground state to a virtual state. The second beam, the Stokes beam, stimulates a transition from

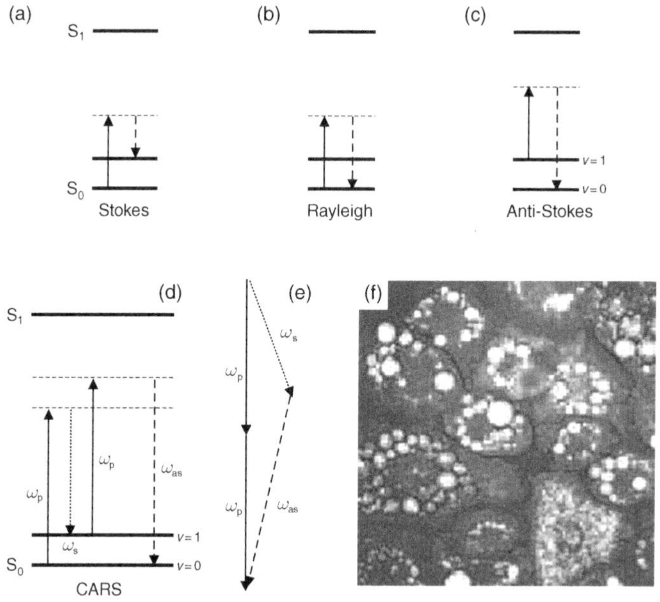

Fig. 5.8 Energy diagram of spontaneous Raman scattering and coherent anti-Stokes Raman scattering (CARS). Light scattering from a molecule can result in (a) Stokes-shifted photons, (b) Rayleigh scattering, or (c) anti-Stokes emission. (d) CARS is a four-wave mixing process using two tunable (pulsed) lasers at ω_p and ω_S. If the difference in frequency between the two lasers hits the energy of a vibration, the CARS signal ω_{as} is enhanced and emitted preferentially into a direction determined by the phase-matching condition (e). (f) An image of fibroplast cells stimulated to produce lipids. The lipid droplets can be visualized using CARS when tuning to the aliphatic C—H vibration. The 100 μm $\times$ 100 image was taken in 2.7 s. Image courtesy of X. S. Xie, Harvard University.

the virtual state to one of the vibrational states of the electronic ground state, similarly to Stokes Raman scattering. This procedure diminishes the intensity I_p of the pump beam and increases the intensity I_s of the Stokes beam. Both intensity changes, ΔI_p and ΔI_s, are small but measurable. Typically, a high-frequency modulation (MHz regime) is applied to either the pump or the Stokes beam, and the resulting intensity modulation on the respective other beam is detected using a lock-in amplifier. The observed intensity changes depend linearly on the number of molecules in the focal region, on their Raman cross-sections, as well as on the product of the two intensities. The advantage of SRS over CARS is that there is no non-resonant background signal. Therefore SRS allows highly specific and chemically sensitive, as well as quantitative, optical microscopy with three-dimensional sectioning capability.

5.2.2 The solid immersion lens

According to Eq. (5.7) a higher numerical aperture (NA) leads to better spatial resolution. Solid immersion lenses have been put forward to optimize the NA available in a microscope. A solid immersion lens (SIL) can be viewed as a variant of an oil-immersion microscope objective. It was introduced in 1990 for optical microscopy [17] and applied in 1994 for optical recording [18]. As shown in Fig. 5.9, a SIL produces a diffraction-limited, focused light spot directly at the SIL/object interface. The resulting spotsize scales as λ/n, where n can be as large as 3.4 when using SILs made out of gallium phosphate (GaP). Such a reduction in the focused spotsize has led to advances in optical-disk storage schemes with fast read-out rates for addressing media with very high bit density [18]. The prospect of using such lenses in combination with a shorter-wavelength blue semiconductor laser diode makes SIL techniques potentially very attractive not only for data-storage devices but also in the area of high-light-throughput super-resolution optical microscopy and spectroscopy with high sensitivity.

The SIL is a solid plano-convex lens of high refractive index that provides an optimum focus for a Gaussian beam. There are two configurations with a semispherical lens that achieve diffraction-limited performance. One focus exists at the center of the sphere, with incoming rays perpendicular to the surface and is generally termed a SIL (cf. Fig. 5.9(a)). Also, a second focus exists at a set of aplanatic points some distance below the center of the sphere; rays from this focus are refracted at the spherical surface. This type is generally referred to as a super-SIL [18], or Weierstrass optic (see Fig. 5.9(b)). While the super-SIL configuration has a greater magnification ($\propto n^2$ versus n) and increased numerical aperture, it suffers from strong chromatic aberration. The applications of SIL microscopy fall into two categories: surface and subsurface imaging [19]. In the latter, the SIL (or super-SIL) is used to image objects below the lens and into the sample under study. In this sort of subsurface imaging, a good match in index between the lens and substrate must be maintained.

The principle of subsurface imaging is schematically shown in Fig. 5.10. Without the SIL, most of the light rays emanating from a subsurface structure would undergo total internal reflection (TIR) at the surface of the sample. The remaining propagating rays would

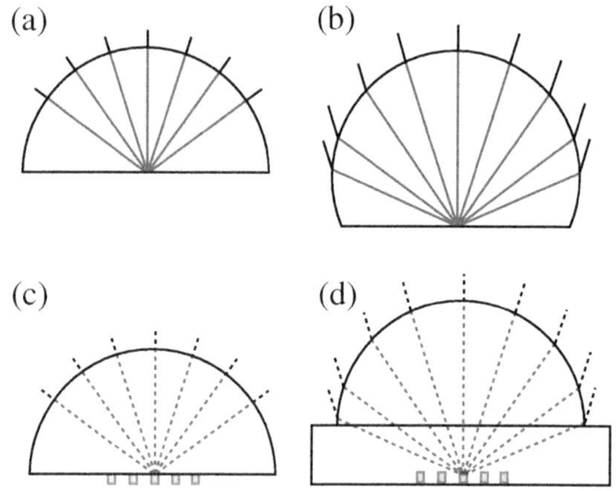

Fig. 5.9 Solid immersion lens (SIL) configurations. In (a), a hemispherical lens increases resolution by $\sim n$. (b) A Weierstrass optic, or super-SIL, has a resolution increase of $\sim n^2$. There are two types of imaging modes, surface SIL microscopy (c) and subsurface SIL microscopy (d).

be confined to a narrow cone around the surface normal, thereby drastically reducing the numerical aperture. By placing an index-matched SIL on the surface of the device, the numerical aperture can be considerably increased. This type of SIL is therefore referred to as a *numerical-aperture-increasing lens* (NAIL) [19]. The dimensions of the SIL have to be adjusted to the depth X of the subsurface structure to be imaged (cf. Fig. 5.10). The vertical thickness D of the lens has to satisfy

$$D = R(1 + 1/n) - X, \tag{5.8}$$

which is the same design condition as encountered in Weierstrass-type SILs. Equation (5.8) ensures that the subsurface object plane coincides with the aplanatic points of the NAIL's spherical surface, which satisfies the sine condition yielding spherical aberration-free or stigmatic imaging.

The addition of a NAIL to a standard microscope increases the NA by a factor of n^2, up to NA $= n$. As an example, Figs. 5.10(c) and (d) demonstrate how a NAIL improves resolution well beyond the state of the art in through-the-substrate imaging of silicon circuits [20]. Image (c) was obtained using a 100$\times$ objective with NA $= 0.5$, whereas image (b) was recorded with a 10$\times$ objective (NA $= 0.25$) and a NAIL. The resulting NA is 3.3. At a wavelength of $\lambda = 1$ μm, the resolution can be as good as 150 nm. Ünlü and coworkers applied the NAIL technique to thermal subsurface imaging, which makes sample illumination unnecessary [19]. In this case, the emitted infrared radiation originates from heating due to electric currents.

Figure 5.11(a) shows a schematic representation of a NAIL confocal microscope. The NAIL is in fixed contact with the sample surface. To obtain an image, the sample and the NAIL are raster scanned using piezoelectric transducers. However, in applications such as data storage and photolithography it is desirable to retain the ability to alter relative positioning of the lens and surface. In order not to sacrifice the NA and not to introduce

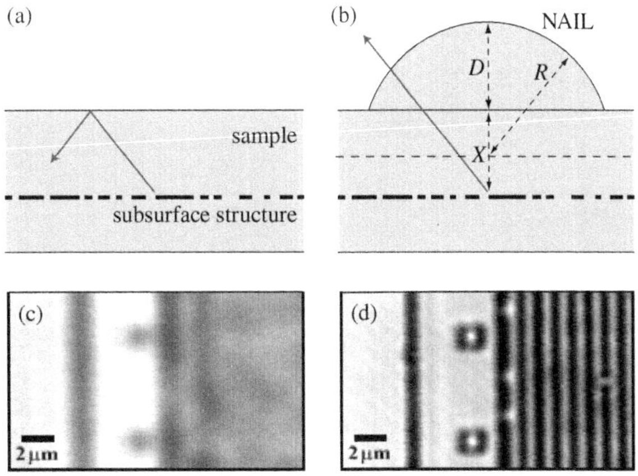

Fig. 5.10 Subsurface imaging using a numerical-aperture-increasing lens (NAIL). (a) In a high-index material, light rays emanating from subsurface structures can undergo total internal reflection, thereby reducing the NA of an imaging system. (b) Addition of a SIL enlarges the NA up to NA $= n$. Images (c) and (d) show a comparison of images of an electronic circuit in silicon taken with, (d), and without, (c), NAIL. From [20] with permission (© 2002 IEEE).

unwanted abberations, the end-face of the SIL must be kept parallel and in close proximity to the sample surface. Naturally, this demands a SIL with small dimensions or a cone-shaped SIL that guarantees that the closest point to the surface is the focal spot. Two approaches have been put forward to control the distance between SIL and surface. The first is based on a cantilever as used in atomic force microscopy (AFM) [21]. The AFM tip is replaced by a miniature conically shaped SIL that is illuminated from the top, Fig. 5.11(b). This combined AFM–SIL technique has successfully been applied to microscopy and photolithography with spatial resolutions on the order of 150 nm [21, 22]. Another approach for controlling the SIL–sample distance is based on a flying head [18]. Rotating the sample at high speeds relative to the stationary SIL results in an air-bearing that keeps the SIL–surface distance at a few tenths of a nanometer (see Fig. 5.11(c)). This approach was originally developed by the IBM company as part of a SIL-based magneto-optical recording system.

An obvious extension of SIL techniques is their marriage with concepts developed in near-field optical microscopy. For example, it has been proposed that one could microfabricate a tiny aperture in the end-face of a SIL [23], implant into the end-face a tiny metal structure acting as a local field enhancer [20], or deposit onto the sides of a conical SIL a bowtie antenna consisting of two electrodes with a small gap [24].

5.2.3 Localization microscopy

The possibility of localizing single emitters on the basis of determining the centroids of their point-spread functions has been discussed in Section 4.5. Here we discuss how this ability can be exploited for super-resolution imaging. Let's first assume that we have densely labeled a structure of interest with a certain type of point-like fluorescent emitter.

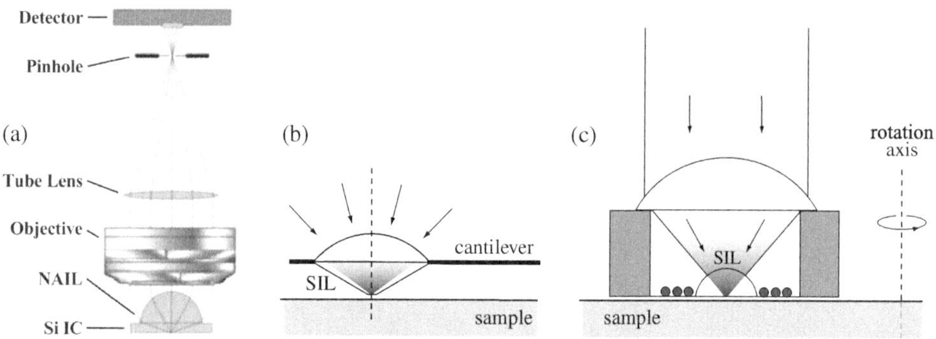

Fig. 5.11 SIL technology in three different instruments: (a) numerical-aperture-increasing lens microscopy with SIL in contact with sample, (b) SIL microscopy using an AFM cantilever for distance control, (c) a flying-head configuration based on a rotating sample surface for magneto-optical data storage.

Let us further assume that the fluorescence emitter can be switched from a bright state (A) to a dark state (B) by some external stimulus, e.g. by means of photons of a certain energy. For the reverse transition B $\rightarrow$ A we assume that it is either spontaneous, albeit with a sufficiently slow rate, or also occurs only upon external stimulation. Since only a few molecules in state A will be imaged in the presence of a large amount of molecules in state B, the contrast in emission intensity between A and B must be large.

In the first step we assume that all emitters present in the sample are in the dark state (B). Then, an external stimulus is uniformly applied to the system in a low dose. This will result in a sparse subset of emitters getting switched into the bright state (A) in a stochastic manner but with a spatially uniform probability. Now that some of the emitters have been turned bright, their fluorescence is imaged onto a two-dimensional detector, e.g. a CCD. To optimize the speed of image acquisition, the stimulation dose may be optimized in such a way that the distance between individual emitters becomes as small as possible but overlapping diffraction-limited image patterns still remain rare. An image of this first subset of emitters is then recorded and stored for further analysis. Since the image contains only isolated spots, the position of each emitter can be determined with high accuracy as explained in Section 4.5. The required acquisition time depends on many parameters of the experimental set-up and the sample, but it is estimated that a total of 500 photons need to be collected per emitter to achieve a position accuracy of 20 nm [25], roughly one order of magnitude beyond the diffraction limit (c.f. Fig. 4.15). A way of representing the result of a localization of a single emitter is to construct a two-dimensional probability density map in which each emitter is represented by a two-dimensional Gaussian probability distribution having its maximum at the spatial coordinate that was found by the least-square fitting algorithm and a full width at half-maximum determined by the uncertainty of this position. A super-resolved image is recorded when the accumulated probability density function appears to be sufficiently densely populated according to Nyquist sampling theory. The process is stopped if the required structural detail is obtained or if no further emitters can be switched to the bright state (A) because they have all been photobleached. The normalized total probability density function that is finally obtained is a continuous function describing the probability of finding a certain number of emitters at any point of the image normalized by the total number of localizations. Assuming uniform labeling of the structure of interest and uniform activation, the total probability density function therefore represents a super-resolved two-dimensional image of the structure of interest (see Fig. 5.12).

The different experimental realizations of the outlined technique mostly vary in the type of single emitter that is used and which external stimulus is needed for photo-switching. One method uses cyanine dyes, which are photo-switched from A to B by the excitation laser which also excites the fluorescence signal. This technique is called stochastic optical reconstruction microscopy (STORM) [27]. Another implementation, known as photoactivated localization microscopy (PALM) [28], uses photoactivatable fluorescent proteins. It should be pointed out that photo-switching to a dark state (B) can be viewed as a transition that is saturable at very low intensities and as such could also be used in STED-like imaging [29].

While the achievements of localization microscopy are impressive, the basic method described so far is applicable only to two-dimensional imaging; that is, it has been assumed that all the emitters are located in a single plane. One possibility to extend localization microscopy to three dimensions is by making use of the fact that the point-spread function of a single-dipole emitter is a function of all three coordinates in the image space as discussed in Section 4.1. Thus, defocusing techniques can be applied to obtain axial resolution [30]. A method to differentiate between positive and negative defocus is based

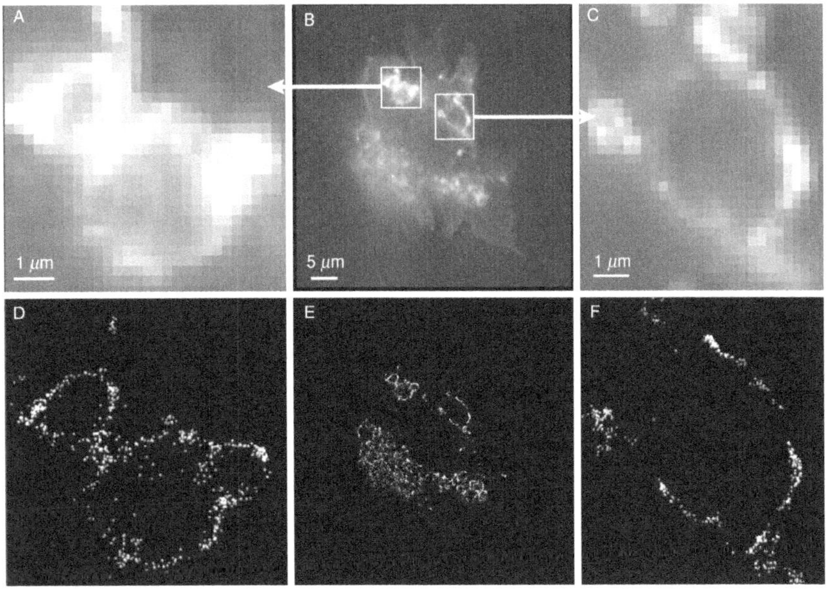

Fig. 5.12 Wide-field fluorescence (B) and PALM images (E) of the protein hemagglutinin tagged with photoactivatable green fluorescent protein (PA-GFP) in a fixed HAb2 fibroblast. Zoom-ins of selected regions to show agreement with wide-field fluorescence, (A) and (C), and illustrate the improvement in resolution achieved by PALM, (D) and (F). Note that the contrast was adjusted in (E) for visualization. The pairs of images (A) and (D), (B) and (E), and (C) and (F) have the same scale. From Ref. [26].

on a cylindrical lens that is introduced into the detection path. This lens induces slight astigmatism, which results in elliptical spot shapes. The ellipticity of the spot increases for increasing defocus. Discrimination between positive and negative defocus is possible since the orientation of the main axis of the elliptical spots turns by 90° upon crossing zero defocus [31].

The compatibility with standard biological fluorescence imaging modalities, such as TIR fluorescence and multi-color imaging, as well as three-dimensional image reconstruction clearly is among the main advantages of PALM and STORM. However, it should be noted that temporal resolution is limited due to the sequential imaging procedure. Furthermore, labeling of structures of interest has to be very dense and homogeneous and, furthermore, the size of intermediate tags or of the labels themselves becomes an issue if the localization accuracy is pushed to the few-nanometer regime.

5.3 Near-field excitation microscopy

In this section we discuss near-field microscopy techniques that are dominated by the **ST** term of the Born series (5.6), that is, incident light first interacts with the probe (T) and then with the sample (S). In general, spatial resolution in optical imaging depends on the bandwidth of transverse spatial frequencies $\Delta k_{\|}$. The numerical aperture (NA) of the optical system limits this bandwidth to $\Delta k_{\|} = [-\text{NA}\,\omega/c \ldots \text{NA}\,\omega/c]$. The NA, as we saw, can be maximized by using a large index of refraction (n) or by increasing the focusing angle. In the best case, NA $= n$, which imposes a strict resolution limit. However, as discussed in Section 4.6, the considerations leading to this resolution limit ignore spatial frequencies associated with evanescent waves. In fact, if evanescent waves are taken into account, the bandwidth of spatial frequencies is in principle unlimited and resolution can be enhanced, in principle, arbitrarily. In this section, we consider optical microscopy with a near-field excitation source, i.e. a source with evanescent field components. The near-field interacts with the sample and the scattered light resulting from this interaction is recorded with standard far-field collection optics, see Fig. 5.1(b). While Section 4.6 provided the necessary theoretical background, this section concentrates on experimental issues. The near-field source is commonly referred to as an "optical probe."

5.3.1 Aperture scanning near-field optical microscopy

The light path of an aperture-type scanning near-field optical microscope differs from a confocal set-up only in that the excitation beam is replaced by the field emanating from a tiny aperture placed near the sample surface, Fig. 5.13 (cf. Fig. 4.17). Most commonly, apertures are formed by coating the sides of a sharply pointed optical fiber with a metal. The uncoated apex of the pointed fiber represents an aperture. The optical properties of aperture probes will be discussed in more detail in Chapter 6. The scattered light originating from the interaction between the near-field of the aperture and the sample surface is recorded with the same scheme as employed in confocal microscopy. The possibility of easily switching back and forth between near-field and far-field illumination modes is an advantage of the similarity between the two techniques.

Since in aperture-type near-field optical microscopy we now have two separate elements for illumination and detection, the two elements must finally share the same optical axis. This requires some means of adjustment for the lateral position of the optical probe. If the sample is scanned, the optical path does not change during image acquisition. This guarantees e.g. the same collection efficiency throughout the image. If probe scanning is required, the back-projected image of the detection aperture has to be large enough to accommodate the whole scan range of the probe.

When a perfect aperture probe is used it is in principle not necessary to use confocal detection optics. However, it turns out that aperture probes are hardly ever as perfect as desired. Pinholes in the metal coating or spurious light escaping from the uncoated upper parts of a probe may contribute to a significant background signal. Limiting the detection

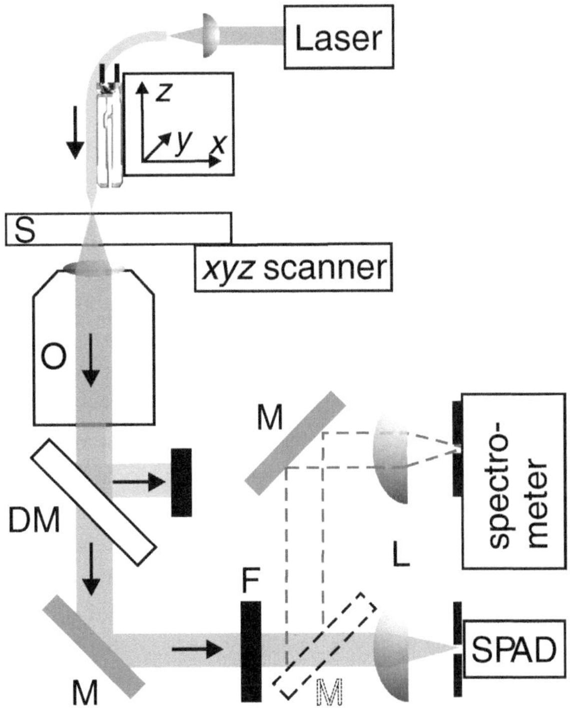

Fig. 5.13 Typical set-up for a near-field optical microscope operating in illumination mode. Note the similarity to the confocal set-up in Fig. 5.3. Laser light is injected into an optical fiber that holds an optical probe at its far end. The probe is held within near-field distance of the sample using e.g. a tuning-fork shear-force feedback (see Chapter 7). The light interacts with the sample and is collected by a microscope objective that is aligned with respect to the fiber axis. In the case of fluorescence imaging, a dichroic mirror reflects most of the excitation light. Residual excitation light is removed by additional filters and the redshifted fluorescence is focused onto a detector or spectrometer. M, mirror; L, lens; DM, dichroic mirror. The dashed mirror can be flipped in and out of the beam path.

of light to the finite confocal volume can improve this problem. For larger apertures the resolution of near-field microscopy can be influenced by the numerical aperture of the collection optics. A large numerical aperture optimizes the collection efficiency, which is important in fluorescence applications. For pure absorption and scattering contrast, light collected below and above the critical angle (allowed and forbidden light, respectively, see Chapter 10) can show inverted contrast [32]. For such applications, high numerical apertures have to be used with care.

5.4 Near-field detection microscopy

In the previous section, the sample was excited locally using a near-field source and the light scattered or emitted by the sample was collected with standard far-field optics. In this section, we consider the reverse situation, i.e. the sample is excited from the far-field and the response is detected locally using a near-field optical probe. The interaction is therefore best described by the **TS** term of the Born series (5.6).

5.4.1 Scanning tunneling optical microscopy

In photon scanning tunneling microscopy (PSTM) [33, 34] a laser beam undergoes total internal reflection at the surface of the sample-support, usually a prism or a hemisphere (c.f. Fig. 5.14). The resulting evanescent surface wave has a typical decay length on the order of 100 nm (see Chapter 2). A bare tapered glass fiber is dipped into this evanescent field to locally couple some of the light into the probe, where it is converted into propagating modes that are guided towards a detector. This conversion is in analogy to the frustrated total internal reflection discussed in Chapter 2. The preparation of sharply pointed fiber probes is described in Chapter 6.

Using a bare fiber probe has both advantages and disadvantages. Counting as an advantage is the fact that a dielectric probe perturbs the field distribution much less than would any kind of metallized probe, justifying the fact that the Born series, Eq. (5.6), may be terminated after the **TS** term. On the other hand, the spatial confinement of the effective area out of which the dielectric probe is collecting light is not very small and not well defined. Since the probe is not a point-like scatterer, the collection efficiency can depend in a complicated way on the specific three-dimensional structure of the probe. Nevertheless, for samples that predominantly exhibit evanescent fields PSTM can resolve field distributions with features down to about 100 nm. An illustrative example is shown in Fig. 2.5. Here, PSTM was used to map a purely evanescent standing wave obtained by the interference of two equal-amplitude counterpropagating evanescent waves obtained by total internal reflection. The observed modulation depth provides information about the effective size of the probe.

Notice that bare fiber probes can generate severe artifacts when imaging scattering samples. These artifacts originate from the fact that fields are most efficiently coupled into the fiber along the probe shaft rather than at the tip apex (cf. Chapter 6).

Recording amplitude and phase of field distributions

A unique feature of photon tunneling microscopy is the possibility to measure not only the time-averaged intensity in the near-field but also its amplitude and phase [35]. These measurements can even be made in a time-resolved manner by employing heterodyne interferometry [36]. The experimental set-up for this type of measurement is shown in Fig. 5.15. The light frequency ω_0 in the reference branch is shifted by acousto-optic modulation by an amount $\delta\omega$. The signal recorded via the fiber probe and the reference field can be described as [35]

$$\mathbf{E}_S(x,y) = \mathbf{A}_S(x,y)\exp\big[\mathrm{i}(\omega_0 t + \phi_S(x,y) + \beta_S)\big]\,, \tag{5.9}$$

$$\mathbf{E}_R = \mathbf{A}_R \exp[\mathrm{i}(\omega_0 t + \delta\omega\, t + \beta_R)]\,. \tag{5.10}$$

Here, $\mathbf{A}_S(x,y)$ and $\mathbf{A}_R$ are the real amplitudes of the signal and the reference field, respectively. $\phi_S(x,y)$ is the relative phase of the optical signal at the sample. Both the signal amplitude and the phase depend on the position of the fiber probe. The factors β_S and β_R are constant phase differences due to the different optical paths in the reference and signal branches. The sampled field is then made to interfere with the reference field and the result is recorded by a photodetector. The resulting signal becomes

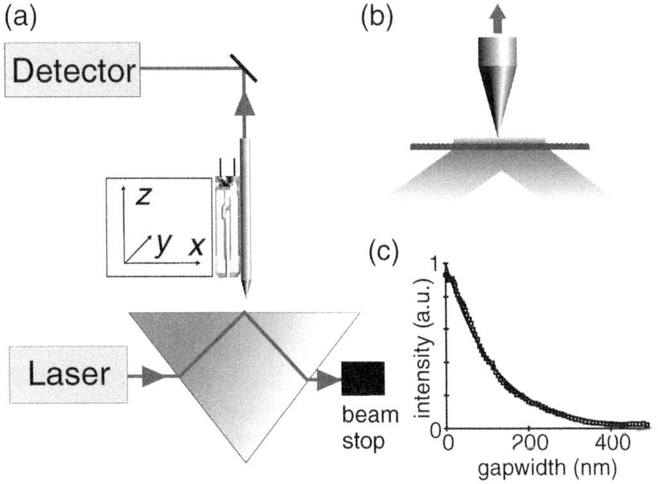

Fig. 5.14 Photon scanning tunneling optical microscopy (PSTM). (a) Typical set-up. A transparent sample on top of a prism is irradiated by total internal reflection. (b) Close-up of the gap region, showing a dielectric probe dipping into the evanescent field above the sample. (c) Exponential decay with increasing gapwidth of the optical signal guided to the detector recorded by approaching a flat air/glass interface with a tapered glass-fiber probe.

$$I = |\mathbf{A}_S(x,y)|^2 + |\mathbf{A}_R|^2 + 2\mathbf{A}_R \cdot \mathbf{A}_S(x,y)\cos\left[-\delta\omega t + \phi_S(x,y) + \beta_S - \beta_R\right]. \tag{5.11}$$

This signal has a DC offset and an oscillating component at $\delta\omega$. The amplitude and phase of this component contain the relevant information. They can be extracted by a dual-output lock-in amplifier locked at the frequency $\delta\omega$. For pulsed excitation, interference can occur only if signal and reference pulses arrive at the detector at the same time. In this way, by varying the delay via the delay line in Fig. 5.15, the propagation of a pulse through a structure of interest can be monitored [36].

As an application example, Fig. 5.16 shows a temporal snap shot (one fixed position of the delay line) of the near-field intensity distribution of a light pulse passing through a photonic crystal waveguide coupler. Figure 5.16(a) shows the structure of the coupler, which in the center consists of two parallel line-defect waveguides that are close enough to allow for electromagnetic coupling. Figure 5.16(b) shows the optical signal, which consists of the product of the sine of the phase and the local field amplitude. The visualized behavior of the pulse shows in direct space that the structure performs very well as a directional coupler. For more details and movies of the propagating pulse see [37].

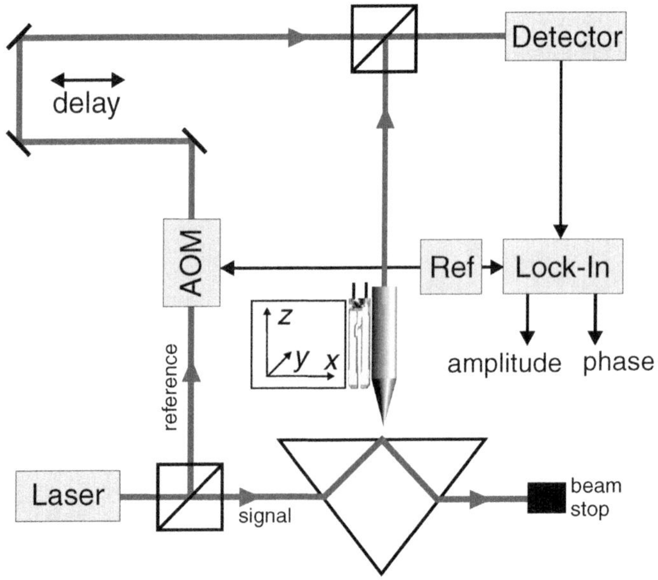

Fig. 5.15 Photon tunneling microscopy combined with time-resolved heterodyne interferometry. The light from a fixed-frequency laser source is divided into a reference and a signal branch. In the reference branch the frequency is shifted by means of an acousto-optic modulator (AOM). Furthermore, in the reference branch there is a delay line for time-resolved experiments. The signal branch undergoes total internal reflection inside a prism and provides evanescent field illumination at a structure of interest. An evanescent field can also be created by coupling the signal branch into a waveguide. A sharp fiber probes the evanescent field above the sample and directs the sampled light to a beamsplitter, where the sampled light interferes with the reference field. The resulting signal is analyzed with a lock-in amplifier.

5.4.2 Field-enhanced near-field microscopy with crossed polarization

Imaging of the near-field intensity distribution of strongly scattering samples, such as metallic nanoparticles exhibiting plasmon resonances (see Chapter 12) requires both very high spatial resolution due to the small effective wavelengths involved and the ability to suppress the detection of scattered homogeneous waves. This is possible using field-enhanced scanning near-field optical microscopy with polarization control [38]. Figure 5.17 shows the principle of the method. The sample is illuminated by a moderately focused s-polarized beam of light, which allows one to excite plasmon resonances of isolated metal nanoparticles on the sample. The optical probe, a sharply pointed probe, is only weakly excited by this polarization in the first place since it exhibits a strong polarizability only along its main axis (see Section 6.5). However, the interaction of the incident radiation with strongly scattering sample features gives rise to "polarization scrambling" and hence to localized fields that are polarized along the probe axis. These field components can now be efficiently scattered by the optical probe and recorded in the far field. In order to suppress multiple scattering between probe and sample, the scattered light is again sent through a polarizer, which selects the field component parallel to the probe axis. The detection scheme is completed by mixing the detected signal with a reference beam in order to determine also the phase of the scattered field and to boost the signal by using lock-in detection referenced to the oscillation frequency of the optical probe. The use of the cross-polarization scheme terminates the Born series after the **TS** term. This near-field imaging technique makes it possible to record amplitude and phase distributions near nanoplasmonic structures, such as the gold discs shown in Fig. 5.17(b). Note that reversing the polarization directions of illumination and detection converts this imaging technique into the **ST** mode described before.

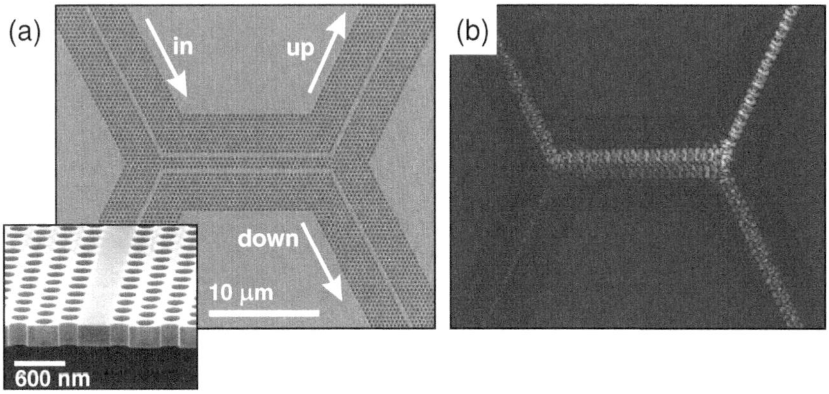

Fig. 5.16 Light-pulse propagation through a photonic crystal waveguide coupler. (a) Scanning electron microscopy overview image of the photonic crystal and the light-guiding line defects. Inset: zoom showing the structure of the photonic crystal membrane and the line defect. (b) Snapshot of an optical pulse recorded while it passes through the junction. The image shows the interferometric signal (field amplitude multiplied by the sine of the phase). A fiber probe with a metal coating and a rather large aperture was used to optimize background suppression. Adapted from [37].

5.5 Near-field excitation and detection microscopy

We now consider configurations of near-field microscopy in which multiple scattering between probe and sample occurs. The leading term in the Born scattering series is the **TST** term. Accordingly, excitation of the sample occurs predominantly via the optical probe's enhanced near-field, i.e. sample excitation by external irradiation is comparatively weak. Furthermore, the emission or scattering of light from the sample into the far-field is also weak. However, the presence of the optical probe in close proximity to the sample helps to scatter the localized sample fields towards the detector. In this mode, the optical probe acts like an optical antenna.

5.5.1 Field-enhanced near-field microscopy

Aperture-type near-field microscopy is limited in resolution because the effective diameter of an aperture cannot be smaller than twice the skin depth of the metal used for coating the glass taper. The skin depth is between 6 and 10 nm for good metals at optical frequencies. As a consequence, even if the physical aperture size is zero, there exists an effective aperture of diameter about 20 nm. It is not at all straightforward to then achieve such a resolution in an experiment because for apertures of such a small size the transmission becomes exceedingly low, as will be discussed in Chapter 6. When working with aperture probes on a routine basis, for signal-to-noise reasons, aperture diameters are usually kept between 50 and 100 nm unless the taper angle of the pointed probe can be drastically increased (see Chapter 6).

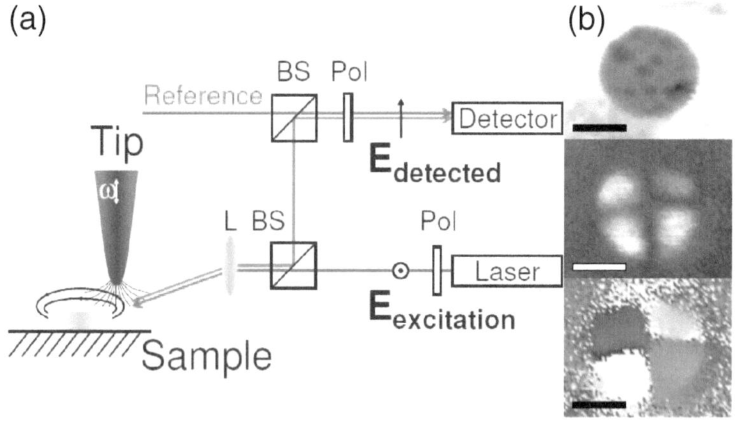

Fig. 5.17 Field-enhanced scanning near-field microcopy with polarization control. (a) Overview of the set-up showing the beam path with respective polarization as well as the probe-sample interaction region. Depolarization effects at the sample lead to a polarization of the probe along its main axis. The sensitivity is enhanced by using both a reference field and lock-in detection. (b) Images of gold disks on glass. Top, AFM topography; middle, measured field amplitude; bottom, the phase of the field. From [38].

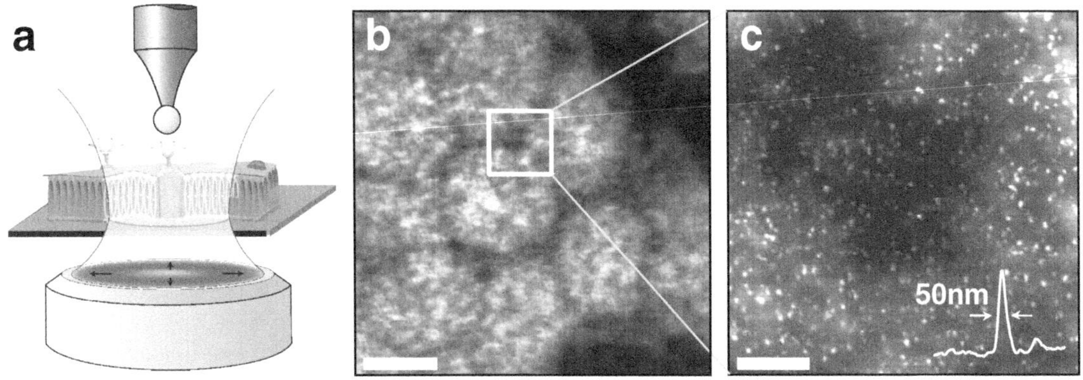

Fig. 5.18 Field-enhanced near-field fluorescence imaging. (a) Sketch of the set-up. (b) Far-field confocal fluorescence image showing erythrocyte membranes with fluorescently labeled calcium-ion-channel proteins. The protein density is too high for individual proteins to be resolved confocally. Scale bar: 5 μm. (c) Near-field fluorescence image of the area indicated in (b) showing individually resolved proteins. The image was acquired with a gold nanoparticle antenna of size $\sim$60 nm irradiated by a tightly focused radially polarized laser beam. The inset shows a cross-section through one of the fluorescence spots, indicating a resolution of $\sim$50 nm. Scale bar: 1 μm. Adapted with permission from [39].

To go beyond the light confinement that is achievable with aperture probes, one can rely on the fact that an optical near-field is created by *any* irradiated small material structure. This near-field is localized to the surface of the material and, depending on the material properties, the near-field intensity can be enhanced over the intensity of the irradiating light. The goal, of course, is to find specific structures that yield particularly strong near-field confinement and enhancement. One possibility is to exploit laser-irradiated metal particles and sharp, tip-like metal structures that provide "field-line crowding" (the lightning-rod effect). Another possibility is to take advantage of geometry-dependent plasmon resonances that occur for high-enough frequencies near or in the optical regime. These plasmon resonances are associated with strong field enhancement and can be employed for the realization of efficient near-field probes. Plasmons will be discussed in more detail in Chapter 12.

Using a resonant probe consisting of a single spherical gold particle attached to a dielectric fiber tip Hoeppener has [39] have shown that it is possible to image single fluorescently labeled proteins in their natural environment. Figure 5.18(a) shows a sketch of the set-up. Pure confocal imaging cannot resolve individual proteins as shown in the overview scan Fig. 5.18(b). The resonant spherical gold particle enhances the fluorescence of nearby single emitters (see also Fig. 9.21) by a factor of 8–10 (see Section 13.4). As long as the surface density of labeled structures is not too high, it is possible to obtain high-resolution imaging of individual labeled proteins at the sample surface on top of a background of fluorescence due to the external illumination (Fig. 5.18(c)).

The background due to external illumination generally causes a deterioration of the signal-to-noise ratio. Although the intensity associated with the external irradiation is weak, the irradiated sample area is much larger than the area associated with the confined near-field. To discriminate the signal generated by the near-field interaction from the signal generated by the far-field irradiation, nonlinear interactions such as two-photon excitation or sum-frequency generation can be employed.

For a diffraction-limited excitation spot, the ratio between the areas associated with external excitation and with near-field excitation is on the order of 10^3. Hence, assuming a uniform surface coverage of molecules, the near-field intensity has to be enhanced by a factor of at least 10^3 in order to generate a near-field signal that is as strong as the signal associated with far-field irradiation. On the other hand, for a second-order nonlinear process, which scales with the square of the excitation intensity, the required enhancement factor is only $\sqrt{10^3} \approx 32$. Of course, for very low surface coverage the problem of near-field vs. far-field discrimination is less important. With only a single species or an isolated cluster in the far-field illumination focus, such background can even become negligible.

The use of nonlinear optical processes can also pose problems because new sources of background may appear. Prominent examples are broad-band photoluminescence [40] and second-harmonic generation [41] at increased illumination levels. Being disturbing effects in luminescence measurements, both effects can be exploited, e.g. to generate local light sources for spectroscopy or lithography.

Another way to solve the background problem was demonstrated by Frey *et al.* [42]. Tips can be grown on the end-faces of aperture probes. Excitation through the aperture instead of using a far-field illumination spot drastically reduces the far-field background.

Field-enhanced scanning near-field optical microscopy has also been combined with types of vibrational spectroscopy such as Raman scattering [43] and CARS [44]. In this context the method is generally referred to as tip-enhanced Raman scattering (TERS). Since in the presence of a field-enhancing structure not only the excitation field but also the Raman-scattered radiation is enhanced according to the **TST** character of the interaction, usually the Raman signal is assumed to scale with the fourth power of the local field strength [45]. As an example, Fig. 5.19 shows near-field Raman scattering images of a sample of carbon nanotubes [46]. Carbon nanotubes possess comparatively large Raman scattering cross-sections and are easily imaged at low sample coverage. The Raman image in Fig. 5.19 was obtained by integrating over a narrow spectral band centered around the G band at $\nu = 1580\ \mathrm{cm}^{-1}$.

Modulation techniques

Modulation techniques are used to discriminate the near-field signal generated near the probe apex against the background signal associated with the external irradiation of the sample. Most commonly, the distance between probe and sample is modulated and the optical signal is detected at the same modulation frequency, or at higher harmonics, using lock-in amplifiers. Modulation techniques are mostly applied to Rayleigh-scattered light at the same frequency as the external excitation. The excitation field induces a dipole in the probe tip, which itself induces an image dipole in the sample. The signal that is observed is the light scattered by the effective dipole emerging from the combination of probe and sample dipoles, again highlighting the **TST** character of the interaction. Using a model that replaces the probe by a spherical particle above a plane interface, the following effective polarizability of the coupled probe–sample system can be derived:

$$\alpha_{\text{eff}} = \frac{\alpha(1+\beta)}{1-\alpha\beta/[16\pi(a+z)^3]}, \tag{5.12}$$

where $\alpha = 4\pi a^3(\varepsilon_{\text{probe}}-1)/(\varepsilon_{\text{probe}}+2)$, $\beta = (\varepsilon_{\text{sample}}-1)/(\varepsilon_{\text{sample}}+1)$, a is the radius of curvature of the probe tip, and z is the gapwidth between probe and sample [47]. For a small particle, the scattered field amplitude is proportional to the polarizability α_{eff}. Therefore, changing the wavelength of illumination will lead to changes in the scattering efficiency because the values of the dielectric constants of the sample, $\varepsilon_{\text{sample}}$, and the probe, $\varepsilon_{\text{probe}}$, will be subject to change. This type of spectroscopy allows one to distinguish between different materials if the probe's response is flat in the spectral region of interest.

Usually it is found that detection of the optical signal at the fundamental probe oscillation frequency is not very favorable since scattering from the probe shaft can also contribute a modulation of the signal. This problem can be solved by demodulation at higher harmonic frequencies of the fundamental probe oscillation frequency. Since the probe–sample distance dependence of the near-field optical signal is strongly nonlinear (see e.g. Eq. (5.12)), it will introduce higher harmonics in the detected signal. These higher harmonics can be extracted by using lock-in detection in combination with heterodyne or homodyne interferometry in a similar way to that described in Section 5.4.1. Figure 5.20 shows the set-ups used in this context as well as the effect of demodulation at higher harmonics. Exploiting

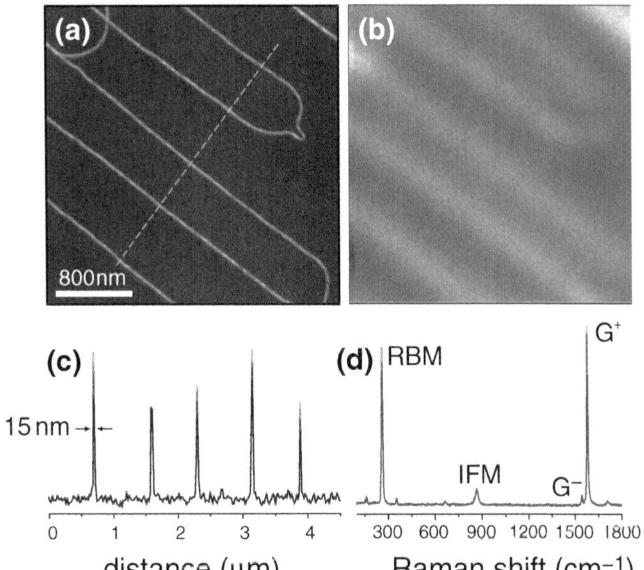

Fig. 5.19 Tip-enhanced Raman scattering (TERS) of a single-walled carbon nanotube. (a) Near-field image representing the spatial distribution of the G-band intensity. (b) Corresponding confocal Raman image. (c) Cross-section along the line in (a) indicating a resolution of 15 nm and a high signal-to-noise ratio. (d) Near-field Raman spectrum showing characteristic vibrational bands. Adapted with permission from [46].

higher harmonics, the near-field can be extracted more specifically. The possible order of the harmonics to be used is, however, limited by the measurement noise, which is usually the shot-noise of the detected signal. Detecting at the third harmonic seems to be a good compromise between good background suppression and tolerable noise. Figure 5.20 demonstrates the effect of demodulation at the third harmonic on the image quality. The set-up of Fig. 5.20(a) is used to image a latex sphere projection pattern. The topography is shown in Fig. 5.20(c). Figures 5.20(d) and (e) show the optical signals demodulated at the fundamental frequency and at the third harmonic, respectively. The third-harmonic picture is much clearer since far-field contributions are better suppressed. This can also be seen by comparing the respective approach curves beneath the optical images [47].

Modulation techniques have also been implemented for discrete signals, such as streams of single photons. In photon time-stamping, for example, the arrival times of individual photons (the so-called time-stamps) are recorded and only those photons that fall into a predefined time-window are retained [48]. Typically, only those photons that arrive during a short period starting before and ending after the probe reaches its shortest distance to

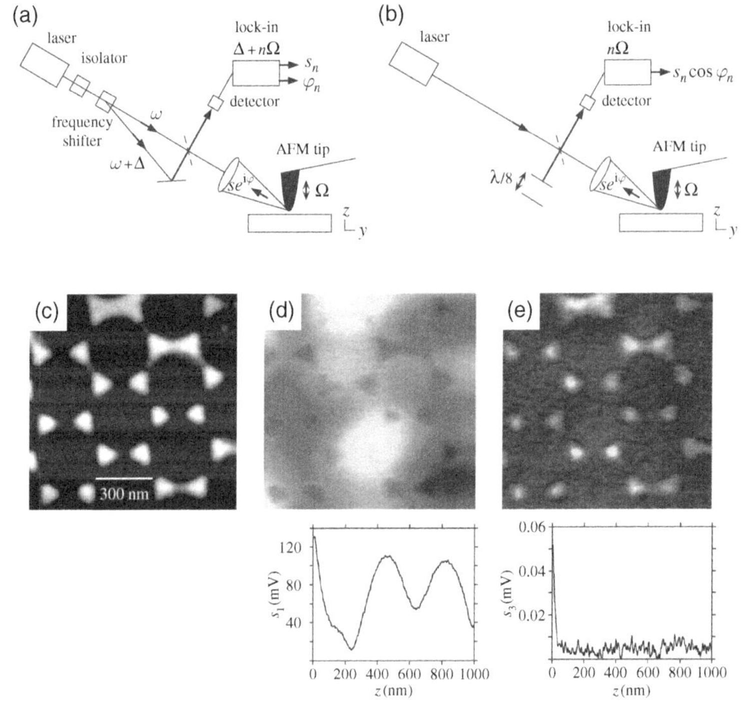

Fig. 5.20 Set-ups for scattering-type near-field microscopy using heterodyne (a) and homodyne (b) detection. (c) Topography of a latex sphere projection pattern. (d) Upper panel: scattered light image at the fundamental oscillation frequency of the cantilever. Lower panel: approach curve showing strong interference fringes due to remaining far-field contributions. (e) Upper panel: scattered light image at the third harmonic of the cantilever. Lower panel: approach curve recorded on the third harmonic of the cantilever oscillation frequency, showing a clean near-field signal. From [47].

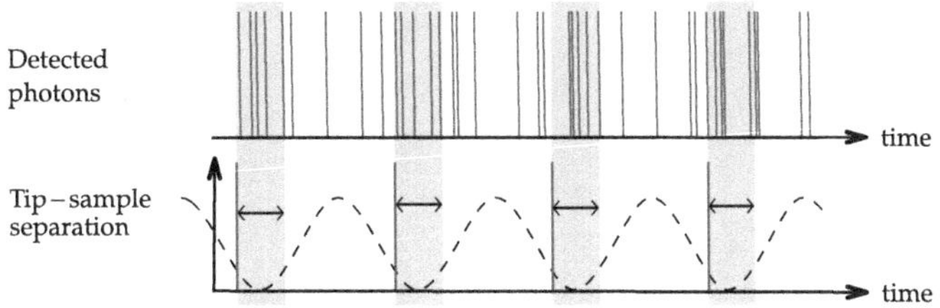

Fig. 5.21 Correlation of photon arrival times with the vertical oscillation of the near-field probe. In time-stamping only photons that fall into periodic time-windows (shaded areas) with preset widths (arrows) are counted.

the sample surface are considered. Besides better sensitivity, a further advantage of this method is that different analysis techniques can be applied to the raw data, depending on the signal properties that are to be extracted. Figure 5.21 illustrates the relation between the time-stamps of optical and probe-position data.

It was also shown that discrete voltage pulses as they occur in single-photon counting can be directly fed into a lock-in amplifier referenced to the frequency of the probe–sample distance modulation. The technique even works for low count rates down to the single-emitter level. This possibility has been exploited to image fluorescent quantum dots at high resolution [49] as well as membrane proteins labeled with single molecules [50].

5.5.2 Double-passage near-field microscopy

In this section we briefly discuss configurations of near-field microscopy which are dominated by the **TST** term of the interaction series, but do not require external far-field illumination of the probe. A first configuration falling into this category is shown in Fig. 5.22. In this microscope a fiber probe or an aperture probe is used to excite the sample and, at the same time, to collect the optical response. In the case of a bare fiber probe, light has to pass through the probe twice and hence the resolution is improved compared with configurations that use fiber probes only for illumination. Resolutions of about 150 nm at a wavelength of 633 nm have been demonstrated using fiber probes both for excitation and for collection [51]. Aperture-type probes are more difficult to be used in the "double-passage" configuration, because of signal-to-noise limitations. Light throughput through a subwavelength aperture is very small and if light has to pass twice the throughput is even lower (cf. Chapter 6). Nevertheless, the throughput can be optimized by making use of metal-coated fibers with large taper angles or probes with a double taper. In fact, Hosaka and Saiki have demonstrated single-molecule imaging with resolution $\approx$20 nm using "double-passage" through aperture probes [52]. Near-field microscopy in the "double-passage" configuration is attractive because of its numerous conceivable technical applications to non-transparent samples, including data storage. To overcome the limitation of low throughput, a combination with local field enhancement is desirable.

An early version of a near-field microscope working in the "double-passage" mode was devised by the pioneers of near-field optics, U. Fischer and D. W. Pohl, in 1988 [53].

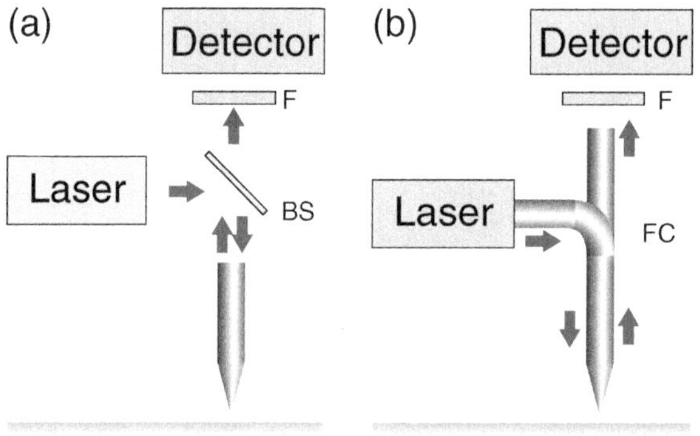

Fig. 5.22 Concept of near-field microscopy in the "double-passage" mode. The probe is used both for excitation and for collection. Implementation with (a) an external beamsplitter and (b) a y-shaped fiber coupler. F, filter; BS, beamsplitter; FC, fiber coupler.

A sketch is shown in Fig. 5.23. A subwavelength aperture in a metal screen is illuminated by a waveguide mode supported by a glass slab. Light scattered at the aperture is recorded as a function of the aperture–sample distance and as a function of the lateral scan coordinates [53]. The scattering strength depends on the local effective index of refraction in the vicinity of the aperture. High-resolution optical images were obtained using this type of microscopy.

5.6 Conclusion

The interaction of an optical probe and a sample can be described using the intuitive picture of multiple consecutive scattering events. As a result near-field optical microscopes can be classified according to which term of this interaction series dominates over the others. It is not always possible to extract a single term from the interaction series and in exotic situations the series even fails to converge. In general, what is measured in near-field microscopy is *not* the field of the sample but the interaction between probe and sample.

Problems

5.1 Interaction series. Derive Eq. (5.6) by explicit calculation and rearrangement of the resulting terms.

5.2 Surface-enhanced spectroscopy. Using Ref. [45] discuss why the enhancement of Raman scattering near nanostructure is proportional to the fourth power of the field-enhancement factor. Does the same scaling also hold for other spectroscopic signals?

5.3 Use the formalism of Section 3.6 to determine the diameter of the on-axis phase plate that should be used in STED microscopy in order to exactly cancel out the total field

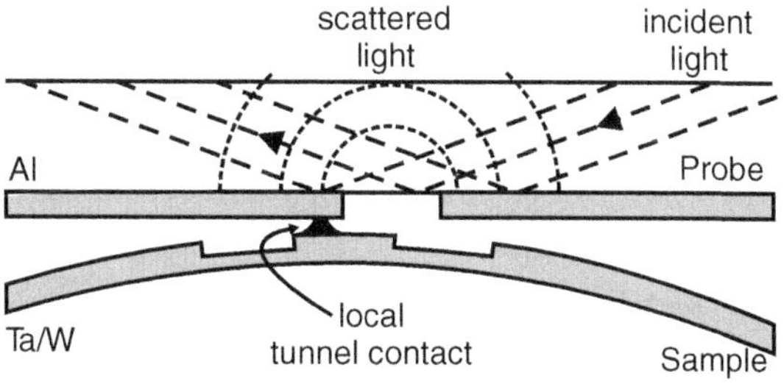

Fig. 5.23 Reflection-mode near-field microscopy. A subwavelength aperture is illuminated by a waveguide mode in a glass slab. The scattering from the aperture is recorded as a function of the local environment of the aperture. From [53].

in the geometrical focus. Discuss why it is important to really achieve zero field with a high degree of accuracy.

5.4 Derive Eq. (5.12) assuming a small spherical particle above a plane interface. The particle is treated as a single dipole which induces an image dipole in the substrate.

5.5 Imaging artifacts in localization microscopy. Consider a molecule located at the geometrical focus of an objective with numerical aperture NA. The fluorescence of the molecule is imaged onto a CCD, from which we determine the center position with an accuracy of 10 nm. The imaging system is characterized by the transverse magnification M.

We now place a nanoparticle (radius a_0, dielectric constant ϵ) to the side of the molecule. The distance between molecule and nanoparticle (center-to-center) is d. We end up with the emission from two coherent dipoles, the molecule and the dipole induced in the nanoparticle.

Calculate the effects of a a_0, d, and ϵ on the measured center position and its accuracy. What conclusions do you draw?

References

[1] J. Sun, P. S. Carney, and J. C. Schotland "Strong tip effects in near-field scanning optical tomography," *J. Appl. Phys.* **102**, 103103 (2007).

[2] M. Born and E. Wolf, *Principles of Optics*, 7th edn. New York: Cambridge University Press (1999).

[3] S. W. Hell, M. Schrader, P. E. Hänninen, and E. Soini, "Resolving fluorescence beads at 100–200 nm axial distance with a two photon 4π-microscope operated in the near infrared," *Opt. Commun.* **120**, 129–133 (1995).

[4] R. H. Webb, "Confocal optical microscopy," *Rep. Prog. Phys.* **59**, 427–471 (1996).

[5] J. B. Pawley (ed.), *Handbook of Biological Confocal Microscopy*, 2nd edn. New York: Plenum Press (1995).

[6] S. W. Hell and E. H. K. Stelzer, "Properties of a 4π-confocal fluorescence microscope," *J. Opt. Soc. Am. A* **9**, 2159–2166 (1992).

[7] S. Lindek, R. Pick, and E. H. K. Stelzer, "Confocal theta microscope with three objective lenses," *Rev. Sci. Instrum.* **65**, 3367–3372 (1994).
[8] S. W. Hell, *Increasing the Resolution of Far-Field Fluorescence Microscopy by Point-Spread-Function Engineering*. New York: Plenum Press (1997).
[9] M. Göppert-Mayer, "Über die Wahrscheinlichkeit des Zusammenwirkens zweier Lichtquanten in einem Elementarakt," *Naturwissenschaften* **17**, 932 (1929).
[10] W. Kaiser and C. G. B. Garret, "Two-photon excitation in CaF_2:Eu^{2+}," *Phys. Rev. Lett.* **7**, 229–231 (1961).
[11] W. Denk, J. H. Strickler, and W. W. Webb, "2-Photon laser scanning fluorescence microscopy," *Science* **248**, 73–76 (1990).
[12] P. S. Dittrich and P. Schwille, "Photobleaching and stabilization of fluorophores used for single-molecule analysis with one- and two-photon excitation," *Appl. Phys. B* **73**, 829–837 (2001).
[13] S. W. Hell, "Towards fluorescence nanoscopy," *Nature Biotechnol.* **21**, 1347–1355 (2003).
[14] P. D. Maker and R. W. Terhune, "Study of optical effects due to an induced polarization third order in the electric field strength," *Phys. Rev. A* **137**, 801–818 (1965).
[15] A. Zumbusch, G. R. Holtom, and X. S. Xie, "Three-dimensional vibrational imaging by coherent anti-Stokes Raman scattering," *Phys. Rev. Lett.* **82**, 4142–4145 (1999).
[16] C. W. Freudiger, W. Min, B. G. Saar *et al.*, "Label-free biomedical imaging with high sensitivity by stimulated Raman scattering microscopy," *Science* **322**, 1857–1861 (2008).
[17] S. M. Mansfield and G. S. Kino, "Solid immersion microscope," *Appl. Phys. Lett.* **77**, 2615–2616 (1990).
[18] B. D. Terris, H. J. Mamin, and D. Rugar, "Near-field optical data storage," *Appl. Phys. Lett.* **68**, 141–143 (1996).
[19] S. B. Ippolito, B. B. Goldberg, and M. S. Ünlü, "High spatial resolution subsurface microscopy," *Appl. Phys. Lett.* **78**, 4071–4073 (2001).
[20] B. B. Goldberg, S. B. Ippolito, L. Novotny, Z. Liu, and M. S. Ünlü, "Immersion lens microscopy of nanostructures and quantum dots," *IEEE J. Selected Topics Quantum Electron.* **8**, 1051–1059 (2002).
[21] L. P. Ghislain and V. B. Elings, "Near-field scanning solid immersion microscope," *Appl. Phys. Lett.* **72**, 2779–2781 (1998).
[22] L. P. Ghislain, V. B. Elings, K. B. Crozier, *et al.*, "Near-field photolithography with a solid immersion lens," *Appl. Phys. Lett.* **74**, 501–503 (1999).
[23] T. D. Milster, F. Akhavan, M. Bailey, *et al.*, "Super-resolution by combination of a solid immersion lens and an aperture," *Jap. J. Appl. Phys.* **40**, 1778–1782 (2001).
[24] J. N. Farahani, H. J. Eisler, D. W. Pohl, and B. Hecht, "Single quantum dot coupled to a scanning optical antenna: a tunable super emitter," *Phys. Rev. Lett.* **95**, 017402 (2005).
[25] S. T. Hess, T. P. K. Girirajan, and M. D. Mason "Ultra-high resolution imaging by fluorescence photoactivation localization microscopy," *Biophys. J.* **91**, 4258–4272 (2006).

[26] T. J. Gould and S. T. Hess "Nanoscale biological fluorescence imaging: breaking the diffraction barrier," *Methods Cell Bio.* **89**, 329–358 (2008). With permission from Elsevier.

[27] M. J. Rust, M. Bates, and X. Zhuang, "Sub-diffraction-limit imaging by stochastic optical reconstruction microscopy (STORM)," *Nature Methods* **3**, 793–795 (2006).

[28] E. Betzig, G. H. Patterson, R. Sougrat, *et al.*, "Imaging intracellular fluorescent proteins at nanometer resolution," *Science* **313**, 1642–1645 (2006).

[29] M. Hofmann, C. Eggeling, S. Jakobs, and S. W. Hell, "Breaking the diffraction barrier in fluorescence microscopy at low light intensities by using reversibly photoswitchable proteins," *Proc. Nat. Acad. Sci.* **102**, 17565–17569 (2005).

[30] M. F. Juette, T. J. Gould, M. D. Lessard, *et al.*, "Three-dimensional sub-100 nm resolution fluorescence microscopy of thick samples," *Nature Methods* **5**, 527–529 (2008).

[31] B. Huang, W. Wang, M. Bates, and X. Zhuang, "Three-dimensional super-resolution imaging by stochastic optical reconstruction microscopy," *Science* **319**, 810–813 (2008).

[32] B. Hecht, H. Bielefeldt, D. W. Pohl, L. Novotny, and H. Heinzelmann, "Influence of detection conditions on near-field optical imaging," *J. Appl. Phys.* **84**, 5873–5882 (1998).

[33] D. Courjon, K. Sarayeddine, and M. Spajer, "Scanning tunneling optical microscopy," *Opt. Commun.* **71**, 23–28 (1989).

[34] R. C. Reddick, R. J. Warmack, D. W. Chilcott, S. L. Sharp, and T. L. Ferrell, "Photon scanning tunneling microscopy," *Rev. Sci. Instrum.* **61**, 3669–3677 (1990).

[35] M. L. M. Balistreri, J. P. Korterik, L. Kuipers, and N. F. van Hulst, "Phase mapping of optical fields in integrated optical waveguide structures," *J. Lightwave Technol.* **19**, 1169–1176 (2001).

[36] M. L. M. Balistreri, H. Gersen, J. P. Korterik, L. Kuipers, and N. F. van Hulst, "Tracking femtosecond laser pulses in space and time," *Science* **294**, 1080–1082 (2001).

[37] R. J. Engelen, Y. Sugimoto, H. Gersen, *et al.*, "Ultrafast evolution of photonic eigenstates in **k**-space," *Nature Phys.* **3**, 401–405 (2007). Reprinted by permission from Macmillan Publisher Ltd.

[38] R. Esteban, R. Vogelgesang, J. Dorfmüller, *et al.*, "Direct near-field optical imaging of higher order plasmonic resonances," *Nano Lett.* **8**, 3155–3159 (2008).

[39] C. Hoeppener and L. Novotny "Antenna-based optical imaging of single Ca^{2+} transmembrane proteins in liquids," *Nano Lett.* **8**, 642–646 (2008). Copyright 2008 American Chemical Society.

[40] M. R. Beversluis, A. Bouhelier, and L. Novotny, "Continuum generation from single gold nanostructures through near-field mediated intraband transitions," *Phys. Rev. B* **68**, 115433 (2003).

[41] A. Bouhelier, M. Beversluis, A. Hartschuh, and L. Novotny, "Near-field second-harmonic generation induced by local field enhancement," *Phys. Rev. Lett.* **90**, 013903 (2003).

[42] H. G. Frey, F. Keilmann, A. Kriele, and R. Guckenberger, "Enhancing the resolution of scanning near-field optical microscopy by a metal tip grown on an aperture probe," *Appl. Phys. Lett.* **81**, 5030–5032 (2002).

[43] R. M. Stockle, Y. D. Suh, V. Deckert, and R. Zenobi, "Nanoscale chemical analysis by tip-enhanced Raman spectroscopy," *Chem. Phys. Lett.* **318**, 131–136 (2000).

[44] T. Ichimura, N. Hayazawa, M. Hashimoto, Y. Inouye, and S. Kawata, "Tip-enhanced coherent anti-Stokes Raman scattering for vibrational nanoimaging," *Phys. Rev. Lett.* **92**, 220801 (2004).

[45] H. Metiu, "Surface enhanced spectroscopy," *Prog. Surf. Sci.* **17**, 153–320 (1984).

[46] L. G. Cancado, A. Jorio, A. Hartschuh, *et al.*, "Mechanism of near-field Raman enhancement in one-dimensional systems," *Phys. Rev. Lett.* **103**, 186101 (2009). Copyright 2009 American Physical Society.

[47] F. Keilmann and R. Hillenbrand, "Near-field microscopy by elastic light scattering from a tip," *Phil. Trans. Roy. Soc. Lond. A* **362**, 787–805 (2004).

[48] T. J. Yang, G. A. Lessard, and S. R. Quake, "An apertureless near-field microscope for fluorescence imaging," *Appl. Phys. Lett.* **76**, 378–380 (2000).

[49] C. Xie, C. Mu, J. R. Cox, and J. M. Gerton "Tip-enhanced fluorescence microscopy of high-density samples," *Appl. Phys. Lett.* **89**, 143117 (2006).

[50] C. Hoeppener, R. Beams and L. Novotny "Background suppression in near- field optical imaging," *Nano Lett.* **9**, 903–908 (2009).

[51] Ch. Adelmann, J. Hetzler, G. Scheiber, *et al.*, "Experiments on the depolarization near-field scanning optical microscope," *Appl. Phys. Lett.* **74**, 179–181 (1999).

[52] N. Hosaka and T. Saiki, "Near-field fluorescence imaging of single molecules with a resolution in the range of 10 nm," *J. Microsc.* **202**, 362–364 (2001).

[53] U. Ch. Fischer, U. T. Dürig, and D. W. Pohl, "Near-field optical scanning microscopy in reflection," *Appl. Phys. Lett.* **52**, 249–251 (1988).

6 Localization of light with near-field probes

Near-field optical probes, such as laser-irradiated apertures or metal tips, are the key components of the near-field microscopes discussed in the previous chapter. No matter in which configuration a probe is used, the achievable resolution depends on how well the probe is able to confine the optical energy. This chapter discusses light propagation and light confinement in different probes. Fundamental properties are discussed and an overview of fabrication methods is provided. The most common optical probes are (1) uncoated tapered glass fibers, (2) aperture probes, and (3) pointed metal/semiconductor structures and resonant-particle probes. The reciprocity theorem of electromagnetism states that a signal remains unchanged upon exchange of source and detector (see Chapter 2.13). We therefore consider all probes as localized sources of light.

6.1 Light propagation in a conical transparent dielectric probe

Transparent dielectric probes can be modeled as infinitely long glass rods with a conical and pointed end. The analytically known HE_{11} waveguide mode, incident from the infinite cylindrical glass rod and polarized in the x-direction, excites the field in the conical probe. For weakly guiding fibers, the modes are usually designated LP (linearly polarized). In this case, the fundamental LP_{01} mode corresponds to the HE_{11} mode. The tapered, conical part of the probe may be represented as a series of disks with decreasing diameters and infinitesimal thicknesses. At each intersection, the HE_{11} field distribution adapts to the distribution appropriate for the next slimmer section. This is possible without limit because the fundamental mode HE_{11} has no cut-off [1]. With each step, however, part of the radiation is reflected, and the transmitted HE_{11} mode becomes less confined as the field extends more and more into the surrounding medium (air). One hence expects high throughput but poor confinement for this type of probe.

The calculated field distribution in Fig. 6.1 qualitatively supports the expected behavior but reveals some interesting additional features: the superposition of incident and reflected light leads to an intensity maximum at a diameter of approximately half the internal wavelength. Further down the cone, the light penetrates the sides of the probe so that at the tip apex there is an intensity minimum. Thus, the fiber probe is not a local illumination source and one can expect that the best field confinement is on the order of $\lambda/(2n_{\text{tip}})$, with n_{tip} being the refractive index of the fiber.

To understand the efficiency of the fiber probe in the *collection mode* we apply time-reversal to the illumination-mode configuration. The essence is as follows: in illumination mode, the HE_{11} mode propagating in the fiber is converted into radiation near the end of the tip. The radiation field can be decomposed into plane waves and evanescent waves propagating/decaying into various directions with different magnitudes and polarizations (angular spectrum, see Section 2.15). Reversing the propagation directions of all plane waves and evanescent waves will excite in the fiber probe a HE_{11} mode with the same magnitude as was used in the illumination mode. Hence, at first glance it seems that high resolution cannot be achieved with a fiber probe in collection mode. However, as long as the fields to be probed are purely evanescent, such as along a waveguide structure, the fiber probe will collect only the evanescent modes available and the recorded images will represent the local field distribution. But if the sample contains scatterers that convert the evanescent modes into propagating modes, then there is a good chance that the measured signal is dominated by radiation that is coupled into the probe along the tip shaft and images become obscured. Therefore, the fiber probe turns out to be an unfavorable near-field probe for radiating structures.

6.2 Fabrication of transparent dielectric probes

Transparent dielectric probes are often used for the fabrication of more complex probes, e.g. aperture probes. Transparent dielectric probes can be produced by tapering of optical fibers, yielding conical shapes, by suitable breaking of glass slides to produce tetrahedral tips, by polymer molding processes, or by silicon (nitride or oxide) microfabrication

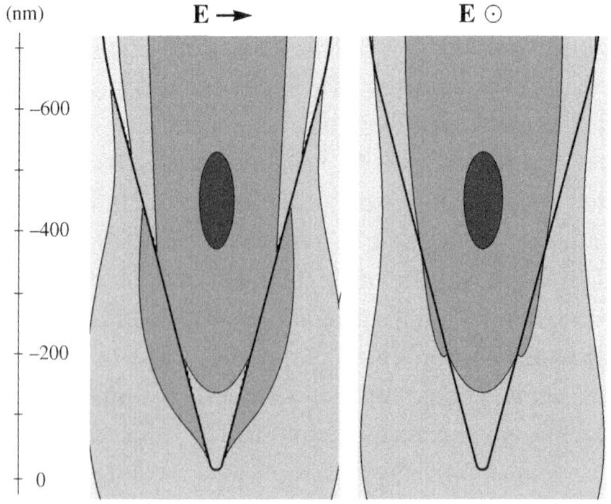

Fig. 6.1 Contours of constant power density on two perpendicular planes through the center of a transparent dielectric probe (with a factor of 3 difference between adjacent lines). The fields are excited by the HE_{11} mode (polarization indicated by symbols) incident from the upper cylindrical part, with $\lambda = 488$ nm and $\varepsilon_{tip} = 2.16$.

techniques. Probes at the end of glass fibers have the distinct advantage that the coupling of light into the taper region can be done easily by exciting the guided modes in the fiber at the far fiber end. In the following we discuss the most important methods that can be used to create sharp transparent dielectric probes.

6.2.1 Tapered optical fibers

Tapering of optical fibers can be achieved by chemical etching, or by local heating and subsequent pulling. Here we compare the results of different etching and pulling techniques and discuss their respective features, advantages, and disadvantages.

Etching

Chemical etching of glass fibers has the potential for batch fabrication of a large number of identical probes. Initially, etching of glass fibers was performed using Turner's method [3, 4]. Here, fibers with their plastic coating stripped off are dipped into a 40% HF solution. A thin overlayer of an organic solvent is usually added (i) to control the height of the meniscus of the HF forming at the glass fiber and (ii) to prevent dangerous vapors escaping from the etching vessel. By using different organic overlayers the opening angle of the resulting conical tapers can be tuned [4]. Large taper angles are of interest because, as we shall see, they result in high-throughput optical probes. Taper formation in the Turner method takes place because the height of the meniscus is a function of the diameter of the remaining cylindrical fiber. The initial meniscus height depends on the type of organic overlayer. Since the fiber diameter shrinks during etching, the meniscus height is reduced, so preventing higher parts of the fiber from being etched further. Finally, if the fiber diameter approaches zero the etching process in principle should be self-terminating.

The Turner method has some important drawbacks. (i) The process is not really self-terminating. Diffusion of the small HF molecules into the organic solvent overlayer degrades the tip if it is not removed immediately after it has formed. (ii) The surface of the conical taper is usually rather rough. This roughness is most probably due to the fact that the meniscus of HF does not move continuously and smoothly during etching but rather jumps from one stable position to the next. This results in a faceted, rather rough surface structure, which can pose problems in later processing steps, e.g. resulting in mediocre opacity of metal coatings.

This roughness problem can be overcome by applying the so-called tube-etching method [5]. Here, the fibers are dipped into the HF solution with an organic solvent overlayer (*p*-xylene or iso-octane) *without* stripping off their plastic coating. The plastic coatings of standard optical fibers are chemically stable against HF. Figure 6.2 shows schematically the progress of the etching process for (a) HF impermeable and (b) permeable cladding. The insets show photographs of the etched fibers *in situ*. Both types of cladding result in different pathways for tip formation. For more details the reader is referred to the original publication [5]. Figure 6.3 shows typical results for fiber tips etched by the different techniques. Note the difference in roughness between Turner and tube-etched probes.

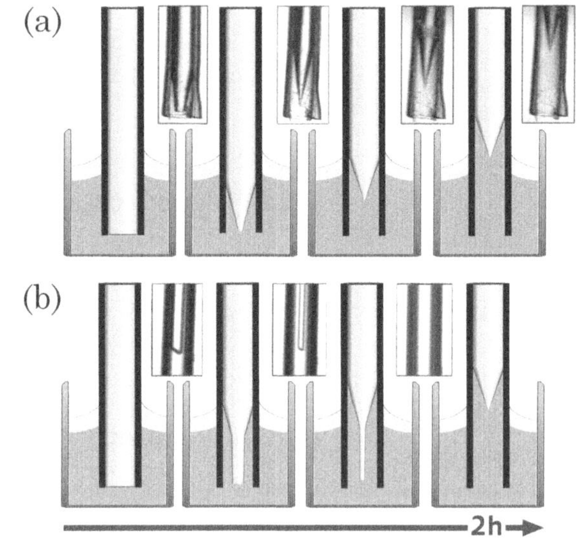

Fig. 6.2 Time evolution of the tube-etching process. The insets show *in situ* video frames of the etching process. Cleaved fibers are dipped into a 40% HF solution with an organic overlayer (*p*-xylene or iso-octane). The etching proceeds along different pathways depending on whether or not the polymer fiber cladding is permeable to HF. In the case of a non-permeable cladding the tip forms at the end of the fiber and keeps its shape while shortening inside the tube (a). In the case of a permeable cladding the tip forms at the meniscus between HF and the organic overlayer (b). From [5].

Besides the Turner and the tube-etching methods there are other etching methods that result in sharp tips. A prominent method is based on dipping cleaved fibers into a buffered HF solution consisting of a mixture with volume ratio $NH_4F:HF:H_2O = X:1:1$, where X denotes a variable volume [6]. In general, mixtures with $X > 1$ are used. The opening angle of the tips monotonically decreases for increasing X and tends to a stationary value for $X > 6$. The magnitude of the stationary angle depends strongly on the Ge concentration in the fiber core. It varies between 100° and 20° for doping ratios of 3.6 and 23 mol%, respectively. The method relies on the fact that in such a solution Ge-rich parts of optical fibers are etched at a lower rate. Since the core of a suitable fiber is doped with Ge, the core starts protruding from an otherwise flat fiber. Figure 6.4 shows the typical shape of fiber probes created by Ohtsu's method. The fiber is flat apart from a short and sharp protrusion sitting on the fiber core. For the method to work, the Ge concentration in the core has to have a suitable profile, which is not the case for all types of standard commercial single-mode fibers. More involved techniques have been applied to achieve tapers with discontinuous opening angles, so-called multiple tapers [7].

Heating and pulling

Another successful method to produce tapered optical fibers is by locally heating a stripped fiber and subsequently pulling it apart. The technology used here was originally developed for electrophysiology studies of cells using the patch-clamp technique. The patch-clamp technique was developed in the 1970s by Erwin Neher and Bert Sakmann [9], for which

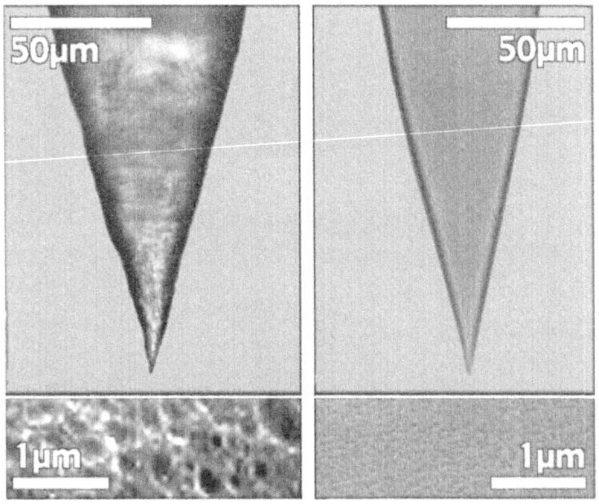

Fig. 6.3 Etched fiber probes. Left: Turner's method. Right: tube-etched probe. The upper panels show optical images obtained with a conventional optical microscope. The lower panel shows higher-resolution scanning electron micrographs of the surface roughness of the tips sputtered with 3 nm of platinum at 77 K. From [5].

they were awarded the 1991 Nobel prize in medicine. Micropipettes for patch-clamp experiments are produced from quartz capillaries by local heating and pulling. The overall shape and the apex diameter of heat-pulled pipettes depend on many parameters, including the pulling speed, the size of the heated area, and the heating time.

For applications in nano-optics, as mentioned before, tapered optical fibers should exhibit a short and robust taper region with a large opening angle at the apex. In order to achieve this goal, the length of the heated area of the fiber should be smaller than or equal to the fiber diameter. In order to achieve a symmetric tip shape, the temperature distribution in the glass should have cylindrical symmetry. Also, heating of the glass should be moderate because a certain minimum viscosity of the glass before pulling is necessary in order to achieve short enough tips. Too low a viscosity leads to the formation of thin filaments upon pulling. In many labs CO_2 lasers at a wavelength of 10.6 μm are used to heat the glass, which at this wavelength is a very efficient absorber. Alternatively, a perforated heating foil or a heating coil can be used. Figure 6.5 shows a typical set-up for heating and pulling of fibers. There exist commercial pipette pullers that can be used to pull optical fibers since they provide control over the magnitude and timing of all relevant process parameters. A detailed study on how to adapt a pipette puller for fiber pulling can be found e.g. in Ref. [10].

Close inspection of fiber tips by scanning electron microscopy reveals that pulled tips tend to show a flat plateau at the apex. The diameter of the plateau is a function of the pulling parameters. A probable explanation for the occurrence of the plateau is that there is brittle rupture once the diameter of the glass filament has become very small and then cooling is very effective. This would imply that the diameter of the plateau should scale with the heating energy applied to the fiber. This was actually observed. Figure 6.6 shows a series of pulled tips with decreasing heating power. There is also a distinct correlation between the opening angle and the heating energy supplied. The angle becomes larger as less heating energy is supplied. Unfortunately, concomitantly the diameter of the flat facet at the apex increases, as can be seen in the insets of Fig. 6.6.

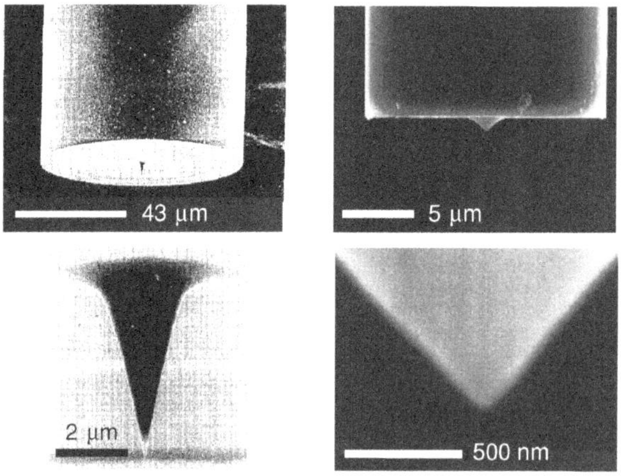

Fig. 6.4 Scanning electron microscopy images of fiber probes etched by Ohtsu's method. Left: highly Ge-doped special fiber. From [6] with permission. Right: commercial fiber. From [8].

It is important to note that tapers created by etching and by pulling are not completely identical. Some groups report problems with pulled probes when polarization of light is an issue. Stress relaxation over time probably creates a time-dependent polarization behavior of pulled probes [11]. Also, for pulled probes the refractive-index profile in the taper is changed since both the fiber core and the cladding are affected by the heating and pulling. On the other hand, the tapers of pulled fibers show very little surface roughness, which is favorable for subsequent processing, such as metal coating.

While the shape of tapered fibers can be accurately determined in scanning electron microscopes, the optical properties, e.g. the effective optical diameter, are more difficult to assess experimentally in a standard way. We point the interested reader to a method that relies on imaging a pattern of standing evanescent waves [12]. By comparing the measured with the expected fringe contrast using a simple model for the probe's collection function, one can estimate the effective optical diameter of a given probe (see Problem 6.1). It is found that for pulled glass-fiber probes this effective diameter is about 50–150 nm.

An alternative to a tapered optical fiber is the so-called tetrahedral probe [13] which can be obtained by cleaving a rectangular slab of glass twice at an angle. The result is a fragment with triangular cross-section. The fragment can be produced from a 170 μm-thick cover slip, so that the overall size of the fragment is rather small. In order to couple in light that is focused to the tip a coupling prism has to be used. A particular feature of tetrahedral probes is that they are not rotationally symmetric, which, after metal coating and aperture formation, can lead to interesting field distributions [14].

6.3 Aperture probes

Probes based on metal-coated dielectrics with a transparent spot at the apex are referred to as aperture probes. The metal coating basically prevents the fields from leaking through

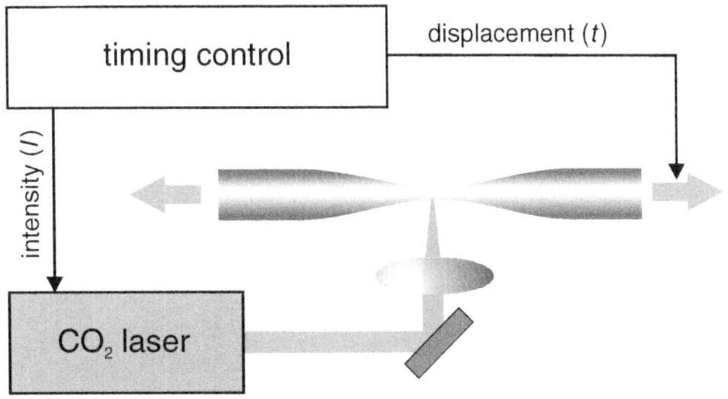

Fig. 6.5 Sketch of a typical set-up for pulling of optical fibers using a CO_2 laser. The laser is focused onto the fiber. For heating, a laser pulse of duration some milliseconds is applied. The pulling starts after the laser pulse and follows a distinct velocity profile. See [10] for details.

the sides of the probe (cf. Fig. 6.1). The most common example is a tapered optical fiber coated with a metal, most often aluminum. In order to understand the light propagation in such a probe we note that it can be viewed as a hollow metal waveguide filled with a transparent dielectric. Towards the probe apex, the diameter of the waveguide is constantly decreasing. The mode structure in a tapered hollow waveguide changes as a function of the characteristic dimension of the transparent core [15]. For large diameters of the transparent dielectric core there will be a number of guided modes in the waveguide. These run into cut-off one after the other as the diameter decreases on approaching the apex. Finally, at a well-defined diameter even the last guided mode runs into cut-off. For smaller diameters of the dielectric core the energy in the core decays exponentially towards the apex because the propagation constants of all modes become purely imaginary. This situation is visualized in Fig. 6.7. The mode cut-off is essentially the reason for the low light throughput of aperture probes. The low light throughput of metal-coated dielectric waveguides is the price for their superior light confinement.

The behavior described above determines some of the design goals and limitations of aperture probes. (i) The larger the opening angle of the tapered structure, and the higher the refractive index of the dielectric core, the better the light transmission of the probe will be. This is because the final cut-off diameter approaches the probe apex [17]. (ii) In the region of cut-off, about two thirds of the incoming energy will be absorbed in the metal layer. This can result in significant heating of the metal coating in this region, which, as a consequence, might be destroyed. The maximum power that can be sent down such a probe is therefore limited. Improving the heat dissipation in the relevant region or increasing the thermal stability of the coating can increase this destruction threshold [18]. These effects will be analyzed in more detail in the following section.

6.3.1 Power transmission through aperture probes

Figure 6.8 shows the calculated power density inside an aperture probe. The probe is excited by the analytically known cylindrical HE_{11} waveguide mode at a wavelength of $\lambda =$

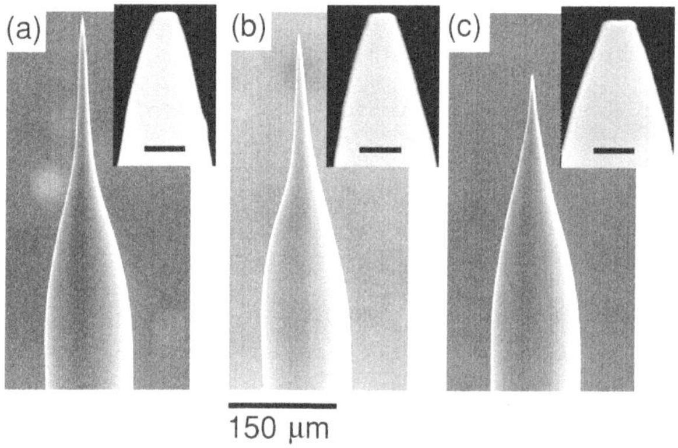

Fig. 6.6 Scanning electron microscopy images of pulled glass fibers sputtered with 20 nm of gold. The insets show magnifications of the respective tip apex. There is a trend that the shorter the tip and therefore the larger the opening angle is, the more pronounced the plateau that occurs at the apex will be. This plateau defines the smallest possible aperture that can be achieved after metal coating.

488 nm. At this wavelength the dielectric constants of the dielectric core and the aluminum coating are $\varepsilon_{\mathrm{core}} = 2.16$ and $\varepsilon_{\mathrm{coat}} = -34.5 + 8.5\mathrm{i}$, respectively.[1] The corresponding skin depth is 6.5 nm. The core has a diameter of 250 nm at the upper cylindrical part and is tapered towards the aperture end.

In the cylindrical part the HE_{11} mode is still in the propagating regime, i.e. its propagation constant has a negligibly small imaginary part. As the core radius becomes smaller, the modes of the tapered part become evanescent and the field decays extremely fast, faster than exponentially, towards the aperture. Since roughly a third of the incident power is reflected backwards this leads to a standing-wave pattern at the upper part of the probe. To the sides of the core the field penetrates into the aluminum coating, where roughly two thirds of the incident power is dissipated into heat.

The fast power decay inside the aperture probe can be well explained by a mode-matching analysis. In this approach, the tapered part of the probe is subdivided into small cylindrical waveguide pieces as shown in Fig. 6.9. For a lossy waveguide the propagation constant k_z can be written as

$$k_z = \beta + \mathrm{i}\alpha, \tag{6.1}$$

where β is the phase constant and α the attenuation constant. According to waveguide theory, the power loss in the nth waveguide section is

$$P_{\mathrm{loss}}(n\,\mathrm{d}z) = P(n\,\mathrm{d}z)(1 - \mathrm{e}^{-2\alpha_{11}(n\,\mathrm{d}z)\mathrm{d}z}), \tag{6.2}$$

where $P(n\,\mathrm{d}z)$ is the incident power and $\alpha_{11}(n\,\mathrm{d}z)$ the attenuation constant of the HE_{11} mode in the nth waveguide section. α_{11} depends on the diameter of the waveguide section,

[1] The complex dielectric function of aluminum for visible wavelengths can be well described by a plasma dispersion law (see Chapter 12), $\varepsilon(\omega) = 1 - \omega_{\mathrm{p}}^2 \cdot \left(\omega^2 + \mathrm{i}\gamma\omega\right)^{-1}$, where a plasma frequency of $\omega_{\mathrm{p}} = 15.565\ \mathrm{eV}/\hbar$ and a damping constant of $\gamma = 0.608\ \mathrm{eV}/\hbar$ yield a good approximation for the dielectric function [15].

on the wavelength and on the material properties. A more detailed discussion on lossy waveguide modes can be found in Ref. [19]. On summing Eq. (6.2) over all waveguide sections, using

$$P([n+1]\mathrm{d}z) = P(n\,\mathrm{d}z) - P_{\mathrm{loss}}(n\,\mathrm{d}z), \tag{6.3}$$

and taking the limit $\mathrm{d}z \to 0$ we obtain the power distribution

$$P(z) = P(z_0)\,\mathrm{e}^{-2\int_{z_0}^{z} \alpha_{11}(z)\mathrm{d}z}. \tag{6.4}$$

In Fig. 6.10 this formula is compared with the computationally determined power along the probe axis (curve a). The power in the probe can also be plotted against the core diameter D using the geometrical relationship

$$z = -\frac{D - D_{\mathrm{a}}}{2\tan\delta}, \tag{6.5}$$

where δ is the half-cone angle and D_{a} the diameter of the aperture. Note that $z_0 \leq z \leq 0$ for the coordinates chosen in Fig. 6.9. The asymptotic values of $P(z)$ are indicated by curves d and e, which describe the decay of the HE_{11} mode in the cylindrical part of the aperture probe and the decay of a wave inside bulk aluminum, respectively. Since the presence of the aperture has hardly any influence on $P(z)$ the curve may be applied in good agreement to any D_{a}. The power transmission of aperture probes with $D_{\mathrm{a}} = 100$ nm, 50 nm and 20 nm therefore is approximately 10^{-3}, 10^{-6}, and 2×10^{-12}, respectively. The steep decay of the transmission curve (see Fig. 6.10) indicates that for the chosen cone angle it is very unfavorable to decrease the aperture size considerably below 50–100 nm, which is actually the diameter most commonly used for aperture probes.

For an aperture probe with a thick (infinite) coating, Fig. 6.11 shows α and β for the HE_{11} mode as functions of z and D. The transition from the propagating to the evanescent region occurs at $D \approx 160$ nm. The agreement of the computed decay (curve α/k_0 in

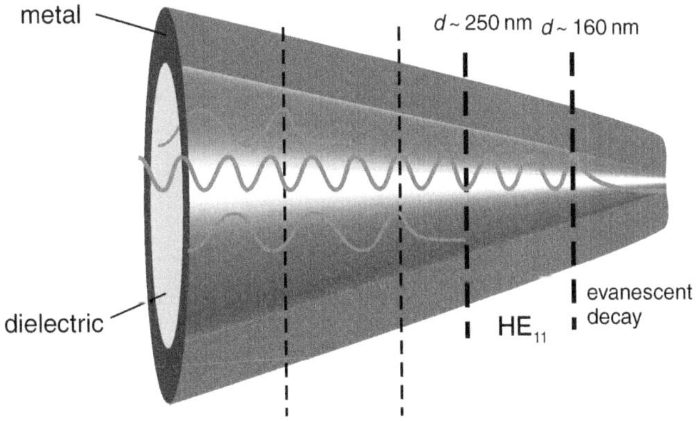

Fig. 6.7 Illustration of the successive cut-off of guided modes and exponential decay of the fields towards the aperture in a tapered, metal-coated waveguide. Adapted from [16].

Fig 6.11) and the power decay obtained by Eq. (6.4) is dependent on the lower integration limit z_0. Excellent fits are obtained if z_0 is chosen to be in the evanescent region of the HE_{11} mode, where $\alpha_{11}(z)$ is well described by an exponential function

$$\alpha_{11}(D) = \mathrm{Im}\{n_{\mathrm{coat}}\} k_0 \mathrm{e}^{-AD}, \tag{6.6}$$

where n_{coat} is the index of refraction of the metal coating, $k_0 = 2\pi/\lambda$ is the propagation constant in free space, and A is a constant determined to be 0.016 nm^{-1} in the present example (cf. Fig. 6.11). If Eq. (6.6) is inserted into Eq. (6.4) and the integration in the exponent is carried out, we arrive at

$$P(z) = P(z_0)\exp[a - b(\mathrm{e}^{2Az\tan\delta})] \tag{6.7}$$

with the two constants

$$a = \frac{\mathrm{Im}\{n_{\mathrm{coat}}\} k_0}{A\tan\delta}\mathrm{e}^{-AD_0}, \qquad b = \frac{\mathrm{Im}\{n_{\mathrm{coat}}\} k_0}{A\tan\delta}\mathrm{e}^{-AD_a},$$

where D_0 is the core diameter at $z = z_0$. The analysis above is valid for a δ that is not too large since reflections in the probe were neglected. This also explains the deviation of curve b in Fig. 6.10, where z_0 was chosen to be in the propagating region of the probe.

The mode-matching analysis outlined above can be simplified if a perfectly conducting metal coating is assumed. In this case, the propagation constant k_z of the lowest-order TE_{11} mode can be calculated as

$$k_z(D) = \sqrt{\varepsilon_{\mathrm{core}} k_0^2 - (3.68236/D)^2}, \tag{6.8}$$

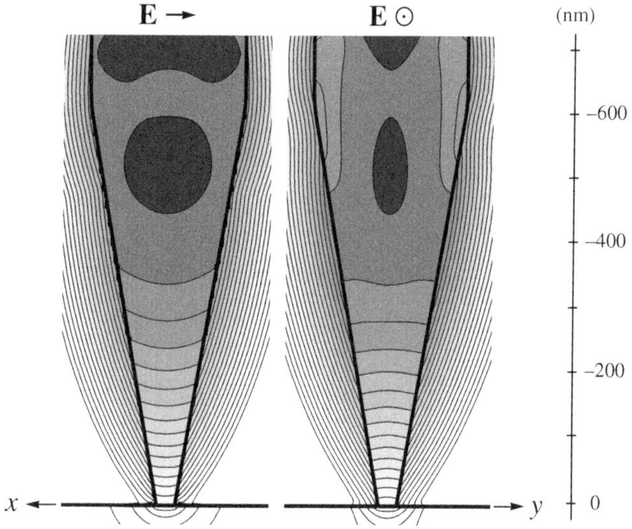

Fig. 6.8 Contours of constant power density on two perpendicular planes, parallel and perpendicular to the electric field, through the center of an aperture probe with infinitely thick coating (with a factor of 3 difference between adjacent lines). The field is excited by the HE_{11} mode incident from the cylindrical part.

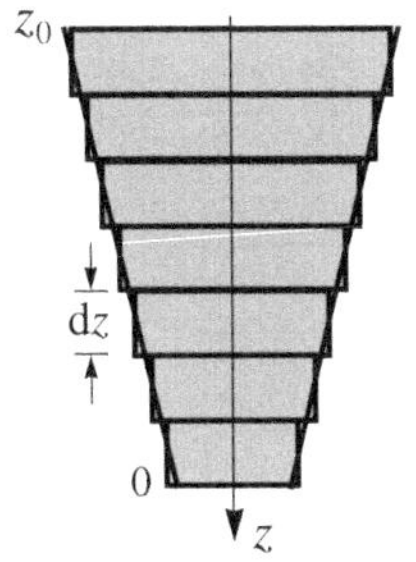

Fig. 6.9 Mode-matching approximation for the power $P(z)$ in the aperture probe. In each waveguide section the attenuation of the HE_{11} mode is calculated analytically. The contributions of all sections are added and the limit $dz \rightarrow 0$ is applied.

with $\varepsilon_{\rm core}$ being the dielectric constant of the core. For large core diameters D the propagation constant is real and the TE_{11} mode propagates without attenuation. However, for diameters $D < 0.586\lambda\sqrt{\varepsilon_{\rm core}}$ the propagation constant becomes purely imaginary and the waveguide mode decays exponentially in the z-direction. Therefore, in the attenuated regime, we can write

$$\alpha_{11}(D) = \sqrt{(3.68236/D)^2 - \varepsilon_{\rm core}k_0^2}, \tag{6.9}$$

which can be inserted into Eq. (6.4). A similar analysis has been carried out by Knoll and Keilmann for a perfectly conducting aperture probe with a square cross-section [20].

The throughput of the aperture probe also depends strongly on the taper angle. As the half-cone angle δ is increased the spotsize will become larger because more and more radiation penetrates through the edges of the aperture. Surprisingly, the spotsize remains almost constant over a large range of δ and increases rapidly for $\delta > 50°$ [21]. However, as shown in Fig. 6.12 the power transmission behaves very differently. A strong variation is observed in the range between 10° and 30°. The data points in the figure are calculated by three-dimensional computations for a probe with an aperture diameter of 20 nm and excitation at $\lambda = 488$ nm. The solid line, on the other hand, is calculated according to mode-matching theory, i.e. by using Eqs. (6.4)–(6.7). The analysis leads to

$$\frac{P_{\rm out}}{P_{\rm in}} \propto {\rm e}^{-B\cot\delta}, \tag{6.10}$$

with B being a constant. While the above theory leads to a value of $B = 3.1$, the best fit to the numerical results is found for $B = 3.6$. Figure 6.12 shows that the agreement is excellent for $10° < \delta < 50°$. The deviation above 50° is mainly due to neglected reflections in the mode-matching model. Changing the taper angle from 10° to 45° increases the power throughput by *nine* orders of magnitude while the spotsize remains almost unaffected. Thus, methods that produce sharp fiber tips with large taper angles are of utmost importance.

However, the cut-off of light propagation in hollow metal waveguides is not always disadvantageous. For example, the rapidly decaying field inside a waveguide with diameter below cut-off has been used for single-molecule studies [22]. Such *zero-mode waveguides* typically consist of holes of diameter ~70 nm fabricated into a metal film ~100 nm thick deposited on a glass substrate. When the film is irradiated from the glass side, the field

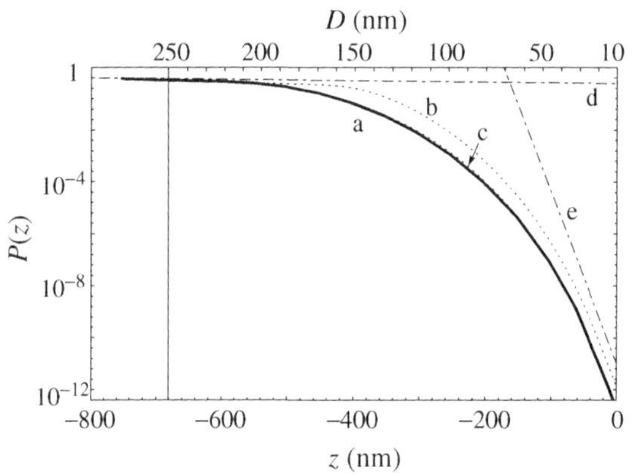

Fig. 6.10 Power decay in an infinitely coated aperture probe as a function of distance from the aperture z and of the core diameter D. Curve a, computed decay; curve b, mode-matching approximation with $z_0 = -600$ nm; curve c, mode-matching approximation with $z_0 = -400$ nm; curve d, decay of the HE_{11} mode in the cylindrical part of the probe; curve e, decay of a wave inside bulk aluminum. The vertical line indicates the transition from the cylindrical to the tapered part of the probe.

inside the holes decays exponentially, which defines very high field confinement. Typically, the observation volume within a single zero-mode waveguide is as low as ~20 zeptoliters. The interior of the zero-mode waveguide can be functionalized in order to capture target molecules from a solution. The small observation volume ensures that single molecules can be detected and monitored with high sensitivity even at very high analyte concentrations.

6.3.2 Field distribution near small apertures

To understand light–matter interactions near aperture probes we need a model for the field distribution near subwavelength-sized apertures. In classical optics, the Kirchhoff approximation is often applied to study the diffraction of light by an aperture in an infinitely thin, perfectly conducting screen. The Kirchhoff approximation assumes that the field inside the aperture is the same as the excitation field in the absence of the aperture. Of course, this assumption fails near the edges of the aperture, and consequently the Kirchhoff approximation becomes inaccurate for small apertures. For an aperture considerably smaller than the wavelength of the exciting radiation it is natural to consider the fields in the electrostatic limit. Unfortunately, for a wave at normal incidence the fields in the electrostatic limit become identically zero because the exciting electric field consisting of a superposition of incident and reflected waves disappears at the surface of the metal screen. Therefore, the electric field has to be calculated by using a first-order perturbative approach. On the other hand, it is possible to derive a solution of the magnetostatic problem.

In 1944 Bethe derived an analytical solution for the electromagnetic field near a small aperture [23]. He also showed that in the far-field the emission of the aperture is equal to the radiation of a magnetic and an electric dipole located at the center of the aperture. The electric dipole is excited only if the exciting plane wave is incident from an oblique angle.

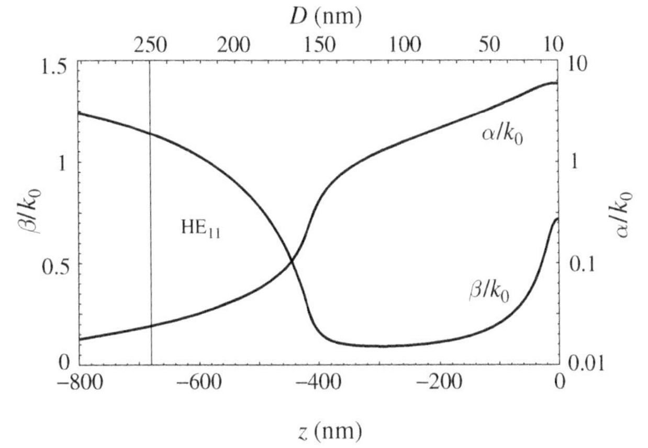

Fig. 6.11 The attenuation constant α_{11} and phase constant β_{11} of the cylindrical HE_{11} mode as functions of the core diameter D. z is the corresponding distance from the aperture. The vertical line indicates the transition from the cylindrical to the tapered part of the aperture probe. From [19].

In 1950 Bouwkamp revealed that the electric field derived by Bethe is discontinuous in the hole, contrary to what is required by the boundary conditions [24].

To derive the correct solution, Bouwkamp first calculated the solution for a disk and then used Babinet's principle to obtain the magnetic currents for the case of the aperture. The solution is derived from an integral equation containing the current distribution function on the disk as an unknown function. The integral equation is then solved using a series-expansion method and making use of the singularity condition at the rim of the disk. This condition states that the electric field component tangential to the edge of the disk must vanish as the square root of the distance from it. Furthermore, the electric field component normal to the edge must become infinite as the inverse square root of the distance from the edge. This boundary condition had already been used by Sommerfeld in the study of diffraction by a semi-infinite metal plate. An alternative approach for solving the fields near a small disk can be found in Ref. [25].

Babinet's principle is equivalent to replacing the electric currents and charges induced in the metal screen by magnetic currents and charges located in the aperture. The magnetic surface current density $\mathbf{K}$ and magnetic charge density η in the aperture give rise to a magnetic vector potential $\mathbf{A}^{(\mathrm{m})}$ and a magnetic scalar potential $\Phi^{(\mathrm{m})}$ as

$$\mathbf{A}^{(\mathrm{m})} = \varepsilon_0 \int \mathbf{K} \frac{\mathrm{e}^{ikR}}{4\pi R}\,\mathrm{d}S, \qquad \Phi^{(\mathrm{m})} = \frac{1}{\mu_0} \int \eta \frac{\mathrm{e}^{ikR}}{4\pi R}\,\mathrm{d}S, \tag{6.11}$$

where $R = |\mathbf{r} - \mathbf{r}'|$ denotes the distance between the source point $\mathbf{r}'$ and the field point $\mathbf{r}$, and the integration runs over the surface of the aperture. Similarly to the electric case, $\mathbf{A}^{(\mathrm{m})}$ and $\Phi^{(\mathrm{m})}$ are related to the electric and magnetic fields as

$$\mathbf{E} = \frac{1}{\varepsilon_0} \nabla \times \mathbf{A}^{(\mathrm{m})}, \qquad \mathbf{H} = \mathrm{i}\omega \mathbf{A}^{(\mathrm{m})} - \nabla \Phi^{(\mathrm{m})} \approx -\nabla \Phi^{(\mathrm{m})}. \tag{6.12}$$

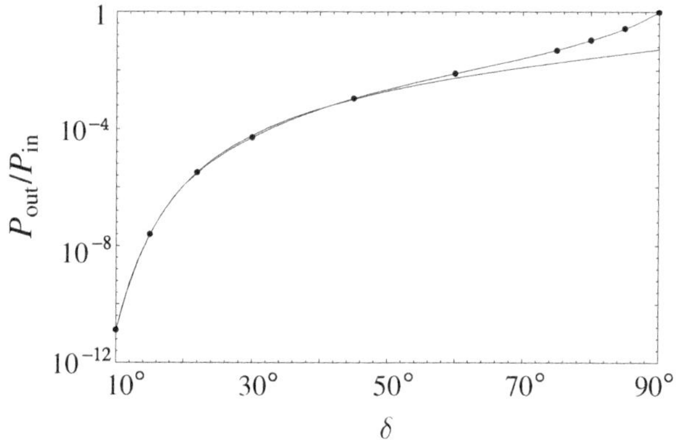

Fig. 6.12 Dependence of power transmission on taper angle (δ is the half-cone angle). The aperture diameter is 20 nm and the wavelength $\lambda = 488$ nm. Changing the taper angle from 10° to 45° increases the power throughput by nine orders of magnitude. Results from three-dimensional computation (points) and according to Eq. (6.10) with a value of $B = 3.6$ (solid line) are shown.

In what follows, we neglect the first term in the expression for $\mathbf{H}$ because it is proportional to $k = \omega/c$ and therefore negligible in the limit of a small aperture a ($ka \ll 1$).

To solve for $\mathbf{A}^{(\mathrm{m})}$ and $\Phi^{(\mathrm{m})}$ it is convenient to introduce oblate-spheroidal coordinates $\mathbf{r} = (u, v, \varphi)$ defined by

$$z = auv, \qquad x = a\sqrt{(1-u^2)(1+v^2)}\cos\varphi, \qquad y = a\sqrt{(1-u^2)(1+v^2)}\sin\varphi, \quad (6.13)$$

where $0 \le u \le 1$, $-\infty \le v \le \infty$, $0 \le \varphi \le 2\pi$. The surfaces $v = 0$ and $u = 0$ correspond to the aperture and the screen, respectively.

Plane wave at normal incidence

For a plane wave at normal incidence, the Laplace equation $\nabla^2\Phi^{(\mathrm{m})} = 0$ yields the solution

$$\Phi^{(\mathrm{m})} = -H_0 \frac{2a}{\pi} P_1^1(u) Q_1^1(iv) \sin\varphi, \qquad (6.14)$$

where P_n^m and Q_n^m are associated Legendre functions of the first and second kind, respectively [26], and E_0 and $H_0 = E_0\sqrt{\varepsilon_0/\mu_0}$ are the magnitudes of the electric and magnetic fields of the incident plane wave polarized in the x-direction ($\varphi = 0$). The solution for the magnetic vector potential $\mathbf{A}^{(\mathrm{m})}$ is much more difficult to derive since it cannot be calculated statically. The expression derived by Bouwkamp reads as

$$\begin{aligned} A_x^{(\mathrm{m})} &= -\varepsilon_0 E_0 \frac{ka^2}{36\pi} P_2^2(u) Q_2^2(iv)\sin(2\varphi), \\ A_y^{(\mathrm{m})} &= \varepsilon_0 E_0 \frac{ka^2}{36\pi}\left[-48 Q_0(iv) + 24 P_2(u) Q_2(iv) + P_2^2(u) Q_2^2(iv)\cos(2\varphi)\right], \end{aligned} \qquad (6.15)$$

and is different from Bethe's previous calculation.

The electric and magnetic fields are now easily derived by substituting $\Phi^{(m)}$ and $\mathbf{A}^{(m)}$ into Eq. (6.12). The electric field becomes

$$
\begin{aligned}
E_x/E_0 &= \mathrm{i}kz - \frac{2}{\pi}\mathrm{i}kau\left[1 + v\arctan v + \frac{1}{3}\frac{1}{u^2+v^2} + \frac{x^2-y^2}{3a^2(u^2+v^2)(1+v^2)^2}\right],\\
E_y/E_0 &= -\frac{4\mathrm{i}kxyu}{3\pi a(u^2+v^2)(1+v^2)^2},\\
E_z/E_0 &= -\frac{4\mathrm{i}kxv}{3\pi(u^2+v^2)(1+v^2)},
\end{aligned}
\tag{6.16}
$$

and the magnetic field turns out to be

$$
\begin{aligned}
H_x/H_0 &= -\frac{4xyv}{\pi a^2(u^2+v^2)(1+v^2)^2},\\
H_y/H_0 &= 1 - \frac{2}{\pi}\left[\arctan v + \frac{v}{u^2+v^2} + \frac{v(x^2-y^2)}{\pi a^2(u^2+v^2)(1+v^2)^2}\right],\\
H_z/H_0 &= -\frac{4ayu}{\pi a^2(u^2+v^2)(1+v^2)}.
\end{aligned}
\tag{6.17}
$$

By evaluating the electric and magnetic fields on the metal screen it is straightforward to solve for the *electric* charge density σ and the *electric* surface current density $\mathbf{I}$ as

$$
\begin{aligned}
\sigma(\rho,\phi) &= \varepsilon_0 E_0 \frac{8\mathrm{i}}{3} ka \frac{a/\rho}{\sqrt{\rho^2/a^2-1}}\cos\phi,\\
\mathbf{I}(\rho,\phi) &= H_0\frac{\mathbf{n}_\rho}{\pi^2}\left[\arctan\left(\sqrt{\rho^2/a^2-1}\right) + \frac{a}{\rho}\sqrt{1-a^2/\rho^2}\right]\cos\phi\\
&\quad - H_0\frac{\mathbf{n}_\phi}{\pi^2}\left[\arctan\left(\sqrt{\rho^2/a^2-1}\right) + \frac{1+a^2/\rho^2}{\sqrt{\rho^2/a^2-1}}\right]\sin\phi.
\end{aligned}
\tag{6.18}
$$

Here, a point on the metal screen is defined by the polar coordinates (ρ,ϕ), and $\mathbf{n}_\rho$ and $\mathbf{n}_\phi$ are the radial and azimuthal unit vectors, respectively. It is important to notice that the current density is independent of the parameter ka, indicating that it is equal to the magnetostatic current for which $\nabla\cdot\mathbf{I} = 0$. On the other hand, the charge density is proportional to ka and therefore cannot be derived from electrostatic considerations. At the edge of the aperture ($\rho = a$) the component of the current normal to the edge vanishes whereas the tangential component of the current and the charge density become infinitely large.

The fields determined above are valid only in the vicinity of the aperture, i.e. within a distance $R \ll a$. To derive expressions for the fields at larger distance one can calculate the spatial spectrum of the fields in the aperture plane and then use the angular spectrum representation to propagate the fields [27]. However, as shown in Problem 3.5 this approach does not correctly reproduce the far-fields because the near-field is correct only up to order ka, whereas the far-field requires orders up to $(ka)^3$. Bouwkamp calculates the fields in the aperture up to order $(ka)^5$ [28]. These fields are sufficiently accurate to be used in an angular spectrum representation that is valid from near-field to far-field.

Bethe and Bouwkamp show that the far-field of a small aperture is equivalent to the far-field of a radiating magnetic dipole located in the aperture and with axis along the

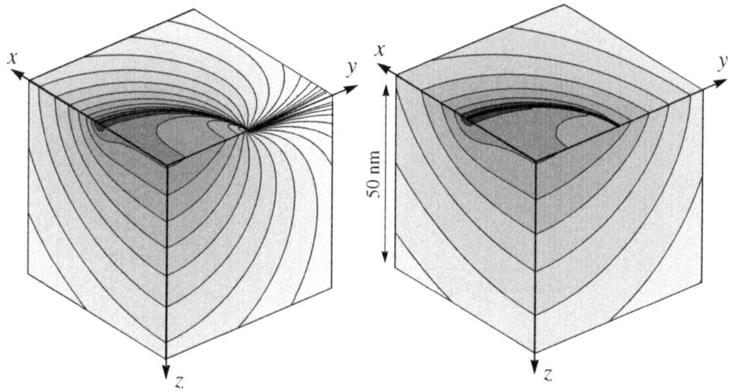

Fig. 6.13 Comparison between Bouwkamp's solution (left) and the fields in front of an aperture probe with aluminum coating ($\lambda = 488$ nm) (right). Contours of constant $|\mathbf{E}|^2$ (with a factor of 2 difference between adjacent lines). The incident polarization is along the x-axis.

negative y-direction, i.e. opposite to the magnetic field vector of the incident plane wave. The magnetic dipole moment $\mathbf{m}$ turns out to be

$$\mathbf{m} = -\frac{8}{3}a_0^3\mathbf{H}_0. \tag{6.19}$$

It scales with the third power of a_0, indicating that the aperture behaves like a three-dimensional polarizable object.

Plane wave at arbitrary incidence

Bouwkamp derived the fields for a small disk irradiated by a plane wave with arbitrary incidence [28]. Using Babinet's principle it is straightforward to translate the solution to the case of an aperture. It turns out that the far-field is no longer equivalent to the radiation of a magnetic dipole alone. Instead, the electric field also induces an electric dipole oriented perpendicular to the plane of the aperture and antiparallel to the driving field component. Thus, the far-field of a small aperture irradiated by an arbitrary plane wave is given by the radiation of an electric dipole and a magnetic dipole with the following moments [23]:

$$\boldsymbol{\mu} = -\frac{4}{3}\varepsilon_0 a_0^3\left[\mathbf{E}_0\cdot\mathbf{n}_z\right]\mathbf{n}_z, \qquad \mathbf{m} = -\frac{8}{3}a_0^3\left[\mathbf{n}_z\times(\mathbf{E}_0\times\mathbf{n}_z)\right], \tag{6.20}$$

with $\mathbf{n}_z$ being the unit vector normal to the plane of the aperture pointing in the direction of propagation.

Bethe–Bouwkamp theory applied to aperture probes

Figure 6.13 compares the near-fields behind the aperture probe and the ideal aperture. The fields look very similar at first glance but there are significant differences. The field of the ideal aperture is singular at the edges in the plane of polarization and zero along the y-axis outside the aperture. This is not the case for an aperture probe with a metal coating of finite conductivity. The Bouwkamp approximation further shows higher confinement of the fields and much higher field gradients, which would lead, if they were real, for instance, to larger forces being exerted on particles next to the aperture. Notice that the infinitely conducting and infinitely thin screen used in the Bethe–Bouwkamp theory is a strong idealization. At optical frequencies, the best metals have skin depths of 6–10 nm, which will enlarge the effective aperture size and smooth out the singular fields at the edges. Furthermore, any realistic metal screen will have a thickness of at least $\lambda/4$. The exciting field of the aperture is therefore given by the waveguide mode in the hole rather than by a plane wave.

An ideal aperture radiates as a coherent superposition of a magnetic and an electric dipole [23]. In the case of an ideal aperture illuminated by a plane wave at normal incidence the electric dipole is not excited. However, the fields in the aperture of a realistic probe are determined by the exciting waveguide mode. A metal coating with finite conductivity always gives rise to an exciting electric field with a net forward component in the plane of the aperture. One therefore might think that a vertical dipole moment must be introduced. However, since such a combination of dipoles leads to an asymmetric far-field, it is not a suitable approximation. Also, the magnetic dipole alone gives no satisfactory correspondence to the radiation of the aperture probe. Obermüller and Karrai proposed an electric and a magnetic dipole, both lying in the plane of the aperture and perpendicular to each other [29]. This configuration fulfills the symmetry requirements for the far-field radiation and is in good agreement with experimental measurements.

6.3.3 Field distribution near aperture probes

Figure 6.14 shows the fields in the aperture region of an aperture probe in vacuum and above a dielectric substrate. The coating is tapered towards the aperture and the final thickness is 70 nm. The aperture diameter is chosen to be 50 nm, and the exciting HE_{11} mode is polarized along the x direction.

Part of the field penetrates the edges of the aperture into the metal thereby increasing the effective width of the aperture. When a dielectric substrate is brought towards the aperture the power transmission through the probe increases. This can be seen in Fig. 6.14 by comparing the contour lines in the probe. Part of the emitted field is scattered around the probe and couples to external surface modes propagating backwards along the coating surface.

External surface modes can also be excited in the forward direction by the field transmitted from the core through the coating. In analogy to cylindrical waveguides they have hardly any attenuation [19]. Most of the energy associated with these modes therefore propagates towards the aperture plane. If the coating chosen is too thin it may happen that the light from the surface of the coating is stronger than the light emitted by the aperture. In this case the field is strongly enhanced at the outer edges of the coating, leading to the field pattern shown in Fig. 6.15 (right panel). To avoid such an unfavorable situation a sufficiently thick coating has to be chosen. A tapered coating could be a reasonable way to reduce the coating thickness near the aperture. It has to be emphasized that surface modes cannot be excited by illumination from outside since they possess propagation constants

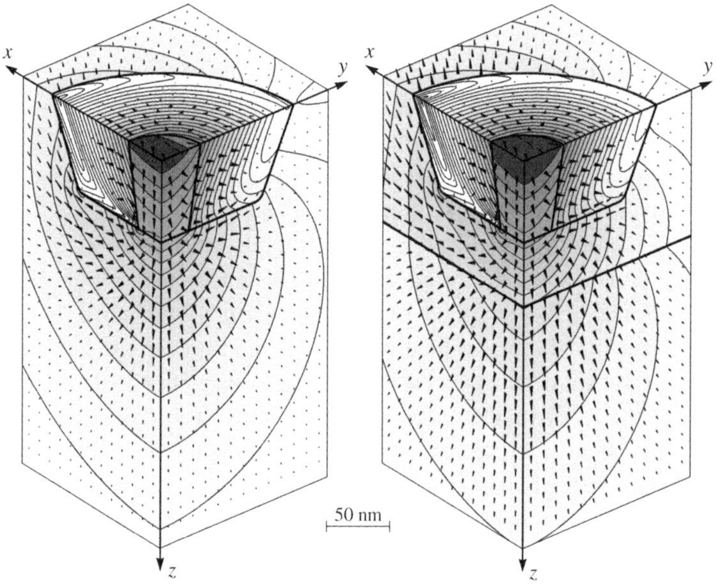

Fig. 6.14 Contours of constant $|\mathbf{E}|^2$ on three perpendicular planes near the foremost end of an aperture probe (with a factor of 2 difference between successive lines). The arrows indicate the time-averaged Poynting vector. The incident polarization is in the plane $y = 0$. The transmission through the probe is increased when a dielectric substrate ($\varepsilon = 2.25$) is brought close (right figure).

that are larger than the k-vector of freely propagating light, being in this regard similar to surface plasmons (see Chapter 12).

The Bethe–Bouwkamp theory has been used by various authors to approximate the near-field of aperture probes. Single-molecule experiments have shown a good qualitative agreement [30] and are the perfect tool with which to analyze the field distribution of a given aperture (see Chapter 9).

6.3.4 Enhancement of transmission and directionality

Ebbesen and coworkers have demonstrated that the transmission through a metal screen with subwavelength-sized holes can be increased if a periodic arrangement of holes is used [31]. The effect originates from the constructive interference of scattered fields at the irradiated surface of the metal screen and thus depends strongly on the excitation wavelength. The periodic arrangement of holes increases the energy density on the surface of the metal screen through the creation of standing surface waves. These surface waves delocalize the energy of the aperture and are responsible for the enhanced transmission. One can view the enhanced transmission of an aperture array as an antenna problem: while the radiation from a single aperture is $P_{\rm rad} \propto |\mathbf{m}|^2$, with $\mathbf{m}$ being the induced magnetic dipole (Eq. 6.19), the radiation from an aperture array can be as high as $P_{\rm rad} \propto N^2|\mathbf{m}|^2$, where N

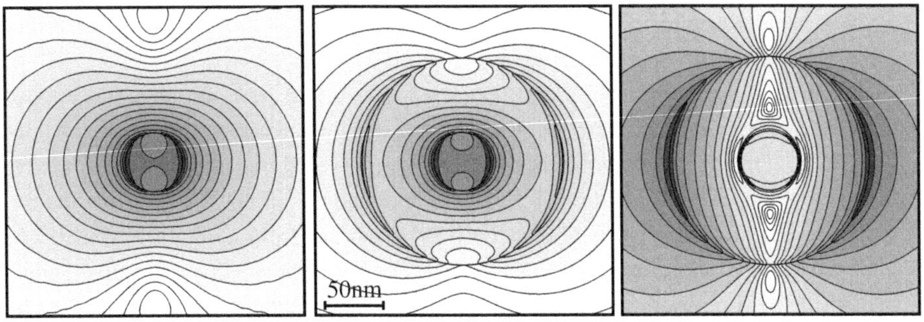

Fig. 6.15 Contours of constant $|\boldsymbol{E}|^2$ (with a factor of $3^{1/2}$ difference between successive lines) in the aperture planes of three aperture probes with different coating thicknesses. Left: infinite coating. Middle: finite coating, the field is dominated by the flux emitted by the aperture. Right: finite coating, the field is dominated by the flux from the outside coating surface.

is the total number of apertures. The multiplication with N^2 is an interference effect that is exploited, for example, in phased-array antennas.

The enhanced transmission in a periodically perforated metal screen was at first ascribed to the creation and interference of surface plasmons until it was pointed out that the same effect persists in an ideal metal that does not support any surface modes. The debate was resolved by realizing that a periodically perforated ideal metal acts as an *effective medium* supporting surface modes that "mimic" surface plasmons encountered on noble-metal surfaces [32]. Thus, even though an ideal metal cannot support any "bound" surface modes, it is the periodic arrangement of holes that helps the ideal metal to mimic a noble metal. Within the effective-medium framework, Pendry and coworkers derived the following dispersion relation for a perforated metal screen [32]:

$$k_{\|}(\omega) = \frac{\omega}{c}\sqrt{1 + \frac{64a^4}{\pi^4 d^4}\frac{\omega^2}{\omega_{\rm pl}^2 - \omega^2}}. \tag{6.21}$$

Here, $k_{\|}$ represents the propagation constant along the surface of the perforated metal screen, c is the vacuum speed of light, a is the hole diameter, and d is the hole spacing. The plasma frequency $\omega_{\rm pl}$ of the effective medium is defined as

$$\omega_{\rm pl} = \frac{\pi c}{a\sqrt{\varepsilon\mu}}, \tag{6.22}$$

with ε and μ being the material constants of the material filling the holes. Equation (6.21) is similar to the familiar dispersion relation of surface plasmons supported by a Drude metal (see Chapter 12). However, whereas for a Drude metal the plasmon resonance ($k_{\|} \to \infty$) occurs at a lower frequency than the plasma frequency, the plasmon resonance for the perforated metal screen is identical with the plasma frequency $\omega_{\rm pl}$. The interesting outcome is that it is possible to simulate real surface plasmons by a perforated metal screen and that the dispersion relation can be tailored by the hole size and the hole periodicity. Notice that the periodicity of the holes implies a periodicity of $2\pi/d$ in the dispersion relation as in the theory of photonic crystals or the electronic theory of semiconductors. This property is not reflected in Eq. (6.21), implying that it is impossible to reach the surface plasmon resonance $k_{\|} \rightarrow \infty$.

In similar experiments, Lezec and coworkers have used a single aperture with a concentric microfabricated grating to delocalize the radiation in the near-zone of the aperture [33]. This delocalization leads to either an increased transmission or improved directionality of the emitted radiation. To better understand this effect, we note that the theory of Bethe and Bouwkamp predicts that the light emerging from a small irradiated aperture propagates in all directions. The smaller the aperture, the stronger the divergence of the radiation will be. A significant portion of the electromagnetic energy does not propagate and stays "attached" to the back surface of the aperture. This energy never reaches a distant observer (see Fig. 6.16(a)). With the help of a concentric grating, Lezec and coworkers convert the non-propagating near-field into propagating fields that can be seen by a distant observer (see Fig. 6.16(b)). Because the grating at the exit plane artificially increases the radiating area it also destroys the light confinement in the near-field, which limits applications in near-field optical microscopy. However, if the grating is placed on the front side of the aperture, the light throughput can be strongly increased. In this case the gratings redirect the radiation hitting the opaque screen towards the aperture, thereby enhancing the light intensity at the aperture.

6.4 Fabrication of aperture probes

In order to create aperture probes in the laboratory, dielectric tips have to be metal-coated. Among all of the metals, aluminum has the smallest skin depth in the visible spectrum [34]. Coating of dielectric tips with aluminum can be done e.g. by thermal evaporation, electron-beam (e-beam)-assisted evaporation or sputtering. Thermal and e-beam evaporation have the advantage of being directed processes. Sputtering, on the other hand, is an isotropic process. All surfaces, even of complex bodies, will be coated

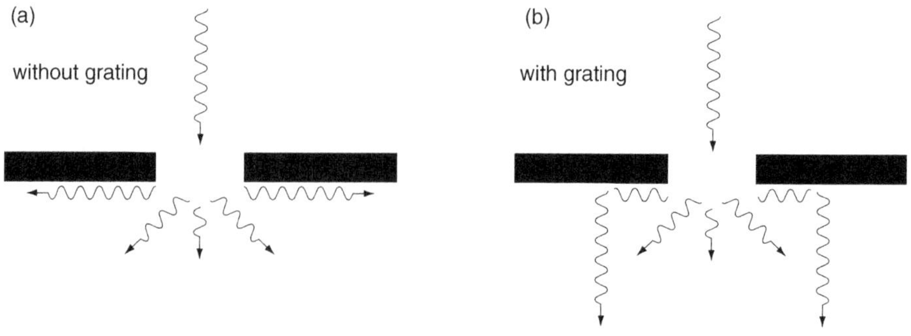

Fig. 6.16 Improving the directionality of light emission by use of a grating fabricated on the exit side of a small aperture. (a) Without the grating radiation diffracts into all directions. (b) The grating delocalizes the near-field and converts it into directional radiation.

at the same rate. The formation of apertures at the apex of fiber tips can be accomplished by exploiting the shadowing effect supported by thermal and e-beam evaporation. In this process, the tips are positioned and oriented such that the stream of metal vapor hits the tip at an angle. At the same time the tips are being rotated. The deposition rate of metal at the tip apex is then much smaller than on the sides, which leads to the formation of an aperture at the apex as illustrated in Fig. 6.17.

Evaporation and sputtering suffer from the tendency of aluminum to form rather large grains. These grains have a typical size of about 100 nm and can be observed very well with a focused ion-beam microscope (see e.g. [35]). Figure 6.18 shows images of apertures formed by focused ion-beam milling and by tip shadowing during evaporation. The presence of grains is manifested by the faceted surface texture of the metal coating. The grain formation in aluminum films is unfavorable for two reasons: (i) leakage of light at grain boundaries and related imperfections can occur, which interferes with the weak wanted emission at the apex; and (ii) the optical apertures are rather ill-defined since the aperture size is usually smaller than the average grain size. Grains, as illustrated in Fig. 6.18(b), also prevent the actual optical aperture from approaching close to the sample because of protruding particles. E-beam evaporation often produces smoother aluminum coatings than does thermal evaporation.

The small amount of light that is emitted by a near-field aperture is a limiting factor in experiments. Therefore one is tempted to increase the input power at the fiber far end. However, aperture probes can be destroyed by too strong an illumination. This happens because of the pronounced energy dissipation in the metal coating, which, as a consequence, is strongly heated. Temperature measurements along a taper of aluminum-coated fiber probes have been performed (see e.g. [36]), and showed that the strongest heating occurs far away from the tip in the upper part of the taper, consistently with the discussion in Section 6.3.1. Here temperatures of several hundred degrees Celsius can be reached for input powers up to 10 mW. For larger input powers the aluminum coating breaks down, which leads to a strong increase of light emission from the structure. Breakdown usually happens either by

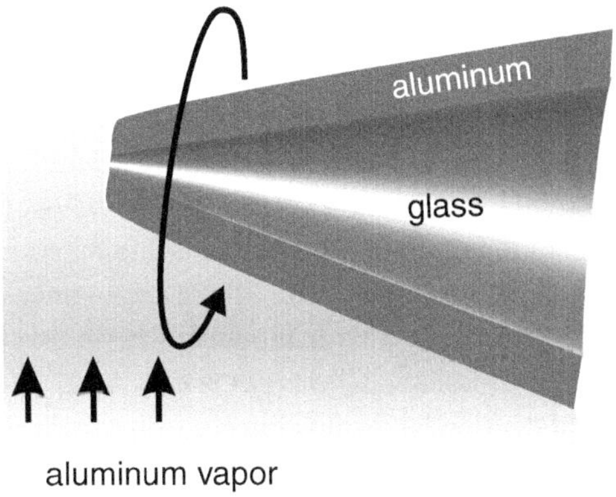

Fig. 6.17 Self-aligned formation of an aperture by thermal evaporation. The evaporation takes place at an angle slightly from behind while the tip is being rotated. Adapted from [16].

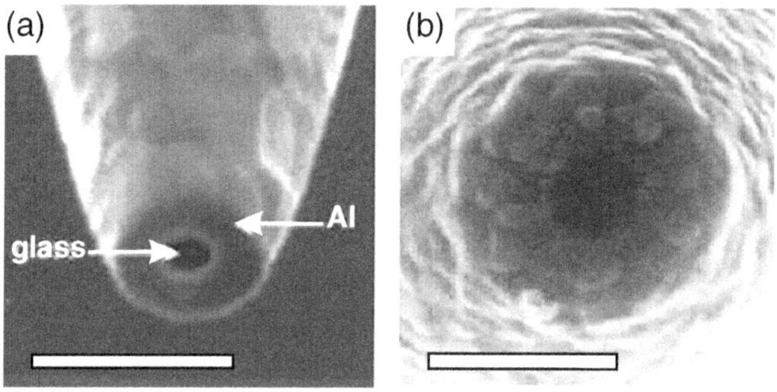

Fig. 6.18 Aluminum-coated aperture probes. (a) Aperture formed by cutting off the apex in a focused ion beam apparatus. Image courtesy of N. F. van Hulst. (b) Aperture formed by thermal evaporation and tip shadowing. From [16]. Scale bars: 300 nm.

straightforward melting of the aluminum layer or by fracture and subsequent rolling up of the metal sheets due to internal stress.

6.4.1 Aperture formation by focused-ion-beam milling

The availability of high-resolution focused-ion beams creates new possibilities for micromachining with nanometer-scale resolution [37]. Current focused-ion-beam (FIB) instruments operate with liquid-metal sources. To ensure a constant supply of ions for the beam, a tungsten coil with a tip [37] is wetted by gallium or indium, which is then field ionized and accelerated towards the sample. Using conventional electromagnetic lenses as in SEM, such an ion beam can be focused down to a diameter of ∼10 nm. At an ion flux of ∼11 pA at 30 kV, aluminum can be locally removed. The ablated material can be chemically analyzed using mass spectrometry [37]. At much lower ion flux (1 pA), the micromachined structure can be inspected with nearly negligible material ablation. Modern FIB microscopes combine an ion column and an electron column in the same machine.

The standard procedure of probe processing by FIB is to cut conventional aluminum-coated probes by slicing them perpendicular to the optical axis [38]. Depending on where the cut is performed, either an existing aperture can be smoothed and improved by removing protruding grains or a closed tip can be opened to any desired aperture radius. An example of the result of such micromachining is shown in Fig. 6.18(a). FIB-treated probes have superior performance since there are no grains to prevent the probe from coming very close to the sample. This is a prerequisite to exploiting the full confinement of the optical near-field. Using single molecules as local field probes, it was found that the optical near-field distribution could be recorded reproducibly and that it very much resembles the fields of a Bethe–Bouwkamp aperture [38]. For conventional non-smoothed apertures such patterns were observed very rarely [30] and could not be reproduced before the advent of FIB-treated optical probes. One challenge that is encountered when using FIB-milled apertures is the adjustment of the aperture plane parallel to the sample surface. Typically, the lateral size of the probe is up to 1 μm and, to ensure high resolution, its aperture has to be placed as close as 5–10 nm from the sample surface.

A particular strength of FIB milling is that it provides the possibility to micromachine prototype structures at the apices of tips that are more complex than simple apertures. This can lead to improved probe structures with very high field confinement and strong enhancement (see Section 6.5).

6.4.2 Alternative aperture-formation schemes

Various other techniques for aperture probe fabrication have been explored. Here we review two electrochemical methods and then a mechanical method.

Electrochemistry is usually performed in liquid environments, which is problematic in its application to micromachining. In the presence of a liquid, in general large areas are wetted and nanometer-scale material processing cannot be achieved. However, there exist solid electrolytes that allow significant transport of metal ions in the solid phase. Such electrolytes have been used to perform controlled all-solid-state electrolysis (CASSE). A prominent solid electrolyte for silver ions is amorphous silver metaphosphate iodide ($AgPO_3$:AgI) [39], which is appreciated for its high ionic conductivity, optical transparency, and ease of fabrication [40]. The aperture formation is induced by carefully contacting a fully silver-covered tapered transparent tip with the flat solid electrolyte and transferring silver ions from the tip to the solid electrolyte by means of an applied voltage until an aperture is formed.

Another electrochemical method that actually involves light-induced corrosion of aluminum was introduced in Ref. [41]. In this approach, an aperture is produced in the metal layer at the probe apex by a simple, one-step, low-power, laser-thermal oxidation process in water. The apex of a tip is locally heated due to the absorption of light from an evanescent field created by total internal reflection at a glass/water interface. Owing to the heating, the passivation layer that normally covers aluminum is dissolved in an aqueous environment.

The formation of small apertures by nanomechanical interaction of a probe with a sample is another means by which to obtain subwavelength apertures. Aperture punching,

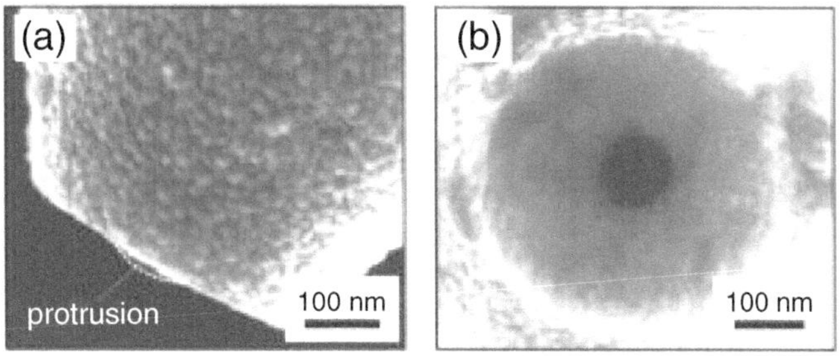

Fig. 6.19 Scanning electron micrographs of (a) a side view and (b) a front view of an aperture with a diameter of 100 nm produced by aperture punching. Adapted from [42].

or, in other words, the opening of a small aperture at the apex of a completely metal-coated dielectric tip by plastic deformation of the metal near the apex, was the method used in the very first near-field optical experiments [43]. This method was later adapted and perfected by other groups [14, 42]. Figure 6.19 shows the result of punching an etched optical fiber sputtered with 200 nm of gold. A circular aperture with a flat rim can be observed.

6.5 Optical antenna probes

Optical near-field probes are optical analogs of classical antennas. Briefly, considering the case of a receiving antenna, electromagnetic energy has to be channeled to the near-field zone of the antenna. Vice versa, the energy has to be released from the near-field zone if the antenna is operated in transmission mode. An antenna is a device that establishes efficient coupling between the near-field and the far-field. Although antenna theory has been developed for the radiofrequency and the microwave range of the electromagnetic spectrum it holds great promise for inspiring new concepts in the optical frequency range [43]. Field enhancement is a natural phenomenon in antenna theory. It occurs because an antenna concentrates electromagnetic energy into a tight space, thereby generating a zone of high energy density. In the context of near-field optics one would like to use this property to create a highly confined light source. A simple type of antenna is a pointed tip acting as a lightning-rod antenna.

Optical antennas will be discussed in depth in Chapter 13.

6.5.1 Solid metal tips

Near-field optical microscopy based on local field enhancement was proposed by Synge as early as in 1928, long before the invention of atomic force microscopy [44]. Since then various related implementations have been demonstrated, most of them using a sharp vibrating tip to locally scatter the near-field at the sample surface. Homodyne or heterodyne detection using lock-in techniques is commonly applied to discriminate the small scattered signal from the tip apex against the background from a diffraction-limited illumination area.

It has been shown that under certain conditions a scattering object can also act as a local light source [45, 46]. As discussed before, this light source is established by the field-enhancement effect, which has similar origins as to the lightning-rod effect in electrostatics. Thus, instead of using an object to scatter the sample's near-field, the object is used to provide a local near-field excitation source to record a local spectroscopic response of the sample. This approach enables simultaneous *spectral* and subdiffraction *spatial* measurements, but it depends sensitively on the magnitude of the field-enhancement factor [47]. The latter is a function of wavelength, material, geometry, and the polarization of the exciting light field. Although theoretical investigations have led to an inconsistent spread of values for the field-enhancement factor, these results are consistent with respect to polarization conditions and local field distributions.

Figure 6.20 shows the field distribution near a sharp gold tip in water irradiated by two different monochromatic plane-wave excitations. In Fig. 6.20(a), a plane wave is incident from the bottom with the polarization perpendicular to the tip axis, whereas in Fig. 6.20(b) the tip is illuminated from the side with the polarization parallel to the tip axis. A striking difference is seen for the two different polarizations: in Fig. 6.20(b), the intensity near the tip end is strongly increased over the illuminating intensity, whereas there is no enhancement beneath the tip in Fig. 6.20(a). This result suggests that it is crucial to have a large component of the excitation field along the axial direction in order to obtain a high field enhancement. Calculations for platinum and tungsten tips show lower enhancements, whereas the field beneath a dielectric tip is reduced compared with the excitation field (cf. Section 6.1).

Figure 6.21 shows the induced surface charge density for the two situations shown in Fig. 6.20. The incident light drives the free electrons in the metal along the direction of polarization. While the charge density is zero inside the metal at any instant of time ($\nabla \cdot \mathbf{E} = 0$), charges accumulate on the surface of the metal. When the incident polarization is perpendicular to the tip axis (Fig. 6.20(a)), diametrically opposed points on the tip surface have opposite charges. As a consequence, the foremost end of the tip remains uncharged. On the other hand, when the incident polarization is parallel to the tip axis (Fig. 6.20(b)), the induced surface charge density is axially symmetric and has the highest amplitude at the end of the tip. In both cases the surface charges form oscillating standing waves (surface plasmons) with wavelengths shorter than the wavelength of the illuminating light, indicating that it is essential to include retardation in the analysis.

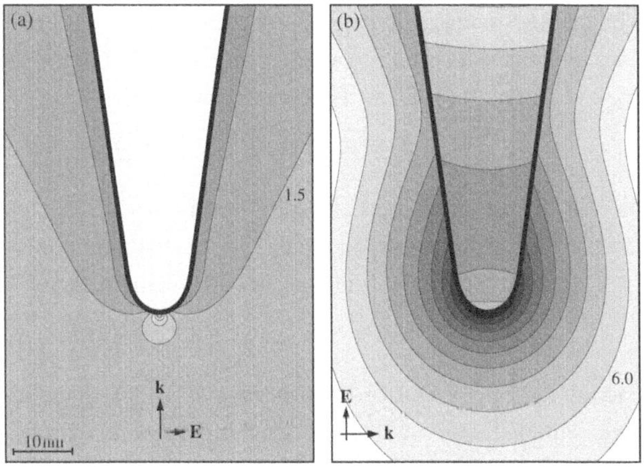

Fig. 6.20 The near-field of a gold tip (5 nm tip radius) in water illuminated by two different monochromatic waves at $\lambda = 810$ nm. The direction and polarization of the incident wave are indicated by the ***k*** and ***E*** vectors. The figures show contours of E^2 (with a factor of 2 difference between successive lines). The fields are almost axially symmetric in the vicinity of the tip.

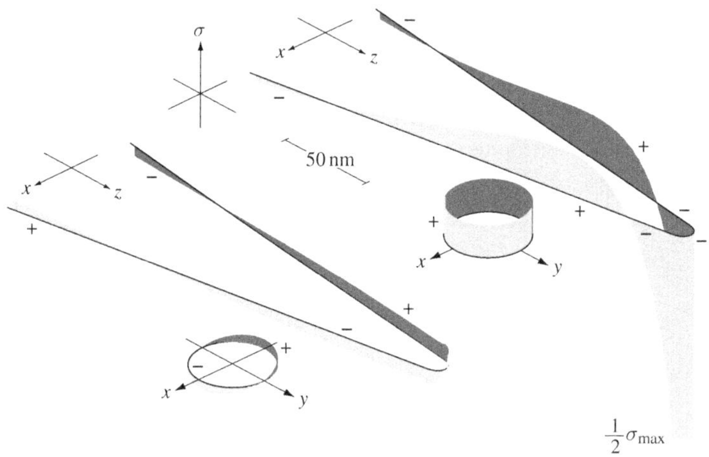

Fig. 6.21 Induced surface charge density σ corresponding to Fig. 6.20(a) (left) and Fig. 6.20(b) (right). The surface charges form an oscillating standing wave in each case. In (a) the surface-charge wave has a node at the end of the tip, whereas in (b) there is a large accumulation of surface charge on the foremost part, which is responsible for the field enhancement.

The magnitude of the field-enhancement factor is crucial for imaging applications. The direct illumination of the sample surface gives rise to a far-field background signal. If we consider an optical interaction that is based on an nth-order nonlinear process and assume that only the sample surface is active, then the far-field background will be proportional to

$$S_{\mathrm{ff}} \sim A I_0^n, \tag{6.23}$$

where A is the illuminated surface area and I_0 is the laser intensity. The signal that we wish to detect and investigate (the near-field signal) is excited by the enhanced field at the tip. If we designate the enhancement factor for the electric field intensity (E^2) by f_{i} then the near-field signal of interest is proportional to

$$S_{\mathrm{nf}} \sim a (f_{\mathrm{i}} I_0)^n, \tag{6.24}$$

where a is a reduced area given by the tip size. If we require that the signal be stronger than the background ($S_{\mathrm{nf}}/S_{\mathrm{ff}} > 1$) and use realistic numbers for the areas ($a = (10\ \mathrm{nm})^2$, $A = (500\ \mathrm{nm})^2$) then we find that an enhancement factor of

$$f_{\mathrm{i}} > \sqrt[n]{2500} \tag{6.25}$$

is required. For a first-order process ($n = 1$), such as scattering or fluorescence, an enhancement factor of three to four orders of magnitude is required, but for a second-order nonlinear process the required enhancement factor is only 50. This is the reason why the first tip-enhanced experiments were performed with two-photon excitation [46]. To maximize the field enhancement various alternative probe shapes and materials have been proposed. It has been found that elongated subwavelength particles exhibit very low radiation damping and therefore provide very high enhancement factors [48, 49]. Even stronger enhancement is found for tetrahedral shapes [46]. It is found that, whatever the magnitude of the enhancement factor is, the field distribution in the vicinity of a sharp tip can be quite accurately described by the fields of an effective dipole $\boldsymbol{p}(\omega)$ located at the center of the tip apex (see Fig. 6.22) and with the magnitude

$$\boldsymbol{p}(\omega) = \begin{bmatrix} \alpha_\perp & 0 & 0 \\ 0 & \alpha_\perp & 0 \\ 0 & 0 & \alpha_\parallel \end{bmatrix} \mathbf{E}_0(\omega), \tag{6.26}$$

where the z-axis coincides with the tip axis, $\mathbf{E}_0$ is the exciting electric field in the absence of the tip, and $\alpha_\perp$ and $\alpha_\parallel$ denote the transverse and longitudinal polarizabilities defined by

$$\alpha_\perp(\omega) = 4\pi\varepsilon_0 r_0^3 \frac{\varepsilon(\omega) - 1}{\varepsilon(\omega) + 2} \tag{6.27}$$

and

$$\alpha_\parallel(\omega) = 2\pi\varepsilon_0 r_0^3 f_{\rm e}(\omega), \tag{6.28}$$

respectively. Here, ε denotes the bulk dielectric constant of the tip, r_0 the tip radius, and $f_{\rm e}$ the complex field-enhancement factor. For a wavelength of $\lambda = 830$ nm, a gold tip with $\varepsilon = -24.9 + 1.57{\rm i}$ and a tip radius of $r_0 = 10$ nm, numerical calculations based on the MMP method lead to $f_{\rm e} = -7.8 + 17.1{\rm i}$. While $\alpha_\perp$ is identical to the polarizability of a small sphere, $\alpha_\parallel$ arises from the requirement that the magnitude of the field produced by $\boldsymbol{p}(\omega)$ at the surface of the tip be equal to the computationally determined field, which we set

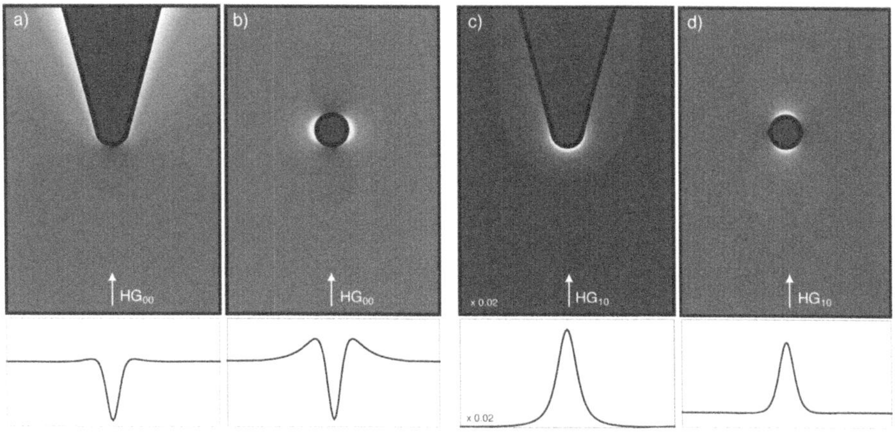

Fig. 6.22 Comparison of the near-fields of a metal tip and a metal sphere. Images (a) and (b) show excitation with an on-axis, focused (NA $= 1.4$) Gaussian beam. Images (c) and (d) show excitation with an on-axis, focused Hermite–Gaussian (1, 0) beam. The strong field enhancement in (c) is due to the longitudinal field of the excitation beam. The cross-sections are evaluated in a plane 1 nm beneath the tip. The results indicate that the field distribution near the tip is well approximated by the dipole fields of a small sphere. However, the field strength for longitudinal excitation (c) is much stronger than the field strength of an irradiated sphere (d). While in (a) and (b) the fields are in-phase, they are 155° out of phase in (c) and (d).

equal to $f_e \mathbf{E}_0$. Once the tip dipole has been determined, the electric field $\mathbf{E}$ in the vicinity of the tip is calculated as

$$\mathbf{E}(\mathbf{r}, \omega) = \mathbf{E}_0(\mathbf{r}, \omega) + \frac{1}{\varepsilon_0} \frac{\omega^2}{c^2} \overleftrightarrow{\mathbf{G}}(\mathbf{r}, \mathbf{r}_0, \omega) \boldsymbol{p}(\omega), \tag{6.29}$$

where $\mathbf{r}_0$ specifies the origin of $\boldsymbol{p}$ and $\overleftrightarrow{\mathbf{G}}$ is the dyadic Green function.

In fluorescence studies, the enhanced field is used to locally excite the sample under investigation to a higher electronic state or band. Image formation is based on the subsequent fluorescence emission. However, the fluorescence can be quenched by the presence of the probe, i.e. the excitation energy can be transferred to the probe and be dissipated through various channels into heat [50] (cf. Problem 8.8). Thus, there is competition between field enhancement and fluorescence quenching (cf. Section 13.4). Whether or not enhanced fluorescence from a molecule placed near a laser-irradiated tip can be observed depends critically on factors such as the tip shape and excitation conditions. Also, not only the magnitude of the field enhancement factor but also its phase play a role.

It has been shown that metal tips are a source of second-harmonic radiation and of broadband luminescence if excited with ultrashort laser pulses. The local second-harmonic generation has been used as a localized photon source for near-field absorption studies [51]. While second-harmonic generation is an instantaneous effect, the lifetime of the tip's broadband luminescence has been measured to be shorter than 4 ps [52].

Fabrication of solid metal tips

Fabrication procedures for sharp metal tips have been established mainly in the context of field ion microscopy [53] and scanning tunneling microscopy (STM) (see e.g. [54]). The actual geometrical shape of the tip is not so important for applications in STM on flat samples as long as there is a foremost atom and there is sufficient conductivity along the tip shaft. On the other hand, in *optical* applications one also cares about the tip's *mesoscopic* structure, i.e. its roughness, cone angle, radius of curvature, and crystallinity. Not all etching techniques yield tips of sufficient "optical" quality. Therefore, FIB milling can be an alternative means by which to produce very well-defined tips [55].

In electrochemical etching, a metal wire is dipped into the etching solution and a voltage is applied between the wire and a counter-electrode immersed into the solution. The surface tension of the solution forms a meniscus around the wire. Etching proceeds most rapidly at the meniscus. After the wire has been etched through, the immersed lower portion of the wire drops down into the supporting vessel. By this time, a tip has been formed at both ends, at the rigidly supported upper portion of the wire and the lower portion that dropped down. By the time of drop-off, the upper tip is still in contact with the solution because of meniscus formation. Therefore, if the etching voltage is not switched off immediately after drop-off, etching will proceed on the upper tip and the sharpness of the tip will be affected. Hence, it is crucial to switch off the etching voltage as soon as drop-off has occurred.

Various electronic schemes have been introduced to control the drop-off event. Most of them use DC etching voltages. However, it has been observed that for certain materials DC etching produces relatively rough tip surfaces. Especially for gold and silver, AC etching is

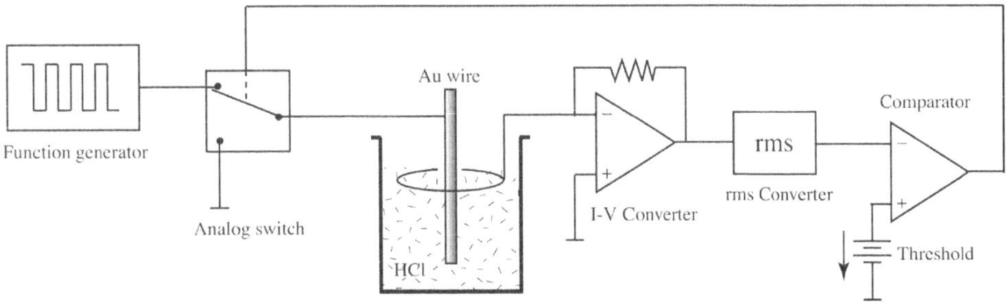

Fig. 6.23 Schematic diagram of an AC etching circuit for gold tips. The etching voltage is automatically switched off after drop-off. The circuit also works for other tip materials if HCl is replaced by a suitable etching solution. See the text for details.

favorable. A schematic diagram of the procedure for fabrication of sharp gold tips is shown in Fig. 6.23. A function generator provides a periodic voltage overlaid with a certain offset. The voltage is sent through an analog switch and applied to a gold wire that is vertically dipped into a solution of hydrochloric acid (HCl) and centered into a circular counter-electrode (Pt) placed just below the surface of the solution. The counter-electrode, held at virtual ground, directs the etching current to a current-to-voltage converter. The resulting voltage is averaged by an r.m.s. converter and then compared with an adjustable threshold voltage by means of a comparator. At the beginning of the etching process, the diameter of the wire and thus the etching current are at a maximum. With ongoing time, the diameter of the wire and the current decrease. The diameter of the wire decreases more rapidly at the meniscus, giving rise to tip formation. When the diameter at the meniscus becomes small enough, the lower portion of the tip drops off and the etching current decreases abruptly. Consequently, the r.m.s. voltage at the input of the comparator drops below the preset voltage threshold and the output of the comparator opens the analog switch, thereby interrupting the etching process. Because of the r.m.s. conversion, the circuit cannot respond faster than the time of 2–10 periods of the waveform provided by the function generator. It turns out that the speed of the circuit is not the limiting factor for achieving good tip quality. The waveform, threshold voltage, concentration of HCl, depth of counter-electrode, and length of wire are factors that are much more important. These factors vary from set-up to set-up and have to be determined empirically. With a good set of parameters one can achieve tip diameters of less than 20 nm with a yield of 50%.

It has to be stressed that the fabricated tips are not monocrystalline, i.e. the metal atoms do not have a periodic arrangement throughout the tip volume. Instead, the tip consists of an arrangement of crystalline grains with sometimes varying lattice configurations. The origin of this grain formation lies in the fabrication process of the original metal wire and has been known since the early days of field-ion microscopy. Because of grain formation it is only a rough approximation to describe the tip's electromagnetic properties by a macroscopic dielectric function $\varepsilon(\omega)$. In fact, it is commonly noticed that the observed field-enhancement factors are weaker than those predicted by calculations and show high variability from tip to tip. This observation is likely to be related to the grain structure of the tips. A quantitative comparison of theory and experiment and the assessment of non-local effects demand the development of single-crystal metal tips or related structures [57].

Resonant probes

While the semi-infinite solid metal probes provide near-field intensity enhancement mainly due to the lightning-rod effect, finite-sized metal particles can support plasmon resonances, which can lead to a resonant enhancement of the near-field intensity. Frey *et al.* [58] pioneered the so-called tip-on-aperture (TOA) probe mainly to reduce the background signal associated with exposure of the sample to the irradiating laser beam. In this approach, a minitip is grown on the end-face of an aperture probe in a scanning electron microcope. Taminiau *et al.* [56] perfected this approach by means of FIB milling and demonstrated antenna-enhanced single-molecule imaging. A resonant aluminum wire of well-defined length was fabricated next to the opening of an aperture probe as shown in Figs. 6.24(a) and (b). The length of the wire is chosen such that the $\lambda/4$ resonance of a single-rod antenna on conducting ground is obtained. Figure 6.24(c) shows the resulting electric field distribution if the tip is illuminated via the adjacent small aperture using the correct polarization that leads to a field maximum at the aperture rim at the position of the minitip (see Section 6.3.2). The electric field maximum at the end is consistent with the presence of a $\lambda/4$ resonance characterized by the current and field distribution sketched in Fig. 6.24(d).

An elegant early demonstration of the principle of a resonant plasmon probe was the experiment by Fischer and Pohl in 1989 [59]. It is shown schematically in Fig. 6.25(a). A 20 nm-thick gold film covers polystyrene beads that are adsorbed on a gold-coated glass substrate. Kretschmann-type illumination is used (see Chapter 12) to launch surface plasmons on the gold film. The surface-plasmon scattering from a selected protrusion (indicated in Fig. 6.25(a)) is recorded as a function of the distance between the scatterer and an approaching glass surface (Fig. 6.25(b)). A peak is observed for p-polarized excitation and for small separations, which is indicative of a surface plasmon resonance. The

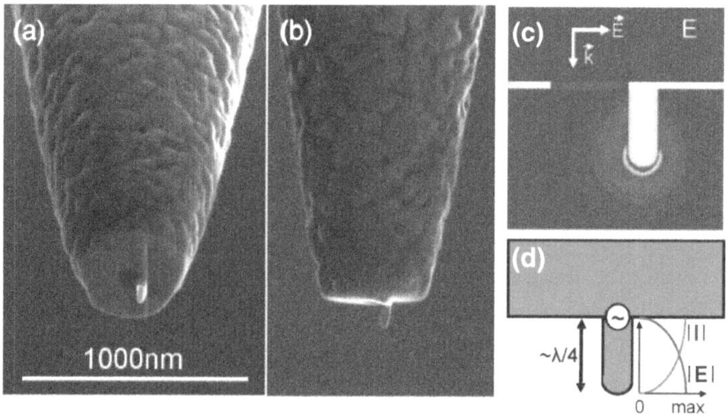

Fig. 6.24 Tip-on-aperture geometry: (a) and (b) SEM images of a tip-on-aperture probe; (c) calculated field distribution and (d) illustration of the current and field-strength of a $\lambda/4$ antenna. Adapted with permission from [56].

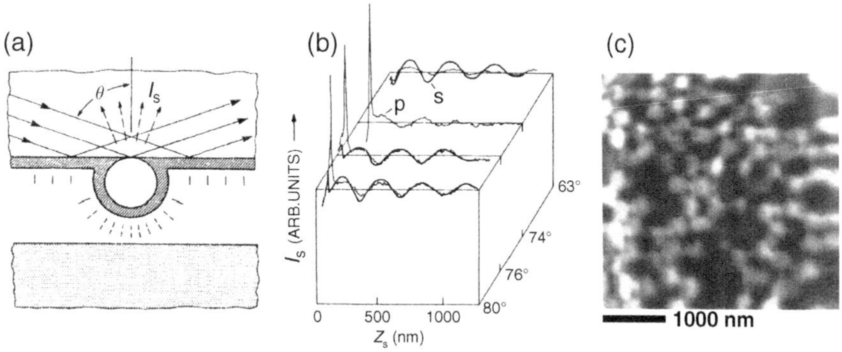

Fig. 6.25 Resonant particle plasmon probe. (a) A polystyrene bead on a flat glass substrate is covered with a 20 nm gold layer and illuminated in the Kretschmann configuration. The scattering of the protrusion is recorded as a sample is approached from the other side. (b) Recorded scattering intensity versus particle–surface distance for both p- and s-polarization. (c) Image recorded in constant-height mode using electron-tunneling feedback. Adapted from [59].

peak is absent for s-polarization, which reinforces the surface-plasmon interpretation. It is evident that the existence of the resonance peak can be used for near-field optical imaging in reflection, i.e. backscattered light is very sensitive to local variations of the dielectric constant near the protrusion. Figure 6.25(c) shows that the technique is able to resolve metal patches on glass with high resolution. A similar approach was adopted later to image magnetic domains on opaque materials [60]. Also, gold-coated dielectric particles, called *nanoshells*, found applications in diverse sensing applications as demonstrated in the work of Halas *et al.* [61].

Another very well-controlled approach that can be used to obtain a resonant probe is the attachment of a spherical or elliptical metal nanoparticle to the apex of a dielectric tip. Following early demonstrations of the feasibility of the approach and applications to scanning near-field optical microscopy [62], and its combination with single-molecule spectroscopy [63, 64], the method is now being used successfully to image fluorescent biological surface structures under physiological conditions [65] and at very high densities [66]. Figure 6.26 shows different realizations of resonant probes. The topic of resonant optical antenna-like probes and optical antennas in general will be picked up again in Chapter 13.

6.6 Conclusion

This chapter provided an overview of the types of probes used in near-field optical microscopy. Besides the theoretical background necessary to understand and correctly apply the respective probe structures we have also discussed fabrication procedures and possible problems that might arise during applications. Many more probe structures and fabrication procedures can be found in the literature. We selected the most important and representative schemes to provide a concise overview.

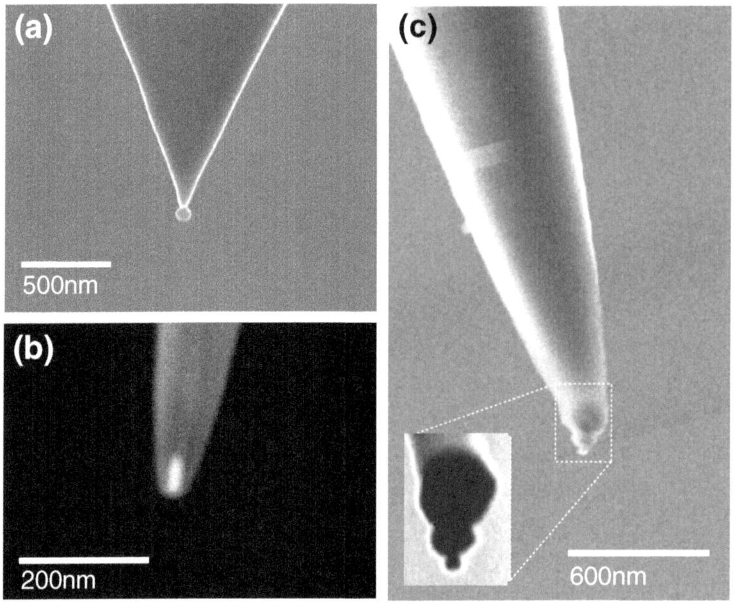

Fig. 6.26 Resonant particle probes. Scanning electron microscopy images of (a) a chemically grafted gold particle (diameter ~80 nm) at the apex of a glass tip, (b) a single gold nano rod taken up by a micropipette, and (c) a glass probe featuring a sequence of attached gold particles of decreasing diameter.

Problems

6.1 Calculate the intensity distribution in a standing evanescent wave above a glass/air interface created by counterpropagating evanescent waves of the same intensity and polarization. Take a line profile perpendicular to the interference fringes and calculate the convolution with a Gaussian of a given full width at half-maximum. How does the full width at half-maximum influence the fringe visibility? Discuss applications to the characterization of glass-fiber probes.

6.2 Calculate the difference in transmission through an aluminum-coated aperture probe and an aperture probe with an infinitely conducting coating. Assume an aperture diameter of 100 nm and a taper angle of $\delta = 10°$.

6.3 Apply Babinet's principle to derive the fields near an ideally conducting disk. Use Bouwkamp's solution and state the fields in the plane of the disk.

6.4 Calculate the second-harmonic generation at a laser-illuminated metal tip. Assume that the fields near the tip are given by Eqs. (6.26)–(6.29) and that second-harmonic generation at the tip originates from a local surface nonlinear polarizability $\chi_{\mathrm{s}}^{(2)}$. The nonlinear surface polarization is determined by the field E_{n} normal to the surface of the tip as

$$P_{\mathrm{n}}^{\mathrm{s}}(\mathbf{r}', 2\omega) = \chi_{nnn}^{\mathrm{s}}(-2\omega; \omega, \omega)\, E_{\mathrm{n}}^{(\mathrm{vac})}(\mathbf{r}', \omega)\, E_{\mathrm{n}}^{(\mathrm{vac})}(\mathbf{r}', \omega), \tag{6.30}$$

where the index n denotes the surface normal, $\mathbf{r}'$ is a point on the surface of the tip, and the superscript (vac) indicates that the fields are evaluated on the vacuum side of the surface. The field at the second-harmonic frequency generated by $\mathbf{P}^s$ is calculated as

$$\mathbf{E}(\mathbf{r}, 2\omega) = \frac{1}{\varepsilon_0} \frac{(2\omega)^2}{c^2} \int_{\text{surface}} \overleftrightarrow{\mathbf{G}}(\mathbf{r}, \mathbf{r}', 2\omega)\, \mathbf{P}^s(\mathbf{r}', 2\omega) \mathrm{d}^2\mathbf{r}'. \tag{6.31}$$

Consider only the near-field of $\overleftrightarrow{\mathbf{G}}$ and assume a semispherical integration surface. Determine an effective tip dipole oscillating at the second-harmonic frequency.

References

[1] D. Marcuse, *Light Transmission Optics*. Malabar, FL: Krieger (1989).

[2] B. Hecht, H. Bielefeldt, D. W. Pohl, L. Novotny, and H. Heinzelmann, "Influence of detection conditions on near-field optical imaging," *J. Appl. Phys.* **84**, 5873–5882 (1998).

[3] D. R. Turner, *Etch Procedure for Optical Fibers*. US patent 4,469,554 (1984).

[4] P. Hoffmann, B. Dutoit, and R.-P. Salathé, "Comparison of mechanically drawn and protection layer chemically etched optical fiber tips," *Ultramicroscopy* **61**, 165–170 (1995).

[5] R. M. Stöckle, C. Fokas, V. Deckert, *et al.*, "High-quality near-field optical probes by tube etching," *Appl. Phys. Lett.* **75**, 160–162 (1999).

[6] T. Pangaribuan, K. Yamada, S. Jiang, H. Ohsawa, and M. Ohtsu, "Reproducible fabrication technique of nanometric tip diameter fiber probe for photon scanning tunneling microscope," *Jap. J. Appl. Phys.* **31**, L1302–L1304 (1992).

[7] T. Yatsui, M. Kourogi, and M. Ohtsu, "Increasing throughput of a near-field optical fiber probe over 1000 times by the use of a triple-tapered structure," *Appl. Phys. Lett.* **73**, 2089–2091 (1998).

[8] S.-K. Eah, W. Jhe, and Y. Arakawa "Nearly diffraction-limited focusing of a fiber axicon microlens," *Rev. Sci. Instrum.* **74**, 4969–4971 (2003).

[9] E. Neher and B. Sakmann, "Noise analysis of drug induced voltage clamp currents in denervated frog muscle fibres," *J. Physiol. (Lond.)* **258**, 705–729 (1976).

[10] G. A. Valaskovic, M. Holton, and G. H. Morrison, "Parameter control, characterization, and optimization in the fabrication of optical fiber near-field probes," *Appl. Opt.* **34**, 1215–1228 (1995).

[11] Ch. Adelmann, J. Hetzler, G. Scheiber, *et al.*, "Experiments on the depolarization near-field scanning optical microscope," *Appl. Phys. Lett.* **74**, 179–181 (1999).

[12] A. J. Meixner, M. A. Bopp, and G. Tarrach, "Direct measurement of standing evanescent waves with a photon scanning tunneling microscope," *Appl. Opt.* **33**, 7995–8000 (1994).

[13] U. Ch. Fischer, J. Koglin, and H. Fuchs, "The tetrahedal tip as a probe for scanning near-field optical microscopy at 30 nm resolution," *J. Microsc.* **176**, 231–237 (1994).

[14] A. Naber, D. Molenda, U. C. Fischer, *et al.*, "Enhanced light confinement in a near-field optical probe with a triangular aperture," *Phys. Rev. Lett.* **89**, 210801 (2002).
[15] L. Novotny and C. Hafner, "Light propagation in a cylindrical waveguide with a complex, metallic dielectric function," *Phys. Rev. E* **50**, 4094–4206 (1994).
[16] B. Hecht, B. Sick, U. P. Wild, *et al.*, "Scanning near-field optical microscopy with aperture probes: fundamentals and applications," *J. Chem. Phys.* **112**, 7761–7774 (2000).
[17] L. Novotny, D. W. Pohl, and B. Hecht, "Scanning near-field optical probe with ultrasmall spot size," *Opt. Lett.* **20**, 970–972 (1995).
[18] R. M. Stöckle, N. Schaller, V. Deckert, C. Fokas, and R. Zenobi, "Brighter near-field optical probes by means of improving the optical destruction threshold," *J. Microsc.* **194**, 378–382 (1999).
[19] L. Novotny and D. W. Pohl, "Light propagation in scanning near-field optical microscopy," in *Photons and Local Probes*, ed. O. Marti and R. Möller. Dordrecht: Kluwer, pp. 21–33 (1995).
[20] B. Knoll and F. Keilmann, "Electromagnetic fields in the cutoff regime of tapered metallic waveguides," *Opt. Commun.* **162**, 177–181 (1999).
[21] M. J. Levene, J. Korlach, S. W. Turner, *et al.*, "Zero-mode waveguides for single-molecule analysis at high concentrations," *Science* **299**, 682–686 (2003).
[22] H. A. Bethe, "Theory of diffraction by small holes," *Phys. Rev.* **66**, 163–182 (1944).
[23] C. J. Bouwkamp, "On Bethe's theory of diffraction by small holes," *Philips Res. Rep.* **5**, 321–332 (1950).
[24] C. T. Tai, "Quasi-static solution for diffraction of a plane electromagnetic wave by a small oblate spheroid," *IRE Trans. Antennas Propag.*, **1**, 13–36 (1952).
[25] M. Abramowitz and I. A. Stegun, *Handbook of Mathematical Functions*. New York: Dover Publications (1974).
[26] D. Van Labeke, D. Barchiesi, and F. Baida, "Optical characterization of nanosources used in scanning near-field optical microscopy," *J. Opt. Soc. Am. A* **12**, 695–703 (1995).
[27] C. J. Bouwkamp, "On the diffraction of electromagnetic waves by small circular disks and holes," *Philips Res. Rep.* **5**, 401–422 (1950).
[28] C. Obermüller and K. Karrai, "Far field characterization of diffracting circular apertures," *Appl. Phys. Lett.* **67**, 3408–3410 (1995).
[29] E. Betzig and R. J. Chichester, "Single molecules observed by near-field scanning optical microscopy," *Science* **262**, 1422–1425 (1993).
[30] T. W. Ebbesen, H. J. Lezec, H. F. Ghaemi, T. Thio, and P. A. Wolff, "Extraordinary optical transmission through sub-wavelength hole arrays," *Nature* **391**, 667–669 (1998).
[31] J. B. Pendry, L. Martin-Moreno, and F. J. Garcia-Vidal, "Mimicking surface plasmons with structured surfaces," *Science* **305**, 847–848 (2004).
[32] H. J. Lezec, A. Degiron, E. Devaux, "Beaming light from a subwavelength aperture," *Science* **297**, 820–822 (2002).

[33] S. Schiller and U. Heisig, *Bedampfungstechnik: Verfahren, Einrichtungen, Anwendungen*, Stuttgart: Wissenschaftliche Verlagsgesellschaft (1975).

[34] D. L. Barr and W. L. Brown, "Contrast formation in focused ion beam images of polycrystalline aluminum," *J. Vac. Sci. Technol. B* **13**, 2580–2583 (1995).

[35] M. Stähelin, M. A. Bopp, G. Tarrach, A. J. Meixner, and I. Zschokke-Gränacher, "Temperature profile of fiber tips used in scanning near-field optical microscopy," *Appl. Phys. Lett.* **68**, 2603–2605 (1996).

[36] J. Orloff, "High-resolution focused ion beams," *Rev. Sci. Instrum.* **64**, 1105–1130 (1993).

[37] J. A. Veerman, M. F. García-Parajó, L. Kuipers, and N. F. van Hulst, "Single molecule mapping of the optical field distribution of probes for near-field microscopy," *J. Microsc.* **194**, 477–482 (1999).

[38] S. Geller (ed.), *Solid Electrolytes*. Berlin: Springer-Verlag (1977).

[39] A. Bouhelier, J. Toquant, H. Tamaru, *et al.*, "Electrolytic formation of nanoapertures for scanning near-field optical microscopy," *Appl. Phys. Lett.* **79**, 683–685 (2001).

[40] D. Haefliger and A. Stemmer, "Subwavelength-sized aperture fabrication in aluminum by a self-terminated corrosion process in the evanescent field," *Appl. Phys. Lett.* **80**, 33973399 (2002).

[41] T. Saiki and K. Matsuda, "Near-field optical fiber probe optimized for illumination-collection hybrid mode operation," *Appl. Phys. Lett.* **74**, 2773–2775 (1999).

[42] D. W. Pohl, W. Denk, and M. Lanz, "Optical stethoscopy: image recording with resolution $\lambda/20$," *Appl. Phys. Lett.* **44**, 651–653 (1984).

[43] D.W. Pohl, "Near field optics seen as an antenna problem," in *Near-Field Optics: Principles and Applications, The Second Asia-Pacific Workshop on Near Field Optics, Beijing, China October 20–23, 1999*, ed. M. Ohtsu and X. Zhu. Singapore: World Scientific, pp. 9–21 (2000).

[44] L. Novotny, "The history of near-field optics," in *Progress in Optics*, vol. 50, ed. E. Wolf. Amsterdam: Elsevier, pp. 137–180 (2007).

[45] J. Wessel, "Surface-enhanced optical microscopy," *J. Opt. Soc. Am. B* **2**, 1538–1540 (1985).

[46] E. J. Sanchez, L. Novotny, and X. S. Xie, "Near-field fluorescence microscopy based on two-photon excitation with metal tips," *Phys. Rev. Lett.* **82**, 4014–4017 (1999).

[47] A. Hartschuh, M. R. Beversluis, A. Bouhelier, and L. Novotny, "Tip-enhanced optical spectroscopy," *Phil. Trans. Roy. Soc. Lond. A* **362**, 807–819 (2004).

[48] Y. C. Martin, H. F. Hamann, and H. K. Wickramasinghe, "Strength of the electric field in apertureless near-field optical microscopy," *J. Appl. Phys.* **89**, 5774–5778 (2001).

[49] C. Sönnichsen, T. Franzl, T. Wilk, G. von Plessen, and J. Feldmann, "Drastic reduction of plasmon damping in gold nanorods," *Phys. Rev. Lett.* **88**, 077402 (2002).

[50] R. X. Bian, R. C. Dunn, X. S. Xie, and P. T. Leung, "Single molecule emission characteristics in near-field microscopy," *Phys. Rev. Lett.* **75**, 4772–4775 (1995).

[51] A. Bouhelier, M. Beversluis, A. Hartschuh, and L. Novotny, "Near-field second harmonic generation excited by local field enhancement," *Phys. Rev. Lett.* **90**, 13903 (2003).

[52] M. R. Beversluis, A. Bouhelier, and L. Novotny, "Continuum generation from single gold nanostructures through near-field mediated intraband transitions," *Phys. Rev. B* **68**, 115433 (2003).
[53] E. W. Müller and T. T. Tsong, *Field Ion Microscopy*. New York: Elsevier (1969).
[54] A. J. Nam, A. Teren, T. A. Lusby, and A. J. Melmed, "Benign making of sharp tips for STM and FIM: Pt, Ir, Au, Pd, and Rh," *J. Vac. Sci. Technol. B* **13**, 1556–1559 (1995).
[55] M. J. Vasile, D. A. Grigg, J. E. Griffith, E. A. Fitzgerald, and P. E. Russell, "Scanning probe tips formed by focused ion beams," *Rev. Sci. Instrum.* **62**, 2167–2171 (1991).
[56] T. H. Taminiau, R. J. Moerland, F. B. Segerink, L. Kuipers, and N. F. van Hulst "$\lambda/4$ Resonance of an optical monopole antenna probed by single molecule fluorescence," *Nano Lett.* **7**, 28–33 (2007). Copyright 2007 American Chemical Society.
[57] J.-S. Huang, V. Callegari, P. Geisler, *et al.*, "Atomically flat single-crystalline gold nanostructures for plasmonic nanocircuitry," *Nature Commun.* **1**, 150 (2010).
[58] H. G. Frey, F. Keilmann, A. Kriele, and R. Guckenberger, "Enhancing the resolution of scanning near-field optical microscopy by a metal tip grown on an aperture probe," *Appl. Phys. Lett.* **81**, 5030–5032 (2002).
[59] U. Ch. Fischer and D. W. Pohl, "Observation of single-particle plasmons by near-field optical microscopy," *Phys. Rev. Lett.* **62**, 458–461 (1989).
[60] T. J. Silva, S. Schultz, and D. Weller, "Scanning near-field optical microscope for the imaging of magnetic domains in optically opaque materials," *Appl. Phys. Lett.* **65**, 658–660 (1994).
[61] S. Lal, S. Link, and N. J. Halas, "Nano-optics from sensing to waveguiding," *Nature Photonics* **1**, 641–648 (2007)
[62] T. Kalkbrenner, M. Ramstein, J. Mlynek, and V. Sandoghdar, "A single gold particle as a probe for apertureless scanning near-field optical microscopy," *J. Microsc.* **202**, 72–76 (2001).
[63] P. Anger, P. Bharadwaj, and L. Novotny "Enhancement and quenching of single molecule fluorescence," *Phys. Rev. Lett.* **96**, 113002 (2006).
[64] S. Kühn, U. Hakanson, L. Rogobete, and V. Sandoghdar "Enhancement of single molecule fluorescence using a gold nanoparticle as an optical nano-antenna," *Phys. Rev. Lett.* **97**, 017402 (2006).
[65] C. Hoeppener and L. Novotny "Antenna-based optical imaging of single Ca^{2+} transmembrane proteins in liquids," *Nano Lett.* **8**, 642–646 (2008).
[66] C. Hoeppener, R. Beams, and L. Novotny "Background suppression in near- field optical imaging," *Nano Lett.* **9**, 903–908 (2009).

7 Probe–sample distance control

In order to measure localized fields one needs to bring a local probe into close proximity to a sample surface. Typically, the probe–sample distance is required to be smaller than the size of lateral field confinement and thus smaller than the spatial resolution to be achieved. An active feedback loop is required in order to maintain a constant distance during the experiment. However, the successful implementation of a feedback loop requires a sufficiently short-ranged interaction between the optical probe and the sample. The dependence of this interaction on the probe–sample distance should be monotonic in order to ensure a unique distance assignment. A typical block diagram of a feedback loop applied to scanning probe microscopy is shown in Fig. 7.1. A piezoelectric element $P(\omega)$ is used to transform an electric signal into a displacement, whilst the interaction measurement $I(\omega)$ takes care of the reverse transformation. The controller $G(\omega)$ is used to optimize the speed of the feedback loop and to ensure stability according to well-established design rules. Most commonly, a so-called PI controller is used, which is a combination of a proportional gain (P) and an integrator stage (I).

Using the (near-field) optical signal itself as a distance-dependent feedback signal seems to be an attractive solution at first glance. However, it turns out that this is problematic. (1) In the presence of a sample of unknown and inhomogeneous composition, unpredictable variations in the near-field distribution give rise to a non-monotonic distance dependence. Such behavior inevitably leads to frequent probe damage. (2) The detected near-field signal is often small and masked by far-field contributions. (3) The decay length of the near fields of optical probes is often too long to serve as a reliable measure for distance changes on the nanometer scale. For these reasons, usually an auxiliary distance feedback is required for the operation of optical probes.

Standard scanning probe techniques basically employ two different types of interactions, i.e. electron tunneling (STM) [1] and interaction forces normal and lateral to the surface (AFM) [2]. Electron tunneling requires a conductive sample. This is a strong limitation in view of the spectroscopic capabilities of optical microscopy that are lost by covering the sample with a metallic layer. Therefore, optical experiments most commonly employ feedback loops based on short-range interaction forces, for example the measurement of lateral shear forces or normal van der Waals forces.

Before we go into more details an important note has to be made. In standard commercial AFMs and STMs the short-ranged interaction used for feedback is also the physical quantity of interest. This is not the case in near-field microscopy with auxiliary feedback. The use of an auxiliary feedback mechanism inherently bears the danger of introducing into the optical signal artifactual changes that are related not to the optical properties of the sample but to changes in the probe–sample distance induced by the auxiliary feedback. These problems and possible solutions are discussed in detail in the final section of this chapter and also in Refs. [3, 4]. In the following we will concentrate lateral shear forces, but similar analysis can be done for normal forces.

7.1 Shear-force methods

The vibration of a probe in a direction parallel to the sample surface is influenced by the proximity of the sample. Typically, the probe is oscillated at the resonance frequency of its mechanical support (vertical beam, tuning fork) and the amplitude, phase, and/or frequency of the oscillation are measured as a function of the probe–sample distance. The interaction range is 1–100 nm, depending on the type of probe and the particular implementation. The nature of this so-called shear force is still under debate. It is accepted that under ambient conditions the effect originates from the interaction with a surface humidity layer. However, the shear force can even be measured under high-vacuum conditions and at ultralow temperatures [5, 6], and thus there must be more fundamental interaction mechanisms such as electromagnetic friction (cf. Section 15.3.2) [7]. Whatever the origin, the distance-dependent shear force is an ideal feedback signal for maintaining an optical probe in close proximity to a sample surface.

7.1.1 Optical fibers as resonating beams

The simplest type of shear-force sensor is the oscillating beam. It consists of a clamped short piece of a glass fiber or a metal rod with a tip at its end. The resonance frequency of the beam depicted in Fig. 7.2 scales with the square of its free length L. This scaling

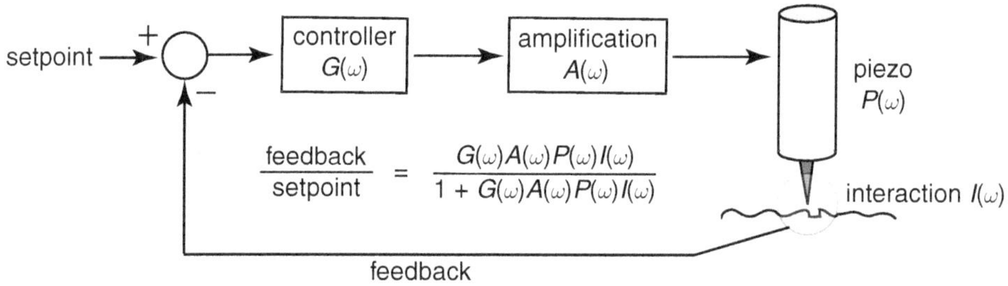

Fig. 7.1 Characteristic diagram of a feedback loop employed in scanning probe microscopy. Ideally, the measured interaction signal corresponds to the externally defined setpoint. The speed and stability of the feedback loop depend on the parameters of the controller $G(\omega)$.

holds for any type of cantilever fixed at one end. The fundamental resonance frequency of an oscillating beam with circular cross-section is calculated as [8]

$$\omega_0 = 1.76\sqrt{\frac{E}{\rho}\frac{R}{L^2}}, \tag{7.1}$$

where E is Young's modulus, ρ the specific mass density, R the radius of the beam, and L the length of the beam. For the example of an optical fiber with radius $R = 125\,\mu\text{m}$ and length $L = 3\,\text{mm}$ we obtain $f_0 = \omega_0/(2\pi) \approx 20$ kHz. A typical quality factor of such a probe in air is about 150. Changing the length of the fiber will strongly change the resonance frequency according to Eq. (7.1).

When the end of the beam starts to interact with a surface the resonance frequency will shift and the oscillation amplitude will drop. This situation is depicted in Figs. 7.3(a) and (b) for a beam that is externally driven at a variable frequency ω. The amplitude and phase of the beam oscillation are shown for two different distances d between the beam-end and the sample surface. Figures 7.3(c) and (d) show the amplitude shift and the phase shift, respectively, as functions of the distance d for the case in which the beam is driven at its resonance frequency $\omega = \omega_0$. The distance range over which the amplitude and phase vary depends on the diameter of the beam, i.e. the tip diameter in the case of a sharp probe. Because of the monotonic behavior of the curves in Figs. 7.3(a) and (b), amplitude and phase are well-suited feedback signals. Usually they are detected with a lock-in amplifier. As will be discussed later on, in high-sensitivity applications that require a high Q-factor (narrow resonances) it is favorable not to drive the beam at a fixed frequency ω. Instead, with a self-oscillating circuit the beam can be vibrated at its distance-dependent resonance frequency [9]. As illustrated in Figs. 7.3(a) and (b), the resonance frequency shifts as the oscillating probe is advanced towards the sample surface and thus the frequency shift $\Delta\omega$ can be used as an alternative feedback signal. A further possibility is to use the Q-factor of the resonance as a feedback signal, which would correspond to operation in constant-dissipation mode. Which type of feedback signal to use depends on the particular type of experiment. In general, complementary information about the probe–sample

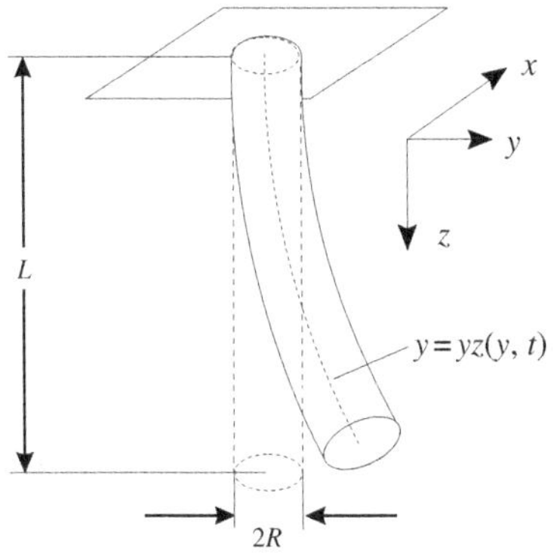

Fig. 7.2 Sketch of a quartz beam of length L used to calculate the resonances of an oscillating fiber probe.

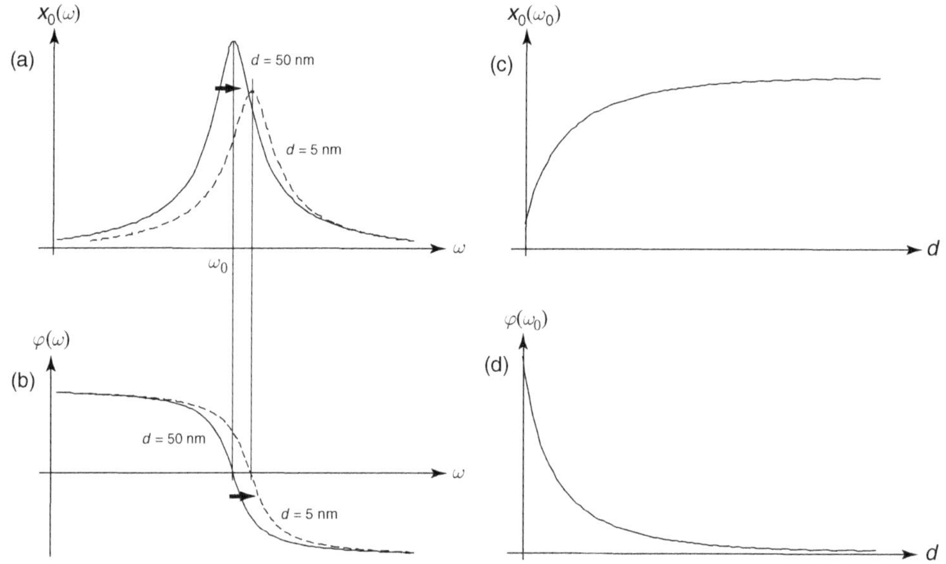

Fig. 7.3 Resonance of a vibrating beam. The amplitude (a) and phase (b) of a beam driven at a frequency ω. As the beam-end starts to interact with a sample surface, the resonance shifts and the amplitude drops. Images (c) and (d) show the amplitude and phase at frequency $\omega = \omega_0$ as functions of the distance between the beam-end (tip) and the surface. The distance range over which the amplitude and phase vary depends on the interaction area (tip sharpness).

interaction can be accessed by recording the amplitude, phase, frequency shift, and Q-factor simultaneously as auxiliary signals, as in standard atomic-force microscopy.

There are several ways of directly detecting the vibration of an oscillating optical probe. The simplest method (see Fig. 7.4(a)) is to project the light emitted or scattered from an optical probe onto a suitably positioned aperture and to detect the transmitted light intensity. The modulation amplitude of the optical signal at the dither frequency of the tip will reflect the amplitude and phase of the tip oscillation [10]. In a near-field optical microscope, this method interferes with the detection path of the optical signal and thus can be influenced by the optical properties of the sample. Therefore, alternative optical detection schemes employing a beam path perpendicular to the optical detection path of the microscope have been developed. An auxiliary laser can be pointed to the probe and the resulting diffraction pattern is detected by a split photodiode (see Fig. 7.4(b)). This scheme works well but it can suffer from mode hopping of the laser diode or drifts in the mechanical set-up leading to changes in the (interference) pattern on the photodiode. Also, it is clear that the motion sensed along the shaft of the probe is not identical to the motion of the tip apex itself. This can be a problem if higher-order oscillation modes of the probe are excited. The same arguments may apply to interferometric detection schemes, e.g. using differential interferometry [11] or a fiber interferometer [12, 13] (see Figs. 7.4(c) and (d)). The latter methods are, however, very sensitive and can detect amplitudes well below 1 nm. However, the direct optical detection of probe oscillation is no longer widely employed because indirect methods, using quartz or piezoceramic sensors, have proven to be favorable in terms of sensitivity and simplicity of implementation.

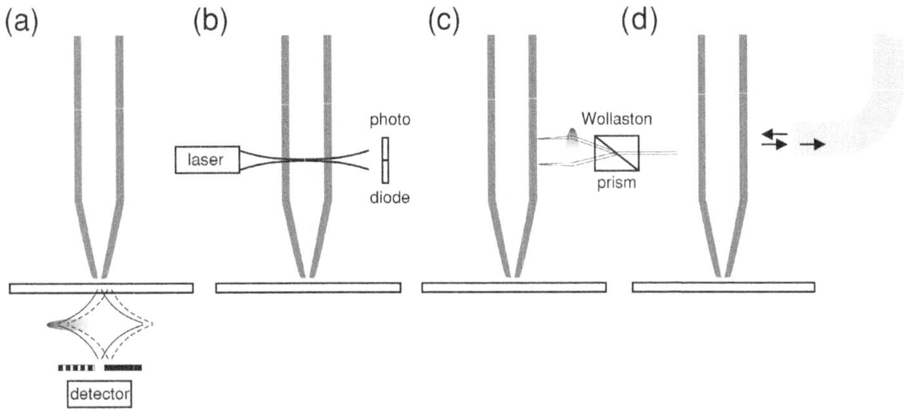

Fig. 7.4 Different methods for the direct detection of the oscillation of an optical probe. (a) Aperture detection scheme: the light emitted or scattered by the probe is focused onto a pinhole. The detected light is modulated at the mechanical resonance frequency of the probe. (b) Laser deflection scheme: an infrared diode laser is scattered or deflected by the fiber probe. The resulting oscillating fringe pattern is directed to a split photodiode. (c) Differential interferometry using a Wollaston prism. (d) Interferometry using a fiber-optic interferometer.

7.1.2 Tuning-fork sensors

When using optical methods to detect the optical probe's lateral vibration there is the danger that the optical detection interferes with the detection of a generally weak near-field optical signal. This is especially important when spectroscopic experiments are performed or photosensitive samples are investigated. Therefore, alternative sensing methods, that do not employ light, have been developed. Many of them are based on measuring changes in the admittance of piezoelectric devices that are related to a change in the resonant behavior upon interaction with the sample of the piezoelectric device itself or an optical probe attached to it. The piezoelectric element can be a piezo plate [14] or tube [15]. However, the most successful and widespread method of shear-force detection today is based on microfabricated quartz tuning forks [16], which were originally developed for use as time standards in quartz watches.

Figure 7.5(a) shows a photograph of a typical quartz tuning fork. It consists of a micromachined quartz element shaped like a tuning fork with electrodes deposited on the surface of the device. At the base, the tuning fork is supported by an epoxy-resin mounting (left side). The overall length of the element without mount is about 5.87 mm. The width is 1.38 mm and the thickness of the element is 220 μm. It has two electric connections that contact the electrodes of the tuning-fork element as sketched in Fig. 7.5(b). For use in clocks and watches, the tuning fork is encapsulated by a metal cap in order to protect it against ambient parameters such as humidity. The metal capsule has to be removed if the tuning fork is to be used as a shear-force sensor. Tuning-fork crystals are fabricated in different sizes and laid out for different resonance frequencies. The most common frequencies are 2^{15} Hz = 32 768 Hz and 100 kHz.

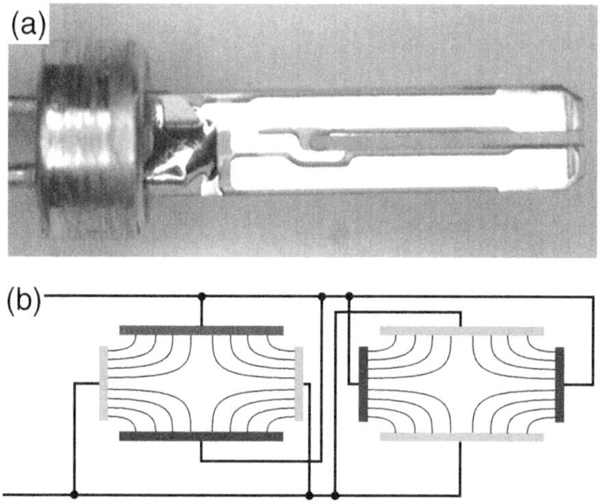

Fig. 7.5 Quartz tuning fork. (a) Enlarged photograph. The dimensions of the quartz element shown are 5870 μm $\times$ 1380 μm $\times$ 220 μm. (b) Connection scheme of a quartz tuning fork for a cut perpendicular to the prongs. Adapted from [17].

The mechanical oscillation of the tuning-fork prongs induces surface charges that are picked up by the electrodes and measured by an external electronic circuit. Hence, the tuning fork acts as a mechanical–electrical converter similar to a piezoceramic actuator. Vice versa, an alternating voltage applied to the tuning-fork electrodes gives rise to a mechanical oscillation of the prongs. The particular electrode layout on the tuning fork ensures that only movements of the prongs against each other can be excited and detected electronically. This is because contraction and dilatation occur perpendicular to the field lines sketched in Fig. 7.5(b). If the tuning-fork oscillation is excited via mechanical coupling to a separate oscillator (e.g. a dither piezo) one has to make sure that the correct mode is excited because otherwise no signal can be detected. The advantages of quartz tuning forks compared with other piezoelectric elements, apart from their small size, are their standardized properties and low price due to large-scale production. The small size allows optical (fiber) probes to be attached to one prong of a fork such that even a weak interaction of the probe apex with the sample will rigidly couple to the motion of the tuning-fork element and influence its oscillation. Figure 7.6 shows a sketch of a typical setting. In this scheme of shear-force detection, the tuning-fork prongs act as oscillating beams and not the probe itself. It is important that the probe itself does not oscillate at the frequency of the tuning fork in order to prevent a coupled-oscillator type of operation. Hence, the length of the probe protruding from the tuning-fork end has to be kept as short as possible. For a tuning fork operating at $\approx$32 kHz with an attached glass-fiber probe, Eq. (7.1) implies that the protruding fiber length needs to be shorter than $\sim$2.3 mm.

7.1.3 The effective-harmonic-oscillator model

For small oscillation amplitudes $x(t)$ of the driven tuning-fork oscillation, the equation of motion for the tuning fork is that of an effective harmonic oscillator:

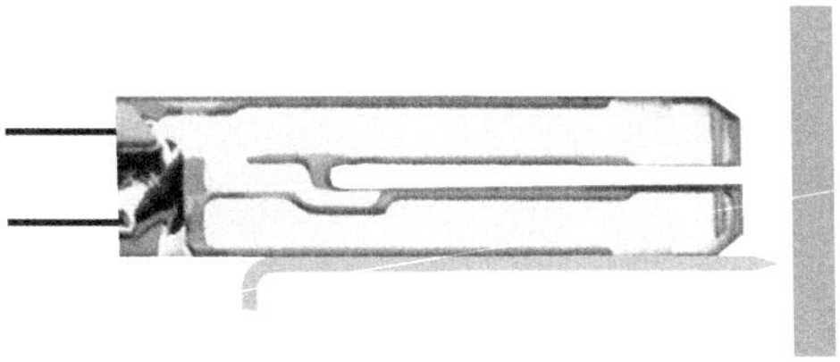

Fig. 7.6 Cartoon of a quartz tuning-fork sensor with attached tapered glass fiber (to scale) showing the relative dimensions of the fiber probe and tuning-fork sensor. Left: sensor. Right: sample.

$$m\ddot{x}(d,t) + m\gamma(d)\dot{x}(d,t) + m\,\omega_0^2(d)\,x(d,t) = F\mathrm{e}^{-\mathrm{i}\omega t}. \tag{7.2}$$

Here, γ is the damping constant, $f_0 = \omega_0/(2\pi)$ the resonance frequency, and F a constant driving force, which is, for example, supplied by an external dither piezo shaking the tuning fork. The parameter d indicates the dependence on probe–sample distance. For ease of notation, the explicit dependence on d will be suppressed. The steady-state solution of Eq. (7.2) is

$$x(t) = \frac{F/m}{\omega_0^2 - \omega^2 - \mathrm{i}\gamma\omega}\mathrm{e}^{-\mathrm{i}\omega t}. \tag{7.3}$$

The amplitude of this oscillation is a Lorentzian lineshape function with a Q-factor

$$Q = \frac{f_0}{\Delta f} = \frac{\omega_0}{\gamma\sqrt{3}}, \tag{7.4}$$

where Δf is the full width at half-maximum of the resonance. Similarly to γ and ω_0, the Q-factor and the oscillation amplitude $x(t)$ depend on the probe–sample distance d (see Fig. 7.3(a)). The Q-factor of a tuning fork is on the order of 10^3–10^4 under ambient conditions and can be several orders of magnitude higher in vacuum. Such a high Q originates from the fact that there is no center-of-mass motion. While one prong moves to the left, the other prong moves to the right, so there is no net mass displacement.

The interaction of the probe with the sample surface affects two types of forces: (1) a dissipative friction force associated with the second term in Eq. (7.2) and (2) a reactive elastic force due to the third term in Eq. (7.2). We will derive expressions for both of these forces and estimate their magnitude. Let us first note that both the damping constant γ and the spring constant $k = m\omega_0^2$ have two different contributions: (1) a static or intrinsic one associated with the physical properties of the tuning fork itself and (2) an interaction mediated contribution due to the probe–sample interaction. An expression for the interaction part of γ can be derived from the oscillation amplitude Eq. (7.3) evaluated at the resonance frequency, i.e.

$$\gamma(d) = \gamma_{\text{stat}} + \gamma_{\text{int}}(d) = \frac{F/m}{\omega_0(d)\,x_0(d)}, \tag{7.5}$$

with x_0 being the oscillation amplitude and γ_{int} the interaction-mediated damping constant. Notice that $\gamma_{\mathrm{int}}(d \to \infty) = 0$, which implies that

$$\gamma_{\mathrm{int}}(d) = \gamma_{\mathrm{stat}} \left[\frac{\omega_0(\infty)\, x_0(\infty)}{\omega_0(d) x_0(d)} - 1 \right]. \tag{7.6}$$

According to the second term in Eq. (7.2), the amplitude of the interaction-induced friction force is calculated as

$$F_{\mathrm{int}}^{\mathrm{friction}}(d) = m\gamma_{\mathrm{int}}(d)\, \omega_0(d)\, x_0(d) = \left[1 - \frac{\omega_0(d) x_0(d)}{\omega_0(\infty) x_0(\infty)} \right] \frac{k_{\mathrm{stat}}\, x_0(\infty)}{\sqrt{3}\, Q(\infty)}, \tag{7.7}$$

where we used Eq. (7.4) and the property $m = k_{\mathrm{stat}}/\omega_0^2(\infty)$. Next, we use the fact that the amplitude x_0 changes faster with distance than does the resonance frequency ω_0, which allows us to drop the dependence on ω_0 in the expression inside the brackets. Furthermore, the voltage V due to the induced surface charge at the surface of the tuning fork is directly proportional to the oscillation amplitude and thus

$$F_{\mathrm{int}}^{\mathrm{friction}}(d) = \left[1 - \frac{V(d)}{V(\infty)} \right] \frac{k_{\mathrm{stat}}}{\sqrt{3}\, Q(\infty)} x_0(\infty). \tag{7.8}$$

This is the key expression for estimating the friction forces in shear-force microscopy. All the parameters in this expression are directly accessible. It can be shown that the ratio x_0/Q is independent of the probe–sample distance d, which supports the hypothesis of a viscous origin of the friction force, i.e. the friction is proportional to the velocity [5]. Thus, as the probe is advanced towards the sample surface a reduction in oscillation amplitude corresponds to a proportional reduction of the quality factor.

Let us now work out the numbers for a realistic situation. The expression in brackets takes on the value of 0.1 if we assume a feedback setpoint corresponding to 90% of the original voltage $V(\infty)$. A 32-kHz tuning fork with spring constant $k_{\mathrm{stat}} = 40\ \mathrm{kN\,m^{-1}}$ can be operated at an oscillation amplitude of $x_0(\infty) = 10$ pm (less than a Bohr radius!), and a typical quality factor with attached tip is $Q(\infty) \approx 1200$. With these parameters, the interaction-induced friction force turns out to be $F_{\mathrm{int}}^{\mathrm{friction}} \approx 20$ pN, which is comparable to AFM measurements obtained using ultrasoft cantilevers.

If a tuning-fork prong with $k_{\mathrm{stat}} = 40\ \mathrm{kN\,m^{-1}}$ is displaced by an amount of $x_0 = 10$ pm a surface charge difference of roughly 1000 electrons is built up between the two electrodes. Typically, the piezo-electromechanical coupling constant is of the order of

$$\alpha = 10\,\mu\mathrm{C\,m^{-1}}. \tag{7.9}$$

The exact value depends on the specific type of tuning fork. For an oscillation with 32 kHz, this corresponds to a current-to-displacement conversion of $2\ \mathrm{A\,m^{-1}}$, which has been confirmed experimentally with a laser-interferometric technique [18]. Using a current-to-voltage conversion with a resistance of 10 MΩ, an oscillation amplitude of $x_0 = 10$ pm gives rise to an oscillating voltage with amplitude $V = 200\,\mu$V. This voltage must be further amplified before it is processed, for example, by a lock-in amplifier. While an oscillation amplitude of 10 pm seems very small, it is nevertheless more than a factor of 20 larger than the thermally induced oscillation amplitude. The latter is calculated with help of the equipartition principle as

$$\frac{1}{2} k_{\mathrm{stat}} x_{\mathrm{rms}}^2 = \frac{1}{2} k_{\mathrm{B}} T, \tag{7.10}$$

where T is the temperature and k_{B} the Boltzmann constant. At room temperature we obtain $x_{\mathrm{rms}} = 0.32$ pm, which corresponds to a peak noise amplitude of 0.45 pm.

Finally, we turn our attention to the elastic force associated with the third term in Eq. (7.2). Similarly to the case of the damping constant, the spring constant k is characterized by a static part and an interaction-induced part. Because the mass m is independent of the probe–sample distance we obtain

$$m = \frac{k_{\mathrm{stat}} + k_{\mathrm{int}}(d)}{\omega_0^2(d)} = \frac{k_{\mathrm{stat}}}{\omega_0^2(\infty)} \rightarrow k_{\mathrm{int}}(d) = k_{\mathrm{stat}} \left[\frac{\omega_0^2(d)}{\omega_0^2(\infty)} - 1 \right]. \tag{7.11}$$

Introducing this relationship into the expression for the amplitude of the interaction-induced elastic force gives

$$F_{\mathrm{int}}^{\mathrm{elastic}}(d) = k_{\mathrm{int}}(d) x_0(d) = \left[\frac{\omega_0^2(d)}{\omega_0^2(\infty)} - 1 \right] k_{\mathrm{stat}} x_0(d). \tag{7.12}$$

As an example, we consider a small frequency shift of 5 Hz and assume that this shift is again associated with a reduction of the oscillation amplitude $x_0(\infty) = 10$ pm to 90%, so that $x_0(d) = 9$ pm. For the same parameters as used before, the elastic force amplitude turns out to be $F_{\mathrm{int}}^{\mathrm{elastic}} \approx 110$ pN, which demonstrates that typically the elastic force is stronger than the friction force. However, as will be discussed later on, measurements of $F_{\mathrm{int}}^{\mathrm{friction}}$ rely on measurements of amplitude variations which are inherently slow for high Q-factors. Therefore, measurements of frequency shifts and thus of $F_{\mathrm{int}}^{\mathrm{elastic}}$ are often a good compromise between sensitivity and speed.

7.1.4 Response time

The higher the Q-factor of a system is, the longer it takes to respond to an external signal. On the other hand, a high Q-factor is a prerequisite for high sensitivity. Thus, short response time and high sensitivity tend to counteract each other and a compromise between the two has to be found. The parameters of a tuning fork used for probe–sample distance control have to be adjusted so that there is sufficient sensitivity to prevent probe or sample damage and the response time is sufficiently short to guarantee reasonable scanning speeds. For example, the use of ductile gold tips as near-field probes demands interaction forces smaller than $\approx$200 pN. The same is true if soft biological tissue is to be imaged. Such small forces require a high Q-factor, which limits the image-acquisition time.

To illustrate the relationship between the Q-factor and the response time, let us consider the amplitude and phase of the complex steady-state solution of the harmonic-oscillator model (cf. Eq. (7.3))

$$x_0 = \frac{F/m}{\sqrt{(\omega_0^2 - \omega^2)^2 + \omega_0^2\omega^2/(3Q^2)}}, \tag{7.13}$$

$$\varphi_0 = \tan^{-1}\left[\frac{\omega_0\omega}{\sqrt{3}Q\,(\omega_0^2 - \omega^2)}\right], \tag{7.14}$$

where we expressed the damping constant in terms of the quality factor using Eq. (7.4). In terms of x_0 and φ_0 the solution can be written as

$$x(t) = x_0 \cos(\omega t + \varphi_0). \tag{7.15}$$

We will now consider what happens if the probe–sample distance d is abruptly changed from one value to another [9]. As an initial condition we assume that the resonance frequency changes instantaneously from ω_0 to ω_0' at the time $t = 0$. The solution is provided by Eq. (7.2), and with the proper boundary conditions we obtain

$$x(t) = x_0' \cos\left(\omega t + \varphi_0'\right) + x_t e^{-\omega_0' t/(2\sqrt{3}Q)} \cos(\omega_t t + \varphi_t). \tag{7.16}$$

The solution consists of a steady-state term (left) and a transient term (right). x_0' and φ_0' are the new steady-state amplitude and phase, respectively. Similarly, x_t and φ_t and ω_t are the corresponding parameters of the transient term. Their exact values follow from the boundary conditions.

Figure 7.7 shows the envelope of typical transient behavior described by Eq. (7.16) for a tuning fork with a typical Q-factor of 2886. Upon a change of distance at $t = 0$ it takes about $2Q$ oscillation cycles to reach the new steady state. The response time of the tuning fork can be defined as

$$\tau = \frac{2\sqrt{3}Q}{\omega_0'} \approx \frac{2\sqrt{3}Q}{\omega_0}, \tag{7.17}$$

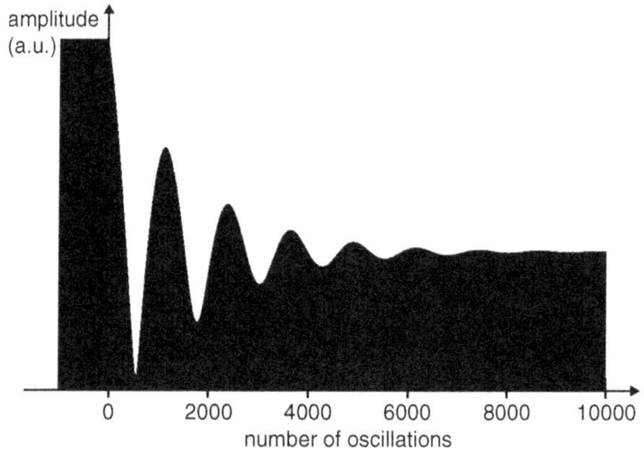

Fig. 7.7 Transient response of an oscillating tuning fork ($Q = 2886$) upon a change in probe–sample distance d at $t = 0$. The step causes a resonance-frequency shift of 16.5 Hz from 33 000 to 33 016.5 Hz following Eq. (7.16). Only after approximately $2Q = 10\,000$ oscillation periods is the new steady state reached. Individual oscillations cannot be resolved – only the envelope is visible.

which is as large as ~300 ms. Thus, the bandwidth of the feedback loop becomes very small and the scanning speeds very slow if the amplitude serves as a feedback signal. To overcome this problem, it was proposed that one could use the resonance-frequency shift as a feedback signal [9]. In a first approximation, the resonance frequency responds instantaneously to a perturbation; however, one has to keep in mind that it takes at least one oscillation period to define a frequency. The frequency shift can be monitored, for example, by using a phase-locked loop (PLL) as in FM demodulators used in radios. However, here also the available bandwidth is not unlimited because of the low-pass filtering used in the process. In other words, one needs a number of oscillation cycles in order to compare the phase to be measured with a reference.

7.1.5 Equivalent electric circuit

So far, we have assumed that the tuning fork is driven by a constant driving force F. This force can be supplied mechanically by an external dither piezo attached in the vicinity of the tuning fork. This type of mechanical excitation is favorable in the sense that the driving circuit is electrically decoupled from the system and hence provides better stability and noise performance. On the other hand, mechanical shaking gives rise to center-of-mass oscillation of the tuning fork that does not correspond to the desired "asymmetric" mode of operation (prongs oscillating out of phase). Consequently, mechanical excitation provides poor coupling to the tuning-fork oscillation. Electrical excitation can be more favorable because of the simplicity of implementation. When using the fully electric operation of a tuning fork, the measurement of the dither motion reduces to a simple impedance $Z(\omega)$ or admittance $Y(\omega)$ measurement.

The admittance of a piezoelectric resonator can be modeled by a Butterworth–Van Dyke equivalent circuit [17] as shown in Fig. 7.8(a). It can be expressed as

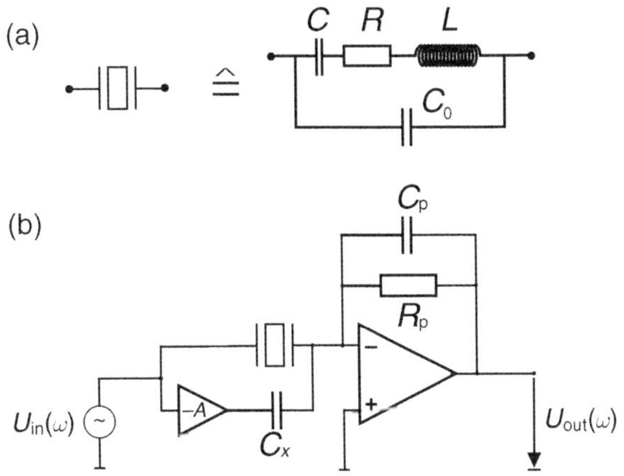

Fig. 7.8 Equivalent electric circuit of a tuning fork and its measurement. (a) Symbol and Butterworth–Van Dyke equivalent circuit. (b) Measurement of the tuning-fork admittance. The capacitor C_x and the gain of $-A$ compensate for the tuning fork's stray capacitance C_0.

$$Y(\omega) = \frac{1}{Z(\omega)} = \frac{1}{R + (\mathrm{i}\omega C)^{-1} + \mathrm{i}\omega L} + \mathrm{i}\omega C_0. \qquad (7.18)$$

Here, the inductance L, the resistance R, and the capacitance C are characteristic values for a certain type of resonator. The parallel capacitance C_0 originates from the pick-up electrodes and external leads connecting to the resonator. Equation (7.18) can be represented by a Nyquist plot (see Fig. 7.9(a)), in which Im(Y) is plotted against Re(Y) parameterized by the frequency ω. The resulting plot, characteristically for a resonator, is a circle in the complex admittance plane that is offset along the imaginary axis by ωC_0. Plotting the absolute value of $Y(\omega)$ as a function of ω using a logarithmic scale yields the resonance curve of the oscillator shown in Fig. 7.9(b). Using the parameters of a typical tuning fork listed in the caption of Fig. 7.9 gives rise to a resonance at 32 765 Hz. The resonance frequency is determined by $f_0 = 1/(2\pi\sqrt{LC})$ and the quality factor by $Q = \sqrt{L/(CR^2)}$. The small negative peak at higher frequencies is a consequence of the stray capacitance C_0, which can be traced back to the offset of the circular admittance locus in Fig. 7.9(a). Increasing C_0 hardly influences the position of the resonance peak, but distorts the shape of the curve by moving the second peak closer to the actual resonance.

A scheme for measuring the admittance is depicted in Fig. 7.8(b). The transfer function of this circuit is determined as

$$\frac{U_{\mathrm{out}}}{U_{\mathrm{in}}}(\omega) = -\frac{R_{\mathrm{p}}}{1 + \mathrm{i}\omega R_{\mathrm{p}} C_{\mathrm{p}}} \frac{\mathrm{i}\omega C + (1 - \omega^2 CL + \mathrm{i}\omega RC)(\mathrm{i}\omega C_0 - A\mathrm{i}\omega C_x)}{1 - \omega^2 CL + \mathrm{i}\omega RC}. \qquad (7.19)$$

It can be seen that by adjusting the variable negative gain $-A$ it is possible to compensate for the influence of C_0, which, if left uncompensated, results in a suboptimal signal-to-noise ratio due to the high- and low-frequency offsets introduced to U_{out}. The first term in Eq. (7.19) corresponds to a low-pass filter due to the feedback resistor's stray resistance. Notice that the current through the tuning fork is directly determined by the applied voltage U_{in}. Thus, following our previous example, an interaction-induced friction force of $F_{\mathrm{int}}^{\mathrm{friction}} = 20$ pN (oscillation amplitude 10 pm) requires an input voltage of $U_{\mathrm{in}} \approx 200\,\mu$V.

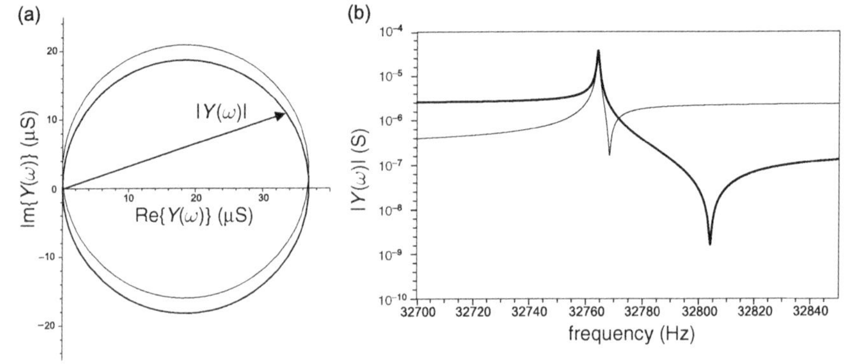

Fig. 7.9 Plots of the admittance $Y(\omega)$. (a) Nyquist plot of the admittance in the complex plane. The values used for the plot are $C_0 = 1.2$ pF, $L = 8.1365$ kH, $R = 27.1$ kΩ, and $C = 2.9$ fF. For the thin curve the stray capacitance was increased by a factor of 10. (b) The absolute value of the admittance as a function of the frequency.

Such a small voltage is difficult to deliver and requires voltage dividers close to the tuning-fork circuit if a reasonable signal-to-noise ratio is to be achieved. From this perspective, mechanical excitation can be favorable over electrical excitation. Finally, it should be noted that the piezo-electromechanical coupling constant α (Eq. (7.9)) can be determined if both the mechanical constants and the equivalent electrical constants of the tuning fork are known [17]. For example, by equating the potential energies $Q^2/(2C) = k_{\text{stat}}x_0^2/2$ and replacing the charge Q by αx_0 one finds

$$\alpha = \sqrt{k_{\text{stat}}C}. \tag{7.20}$$

Similar relationships can be derived by considering the equivalence of kinetic energies.

7.2 Normal-force methods

Using shear-force interactions to control the probe–sample distance has the advantage that any type of probe tip can be used as long as it is shaped approximately like a pencil and is small enough to be attached to a tuning fork. The disadvantage of this configuration is that the spring constant of the probe normal to the surface is very high. This means that a small instability or even the unavoidable small error in probe–sample distance control (as might occur at steep steps in the sample) is immediately translated into a very high normal force acting on the probe apex. Thus, shear-force feedback is a risky operation if there is little information on surface topology. In AFM, this problem is less important since commercial AFM cantilevers have well-defined and rather small spring constants normal to the probing tip. As a consequence, small instabilities result in only small excess forces acting on the probe apex. For these reasons, and with the goal of mass production, integration, and user friendliness in mind, there have been several attempts to integrate optical probes onto AFM cantilevers. In the following, we will discuss two different implementations working in normal-mode operation.

7.2.1 Tuning fork in tapping mode

By using the arrangement shown in Fig. 7.10 a probe attached to a tuning fork can also be operated in the normal-force mode. For optical fibers it was found necessary to break the fiber just above the fixation point on the tuning fork in order to allow free vibration of the prong [19]. Light is then delivered via a second cleaved fiber that is positioned just above the probe fiber. In normal-force operation, the attached fiber probe is allowed to protrude several millimeters beyond the attachment point because the normal motion is not able to excite fiber vibration. For example, the protruding fiber can be dipped into a liquid cell without wetting the tuning fork, which is very favorable for biological imaging. Also, since tuning-fork prongs are very stiff cantilevers they can be used for non-contact AFM in UHV since snap-into-contact effects appear only at very small probe–sample distances [20].

7.2.2 Bent-fiber probes

Cantilevered aperture probes with reasonably soft spring constants can be created by deforming standard fiber probes during the fabrication process using a CO_2 laser. The fiber is aligned parallel to the sample surface with the end of the bent fiber facing the sample perpendicularly. During raster scanning, the vertical motion of the fiber can be read out by standard AFM beam-deflection techniques. Figure 7.11 shows a selection of cantilevered fiber probes found in the literature. Because of their soft spring constants and the good Q-factors, bent-fiber probes have been used for imaging of soft samples under liquids, see e.g. Refs. [21, 22].

7.3 Topographic artifacts

In any type of scanning probe microscopy, image formation relies on recording a strongly distance-dependent physical interaction between probe and sample. The information encoded in the recorded images depends on the tip shape and on the path the tip takes. In AFM, for example, non-ideal tip shapes are an important source of misinterpretations. Blunt tips lead to low-pass-filtered images, i.e. deep and narrow trenches cannot be recorded because the tip does not fit in (see e.g. [23]). In some scanning probe techniques a single tip is capable of measuring several interactions simultaneously. For example, AFM can record force and friction by simultaneously measuring cantilever bending and torsion. However, only one of these measurements can be used as a feedback signal for controlling the probe–sample distance. While the feedback keeps one signal constant, it can introduce artifacts into the other signal. For example, as the shear-force feedback in a near-field optical microscope adjusts for a distance change, the vertical motion of the optical probe can lead to intensity variations that are not related to the optical properties of the sample. In this section we will analyze potential artifacts in near-field optical imaging that arise from the fact that the optical signal is an auxiliary signal not used in the feedback loop.

Let us denote by X the distance-dependent *feedback signal* originating from a specific probe–sample interaction such as shear force or normal force. The respective X image will reflect the piezo movements that were necessary in order to keep X constant during scanning. All other signals are *auxiliary signals* that result from the boundary condition

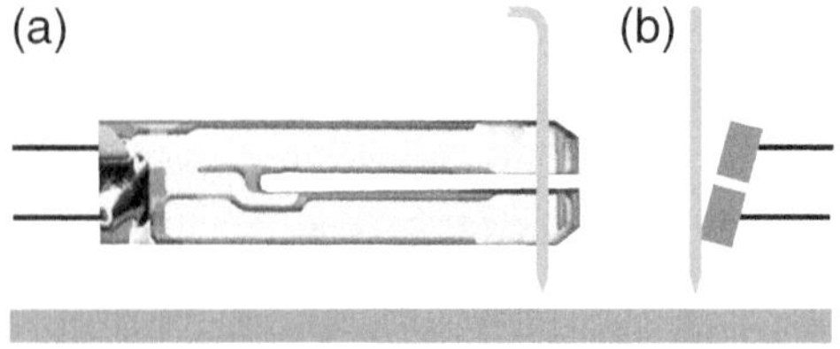

Fig. 7.10 A tuning fork operated in the normal-force mode. The tuning fork is aligned parallel to the sample while the probe is oscillating perpendicularly. (a) Side view. (b) Front view. The fork is tilted slightly in order not to affect the second arm.

X = constant. In principle, any distance-dependent signal can serve as the feedback signal. It must, however, fulfill the following conditions. (1) The probe–sample distance dependence must be short-ranged in order to maintain the probe in close proximity to the sample and thus to guarantee high resolution. (2) The distance dependence must be a piecewise monotonic function in order to guarantee that we have a stable feedback loop. Typically, a near-field optical microscope renders two simultaneously recorded images: (1) a topographic image originating from keeping the shear-force feedback signal constant and (2) an optical near-field image due to spatially varying optical properties of the sample and probe–sample distance variations. The optical image can result, for example, from local variations of sample transmission or from spatially distributed fluorescent centers.

In most cases, the optical interaction is not suitable as a feedback signal because it is neither short-ranged nor monotonically dependent on the probe–sample distance. For example, the optical transmission of an aperture probe near a transparent substrate was discussed in Chapter 6. If the emission is integrated over a large range of angles that also covers angles larger than the critical angle of the substrate, an increase of the transmission for small distances is observed. For larger distances, however, interference undulations render the optical response non-monotonic. Furthermore, the local light transmission could be completely suppressed when the probe is scanned over a metal patch. This would result in a loss of the feedback signal in an unpredictable way. As a consequence, optical signals are recorded under the condition that the shear-force interaction is maintained constant. This condition can be responsible for topographic artifacts in the near-field optical signal.

A representative sample with large topographic variations is depicted in Fig. 7.12. It exhibits uniform optical properties but its topographic features are large compared with the overall shape of the optical probe. From the discussion in Chapter 6 we know that aperture probes have a more or less conical shape with a flat facet at the apex. For the following we assume that the short-range probe–sample distance dependence of the optical signal decreases monotonically. This is reasonable because of the confined and enhanced fields near the aperture. The topography of the sample (S) is assumed to be measured via shear-force feedback and, since the probe's profile is not a delta function, the measured

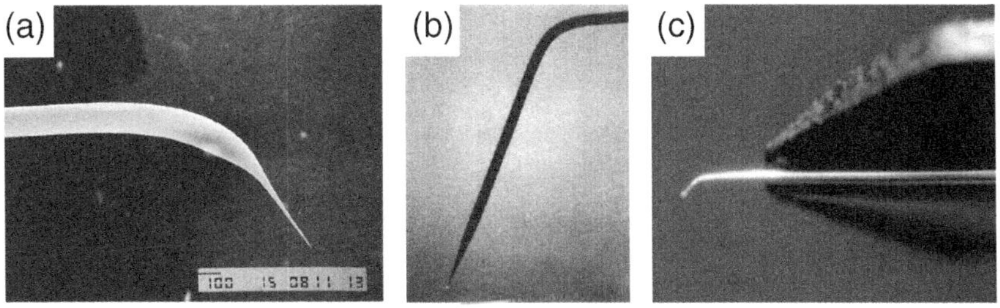

Fig. 7.11 Cantilevered fiber probes. (a) A bent-fiber probe with a mirror facet for beam deflection. The resonance frequency of the cantilevered probe is ~14 kHz and the Q-factor is ~30 in water, which is sufficient to perform near-field imaging on soft samples in liquid cells. From [21]. (b) A different type of bent-fiber probe. The resonances are typically in the range 30–60 kHz, the Q-factors are larger than 100, and the measured spring constants are 300–400 N m^{-1}. From [22]. (c) A commercially available cantilevered fiber probe. Image courtesy of Nanonics Imaging Ltd.

profile (T) will always differ from the actual sample profile S. The "mechanical" point of contact changes during scanning and gives rise to variations in the "optical" probe–sample distance. This distance can be defined as the vertical distance between the center of the aperture and the sample profile S. Since the optical signal is distance-dependent, it will reflect differences between S and T. The resulting optical signal is sketched in Fig. 7.12, trace O. It demonstrates the appearance of features in the optical image that are related purely to a topographic artifact.

A second limiting case is a sample with uniform optical properties with topographic features that are small compared with the overall shape of the probe (see Fig. 7.13). The end of an aperture probe is typically not smooth but exhibits grains that result from the metal-evaporation process (cf. Fig. 6.18). These grains often act as minitips that mediate the shear-force interaction. Here we assume that a single minitip is active. Because of the minitip, the apparent topography T will match the actual topography S very well. The probe produces an excellent high-resolution topographic image. However, while scanning over the small features of the sample S in force feedback, the average distance between the optical probe and the sample surface will change because of the distance-dependent optical signal. This leads to an optical image that contains small features that are highly correlated to the topography. In particular, it is possible that the size of these optical features turns out to be *much smaller* than what could be expected from the available optical resolution, e.g. estimated from the aperture diameter of the optical probe determined independently by SEM.

7.3.1 Phenomenological theory of artifacts

In order to put the discussion on more solid ground, we introduce system signal functions $S_{NFO}(x, y, z)$ and $S_{SF}(x, y, z)$ which represent the optical signal and the distance-dependent feedback signal, respectively [3]. Both signals depend on the coordinates (x, y, z) of the probe relative to the sample. The signal S_{NFO} can represent, for example, the locally transmitted or reflected light, polarized or depolarized components of locally scattered light, or the fluorescence due to local excitation by the near-field probe. S_{NFO} can also be the

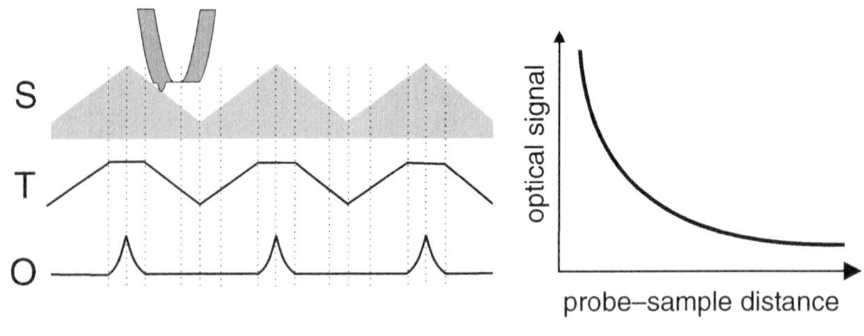

Fig. 7.12 Near-field optical imaging of a sample with large topographic variations. Left panel: S, sample profile; T, apparent topography measured by the probe; O, detected optical signal resulting from the particular probe–sample distance dependence (right panel).

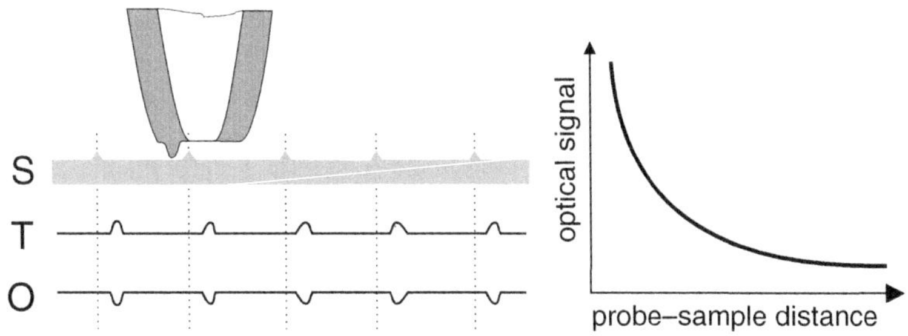

Fig. 7.13 Near-field optical imaging of a sample with small topographic variations. Left panel: S, sample profile; T, apparent topography as measured by the probe; O, detected optical signal resulting from the particular probe–sample distance dependence (right panel).

amplitude or phase of a modulated signal if differential techniques such as probe dithering are employed. Typically, S_{NFO} exhibits a weaker probe–sample distance dependence than that of the feedback signal S_{SF}.

The signals that are actually recorded during an experiment can be derived from $S_{NFO}(x, y, z)$ and $S_{SF}(x, y, z)$ by specifying a path that the probe takes. This path depends on the mode of operation of the microscope. Let these recorded signals be $R_{NFO}(x, y)$ and $R_{SF}(x, y)$, where

$$R_{NFO}(x, y) = S_{NFO}[x, y, z_{scan}(x, y)], \tag{7.21}$$

$$R_{SF}(x, y) = S_{SF}[x, y, z_{scan}(x, y)]. \tag{7.22}$$

Here, $z_{scan}(x, y)$ is the path of the probe. It can be derived from the voltage applied to the distance-controlling piezo element. The relation between the different signals is illustrated in Fig. 7.14.

Constant-height mode

In constant-height mode, the probe is scanned in a plane parallel to the average object surface, resulting in

$$z_{scan} = z_{set}, \tag{7.23}$$

$$R_{NFO}(x, y) = \overline{S}_{NFO}(z_{set}) + \delta S_{NFO}(x, y, z_{set}), \tag{7.24}$$

where we separated a constant background $\overline{S}_{NFO}$ from the signal. Any structure visible in the scan image corresponds to a lateral variation of S_{NFO} originating from *optical* or surface-related properties of the sample.

Constant-gap mode

In constant-gap mode, the feedback forces the probe to follow a path of (nearly) constant probe–sample separation. Consequently,

$$R_{SF}(x, y) = S_{SF}(x, y; z_{scan}) \approx R_{set}, \tag{7.25}$$

$$z_{\text{scan}} = \bar{z} + \delta z(x, y), \tag{7.26}$$

$$R_{\text{NFO}}(x, y) = \bar{S}_{\text{NFO}}(\bar{z}) + \delta S_{\text{NFO}}(x, y, \bar{z}) + \left.\frac{\partial S_{\text{NFO}}}{\partial z}\right|_{\bar{z}} \cdot \delta z. \tag{7.27}$$

In Eq. (7.25) the $\approx$ symbol indicates possible deviations caused by technical limitations of the electromechanical feedback circuit. Such deviations can become significant when the topography undergoes rapid changes and/or the scan speed is too high. Furthermore, $\bar{z}$ is the average z-position of the probe, and $\delta z(x, y)$ describes the variations of the z-position around $\bar{z}$ due to the feedback. It should be emphasized that the following considerations are valid for any path that the probe may take, no matter whether it follows the topography exactly or not.

The signal $R_{\text{NFO}}(x, y)$ in Eq. (7.27) is developed into a power series of δz of which only the first terms are retained. The first two terms render the same signal as is obtained for operation in constant-height mode. However, the third term represents the coupling of the vertical z-motion with the optical signal. It is this term that leads to common artifacts. For the optical properties to dominate, the variations of light intensity in a scan image have to satisfy

$$\delta S_{\text{NFO}}(x, y; \bar{z}) \gg \left.\frac{\partial S_{\text{NFO}}}{\partial z}\right|_{\bar{z}} \cdot \delta z. \tag{7.28}$$

This condition becomes more difficult to achieve the stronger the light confinement of the optical probe is. This is because a laterally confined field decays very rapidly with distance from the probe. Therefore, probe–sample distance variations have a much stronger effect and can easily overshadow any contrast originating from the optical properties of the sample.

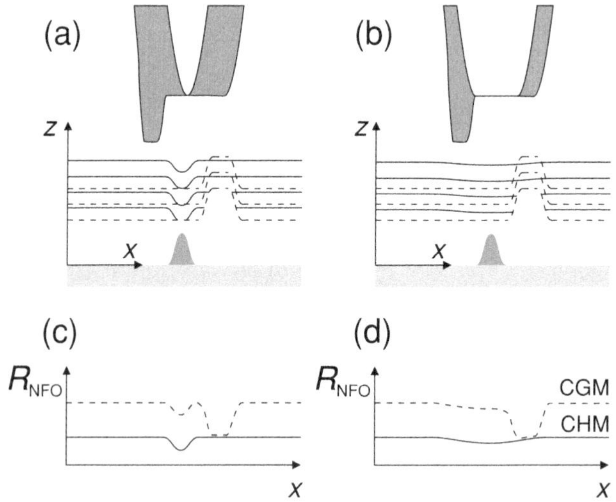

Fig. 7.14 Influence of probe geometry on recorded scan images. The optical signal S_{NFO} is represented by solid lines and the feedback signal S_{SF} by dashed lines. Both scan lines are plotted relative to the center of the aperture probe. The minitip on the rim of the aperture is oversized in order to better visualize the resulting effects. Images (c) and (d) show for cases (a) and (b), respectively, recorded scan lines for different modes of operation: constant-gap mode (CGM) and constant-height mode (CHM). CGM introduces artifacts because of the varying probe–sample distance. From [3].

For two different probes, Figs. 7.14(c) and (d) show the signals that are recorded in constant-height mode and in constant-gap mode. Only the probe with the small aperture provides an optical image that is representative of the sample. The large-aperture probe cannot generate any high-resolution optical image, and for operation in constant-gap mode the scan lines are dominated by the shear-force response specific to the passage over the bump. In Fig. 7.14(c), the true near-field signal can still be recognized, but in Fig. 7.14(d) the CGM trace is not at all related to the optical properties of the sample.

7.3.2 Example of optical artifacts

A simple experiment will serve as an illustration of artifacts originating from different modes of operation. Figure 7.15 shows topographic images and near-field optical transmission-mode images of a so-called Fischer projection pattern [24]. Such patterns are created by evaporating a thin layer of metal onto a closely packed monolayer of latex spheres. The triangular voids between the spheres are filled with metal. After metal evaporation, the latex spheres are washed away in an ultrasonic bath. The result is a sample with periodically arranged triangular patches. These patches, when imaged at close proximity, show strong optical-absorption contrast. The process of using microspheres for creating nanostructured surfaces is also called *nanosphere lithography*.

The same sample was imaged using two different aperture probes: (1) a probe with an aperture on the order of 50 nm (a good probe) and (2) a probe with a large aperture of 200 nm (a bad probe). Because the metal patches are created using spheres of diameter 200 nm, the resulting triangular patches have a characteristic size of about 50 nm, which can be resolved only using the good probe. For both probes, two sets of images

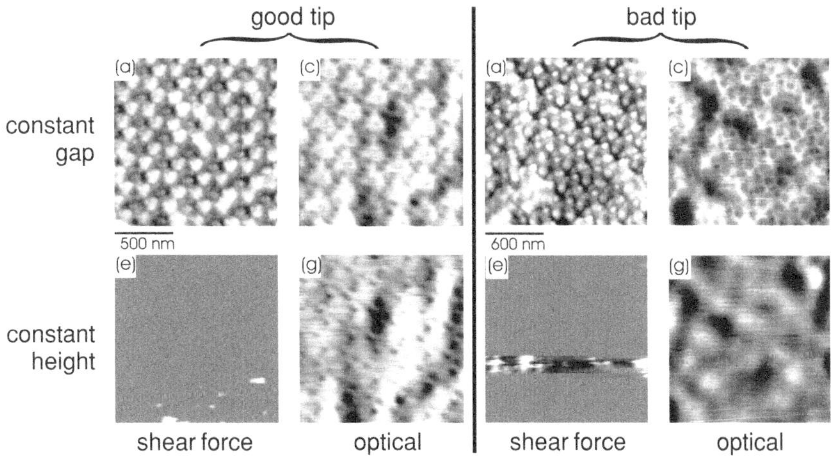

Fig. 7.15 Imaging of a latex-sphere projection pattern in constant-gap mode (upper row) and constant-height mode (lower row) with two different near-field probes, a "good" probe with an aperture of 50 nm (left side) and a "bad" probe with an aperture of 200 nm (right side). The constant-gap topographic and optical images show sharp features for both probes but only the constant-height-mode image for the "good" probe shows optical contrast.

were recorded: one in constant-gap mode, using shear-force feedback, and one in constant-height mode. The left panel shows the results for the good probe: in constant-gap mode the topography of the sample is well reproduced, probably due to a mini-tip on the aperture. The optical image strongly resembles the topographic image. It is difficult to tell how much the optical signal is influenced by the feedback. When the same area is imaged using constant-height mode (lower left row), the topographic signal is constant apart from isolated spots where the feedback becomes activated to prevent the probe from touching the surface (white spots). However, the optical signal completely changes its appearance. The contrast becomes much clearer and the metal patches are well resolved. For the bad probe, we observe an optical image with fine details only in constant-gap mode. As soon as we switch to constant-height mode the optical resolution becomes much worse. This shows clearly that the apparent optical resolution observed in the constant-gap optical image is purely artificially induced by the feedback loop.

7.3.3 Discussion

We have determined that if a force feedback is applied to control the probe–sample distance of samples with structured surfaces it is not possible to record optical images with pure optical contrast. Images recorded in constant-height mode are more likely to reflect the true optical resolution and contrast. Constant-height imaging does not use feedback control while scanning. The probe is raster scanned in a plane parallel to the mean sample surface. The measured optical signal thus cannot be influenced by feedback movements of the tip. Although constant-height images are more likely to represent the optical properties of the sample, they are still subject to misinterpretations because of the varying distance between probe and sample when scanning over structured surfaces. *Real* optical contrast can be expected only if the aperture size is sufficiently small compared with the characteristic size of sample features and the local optical coupling between probe and sample is large. Although the contrast originating from a varying probe–sample distance is a purely topographic effect, it should not be considered an artifact but is rather a property inherent to near-field optical imaging. Since the minimum distance between probe and sample is given by the highest topographic feature within the scan range, high resolution can be expected only for samples with low topography. Features on the bottom of surface depressions will be poorly resolved. In short, only features interacting with the highly localized near-field of the optical probe can be imaged with high resolution.

Image interpretation can be greatly facilitated by taking spectroscopic information into account. For example, fluorescence spectra and Raman spectra provide a highly specific fingerprint for the chemical composition of the sample (e.g. the structure of a carbon nanotube). Thus, near-field optical imaging combined with spectroscopy is able to unambiguously localize certain target molecules. Consequently, the recorded images are artifact-free maps of the spatial distribution of the target species. Despite this clear advantage, varying probe–sample distances will still pose a problem for quantifying the local concentration of the target species.

Problems

7.1 In tip-enhanced microscopy, a pointed gold wire is attached to one arm of a tuning fork. Assume that the wire is cylindrical with a diameter of 100 μm and that the tuning fork's resonance frequency is 32.7 kHz. In order that the attached gold wire follows the oscillation of the tuning fork more or less instantaneously, the resonance frequency of the protruding wire must be at least twice the tuning-fork frequency. Determine the maximum length of the protruding wire.

7.2 With the help of the equipartition principle we determined the thermally activated oscillation x_{rms} of the tuning fork. Here we calculate the spectral force density $S_{\mathrm{F}}(f)$ in units of $\mathrm{N}^2\,\mathrm{Hz}^{-1}$. S_{F} is the spectral noise force that excites the end of a tuning-fork prong to a vibration amplitude x_{rms}. It has a flat frequency dependence (white noise) and can be determined through

$$x_{\mathrm{rms}}^2 = \int_0^\infty S_{\mathrm{F}} \frac{f_0^2/k}{(f_0^2 - f^2) + \mathrm{i} f f_0/Q} \, \mathrm{d}f.$$

Here, the Lorentzian term following S_{F} is the transfer function of the tuning fork.

(1) Determine S_{F} in terms of the spring constant k, the Q-factor Q, the temperature T, and the resonance frequency f_0. Hint: evaluate the integral in the limit $Q \gg 1$ and apply the equipartition theorem.

(2) Use $k = 40\,\mathrm{kN\,m^{-1}}$, $T = 300\,\mathrm{K}$, $f_0 = 32.7\,\mathrm{kHz}$, and $Q = 1000$ to determine the thermal force in a spectral bandwidth of 100 Hz.

7.3 Owing to the typically high Q-factor of a tuning fork it takes a long time for the oscillation amplitude to respond to a sudden change of the feedback signal.

(1) Derive the solution given in Eq. (7.16) for a tuning fork whose frequency changes abruptly from one frequency to another at the time $t = 0$. Determine the values of x_{t}, φ_{t} and ω_{t}.

(2) Repeat the calculation but assume that the driving force F changes abruptly from one value to another at $t = 0$.

(3) Discuss the main difference between the solutions in parts (1) and (2).

References

[1] G. Binnig and H. Rohrer, "Scanning tunneling microscopy," *Helv. Phys. Acta* **55**, 726–735 (1982).

[2] G. Binnig, C. F. Quate, and C. Gerber, "Atomic force microscope," *Phys. Rev. Lett.* **56**, 930–933 (1986).

[3] B. Hecht, H. Bielefeldt, L. Novotny, Y. Inouye, and D. W. Pohl, "Facts and artifacts in near-field optical microscopy," *J. Appl. Phys.* **81**, 2492–2498 (1997).

[4] R. Carminati, A. Madrazo, M. Nieto-Vesperinas, and J.-J. Greffet, "Optical content and resolution of near-field optical images: influence of the operating mode," *J. Appl. Phys.* **82**, 501–509 (1997).

[5] K. Karrai and I. Tiemann, "Interfacial shear force microscopy," *Phys. Rev. B* **62**, 13 174–13 181 (2000).
[6] B. C. Stipe, H. J. Mamin, T. D. Stowe, T. W. Kenny, and D. Rugar, "Noncontact friction and force fluctuations between closely spaced bodies," *Phys. Rev. Lett.* **87**, 96801 (2001).
[7] J. R. Zurita-Sánchez, J.-J. Greffet, and L. Novotny, "Friction forces arising from fluctuating thermal fields," *Phys. Rev. A* **69**, 022902 (2004).
[8] L. D. Landau and E. M. Lifshitz, *Theory of Elasticity*. Oxford: Pergamon (1986).
[9] T. R. Albrecht, P. Grütter, D. Horne, and D. Rugar, "Frequency modulation detection using high-Q cantilevers for enhanced force microscope sensitivity," *J. Appl. Phys.* **69**, 668–673 (1991).
[10] E. Betzig, P. L. Finn, and S. J. Weiner, "Combined shear force and near-field scanning optical microscopy," *Appl. Phys. Lett.* **60**, 2484–2486 (1992).
[11] R. Toledo-Crow, P. C. Yang, Y. Chen, and M. Vaez-Iravani, "Near-field differential scanning optical microscope with atomic force regulation," *Appl. Phys. Lett.* **60**, 2957–2959 (1992).
[12] D. Rugar, H. J. Mamin, and P. Guethner, "Improved fiber-optic interferometer for atomic force microscopy," *Appl. Phys. Lett.* **55**, 2588–2590 (1989).
[13] G. Tarrach, M. A. Bopp, D. Zeisel, and A. J. Meixner, "Design and construction of a versatile scanning near-field optical microscope for fluorescence imaging of single molecules," *Rev. Sci. Instrum.* **66**, 3569–3575 (1995).
[14] J. Barenz, O. Hollricher, and O. Marti, "An easy-to-use non-optical shear-force distance control for near-field optical microscopes," *Rev. Sci. Instrum.* **67**, 1912–1916 (1996).
[15] J. W. P. Hsu, M. Lee, and B. S. Deaver, "A nonoptical tip–sample distance control method for near-field scanning optical microscopy using impedance changes in an electromechanical system," *Rev. Sci. Instrum.* **66**, 3177–3181 (1995).
[16] K. Karrai and R. D. Grober, "Piezoelectric tip–sample distance control for near field optical microscopes," *Appl. Phys. Lett.* **66**, 1842–1844 (1995).
[17] J. Rychen, T. Ihn, P. Studerus, *et al.*, "Operation characteristics of piezoelectric quartz tuning forks in high magnetic fields at liquid helium temperatures," *Rev. Sci. Instrum.* **71**, 1695–1697 (2000).
[18] R. D. Grober, J. Acimovic, J. Schuck, *et al.*, "Fundamental limits to force detection using quartz tuning forks," *Rev. Sci. Instrum.* **71**, 2776–2780 (2000).
[19] A. Naber, H.-J. Maas, K. Razavi, and U. C. Fischer, "Dynamic force distance control suited to various probes for scanning near-field optical microscopy," *Rev. Sci. Instrum.* **70**, 3955–3961 (1999).
[20] F. J. Giessibl, S. Hembacher, H. Bielefeldt, and J. Mannhart, "Subatomic features on the silicon (111)-(7 × 7) surface observed by atomic force microscopy," *Science* **289**, 422–425 (2000).
[21] H. Muramatsu, N. Chiba, K. Homma, *et al.*, "Near-field optical microscopy in liquids," *Appl. Phys. Lett.* **66**, 3245–3247 (1995).

[22] C. E. Talley, G. A. Cooksey, and R. C. Dunn, "High resolution fluorescence imaging with cantilevered near-field fiber optic probes," *Appl. Phys. Lett.* **69**, 3809–3811 (1996).

[23] D. Keller, "Reconstruction of STM and AFM images distorted by finite-size tips," *Surf. Sci.* **253**, 353–364 (1991).

[24] U. Ch. Fischer and H. P. Zingsheim, "Submicroscopic pattern replication with visible light," *J. Vac. Sci. Technol.* **19**, 881–885 (1981).

8 Optical interactions

At the heart of nano-optics are light–matter interactions on the nanometer scale. For example, optically excited single molecules are used to probe local environments and metal nanostructures are exploited for extreme light localization and enhanced sensing. Furthermore, various nanoscale structures are used in near-field optics as local light sources.

The scope of this chapter is to discuss the interactions of light with nanoscale systems. The light–matter interaction depends on many parameters, such as the atomic composition of the materials, their geometry and size, and the frequency and intensity of the radiation field. Nevertheless, there are many issues that can be discussed from a more or less general point of view.

To rigorously understand light–matter interactions we need to invoke quantum electrodynamics (QED). There are many textbooks that provide a good understanding of optical interactions with atoms or molecules, and we especially recommend the books in Refs. [1–3]. Since nanometer-scale structures are often too complex to be solved rigorously by QED, one often needs to stick to classical theory and invoke the results of QED in a phenomenological way.

8.1 The multipole expansion

In this section we consider an arbitrary material system that is small compared with the wavelength of light. We call this material system a *particle*. Although it is small compared with the wavelength, this particle consists of many atoms or molecules. On a macroscopic scale the charge density ρ and current density $\mathbf{j}$ can be treated as continuous functions of position. However, atoms and molecules are made of discrete charges that are spatially separated. Thus, the microscopic structure of matter is not considered in the macroscopic Maxwell equations. The macroscopic fields are local spatial averages over microscopic fields.

In order to derive the potential energy for a microscopic system we have to give up the definitions of the electric displacement $\mathbf{D}$ and the magnetic field $\mathbf{H}$ and consider only the field vectors $\mathbf{E}$ and $\mathbf{B}$ in the empty space between a set of discrete charges q_n. We thus make

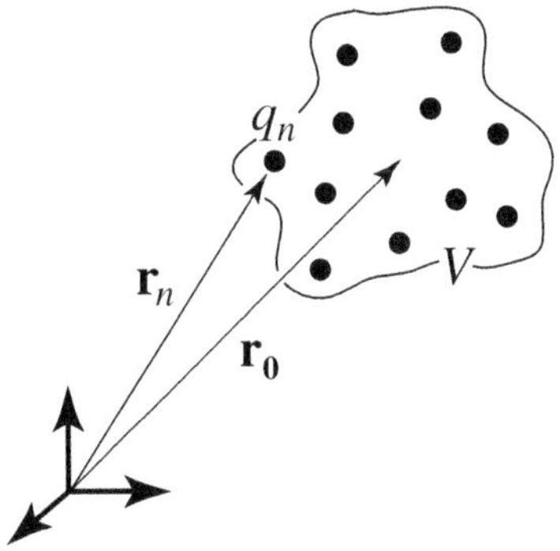

Fig. 8.1 In the microscopic picture, optical radiation interacts with the discrete charges q_n of matter. The collective response of the charges with coordinates $\mathbf{r}_n$ can be described by a multipole expansion with origin $\mathbf{r}_0$.

the replacements $\mathbf{D} = \varepsilon_0\mathbf{E}$ and $\mathbf{B} = \mu_0\mathbf{H}$ in Maxwell's equations (cf. Eqs. (2.1)–(2.4)) and set

$$\rho(\mathbf{r}) = \sum_n q_n\delta[\mathbf{r} - \mathbf{r}_n], \tag{8.1}$$

$$\mathbf{j}(\mathbf{r}) = \sum_n q_n\dot{\mathbf{r}}_n\delta[\mathbf{r} - \mathbf{r}_n], \tag{8.2}$$

where $\mathbf{r}_n$ denotes the position vector of the nth charge and $\dot{\mathbf{r}}_n$ its velocity (Fig. 8.1). The total charge and current of the particle are obtained by a volume integration over ρ and $\mathbf{j}$.

To derive the polarization and magnetization of the charge distribution we consider the total current density as defined in Eq. (2.10),

$$\mathbf{j} = \frac{\mathrm{d}\mathbf{P}}{\mathrm{d}t} + \nabla\times\mathbf{M}. \tag{8.3}$$

We ignored the contribution of the source current $\mathbf{j}_\mathrm{s}$ which generates the incident field $\mathbf{E}_\mathrm{inc}$ since it is not part of the considered particle. Furthermore, we incorporated the conduction current $\mathbf{j}_\mathrm{c}$ into the polarization current. To solve for $\mathbf{P}$ we apply the operator $\nabla\cdot$ to both sides of Eq. (8.3). The last term on the right-hand side vanishes because $\nabla\cdot\nabla\times = 0$, and the term on the left can be related to the time derivative of the charge density through the continuity equation for charge (2.5). We then obtain

$$\rho = -\nabla\cdot\mathbf{P}. \tag{8.4}$$

If the particle is not charge neutral we need to add the net charge density to the right-hand side. Using Eq. (8.1) for the charge density it is possible to solve for $\mathbf{P}$ as [1]

$$\mathbf{P}(\mathbf{r}) = \sum_n q_n\mathbf{r}_n\int_0^1 \delta[\mathbf{r} - s\mathbf{r}_n]\mathrm{d}s. \tag{8.5}$$

Together with the current density in Eq. (8.2) this expression can be introduced into Eq. (8.3). It is then possible to solve for **M** as [1]

$$\mathbf{M}(\mathbf{r}) = \sum_n q_n \mathbf{r}_n \times \dot{\mathbf{r}}_n \int_0^1 s\delta[\mathbf{r} - s\mathbf{r}_n]\mathrm{d}s. \tag{8.6}$$

To calculate the potential energy of the particle in the incident field we first consider fixed charges, i.e. the charge distribution is not induced by the incident field. Instead, the charge distribution is determined by the atomic and interatomic potentials. Of course, the particle is polarizable, but for the moment we consider this to be a secondary effect.

We now consider the interaction between a discrete charge distribution and an electromagnetic field. The incident field in the absence of the charge distribution is denoted as $\mathbf{E}_{\mathrm{inc}}$. The electric potential energy of the *permanent* microscopic charge distribution is determined as [4]

$$V_{\mathrm{E}} = -\int_V \mathbf{P} \cdot \mathbf{E}_{\mathrm{inc}}\, \mathrm{d}V = -\sum_n q_n \int_0^1 \mathbf{r}_n \cdot \mathbf{E}_{\mathrm{inc}}(s\mathbf{r}_n)\mathrm{d}s. \tag{8.7}$$

Next, we expand the electric field $\mathbf{E}_{\mathrm{inc}}$ in a Taylor series with origin at the center of the particle. For convenience we choose this origin at $\mathbf{r} = 0$ and obtain

$$\mathbf{E}_{\mathrm{inc}}(s\mathbf{r}_n) = \mathbf{E}_{\mathrm{inc}}(0) + [s\mathbf{r}_n \cdot \nabla]\mathbf{E}_{\mathrm{inc}}(0) + \frac{1}{2!}[s\mathbf{r}_n \cdot \nabla]^2\mathbf{E}_{\mathrm{inc}}(0) + \cdots . \tag{8.8}$$

This expansion can now be inserted into Eq. (8.7) and the integration over s can be carried out. Then, the electric potential energy expressed in terms of the multipole moments of the charges becomes

$$\begin{aligned} V_{\mathrm{E}} = & -\sum_n q_n\mathbf{r}_n \cdot \mathbf{E}_{\mathrm{inc}}(0) - \sum_n \frac{q_n}{2!}\mathbf{r}_n \cdot [\mathbf{r}_n \cdot \nabla]\mathbf{E}_{\mathrm{inc}}(0) \\ & - \sum_n \frac{q_n}{3!}\mathbf{r}_n \cdot [\mathbf{r}_n \cdot \nabla]^2\mathbf{E}_{\mathrm{inc}}(0) - \cdots . \end{aligned} \tag{8.9}$$

The first term is recognized as the *electric dipole interaction*

$$V_{\mathrm{E}}^{(1)} = -\mathbf{p} \cdot \mathbf{E}_{\mathrm{inc}}(0), \tag{8.10}$$

with the electric dipole moment defined as

$$\mathbf{p} = \sum_n q_n\mathbf{r}_n. \tag{8.11}$$

The next higher term in Eq. (8.9) is the *electric quadrupole interaction*, which can be written as

$$V_{\mathrm{E}}^{(2)} = -[\overleftrightarrow{\mathbf{Q}}\, \nabla] \cdot \mathbf{E}_{\mathrm{inc}}(0), \tag{8.12}$$

with the electric quadrupole moment defined as

$$\overleftrightarrow{\mathbf{Q}} = \frac{1}{2}\sum_n q_n\mathbf{r}_n\mathbf{r}_n, \tag{8.13}$$

where $\mathbf{r}_n\mathbf{r}_n$ denotes the outer product. Therefore, $\overleftrightarrow{\mathbf{Q}}$ becomes a tensor of rank two.[1] Since $\nabla \cdot \mathbf{E}_{\mathrm{inc}} = 0$ we can subtract any multiple of $\nabla \cdot \mathbf{E}_{\mathrm{inc}}$ from Eq. (8.12). We can therefore rewrite Eq. (8.12) as

$$V_{\mathrm{E}}^{(2)} = -\frac{1}{2}[(\overleftrightarrow{\mathbf{Q}} - A\overleftrightarrow{\mathbf{I}})\nabla] \cdot \mathbf{E}_{\mathrm{inc}}(0), \tag{8.14}$$

with an arbitrary constant A, which commonly is chosen as $A = (1/3)\,|\mathbf{r}_n|^2$ because this generates a traceless quadrupole moment. Thus, we can also define the quadrupole moment as

$$\overleftrightarrow{\mathbf{Q}} = \frac{1}{2}\sum_n q_n \left[\mathbf{r}_n\mathbf{r}_n - \frac{\overleftrightarrow{\mathbf{I}}}{3}\,|\mathbf{r}_n|^2\right]. \tag{8.15}$$

We avoid writing down the next higher multipole orders but we note that the rank of every next higher multipole increases by one.

The dipole interaction is determined by the electric field at the center of the charge distribution, whereas the quadrupole interaction is defined by the electric field gradient at the center. Thus, if the electric field is sufficiently homogeneous over the dimensions of the particle, the quadrupole interaction vanishes. This is why in small systems of charge, such as atoms and molecules, often only the dipole interaction is considered. This *dipole approximation* leads to the standard selection rules encountered in optical spectroscopy. However, the dipole approximation is not necessarily sufficient for nanoscale particles because of their larger size compared with that of an atom. Furthermore, if the particle interacts with an optical near-field it will experience strong field gradients. This increases the importance of the quadrupole interaction and modifies the standard selection rules. Thus, the strong field gradients encountered in near-field optics have the potential to excite usually forbidden transitions in larger quantum systems and thus extend the capabilities of optical spectroscopy.

A similar multipole expansion can be performed for the magnetic potential energy V_{M}. The lowest-order term is the *magnetic dipole interaction*

$$V_{\mathrm{M}}^{(1)} = -\mathbf{m}_{\mathrm{d}} \cdot \mathbf{B}_{\mathrm{inc}}(0), \tag{8.16}$$

with the magnetic dipole moment defined as

$$\mathbf{m}_{\mathrm{d}} = \sum_n [q_n/(2m_n)]\mathbf{r}_n \times (m_n\dot{\mathbf{r}}_n). \tag{8.17}$$

The magnetic moment is often expressed in terms of the angular momenta $\mathbf{I}_n = m_n\mathbf{r}_n \times \dot{\mathbf{r}}_n$, where m_n denotes the mass of the nth particle. We avoid deriving higher-order magnetic multipole terms since the procedure is analogous to the electric case.

So far we have considered the polarization and magnetization of a charge distribution that is *not* affected by the incident electromagnetic field. However, it is clear that the incident radiation will act on the charges and displace them from their unperturbed positions.

[1] If we denote the Cartesian components of $\mathbf{r}_n$ by $(x_{n_1}, x_{n_2}, x_{n_3})$ we can write Eq. (8.12) as $V_{\mathrm{E}}^{(2)} = -(1/2)\sum_{i,j}\left[\sum_n q_n x_{n_i} x_{n_j}\right]\left[(\partial/\partial x_i)E_j(0)\right]$.

This gives rise to an induced polarization and magnetization. The interaction of the incident field $\mathbf{E}_{\text{inc}}$ with the particle causes a change $\mathrm{d}\mathbf{P}$ in the polarization $\mathbf{P}$. The change in the electric potential energy $\mathrm{d}V_{\mathrm{E}}$ due to this interaction is

$$\mathrm{d}V_{\mathrm{E}} = -\int_V \mathbf{E}_{\text{inc}} \cdot \mathrm{d}\mathbf{P}\,\mathrm{d}V. \tag{8.18}$$

To calculate the total induced electric potential energy $V_{\mathrm{E,ind}}$ we have to integrate $\mathrm{d}V_{\mathrm{E}}$ over the polarization range $\mathbf{P}_{\mathrm{p}} \ldots \mathbf{P}_{\mathrm{p+i}}$, where $\mathbf{P}_{\mathrm{p}}$ and $\mathbf{P}_{\mathrm{p+i}}$ are the initial and final values of the polarization. We now assume that the interaction between the field and the particle is *linear* so that we can write $\mathbf{P} = \varepsilon_0 \chi \mathbf{E}_{\text{inc}}$. In this case we find for the total differential $\mathrm{d}(\mathbf{P} \cdot \mathbf{E}_{\text{inc}})$

$$\mathrm{d}(\mathbf{P} \cdot \mathbf{E}_{\text{inc}}) = \mathbf{E}_{\text{inc}} \cdot \mathrm{d}\mathbf{P} + \mathbf{P} \cdot \mathrm{d}\mathbf{E}_{\text{inc}} = 2\,\mathbf{E}_{\text{inc}} \cdot \mathrm{d}\mathbf{P}, \tag{8.19}$$

and the induced potential energy becomes

$$V_{\mathrm{E,ind}} = -\frac{1}{2}\int_V \left[\int_{\mathbf{P}_{\mathrm{p}}}^{\mathbf{P}_{\mathrm{p+i}}} \mathrm{d}(\mathbf{P} \cdot \mathbf{E}_{\text{inc}})\right] \mathrm{d}V. \tag{8.20}$$

Using $\mathbf{P}_{\mathrm{p+i}} = \mathbf{P}_{\mathrm{p}} + \mathbf{P}_{\mathrm{i}}$ we finally obtain

$$V_{\mathrm{E,ind}} = -\frac{1}{2}\int_V \mathbf{P}_{\mathrm{i}} \cdot \mathbf{E}_{\text{inc}}\,\mathrm{d}V. \tag{8.21}$$

This result states that the induced potential energy is smaller than the permanent potential energy by a factor 1/2. The other 1/2 portion is related to the work needed to build up the polarization. For $\mathbf{P}_{\mathrm{i}} > 0$ regions of high electric field create an attracting force on polarizable objects, a property that is used in optical trapping (cf. Section 14.4).

A similar derivation can be performed for the induced magnetization $\mathbf{M}_{\mathrm{i}}$ and its associated energy. The interesting outcome is that objects with $\mathbf{M}_{\mathrm{i}} > 0$ are repelled from regions of high magnetic field. This finding underlies the phenomenon of eddy-current damping. However, at optical frequencies induced magnetizations are practically zero.

8.2 The classical particle–field Hamiltonian

So far we have been concerned with the potential energy of a particle in an external electromagnetic field. However, for a fundamental understanding of the interaction of a particle with the electromagnetic field we need to know the total energy of the system consisting of particle and field. This energy remains conserved; the particle can borrow energy from the field (absorption) or it can donate energy to it (emission). The total energy corresponds to the classical Hamiltonian H, which constitutes the Hamiltonian operator $\hat{H}$ encountered in quantum mechanics. For particles consisting of many charges, the Hamiltonian soon becomes a very complex function: it depends on the mutual interaction between the charges, their kinetic energies, and the exchange of energy with the external field.

To understand the interaction between a particle and an electromagnetic field we first consider a single point-like particle with mass m and charge q. Later we generalize the situation to systems consisting of multiple charges and with finite size. The Hamiltonian for a single charge in an electromagnetic field is found by first deriving a Lagrangian function $L(\mathbf{r}, \dot{\mathbf{r}})$ that satisfies the Lagrange–Euler equation

$$\frac{\mathrm{d}}{\mathrm{d}t}\left(\frac{\partial L}{\partial \dot{q}}\right) - \frac{\partial L}{\partial q} = 0, \quad q = x, y, z. \tag{8.22}$$

Here, $\mathbf{q} = (x, y, z)$ and $\dot{\mathbf{q}} = (\dot{x}, \dot{y}, \dot{z})$ denote the coordinates and velocities of the charge, respectively.[2] To determine L, we first consider the (non-relativistic) equation of motion for the charge

$$\mathbf{F} = \frac{\mathrm{d}}{\mathrm{d}t}[m\dot{\mathbf{r}}] = q(\mathbf{E} + \dot{\mathbf{r}} \times \mathbf{B}), \tag{8.23}$$

and replace $\mathbf{E}$ and $\mathbf{B}$ by the vector potential $\mathbf{A}$ and scalar potential ϕ according to

$$\mathbf{E}(\mathbf{r}, t) = -\frac{\partial}{\partial t}\mathbf{A}(\mathbf{r}, t) - \nabla\phi(\mathbf{r}, t), \tag{8.24}$$

$$\mathbf{B}(\mathbf{r}, t) = \nabla \times \mathbf{A}(\mathbf{r}, t). \tag{8.25}$$

Now we consider the vector components of Eq. (8.23) separately. For the x-component we obtain

$$\begin{aligned}\frac{\mathrm{d}}{\mathrm{d}t}[m\dot{x}] &= -q\left[\frac{\partial \phi}{\partial x} + \frac{\partial A_x}{\partial t}\right] + q\left[\dot{y}\left(\frac{\partial A_y}{\partial x} - \frac{\partial A_x}{\partial y}\right) - \dot{z}\left(\frac{\partial A_x}{\partial z} - \frac{\partial A_z}{\partial x}\right)\right]\\ &= \frac{\partial}{\partial x}\left[-q\phi + q\left(A_x\dot{x} + A_y\dot{y} + A_z\dot{z}\right)\right]\\ &\quad - q\left[\frac{\partial A_x}{\partial t} + \dot{x}\frac{\partial A_x}{\partial x} + \dot{y}\frac{\partial A_y}{\partial y} + \dot{z}\frac{\partial A_z}{\partial z}\right].\end{aligned} \tag{8.26}$$

Identifying the last expression in brackets with $\mathrm{d}A_x/\mathrm{d}t$ (total differential) and rearranging terms, the equation above can be written as

$$\frac{\mathrm{d}}{\mathrm{d}t}\left[m\dot{x} + qA_x\right] - \frac{\partial}{\partial x}\left[-q\phi + q\left(A_x\dot{x} + A_y\dot{y} + A_z\dot{z}\right)\right] = 0. \tag{8.27}$$

This equation has almost the form of the Lagrange–Euler equation (8.22). Therefore, we seek a Lagrangian of the form

$$L = -q\phi + q\left(A_x\dot{x} + A_y\dot{y} + A_z\dot{z}\right) + f(x, \dot{x}), \tag{8.28}$$

with $\partial f/\partial x = 0$. With this choice, the first term in Eq. (8.22) leads to

$$\frac{\mathrm{d}}{\mathrm{d}t}\left(\frac{\partial L}{\partial \dot{q}}\right) = \frac{\mathrm{d}}{\mathrm{d}t}\left[qA_x + \frac{\partial f}{\partial \dot{x}}\right]. \tag{8.29}$$

This expression has to be identical with the first term in Eq. (8.27), which leads to $\partial f/\partial \dot{x} = m\dot{x}$. The solution $f(x, \dot{x}) = m\dot{x}^2/2$ can be substituted into Eq. (8.28) and, after generalizing to all degrees of freedom, we finally obtain

$$L = -q\phi + q\left(A_x\dot{x} + A_y\dot{y} + A_z\dot{z}\right) + \frac{1}{2}m\left(\dot{x}^2 + \dot{y}^2 + \dot{z}^2\right), \tag{8.30}$$

[2] It is a convention of the Hamiltonian formalism to designate the generalized coordinates by the symbol q. Here, it should not be confused with the charge q.

which can be written as

$$L = -q\phi + q\mathbf{v}\cdot\mathbf{A} + \frac{m}{2}\mathbf{v}\cdot\mathbf{v}. \tag{8.31}$$

To determine the Hamiltonian H we first calculate the canonical momentum[3] $\mathbf{p} = (p_x, p_y, p_z)$ conjugate to the coordinate $\mathbf{q} = (x, y, z)$ according to $p_i = \partial L/\partial \dot{q}_i$. The canonical momentum turns out to be

$$\mathbf{p} = m\mathbf{v} + q\mathbf{A}, \tag{8.32}$$

which is the sum of mechanical momentum $m\mathbf{v}$ and field momentum $q\mathbf{A}$. According to Hamiltonian mechanics, the Hamiltonian is derived from the Lagrangian according to

$$H(\mathbf{q},\mathbf{p}) = \sum_i \left[p_i\dot{q}_i - L(\mathbf{q},\dot{\mathbf{q}})\right], \tag{8.33}$$

in which all the velocities $\dot{q}_i$ have to be expressed in terms of the coordinates q_i and conjugate momenta p_i. This is easily done by using Eq. (8.32) as $\dot{q}_i = p_i/m - qA_i/m$. Using this substitution in Eqs. (8.30) and (8.33) we finally obtain

$$H = \frac{1}{2m}(\mathbf{p} - q\mathbf{A})^2 + q\phi. \tag{8.34}$$

This is the Hamiltonian of a *free* charge q with mass m in an external electromagnetic field. The first term renders the kinetic mechanical energy and the second term the potential energy of the charge. Notice that the derivation of L and H is independent of gauge, i.e. we did not imply any condition on $\nabla\cdot\mathbf{A}$. Using Hamilton's canonical equations $\dot{q}_i = \partial H/\partial p_i$ and $\dot{p}_i = -\partial H/\partial q_i$ it is straightforward to show that the Hamiltonian in Eq. (8.34) reproduces the equations of motion stated in Eq. (8.23).

The Hamiltonian of Eq. (8.34) is not yet the total Hamiltonian H_{tot} of the system "charge plus field" since we did not include the energy of the electromagnetic field. Furthermore, if the charge is interacting with other charges, as in the case of an atom or a molecule, we must take into account the interaction between the charges. In general, the total Hamiltonian for a system of charges can be written as

$$H_{\text{tot}} = H_{\text{particle}} + H_{\text{rad}} + H_{\text{int}}. \tag{8.35}$$

Here, H_{rad} is the Hamiltonian of the radiation field in the absence of the charges and H_{particle} is the Hamiltonian of the system of charges (particle) in the absence of the electromagnetic field. The interaction between the two systems is described by the interaction Hamiltonian H_{int}. Let us determine the individual contributions.

The particle Hamiltonian H_{particle} is determined by the sum of the kinetic energies $\mathbf{p}_n\cdot\mathbf{p}_n/(2m_n)$ of the N charges and the potential energies $V(\mathbf{r}_m,\mathbf{r}_n)$ between the charges (intramolecular potential), i.e.

$$H_{\text{particle}} = \sum_{n,m}\left[\frac{\mathbf{p}_n\cdot\mathbf{p}_n}{2m_n} + V(\mathbf{r}_m,\mathbf{r}_n)\right], \tag{8.36}$$

[3] Careful, we're using the same symbol for the dipole moment and the canonical momentum!

where the nth particle is specified by its charge q_n, mass m_n, and coordinate $\mathbf{r}_n$. Notice that $V(\mathbf{r}_m, \mathbf{r}_n)$ is determined in the absence of the external radiation field. This term is solely due to the Coulomb interaction between the charges. H_{rad} is defined by integrating the electromagnetic energy density W of the radiation field (Eq. (2.57)) over all space as[4]

$$H_{\text{rad}} = \frac{1}{2} \int \left[\varepsilon_0 E^2 + \mu_0^{-1} B^2 \right] \mathrm{d}V, \tag{8.37}$$

where $E^2 = |\mathbf{E}|^2$ and $B^2 = |\mathbf{B}|^2$. It should be noted that the inclusion of H_{rad} is essential for a rigorous quantum-electrodynamical treatment of light–matter interactions. This term ensures that the system consisting of particles and fields is conservative; it permits the interchange of energy between the atomic states and the states of the radiation field. Spontaneous emission is a direct consequence of the inclusion of H_{rad} and cannot be derived by semiclassical calculations in which H_{rad} is not included. Finally, to determine H_{int} we first consider each charge separately. Each charge contributes to H_{int} a term that can be derived from Eq. (8.34) as

$$H - \frac{\mathbf{p} \cdot \mathbf{p}}{2m} = -\frac{q}{2m} \left[\mathbf{p} \cdot \mathbf{A} + \mathbf{A} \cdot \mathbf{p} \right] + \frac{q^2}{2m} \mathbf{A} \cdot \mathbf{A} + q\phi. \tag{8.38}$$

Here, we subtracted the kinetic energy of the charge from the classical "particle–field" Hamiltonian since this term is already included in H_{particle}. Using $\mathbf{p} \cdot \mathbf{A} = \mathbf{A} \cdot \mathbf{p}$ and then summing the contributions of all N charges in the system we can write H_{int} as[5]

$$H_{\text{int}} = \sum_n \left[-\frac{q_n}{m_n} \mathbf{A}(\mathbf{r}_n, t) \cdot \mathbf{p}_n + \frac{q_n^2}{2m_n} \mathbf{A}(\mathbf{r}_n, t) \cdot \mathbf{A}(\mathbf{r}_n, t) + q_n \phi(\mathbf{r}_n, t) \right]. \tag{8.39}$$

In the next section we will show that H_{int} can be expanded into a multipole series similar to our previous results for V_{E} and V_{M}.

8.2.1 Multipole expansion of the interaction Hamiltonian

The Hamiltonian expressed in terms of the vector potential $\mathbf{A}$ and scalar potential ϕ is not unique. This is caused by the freedom of gauge, i.e. if the potentials are replaced by new potentials $\tilde{\mathbf{A}}$ and $\tilde{\phi}$ according to

$$\mathbf{A} \rightarrow \tilde{\mathbf{A}} + \nabla\chi \quad \text{and} \quad \phi \rightarrow \tilde{\phi} - \partial\chi/\partial t, \tag{8.40}$$

with $\chi(\mathbf{r}, t)$ being an arbitrary gauge function, then Maxwell's equations remain unaffected. This is easily seen by introducing the above substitutions into the definitions of $\mathbf{A}$ and ϕ

[4] This integration leads necessarily to an infinite result, which caused difficulties in the development of the quantum theory of light.

[5] In quantum mechanics, the canonical momentum $\mathbf{p}$ is converted into an operator according to $\mathbf{p} \rightarrow -\mathrm{i}\hbar\nabla$ (Jordan's rule), which also turns H_{int} into an operator. $\mathbf{p}$ and $\mathbf{A}$ commute only if the Coulomb gauge ($\nabla \cdot \mathbf{A} = 0$) is adopted.

(Eqs. (8.24) and (8.25)). To remove the ambiguity caused by the freedom of gauge we need to express H_{int} in terms of the original fields **E** and **B**. To do this, we first expand the electric and magnetic fields in a Taylor series with origin $\mathbf{r} = 0$ (cf. Eq. (8.8)),

$$\mathbf{E}(\mathbf{r}) = \mathbf{E}(0) + [\mathbf{r} \cdot \nabla]\mathbf{E}(0) + \frac{1}{2!}[\mathbf{r} \cdot \nabla]^2\mathbf{E}(0) + \cdots, \tag{8.41}$$

$$\mathbf{B}(\mathbf{r}) = \mathbf{B}(0) + [\mathbf{r} \cdot \nabla]\mathbf{B}(0) + \frac{1}{2!}[\mathbf{r} \cdot \nabla]^2\mathbf{B}(0) + \cdots, \tag{8.42}$$

and introduce these expansions into the definitions for **A** and ϕ (Eqs. (8.24) and (8.25)). The task is now to find an expansion of **A** and ϕ in terms of **E** and **B** such that the left- and right-hand sides of Eqs. (8.24) and (8.25) are identical. These expansions have been determined by Barron and Gray [5] as

$$\phi(\mathbf{r}) = \phi(0) - \sum_{i=0}^{\infty} \frac{\mathbf{r}[\mathbf{r} \cdot \nabla]^i}{(i+1)!} \cdot \mathbf{E}(0), \quad \mathbf{A}(\mathbf{r}) = \sum_{i=0}^{\infty} \frac{[\mathbf{r} \cdot \nabla]^i}{(i+2)i!}\, \mathbf{B}(0) \times \mathbf{r}. \tag{8.43}$$

Their insertion into the expression for H_{int} in Eq. (8.39) leads to the so-called *multipolar interaction Hamiltonian*

$$H_{\text{int}} = q_{\text{tot}}\,\phi(0,t) - \mathbf{p} \cdot \mathbf{E}(0,t) - \mathbf{m} \cdot \mathbf{B}(0,t) - [\overset{\leftrightarrow}{\mathbf{Q}}\,\nabla] \cdot \mathbf{E}(0,t) - \cdots, \tag{8.44}$$

in which we used the following definitions:

$$q_{\text{tot}} = \sum_n q_n, \qquad \mathbf{p} = \sum_n q_n \mathbf{r}_n, \qquad \mathbf{m} = \sum_n \frac{q_n}{2m_n}\mathbf{r}_n \times \tilde{\mathbf{p}}_n, \qquad \overset{\leftrightarrow}{\mathbf{Q}} = \sum_n \frac{q_n}{2}\mathbf{r}_n\mathbf{r}_n, \tag{8.45}$$

where q_{tot} is the total charge of the system, **p** the total electric dipole moment, **m** the total magnetic dipole moment, and $\overset{\leftrightarrow}{\mathbf{Q}}$ the total electric quadrupole moment. If the system of charges is charge neutral, the first term in H_{int} vanishes and we are left with an expansion that looks very much like the former expansion of the potential energy $V_{\text{E}} + V_{\text{M}}$. However, the two expansions are *not* identical! First, the new magnetic dipole moment is defined in terms of the canonical momenta $\tilde{\mathbf{p}}_n$ rather than by the mechanical momenta $m_n\dot{\mathbf{r}}_n$.[6] Second, the expansion of H_{int} contains a term nonlinear in $\mathbf{B}(0,t)$, which is non-existent in the expansion of $V_{\text{E}} + V_{\text{M}}$. The nonlinear term arises from the term $\mathbf{A} \cdot \mathbf{A}$ of the Hamiltonian and is referred to as the diamagnetic term. It reads

$$\sum_n \frac{q_n^2}{8m_n}\,[\mathbf{r}_n \times \mathbf{B}(0,t)]^2. \tag{8.46}$$

Our previous expressions for V_{E} and V_{M} have been derived by neglecting retardation and assuming weak fields. In this limit, the nonlinear term in Eq. (8.46) can be neglected and the canonical momentum can be approximated by the mechanical momentum.

The multipolar interaction Hamiltonian can easily be converted to an operator by simply applying Jordan's rule $\mathbf{p} \to -i\hbar\nabla$ and replacing the fields **E** and **B** by the corresponding

[6] A gauge transformation also transforms the canonical momenta. Therefore, the canonical momenta $\tilde{\mathbf{p}}_n$ are different from the original canonical momenta $\mathbf{p}_n$.

electric and magnetic field operators. However, this is beyond the present scope. Notice that the Hamiltonian H_{int} in Eq. (8.44) is gauge-independent. The gauge affects H_{int} only when the latter is expressed in terms of $\mathbf{A}$ and ϕ, not when it is represented by the original fields $\mathbf{E}$ and $\mathbf{B}$. The first term in the multipolar Hamiltonian of a charge-neutral system is the dipole interaction, which is identical to the corresponding term in V_{E}. In most circumstances, it is sufficiently accurate to reject the higher terms in the multipolar expansion. This is especially true for far-field interactions, for which the magnetic dipole and electric quadrupole interactions are roughly two orders of magnitude weaker than the electric dipole interaction. Therefore, standard selection rules for optical transitions are based on the electric dipole interaction. However, in strongly confined optical fields, as are encountered in near-field optics, higher-order terms in the expansion of H_{int} can become important and the standard selection rules can be violated. Finally, it should be noted that the multipolar form of H_{int} can also be derived from Eq. (8.39) by a unitary transformation [6]. This transformation, commonly referred to as the *Power–Zienau–Woolley transformation*, plays an important role in quantum optics [3].

We have established that to first order any neutral system of charges (particle) that is smaller than the wavelength of the interacting radiation can be viewed as a dipole. In the next section we will consider its radiating properties.

8.3 The radiating electric dipole

The current density due to a distribution of charges q_n with coordinates $\mathbf{r}_n$ and velocities $\dot{\mathbf{r}}_n$ has been given in Eq. (8.2). We can develop this current density in a Taylor series with origin $\mathbf{r}_0$, which is typically at the center of the charge distribution. If we keep only the lowest-order term we find

$$\mathbf{j}(\mathbf{r},t) = \frac{\mathrm{d}}{\mathrm{d}t}\mathbf{p}(t)\delta[\mathbf{r}-\mathbf{r}_0], \tag{8.47}$$

with the dipole moment

$$\mathbf{p}(t) = \sum_n q_n[\mathbf{r}_n(t)-\mathbf{r}_0]. \tag{8.48}$$

The dipole moment is identical with the definition in Eq. (8.11) for which we had $\mathbf{r}_0 = 0$. We assume a harmonic time dependence, which allows us to write the current density as $\mathbf{j}(\mathbf{r},t) = \mathrm{Re}\{\mathbf{j}(\mathbf{r})\exp(-\mathrm{i}\omega t)\}$ and the dipole moment as $\mathbf{p}(t) = \mathrm{Re}\{\mathbf{p}\exp(-\mathrm{i}\omega t)\}$. Equation (8.47) can then be written as

$$\mathbf{j}(\mathbf{r}) = -\mathrm{i}\omega\mathbf{p}\delta[\mathbf{r}-\mathbf{r}_0]. \tag{8.49}$$

Thus, to lowest order, any current density can be thought of as an oscillating dipole with its origin at the center of the charge distribution.

8.3.1 Electric dipole fields in a homogeneous space

In this section we will derive the fields of a dipole representing the current density of a small charge distribution located in a homogeneous, linear, and isotropic space. The fields of the dipole can be derived by considering two oscillating charges q of opposite sign, separated by an infinitesimal vector $\mathbf{ds}$. In this physical picture the dipole moment is given by $\mathbf{p} = q\,\mathbf{ds}$. However, it is more elegant to derive the dipole fields using the Green-function formalism developed in Section 2.12. There, we derived the so-called volume-integral equations (Eqs. (2.89) and (2.90))

$$\mathbf{E}(\mathbf{r}) = \mathbf{E}_0 + \mathrm{i}\omega\mu\mu_0 \int_V \overleftrightarrow{\mathbf{G}}(\mathbf{r},\mathbf{r}')\mathbf{j}(\mathbf{r}')\mathrm{d}V', \tag{8.50}$$

$$\mathbf{H}(\mathbf{r}) = \mathbf{H}_0 + \int_V \left[\nabla \times \overleftrightarrow{\mathbf{G}}(\mathbf{r},\mathbf{r}')\right]\mathbf{j}(\mathbf{r}')\mathrm{d}V'. \tag{8.51}$$

$\overleftrightarrow{\mathbf{G}}$ denotes the dyadic Green's function and $\mathbf{E}_0$ and $\mathbf{H}_0$ are the fields in the absence of the current $\mathbf{j}$. The integration runs over the source volume specified by the coordinate $\mathbf{r}'$. If we introduce the current from Eq. (8.49) into the last two equations and assume that all fields are produced by the dipole we find

$$\mathbf{E}(\mathbf{r}) = \omega^2\mu\mu_0\overleftrightarrow{\mathbf{G}}(\mathbf{r},\mathbf{r}_0)\mathbf{p}, \tag{8.52}$$

$$\mathbf{H}(\mathbf{r}) = -\mathrm{i}\omega\left[\nabla \times \overleftrightarrow{\mathbf{G}}(\mathbf{r},\mathbf{r}_0)\right]\mathbf{p}. \tag{8.53}$$

Hence, the fields of an arbitrarily oriented electric dipole located at $\mathbf{r} = \mathbf{r}_0$ are determined by the Green function $\overleftrightarrow{\mathbf{G}}(\mathbf{r},\mathbf{r}_0)$. As mentioned earlier, each column vector of $\overleftrightarrow{\mathbf{G}}$ specifies the electric field of a dipole whose axis is aligned with one of the coordinate axes. For a homogeneous space, $\overleftrightarrow{\mathbf{G}}$ has been derived as

$$\overleftrightarrow{\mathbf{G}}(\mathbf{r},\mathbf{r}_0) = \left[\overleftrightarrow{\mathbf{I}} + \frac{1}{k^2}\nabla\nabla\right]G(\mathbf{r},\mathbf{r}_0), \qquad G(\mathbf{r},\mathbf{r}_0) = \frac{\exp(\mathrm{i}k|\mathbf{r}-\mathbf{r}_0|)}{4\pi\,|\mathbf{r}-\mathbf{r}_0|}, \tag{8.54}$$

where $\overleftrightarrow{\mathbf{I}}$ is the unit dyad and $G(\mathbf{r},\mathbf{r}_0)$ is the scalar Green function. It is straightforward to calculate $\overleftrightarrow{\mathbf{G}}$ in the major three coordinate systems. In a Cartesian system $\overleftrightarrow{\mathbf{G}}$ can be written as

$$\overleftrightarrow{\mathbf{G}}(\mathbf{r},\mathbf{r}_0) = \frac{\exp(\mathrm{i}kR)}{4\pi R}\left[\left(1 + \frac{\mathrm{i}kR-1}{k^2R^2}\right)\overleftrightarrow{\mathbf{I}} + \frac{3-3\mathrm{i}kR-k^2R^2}{k^2R^2}\frac{\mathbf{RR}}{R^2}\right], \tag{8.55}$$

where R is the absolute value of the vector $\mathbf{R} = \mathbf{r} - \mathbf{r}_0$ and $\mathbf{RR}$ denotes the outer product of $\mathbf{R}$ with itself. Equation (8.55) defines a symmetric 3×3 matrix

$$\overleftrightarrow{\mathbf{G}} = \begin{bmatrix} G_{xx} & G_{xy} & G_{xz} \\ G_{xy} & G_{yy} & G_{yz} \\ G_{xz} & G_{yz} & G_{zz} \end{bmatrix}, \tag{8.56}$$

which, together with Eqs. (8.52) and (8.53), determines the electromagnetic field of an arbitrary electric dipole $\mathbf{p}$ with Cartesian components p_x, p_y, p_z. The tensor $[\nabla \times \overleftrightarrow{\mathbf{G}}]$ can be expressed as

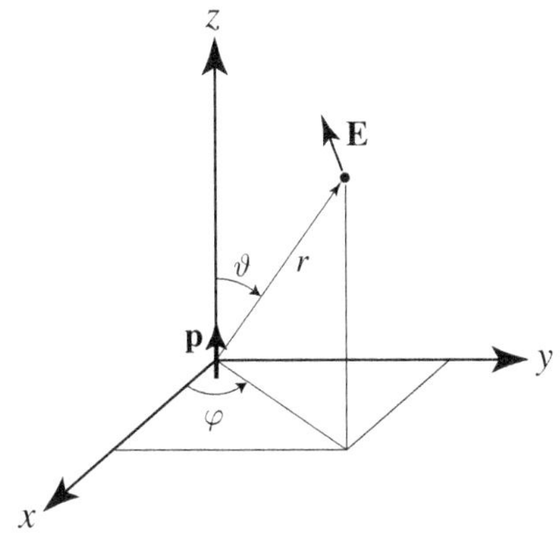

Fig. 8.2 The fields of a dipole are most conveniently represented in a spherical coordinate system (r, ϑ, φ) in which the dipole points along the z-axis ($\vartheta = 0$).

$$\nabla \times \overleftrightarrow{\mathbf{G}}(\mathbf{r}, \mathbf{r}_0) = \frac{\exp(\mathrm{i}kR)}{4\pi R} \frac{k\left(\mathbf{R} \times \overleftrightarrow{\mathbf{I}}\right)}{R} \left(\mathrm{i} - \frac{1}{kR}\right), \tag{8.57}$$

where $\mathbf{R} \times \overleftrightarrow{\mathbf{I}}$ denotes the matrix generated by the cross-product of $\mathbf{R}$ with each column vector of $\overleftrightarrow{\mathbf{I}}$.

The Green function $\overleftrightarrow{\mathbf{G}}$ has terms in $(kR)^{-1}$, $(kR)^{-2}$, and $(kR)^{-3}$. In the *far-field*, for which $R \gg \lambda$, only the terms with $(kR)^{-1}$ survive. On the other hand, the dominant terms in the *near-field*, for which $R \ll \lambda$, are the terms with $(kR)^{-3}$. The terms with $(kR)^{-2}$ dominate the *intermediate-field* at $R \approx \lambda$. To distinguish these three ranges it is convenient to write

$$\overleftrightarrow{\mathbf{G}} = \overleftrightarrow{\mathbf{G}}_{\mathrm{NF}} + \overleftrightarrow{\mathbf{G}}_{\mathrm{IF}} + \overleftrightarrow{\mathbf{G}}_{\mathrm{FF}}, \tag{8.58}$$

where the near-field (G_{NF}), intermediate-field (G_{IF}), and far-field (G_{FF}) Green functions are given by

$$\overleftrightarrow{\mathbf{G}}_{\mathrm{NF}} = \frac{\exp(\mathrm{i}kR)}{4\pi R} \frac{1}{k^2R^2} \left[-\overleftrightarrow{\mathbf{I}} + 3\mathbf{RR}/R^2\right], \tag{8.59}$$

$$\overleftrightarrow{\mathbf{G}}_{\mathrm{IF}} = \frac{\exp(\mathrm{i}kR)}{4\pi R} \frac{\mathrm{i}}{kR} \left[\overleftrightarrow{\mathbf{I}} - 3\mathbf{RR}/R^2\right], \tag{8.60}$$

$$\overleftrightarrow{\mathbf{G}}_{\mathrm{FF}} = \frac{\exp(\mathrm{i}kR)}{4\pi R} \left[\overleftrightarrow{\mathbf{I}} - \mathbf{RR}/R^2\right]. \tag{8.61}$$

Notice that the intermediate-field is 90° out of phase with respect to the near- and far-field.

Because the dipole is located in a homogeneous environment, all three dipole orientations lead to fields that are identical upon suitable frame rotations. We therefore choose a coordinate system with origin at $\mathbf{r} = \mathbf{r}_0$ and a dipole orientation along the dipole axis, i.e. $\mathbf{p} = |\mathbf{p}|\,\mathbf{n}_z$ (see Fig. 8.2). It is most convenient to represent the dipole fields in spherical coordinates $\mathbf{r} = (r, \vartheta, \varphi)$ and in spherical vector components $\mathbf{E} = (E_r, E_\vartheta, E_\varphi)$. In this system the field components E_φ and H_r and H_ϑ are identical to zero and the only non-vanishing field components are

$$E_r = \frac{|\mathbf{p}|\cos\vartheta}{4\pi\varepsilon_0\varepsilon}\frac{\exp(ikr)}{r}k^2\left[\frac{2}{k^2r^2} - \frac{2i}{kr}\right], \tag{8.62}$$

$$E_\vartheta = \frac{|\mathbf{p}|\sin\vartheta}{4\pi\varepsilon_0\varepsilon}\frac{\exp(ikr)}{r}k^2\left[\frac{1}{k^2r^2} - \frac{i}{kr} - 1\right], \tag{8.63}$$

$$H_\varphi = \frac{|\mathbf{p}|\sin\vartheta}{4\pi\varepsilon_0\varepsilon}\frac{\exp(ikr)}{r}k^2\left[-\frac{i}{kr} - 1\right]\sqrt{\frac{\varepsilon_0\varepsilon}{\mu_0\mu}}. \tag{8.64}$$

The fact that E_r has no far-field term ensures that the far-field is purely transverse. Furthermore, since the magnetic field has no terms in $(kr)^{-3}$ the near-field is dominated by the electric field. This justifies a quasi-electrostatic consideration. See Fig. 8.3.

It is instructive to also have a look at the phase of the dipole field since close to the origin it deviates considerably from the familiar phase of a spherical wave $\exp[ikr]$. The phase of the field is defined relative to the oscillation of the dipole p_z. In Fig. 8.4 we plot the phase of the field E_z along the x-axis and along the z-axis (c.f. Fig. 8.2). Interestingly, at the origin the phase of the transverse field is 180° out of phase with the dipole oscillation (Fig. 8.4(a)). The phase of the transverse field then drops to a minimum value at a distance of $x \sim \lambda/5$, after which it increases and then asymptotically approaches the phase of a spherical wave with origin at the dipole (dashed line). On the other hand, the phase of the longitudinal field, shown in Fig. 8.4(b), starts out the same as for the oscillating dipole, but it runs 90° out of phase for distances $z \gg \lambda$. The reason for this behavior is the missing far-field term in the longitudinal field (c.f. Eq. (8.62)). The 90° phase shift is due to the intermediate field represented by the Green function in Eq. (8.60). The same intermediate field is also responsible for the dip near $x \sim \lambda/5$ in Fig. 8.4(a). This phase dip is of relevance for the design of multi-element antennas, such as the Yagi–Uda antennas that will be discussed

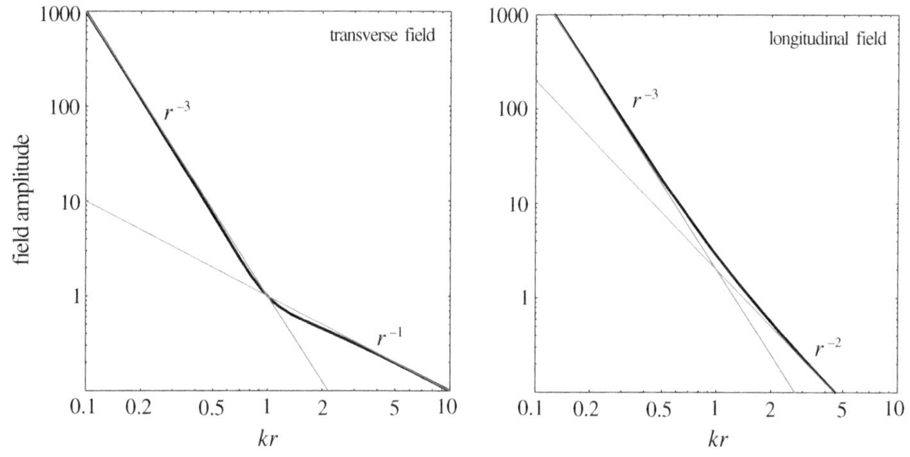

Fig. 8.3 Radial decay of the dipole's transverse and longitudinal fields. The curves correspond to the absolute value of the expressions in brackets of Eqs. (8.62) and (8.63), respectively. While both the transverse and the longitudinal field contribute to the near-field, only the transverse field survives in the far-field. Notice that the intermediate-field with $(kr)^{-2}$ does not really show up for the transverse field. Instead the near-field dominates for $(kr) < 1$ and the far-field for $(kr) > 1$.

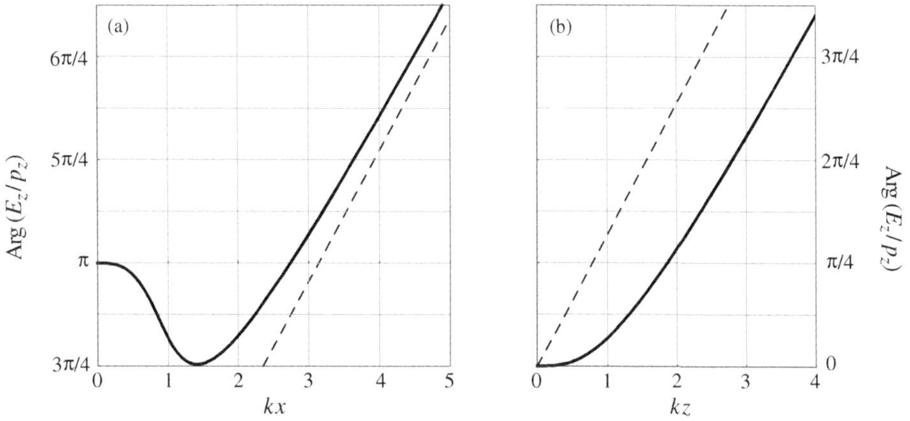

Fig. 8.4 Phase of the electric field near the origin. (a) Phase of the transverse field E_z evaluated along the x-axis. At the origin, the electric field is 180° out of phase with the dipole. The phase drops to a minimum at a distance of $x \sim \lambda/5$. For larger distances, the phase approaches that of a spherical wave $\exp[ikr]$ (dashed line). (b) Phase of the longitudinal field E_z evaluated along the z-axis. At the origin, the electric field is in phase with the dipole. At larger distances, the phase is 90° out of phase with a spherical wave $\exp[ikr]$ (dashed line).

in Chapter 13 (see Problem 13.4). It is important to remember that close to the source the phase of the field *does not* evolve linearly with distance and that the phase can be advanced or delayed by small distance variations.

So far we have considered a dipole that oscillates harmonically in time, i.e. $\mathbf{p}(t) = \mathrm{Re}\{\mathbf{p}\exp(-\mathrm{i}\omega t)\}$. Therefore, the electromagnetic field is monochromatic and oscillates at the same frequency. Although it is possible to generate any time dependence by a superposition of monochromatic fields (Fourier transformation), it is advantageous for ultrafast applications to have the full time dependence available. The fields of a dipole $\mathbf{p}(t)$ with arbitrary time dependence can be derived by using the time-dependent Green function. In a non-dispersive medium it is easier to introduce the explicit time dependence by using the substitutions

$$\exp(\mathrm{i}kr)k^m\mathbf{p} = \exp(\mathrm{i}kr)\left[\frac{\mathrm{i}n}{c}\right]^m(-\mathrm{i}\omega)^m\mathbf{p} \rightarrow \left[\frac{\mathrm{i}n}{c}\right]^m\frac{\mathrm{d}^m}{\mathrm{d}t^m}\mathbf{p}(t-nr/c)\,, \tag{8.65}$$

where n denotes the (dispersion-free) index of refraction[7] and $(t - nr/c)$ is the retarded time. With this substitution, the dipole fields read as

$$E_r(t) = \frac{\cos\vartheta}{4\pi\varepsilon_0\varepsilon}\left[\frac{2}{r^3} + \frac{n}{c}\frac{2}{r^2}\frac{\mathrm{d}}{\mathrm{d}t}\right]|\mathbf{p}(t-nr/c)|\,, \tag{8.66}$$

$$E_\vartheta(t) = -\frac{\sin\vartheta}{4\pi\varepsilon_0\varepsilon}\left[\frac{1}{r^3} + \frac{n}{c}\frac{1}{r^2}\frac{\mathrm{d}}{\mathrm{d}t} + \frac{n^2}{c^2}\frac{1}{r}\frac{\mathrm{d}^2}{\mathrm{d}t^2}\right]|\mathbf{p}(t-nr/c)|\,, \tag{8.67}$$

$$H_\varphi(t) = -\frac{\sin\vartheta}{4\pi\varepsilon_0\varepsilon}\sqrt{\frac{\varepsilon_0\varepsilon}{\mu_0\mu}}\left[\frac{n}{c}\frac{1}{r^2}\frac{\mathrm{d}}{\mathrm{d}t} + \frac{n^2}{c^2}\frac{1}{r}\frac{\mathrm{d}^2}{\mathrm{d}t^2}\right]|\mathbf{p}(t-nr/c)|\,. \tag{8.68}$$

[7] A dispersion-free index of refraction different from unity is an approximation since it violates causality.

We see that the far-field is generated by the acceleration of the charges that constitute the dipole moment. Similarly, the intermediate-field and the near-field are generated by the speed and the position of the charges, respectively.

8.3.2 Dipole radiation

It can be shown (see Problem 8.3) that in the steady state only the far-field of the dipole contributes to the net energy transport (Fig. 8.5). The Poynting vector $\mathbf{S}(t)$ associated with the far-field can be calculated by retaining only the r^{-1} terms in the dipole fields. We obtain

$$\mathbf{S}(t) = \mathbf{E}(t) \times \mathbf{H}(t) = \frac{1}{16\pi^2\varepsilon_0\varepsilon}\,\frac{\sin^2\vartheta}{r^2}\,\frac{n^3}{c^3}\left[\frac{\mathrm{d}^2}{\mathrm{d}t^2}|\mathbf{p}(t-nr/c)|\right]^2 \mathbf{n}_r. \tag{8.69}$$

The radiated power P can be determined by integrating $\mathbf{S}(t)$ over a closed spherical surface as

$$P(t) = \int_{\partial V} \mathbf{S}\cdot\mathbf{n}\,\mathrm{d}a = \frac{1}{4\pi\varepsilon_0\varepsilon}\,\frac{2\,n^3}{3\,c^3}\left[\frac{\mathrm{d}^2|\mathbf{p}(t)|}{\mathrm{d}t^2}\right]^2, \tag{8.70}$$

where we have shrunk the radius of the sphere to zero to get rid of the retarded time. The average radiated power for a harmonically oscillating dipole turns out to be

$$\bar{P} = \frac{|\mathbf{p}|^2}{4\pi\varepsilon_0\varepsilon}\,\frac{n^3\omega^4}{3c^3}, \tag{8.71}$$

which could also have been calculated by integrating the time-averaged Poynting vector $\langle\mathbf{S}\rangle = (1/2)\mathrm{Re}\,\{\mathbf{E}\times\mathbf{H}^*\}$, $\mathbf{E}$ and $\mathbf{H}$ being the dipole's complex field amplitudes given by Eqs. (8.62)–(8.64). We find that the radiated power scales with the fourth power of the frequency. To determine the normalized radiation pattern we calculate the power $\bar{P}(\vartheta,\varphi)$ radiated into an infinitesimal unit solid angle $\mathrm{d}\Omega = \sin\vartheta\,\mathrm{d}\vartheta\,\mathrm{d}\varphi$ and divide by the total radiated power $\bar{P}$,

$$\frac{\bar{P}(\vartheta,\varphi)}{\bar{P}} = \frac{3}{8\pi}\sin^2\vartheta. \tag{8.72}$$

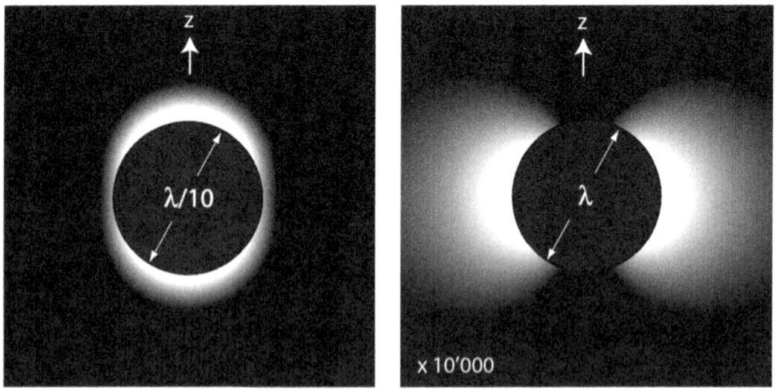

Fig. 8.5 The electric energy density outside a fictitious sphere enclosing a dipole $\boldsymbol{p} = p_z$. Left: close to the dipole's origin the field distribution is elongated along the dipole axis (near-field). Right: at larger distances the field spreads transverse to the dipole axis (far-field).

Most of the energy is radiated perpendicular to the dipole moment and there is no radiation at all in the direction of the dipole.

Although we have considered an arbitrary time dependence for the dipole we will restrict ourselves in the following to the time-harmonic case. It is straightforward to account for dispersion when working with time-harmonic fields, and arbitrary time dependences can be introduced by using Fourier transforms.

8.3.3 Rate of energy dissipation in inhomogeneous environments

According to Poynting's theorem (cf. Eq. (2.58)) the radiated power of any current distribution with a harmonic time dependence in a linear medium has to be identical to the rate of energy dissipation $\mathrm{d}W/\mathrm{d}t$ given by

$$\frac{\mathrm{d}W}{\mathrm{d}t} = -\frac{1}{2}\int_V \mathrm{Re}\{\mathbf{j}^* \cdot \mathbf{E}\}\mathrm{d}V, \tag{8.73}$$

with V being the source volume and $\mathbf{j}$ representing both sources and energy sinks. If we introduce the dipole's current density from Eq. (8.49) we obtain the important result

$$\frac{\mathrm{d}W}{\mathrm{d}t} = \frac{\omega}{2}\,\mathrm{Im}\big\{\mathbf{p}^* \cdot \mathbf{E}(\mathbf{r}_0)\big\} \tag{8.74}$$

where the field $\mathbf{E}$ is evaluated at the dipole's origin $\mathbf{r}_0$. This equation can be rewritten in terms of the Green function by using Eq. (8.52) as

$$\frac{\mathrm{d}W}{\mathrm{d}t} = \frac{\omega^3|\mathbf{p}|^2}{2c^2\varepsilon_0\varepsilon}\left[\mathbf{n}_\mathrm{p} \cdot \mathrm{Im}\left\{\overleftrightarrow{\mathbf{G}}(\mathbf{r}_0, \mathbf{r}_0; \omega)\right\} \cdot \mathbf{n}_\mathrm{p}\right], \tag{8.75}$$

with $\mathbf{n}_\mathrm{p}$ being the unit vector in the direction of the dipole moment. At first sight it does not seem possible to evaluate Eq. (8.74) since $\exp(\mathrm{i}kR)/R$ appears to be infinite at $\mathbf{r} = \mathbf{r}_0$. As we shall see, this is not the case. We first note that due to the dot product between $\mathbf{p}$ and $\mathbf{E}$ we need only evaluate the component of $\mathbf{E}$ in the direction of $\mathbf{p}$. Choosing $\mathbf{p} = |\mathbf{p}|\,\mathbf{n}_z$, we calculate E_z as

$$E_z = \frac{|\mathbf{p}|}{4\pi\,\varepsilon_0\,\varepsilon}\frac{\mathrm{e}^{\mathrm{i}kR}}{R}\left[k^2\sin^2\vartheta + \frac{1}{R^2}(3\cos^2\vartheta - 1) - \frac{\mathrm{i}k}{R}(3\cos^2\vartheta - 1)\right]. \tag{8.76}$$

Since the interesting part is the field at the origin of the dipole, the exponential term is expanded into a series $[\exp(\mathrm{i}kR) = 1 + \mathrm{i}kR + (1/2)(\mathrm{i}kR)^2 + (1/6)(\mathrm{i}kR)^3 + \cdots]$ and the limiting case $R \to 0$ is considered. Thus,

$$\begin{aligned}\frac{\mathrm{d}W}{\mathrm{d}t} &= \lim_{R\to 0}\frac{\omega}{2}\,|\mathbf{p}|\;\mathrm{Im}\{E_z\} = \frac{\omega\,|\mathbf{p}|^2}{8\pi\,\varepsilon_0\varepsilon}\lim_{R\to 0}\left\{\frac{2}{3}k^3 + R^2(\cdots) + \cdots\right\}\\ &= \frac{|\mathbf{p}|^2}{12\pi}\frac{\omega}{\varepsilon_0\varepsilon}k^3,\end{aligned} \tag{8.77}$$

which is identical with Eq. (8.71). Thus, Eq. (8.74) leads to the correct result despite the apparent singularity at $R = 0$.

The importance of Eq. (8.74) becomes obvious if we consider an emitting dipole in an inhomogeneous environment, such as an atom in a cavity or a molecule in a superlattice. The rate at which energy is released can still be calculated by integrating the Poynting vector over a surface enclosing the dipole emitter. However, to do this, we need to know the electromagnetic field everywhere on the enclosing surface. Because of the inhomogeneous environment, this field is not equal to the dipole field alone! Instead, it is the self-consistent field, i.e. the field $\mathbf{E}$ generated by the superposition of the dipole field $\mathbf{E}_0$ and the scattered field $\mathbf{E}_s$ from the environment. Thus, to determine the energy dissipated by the dipole we first need to determine the electromagnetic field everywhere on the enclosing surface. However, by using Eq. (8.74) we can do the same job by evaluating only the total field at the dipole's origin $\mathbf{r}_0$. It is convenient to decompose the electric field at the dipole's position as

$$\mathbf{E}(\mathbf{r}_0) = \mathbf{E}_0(\mathbf{r}_0) + \mathbf{E}_s(\mathbf{r}_0), \tag{8.78}$$

where $\mathbf{E}_0$ and $\mathbf{E}_s$ are the primary dipole field and the scattered field, respectively. Introducing Eq. (8.78) into Eq. (8.74) allows us to split the rate of energy dissipation $P = \mathrm{d}W/\mathrm{d}t$ into two parts. The contribution of $\mathbf{E}_0$ has been determined in Eq. (8.71) and Eq. (8.77) as

$$P_0 = \frac{|\mathbf{p}|^2}{12\pi} \frac{\omega}{\varepsilon_0 \varepsilon} k^3, \tag{8.79}$$

which allows us to write for the normalized rate of energy dissipation

$$\frac{P}{P_0} = 1 + \frac{6\pi\varepsilon_0\varepsilon}{|\mathbf{p}|^2} \frac{1}{k^3} \operatorname{Im}\{\mathbf{p}^* \cdot \mathbf{E}_s(\mathbf{r}_0)\}. \tag{8.80}$$

Thus, the change of energy dissipation depends on the *secondary field* of the dipole. This field corresponds to the dipole's own field emitted at an earlier time. It arrives at the position of the dipole after it has been scattered in the environment.

8.3.4 Radiation reaction

An oscillating charge produces electromagnetic radiation. This radiation not only dissipates the energy of the oscillator but also influences the motion of the charge. This back-action is called radiation damping or radiation reaction. With the inclusion of the reaction force $\mathbf{F}_r$ the equation of motion for an undriven harmonic oscillator becomes

$$m\ddot{\mathbf{r}} + \omega_0^2 m\mathbf{r} = \mathbf{F}_r, \tag{8.81}$$

where $\omega_0^2 m$ is the linear spring constant. According to Eq. (8.70) the average rate of energy dissipation is

$$P(t) = \frac{1}{4\pi\varepsilon_0} \frac{2}{3c^3} \left[\frac{\mathrm{d}^2|\mathbf{p}(t)|}{\mathrm{d}t^2} \right]^2 = \frac{q^2(\ddot{\mathbf{r}} \cdot \ddot{\mathbf{r}})}{6\pi\varepsilon_0 c^3}. \tag{8.82}$$

Integrated over a certain time period $T = [t_1 \dots t_2]$, this term must be equal to the work exerted on the oscillating charge by the radiation reaction force. Thus,

$$\int_{t_1}^{t_2} \left[\mathbf{F}_{\mathrm{r}} \cdot \dot{\mathbf{r}} + \frac{q^2 (\ddot{\mathbf{r}} \cdot \ddot{\mathbf{r}})}{6\pi \varepsilon_0 c^3} \right] \mathrm{d}t = 0. \tag{8.83}$$

After integrating the second term by parts we obtain

$$\int_{t_1}^{t_2} \left[\mathbf{F}_{\mathrm{r}} \cdot \dot{\mathbf{r}} - \frac{q^2 (\dot{\mathbf{r}} \cdot \dddot{\mathbf{r}})}{6\pi \varepsilon_0 c^3} \right] \mathrm{d}t + \left. \frac{q^2 (\ddot{\mathbf{r}} \cdot \dot{\mathbf{r}})}{6\pi \varepsilon_0 c^3} \right|_{t_1}^{t_2} = 0. \tag{8.84}$$

Assuming that $\mathbf{r}$ is time-harmonic, then the integrated term is zero if $(t_2 - t_1)$ is chosen to be a multiple of the oscillation period. Consequently the remaining integrand has to vanish, i.e.

$$\mathbf{F}_{\mathrm{r}} = \frac{q^2 \dddot{\mathbf{r}}}{6\pi \varepsilon_0 c^3}, \tag{8.85}$$

which is the *Abraham–Lorentz formula* for the radiation reaction force. The equation of motion (8.81) now becomes

$$\ddot{\mathbf{r}} - \frac{q^2}{6\pi \varepsilon_0 c^3 m} \dddot{\mathbf{r}} + \omega_0^2 \mathbf{r} = 0. \tag{8.86}$$

Assuming that the damping introduced by the radiation reaction force is negligible, the solution becomes $\mathbf{r}(t) = \mathbf{r}_0 \exp(-\mathrm{i}\omega_0 t)$ and hence $\dddot{\mathbf{r}} = -\omega_0^2 \dot{\mathbf{r}}$. Thus, for small damping, we obtain

$$\ddot{\mathbf{r}} + \gamma_0 \dot{\mathbf{r}} + \omega_0^2 \mathbf{r} = 0, \quad \gamma_0 = \frac{1}{4\pi \varepsilon_0} \frac{2q^2 \omega_0^2}{3c^3 m}. \tag{8.87}$$

This equation corresponds to an undriven Lorentzian atom model with transition frequency ω_0 and linewidth γ_0. A more rigorous derivation shows that radiation reaction affects not only the damping of the oscillator due to radiation but also the oscillator's effective mass. This additional mass contribution is called the *electromagnetic mass* and it is the source of many controversies [7].

Owing to radiation damping the undriven oscillator will ultimately come to rest. However, the oscillator interacts with the vacuum field that keeps the oscillator alive. Consequently, a driving term accounting for the fluctuating vacuum field $\mathbf{E}_0$ has to be added to the right-hand side of Eq. (8.87). The fluctuating vacuum field compensates for the dissipation of the oscillator. Such fluctuation–dissipation relations will be discussed in Chapter 15. In short, to preserve an equilibrium between the oscillator and the vacuum, the vacuum must give rise to fluctuations if it takes energy from the oscillator (radiation damping). It can be shown that spontaneous emission is the result of both radiation reaction *and* vacuum fluctuations [7].

Finally, let us remark that radiation reaction is an important ingredient in obtaining the correct result for the *optical theorem* in the dipole limit [8], i.e. for a particle that is described by a polarizability α. In this limit, an incident field polarizes the particle and induces a dipole moment $\mathbf{p}$, which in turn radiates a scattered field. According to the optical theorem, the extinct power (the sum of scattered and absorbed power) can be expressed by the field scattered in the forward direction. However, it turns out that in the dipole limit the extinct power is identical with the absorbed power and hence light scattering is not taken into account! The solution to this dilemma is provided by the radiation reaction term in Eq. (8.85) and is analyzed in more detail in Problem 8.5. In short, the particle interacts not only with the external driving field but also with its own field, causing a phase-lag between the induced dipole oscillation and the driving electric field oscillation. This phase-lag recovers the optical theorem and is responsible for light scattering in the dipole limit.

8.4 Spontaneous decay

Before Purcell's analysis in 1946, spontaneous emission was considered a radiative intrinsic property of atoms or molecules [9]. Purcell's work predicted that the spontaneous transition rate of a nuclear magnetic moment coupled to a resonant electronic device can be enhanced compared with the free-space transition rate. Thus, it can be inferred that the environment in which an atom is embedded modifies the radiative properties of the atom. In order to experimentally observe this effect a physical device with dimensions on the order of the emission wavelength λ is needed. Since most of the atomic transitions occur in or near the visible spectral range, the modification of spontaneous decay was not an obvious fact. In 1966, Drexhage investigated the effect of planar interfaces on the spontaneous decay rate of molecules [10], and the enhancement of the atomic decay rate in a cavity was later verified by Goy *et al.* [11]. However, it was also observed that the decay of excited atoms can be inhibited by a cavity [12]. Since then, the modification of the spontaneous decay rate of an atom or molecule has been investigated in various environments, including photonic crystals and optical antennas [13–16]. Recently, it was also demonstrated that non-radiative energy transfer between adjacent molecules (Förster transfer) can be modified by an inhomogeneous environment [17].

In the theory of atom–field interactions there are two physically distinct regimes, namely the strong and weak coupling regimes. The two regimes are distinguished on the basis of the atom–field coupling constant, which is estimated as

$$\kappa = \frac{p_{ij}}{\hbar}\sqrt{\frac{\hbar\omega_0}{2\varepsilon_0 V}}, \tag{8.88}$$

where ω_0 is the atomic transition frequency, p_{ij} the dipole matrix element, and V the volume of the cavity. Strong coupling satisfies the condition $\kappa \gg \gamma$, γ being the spontaneous decay rate inside the cavity. In the strong-coupling regime, the emission spectrum of an atom inside a cavity with a high quality factor ($Q \rightarrow \infty$) exhibits two distinct peaks, which are the result of mode splitting [18, 19]. In the weak-coupling regime ($\kappa \ll \gamma$) it has been shown that QED and classical theory give the same results for the *modification* of the spontaneous-emission decay rate. Classically, the modification of the spontaneous decay rate is generated by back-action, namely the interaction of the atom with its own "retarded" field. The latter is the field that arrives back at the atom after having been scattered in the environment. On the other hand, in the QED picture the decay rate is stimulated by vacuum field fluctuations, the latter being a function of the environment.

8.4.1 QED of spontaneous decay

In this section we derive the spontaneous emission rate γ for a two-level quantum system ("atom" for short) located at $\mathbf{r} = \mathbf{r}_0$. Spontaneous decay is a purely quantum effect and requires a QED treatment. This section is intended to put classical treatments into the proper context. We consider the combined "*atom plus field*" states and calculate the transitions from an initial state $|i\rangle$ with energy E_i to a set of final states $|f\rangle$ with identical energies E_f (see Fig. 8.6). The final states differ only by the mode $\mathbf{k}$ of the radiation field.[8] The derivation presented here is based on the Heisenberg picture. An equivalent derivation is presented in Appendix B.

According to Fermi's golden rule γ is given by [2]

$$\gamma = \frac{2\pi}{\hbar^2} \sum_f \left| \langle f | \hat{H}_\mathrm{I} | i \rangle \right|^2 \delta(\omega_i - \omega_f), \tag{8.89}$$

where $\hat{H}_\mathrm{I} = -\hat{\mathbf{p}}\cdot\hat{\mathbf{E}}$ is the interaction Hamiltonian in the dipole approximation, with $\hat{\mathbf{E}}$ being the vacuum electric field operator. It has to be emphasized that $\hbar\omega_i$ and $\hbar\omega_f$ are the energies of the combined "atom plus field" system. The initial energy is solely determined by the atom, that is $\hbar\omega_i = E_\mathrm{e}$, where E_e denotes the atom's excited-state energy. On the other hand, the final energy is determined by the atom and the field as $\hbar\omega_f = E_g + \hbar\omega_0$, with E_f being the atom's ground-state energy and $\hbar\omega_0$ being the quantum of electromagnetic

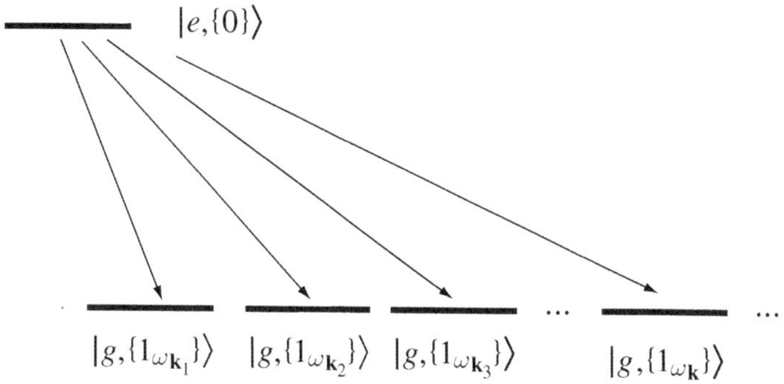

Fig. 8.6 Transition from an initial state $|i\rangle = |e,\{0\}\rangle$ to a set of final states $|f\rangle = |g,\{1_{\omega_k}\}\rangle$. The states are products of atomic states ($|e\rangle$ or $|g\rangle$) and single-photon states ($|\{0\}\rangle$ or $|\{1_{\omega_k}\}\rangle$). The energy difference between the atom's excited and ground states is $E_\mathrm{e} - E_\mathrm{g} = \hbar\omega_0$, with $\hbar\omega_0$ being the photon energy. The number of distinct final single-photon states is defined by the partial local density of states $\rho_\mathrm{p}(\mathbf{r}_0, \omega_0)$, with $\mathbf{r}_0$ being the origin of the two-level system.

[8] $\mathbf{k}$ is not to be confused with the wavevector. It is a label denoting a specific mode, which in turn is characterized by the polarization vector and the wavevector.

energy (photon). Thus the delta function in Eq. (8.89) is a statement of energy conservation; that is, $E_e - E_g = \hbar\omega_0$. Using the expression for $\hat{H}_I$ we can substitute as follows:[9]

$$\left|\langle f|\hat{H}_I|i\rangle\right|^2 = \langle f|\hat{\mathbf{p}}\cdot\hat{\mathbf{E}}|i\rangle^* \langle f|\hat{\mathbf{p}}\cdot\hat{\mathbf{E}}|i\rangle = \langle i|\hat{\mathbf{p}}\cdot\hat{\mathbf{E}}|f\rangle\langle f|\hat{\mathbf{p}}\cdot\hat{\mathbf{E}}|i\rangle. \tag{8.90}$$

Let us represent the electric field operator $\hat{\mathbf{E}}$ at $\mathbf{r} = \mathbf{r}_0$ as [2]

$$\hat{\mathbf{E}} = \sum_{\mathbf{k}} \left[\mathbf{E}_{\mathbf{k}}^+ \hat{a}_{\mathbf{k}}(t) + \mathbf{E}_{\mathbf{k}}^- \hat{a}_{\mathbf{k}}^\dagger(t)\right], \tag{8.91}$$

where

$$\hat{a}_{\mathbf{k}}^\dagger(t) = \hat{a}_{\mathbf{k}}^\dagger(0)\exp(\mathrm{i}\omega_{\mathbf{k}}t), \qquad \hat{a}_{\mathbf{k}}(t) = \hat{a}_{\mathbf{k}}(0)\exp(-\mathrm{i}\omega_{\mathbf{k}}t). \tag{8.92}$$

Here, $\hat{a}_{\mathbf{k}}(0)$ and $\hat{a}_{\mathbf{k}}^\dagger(0)$ are the annihilation and creation operators, respectively. The sum over $\mathbf{k}$ refers to summation over all modes. $\omega_{\mathbf{k}}$ denotes the frequency of mode $\mathbf{k}$. The spatially dependent complex fields $\mathbf{E}_{\mathbf{k}}^+ = (\mathbf{E}_{\mathbf{k}}^-)^*$ are the positive and negative frequency parts of the complex field $\mathbf{E}_{\mathbf{k}}$. For a two-level atomic system with the ground state $|g\rangle$ and the excited state $|e\rangle$, the dipole moment operator $\hat{\mathbf{p}}$ can be written as

$$\hat{\mathbf{p}} = \mathbf{p}\left[\hat{r}^+ + \hat{r}\right], \text{ with } \hat{r}^+ = |e\rangle\langle g| \text{ and } \hat{r} = |g\rangle\langle e|. \tag{8.93}$$

In this notation, $\mathbf{p}$ is simply the transition dipole moment, which is assumed to be real, i.e. $\langle g|\hat{\mathbf{p}}|e\rangle = \langle e|\hat{\mathbf{p}}|g\rangle$. Using the expressions for $\hat{\mathbf{E}}$ and $\hat{\mathbf{p}}$, the interaction Hamiltonian takes on the form

$$-\hat{\mathbf{p}}\cdot\hat{\mathbf{E}} = -\sum_{\mathbf{k}} \mathbf{p}\cdot\left[\mathbf{E}_{\mathbf{k}}^+\hat{r}^+\hat{a}_{\mathbf{k}}(t) + \mathbf{E}_{\mathbf{k}}^-\hat{r}\hat{a}_{\mathbf{k}}^\dagger(t) + \mathbf{E}_{\mathbf{k}}^+\hat{r}\hat{a}_{\mathbf{k}}(t) + \mathbf{E}_{\mathbf{k}}^-\hat{r}^+\hat{a}_{\mathbf{k}}^\dagger(t)\right]. \tag{8.94}$$

We now define the initial and final state of the combined system "*field plus atom*" as (see Fig. 8.6)

$$|i\rangle = |e, \{0\}\rangle = |e\rangle\,|\{0\}\rangle \tag{8.95}$$

$$|f\rangle = \left|g, \{1_{\omega_{\mathbf{k}'}}\}\right\rangle = |g\rangle\left|\{1_{\omega_{\mathbf{k}'}}\}\right\rangle, \tag{8.96}$$

respectively. Here, $|\{0\}\rangle$ denotes the zero-photon state, and $\left|\{1_{\omega_{\mathbf{k}'}}\}\right\rangle$ designates the one-photon state associated with mode $\mathbf{k}'$ and frequency $\omega_0 = (E_e - E_g)/\hbar$. Thus, the final states in Eq. (8.89) are associated with the different modes $\mathbf{k}'$. Operating with $\hat{\mathbf{p}}\cdot\hat{\mathbf{E}}$ on state $|i\rangle$ leads to

$$\hat{\mathbf{p}}\cdot\hat{\mathbf{E}}\,|i\rangle = \mathbf{p}\cdot\sum_{\mathbf{k}} \mathbf{E}_{\mathbf{k}}^- \mathrm{e}^{\mathrm{i}\omega_{\mathbf{k}}t}\left|g, \{1_{\omega_{\mathbf{k}}}\}\right\rangle, \tag{8.97}$$

where we used $\hat{a}_{\mathbf{k}}^\dagger(0)\,|\{0\}\rangle = \left|\{1_{\omega_{\mathbf{k}}}\}\right\rangle$. Operating with $\langle f|$ gives

$$\langle f|\,\hat{\mathbf{p}}\cdot\hat{\mathbf{E}}\,|i\rangle = \mathbf{p}\cdot\sum_{\mathbf{k}} \mathbf{E}_{\mathbf{k}}^- \mathrm{e}^{\mathrm{i}\omega_{\mathbf{k}}t}\langle g, \{1_{\omega_{\mathbf{k}'}}\}\left|g, \{1_{\omega_{\mathbf{k}}}\}\right\rangle, \tag{8.98}$$

where we used $\hat{a}_{\mathbf{k}}(0)\left|\{1_{\omega_{\mathbf{k}}}\}\right\rangle = \{0\}$. A similar procedure leads to

$$\langle i|\,\hat{\mathbf{p}}\cdot\hat{\mathbf{E}}\,|f\rangle = \mathbf{p}\cdot\sum_{\mathbf{k}} \mathbf{E}_{\mathbf{k}}^+ \mathrm{e}^{-\mathrm{i}\omega_{\mathbf{k}}t}\langle g, \{1_{\omega_{\mathbf{k}}}\}\,|g, \{1_{\omega_{\mathbf{k}'}}\}\rangle. \tag{8.99}$$

[9] Remember that $\hat{\mathbf{p}}$ is the dipole operator, *not* the momentum operator.

The matrix elements can now be introduced into Eqs. (8.90) and (8.89). On expressing the sum over the final states as a sum over the modes $\mathbf{k}'$, the transition rate becomes

$$\gamma = \frac{2\pi}{\hbar^2}\sum_{\mathbf{k}}\sum_{\mathbf{k}''}\left[\mathbf{p}\cdot\mathbf{E}^+_{\mathbf{k}''}\mathbf{E}^-_{\mathbf{k}}\cdot\mathbf{p}\right]\mathrm{e}^{\mathrm{i}(\omega_{\mathbf{k}}-\omega_{\mathbf{k}''})t}$$
$$\times\sum_{\mathbf{k}'}\langle g,\{1_{\omega_{\mathbf{k}''}}\}\,|g,\{1_{\omega_{\mathbf{k}'}}\}\rangle\langle g,\{1_{\omega_{\mathbf{k}'}}\}\,|g,\{1_{\omega_{\mathbf{k}}}\}\rangle\,\delta(\omega_{\mathbf{k}'}-\omega_0). \tag{8.100}$$

Because of orthogonality, the only non-vanishing terms are those for which $\mathbf{k}'=\mathbf{k}''=\mathbf{k}$, which leads to the simple expression

$$\gamma = \frac{2\pi}{\hbar^2}\sum_{\mathbf{k}}\left[\mathbf{p}\cdot(\mathbf{E}^+_{\mathbf{k}}\mathbf{E}^-_{\mathbf{k}})\cdot\mathbf{p}\right]\delta(\omega_{\mathbf{k}}-\omega_0). \tag{8.101}$$

Here, $\mathbf{E}^+_{\mathbf{k}}\mathbf{E}^-_{\mathbf{k}}$ denotes the outer product, i.e. the result is a 3×3 matrix. For later purposes it is convenient to rewrite this expression in terms of normal modes $\mathbf{u}_{\mathbf{k}}$ defined as

$$\mathbf{E}^+_{\mathbf{k}} = \sqrt{\frac{\hbar\omega_{\mathbf{k}}}{2\varepsilon_0}}\,\mathbf{u}_{\mathbf{k}},\qquad \mathbf{E}^-_{\mathbf{k}} = \sqrt{\frac{\hbar\omega_{\mathbf{k}}}{2\varepsilon_0}}\,\mathbf{u}^*_{\mathbf{k}}. \tag{8.102}$$

Because the delta function imposes $\omega_{\mathbf{k}}=\omega_0$ the decay rate can be written as

$$\gamma = \frac{\pi\omega}{3\hbar\varepsilon_0}\,|\mathbf{p}|^2\,\rho_{\mathrm{p}}(\mathbf{r}_0,\omega_0),\qquad \rho_{\mathrm{p}}(\mathbf{r}_0,\omega_0) = 3\sum_{\mathbf{k}}\left[\mathbf{n}_{\mathrm{p}}\cdot(\mathbf{u}_{\mathbf{k}}\mathbf{u}^*_{\mathbf{k}})\cdot\mathbf{n}_{\mathrm{p}}\right]\delta(\omega_{\mathbf{k}}-\omega_0), \tag{8.103}$$

where we introduced the *partial local density of states* $\rho_{\mathrm{p}}(\mathbf{r}_0,\omega_0)$, which will be discussed in the next section. The dipole moment has been decomposed as $\mathbf{p}=|\mathbf{p}|\mathbf{n}_{\mathrm{p}}$, with $\mathbf{n}_{\mathrm{p}}$ being the unit vector in the direction of $\mathbf{p}$. The above equation for γ is our main result. The delta function in the expression suggests that we need to integrate over a finite distribution of final frequencies. However, even for a single final frequency, the apparent singularity introduced through $\delta(\omega_{\mathbf{k}}-\omega_0)$ is compensated for by the normal modes, whose magnitude tends to zero for a sufficiently large mode volume. In any case, it is convenient to get rid of these singularities by representing $\rho_{\mathrm{p}}(\mathbf{r}_0,\omega_0)$ in terms of the Green function instead of normal modes.

8.4.2 Spontaneous decay and Green's dyadics

We aim to derive an important relationship between the normal modes $\mathbf{u}_{\mathbf{k}}$ and the dyadic Green function $\overleftrightarrow{\mathbf{G}}$. Subsequently, this relationship is used to express the spontaneous decay rate γ and to establish an elegant expression for the local density of states. While we suppressed the explicit position dependence of $\mathbf{u}_{\mathbf{k}}$ in the previous section for notational convenience, it is essential in the current context to carry all the arguments. The normal modes defined in the previous section satisfy the wave equation

$$\nabla\times\nabla\times\mathbf{u}_{\mathbf{k}}(\mathbf{r},\omega_{\mathbf{k}}) - \frac{\omega_{\mathbf{k}}^2}{c^2}\mathbf{u}_{\mathbf{k}}(\mathbf{r},\omega_{\mathbf{k}}) = 0 \tag{8.104}$$

and they fulfill the orthogonality relation

$$\int \mathbf{u_k}(\mathbf{r}, \omega_\mathbf{k}) \cdot \mathbf{u}^*_{\mathbf{k}'}(\mathbf{r}, \omega_{\mathbf{k}'}) \mathrm{d}^3\mathbf{r} = \delta_{\mathbf{kk}'}, \tag{8.105}$$

where the integration runs over the entire mode volume. $\delta_{\mathbf{kk}'}$ is the Kronecker delta and $\overleftrightarrow{\mathbf{I}}$ the unit dyad. We now expand the Green function $\overleftrightarrow{\mathbf{G}}$ in terms of the normal modes as

$$\overleftrightarrow{\mathbf{G}}(\mathbf{r}, \mathbf{r}'; \omega) = \sum_\mathbf{k} \mathbf{A_k}(\mathbf{r}', \omega) \mathbf{u_k}(\mathbf{r}, \omega_\mathbf{k}), \tag{8.106}$$

where the vectorial expansion coefficients $\mathbf{A_k}$ have yet to be determined.

We recall the definition of the Green function (cf. Eq. (2.87))

$$\nabla \times \nabla \times \overleftrightarrow{\mathbf{G}}(\mathbf{r}, \mathbf{r}'; \omega) - \frac{\omega^2}{c^2} \overleftrightarrow{\mathbf{G}}(\mathbf{r}, \mathbf{r}'; \omega) = \overleftrightarrow{\mathbf{I}}\, \delta(\mathbf{r}-\mathbf{r}') \,. \tag{8.107}$$

To determine the coefficients $\mathbf{A_k}$ we substitute the expansion for $\overleftrightarrow{\mathbf{G}}$ and obtain

$$\sum_\mathbf{k} \mathbf{A_k}(\mathbf{r}', \omega) \left[\nabla \times \nabla \times \mathbf{u_k}(\mathbf{r}, \omega_\mathbf{k}) - \frac{\omega^2}{c^2} \mathbf{u_k}(\mathbf{r}, \omega_\mathbf{k}) \right] = \overleftrightarrow{\mathbf{I}} \delta(\mathbf{r} - \mathbf{r}'). \tag{8.108}$$

Using Eq. (8.104) we can rewrite the latter as

$$\sum_\mathbf{k} \mathbf{A_k}(\mathbf{r}', \omega) \left[\frac{\omega_\mathbf{k}^2}{c^2} - \frac{\omega^2}{c^2} \right] \mathbf{u_k}(\mathbf{r}, \omega_\mathbf{k}) = \overleftrightarrow{\mathbf{I}}\, \delta(\mathbf{r} - \mathbf{r}'). \tag{8.109}$$

Multiplying on both sides by $\mathbf{u}^*_{\mathbf{k}'}$, integrating over the mode volume, and making use of the orthogonality relation leads to

$$\mathbf{A}_{\mathbf{k}'}(\mathbf{r}', \omega) \left[\frac{\omega_{\mathbf{k}'}^2}{c^2} - \frac{\omega^2}{c^2} \right] = \mathbf{u}^*_{\mathbf{k}'}(\mathbf{r}', \omega_\mathbf{k}). \tag{8.110}$$

Substituting this expression back into Eq. (8.106) leads to the desired expansion for $\overleftrightarrow{\mathbf{G}}$ in terms of the normal modes:

$$\overleftrightarrow{\mathbf{G}}(\mathbf{r}, \mathbf{r}'; \omega) = \sum_\mathbf{k} c^2 \frac{\mathbf{u}^*_\mathbf{k}(\mathbf{r}', \omega_\mathbf{k}) \mathbf{u_k}(\mathbf{r}, \omega_\mathbf{k})}{\omega_\mathbf{k}^2 - \omega^2}. \tag{8.111}$$

To proceed we make use of the following mathematical identity which can be easily proved by complex contour integration:

$$\lim_{\eta \to 0} \operatorname{Im} \left\{ \frac{1}{\omega_\mathbf{k}^2 - (\omega + \mathrm{i}\eta)^2} \right\} = \frac{\pi}{2\omega_\mathbf{k}} \left[\delta(\omega - \omega_\mathbf{k}) - \delta(\omega + \omega_\mathbf{k}) \right]. \tag{8.112}$$

Multiplying on both sides by $\mathbf{u}^*_\mathbf{k}(\mathbf{r}, \omega_\mathbf{k}) \mathbf{u_k}(\mathbf{r}, \omega_\mathbf{k})$ and summing over all $\mathbf{k}$ yields

$$\operatorname{Im} \left\{ \lim_{\eta \to 0} \sum_\mathbf{k} \frac{\mathbf{u}^*_\mathbf{k}(\mathbf{r}, \omega_\mathbf{k}) \mathbf{u_k}(\mathbf{r}, \omega_\mathbf{k})}{\omega_\mathbf{k}^2 - (\omega + \mathrm{i}\eta)^2} \right\} = \frac{\pi}{2} \sum_\mathbf{k} \frac{1}{\omega_\mathbf{k}} \mathbf{u}^*_\mathbf{k}(\mathbf{r}, \omega_\mathbf{k}) \mathbf{u_k}(\mathbf{r}, \omega_\mathbf{k}) \delta(\omega - \omega_\mathbf{k}), \tag{8.113}$$

where we dropped the term $\delta(\omega + \omega_{\mathbf{k}})$ because we are concerned only with positive frequencies. By comparison with Eq. (8.111), the expression in brackets on the left-hand side can be identified with the Green function evaluated at its origin $\mathbf{r} = \mathbf{r}'$. Furthermore, the delta function on the right-hand side restricts all values of $\omega_{\mathbf{k}}$ to ω, which allows us to move the first factor out of the sum. We therefore obtain the important relationship

$$\mathrm{Im}\left\{\overleftrightarrow{\mathbf{G}}(\mathbf{r},\mathbf{r};\omega)\right\} = \frac{\pi c^2}{2\omega}\sum_{\mathbf{k}} \mathbf{u}^*_{\mathbf{k}}(\mathbf{r},\omega_{\mathbf{k}})\mathbf{u}_{\mathbf{k}}(\mathbf{r},\omega_{\mathbf{k}})\delta(\omega-\omega_{\mathbf{k}}). \tag{8.114}$$

We now set $\mathbf{r} = \mathbf{r}_0$ and $\omega = \omega_0$ and rewrite the decay rate γ and the partial local density of states ρ_{p} in Eq. (8.103) as

$$\gamma = \frac{\pi\omega_0}{3\hbar\varepsilon_0}\left|\mathbf{p}\right|^2 \rho_{\mathrm{p}}(\mathbf{r}_0,\omega_0), \qquad \rho_{\mathrm{p}}(\mathbf{r}_0,\omega_0) = \frac{6\omega_0}{\pi c^2}\left[\mathbf{n}_{\mathrm{p}}\cdot \mathrm{Im}\left\{\overleftrightarrow{\mathbf{G}}(\mathbf{r}_0,\mathbf{r}_0;\omega_0)\right\}\cdot\mathbf{n}_{\mathrm{p}}\right]. \tag{8.115}$$

This formula is the main result of this section. It allows us to calculate the spontaneous decay rate of a two-level quantum system in an arbitrary reference system. All that is needed is knowledge of the Green dyadic for the reference system. The Green dyadic is evaluated at its origin, which corresponds to the location of the atomic system. From a classical viewpoint this is equivalent to the electric field previously emitted by the quantum system and now arriving back at its origin. The mathematical analogy of the quantum and the classical treatments now becomes obvious on comparing Eq. (8.115) and Eq. (8.75). The latter is the classical equation for energy dissipation based on Poynting's theorem.

Notice, however, that the dipoles in Eqs. (8.115) and (8.75) are *not* the same! In Eq. (8.75), $\mathbf{p}$ denotes a classical dipole, whereas in Eq. (8.115) it represents the matrix element $\langle e|\hat{\mathbf{p}}|g\rangle$. A comparison of Eq. (8.75) with Eq. (8.115) yields

$$\frac{\gamma}{\gamma_0} = \frac{P}{P_0}, \tag{8.116}$$

where γ_0 and P_0 refer to the free-space values, or to any other known reference values. The normalization by γ_0 and P_0 eliminates the dependence on $\mathbf{p}$ and establishes a safe link between quantum and classical formalisms. While the left-hand side of Eq. (8.116) represents the quantum-mechanical picture of spontaneous emission, the right-hand side corresponds to the classical formalism of dipole radiation. Because of the relationship of Eq. (8.116) we are able to classically calculate the relative change of the spontaneous decay rate of a quantum emitter. This relationship is valid as long as $\langle e|\hat{\mathbf{p}}|g\rangle$ is not affected by the environment, i.e. the quantum orbitals remain unaffected.

In Eq. (8.115) we have expressed γ in terms of the partial local density of states ρ_{p}, which corresponds to the number of modes per unit volume and frequency, at the origin $\mathbf{r}$ of the (point-like) quantum system, into which a photon with energy $\hbar\omega_0$ can be released during the spontaneous decay process. In the next section we discuss some important aspects of ρ_{p}.

8.4.3 Local density of states

In situations where the transitions of the quantum system have no fixed dipole axis $\mathbf{n}_p$ and the medium is isotropic and homogeneous, the decay rate is averaged over the various orientations, leading to (see Problem 8.6)

$$\left\langle \mathbf{n}_p \cdot \mathrm{Im}\left\{ \overleftrightarrow{\mathbf{G}}(\mathbf{r}_0, \mathbf{r}_0; \omega_0) \right\} \cdot \mathbf{n}_p \right\rangle = \frac{1}{3}\, \mathrm{Im}\left\{ \mathrm{Tr}[\overleftrightarrow{\mathbf{G}}(\mathbf{r}_0, \mathbf{r}_0; \omega_0)] \right\}. \tag{8.117}$$

Substituting into Eq. (8.115), we find that in this case the partial local density of states ρ_p becomes identical with the *total local density of states* ρ defined as

$$\rho(\mathbf{r}_0, \omega_0) = \frac{2\omega_0}{\pi c^2}\, \mathrm{Im}\left\{ \mathrm{Tr}[\overleftrightarrow{\mathbf{G}}(\mathbf{r}_0, \mathbf{r}_0; \omega_0)] \right\} = \sum_{\mathbf{k}} |\mathbf{u}_{\mathbf{k}}|^2\, \delta(\omega_{\mathbf{k}} - \omega_0), \tag{8.118}$$

where Tr[...] denotes the trace of the tensor in brackets. ρ corresponds to the total number of electromagnetic modes per unit volume and unit frequency at a given location $\mathbf{r}_0$. In practice, ρ has little significance because any detector or measurement relies on the translation of charge carriers from one point to another. On defining the axis between these points as $\mathbf{n}_p$, it is obvious that ρ_p is of much greater practical significance since it also enters the well-known formula for spontaneous decay.

As shown earlier in Section 8.3.3, the imaginary part of $\overleftrightarrow{\mathbf{G}}$ evaluated at its origin is not singular. For example, in free space ($\overleftrightarrow{\mathbf{G}}=\overleftrightarrow{\mathbf{G}}_0$) we have (see Problem 8.7)

$$\left[\mathbf{n}_p \cdot \mathrm{Im}\left\{ \overleftrightarrow{\mathbf{G}}_0(\mathbf{r}_0, \mathbf{r}_0; \omega_0) \right\} \cdot \mathbf{n}_p \right] = \frac{1}{3}\, \mathrm{Im}\left\{ \mathrm{Tr}[\overleftrightarrow{\mathbf{G}}_0(\mathbf{r}_0, \mathbf{r}_0; \omega_0)] \right\} = \frac{\omega_0}{6\pi c}, \tag{8.119}$$

where no orientational averaging has been performed. It is the symmetric form of $\overleftrightarrow{\mathbf{G}}_0$ that leads to this simple expression. Thus, ρ and ρ_p take on the well-known value of

$$\rho_0 = \frac{\omega_0^2}{\pi^2 c^3}, \tag{8.120}$$

which is the density of electromagnetic modes as encountered in blackbody radiation. The free-space spontaneous decay rate turns out to be

$$\gamma_0 = \frac{\omega_0^3 |\mathbf{p}|^2}{3\pi \varepsilon_0 \hbar c^3}, \tag{8.121}$$

where $\mathbf{p} = \langle g|\hat{\mathbf{p}}|e\rangle$ denotes the transition dipole matrix element.

To summarize, the spontaneous decay rate is proportional to the partial local density of states, which depends on the transition dipole defined by the two atomic states involved in the transition. Only in homogeneous environments or after orientational averaging can ρ_p be replaced by the total local density of states.

8.5 Classical lifetimes and decay rates

We now derive the classical picture of spontaneous decay by considering an undriven harmonically oscillating dipole. As the dipole oscillates it radiates energy according to Eq. (8.70). As a consequence, the dipole dissipates its energy into radiation and its dipole moment decreases. We are interested in calculating the time τ after which the dipole's energy has decreased to $1/e$ of its initial value (Fig. 8.7).

8.5.1 Radiation in homogeneous environments

The equation of motion for an undriven harmonically oscillating dipole is (cf. Eq. (8.87))

$$\frac{d^2}{dt^2}\mathbf{p}(t) + \gamma_0 \frac{d}{dt}\mathbf{p}(t) + \omega_0^2 \mathbf{p}(t) = 0. \tag{8.122}$$

The natural frequency of the oscillator is ω_0 and its damping constant is γ_0. The solution for $\mathbf{p}$ is

$$\mathbf{p}(t) = \mathrm{Re}\left\{\mathbf{p}_0 e^{-i\omega_0\sqrt{1-\gamma_0^2/(4\omega_0^2)}\,t}\, e^{\gamma_0 t/2}\right\}. \tag{8.123}$$

Because of losses introduced through γ_0 the dipole forms a non-conservative system. The damping rate not only attenuates the dipole strength but also produces a shift in resonance frequency. In order to be able to define an average dipole energy $\bar{W}$ at any instant of time, we have to make sure that the oscillation amplitude stays constant over one period of oscillation. In other words, we require

$$\gamma_0 \ll \omega_0. \tag{8.124}$$

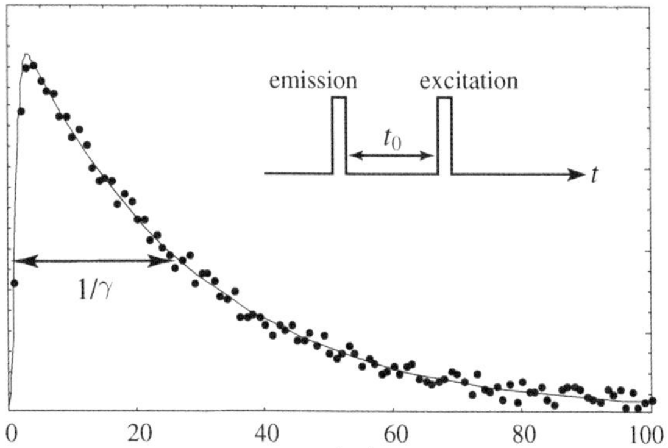

Fig. 8.7 Radiative decay rate γ of the $2P_{1/2}$ state of Li. The time interval t_0 between an excitation pulse and the subsequent photon count is measured and plotted in a histogram. The $1/e$ width of the exponential distribution corresponds to the lifetime $\tau = 1/\gamma = 27.1$ ns. For $t_0 \rightarrow 0$ the distribution falls to zero because of the finite response time of the photon detector.

The average energy of a harmonic oscillator is the sum of the average kinetic and potential energy. At time t this average energy reads as[10]

$$\bar{W}(t) = \frac{m}{2q^2}\left[\omega_0^2 p^2(t) + \dot{p}^2(t)\right] = \frac{m\omega_0^2}{2q^2}|\mathbf{p}_0|^2 \mathrm{e}^{-\gamma_0 t}, \tag{8.125}$$

where m is the mass of the particle with charge q. The *lifetime* τ_0 of the oscillator is defined as the time after which the energy has decayed to $1/\mathrm{e}$ of its initial value at $t = 0$. We simply find

$$\tau_0 = 1/\gamma_0. \tag{8.126}$$

We now turn to the rate of energy loss due to radiation. The average radiated power P_0 in free space at time t is (c.f. Eq. (8.71))

$$P_0(t) = \frac{|\mathbf{p}(t)|^2}{4\pi\varepsilon_0}\frac{\omega_0^4}{3c^3}. \tag{8.127}$$

Energy conservation requires that the decrease in oscillator energy must equal the energy losses, i.e.

$$\bar{W}(t=0) - \bar{W}(t) = q_\mathrm{i}\int_0^t P_0(t')\mathrm{d}t', \tag{8.128}$$

where we introduced the so-called *intrinsic quantum yield* q_i. This parameter has a value between zero and one and indicates the fraction of the energy loss associated with radiation. For $q_\mathrm{i} = 1$, all of the oscillator's dissipated energy is transformed to radiation. It is now straightforward to solve for the decay rate. We introduce Eqs. (8.125) and (8.127) into the last equation and obtain

$$\gamma_0 = q_\mathrm{i}\frac{1}{4\pi\varepsilon_0}\frac{2q^2\omega_0^2}{3mc^3}, \tag{8.129}$$

which (besides q_i) is identical with Eq. (8.87). γ_0 is the classical expression for the atomic decay rate and, through Eq. (8.126), also for the atomic lifetime. It depends on the oscillation frequency and the particle's mass and charge. The higher the index of refraction of the surrounding medium is, the shorter the lifetime of the oscillator will be. γ_0 can easily be generalized to multiple-particle systems by summing over the individual charges q_n and masses m_n. At optical wavelengths we obtain a value for the decay rate of $\gamma_0 \approx 2 \times 10^{-8}\omega_0$, which is in the MHz regime. The quantum-mechanical analog of the decay rate (cf. Eq. (8.121)) can be arrived at by replacing the oscillator's initial average energy $m\omega_0^2|\mathbf{p}_0|^2/(2q^2)$ by the energy quantum $\hbar\omega_0$. At the same time, the classical dipole moment has to be associated with half of the transition dipole matrix element.[11]

In the treatments so far, we have assumed that the atom is locally surrounded by vacuum ($n = 1$). For an atom placed in a dielectric medium there are two corrections that need to be performed: (1) the bulk dielectric behavior has to be accounted for by a dielectric constant

[10] This is easily derived by setting $p = qx$, $\omega_0^2 = c/m$ and using the expressions $m\dot{x}^2/2$ and $cx^2/2$ for the kinetic and potential energy, respectively.

[11] The factor 1/2 in the substitution $\mathbf{p}_0 \rightarrow (1/2)\langle g|\hat{\mathbf{p}}|e\rangle$ is due to the fact that the Fourier transform of the classical dipole moment spans over positive and negative frequencies.

and (2) the local field at the dipole's position has to be corrected. The latter arises from the depolarization of the dipole's microscopic environment, which influences the dipole's emission properties. The resulting correction is similar to the Clausius–Mossotti relation, but more sophisticated models have been put forward recently.

The Lorentzian lineshape function

Spontaneous emission is well represented by an undriven harmonic oscillator. Although the oscillator acquires its energy through an exciting local field, the phases of excitation and emission are uncorrelated. Therefore, we can envision spontaneous emission as the radiation emitted by an undriven harmonic oscillator whose dipole moment is restored by the local field whenever the oscillator has lost its energy to the radiation field. The spectrum of spontaneous emission by a single atomic system is well described by the spectrum of the emitted radiation of an undriven harmonic oscillator. In free space, the electric far-field of a radiating dipole is calculated as (cf. Eq. (8.67))

$$E_\vartheta(t) = \frac{\sin\vartheta}{4\pi\varepsilon_0}\frac{1}{c^2}\frac{1}{r}\frac{\mathrm{d}^2}{\mathrm{d}t^2}\left|\mathbf{p}(t-r/c)\right|, \tag{8.130}$$

where r is the distance between the observation point and the dipole origin. The spectrum $\hat{E}_\vartheta(\omega)$ can be calculated as (cf. Eq. (2.17))

$$\hat{E}_\vartheta(\omega) = \frac{1}{2\pi}\int_{r/c}^{\infty} E_\vartheta(t)\mathrm{e}^{\mathrm{i}\omega t}\,\mathrm{d}t. \tag{8.131}$$

Here we set the lower integration limit to $t = r/c$ because the dipole starts emitting at $t = 0$ and it takes the time $t = r/c$ for the radiation to propagate to the observation point. Therefore $E_\vartheta(t < r/c) = 0$. On inserting the solution for the dipole moment from Eq. (8.123) and making use of $\gamma_0 \ll \omega_0$ we obtain after integration

$$\hat{E}_\vartheta(\omega) = \frac{1}{2\pi}\frac{|\mathbf{p}|\sin\vartheta\,\omega_0^2}{8\pi\varepsilon_0 c^2 r}\left[\frac{\exp(\mathrm{i}\omega r/c)}{\mathrm{i}(\omega+\omega_0)-\gamma_0/2}+\frac{\exp(\mathrm{i}\omega r/c)}{\mathrm{i}(\omega-\omega_0)-\gamma_0/2}\right]. \tag{8.132}$$

The energy radiated into the unit solid angle $\mathrm{d}\Omega = \sin\vartheta\,\mathrm{d}\vartheta\,\mathrm{d}\varphi$ is calculated as

$$\begin{aligned}\frac{\mathrm{d}W}{\mathrm{d}\Omega} &= \int_{-\infty}^{\infty} I(\mathbf{r},t)r^2\,\mathrm{d}t = r^2\sqrt{\frac{\varepsilon_0}{\mu_0}}\int_{-\infty}^{\infty}|E_\vartheta(t)|^2\,\mathrm{d}t \\ &= 4\pi r^2\sqrt{\frac{\varepsilon_0}{\mu_0}}\int_0^{\infty}|\hat{E}_\vartheta(\omega)|^2\,\mathrm{d}\omega,\end{aligned} \tag{8.133}$$

where we applied Parseval's theorem and used the definition of the intensity $I = \sqrt{\varepsilon_0/\mu_0}\,|E_\vartheta|^2$ of the emitted radiation. The total energy per unit solid angle $\mathrm{d}\Omega$ and per unit frequency interval $\mathrm{d}\omega$ can now be expressed as

$$\frac{\mathrm{d}W}{\mathrm{d}\Omega\,\mathrm{d}\omega} = \frac{1}{4\pi\varepsilon_0}\frac{|\mathbf{p}|^2\sin^2\!\vartheta\,\omega_0^2}{4\pi^2c^3\gamma_0^2}\left[\frac{\gamma_0^2/4}{(\omega-\omega_0)^2+\gamma_0^2/4}\right]. \tag{8.134}$$

The spectral shape of this function is determined by the expression in the brackets known as the *Lorentzian lineshape function*. The function is shown in Fig. 8.8. The width of the

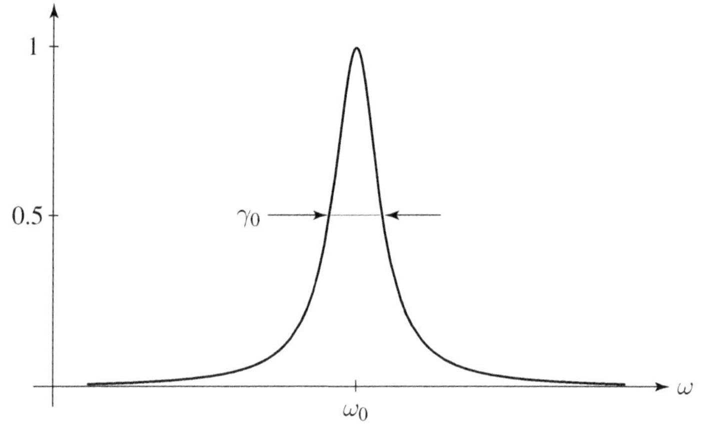

Fig. 8.8 The Lorentzian lineshape function as defined by the expression in brackets in Eq. (8.134).

curve measured at half its maximum height is $\Delta\omega = \gamma_0$, which is called the "radiative linewidth." Thus, the decay rate of an atomic system is identical with the linewidth of its emitted radiation. This correspondence between decay rate and linewidth also follows from Heisenberg's uncertainty principle $\Delta E\, \Delta t \approx \hbar$. The energy uncertainty is defined by the linewidth as $\Delta E = \hbar\, \Delta\omega$ and the average time available to measure the excited state is $\Delta t \approx 1/\gamma_0$ (cf. Eq. (8.125)). Thus, we obtain $\Delta\omega \approx \gamma_0$, in agreement with the Lorentzian lineshape function. For atomic transitions at optical frequencies and with typical lifetimes of $\tau = 10\,\text{ns}$ the radiative linewidth corresponds to a wavelength range of $\Delta\lambda \approx 2 \times 10^{-3}\,\text{nm}$.

Integrating the lineshape function over the entire spectral range yields a value of $\pi\gamma_0/2$. Integrating Eq. (8.134) over all frequencies and all directions leads to the totally radiated energy

$$W = \frac{|\mathbf{p}|^2}{4\pi\varepsilon_0} \frac{\omega_0^4}{3c^3\gamma_0}. \tag{8.135}$$

This value is equal to the average power $\bar{P}$ radiated by a driven harmonic oscillator divided by the linewidth γ_0 (cf. Eq. (8.71)).

The Fano lineshape function

In many experiments the Lorentzian response is superimposed on a broadband background field. The interference of the two contributions leads to a Fano lineshape function. For example, in light-scattering experiments the scattered field from a resonant particle or molecule often interferes with the broadband excitation field of the source. Another example is the coupling of a spectrally broad dipole resonance (bright mode) of a metal nanostructure and a spectrally narrow quadrupole resonance (dark mode) of the same nanostructure [20]. The situation is analogous to a discrete quantum-mechanical system that interacts with a continuum of states, which was analyzed in 1935 by Ugo Fano [21].

To derive the Fano lineshape function we rewrite the spectrum of the Lorentzian field in Eq. (8.132) as

$$\hat{E}_1 \frac{\gamma_0/2}{-\mathrm{i}(\omega - \omega_0) + \gamma_0/2}, \tag{8.136}$$

where we have dropped the first term containing $(\omega + \omega_0)$ in the denominator because it is much smaller in the region of interest $\omega \sim \omega_0$. This is equivalent to the so-called rotating-wave approximation. The amplitude $\hat{E}_1$ accounts for all the remaining factors in Eq. (8.132).

The Lorentzian field is now superimposed on a background field whose spectrum does not vary over the bandwidth γ_0 of the Lorentzian spectrum. The spectral lineshape function of the two interfering fields becomes

$$\left|\hat{E}(\omega)\right|^2 = \left|\frac{\hat{E}_1 \gamma_0/2}{-\mathrm{i}(\omega - \omega_0) + \gamma_0/2} + \hat{E}_2\right|^2, \tag{8.137}$$

with $\hat{E}_2$ representing the broadband background field. The above simplifies to

$$\left|\hat{E}(\omega)\right|^2 = \left|\hat{E}_1\right|^2 \frac{\left|\gamma_0/2 - (\hat{E}_2/\hat{E}_1)\left[\mathrm{i}(\omega - \omega_0) - \gamma_0/2\right]\right|^2}{(\omega - \omega_0)^2 + \gamma_0^2/4}, \tag{8.138}$$

and is illustrated in Fig. 8.9. The first term in the brackets of the numerator in Eq. (8.138) originates from the Lorentz field, whereas the latter term is due to the background field. It is the interference between these two fields that leads to the characteristic Fano lineshape. Fano lineshapes are often encountered in coherent-scattering experiments, such as Rayleigh scattering [22] and surface-enhanced infrared absorption (SEIRA) [23]. In general, Fano interference is the result of energy transfer from an initial to a final state via two indistinguishable paths.

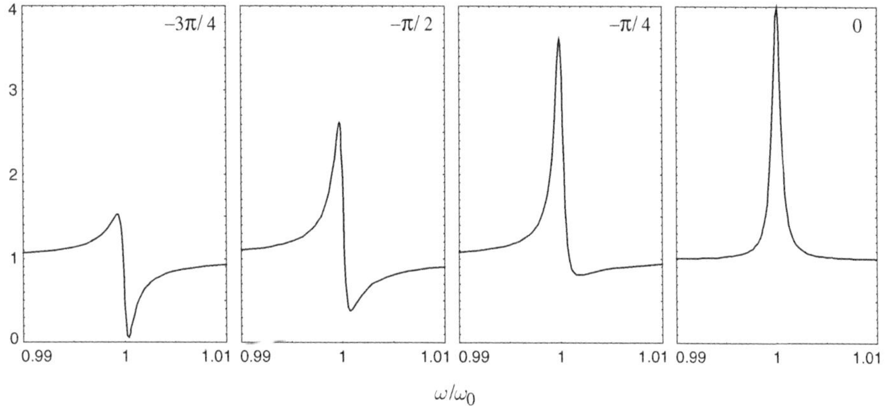

Fig. 8.9 The Fano lineshape function for different phase differences between the Lorentz field and the background field. $\hat{E}_2/\hat{E}_1 = \exp(\mathrm{i}\phi)$, where the value of ϕ is indicated in the individual figures.

8.5.2 Radiation in inhomogeneous environments

In an inhomogeneous environment, a harmonically oscillating dipole left to itself will experience its own field as a driving force. This driving field is the field that arrives back at the oscillator after it has been scattered in the environment. The equation of motion is

$$\frac{\mathrm{d}^2}{\mathrm{d}t^2}\mathbf{p}(t) + \gamma_0 \frac{\mathrm{d}}{\mathrm{d}t}\mathbf{p}(t) + \omega_0^2 \mathbf{p}(t) = \frac{q^2}{m}\,\mathbf{E}_{\mathrm{s}}(t), \tag{8.139}$$

with $\mathbf{E}_{\mathrm{s}}$ being the secondary local field. We expect that the interaction with $\mathbf{E}_{\mathrm{s}}$ will cause a shift of the resonance frequency and a modification of the decay rate. Therefore, we use the following trial solutions for the dipole moment and the driving field

$$\mathbf{p}(t) = \mathrm{Re}\left\{\mathbf{p}_0\, \mathrm{e}^{-\mathrm{i}\omega t}\mathrm{e}^{-\gamma t/2}\right\}, \quad \mathbf{E}_{\mathrm{s}}(t) = \mathrm{Re}\left\{\mathbf{E}_0\, \mathrm{e}^{-\mathrm{i}\omega t}\mathrm{e}^{-\gamma t/2}\right\}. \tag{8.140}$$

γ and ω are the new decay rate and resonance frequency, respectively. The two trial solutions can be inserted into Eq. (8.139). As before, we assume that γ is much smaller than ω (cf. Eq. (8.124)), which allows us to reject terms in γ^2. Furthermore, we assume that the interaction with the field $\mathbf{E}_{\mathrm{s}}$ is weak. In this limit the last term on the left-hand side of Eq. (8.139) is always larger than the driving term on the right-hand side. Using the expression for γ_0 from Eq. (8.129) we obtain

$$\frac{\gamma}{\gamma_0} = 1 + q_{\mathrm{i}}\, \frac{6\pi\varepsilon_0}{|\mathbf{p}_0|^2}\frac{1}{k^3}\, \mathrm{Im}\{\mathbf{p}_0^* \cdot \mathbf{E}_{\mathrm{s}}(\mathbf{r}_0)\}. \tag{8.141}$$

Since $\mathbf{E}_{\mathrm{s}}$ is proportional to $\mathbf{p}_0$, the dependence on the magnitude of the dipole moment cancels out. Besides the introduction of q_{i}, Eq. (8.141) is identical with Eq. (8.80) for the rate of energy dissipation in inhomogeneous environments. Thus, for $q_{\mathrm{i}} = 1$ we find

$$\frac{\gamma}{\gamma_0} = \frac{P}{P_0}, \tag{8.142}$$

in analogy to Eq. (8.116) derived earlier. In Section 8.6.2 we will use this equation to derive energy transfer between two molecules.

Equation (8.141) can be adapted to describe the (normalized) spontaneous emission rate of a quantum system. In this case the classical dipole represents (one half of) the quantum-mechanical transition dipole matrix element from the excited to the ground state. The decay rate of the excited state is equal to the spontaneous emission rate $P/(\hbar\omega)$, where $\hbar\omega$ is the photon energy. Equation (8.141) provides a simple means to calculate lifetime variations of atomic systems in arbitrary environments. In fact, this formula has been used by different authors to describe fluorescence quenching near planar interfaces and the agreement achieved with experiment is excellent (see Fig. 8.10).

8.5.3 Frequency shifts

The inhomogeneous environment not only influences the lifetime of the oscillating dipole but also causes a frequency shift $\Delta\omega = \omega - \omega_0$ of the emitted light. An expression for $\Delta\omega$

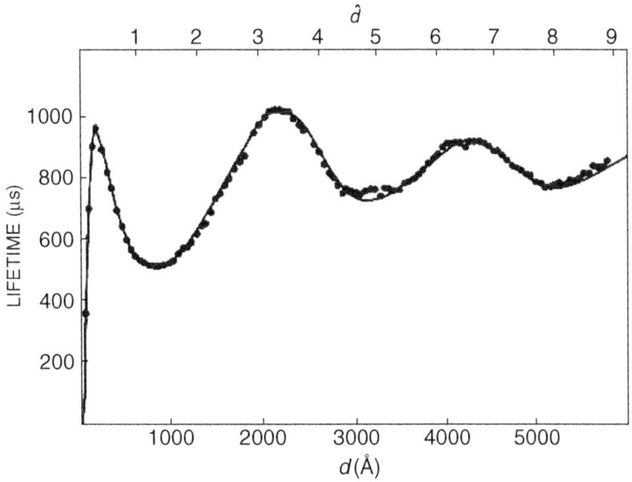

Fig. 8.10 Comparison of classical theory (curve) and experimental data (points) of molecular lifetimes in inhomogeneous environments. In the experiment, a layer of Eu^{3+} ions is held by fatty-acid spacers of variable thickness close to a silver surface (data after Drexhage [24]). The calculated curve is due to Chance *et al.* [25].

can be derived by inserting Eq. (8.140) into Eq. (8.139). The resulting expression for $\Delta\omega$ reads as

$$\Delta\omega = \omega \left[1 - \sqrt{1 - \frac{1}{\omega^2}\left[\frac{q^2}{m\,|\mathbf{p}_0|^2}\,\mathrm{Re}\{\mathbf{p}_0^* \cdot \mathbf{E}_s\} + \frac{\gamma\,\gamma_0}{2} - \frac{\gamma\,\gamma}{4} \right]} \right]. \tag{8.143}$$

After expanding the square root to first order and neglecting the quadratic terms in γ, the expression for the normalized frequency shift reduces to

$$\frac{\Delta\omega}{\gamma_0} = q_{\mathrm{i}}\,\frac{3\pi\varepsilon_0}{|\mathbf{p}_0|^2}\,\frac{1}{k^3}\,\mathrm{Re}\{\mathbf{p}_0^* \cdot \mathbf{E}_s\}. \tag{8.144}$$

The frequency shift is very small, in the range of the radiative linewidth.

For molecules close to planar interfaces, the frequency shift varies as h^{-3}, h being the height of the molecule, and reaches its maximum near the surface plasmon frequency. The dependence on h^{-3} suggests that observation of the frequency shift should be possible for small h. Yet this is not the case because for small h the linewidth also increases. A shift in the range of $\Delta\lambda \approx 20\,\mathrm{nm}$ was experimentally observed for small dipolar scatterers (silver islands) close to a silver layer [26]. In this configuration the dipolar scatterers were excited close to their resonance frequency, leading to a highly enhanced polarizability. At cryogenic temperatures, the vibrational broadening of the emission spectrum of single molecules is frozen out and the linewidths become very narrow, allowing frequency shifts to be observed.

Notice again that, since $\mathbf{E}_s$ is proportional to $\mathbf{p}_0$, the dependence on the magnitude of the dipole moment in Eq. (8.144) cancels out.

8.6 Dipole–dipole interactions and energy transfer

So far we have discussed the interaction of a nanoscale system with its local environment. In this section we shall focus on the interaction between two systems (atoms, molecules, quantum dots, ...) that we refer to as "particles." These considerations are important for the understanding of delocalized excitations (excitons), energy transfer between particles, and collective phenomena. We shall assume that the internal structure of a particle is not affected by the interactions. Therefore, processes such as electron transfer and molecular binding are not considered, and the interested reader is referred to texts on physical chemistry [27].

8.6.1 Multipole expansion of the Coulombic interaction

Let us consider two separate particles A and B represented by the charge densities ρ_A and ρ_B, respectively. For simplicity, we consider only non-retarded interactions. In this case, the Coulomb interaction energy between the systems A and B reads as

$$V_{AB} = \frac{1}{4\pi\varepsilon_0} \int\int \frac{\rho_A(\mathbf{r}')\rho_B(\mathbf{r}'')}{|\mathbf{r}' - \mathbf{r}''|} \, dV' \, dV''. \tag{8.145}$$

If we assume that the extent of the charge distributions ρ_A and ρ_B is much smaller than their separation R, we may expand V_{AB} in a multipole series with respect to the center-of-mass coordinates $\mathbf{r}_A$ and $\mathbf{r}_B$ (cf. Fig. 8.11). The first few multipole moments of the charge

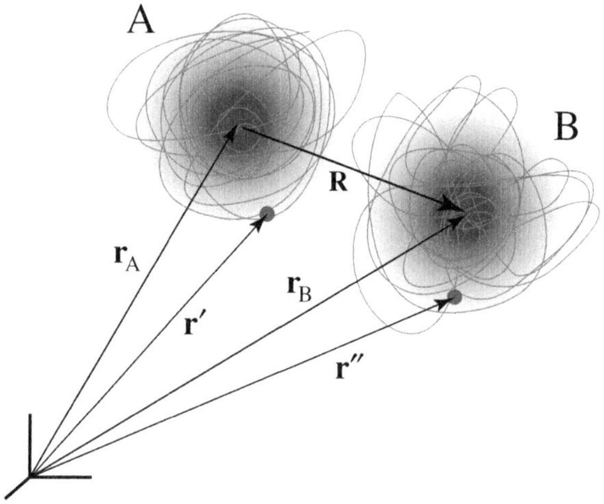

Fig. 8.11 Interaction between two particles A and B, which are represented by their charge distributions.

distribution ρ_A are determined as

$$q_A = \int \rho_A(\mathbf{r}')\mathrm{d}V', \tag{8.146}$$

$$\mathbf{p}_A = \int \rho_A(\mathbf{r}')(\mathbf{r}' - \mathbf{r}_A)\mathrm{d}V', \tag{8.147}$$

$$\overleftrightarrow{\mathbf{Q}}_A = \int \rho_A(\mathbf{r}')\frac{1}{2}\left[(\mathbf{r}' - \mathbf{r}_A)(\mathbf{r}' - \mathbf{r}_A) - \frac{\overleftrightarrow{\mathbf{I}}}{3}\left|\mathbf{r}' - \mathbf{r}_A\right|^2\right]\mathrm{d}V', \tag{8.148}$$

and similar expressions hold for the charge distribution ρ_B. With these multipole moments we can express the interaction potential as

$$V_{AB}(\mathbf{R}) = \frac{1}{4\pi\varepsilon_0}\left[\frac{q_A q_B}{R} + \frac{q_A\,\mathbf{p}_B\cdot\mathbf{R}}{R^3} - \frac{q_B\,\mathbf{p}_A\cdot\mathbf{R}}{R^3} + \frac{R^2\,\mathbf{p}_A\cdot\mathbf{p}_B - 3\,(\mathbf{p}_A\cdot\mathbf{R})(\mathbf{p}_B\cdot\mathbf{R})}{R^5} + \cdots\right], \tag{8.149}$$

where $\mathbf{R} = \mathbf{r}_B - \mathbf{r}_A$. The first term in the expansion is the charge–charge interaction. It is non-zero only if both particles A and B carry a net charge. Charge–charge interactions span long distances, since the distance dependence is only R^{-1}. The next two terms are charge–dipole interactions. They require that at least one particle carries a net charge. These interactions decay as R^{-2} and are therefore of shorter range than the charge–charge interaction. Finally, the fourth term is the dipole–dipole interaction. It is the most important interaction among neutral particles. This term gives rise to van der Waals forces and to Förster energy transfer. The dipole–dipole interaction decays as R^{-3} and depends strongly on the dipole orientations. The next higher expansion terms are the quadrupole–charge, quadrupole–dipole, and quadrupole–quadrupole interactions. These are usually of much shorter range and therefore we do not list them explicitly. It has to be emphasized that the potential V_{AB} accounts only for interactions mediated by the near-field of the two dipoles. Inclusion of the intermediate field and the far-field gives rise to additional terms. We will include these terms in the derivation of energy transfer between particles.

8.6.2 Energy transfer between two particles

Energy transfer between particles is a photophysical process encountered in various systems. Probably the most important example is radiationless energy transfer between light-harvesting proteins in photosynthetic membranes [28]. In these systems, the optical energy absorbed by chlorophyll molecules has to be channeled over longer distances to a protein called the reaction center. This protein uses the energy in order to perform a charge separation across the membrane surface. Energy transfer is also observed between closely arranged semiconductor nanoparticles [29] and it is the basis for Förster energy-transfer (FRET) studies of biological processes [30].

Energy transfer between individual particles can be understood within the same quasi-classical framework as developed in Section 8.5. The system to be analyzed is shown in

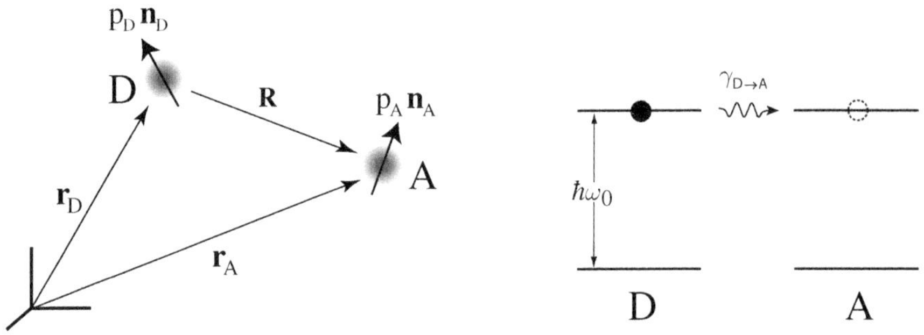

Fig. 8.12 Energy transfer between two particles D (donor) and A (acceptor). Initially, the donor is in its excited state, whereas the acceptor is in its ground state. The transition rate $\gamma_{D\to A}$ depends on the relative orientation of the transition dipole moments and the distance R between donor and acceptor.

Fig. 8.12. Two neutral particles D (donor) and A (acceptor) are characterized by a set of discrete energy levels. We assume that initially the donor resides in an excited state with energy $E_D = \hbar\omega_0$. We are interested in calculating the rate $\gamma_{D\to A}$ of energy transfer from donor to acceptor. The transition dipole moments of donor and acceptor are denoted $\mathbf{p}_D$ and $\mathbf{p}_A$, respectively, and $\mathbf{R}$ is the vector from donor to acceptor. The corresponding unit vectors are $\mathbf{n}_D$, $\mathbf{n}_A$, and $\mathbf{n}_R$, respectively. Our starting point is Eq. (8.116), which connects the quantum-mechanical picture with the classical picture. In the current context this equation reads as

$$\frac{\gamma_{D\to A}}{\gamma_0} = \frac{P_{D\to A}}{P_0}. \tag{8.150}$$

Here, $\gamma_{D\to A}$ is the rate of energy transfer from donor to acceptor and γ_0 is the donor's decay rate in the absence of the acceptor (cf. Eq. (8.129)). Similarly, $P_{D\to A}$ is the donor's energy per unit time absorbed by the acceptor, and P_0 is the energy per unit time released from the donor in the absence of the acceptor. P_0 can be written as (cf. Eq. (8.71))

$$P_0 = \frac{|\mathbf{p}_D|^2 n(\omega_0)}{12\pi\varepsilon_0 c^3}\omega_0^4. \tag{8.151}$$

Classically, we envision the donor to be a dipole radiating at the frequency ω_0 and the acceptor to be an absorber at ω_0. Both systems are embedded in a medium with index of refraction $n(\omega_0)$. Since the expressions for γ_0 and P_0 are known, we need only determine $P_{D\to A}$.

According to Poynting's theorem the power transferred from donor to acceptor is (cf. Eq. (8.73))

$$P_{D\to A} = -\frac{1}{2}\int_{V_A} \mathrm{Re}\{\mathbf{j}_A^* \cdot \mathbf{E}_D\}\mathrm{d}V. \tag{8.152}$$

Here, $\mathbf{j}_A$ is the current density associated with the charges of the acceptor and $\mathbf{E}_D$ is the electric field generated by the donor. In the dipole approximation, the current density reads as $\mathbf{j}_A = -i\omega_0\mathbf{p}_A\delta(\mathbf{r}-\mathbf{r}_A)$ and Eq. (8.152) reduces to

$$P_{D\to A} = \frac{\omega_0}{2}\,\mathrm{Im}\{\mathbf{p}_A^* \cdot \mathbf{E}_D(\mathbf{r}_A)\}. \tag{8.153}$$

It is important to realize that the acceptor's dipole moment $\mathbf{p}_A$ is not a permanent dipole moment. Instead, it is a dipole moment induced by the donor's field (c.f. Appendix A). In the linear regime we may write

$$\mathbf{p}_A = \overleftrightarrow{\alpha}_A \mathbf{E}_D(\mathbf{r}_A), \tag{8.154}$$

where $\overleftrightarrow{\alpha}_A$ is the acceptor's polarizability tensor. The dipole moment can now be substituted into Eq. (8.153) and, if we assume that the acceptor can be polarized only in the direction of a fixed axis given by the unit vector $\mathbf{n}_A$ in the direction of $\mathbf{p}_A$, i.e. $\overleftrightarrow{\alpha}_A = \alpha_A\mathbf{n}_A\mathbf{n}_A$, the power transferred from donor to acceptor can be written as

$$P_{D\to A} = \frac{\omega_0}{2}\,\mathrm{Im}\{\alpha_A\}\left|\mathbf{n}_A \cdot \mathbf{E}_D(\mathbf{r}_A)\right|^2. \tag{8.155}$$

This result demonstrates that energy absorption is associated with the imaginary part of the polarizability. Furthermore, because $\mathbf{p}_A$ is an induced dipole, the absorption rate scales with the square of the electric field projected onto the dipole axis. It is convenient to express the polarizability in terms of the absorption cross-section σ defined as

$$\sigma(\omega_0) = \frac{\langle P(\omega_0)\rangle}{I(\omega_0)}, \tag{8.156}$$

where $\langle P\rangle$ is the power absorbed by the acceptor averaged over all absorption dipole orientations, and I_0 is the incident intensity. In terms of the electric field $\mathbf{E}_D$, the absorption cross-section can be expressed as[12]

$$\sigma(\omega_0) = \frac{(\omega_0/2)\mathrm{Im}\{\alpha(\omega_0)\}\langle\left|\mathbf{n}_p \cdot \mathbf{E}_D\right|^2\rangle}{(1/2)(\varepsilon_0/\mu_0)^{1/2} n(\omega_0)\,|\mathbf{E}_D|^2} = \frac{\omega_0}{3}\sqrt{\frac{\mu_0}{\varepsilon_0}}\,\frac{\mathrm{Im}\{\alpha(\omega_0)\}}{n(\omega_0)}. \tag{8.157}$$

Here, we used the orientational average of $\langle\left|\mathbf{n}_p \cdot \mathbf{E}_D\right|^2\rangle$, which is calculated as

$$\langle\left|\mathbf{n}_p \cdot \mathbf{E}_D\right|^2\rangle = \frac{|\mathbf{E}_D|^2}{4\pi}\int_0^{2\pi}\!\!\int_0^{\pi}\left[\cos^2\theta\right]\sin\theta\,\mathrm{d}\theta\,\mathrm{d}\phi = \frac{1}{3}\,|\mathbf{E}_D|^2, \tag{8.158}$$

where θ is the angle enclosed by the dipole axis and the electric field vector. Thus, in terms of the absorption cross-section, the power transferred from donor to acceptor can be written as

$$P_{D\to A} = \frac{3}{2}\sqrt{\frac{\varepsilon_0}{\mu_0}}\,n(\omega_0)\sigma_A(\omega_0)\left|\mathbf{n}_A \cdot \mathbf{E}_D(\mathbf{r}_A)\right|^2. \tag{8.159}$$

12 Notice that the replacement of the polarizability α by the absorption cross-section σ is not strictly valid in the present context because σ is defined for homogeneous plane-wave excitation. Here, we perform this substitution in order to be consistent with the literature.

The donor's field $\mathbf{E}_\mathrm{D}$ evaluated at the origin of the acceptor $\mathbf{r}_\mathrm{A}$ can be expressed in terms of the free-space Green function $\overleftrightarrow{\mathbf{G}}$ as (cf. Eq. (8.52))

$$\mathbf{E}_\mathrm{D}(\mathbf{r}_\mathrm{A}) = \omega_0^2\,\mu_0 \overleftrightarrow{\mathbf{G}}(\mathbf{r}_\mathrm{D}, \mathbf{r}_\mathrm{A})\mathbf{p}_\mathrm{D}. \tag{8.160}$$

The donor's dipole moment can be represented as $\mathbf{p}_\mathrm{D} = |\mathbf{p}_\mathrm{D}|\,\mathbf{n}_\mathrm{D}$ and the frequency dependence can be substituted as $k = (\omega_0/c)n(\omega_0)$. Furthermore, for later convenience we define the function

$$T(\omega_0) = 16\pi^2 k^4 R^6 \left|\mathbf{n}_\mathrm{A} \cdot \overleftrightarrow{\mathbf{G}}(\mathbf{r}_\mathrm{D}, \mathbf{r}_\mathrm{A})\mathbf{n}_\mathrm{D}\right|^2, \tag{8.161}$$

where $R = |\mathbf{r}_\mathrm{D} - \mathbf{r}_\mathrm{A}|$ is the distance between donor and acceptor. On using Eqs. (8.159)–(8.161) together with Eq. (8.151) in the original equation (8.150), we obtain for the normalized transfer rate from donor to acceptor

$$\frac{\gamma_{\mathrm{D}\to\mathrm{A}}}{\gamma_0} = \frac{9c^4}{8\pi R^6}\,\frac{\sigma_\mathrm{A}(\omega_0)}{n^4(\omega_0)\,\omega_0^4}\,T(\omega_0). \tag{8.162}$$

In terms of the Dirac delta function this equation can be rewritten as

$$\frac{\gamma_{\mathrm{D}\to\mathrm{A}}}{\gamma_0} = \frac{9c^4}{8\pi R^6}\int_0^\infty \frac{\delta(\omega-\omega_0)\sigma_\mathrm{A}(\omega)}{n^4(\omega)\omega^4}\,T(\omega)\mathrm{d}\omega. \tag{8.163}$$

We notice that the normalized frequency distribution of the donor emission is given by

$$\int_0^\infty \delta(\omega-\omega_0)\mathrm{d}\omega = 1. \tag{8.164}$$

Since the donor emits over a range of frequencies we need to generalize the distribution as

$$\int_0^\infty f_\mathrm{D}(\omega)\mathrm{d}\omega = 1, \tag{8.165}$$

with $f_\mathrm{D}(\omega)$ being the donor's normalized emission spectrum in a medium with index $n(\omega)$. Thus, we finally obtain the important result

$$\frac{\gamma_{\mathrm{D}\to\mathrm{A}}}{\gamma_0} = \frac{9c^4}{8\pi R^6}\int_0^\infty \frac{f_\mathrm{D}(\omega)\sigma_\mathrm{A}(\omega)}{n^4(\omega)\omega^4}\,T(\omega)\mathrm{d}\omega. \tag{8.166}$$

The transfer rate from donor to acceptor depends on the spectral overlap of the donor's emission spectrum f_D and the acceptor's absorption cross-section. Notice that f_D has units of ω^{-1}, whereas the units of σ_A are m^2. In order to understand the orientation dependence and the distance dependence of the transfer rate we need to evaluate the function $T(\omega)$. Using the definition in Eq. (8.161) and inserting the free-space dyadic Green function from Eq. (8.55), we obtain

$$\begin{aligned} T(\omega) = {} & (1 - k^2R^2 + k^4R^4)(\mathbf{n}_\mathrm{A}\cdot\mathbf{n}_\mathrm{D})^2 \\ & + (9 + 3k^2R^2 + k^4R^4)(\mathbf{n}_\mathrm{R}\cdot\mathbf{n}_\mathrm{D})^2(\mathbf{n}_\mathrm{R}\cdot\mathbf{n}_\mathrm{A})^2 \\ & + (-6 + 2k^2R^2 - 2k^4R^4)(\mathbf{n}_\mathrm{A}\cdot\mathbf{n}_\mathrm{D})(\mathbf{n}_\mathrm{R}\cdot\mathbf{n}_\mathrm{D})(\mathbf{n}_\mathrm{R}\cdot\mathbf{n}_\mathrm{A}), \end{aligned} \tag{8.167}$$

where $\mathbf{n}_R$ is the unit vector pointing from donor to acceptor. $T(\omega)$ and Eq. (8.166) determine the rate of energy transfer from donor to acceptor for arbitrary dipole orientation and arbitrary separations. Figure 8.13 shows the normalized distance dependence of $T(\omega)$ for three different relative orientations of $\mathbf{n}_D$ and $\mathbf{n}_A$. At short distances R, $T(\omega)$ is constant and the transfer rate in Eq. (8.166) decays as R^{-6}. For large distances R, $T(\omega)$ scales in most cases as R^{-4} and the transfer rate decays as R^{-2}.

In many situations the dipole orientations are not known and the transfer rate $\gamma_{D\rightarrow A}$ has to be determined by taking a statistical average over many donor–acceptor pairs. The same applies to a single donor–acceptor pair subject to random rotational diffusion. We therefore replace $T(\omega)$ by its orientational average $\langle T(\omega)\rangle$. The calculation is similar to the procedure encountered before (cf. Eq. (8.158)) and gives

$$\langle T(\omega)\rangle = \frac{2}{3} + \frac{2}{9}k^2R^2 + \frac{2}{9}k^4R^4. \tag{8.168}$$

The transfer rate decays very rapidly with distance between donor and acceptor. Therefore, only distances $R \ll 1/k$, where $k = 2\pi n(\omega)/\lambda$, are experimentally significant and the terms scaling with R^2 and R^4 in $T(\omega)$ can be neglected. In this limit, $T(\omega)$ is commonly denoted as κ^2 and the transfer rate can be expressed as

$$\frac{\gamma_{D\rightarrow A}}{\gamma_0} = \left[\frac{R_0}{R}\right]^6, \quad R_0^6 = \frac{9c^4\kappa^2}{8\pi}\int_0^\infty \frac{f_D(\omega)\sigma_A(\omega)}{n^4(\omega)\,\omega^4}\,\mathrm{d}\omega \tag{8.169}$$

where κ^2 is given by

$$\kappa^2 = \left[\mathbf{n}_A\cdot\mathbf{n}_D \;-\; 3(\mathbf{n}_R\cdot\mathbf{n}_D)(\mathbf{n}_R\cdot\mathbf{n}_A)\right]^2. \tag{8.170}$$

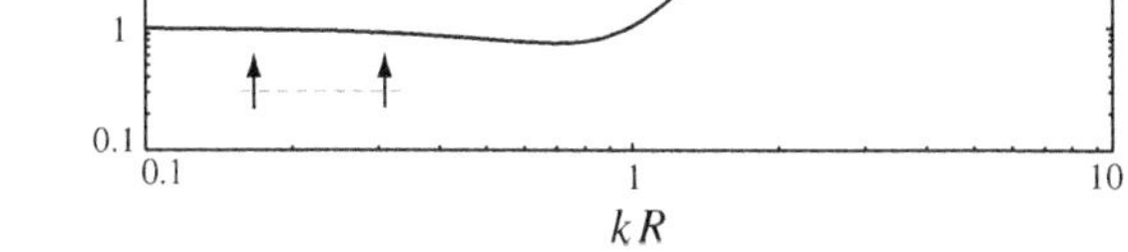

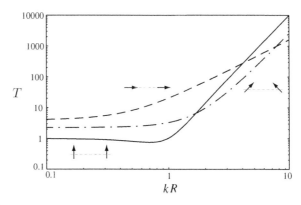

Fig. 8.13 Dependence of the function $T(\omega)$ on the distance R between donor and acceptor for different dipole orientations. In all cases, the short-distance behavior ($kR \ll 1$) is constant. Therefore the short-distance transfer rate $\gamma_{D\rightarrow A}$ scales as R^{-6}. The long-distance behavior ($kR \gg 1$) depends on the relative orientation of donor and acceptor. If the dipoles are aligned, $T(\omega)$ scales as $(kR)^2$ and $\gamma_{D\rightarrow A}$ decays as R^{-4}. In all other cases, the long-distance behavior of $T(\omega)$ shows a $(kR)^4$ dependence and $\gamma_{D\rightarrow A}$ decays as $(kR)^{-2}$.

The process described by Eq. (8.169) is known as Förster energy transfer. It is named after Th. Förster, who first derived this formula in 1946 in a slightly different form [31]. The quantity R_0 is called the Förster radius and it indicates the efficiency of energy transfer between donor and acceptor. For $R = R_0$ the transfer rate $\gamma_{\mathrm{D}\to\mathrm{A}}$ is equal to the decay rate γ_0 of the donor in the absence of the acceptor. R_0 is typically within the range 2–9 nm [32]. Notice that the refractive index $n(\omega)$ of the environment (solvent) is included in the definition of R_0. The Förster radius therefore has different values for different solvents. The literature is not consistent about the usage of $n(\omega)$ in R_0. A discussion can be found in Ref. [33]. The factor κ^2 has a value in the range $\kappa^2 = [0 \ldots 4]$. The relative orientation of donor and acceptor is often not known and the orientational average

$$\langle \kappa^2 \rangle = \frac{2}{3} \tag{8.171}$$

is adopted for κ^2.

In the limit of Förster energy transfer only the non-radiative near-field term in Eq. (8.168) is retained. For distances $kR \gg 1$ the transfer becomes radiative and scales with R^{-2}. In this limit we retain only the last term in Eq. (8.168). The result is identical with the quantum-electrodynamical calculation by Andrews and Juzeliunas [34]. In the radiative limit the donor emits a photon and the acceptor absorbs the same photon. However, the probability of such an event is extremely small. Besides the R^{-6} and the R^{-2} terms we also find an intermediate term that scales as R^{-4}. The inclusion of this term is important for distances $kR \approx 1$.

Recently, it has been demonstrated that the energy-transfer rate is modified in an inhomogeneous environment such as in a microcavity [35, 36]. This modification follows directly from the formalism outlined in this section: the inhomogeneous environment has to be accounted for by a modified Green function $\overleftrightarrow{\mathbf{G}}$ that alters not only the donor's decay rate γ_0 but also the transfer rate $\gamma_{\mathrm{D}\to\mathrm{A}}$ through Eq. (8.161). Using the formalism developed here, it is possible to calculate the energy transfer in an arbitrary environment.

Example: energy transfer (FRET) between two molecules

In order to illustrate the derived formulas for energy transfer we shall calculate the fluorescence from a donor molecule and an acceptor molecule with fixed separation, e.g. two molecules attached to specific sites of a protein. Such a configuration is encountered in studies of protein folding and molecular dynamics [37]. For the current example we choose *fluorescein* as the donor molecule and *Alexa Fluor 532* as the acceptor molecule. At room temperatures the emission and absorption spectra of donor and acceptor can be well fitted by a superposition of Gaussian distribution functions of the form

$$\sum_{n=1}^{N} A_n \mathrm{e}^{-(\lambda-\lambda_n)^2/\Delta\lambda_n^2}. \tag{8.172}$$

For the two dye molecules we obtain good fits with only two Gaussians ($N = 2$). The parameters for the donor emission spectrum f_D are $A_1 = 2.52$ fs, $\lambda_1 = 512.3$ nm, $\Delta\lambda_1 = 16.5$ nm; $A_2 = 1.15$ fs, $\lambda_2 = 541.7$ nm, $\Delta\lambda_2 = 35.6$ nm, and those for the acceptor absorption spectrum σ_A are $A_1 = 0.021$ nm^2, $\lambda_1 = 535.8$ nm, $\Delta\lambda_1 = 15.4$ nm; $A_2 = 0.013$ nm^2, $\lambda_2 = 514.9$ nm, $\Delta\lambda_2 = 36.9$ nm. The fitted absorption and emission spectra are shown in Fig. 8.14. The third panel in Fig. 8.14 shows the overlap of the donor emission spectrum and the acceptor absorption spectrum. In order to calculate the transfer rate we adopt the orientational average of κ^2 from Eq. (8.168). For the index of refraction we choose $n = 1.33$ (water) and we ignore any dispersion effects. Thus, the Förster radius is calculated as

$$R_0 = \left[\frac{3c}{32\pi^4 n^4} \int_0^\infty f_D(\lambda)\sigma_A(\lambda)\,\lambda^2\,\mathrm{d}\lambda \right]^{1/6} = 6.3\,\mathrm{nm}, \tag{8.173}$$

where we substituted ω by $2\pi c/\lambda$.[13] In air ($n = 1$) the Förster radius would be $R_0 = 7.6$ nm, which indicates that the local medium has a strong influence on the transfer rate.

In order to experimentally measure energy transfer the donor molecule has to be promoted into its excited state. We choose an excitation wavelength of $\lambda_{exc} = 488$ nm, which is close to the peak of fluorescein's absorption of $\lambda = 490$ nm. At λ_{exc} the acceptor absorption is a factor of 4 lower than the donor absorption. The non-zero absorption cross-section of the acceptor will lead to a background acceptor fluorescence signal. With the help of spectral filtering it is possible to experimentally separate the fluorescence emission from donor and acceptor. Energy transfer from donor to acceptor is then observed as a decrease of the donor's fluorescence intensity and as an increase of the acceptor's fluorescence intensity. The energy-transfer efficiency E is usually defined as the relative change of the donor's fluorescence emission:

$$E = \frac{P_{D\to A}}{P_0 + P_{D\to A}} = \frac{1}{1 + \gamma_0/\gamma_{D\to A}} = \frac{1}{1 + (R/R_0)^6}. \tag{8.174}$$

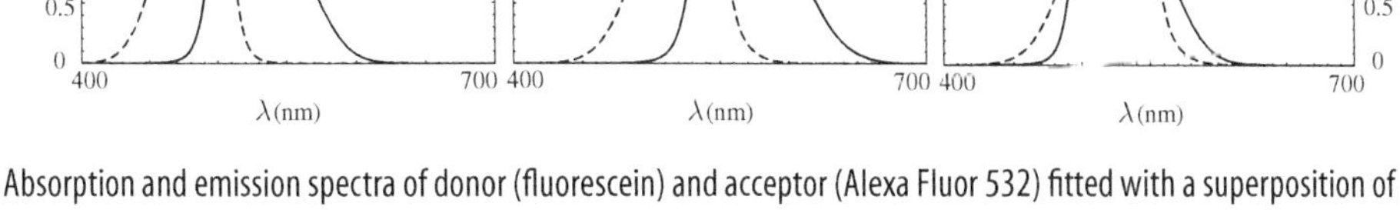

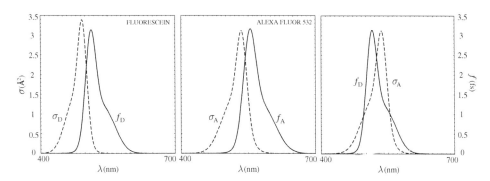

Fig. 8.14 Absorption and emission spectra of donor (fluorescein) and acceptor (Alexa Fluor 532) fitted with a superposition of two Gaussian distribution functions. The panel on the right shows the overlap between f_D and σ_A, which determines the value of the Förster radius.

[13] Notice that in the λ-representation the emission spectrum needs to be normalized as $2\pi c \int_0^\infty f_D(\lambda)/\lambda^2\,\mathrm{d}\lambda = 1$.

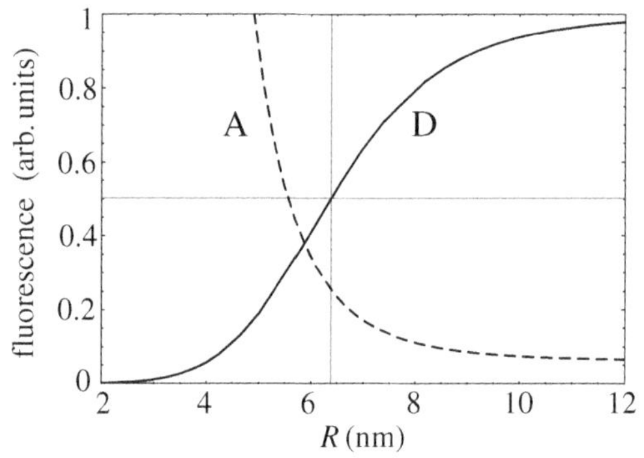

Fig. 8.15 Fluorescence intensity of donor and acceptor as a function of their separation R. The donor emission halves at the distance $R = R_0$. The acceptor fluorescence increases as R^{-6} and saturates at a value determined by the acceptor's excited-state lifetime.

Figure 8.15 illustrates the change of donor and acceptor fluorescence as a function of their separation R. It is assumed that the absorption cross-section of the acceptor is sufficiently small at the excitation wavelength λ_{exc}. At the distance $R = R_0$ the emission of the donor halves. The fluorescence intensity of the acceptor increases as R^{-6} and saturates at a level determined by the lifetime of the acceptor's excited state. See also Fig. 8.16.

In single-molecule experiments it is important to know the orientation of donor and acceptor. Depending on the relative orientation, the value of κ^2 can vary in the range $\kappa^2 = [0\ldots4]$. It is common practice to adopt the averaged value of $\kappa^2 = 2/3$. However, in some situations this might affect the conclusions drawn on the basis of experimental data.

8.7 Strong coupling (delocalized excitations)

The theory of Förster energy transfer assumes that the transfer rate from donor to acceptor is smaller than the vibrational relaxation rate. This ensures that once the energy has been transferred to the acceptor, there is little chance of a backtransfer to the donor. However, if the dipole–dipole interaction energy is larger than the energy associated with vibrational broadening of the electronic excited states, a delocalized excitation of donor and acceptor is more probable. In this so-called *strong-coupling* regime it is not possible to distinguish between donor and acceptor and one must view the pair as one system, i.e. the excitation becomes delocalized over the pair of particles. In this section we discuss strong coupling between a pair of particles A and B, but the analysis can be extended to larger systems such as *J-aggregates*, which are chains of strongly coupled molecules. A characteristic feature of the strong-coupling regime is energy-level splitting, a property that can be well understood from a classical perspective. Therefore, we first discuss the coupling of two harmonic oscillators as illustrated in Fig. 8.17 before diving into a more rigorous analysis.

Optical interactions

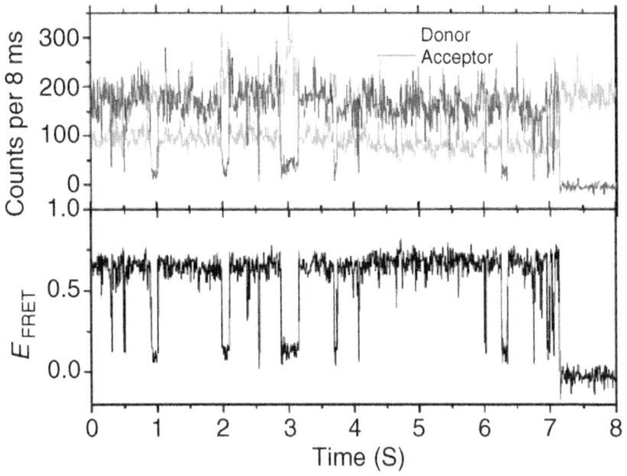

Fig. 8.16 Time trajectory of donor and acceptor fluorescence and corresponding FRET efficiency for a donor–acceptor pair attached to a four-way DNA (Holliday) junction. The data indicate that the DNA structure is switching back and forth between two conformations. Reprinted with permission from Macmillan Publishers Ltd. [38].

8.7.1 Coupled oscillators

In the absence of coupling ($\kappa = 0$) the two oscillators shown in Fig. 8.17 have eigenfrequencies $\omega_A^0 = \sqrt{k_A/m_A}$ and $\omega_B^0 = \sqrt{k_B/m_B}$, respectively. In the presence of coupling ($\kappa \neq 0$) the equations of motion become

$$\begin{aligned} m_A \ddot{x}_A + k_A x_A + \kappa(x_A - x_B) &= 0, \\ m_B \ddot{x}_B + k_B x_B - \kappa(x_A - x_B) &= 0. \end{aligned} \tag{8.175}$$

We seek homogeneous solutions of the form $x_i(t) = x_i^0 \exp(-i\omega_\pm t)$, where $\omega_\pm$ are the new eigenfrequencies. We insert this ansatz into Eq. (8.175) and obtain two coupled linear equations for x_A^0 and x_B^0, which can be written in matrix form as $\overleftrightarrow{\mathbf{M}}\,[x_A^0, x_B^0]^T = 0$. Non-trivial solutions for this homogeneous system of equations exist only if $\det[\overleftrightarrow{\mathbf{M}}] = 0$. The resulting characteristic equation yields

$$\omega_\pm^2 = \frac{1}{2}\left[\omega_A^2 + \omega_B^2 \pm \sqrt{(\omega_A^2 - \omega_B^2)^2 + 4\Gamma^2\omega_A\omega_B}\right], \tag{8.176}$$

where $\omega_A = \sqrt{(k_A + \kappa)/m_A}$, $\omega_B = \sqrt{(k_B + \kappa)/m_B}$, and

$$\Gamma = \frac{\sqrt{\kappa/m_A}\,\sqrt{\kappa/m_B}}{\sqrt{\omega_A\,\omega_B}}. \tag{8.177}$$

In analogy to the dressed-atom picture [39], the eigenfrequencies $\omega_\pm$ can be associated with dressed states, that is, the oscillator frequencies of systems A and B in the presence of mutual coupling. To illustrate the solutions defined by Eq. (8.176) we set $k_A = k_0$, $k_B = k_0 + \Delta k$, and $m_A = m_B = m_0$. Figure 8.18(a) shows the frequencies of the two

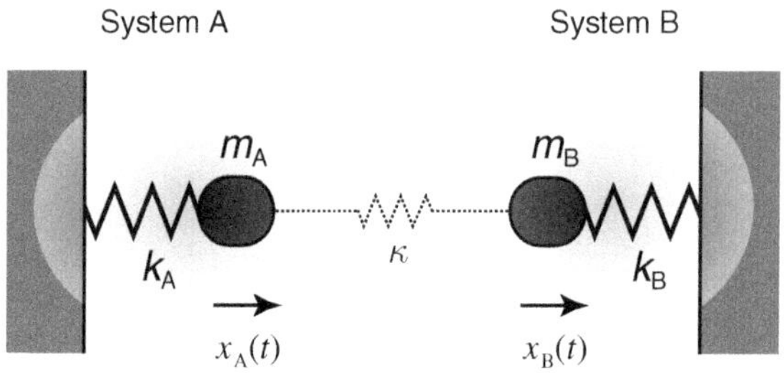

Fig. 8.17 The strong-coupling regime illustrated by mechanical oscillators. The coupling κ of the two oscillators (mass m_i, spring constant k_i) leads to a shift of the eigenfrequencies and a characteristic frequency splitting.

oscillators in the absence of coupling ($\kappa = 0$). As Δk is increased from $-k_0$ to k_0 the frequency of oscillator B increases from zero to $\sqrt{2}\,\omega_0$, whereas the frequency of oscillator B stays constant. The two curves intersect at $\Delta k = 0$. Once a coupling has been introduced the two curves no longer intersect. Instead, as shown in Fig. 8.18(b), there is a characteristic *anti-crossing* with a frequency splitting of

$$\omega_+ - \omega_- = \Gamma. \tag{8.178}$$

Anti-crossing is a characteristic fingerprint of strong coupling. Since $\Gamma \propto \kappa$, the splitting increases with increasing coupling strength.

Note that we have ignored damping in the analysis of the coupled oscillators. Damping can be readily introduced by adding frictional terms $\gamma_A \dot{x}_A$ and $\gamma_B \dot{x}_B$ to the equations of motion of the coupled oscillators, Eq. (8.175). The introduction of damping gives rise to complex frequency eigenvalues, the imaginary parts of which represent the linewidths. The latter give rise to a "smearing out" of the curves shown in Fig. 8.18, and for very strong damping it is no longer possible to discern the frequency splitting $\omega_+ - \omega_-$. Therefore, in order to observe *strong coupling* the frequency splitting needs to be larger than the sum of the linewidths,

$$\frac{\Gamma}{\gamma_A/m_A + \gamma_B/m_B} > 1. \tag{8.179}$$

In other words, the dissipation in each system needs to be smaller than the coupling strength.

Coupled mechanical oscillators are a generic model system for many physical systems, including atoms in external fields [40], coupled quantum dots [41], and cavity optomechanics [42]. Although our analysis is purely classical, a quantum-mechanical analysis yields the same result for the frequency splitting in Eq. (8.176) [43]. The coupling between energy states gives rise to avoided level crossings. As an illustration, Fig. 8.19 shows the Stark structure of the $|m| = 1$ energy states of sodium atoms. The coupling between energy states gives rise to avoided level crossings. The coupled-oscillator picture can be readily extended by adding external forces $F_A(t)$ and $F_B(t)$ acting on the masses m_A and m_B to account for externally driven systems, as in the case of electromagnetically induced transparency (EIT) [44].

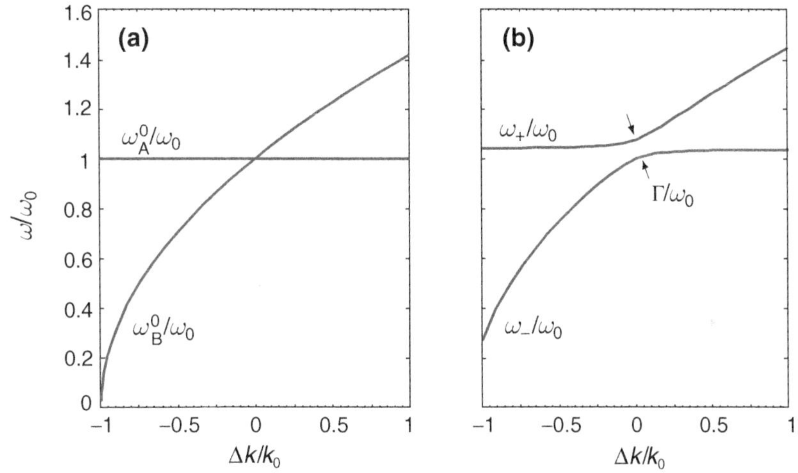

Fig. 8.18 Frequency splitting of coupled oscillators. (a) Eigenfrequencies of two uncoupled oscillators ($\kappa = 0$) with equal mass and spring constants k_0 and $k_0 + \Delta k$. (b) Frequency anti-crossing due to coupling of strength $\kappa = 0.08k_0$. The frequency splitting $\omega_+ - \omega_-$ scales linearly with the coupling strength κ.

8.7.2 Adiabatic and diabatic transitions

We now investigate what happens if one of the oscillator parameters changes as a function of time. For example, to record the curves shown in Fig. 8.18 we need to tune Δk from an initial value of $-k_0$ to a final value k_0. Thus, Δk becomes a function of time, a fact that we have ignored in the analysis above. We have assumed that Δk is tuned so slowly that for every measurement window Δt the system parameters can be regarded as constant. Thus, if we initially have $\Delta k = -k_0$ the coupled system oscillates at frequency ω_- (the bottom curve in Fig. 8.18(b)), and the system will follow the same curve as we slowly increase Δk. We can fine-tune the oscillation frequency by adjusting Δk. The same applies if we initially start with frequency ω_+ (the top curve in Fig. 8.18(b)). In both cases the anti-crossing region is passed by staying on the same branch. This scenario is referred to as an *adiabatic transition* and is illustrated in Fig. 8.20(a).

In the adiabatic limit it is possible to transfer the energy from one oscillator to the other by slowly tuning the coupled system through resonance. To see this effect, we introduce normal coordinates (x_+, x_-) defined by

$$x_A(t) = x_+(t)\sin\beta + x_-(t)\cos\beta(m_2/m_1),$$
$$x_B(t) = x_+(t)\cos\beta - x_-(t)\sin\beta, \tag{8.180}$$

where β is given by $\tan\beta = (\omega_B^2 - \omega_+^2)/(\kappa/m_B) = -(\omega_A^2 - \omega_-^2)/(\kappa/m_B)$. We substitute these expressions for x_A and x_B into Eqs. (8.175) and obtain

$$\ddot{x}_+(t) + \omega_+^2 x_+(t) = 0,$$
$$\ddot{x}_-(t) + \omega_-^2 x_-(t) = 0, \tag{8.181}$$

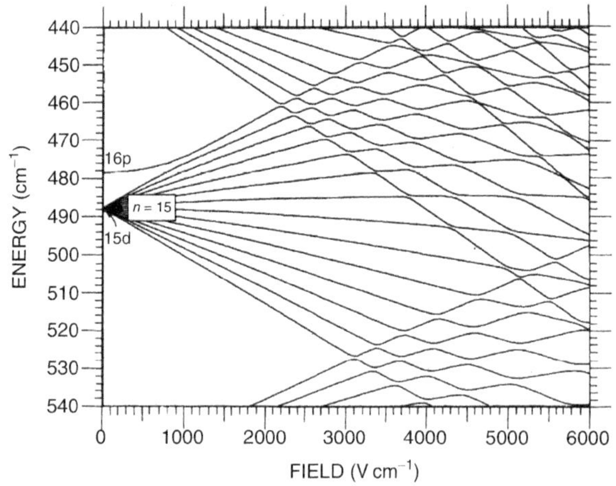

Fig. 8.19 Anti-crossing of $|m| = 1$ energy levels of Na atoms. The energy levels are tuned by an external electric field (the Stark effect). The coupling between the atomic states gives rise to avoided crossings. From [40].

which represent two independent harmonic oscillators oscillating at the eigenfrequencies $\omega_\pm$ defined by Eq. (8.176). In other words, there are two sets of coordinates, x_+ and x_-, which oscillate independently of each other.

Now imagine that Δk is slowly tuned in time from the initial value of $\Delta k = -k_0$ through resonance to a value of $\Delta k = k_0$. According to Fig. 8.18, at the initial time we have $\omega_A - \omega_- \gg \Gamma$ and therefore $\beta \sim -\pi/2$. If we use this value in Eq. (8.180) and assume that initially only oscillator A is active, we find that all the energy is associated with normal mode x_+, that is, $x_- = 0$. Once Δk has been tuned past resonance to k_0 we have $\omega_A - \omega_- \ll \Gamma$ and $\beta \approx 0$. According to Eqs. (8.180) the energy of mode x_+ now coincides with oscillator B, and hence the energy is transferred from oscillator A to oscillator B as the system is slowly tuned through resonance. If Δk changes with time, then also k_B, ω_B, and the eigenfrequencies $\omega_\pm$ become time-dependent. Assuming a slowly varying $\omega_\pm(\mathrm{t})$ in Eq. (8.181) we find

$$x_\pm(\mathrm{t}) = x_\pm(t_\mathrm{i})\mathrm{Re}\left\{\exp\left[\mathrm{i}\int_{t_\mathrm{i}}^{t}\omega_\pm(t')\mathrm{d}t'\right]\right\}, \tag{8.182}$$

where we used the ansatz $x_\pm(t) = x_\pm(t_i)\exp[\mathrm{i}f(t)]$ and $\mathrm{d}^2f/\mathrm{d}t^2 \ll (\mathrm{d}f/\mathrm{d}t)^2$. Equation (8.182) describes the *adiabatic* evolution of the normal modes.

We now analyze what happens if we change Δk more rapidly. We use the ansatz

$$x_A(t) = x_0 c_A(t)\exp(\mathrm{i}\omega_A t), \qquad x_B(t) = x_0 c_B(t)\exp(\mathrm{i}\omega_A t), \tag{8.183}$$

and assume that initially only oscillator A is active; that is, $c_B(-\infty) = 0$. The amplitude x_0 is a normalization constant that ensures $|c_A|^2 + |c_B|^2 = 1$. We substitute these expressions for x_A and x_B into Eqs. (8.175) and obtain the following coupled differential equations for c_A and c_B:

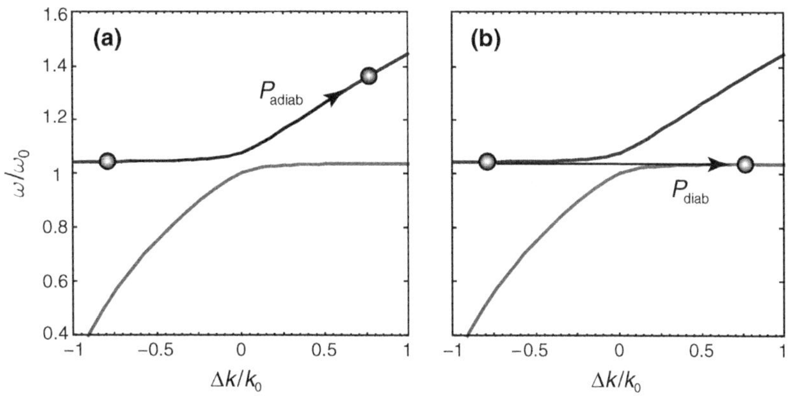

Fig. 8.20 Adiabatic and diabatic transitions caused by a time-varying Δk. (a) In the adiabatic case, the dynamics of the system is not affected by the time dependence of Δk and the system evolves along the eigenmodes $\omega_{\pm}$. (b) In the diabatic case, the time dependence of Δk gives rise to a "level crossing," namely a transition from one eigenmode to the other.

$$\ddot{c}_{\mathrm{A}} + 2\mathrm{i}\omega_{\mathrm{A}}\dot{c}_{\mathrm{A}} = (\kappa/m_{\mathrm{A}})c_{\mathrm{B}}, \tag{8.184}$$
$$\ddot{c}_{\mathrm{B}} + 2\mathrm{i}\omega_{\mathrm{A}}\dot{c}_{\mathrm{B}} + \left[\omega_{\mathrm{B}}^2(t) - \omega_{\mathrm{A}}^2\right]c_{\mathrm{B}} = (\kappa/m_{\mathrm{B}})c_{\mathrm{A}},$$

where we emphasized the time dependence of ω_{B}. For weak coupling between the two oscillators, the amplitudes $c_{\mathrm{A}}(t)$ and $c_{\mathrm{B}}(t)$ in Eq. (8.183) vary much more slowly with time than does the oscillatory term $\exp(\mathrm{i}\omega_{\mathrm{A}}t)$. Therefore, $\ddot{c}_{\mathrm{A}} \ll \mathrm{i}\omega_{\mathrm{A}}\dot{c}_{\mathrm{A}}$ and $\ddot{c}_{\mathrm{B}} \ll \mathrm{i}\omega_{\mathrm{A}}\dot{c}_{\mathrm{B}}$, which allows us to drop the second-order derivatives in Eq. (8.184) to obtain

$$2\mathrm{i}\omega_{\mathrm{A}}\dot{c}_{\mathrm{A}} = (\kappa/m_{\mathrm{A}})c_{\mathrm{B}}, \tag{8.185}$$
$$2\mathrm{i}\omega_{\mathrm{A}}\dot{c}_{\mathrm{B}} + \left[\omega_{\mathrm{B}}^2(t) - \omega_{\mathrm{A}}^2\right]c_{\mathrm{B}} = (\kappa/m_{\mathrm{B}})c_{\mathrm{A}}. \tag{8.186}$$

From Eq. (8.185) we find c_{B} and $\dot{c}_{\mathrm{B}}$ (by taking a derivative), and substitute the results into Eq. (8.186) to find, after some algebra,

$$\ddot{c}_{\mathrm{A}} - \mathrm{i}\dot{c}_{\mathrm{A}}\left[\frac{\omega_{\mathrm{B}}^2(t) - \omega_{\mathrm{A}}^2}{2\omega_{\mathrm{A}}}\right] + c_{\mathrm{A}}\frac{\kappa^2/(m_{\mathrm{A}}m_{\mathrm{B}})}{4\omega_{\mathrm{A}}^2} = 0. \tag{8.187}$$

The time dependence of ω_{B} makes Eq. (8.187) nonlinear. During the time interval of interest, close to the anti-crossing region, $\omega_{\mathrm{B}}(t) \approx \omega_{\mathrm{A}}$, and hence $[\omega_{\mathrm{B}}^2(t) - \omega_{\mathrm{A}}^2]/(2\omega_{\mathrm{A}}) \approx \omega_{\mathrm{B}}(t) - \omega_{\mathrm{A}}$. For the same reason we can set $\Gamma^2 \approx \kappa^2/(m_{\mathrm{A}}m_{\mathrm{B}}\omega_{\mathrm{A}}^2)$ (c.f. Eq. (8.177)). Finally, we assume that near the anti-crossing region the frequency difference of oscillators A and B changes linearly in time, that is,

$$\omega_{\mathrm{B}}(t) - \omega_{\mathrm{A}} = \alpha t. \tag{8.188}$$

According to Eq. (8.188), the anti-crossing region is passed at time $t \approx 0$, and the frequency difference is negative for $t < 0$ and positive for $t > 0$ (see Fig. 8.18). With these approximations, Eq. (8.187) becomes

$$\ddot{c}_{\mathrm{A}} - \mathrm{i}\dot{c}_{\mathrm{A}}\alpha t + c_{\mathrm{A}}\Gamma^2/4 = 0. \tag{8.189}$$

Despite the approximations we have made there is no analytical solution of Eq. (8.189) for $c_A(t)$. However, we are not interested in the temporal behavior of c_A but rather in the value which c_A assumes after the anti-crossing regime has long passed. By using contour integration we find that the solution for $c_A(t \to \infty)$ is [45]

$$c_A(\infty) = \exp\left[-\frac{\pi}{4}\,\Gamma^2/\alpha\right]. \tag{8.190}$$

Because the energy of oscillator A is $E_A \propto |c_A|^2$, the probability of level crossing is

$$P_{\text{diab}} = \exp\left[-\frac{\pi}{2}\,\Gamma^2/\alpha\right], \tag{8.191}$$

which is also referred to as a *diabatic transition*, a transition involving loss or gain.

Equation (8.191) is the classical analog of the Landau–Zener formula in quantum mechanics [46, 47]. As illustrated in Fig. 8.20(b), Eq. (8.191) defines the probability that the energy of oscillator A remains the same after transitioning through the anti-crossing region. P_{diab} is the probability for passing through the anti-crossing region by switching branches, that is, for jumping from one eigenmode to the other (see Fig. 8.20(b)). Consequently, the probability of an adiabatic transition is $P_{\text{adiab}} = 1 - P_{\text{diab}}$.

The probability of a diabatic transition depends on the frequency splitting, $\Gamma = \omega_+ - \omega_-$, and on the time $\tau \sim \Gamma/\alpha$ that it takes to transition through the anti-crossing region. A diabatic transition is likely for $\Gamma\tau \ll 1$, which corresponds to a rapid transition through the anti-crossing region. In contrast, for a slow transition ($\Gamma\tau \gg 1$) an adiabatic transition is more probable. Note that the product $\Gamma\tau$ has analogies with the time–energy uncertainty principle. For times $\tau \ll 1/\Gamma$ the energy uncertainty becomes larger than the level splitting, thereby "closing up" the anti-crossing region and making diabatic transitions possible.

In our example, we can control which branch (eigenmode) we end up with by setting the speed at which $\Delta k(t)$ changes. Figure 8.21(b) shows computed results for $|c_A(t)|^2$ for two time dependences of Δk. In both cases Δk changes from $-k_0$ to k_0, but the speed of this change differs. Figure 8.21(a) shows the corresponding time dependence of the frequency shifts. For the computations of c_A we assumed that only oscillator A is active initially; that is, $c_A(-\infty) = 1$ and $c_B(-\infty) = 0$. As time evolves we observe small oscillations in c_A and then an abrupt change in the transition region. This change is followed by a slowly damped oscillation. The two limiting values for the diabatic probability P_{diab} are indicated in Fig. 8.21(b). Although one of the curves represents a nearly adiabatic transition ($P_{\text{diab}} \sim 0$), the other represents a nearly diabatic transition ($P_{\text{diab}} \sim 1$). Situations in between can be selected by adjusting the speed at which $\Delta k(t)$ changes.

Let us now verify the transition probabilities using the Landau–Zener formula. A linear approximation to the curves in Fig. 8.21(a) yields the slopes: $\alpha_1 = 0.003\omega_A^2$ and $\alpha_2 = 0.075\omega_A^2$. The level splitting of the two oscillators is $\Gamma = (\kappa/m_0)/\omega_A$, with $\omega_A^2 = (k_0 + \kappa)/m_0$. If we use $\kappa = 0.08k_0$ and substitute the expressions for Γ and α into Eq. (8.191), we find $P_{\text{diab}}(\alpha_1) = 0.06$ and $P_{\text{diab}}(\alpha_2) = 0.89$, in agreement with the computed results in Fig. 8.21(b).

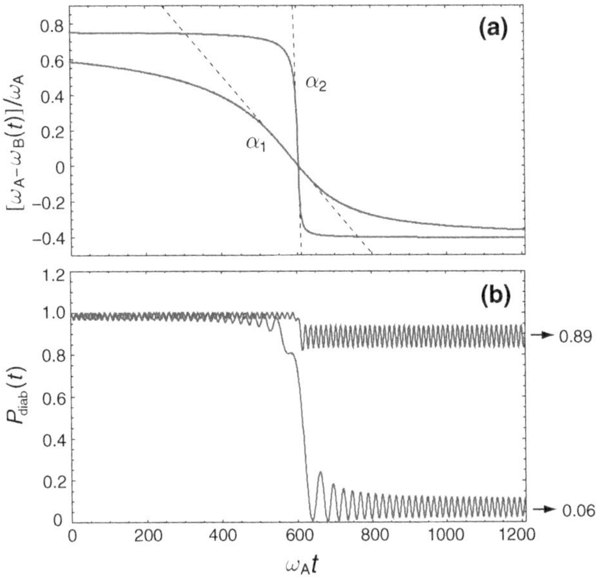

Fig. 8.21 The diabatic transition probability computed for two different time dependences of $\Delta k(t)$. (a) The time dependence of the frequency shift. The linear approximation Eq. (8.188) yields $\alpha_1 = 0.003\omega_A^2$ and $\alpha_2 = 0.075\omega_A^2$. (b) The transition probability $P_{diab}(t) = c_A(t)c_A^*(t)$ computed for the two functions in (a). For a frequency difference that changes slowly (curve labeled α_1) we obtain a nearly adiabatic transition with a limiting value of $P_{diab} = 0.06$, whereas a rapidly changing frequency difference (curve labeled α_2) results in a nearly diabatic transition with a limiting value of $P_{diab} = 0.89$. $k_A = k_0$, $k_B = k_0 + \Delta k$, $m_A = m_B = m_0$, and $\kappa = 0.08k_0$.

The Landau–Zener formula is an important result because it finds application in a wide range of problems. It allows us to readily calculate transitions between eigenmodes of two coupled systems. For example, consider light propagation along a dielectric fiber with a thin metal coating. There are two sets of modes, namely those propagating inside the dielectric (waveguide modes) and those propagating on the outside of the metal coating (surface modes). Let's assume that only the fundamental, radially polarized waveguide mode is excited and that the radius R of the dielectric fiber is steadily reduced. It can be shown that for a certain fiber radius R the propagation constants of the waveguide mode and the surface mode become equal, that is they cross. But because of the finite metal coating the two sets of modes interact and give rise to an anti-crossing region. If R is slowly reduced (small taper angle) we end up with an adiabatic transition and the energy of the waveguide mode is transferred to the surface mode. On the other hand, a diabatic transition is obtained if R is rapidly reduced (large taper angle) and the energy remains with the waveguide mode. This scenario has been discussed in the context of plasmon focusing [48].

Notice that the result in Eq. (8.191) can be extended to a quantum-mechanical system with eigenstates $|1\rangle$ and $|2\rangle$ by using the substitutions $\hbar\Gamma/2 \rightarrow |\langle 1|\hat{H}_{int}|2\rangle|$ and $\hbar\alpha \rightarrow \mathrm{d}[E_1(t) - E_2(t)]/\mathrm{d}t$, where $\hat{H}_{int}$ is the interaction Hamiltonian and E_i are the energy eigenvalues. These substitutions yield the original quantum Landau–Zener formula.

8.7.3 Coupled two-level systems

We next discuss the coupling of two particles from a quantum-mechanical perspective. We consider two particles A and B, which are represented by two-level systems. In the absence of any interactions between the two particles, the ground state and energy eigenvalue of A are denoted as $|A\rangle$ and $E_{\rm A}$, respectively, and for the excited state we use $|A^\star\rangle$ and $E_{\rm A}^\star$ (see Fig. 8.22). Similar notation is used for the eigenstates and eigenvalues of B. To solve the coupled system we define four states $|AB\rangle$, $|A^\star B\rangle$, $|AB^\star\rangle$, $|A^\star B^\star\rangle$ that satisfy the Schrödinger equation of the uncoupled system,

$$[\hat{H}_{\rm A} + \hat{H}_{\rm B}]\,|\phi_n\rangle = e_n\,|\phi_n\rangle\,. \tag{8.192}$$

Here, $|\phi_n\rangle$ is any of the defined four states and E_n denotes the eigenvalue associated with this state, i.e. $e_n \in [(E_{\rm A} + E_{\rm B}), (E_{\rm A}^\star + E_{\rm B}), (E_{\rm A} + E_{\rm B}^\star), (E_{\rm A}^\star + E_{\rm B}^\star)]$. After introducing interaction terms between the four states, the Schrödinger equation of the coupled system becomes

$$[\hat{H}_{\rm A} + \hat{H}_{\rm B} + \hat{V}_{\rm int}]\,|\Phi_n\rangle = E_n\,|\Phi_n\rangle\,, \tag{8.193}$$

where $\hat{V}_{\rm int}$ is the interaction Hamiltonian, $|\Phi_n\rangle$ are the new eigenstates and E_n the new eigenvalues. To determine the eigenstates $E_n = \langle\Phi_n|\,\hat{H}_{\rm A} + \hat{H}_{\rm B} + V_{\rm int}\,|\Phi_n\rangle$ we can now expand the new eigenstates in terms of the old eigenstates as

$$|\Phi_n\rangle = a_n\,|AB\rangle + b_n\,\big|A^\star B\big\rangle + c_n\,\big|AB^\star\big\rangle + d_n\,\big|A^\star B^\star\big\rangle\,, \tag{8.194}$$

and diagonalize the Hamiltonian $[\hat{H}_{\rm A} + \hat{H}_{\rm B} + \hat{V}_{\rm int}]$ using standard procedures.

The problem with this rigorous approach is the lack of information on the coupling terms that lead to $V_{\rm int}$. These terms are defined by the combined system of particles A and B. They only approximately correspond to the interparticle interaction potentials $V_{\rm AB}$ in Eq. (8.149). $V_{\rm int}$ could be determined by first rigorously solving the Schrödinger equation of the combined system of particles and then trying to decouple the unperturbed Hamiltonians from the system. But this is a difficult task. To better understand this subtle point let us consider a system made of two electrons, two neutrons, and two protons. The resulting

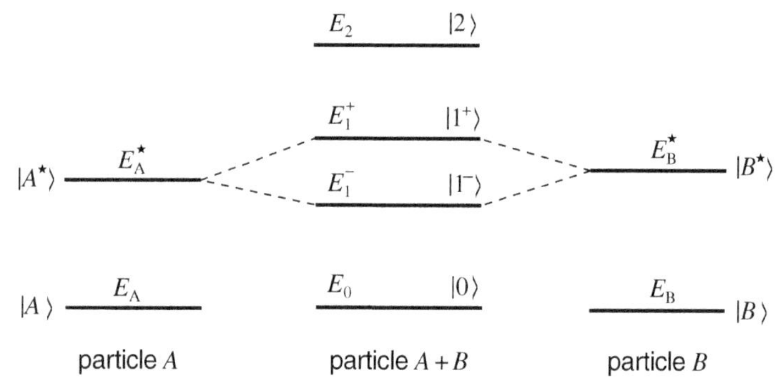

Fig. 8.22 Coherent interaction between two particles A and B. In the resonant limit, the excitation becomes delocalized over the two particles.

element is known as helium (^{4}He). If the particles are equally divided into two separate systems, we obtain two deuterium atoms (D). It is quite evident that it is challenging to view ^{4}He as two interacting deuterium atoms.

Since the exact parameters of a coupled system are not known a priori it is favorable to describe the interaction between two particles in terms of their unperturbed parameters. In this picture, the interaction between two particles can be viewed as a perturbation of the two isolated particles. In particular, if we assume that the dipole moments of particle A and particle B are known, the interaction between the two particles can be described in terms of the interparticle interaction potentials V_{AB} in Eq. (8.149). Assuming that the two particles are charge-neutral, the leading term in the interaction becomes the dipole–dipole interaction.

Using first-order non-degenerate perturbation theory [49], we write for the states of the ground state $|0\rangle$ and the doubly excited state $|A^{\star}B^{\star}\rangle$ of the coupled system

$$|0\rangle = |AB\rangle \tag{8.195}$$
$$|2\rangle = \left|A^{\star}B^{\star}\right\rangle \tag{8.196}$$

and obtain the following first-order-corrected energy eigenstates

$$E_0 = E_{\mathrm{A}} + E_{\mathrm{B}} + \langle 0|V_{\mathrm{AB}}\,|0\rangle, \tag{8.197}$$
$$E_2 = E_{\mathrm{A}^{\star}} + E_{\mathrm{B}^{\star}} + \langle 2|V_{\mathrm{AB}}|2\rangle. \tag{8.198}$$

Thus, the coupling between A and B gives rise to a displacement of the ground-state energy and the energy of the doubly excited state. For the singly excited states $\left|1^{+}\right\rangle$ and $\left|1^{-}\right\rangle$ we cannot proceed in the same way. If the particles A and B were identical, the unperturbed singly excited states $|A^{\star}B\rangle$ and $|AB^{\star}\rangle$ would be degenerate. Even if the two particles are not identical, non-degenerate perturbation theory can be applied only if the energy separation of the unperturbed states, $\Delta E = \left|(E^{\star}_{\mathrm{A}} + E_{\mathrm{B}}) - (E_{\mathrm{A}} + E^{\star}_{\mathrm{B}})\right|$, is much bigger than the strength of the perturbations $\langle A^{\star}B|\,V_{\mathrm{AB}}\,|AB^{\star}\rangle$ and $\langle AB^{\star}|\,V_{\mathrm{AB}}\,|A^{\star}B\rangle$. If this is not the case, degenerate perturbation theory has to be applied even to a non-degenerate system. Therefore, we define the states $\left|1^{+}\right\rangle$ and $\left|1^{-}\right\rangle$ of the coupled system as a linear combination of the unperturbed states

$$\left|1^{+}\right\rangle = \cos\alpha\left|A^{\star}B\right\rangle + \sin\alpha\left|AB^{\star}\right\rangle, \tag{8.199}$$
$$\left|1^{-}\right\rangle = \sin\alpha\left|A^{\star}B\right\rangle - \cos\alpha\left|AB^{\star}\right\rangle, \tag{8.200}$$

where α is an arbitrary coefficient to be determined later. The states $\left|1^{+}\right\rangle$ and $\left|1^{-}\right\rangle$ have to satisfy the Schrödinger equations

$$[\hat{H}_{\mathrm{A}} + \hat{H}_{\mathrm{B}} + V_{\mathrm{AB}}]\left|1^{+}\right\rangle = E_1^{+}\left|1^{+}\right\rangle, \tag{8.201}$$
$$[\hat{H}_{\mathrm{A}} + \hat{H}_{\mathrm{B}} + V_{\mathrm{AB}}]\left|1^{-}\right\rangle = E_1^{-}\left|1^{-}\right\rangle. \tag{8.202}$$

To facilitate notation we introduce the following abbreviations

$$W_{\mathrm{A^{\star}B}} = \left\langle A^{\star}B\middle|V_{\mathrm{AB}}\middle|A^{\star}B\right\rangle, \qquad W_{\mathrm{AB^{\star}}} = \left\langle AB^{\star}\middle|V_{\mathrm{AB}}\middle|AB^{\star}\right\rangle. \tag{8.203}$$

On inserting $|1^+\rangle$ from Eq. (8.199) into Eq. (8.201) and operating from the left with $\langle 1^+|$ we obtain

$$\begin{aligned} E_1^+ =& \sin^2\alpha\,[E_A + E_{B^\star} + W_{AB^\star}] + \cos^2\alpha\,[E_{A^\star} + E_B + W_{A^\star B}] \\ &+ 2\sin\alpha\cos\alpha\,\mathrm{Re}\left\{\langle A^\star B|V_{AB}|AB^\star\rangle\right\}. \end{aligned} \tag{8.204}$$

We made use of the fact that $\hat{H}_A$ operates only on the states of particle A and $\hat{H}_B$ operates only on the states of particle B. We also applied the orthogonality relations $\langle A|A\rangle = 1$, $\langle A^\star|A\rangle = 0$, $\langle B|B\rangle = 1$, and $\langle B^\star|B\rangle = 0$. Furthermore, since V_{AB} is a Hermitian operator we have $\left[\langle AB^\star|\, V_{AB}\, |A^\star B\rangle\right]^\star = \langle A^\star B|\, V_{AB}\, |AB^\star\rangle$, with $[\ldots]^\star$ denoting the complex conjugate. The energy E^- is derived in a similar way as

$$\begin{aligned} E_1^- =& \cos^2\alpha[E_A + E_{B^\star} - W_{AB^\star}] + \sin^2\alpha[E_{A^\star} + E_B + W_{A^\star B}] \\ &- 2\sin\alpha\cos\alpha\,\mathrm{Re}\left\{\langle A^\star B|\, V_{AB}\, |AB^\star\rangle\right\}. \end{aligned} \tag{8.205}$$

The energy levels E^+ and E^- depend on the coefficient α, which can be determined by requiring orthogonality between the states $|1^+\rangle$ and $|1^-\rangle$. Operating with $\langle 1^-|$ on Eq. (8.201) and making use of $\langle 1^-|1^+\rangle = 0$ leads to the condition

$$\langle 1^-|\hat{H}_A + \hat{H}_B + V_{AB}|1^+\rangle = 0, \tag{8.206}$$

from which we derive

$$\tan(2\alpha) = \frac{2\,\mathrm{Re}\{\langle A^\star B|V_{AB}|AB^\star\rangle\}}{[E_{A^\star} + E_B + W_{A^\star B}] - [E_A + E_{B^\star} + W_{AB^\star}]}. \tag{8.207}$$

The coefficient α can have any value in the range $[0\ldots\pi/2]$ depending on the strength of interaction between particles A and B. A better insight is gained by considering the two limiting cases of $\alpha = 0$ ($\alpha = \pi/2$) and $\alpha = \pi/4$.

For $\alpha = 0$, the singly excited states reduce to $|1^+\rangle = |A^\star B\rangle$ and $|1^-\rangle = -|AB^\star\rangle$. Thus, in state $|1^+\rangle$ the excitation is entirely localized on particle A, whereas in state $|1^-\rangle$ the excitation is localized on particle B. The energy eigenvalues become

$$E_1^+ = [E_{A^\star} + E_B + W_{A^\star B}] \qquad (\alpha = 0), \tag{8.208}$$

$$E_1^- = [E_A + E_{B^\star} + W_{A^\star B}] \qquad (\alpha = 0). \tag{8.209}$$

There is no energy-level splitting if A and B are identical particles. The interaction gives rise only to a level shift by an amount $W_{A^\star B}$.

The situation is similar for the case $\alpha = \pi/2$, for which the roles of A and B are simply reversed. The singly excited states become $|1^+\rangle = |AB^\star\rangle$ and $|1^-\rangle = |A^\star B\rangle$ and the energy eigenvalues are

$$E_1^+ = [E_A + E_{B^\star} + W_{AB^\star}] \qquad (\alpha = \pi/2), \tag{8.210}$$

$$E_1^- = [E_{A^\star} + E_B + W_{AB^\star}] \qquad (\alpha = \pi/2). \tag{8.211}$$

If the nominator of Eq. (8.207) goes to infinity or if the denominator goes to zero we obtain the limiting case of $\alpha = \pi/4$. For this so-called resonant case the excitation is distributed equally over both particles and the energy eigenvalues read as

$$E_1^+ = \frac{1}{2}\left[E_A + E_{A^\star} + E_B + E_{B^\star} + W_{AB^\star} + W_{A^\star B}\right] + \mathrm{Re}\left\{\langle A^\star B|V_{AB}|AB^\star\rangle\right\} \quad (\alpha = \pi/4), \tag{8.212}$$

$$E_1^- = \frac{1}{2}\left[E_A + E_{A^\star} + E_B + E_{B^\star} + W_{AB^\star} + W_{A^\star B}\right] - \mathrm{Re}\left\{\langle A^\star B|V_{AB}|AB^\star\rangle\right\} \quad (\alpha = \pi/4). \tag{8.213}$$

This delocalized excitation is also known as an *exciton* and the regime for which $\alpha \approx \pi/4$ is called the *strong-coupling* regime. Strong coupling is always achieved if particles A and B are identical, if they interact, and if there are no losses in the system. In general, we have to require that

$$\mathrm{Re}\left\{\langle A^\star B|\, V_{AB}\, |AB^\star\rangle\right\} \gg \frac{1}{2}\left(\left[E_{A^\star} + E_B + W_{A^\star B}\right] - \left[E_A + E_{B^\star} + W_{AB^\star}\right]\right). \tag{8.214}$$

Our analysis shows that the interaction between two identical particles leads to a level splitting of the singly excited states. In the case of many interacting particles, the multiple splitting of the singly excited states will lead to an energy band (exciton band). Delocalized excitation is based on a coherent superposition of states. The time required to establish this coherence is on the order of $\tau_c = h/V_{AB}$. Vibronic relaxation can easily destroy the coherence within a few picoseconds. As a result, the excitation becomes localized, and incoherent energy transfer between particles (Förster energy transfer) becomes more probable. In general, strong coupling in a system can be established only if vibrational relaxation times τ_{vib} are longer than τ_c.

As an illustration of strong coupling, Fig. 8.23 shows the level splitting of two InAs quantum dots separated by a GaAs barrier of variable thickness. The peaks correspond to

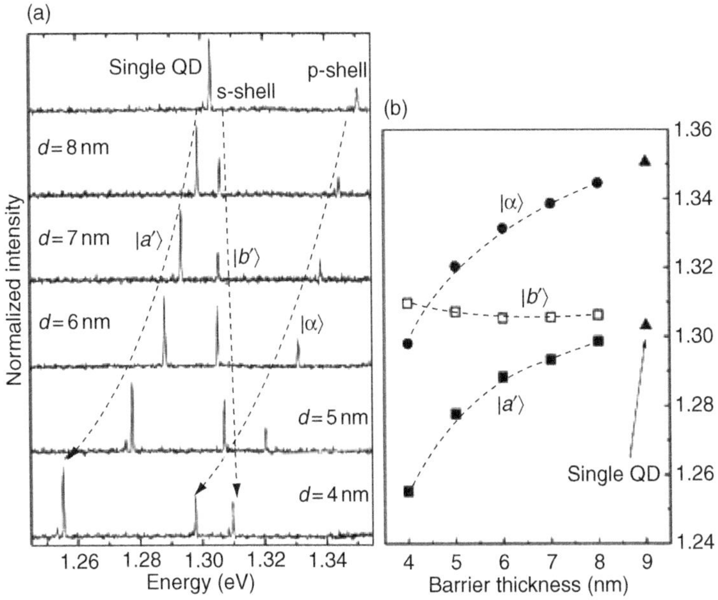

Fig. 8.23 Level splitting of two InAs quantum dots separated by a GaAs barrier. (a) Emission spectra at $T \approx 60$ K for varying dot separation d. (b) Corresponding energy-level diagram. From [41].

the emission of the ground state exciton (s-shell) and first excited state exciton (p-shell). At large separations, only one ground state emission line is seen, but as the barrier thickness is decreased the emission line splits. The same is true for the first excited state exciton although only the lower energy level is shown. In these experiments, low excitation powers were used to prevent the excitation of multiexcitons.

8.7.4 Entanglement

The concept of entanglement is of key importance in the context of quantum information theory. The term is adapted from the German "*verschränkter Zustand*" and was first introduced by Schrödinger [50]. It refers to a combined state of two systems (e.g. the singly excited states encountered in the previous section) that cannot be written as a product of the individual states. More qualitatively, entanglement refers to the degree of "quantum memory" in a system. There exist different definitions for the degree of entanglement, but we restrict the discussion to the so-called *Schmidt decomposition* applicable to pure states [51].

Entanglement refers to a joint property of *two* systems A and B, a so-called bipartite system. Each system is characterized by its eigenstates, i.e. $|A_n\rangle$ and $|B_m\rangle$, with $n = 1, 2, \ldots, N$ and $m = 1, 2, \ldots M$. A and B are called *qubits* if $N = M = 2$. The combined system of $\mathrm{A} + \mathrm{B}$ has its own eigenstates $|\Psi_i\rangle$, which can be arbitrarily superimposed and which define the density matrix

$$\hat{\rho} = |\Psi\rangle\langle\Psi|. \tag{8.215}$$

Because $|\Psi\rangle$ can be expressed in terms of $|A_n\rangle$ and $|B_m\rangle$ we define the *reduced* density matrices $\hat{\rho}_{\mathrm{A}}$ and $\hat{\rho}_{\mathrm{B}}$ as

$$\begin{aligned} \hat{\rho}_{\mathrm{A}} &= \mathrm{Tr}_{\mathrm{B}}\left[\hat{\rho}\right] = \sum_m \langle B_m \left|\hat{\rho}\right| B_m\rangle, \\ \hat{\rho}_{\mathrm{B}} &= \mathrm{Tr}_{\mathrm{A}}\left[\hat{\rho}\right] = \sum_n \langle A_n \left|\hat{\rho}\right| A_n\rangle, \end{aligned} \tag{8.216}$$

where Tr stands for trace. A given (normalized) state $|\Psi\rangle$ is called *separable* if all but one of the eigenvalues λ_i of the reduced density matrix $\hat{\rho}_{\mathrm{A}}$ are zero. It can be shown that $\hat{\rho}_{\mathrm{B}}$ has the same eigenvalues and so it suffices to consider only one of the reduced matrices. Notice that the sum of all λ_i is equal to unity. If $|\Psi\rangle$ is *not* separable, the state is called entangled and the degree of entanglement is defined by the *Grobe–Rzazewski–Eberly number* [52]

$$K = \left[\sum_i \lambda_i^2\right]^{-1}, \tag{8.217}$$

which is always larger than or equal to unity and smaller than or equal to the total number of non-zero eigenvalues.

As an example, let us discuss the state

$$\left|1^+\right\rangle = \cos\alpha\left|A^\star B\right\rangle + \sin\alpha\left|AB^\star\right\rangle \tag{8.218}$$

encountered in the previous section. This is a state of the combined system of A ($|A\rangle$, $|A^\star\rangle$) and B ($|B\rangle$, $|B^\star\rangle$). Thus, $N = M = 2$. The density matrix $\hat{\rho}$ is calculated as

$$\begin{aligned}\hat{\rho} &= \left[\cos\alpha\left|A^\star B\right\rangle + \sin\alpha\left|AB^\star\right\rangle\right]\left[(\cos\alpha)^*\left\langle A^\star B\right| + (\sin\alpha)^*\left\langle AB^\star\right|\right] \\ &= \cos^2\alpha\left|A^\star B\right\rangle\left\langle A^\star B\right| + \sin^2\alpha\left|AB^\star\right\rangle\left\langle AB^\star\right| \\ &\quad + \sin\alpha\cos\alpha\left|A^\star B\right\rangle\left\langle AB^\star\right| + \sin\alpha\cos\alpha\left|AB^\star\right\rangle\left\langle A^\star B\right|,\end{aligned} \tag{8.219}$$

and the reduced density matrix $\hat{\rho}_A$ becomes

$$\hat{\rho}_A = \cos^2\alpha\left|A^\star\right\rangle\left\langle A^\star\right| + \sin^2\alpha|A\rangle\langle A| = \begin{bmatrix}\cos^2\alpha & 0 \\ 0 & \sin^2\alpha\end{bmatrix}, \tag{8.220}$$

where we have made use of the orthonormality of $|B\rangle$ and $|B^\star\rangle$. Because the off-diagonal elements are zero, the eigenvalues are $\lambda_1 = \sin^2\alpha$ and $\lambda_2 = \cos^2\alpha$, and the Grobe–Rzazewski–Eberly number becomes

$$K = \frac{1}{\sin^4\alpha + \cos^4\alpha}. \tag{8.221}$$

Thus, the state $\left|1^+\right\rangle$ is separable if $\alpha = 0$ or $\alpha = \pi/2$. For angles in between the state is entangled. For $\alpha = \pi/4$ the state is maximally entangled ($K = 2$) and is called a *Bell state*. This is consistent with our discussion in the previous section, where we determined that for $\alpha = \pi/4$ the excitation is equally distributed over the two particles (the resonant case) and that the strongest coupling is achieved for this case. Finally, it should be noted that the Schmidt decomposition works only for pure states and that other procedures have to be applied to mixed states.

Problems

8.1 Derive the potential energy V for a system of two charges q, $-q$ in an external field $\mathbf{E}, \mathbf{H}$. The charges are separated by a vector $\mathbf{s}$ with $s = |\mathbf{s}| \ll \lambda$. Calculate first the force $\mathbf{F} = (m_1 + m_2)\ddot{\mathbf{r}}$ acting on the two charges and expand $\mathbf{F}$ in a Taylor series with origin $\mathbf{r}$ at the center of the two charges. Retain only the lowest order in the expansion. Then derive V for the two cases of
(1) a permanent dipole moment $\mathbf{p}$ and
(2) an induced dipole moment $\mathbf{p} = \alpha\mathbf{E}$.

8.2 Derive the far-field Green function $\overleftrightarrow{\mathbf{G}}_{FF}$ in spherical coordinates and Cartesian vector components. Calculate the radiation pattern $P(\vartheta, \varphi)/P$ for a dipole $\mathbf{p}$ that encloses an angle α with the z-axis.

8.3 Prove that the near-field and intermediate-field terms of a dipole in free space do not contribute to radiation.

8.4 Calculate the interaction energy between two dipoles given by $V = -\mathbf{p}_1 \cdot \mathbf{E}_2(\mathbf{r}_1) - \mathbf{p}_2 \cdot \mathbf{E}_1(\mathbf{r}_2)$. $\mathbf{E}_1(\mathbf{r}_2)$ is the field of dipole $\mathbf{p}_1$ evaluated at the position $\mathbf{r}_2$ of dipole $\mathbf{p}_2$. Similarily, $\mathbf{E}_2(\mathbf{r}_1)$ is the field of dipole $\mathbf{p}_2$ evaluated at the position $\mathbf{r}_1$ of dipole $\mathbf{p}_1$. Separate the near-field, intermediate-field, and far-field interactions.

8.5 In Section 8.3.4 it was pointed out that radiation reaction is necessary in order to account for light scattering from particles that are treated in the dipole limit. In this exercise we derive a correction for the particle polarizability α in order to be consistent with the optical theorem.

(1) The radiation reaction force $\mathbf{F}_r$ defines a self-field $\mathbf{E}_{self}$ according to $\mathbf{F}_r = q\mathbf{E}_{self}$. Express Eq. (8.85) in terms of the dipole moment $\mathbf{p} = q\mathbf{r}$ and represent the associated self-field in the frequency domain, i.e. find $\mathbf{E}_{self}(\omega)$.

(2) The dipole $\mathbf{p}$ is induced by the local field consisting of the external field $\mathbf{E}_0$ and the self-field $\mathbf{E}_{self}$ according to $\mathbf{p} = \alpha(\omega)(\mathbf{E}_0 + \mathbf{E}_{self})$. Substitute $\mathbf{E}_{self}$ from (1) and arrange terms to get $\mathbf{p} = \alpha_{eff}(\omega)\mathbf{E}_0$. Show that the effective polarizability is given by

$$\alpha_{eff}(\omega) = \frac{\alpha(\omega)}{1 - \mathrm{i}[k^3/(6\pi\varepsilon_0)]\,\alpha(\omega)}. \tag{8.222}$$

When applied to the optical theorem, the first term in the series expansion of α_{eff} leads to absorption, whereas the second term defines scattering. The inconsistency of the optical theorem in the dipole limit is also discussed in Problem 16.4.

8.6 The partial local density of states ρ_p depends on the orientation of the unit vector $\mathbf{n}_p$. Show that, on averaging $\mathbf{n}_p$ over all orientations, ρ_p becomes identical with the total density of states ρ. It suffices to show that

$$\left\langle \mathbf{n}_p \cdot \mathrm{Im}\left\{\overleftrightarrow{\mathbf{G}}\right\} \cdot \mathbf{n}_p \right\rangle = \frac{1}{3}\,\mathrm{Im}\left\{\mathrm{Tr}[\overleftrightarrow{\mathbf{G}}]\right\}.$$

8.7 In free space, the partial local density of states ρ_p is identical to the total density of states ρ. To show this, prove that

$$\left[\mathbf{n}_p \cdot \mathrm{Im}\left\{\overleftrightarrow{\mathbf{G}}_0\right\} \cdot \mathbf{n}_p\right] = \frac{1}{3}\,\mathrm{Im}\left\{\mathrm{Tr}[\overleftrightarrow{\mathbf{G}}_0]\right\},$$

where $\overleftrightarrow{\mathbf{G}}_0$ is the free-space dyadic.

8.8 A molecule with emission dipole moment in the direction of the x-axis is scanned in the (x, y) plane. A spherical gold particle ($\varepsilon = -7.6 + 1.7\mathrm{i}$) with radius $r_0 = 10$ nm is placed above the (x, y) plane. The emission wavelength is $\lambda = 575$ nm (DiI molecule). The center of the particle is located at the fixed position $(x, y, z) = (0, 0, 20)$ nm.

(1) Calculate the normalized decay rate γ/γ_0 as a function of x, y. Neglect retardation effects and draw a contour plot. What is the minimum value of γ/γ_0? How does the quenching rate scale with the sphere radius r_0?

(2) Repeat the calculation for a dipole oriented in the direction of the z-axis.

8.9 Two molecules, fluorescein (donor) and Alexa Green 532 (acceptor), are located in a plane centered between two perfectly conducting surfaces separated by the distance d. The emission spectrum of the donor (f_D) and the absorption spectrum of the acceptor (σ_A) are approximated by a superposition of two Gaussian distribution functions. Use the fit parameters from Section 8.6.2.

(1) Determine the Green function for this configuration.
(2) Calculate the decay rate γ_0 of the donor in the absence of the acceptor.
(3) Determine the transfer rate $\gamma_{D \to A}$ as a function of the separation R between donor and acceptor. Assume random dipole orientations.
(4) Plot the Förster radius R_0 as a function of the separation d.

8.10 Prove Eq. (8.207) by following the steps in Section 8.7.

8.11 Consider the state

$$|\Psi\rangle = \beta_1 \left|1^+\right\rangle + \beta_2 \left|1^-\right\rangle,$$

where $\left|1^+\right\rangle$ and $\left|1^-\right\rangle$ are defined by Eqs. (8.199) and (8.200), respectively. Assume that $\left|1^+\right\rangle$ and $\left|1^-\right\rangle$ are maximally entangled states ($\alpha = \pi/4$) and investigate the separability of $|\Psi\rangle$ as a function of β_1 and β_2. Can superpositions of entangled states be unentangled? Determine the Grobe–Rzazewski–Eberly number.

8.12 Systems A and B are three-level systems with the states $|-1\rangle$, $|0\rangle$, and $|1\rangle$. Determine the combined, maximally entangled state(s).

References

[1] D. P. Craig and T. Thirunamachandran, *Molecular Quantum Electrodynamics*. Mineola, NY: Dover Publications (1998).

[2] R. Loudon, *The Quantum Theory of Light*, 2nd edn. Oxford: Oxford University Press (1983).

[3] C. Cohen-Tannoudji, J. Dupond-Roc, and G. Grynberg, *Photons and Atoms*. New York: John Wiley & Sons (1997).

[4] J. A. Stratton, *Electromagnetic Theory*. New York: McGraw-Hill (1941).

[5] L. D. Barron and C. G. Gray, "The multipole interaction Hamiltonian for time dependent fields," *J. Phys. A* **6**, 50–61 (1973).

[6] R. G. Woolley, "A comment on 'The multipole interaction Hamiltonian for time dependent fields'," *J. Phys. B* **6**, L97–L99 (1973).

[7] P. W. Milonni, *The Quantum Vacuum*. San Diego, CA: Academic Press (1994).

[8] H. C. van de Hulst, *Light Scattering by Small Particles*. Mineola, NY: Dover Publications (1981).

[9] E. M. Purcell, "Spontaneous emission probabilities at radio frequencies," *Phys. Rev.* **69**, 681 (1946).

[10] K. H. Drexhage, M. Fleck, F. P. Schäfer, and W. Sperling, "Beeinflussung der Fluoreszenz eines Europium-chelates durch einen Spiegel," *Ber. Bunsenges. Phys. Chem.* **20**, 1176 (1966).

[11] P. Goy, J. M. Raimond, M. Gross, and S. Haroche, "Observation of cavity-enhanced single-atom spontaneous emission," *Phys. Rev. Lett.* **50**, 1903–1906 (1983).

[12] D. Kleppner, "Inhibited spontaneous emission," *Phys. Rev. Lett.* **47**, 233–236 (1981).

[13] E. Yablonovitch, "Inhibited spontaneous emission in solid-state physics and electronics," *Phys. Rev. Lett.* **58**, 2059–2062 (1987).

[14] S. John, "Strong localization of photons in certain disordered dielectric superlattices," *Phys. Rev. Lett.* **58**, 2486–2489 (1987).

[15] J. D. Joannopoulos, P. R. Villeneuve, and S. Fan, "Photonic crystals: putting a new twist on light," *Nature* **386**, 143–149 (1997).

[16] S. Kühn, U. Hakanson, L. Rogobete, and V. Sandoghdar, "Enhancement of single-molecule fluorescence using a gold nanoparticle as an optical nanoantenna," *Phys. Rev. Lett.* **97**, 017402 (2006).

[17] P. Andrew and W. L. Barnes, "Forster energy transfer in an optical microcavity," *Science* **290**, 785–788 (2000).

[18] J. J. Sánchez-Mondragón, N. B. Narozhny, and J. H. Eberly, "Theory of spontaneous-emission line shape in an ideal cavity," *Phys. Rev. Lett.* **51**, 550–553 (1983).

[19] G. S. Agarwal, "Spectroscopy of strongly coupled atom–cavity systems: a topical review," *J. Mod. Opt.* **45**, 449–470 (1998).

[20] B. Luk'yanchuk, N. I. Zheludev, S. A. Maier, *et al.*, "The Fano resonance in plasmonic nanostructures and metamaterials," *Nature Mater.* **9**, 707–715 (2010).

[21] U. Fano, "Sullo spettro di assorbimento dei gas nobili presso il limite dello spettro d'arco," *Nuovo Cimento* **12**, 154–161 (1935). English translation: "On the absorption spectrum of noble gases at the arc spectrum limit," arXiv:cond-mat/0502210 v1 by G. Pupillo, A. Zannoni, and C. W. Clark (2005).

[22] B. Lounis and C. Cohen-Tannoudji, "Coherent population trapping and Fano profiles," *J. Physique II* **2**, 579–592 (1992).

[23] O. Krauth, G. Fahsold, N. Magg, and A. Pucci, "Anomalous infrared transmission of adsorbates on ultrathin metal films: Fano effect near the percolation threshold," *J. Chem. Phys.* **113**, 6330–6333 (2000).

[24] K. H. Drexhage, "Influence of a dielectric interface on fluorescent decay time," *J. Lumin.* **1–2**, 693–701 (1970).

[25] R. R. Chance, A. Prock, and R. Silbey, "Molecular fluorescence and energy transfer near interfaces," in *Advances in Chemical Physics*, vol. 37, ed. I. Prigogine and S. A. Rice. New York: Wiley, pp. 1–65 (1978).

[26] W. R. Holland and D. G. Hall, "Frequency shifts of an electric-dipole resonance near a conducting surface," *Phys. Rev. Lett.* **52**, 1041–1044 (1984).

[27] See, for example, H. Haken, W. D. Brewer, and H. C. Wolf, *Molecular Physics and Elements of Quantum Chemistry*. Berlin: Springer-Verlag (1995).

[28] See, for example, H. van Amerongen, L. Valkunas, and R. van Grondelle, *Photosynthetic Excitons*. Singapore: World Scientific (2000).

[29] C. R. Kagan, C. B. Murray, M. Nirmal, and M. G. Bawendi, "Electronic energy transfer in CdSe quantum dot solids," *Phys. Rev. Lett.* **76**, 1517–1520 (1996).

[30] S. Weiss, "Fluorescence spectroscopy of single biomolecules," *Science* **283**, 1676–1683 (1999).

[31] Th. Förster, "Energiewanderung und Fluoreszenz," *Naturwissenschaften* **33**, 166–175 (1946); Th. Förster, "Zwischenmolekulare Energiewanderung und Fluoreszenz," *Ann. Phys. (Leipzig)* **2**, 55–75 (1948). An English translation of Förster's original work is provided by R. S. Knox, "Intermolecular energy migration and fluorescence,"

in *Biological Physics*, ed. E. Mielczarek, R. S. Knox, and E. Greenbaum. New York: American Institute of Physics, pp. 148–160 (1993).

[32] P. Wu and L. Brand, "Resonance energy transfer," *Anal. Biochem.* **218**, 1–13 (1994).

[33] R. S. Knox and H. van Amerongen, "Refractive index dependence of the Förster resonance excitation transfer rate," *J. Phys. Chem. B* **106**, 5289–5293 (2002).

[34] D. L. Andrews and G. Juzeliunas, "Intermolecular energy transfer: radiation effects," *J. Chem. Phys.* **96**, 6606–6612 (1992).

[35] P. Andrew and W. L. Barnes, "Förster energy transfer in an optical microcavity," *Science* **290**, 785–788 (2000).

[36] C. E. Finlayson, D. S. Ginger, and N. C. Greenham, "Enhanced Förster energy transfer in organic/inorganic bilayer optical microcavities," *Chem. Phys. Lett.* **338**, 83–87 (2001).

[37] P. R. Selvin, "The renaissance of fluorescence resonance energy transfer," *Nature Struct. Biol.* **7**, 730–734 (2000).

[38] S. A. McKinney, A. C. Declais, D. M. J. Lilley, and T. Ha, "Structural dynamics of individual Holliday junctions," *Nature Struct. Biol.* **10**, 93–97 (2003).

[39] C. Cohen-Tannoudji, J. Dupont-Roc, and G. Grynberg, *Atom–Photon Interactions*. Weinheim: Wiley-VCH Verlag, Chapter VI (2004).

[40] M. L. Zimmerman, M. G. Littman, M. M. Kash, and D. Kleppner, "Stark structure of the Rydberg states of alkali-metal atoms," *Phys. Rev. A* **20**, 2251–2275 (1979).

[41] M. Bayer, P. Hawrylak, K. Hinzer, *et al.*, "Coupling and entangling of quantum dot states in quantum dot molecules," *Science* **291**, 451–453 (2001).

[42] T.J. Kippenberg and K. J. Vahala, "Cavity optomechanics: back-action at the mesoscale," *Science* **321**, 1172–1176 (2008).

[43] A. R. Bosco de Magalhaes, C. H. d'Avila Fonseca, and M. C. Nemes, "Classical and quantum coupled oscillators: symplectic structure," *Phys. Scripta* **74**, 472–480 (2006).

[44] C. L. Garrido Alzar, M. A. G. Martinez, and P. Nussenzveig, "Classical analog of electromagnetically induced transparency," *Am J. Phys.* **70**, 37–41 (2002).

[45] C. Wittig, "The Landau–Zener formula," *J. Phys. Chem. B* **109**, 8428–8430 (2005).

[46] L. Landau, "Zur Theorie der Energieübertragung. II," *Phys. Z. Sowjetunion* **2**, 46–51 (1932).

[47] C. Zener, "Non-adiabatic crossing of energy levels," *Proc. Roy. Soc. A* **137**, 692–702 (1932).

[48] A. Bouhelier, J. Renger, M. Beversluis, and L. Novotny, "Plasmon coupled tip-enhanced near-field microscopy," *J. Microsc.* **210**, 220–224 (2003).

[49] See, for example, D. J. Griffiths, *Introduction to Quantum Mechanics*. Upper Saddle River, NJ: Prentice Hall (1994).

[50] E. Schrödinger, "Die gegenwärtige Situation in der Quantenmechanik," *Naturwissenschaften* **23**, 807–812 (1935).

[51] A. Ekert and P. L. Knight, "Entangled quantum systems and the Schmidt decomposition," *Am. J. Phys.* **63**, 415–423 (1995).

[52] R. Grobe, K. Rzazewski, and J. H. Eberly, "Measure of electron–electron correlation in atomic physics," *J. Phys. B* **27**, L503–L508 (1994).

9 Quantum emitters

The interaction of light with nanoscale structures is at the core of nano-optics. As the structures become smaller and smaller the laws of quantum mechanics will become apparent. In this limit, the discrete nature of atomic states gives rise to resonant light–matter interactions. In atoms, molecules, and nanoparticles, such as semiconductor nanocrystals and other "quantum confined" systems, these resonances occur when the photon energy matches the energy difference of discrete internal (electronic) energy levels. Owing to the resonant character, light–matter interaction can often be approximated by treating these quantum emitters as effective two-level systems, i.e. by considering only those two (electronic) levels whose difference in energy is close to the interacting photon energy $\hbar\omega_0$.

In this chapter we discuss quantum emitters that are used in optical experiments. We will discuss their use as single-photon sources and analyze their photon statistics. While the radiative properties of quantum emitters have been discussed in Chapter 8, this chapter focuses on the properties of the quantum emitters themselves. We adopt a rather practical perspective since more detailed accounts can be found elsewhere (see e.g. [1–4]).

9.1 Types of quantum emitters

The possibility of detecting single quantum emitters optically relies mostly on the fact that redshifted emission can be very efficiently discriminated against excitation light [5, 6]. This opens the road for experiments in which the properties of these emitters are studied or in which they are used as discrete light sources. We will now introduce three important classes of single emitters: organic dye molecules, semiconductor quantum dots, and impurity centers in wide-bandgap semiconductors, in particular diamond.

9.1.1 Fluorescent molecules

For an organic molecule, the lowest-energy electronic transition appears between the highest occupied molecular orbital (HOMO) and the lowest unoccupied molecular orbital (LUMO). Higher unoccupied molecular orbitals can be taken into account if necessary. In addition to the electronic energy levels, multi-atomic particles, such as molecules, have vibrational degrees of freedom. For molecules, all of the electronic states involved in the interaction of a molecule with light have a manifold of (harmonic-oscillator-like) vibrational states superimposed. Since the nuclei are much more massive than the electrons, the latter are considered to follow the (vibrational) motion of the nuclei instantaneously. Within this so-called adiabatic or Born–Oppenheimer approximation, the electronic and the vibrational wavefunctions can be separated and the total wavefunction may be written as a product of a purely electronic and a purely vibrational wavefunction. At ambient temperature the thermal energy is small compared with the separation between vibrational states. Thus excitation of a molecule usually starts from the electronic ground state with no vibrational quanta excited (see Fig. 9.1).

Excitation of fluorescent molecules

Excitation of the molecule can be resonant, into the vibrational ground state of the LUMO, or it can be non-resonant, involving higher vibrational modes of the LUMO. Vibrational relaxation then causes a fast decay cascade, which, for good chromophores,[1] ends in the

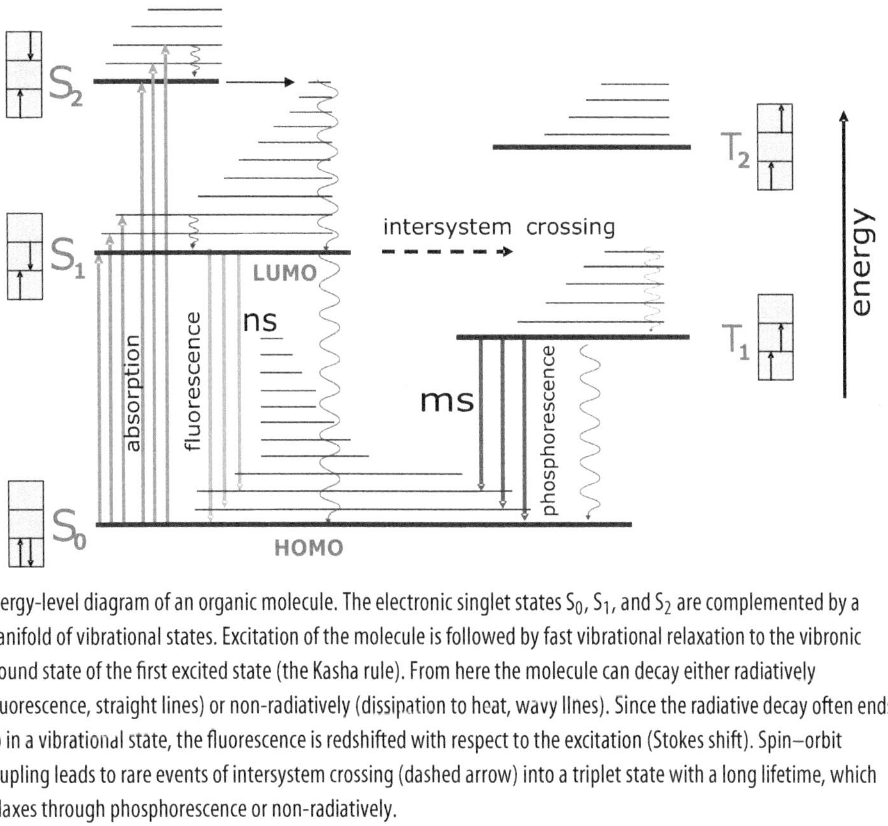

Fig. 9.1 Energy-level diagram of an organic molecule. The electronic singlet states S_0, S_1, and S_2 are complemented by a manifold of vibrational states. Excitation of the molecule is followed by fast vibrational relaxation to the vibronic ground state of the first excited state (the Kasha rule). From here the molecule can decay either radiatively (fluorescence, straight lines) or non-radiatively (dissipation to heat, wavy lines). Since the radiative decay often ends up in a vibrational state, the fluorescence is redshifted with respect to the excitation (Stokes shift). Spin–orbit coupling leads to rare events of intersystem crossing (dashed arrow) into a triplet state with a long lifetime, which relaxes through phosphorescence or non-radiatively.

[1] For inefficient chromophores, non-fluorescent molecules, or molecules strongly coupling to the environment (e.g. phonons), the collisional deactivation continues to the ground state.

vibrational ground state of the LUMO.[2] For a fluorescent molecule the lifetime of this excited state is on the order of 1–10 ns. For resonant pumping, coherence between the pump and the emitted light can be conserved only if the molecule is sufficiently isolated from its environment that environmental dephasing due to collisions or phonon scattering becomes small. Isolated atoms or molecules in particle beams or traps and molecules embedded in crystalline matrices at cryogenic temperatures can show coherence between resonant excitation light and the zero-phonon emission line [5], leading to extreme peak absorption cross-sections and to Rabi oscillations (see Appendix A). Note that even if a molecule is excited resonantly, besides the resonant zero-phonon radiative decay to the LUMO, non-resonant radiative relaxation also occurs (redshifted fluorescence), which leaves the molecule initially in one of the higher vibrational states of the HOMO. This state also relaxes fast by the same process as discussed before, called internal conversion, to the vibrational ground state of the HOMO.

The strength of the HOMO–LUMO transition is determined by a transition matrix element. In the dipole approximation this is the matrix element of the dipole operator between the HOMO and the LUMO wavefunctions supplemented by corresponding vibronic states. This matrix element is called the absorption dipole moment of the molecule (see Appendix A). The dipole approximation assumes that the exciting electric field is constant over the dimensions of the molecule. In nano-optics this is not always the case, and corrections to the dipole approximation, especially for larger quantum systems, might become necessary. Those corrections can result in modified selection rules for optical transitions [7].

The molecular wavefunctions emerge as the result of interacting atomic wavefunctions. Since the atoms have a fixed position within the molecular structure, the direction of the dipole moment vector is fixed with respect to the molecular structure. Degeneracies are observed only for highly symmetric molecules. For the molecules consisting of interconnected aromatic rings and the linear polyenes of Fig. 9.2 the absorption dipole moment approximately points along the long axis of the structure, although this is not a general rule. The emission dipole moment typically points in the same direction as the absorption dipole. Exceptions to this rule may occur if the geometry of the molecule changes significantly between the electronic ground state and the excited state. With increasing length of the aromatic or conjugated system, the absorption of a molecule shifts to the red. This behavior resembles that of a quantum-mechanical particle in a box system in which the level splitting decreases for increasing box length (see Fig. 9.2).

Relaxation of flurescent molecules

Radiative relaxation from the LUMO is called *fluorescence*. But relaxation can also occur non-radiatively via vibrations or collisions that ultimately lead to the generation of heat. The ratio of the radiative decay rate γ_{r} and the total decay rate $(\gamma_{\mathrm{r}} + \gamma_{\mathrm{nr}})$ is denoted the internal quantum efficiency

$$q_{\mathrm{i}} = \frac{\gamma_{\mathrm{r}}}{\gamma_{\mathrm{r}} + \gamma_{\mathrm{nr}}}, \tag{9.1}$$

[2] This is the so-called Kasha rule.

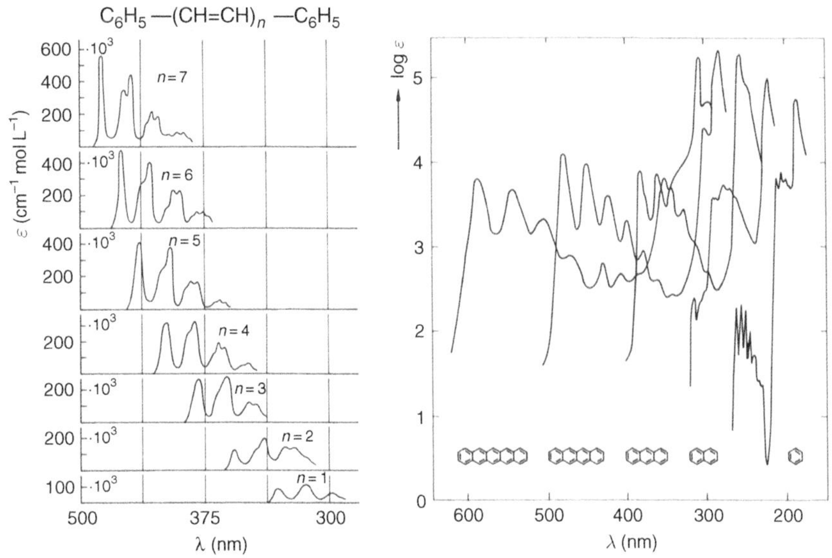

Fig. 9.2 Characteristic absorption spectra of fluorescent molecules. Left: Linear polyenes featuring a conjugated carbon chain on which delocalized electrons exist. Right: aromatic molecules. Electrons are delocalized over the aromatic system. Increasing the length of the conjugated chain or the aromatic system shifts the absorption to the red spectral region. From [8].

where γ_{nr} is the non-radiative decay rate. If the radiative decay of the LUMO prevails, the corresponding lifetime is typically on the order of some nanoseconds. The emission spectrum consists of a sum of Lorentzians (see Section 8.6.2), the so-called vibrational progression, corresponding to the different decay pathways into ground-state vibronic levels (see Fig. 9.1). At ambient temperatures dephasing is strong and leads to additional line broadening such that the vibrational progression becomes washed out. However, vibrational bands become discernable at low temperatures. For a molecule, the probability of decaying into a vibrational state of the HOMO is determined by the overlap integrals of the respective LUMO vibrational state wavefunction and the HOMO vibrational state wavefunctions. These overlap integrals are known as Frank–Condon factors. Their relative magnitude determines the shape of the fluorescence spectrum [5, 6].

Not all molecules fluoresce efficiently. *Radiative* decay occurs only for a special class of molecules that exhibit a low density of (vibronic) states (of the HOMO) at the LUMO energy. Under such circumstances a non-radiative decay via the HOMO vibrational manifold is not likely to occur. Particularly efficient fluorescence is observed for small and rigid aromatic or conjugated molecules, called dye molecules or fluorophores. The same principles hold for other quantum objects in the sense that the more degrees of freedom they have, the lower the probability of radiative decay will be.

Owing to the non-negligible spin–orbit coupling in molecules (which is particularly important for heavy elements) there is a finite torque acting on the spin of the electron in the excited state. This results in a small but significant probability that the spin of the excited electron is flipped upon excitation or radiative decay. This process is known as *intersystem crossing* and, for a good chromophore, it typically occurs at a rate γ_{ISC} much smaller than the excited state's decay rate. If a spin flip happens, the total electronic spin of the molecule changes from 0 to 1. Spin 1 has three possible orientations in an external magnetic field, leading to a triplet of eigenstates. This is the reason for calling a state with spin = 1 a triplet state as opposed to a singlet state for spin = 0. The energy of the electron in the triplet state is usually below the singlet excited-state energy because the exchange interaction between the parallel spins increases the average distance between the electrons in accordance with Hund's rule. The increased average distance leads to a smaller Coulomb repulsion. Once a molecule has undergone intersystem crossing into the triplet state it can only decay into a singlet ground state. However, this is a spin-forbidden transition. Triplet states therefore have a very long lifetime on the order of milliseconds.

Because of triplet-state excursions, the time-trace of fluorescence emission of a molecule shows a characteristic effect: the relatively high count rates associated with singlet–singlet transitions are interrupted by dark periods of duration a few milliseconds corresponding to the triplet-state lifetime. This *fluorescence-blinking* behavior can easily be observed when studying single molecules, and we will analyze it quantitatively later on in this chapter.

Frequently, blinking on longer timescales is also observed. Dark periods that are much longer than the typical triplet-state lifetime are mostly attributed to fluctuating local environments and transient interactions with other chemical species such as oxygen. Finally, a molecule eventually ceases to fluoresce completely. This so-called photobleaching event is often due to chemical reactions with singlet oxygen: a molecule residing in its triplet state can efficiently generate singlet oxygen in its immediate environment by triplet–triplet annihilation.[3] This reactive singlet oxygen then attacks and interrupts the conjugated or aromatic system of the molecule [9].

9.1.2 Semiconductor quantum dots

The use of colloidally dispersed pigment particles for producing colorful effects has been known since ancient times. In the early 1980s experiments with colloidal solutions of semiconductor nanocrystals were performed with applications in solar energy conversion and photocatalysis in mind. It was found that colloidal solutions of the same semiconductor showed striking changes in colors when the size of the nanocrystals was varied. This observation can be attributed to the so-called quantum confinement effect. The excitons in semiconductors,[4] i.e. bound electron–hole pairs, are described by a hydrogen-like Hamiltonian

$$\hat{H} = -\frac{\hbar^2}{2m_{\mathrm{h}}}\nabla_{\mathrm{h}}^2 - \frac{\hbar^2}{2m_{\mathrm{e}}}\nabla_{\mathrm{e}}^2 - \frac{e^2}{\varepsilon\,|r_{\mathrm{e}} - r_{\mathrm{h}}|}, \tag{9.2}$$

where m_{e} and m_{h} are the effective masses of the electron and the hole, respectively, and ε is the dielectric constant of the semiconductor [10]. The subscripts e and h denote the electron and the hole, respectively. Once the size of a nanocrystal has approached the limit of the

[3] The ground state of molecular oxygen is a triplet state.

[4] In solid-state physics, the term "exciton" denotes a bound electron–hole pair. In physical chemistry, however, the term "exciton" is used also for a strongly coupled system (Section 8.7.3).

Bohr radius of an exciton (see Problem 9.1), the states of the exciton shift to higher energy as the confinement energy increases. In semiconductors, due to the small effective masses of the electrons and holes, the Bohr radius can be on the order of 10 nm, which means that quantum confinement in semiconductor nanocrystals becomes prominent at length scales 10–100 times larger than the characteristic size of atoms or fluorescent molecules. The confinement energy arises from the fact that according to the Heisenberg uncertainty principle the momentum of a particle increases if its position becomes better defined. In the limit of small particles, the strongly screened Coulomb interaction between the electron and the hole, the last term in Eq. (9.2), can be completely neglected. Both the electron and the hole can consequently be described by a particle-in-a-box model, which leads to discrete energy levels that shift to higher energies as the box is made smaller. Therefore, for a semiconductor like CdSe with a bandgap in the infrared, this yields luminescence in the visible if sufficiently small particles ($\approx$3 nm) are prepared. The quantum efficiencies for radiative decay of the confined excitons are rather high because both the electron and the hole are confined to a nanometer-sized volume inside the dot. This property renders quantum dots extremely interesting for optoelectronic applications.

In order to fully understand the structure of the electronic states in a semiconductor nanocrystal several refinements to the particle-in-a-box model have to be considered. Most importantly, for larger particles the Coulomb interaction becomes significant and reduces the energy of the exciton. Other effects, like crystal-field splitting and the asymmetry of the particles as well as an exchange interaction between electrons and holes also have to be taken into account [11].

For metal nanoclusters, e.g. made from gold, confinement of the free electrons to dimensions of a few nanometers does not lead to notable quantum confinement effects. This is because the Fermi energy for conductors lies at the center of the conduction band and, upon shrinking the clusters, quantization effects start to become prominent at the band edges first. However, if the confinement reaches the level of the Fermi wavelength of the free electrons ($\approx$0.7 nm), discrete, quantum-confined electronic transitions appear as demonstrated in Ref. [12]. Here, chemically prepared gold nanoclusters are shown to luminesce with comparatively high quantum yield with a spectrum that varies with the gold cluster size.

Surface passivation

Since in nanoparticles the number of surface atoms is comparable to the number of bulk atoms, the properties of the surface strongly influence the electronic structure of a nanoparticle. For semiconductor nanocrystals, it is found that for "naked" particles surface defects created by chemical reactions or surface reconstruction drastically reduce the luminescence quantum yield, since any surface defect will lead to allowed electronic states in the bandgap. Consequently, non-radiative relaxation pathways involving trap states become predominant. This leads to a strong reduction of the quantum yield for visible light emission. In order to avoid surface defects, nanocrystals are usually capped with a protective layer of a second, larger-bandgap semiconductor with very similar lattice constants. Such a material grows epitaxially over the core such that the chemical composition changes abruptly within one atomic layer. The resulting structures are designated as Type I. If the capping layer has a lower bandgap, then the electrons are preferentially located in the outer shell with less confinement. These structures are then designated as Type II, typically showing optical response in the infrared spectral region. For CdSe nanocrystals usually a high-bandgap ZnS capping layer is applied. With this protective shell, it is possible to functionalize the particles by applying suitable surface chemistry without interfering with the optical properties. The overall structure of a typical semiconductor nanocrystal is shown in Fig. 9.3. The implementation of such a complex architecture at the nanometer scale paved the way for widespread application of semiconductor nanocrystals as fluorescent markers in the life sciences and in optoelectronic applications.

Another way to produce semiconductor quantum dots, which is different from the wet-chemistry approach, is to exploit self-assembly during epitaxial growth of semiconductor heterostructures. Here, the most common way to produce quantum dots is the so-called Stranski–Krastanow (SK) method. In 1937, Stranski and Krastanow proposed that island formation could take place on an epitaxially grown surface [13]. For example, on depositing a material with a slightly larger lattice constant, e.g. InAs, onto a GaAs surface, the lattice mismatch ($\approx 7\%$ in this case) introduces strain. The first few layers of InAs form a pseudomorphic two-dimensional layer, the so-called wetting layer. If more material is deposited, the two-dimensional growth is no longer energetically favorable and the material deposited in excess after the wetting layer has formed organizes itself into three-dimensional islands as sketched in Fig. 9.4. These islands are usually called self-assembled quantum dots. The size and the density of these quantum dots can be controlled by the growth parameters. To complete the structure, the quantum dots have to be embedded in a suitable capping layer, similarly to the case of colloidal nanocrystals. The capping-layer material and growth parameters also have to be carefully chosen in order to end up with defect-free quantum dots with high luminescence quantum yield.

Excitation of quantum dots

The absorption spectrum of semiconductor nanocrystals is characterized by increasing absorption strength towards shorter wavelengths. This behavior originates from the electronic density of states, which increases towards the center of the semiconductor

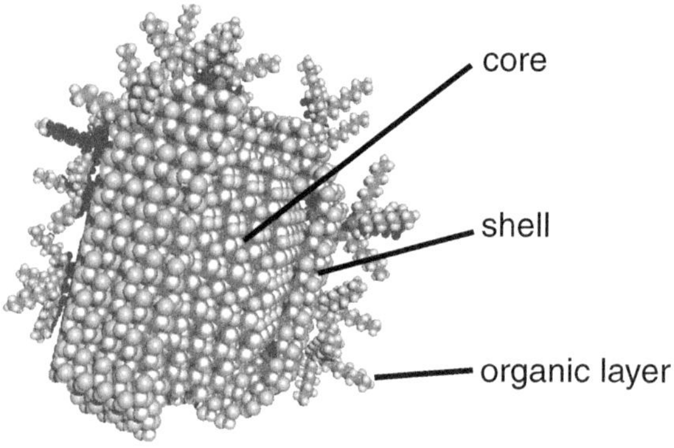

Fig. 9.3 Structure of a typical colloidal semiconductor nanocrystal. Courtesy of Hans Eisler.

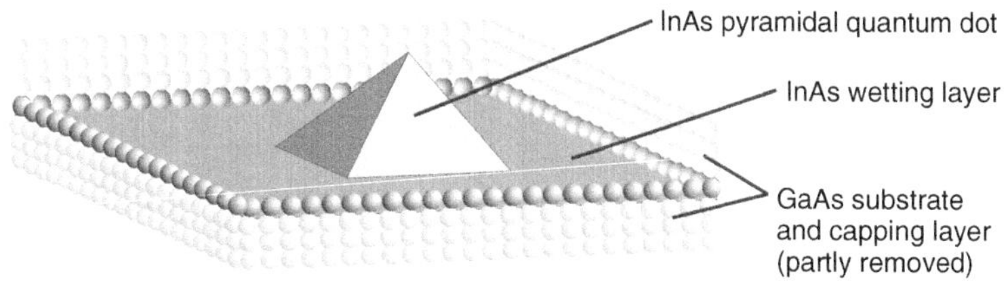

Fig. 9.4 Sketch of an InAs pyramidal quantum dot. InAs pyramidal quantum dots are formed by self-assembly during epitaxial growth. The capping layer is not shown.

conduction band. The very broad absorption spectrum allows different-sized nanocrystals to be excited by a single blue-light source as illustrated in Fig 9.5(b). Similarly to fluorescent molecules, semiconductor nanocrystals excited with excess energy first relax by fast internal conversion to the lowest-energy excitonic state, from which recombination occurs by photon emission. Differently from the case of molecules, multiple excitons can be excited in the same quantum dot at higher excitation powers. The energy necessary to excite a second exciton is lowered by the presence of the first exciton due to Coulomb interactions between the charges involved. Figure 9.5 shows the excitation and emission spectra of a range of CdSe nanocrystals of varying size. Apart from some fine structure near the band edge due to the low density of states, the increasing absorption for blue excitation can be clearly observed independently of the particle size. For the emitted light, a shift of the emission (dashed curve) towards the blue spectral region can be observed as the particle size is reduced.

Because of the symmetry of a nanocrystal, its dipole moment is degenerate. CdSe nanocrystals are slightly elongated in the direction of the crystal axis (the "dark axis"). Figure 9.6(a) sketches the orientation of the so-called "bright plane" and the "dark axis" within the nanocrystal.

It is observed that, no matter what the polarization direction of the excitation light relative to the crystal axis is, photon emission from a CdSe nanocrystal always originates from a transition dipole oriented in the bright plane. There are no transitions along the crystal axis and hence this axis is referred to as the dark axis. However, the nanocrystal can be excited along the dark axis. In this case, there is a 90° difference between the absorption dipole and the emission dipole orientation. Because of the degenerate dipole moment in the bright plane, the emission direction in this plane is arbitrary, unless an external perturbation is applied. For a sample with randomly oriented nanocrystals, the emission of the tilted bright plane in Fig. 9.6(b) gives rise to an anisotropy in the polarization of the emitted light as depicted in Fig. 9.6(c). This anisotropy can be exploited to determine the three-dimensional orientation of the dark axis of the nanocrystal (see Fig. 9.6(d)) [14], which can be of interest in various particle-tracking applications in which semiconductor nanocrystals are used as markers.

Coherent control of excitons

It has been demonstrated that an exciton in a semiconductor quantum dot can act as a qubit, the unit of quantum information [16]. In such experiments, short laser pulses are

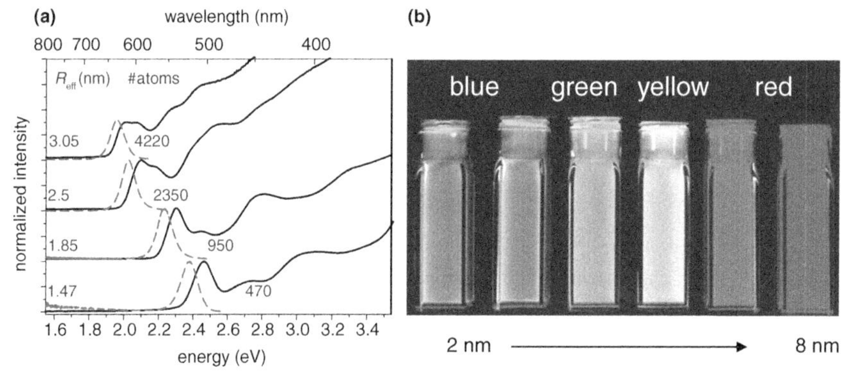

Fig. 9.5 Spectral response of semiconductor nanocrystals with different sizes. (a) Emission spectra (dashed) and excitation spectra. (b) Emission from a series of nanocrystal solutions with increasing particle size excited simultaneously by a UV lamp. Courtesy of Hans Eisler.

used to coherently manipulate the exciton state. A pump pulse prepares an exciton in a well-defined superposition of the ground state, $|00\rangle$, and the excited state, $|10\rangle$. A weak probe pulse reads out the population of the excited state. Changing the pulse area and keeping the delay fixed gives rise to an oscillatory behavior of the excited-state population as a function of the pulse area (excitation power). These oscillations are known as Rabi oscillations (see Appendix A). To realize a quantum logic gate with a single quantum dot, it is necessary to excite two or more interacting excitons within the same dot. In the case of two excitons it is observed that the Coulomb interaction between the two excitons lowers the total energy of the biexciton state with respect to the case of two independent excitons. The resulting energy diagram is shown in Fig. 9.7(a), where the binding energy is denoted as Δ. Note that the two excitons that can be excited in the quantum dot can be distinguished by their polarizations. Inspection of the resulting four-level scheme suggests that it is possible to realize a universal controlled-rotation quantum logic gate for which the target bit (the second exciton) is rotated through a π phase shift, e.g. from state $|01\rangle$ to state $|11\rangle$ or vice versa, if and only if the control bit (the first exciton) is in the excited state $|01\rangle$. The definition of the states is shown in Fig. 9.7(b). Such an experiment requires a two-color excitation scheme since the transition energies of the single exciton and the biexciton differ by the binding energy Δ. A first pulse (tuned to the single exciton transition) is used to excite a single exciton. Now one can apply a so-called operational pulse, which is tuned to one of the biexciton transitions. The truth table of the quantum logic gate (controlled rotation, CROT) can now be mapped out using a π-pulse tuned to e.g. the $|10\rangle$–$|11\rangle$ transition. If the input is $|00\rangle$, the operational pulse is off-resonant and the output will again be $|00\rangle$. If the input is $|10\rangle$ then the π-pulse creates $-|11\rangle$. If the input is already $|11\rangle$ it is transferred to $|10\rangle$ by stimulated emission. The basic operation of the CROT gate is shown in Fig. 9.7(c) by demonstrating Rabi flopping of the second exciton. This shows that the state of the biexciton can be prepared in any superposition of $|10\rangle$ and $|11\rangle$ by varying the pulse duration. Note that while it is possible to perform quantum logic computations with qubits encoded in the excitonic degrees of freedom of a quantum dot the computational

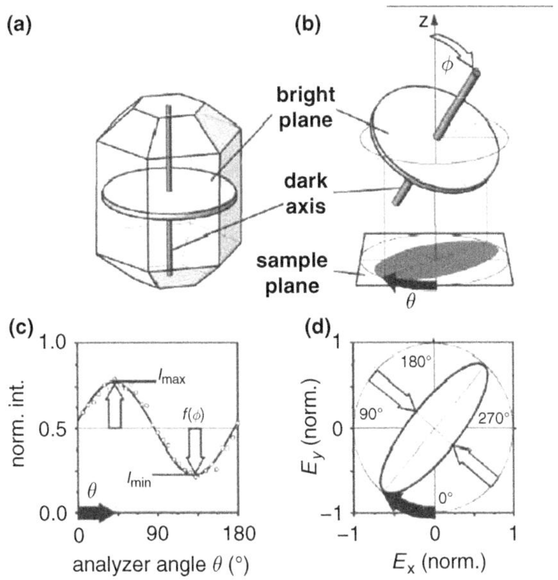

Fig. 9.6 The degenerate emission dipole moment of a semiconductor nanocrystal. (a) Orientation of the bright plane and the dark axis within a CdSe nanocrystal. (b) If the crystal axis is tilted with respect to the normal of a supporting surface, the projection of the bright plane into this surface changes. (c) Light emitted by such a tilted nanocrystal is therefore only partially polarized. (d) From the absolute orientation and the ellipticity of the polarization ellipse, the three-dimensional orientation of the dark axis can be determined. From [15].

time window is limited by the short decoherence time of $\approx$100 ps. Considerably longer times can be achieved with charged quantum dots.

9.1.3 Color centers in diamond

A third class of quantum emitters consists of fluorescent defect centers in wide-bandgap semiconductors. Diamond offers the largest bandgap (5.5 eV) of all known materials and is therefore transparent from deep-ultraviolet up to infrared wavelengths. Diamond hosts more than a hundred known luminescent defect centers, many of which have been characterized by optical spectroscopy [17, 18]. They are the reason for the different shades of color found in natural diamonds. A prominent impurity-related defect center in diamond is the nitrogen-vacancy (NV) center. It consist of a single substitutional nitrogen atom (N) and an empty site, i.e. a vacancy, at a nearest-neighbor lattice position as sketched in Fig. 9.8(a). Such NV centers can be created artificially in nitrogen-rich type Ib diamond samples by inducing electron-irradiation damage [21].

The NV center contains two unbound electrons originating from the substitutional nitrogen. The two unbound electrons form a lone pair. Furthermore, there are three more unbound carbon electrons. Two of those form a quasi-bond while one remains unbound. The NV center therefore can efficiently trap an additional electron (originating from nearby

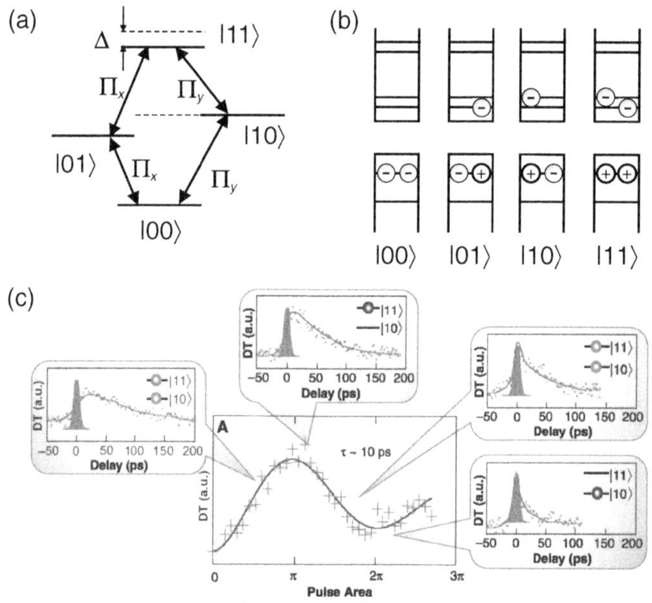

Fig. 9.7 Coherent control of the state of a biexciton. (a) The energy-level scheme. (b) Corresponding exciton transitions encoding two qubits. (c) Biexciton Rabi oscillation, demonstrating the possibility of being able to realize a controlled-rotation quantum logic gate using the biexciton transitions shown in (a). Adapted from [16].

electron donors, e.g. nitrogen impurities), which turns it into the abundant NV^- center as opposed to the neutral NV^0 [22]. Under both conditions NV centers show a bright red fluorescence at room temperature with a zero-phonon line and a vibronic progression. The NV^- and NV^0 centers exhibit zero-phonon lines at 637 nm and 575 nm, respectively, which allows them to be easily distinguished. Figure 9.8(b) shows a typical NV^- luminescence spectrum. Other defect centers in diamond also exist and have promising properties, e.g. a nickel–nitrogen complex (NE8), which at room temperature exhibits very pronounced and narrow zero-phonon-line emission at around 800 nm and only very weak phonon progression.

Because of the additional trapped electron and the observed bright luminescence, the zero-phonon line of the NV^- center is thought to be due to an allowed transition between a triplet ground state (^3A) and a triplet excited state (^3E). The excited-state lifetime is about 11 ns in bulk diamond and the quantum yield of the transition is about 0.7. These properties make it possible to easily detect single NV^- centers in fluorescence experiments with a saturation count rate of 10^7 photons per second [21] (see also Section 9.3). Furthermore, it is found that NV centers in bulk diamond do not show blinking or photo-bleaching. There are, however, indications that NV centers in very small diamond nanoparticles can show blinking related to interaction of the NV center with the nearby surface [23].

Optically detected magnetic resonance (ODMR) in diamond NV centers

In the absence of an external magnetic field the ground state is split into a nearly degenerate doublet X, Y ($m_S = \pm 1$) and the spin sublevel Z ($m_S = 0$) by 2.88 GHz [24]. For

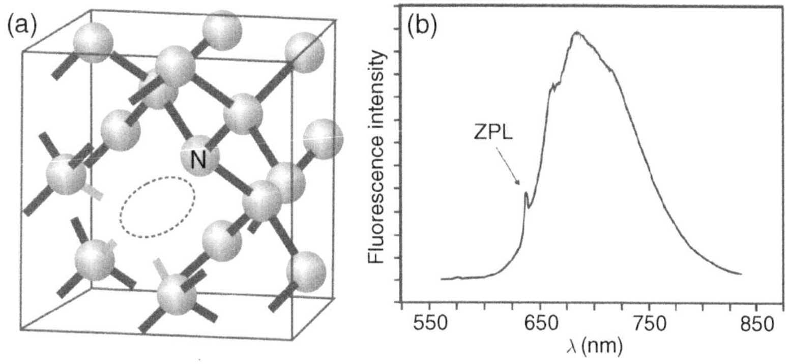

Fig. 9.8 The NV center in diamond. (a) Structure of the NV center, showing the substitutional nitrogen atom (N) and an empty site (dotted circle). (b) The luminescence spectrum of the NV^- center. ZPL, zero-phonon line. Adapted from [19].

temperatures above 1 K the ground-state levels are nearly equally populated. By means of optical excitation a non-Boltzmann spin polarization can be created in the ^3A ground state. While the x, y spin states ($m_S = \pm 1$) become depopulated, eventually more that 80% of the total population accumulates in the z state ($m_S = 0$). The explanation for this effect is based on the greatly different intersystem crossing rates exhibited by the spin levels of the ^3E state. Experimental evidence indicates that x', y' spin states ($m_S = \pm 1$) exhibit a significantly higher intersystem crossing rate than that of the z' sublevel. After relaxation within the singlet manifold the cross-over into the triplet ground state occurs predominantly to the z state ($m_S = 0$). Since the relaxation within the singlet manifold occurs non-radiatively and ends up in a long-lived state, NV centers in z spin sublevels have a significantly higher saturation emission rate R_∞ (see Eq. (9.16)). This effect can be used to optically read out the spin polarization of the NV center. If during optical excitation of the NV center a resonant microwave field is applied, the spin sublevels become equally populated. As a consequence the fluorescence rate of the NV center decreases drastically. The intensity change for single NV centers can be as high as 30%, as depicted in Fig. 9.9(b). The spin sublevels in NV^- centers can be coherently manipulated using a sequence of microwave pulses. The observed relaxation times are on the order of hundreds of microseconds. The electronic spin states of NV^- centers can also be coupled to nuclear spins via the hyperfine interaction. Fluorescence provides a simple means by which to read out the spin quantum states. Such NV centers are therefore promising candidates for solid-state quantum information processing. Furthermore, the spin levels are sensitive to external magnetic fields which can be exploited for high-resolution magnetometry [21].

Stimulated-emission-depletion microscopy of NV centers in diamond

The nearly perfect photostability of single NV centers in bulk diamond and their well-known spectral properties, i.e. the strong phonon progression, make them ideal candidates for high-resolution far-field imaging by means of STED (see Section 5.2.1). Indeed, the fluorescence-depletion curve of NV centers in diamond exhibits an exceptionally sharp decay due to a small stimulated-emission saturation intensity. It is therefore expected that very high spatial resolution can be achieved using a donut-shaped STED beam with a deep

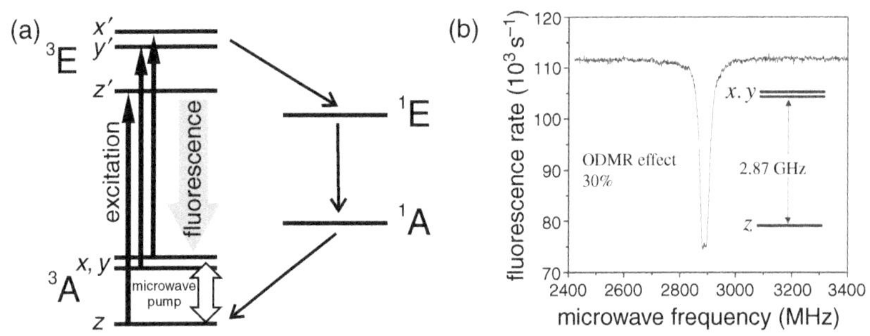

Fig. 9.9 Interaction of light and microwaves with NV centers. (a) The energy-level scheme of NV^- centers in diamond. Relaxation within the singlet manifold is mostly non-radiative (thin arrows) [25]. (b) An optically detected magnetic resonance trace showing the fluorescence rate as a function of the applied microwave frequency. Once the magnetic resonance frequency has been hit, the fluorescence drops because of enhanced intersystem crossing. From [21].

zero at its center. Figure 9.10(a) shows a combined confocal and STED image of a single NV center in diamond. Close to the center of the diffraction-limited spot, the donut-shaped STED beam is switched on. The STED beam extinguishes the fluorescence everywhere but in the close vicinity of the donut zero. Figure 9.10(b) shows a vertical cut through the center of Fig. 9.10(a), revealing both the diffraction-limited width of the confocal spot and the 8 nm full width at half-maximum of the STED spot.

9.2 The absorption cross-section

Absorption of light by a quantum emitter can be characterized by a frequency-dependent absorption cross-section. For weak excitation, the rate at which a two-level system is excited is proportional to the absolute square of the projection of the exciting electric field **E** on the absorption dipole moment **p** (see Appendix A). In this regime, the power absorbed by the system is given by (cf. Chapter 8)

$$P = \frac{\omega}{2}\,\mathrm{Im}\{\alpha\}\left|\mathbf{n_p}\cdot\mathbf{E}\right|^2, \tag{9.3}$$

where $\mathbf{n_p}$ is the unit vector in the direction of **p** and α is the polarizability. To define the absorption cross-section σ and to show its relevance for macroscopic measurements on ensembles of absorbers, we first average the dipole orientation over all directions and then assume that the local field **E** originates from a single incident *plane wave*.[5] In this case, the field **E** can be expressed by the intensity I of the plane wave, which allows us to define the absorption cross-section as

$$\sigma(\omega) = \frac{\langle P(\omega)\rangle}{I(\omega)} = \frac{\omega}{3}\sqrt{\frac{\mu_0}{\varepsilon_0}}\,\frac{\mathrm{Im}\{\alpha(\omega)\}}{n(\omega)}, \tag{9.4}$$

[5] The concept of "cross-section" is strictly valid only for single-mode (plane-wave) excitation.

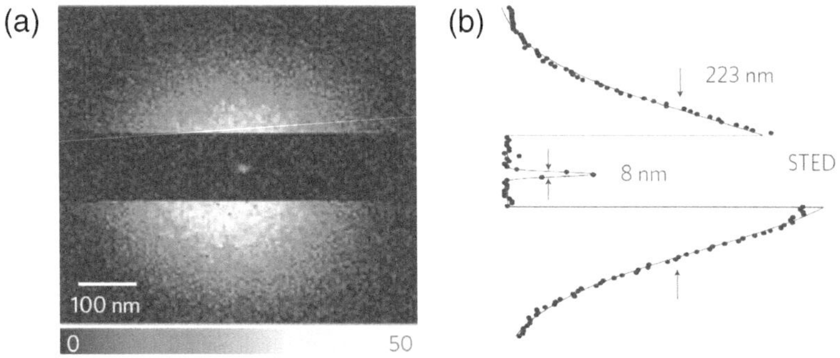

Fig. 9.10 STED microscopy of a single NV center in diamond. (a) A confocal/STED image of a single NV center in diamond. Close to the center the donut-shaped STED beam is unblocked, switching off the fluorescence everywhere but at the position of the donut's zero-intensity point. (b) A vertical cross section through the center of (a), revealing the 8 nm full width at half-maximum of the STED point-spread function as well as the diffraction-limited diameter of the confocal spot. From [26].

with n being the index of refraction of the surrounding medium and $\langle P \rangle$ the power absorbed by the molecule as an average over the random orientations of dipoles in the ensemble. Consider now an excitation beam with intensity I propagating in the direction of z through a dilute sample of randomly oriented molecules. After propagating an infinitesimal distance $\mathrm{d}z$ the laser intensity will be attenuated by an amount

$$I(z) - I(z + \mathrm{d}z) = -\frac{N}{V}\langle P(z) \rangle \mathrm{d}z, \tag{9.5}$$

where N/V is the volume concentration of the absorbers and $\langle P \rangle$ is related to σ and $I(z)$ by Eq. (9.3). In the limit $\mathrm{d}z \to 0$ we obtain

$$I(z) = I_0 \mathrm{e}^{-(N/V)\sigma z}, \tag{9.6}$$

with $I_0 = I(z = 0)$ (the Lambert–Beer law). σ has the units of area per photon, which justifies its designation as the absorption cross-section. According to Eq. (9.6), the absorption cross-section can be determined by an ensemble measurement, i.e. by measuring the attenuation of a laser beam as it propagates through a sample with a dilute concentration of absorbers.

Most commonly, the absorption is measured in terms of the *molar extinction coefficient* $\varepsilon(\lambda)$ according to

$$I(z, \lambda) = I_0 10^{-\varepsilon(\lambda)[M]z}, \tag{9.7}$$

where $[M]$ is the concentration of absorbers in moles per liter and z is the thickness of the absorbing layer in centimeters.

It is easy to see that the cross-section can be calculated from the extinction coefficient as $\sigma = 1000 \ln 10\, \varepsilon / N_{\mathrm{A}}$, with N_{A} being Avogadro's constant. Typical measured values of ε, e.g. for good laser dyes at room temperature, are around 200 000 l mol^{-1} cm^{-1}, which corresponds to a cross-section of 8×10^{-16} cm^2, i.e. a circle of radius 0.16 nm. This is a dimension that roughly coincides with the geometrical area of the aromatic or conjugated system of a small dye molecule. For semiconductor quantum dots, the absorption cross-section is correspondingly higher because of their greater geometrical size. This coincidence suggests that every photon passing the molecule within the area of σ gets absorbed by the molecule. Of course, this is a naive picture, which from the point of view of quantum mechanics cannot be true because of the uncertainty relation that does not allow the photon to be localized. So what is the physical meaning of the absorption cross-section? From a purely classical point of view, the field of the incoming plane wave is modified by the field scattered off the molecule being represented by a point dipole. The emitted dipole field and the exciting plane wave interfere and give rise to a resulting energy flow which, within an area defined by σ, is directed towards the dipole. This leads to an increase of its apparent area way beyond its geometrical size [27].

The spectral shape of the absorption cross-section $\sigma(\omega)$ is a Lorentzian with a width determined by the degree of dephasing between excitation and emission (see [1], p. 780). Almost full coherence between excitation and emission can be established at cryogenic temperatures. Under these conditions, the peak absorption cross-section of an isolated quantum emitter approaches the limit of (see Problem 9.5)

$$\sigma_{\max} = \frac{3\lambda^2}{2\pi}. \tag{9.8}$$

This is huge compared with the physical size of the quantum emitter! σ directly follows from the free-space spontaneous decay rate γ_0 (the natural linewidth) and can be further increased by reducing the local density of states (LDOS) (see Section 8.4.3). At ambient temperatures, or for systems that interact with a dissipative environment, due to dephasing events, $\sigma(\omega)$ broadens and the peak absorption cross-section becomes weaker, until it finally reaches the geometry-limited values for molecules in solutions or quantum dots under ambient conditions. For ambient temperatures the absorption cross-section can be represented as [4]

$$\sigma = \frac{3\lambda^2}{2\pi}\frac{\gamma_0}{\gamma}, \tag{9.9}$$

where γ_0 is the homogeneous linewidth and γ the inhomogeneous linewidth. Typically, $\gamma_0/\gamma \approx 10^{-6}$ and hence $\sigma \approx 0.3\ \text{nm}^2$ for optical frequencies.

9.3 Single-photon emission by three-level systems

We continue our analysis by studying the emission from single emitters. In order to do so we simplify the Jablonski diagram of Fig. 9.1 to its bare bones by neglecting the very fast relaxation within the vibrational manifold. We then end up with a system of three levels: the singlet ground state, the singlet first excited state, and the triplet state, denoted by 1, 2, and 3 as indicated in Fig. 9.11. These three levels are interconnected by excitation and relaxation rates according to the processes that we have just described. Taking into account these rates, we can formulate a system of differential equations for the change of the population p_i, $i = \{1, 2, 3\}$, of each level:

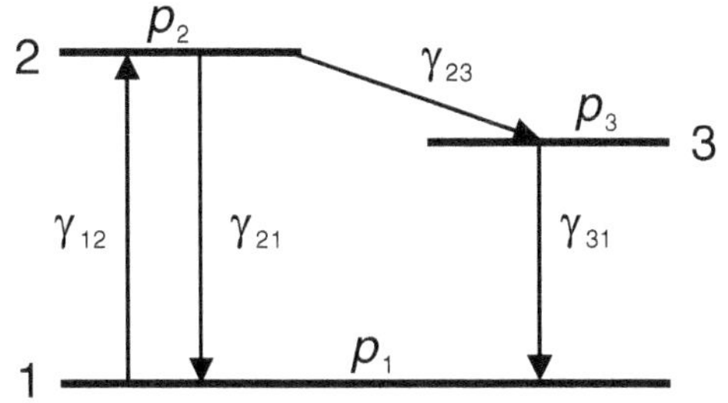

Fig. 9.11 A quantum emitter approximated by a system of three levels. A third level is taken into account in order to accommodate transitions to triplet or dark states.

$$\dot{p}_1 = -\gamma_{12}p_1 + (\gamma_r + \gamma_{nr})p_2 + \gamma_{31}p_3, \tag{9.10}$$
$$\dot{p}_2 = \gamma_{12}p_1 - (\gamma_r + \gamma_{nr} + \gamma_{23})p_2, \tag{9.11}$$
$$\dot{p}_3 = \gamma_{23}p_2 - \gamma_{31}p_3, \tag{9.12}$$
$$1 = p_1 + p_2 + p_3. \tag{9.13}$$

The last equation ensures that the emitter is in one of the three states at any time. The de-excitation rate γ_{21} is divided into a radiative contribution γ_r and a non-radiative contribution γ_{nr} such that $\gamma_{21} = \gamma_r + \gamma_{nr}$. We should note that introducing the population of a state, more precisely the probability that a certain state is occupied, p_i, makes sense only if we assume that we are either describing an ensemble of identical quantum emitters or observing the same quantum emitter many times under identical conditions. Also, by using rate equations we assume that coherence is lost in the excitation–relaxation cycle, e.g. due to dissipative coupling to vibrations. This is a very good approximation at room temperature and for non-resonant or broadband excitation [4]. At cryogenic temperatures with resonant excitation, or for isolated atoms or ions, the full quantum master equation must be considered. This approach also includes coherent effects that show up e.g. as Rabi oscillations (stimulated emission) between the populations of ground and excited states, but are not included in the present discussion (see Appendix A).

9.3.1 Steady-state analysis

Let us first consider the steady-state solution of Eqs. (9.10)–(9.13). We assume that in the steady state the populations are constant in time and consequently their time derivatives can be set to zero. This leads to a set of four equations for the equilibrium populations p_i, $i = \{1, 2, 3\}$.[6] We are interested in the rate R at which the system emits photons. This rate is given by

$$R = p_2\gamma_r, \tag{9.14}$$

which means that we have to determine the population of the excited state and multiply it by the radiative decay rate γ_r. If we solve for the population p_2 (see Problem 9.3) we end up with the following relation:

[6] Inspection of the four equations shows that two of them are linearly dependent.

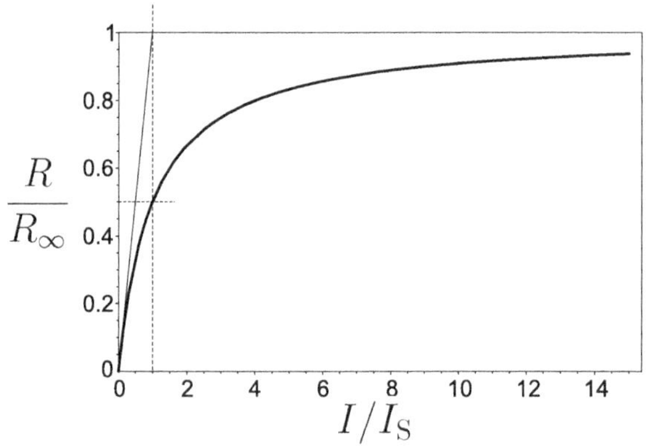

Fig. 9.12 Saturation of the emission rate of a single molecule as a function of the excitation intensity (solid line). The straight lines indicate the slope at zero intensity (solid line), which is equal to R_∞/I_S, as well as the values $I_S = 1$ and $R_\infty = 1/2$ (dashed lines).

$$R(I) = R_\infty \frac{I/I_S}{1 + I/I_S}, \tag{9.15}$$

where I is the intensity of the exciting light entering via the relation $\gamma_{12} = P/(\hbar\omega)$ and the expression for P in Eqs. (9.3) and (9.4). The constants R_∞ and I_S are defined as

$$\begin{aligned} R_\infty &= \gamma_r \left(1 + \frac{\gamma_{23}}{\gamma_{31}}\right)^{-1}, \\ I_S &= \frac{\gamma_r + \gamma_{nr} + \gamma_{23}}{\sigma(1 + \gamma_{23}/\gamma_{31})} \hbar\omega. \end{aligned} \tag{9.16}$$

Equation (9.15) describes the saturation behavior of the emission rate that is visualized in Fig. 9.12. This kind of saturation behavior is expected since the excited state has a finite lifetime, which limits the average time between two photons to a finite value. The saturation behavior is characterized by the two parameters R_∞ and I_S. The first describes the emission rate at infinitely strong excitation intensities and the second is the intensity at which the emission rate equals $R_\infty/2$ (see also Fig. 9.12). Typical values for R_∞ and I_S for a single dye molecule at room temperature are $R_\infty = 6 \times 10^6$ s^{-1} and $I_S = 7.5 \times 10^{21}$ photons s$^{-1} \approx 3$ kW cm^{-2} at wavelength 500 nm. Taking into account a collection and detection efficiency of about 15%, we can expect a photon count-rate of roughly 10^6 photons s^{-1} to be detected from a single dye molecule under saturation. Typically, a moderate excitation power of 1 μW focused onto a spot of diameter 250 nm, e.g. in a confocal microscope or a near-field microscope (see Chapter 5), is sufficient to saturate a molecule.

9.3.2 Time-dependent analysis

Now that we understand the steady-state emission of a single emitter characterized by a three-level system we can analyze the time dependence of the populations. This will give us some insight into the statistical properties of the light emitted by a single emitter.

Specifically, we will show that the light exhibits a striking non-classical behavior, which means that the radiation emitted by a single emitter cannot be characterized by a continuous electromagnetic field. Instead, quantized fields are necessary for a correct description. This does not affect the results obtained in Chapter 8, where a single emitter is modeled as a classical dipole and the statistics of the emitted radiation was not considered. Averaged over many photons, we naturally retain the classical description.

The light emitted by a light source can be characterized by the way it fluctuates. The deeper reason for this fact is provided by the fluctuation–dissipation theorem which, as discussed in Chapter 15, connects the fluctuations of a source characterized by an autocorrelation function to the emission spectrum of the source.

The normalized second-order autocorrelation function of an optical field, also called the intensity autocorrelation function, is defined as

$$g^{(2)}(\tau) = \frac{\langle I(t)I(t+\tau)\rangle}{\langle I(t)\rangle^2}, \tag{9.17}$$

where $\langle\,\rangle$ denotes the time average. $g^{(2)}(\tau)$ describes how the probability of measuring an intensity I at time $t+\tau$ depends on the value of the intensity at time t. In the language of single-photon detection events, $g^{(2)}(\tau)$ is the probability of detecting a photon at time $t+\tau$, provided that there was a photon at time t, normalized by the average photon detection rate. It can be shown generally [4] that $g^{(2)}(\tau)$ must fulfill certain relations if the intensity I is a *classical* variable. These are

$$\begin{aligned} g^{(2)}(0) &\geq 1, \\ g^{(2)}(\tau) &\leq g^{(2)}(0). \end{aligned} \tag{9.18}$$

The resulting typical shape of $g^{(2)}(\tau)$ in the classical limit is shown in Fig. 9.13(a). It is characteristic for the so-called bunching behavior of the light intensity. While the continuous *field amplitude* fluctuates around zero, the respective intensity fluctuations are characterized by "bunches" separated by intensity zeros. This effect is illustrated in Fig. 9.13(b).

While bunching behavior is characteristic for a classical light source, a single quantum emitter is characterized by antibunching, meaning that photons are emitted one after

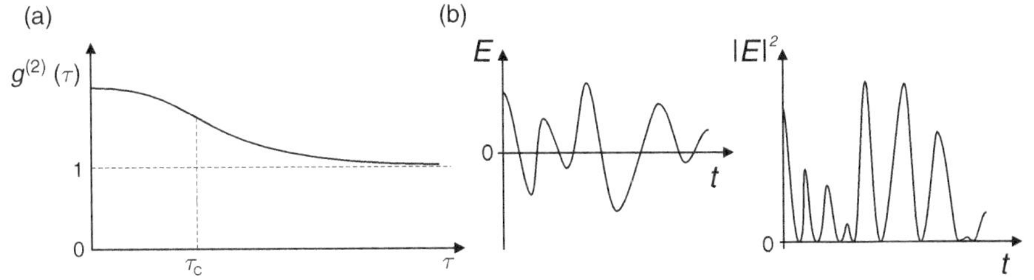

Fig. 9.13 The characteristic shape of the second-order autocorrelation function valid for classical light, showing a bunching behavior for short times (a). The bunching behavior is caused by statistical fluctuations of the classical field amplitude (b), which translate into intensity fluctuations separated by intensity zeros (c).

another, separated by some finite characteristic time. This is not difficult to understand since once a photon has been emitted the molecule has to be excited again, which takes a characteristic time γ_{12}^{-1}. Then it must decay to the ground state, which takes a time γ_r^{-1}. As a consequence, two consecutive photons will on average be separated by a finite time given by $(\gamma_{12} + \gamma_r)^{-1}$. The corresponding intensity autocorrelation function features a dip at $\tau = 0$, which means that the probability of emission of two photons at the same time vanishes. Since this dip violates the conditions of Eq. (9.18), the light emitted by a single quantum emitter is designated "non-classical" light. The generation of non-classical light is of importance for the field of quantum information [28].

We can calculate $g^{(2)}(\tau)$ for the three-level system using the relation Eq. (9.17) for $t = 0$, which is no limitation for a stationary process. For $t = 0$ we prepare the emitter in the ground state.[7] According to [29], Eq. (9.17) can be rewritten as

$$g^{(2)}(\tau) = \frac{\langle I(t)\rangle J(\tau)}{\langle I(t)\rangle^2} = \frac{J(\tau)}{\langle I(t)\rangle}. \tag{9.19}$$

Here, $J(\tau)$ is the probability of recording another photon after a time τ, provided that a photon was recorded at $t = 0$. $J(\tau)$ should not be confused with the probability $K(\tau)$ of recording the *next* photon after a time τ, provided that a photon was recorded at $t = 0$. Experimentally, $K(\tau)$ is determined from the distribution of interphoton times which is measured in start–stop experiments. For sufficiently small τ, J and K are indistinguishable. For longer times J and K are related via their Laplace transforms $\tilde{J}$ and $\tilde{K}$ as [29]

$$\tilde{J} = \frac{\tilde{K}}{1 - \tilde{K}}. \tag{9.20}$$

$J(\tau)$ can be expressed as $J(\tau) = \eta\gamma_r p_2(\tau)$, where η is the collection efficiency of the detection system and $p_2(\tau)$ is the time-dependent solution for the population of level 2 with the initial condition $p_2(0) = 0$. The steady-state count-rate reads as $\langle I(t)\rangle = \eta\gamma_r p_2(\infty)$. We therefore write

$$g^{(2)}(\tau) = \frac{\eta\gamma_r p_2(\tau)}{\eta\gamma_r p_2(\infty)} = \frac{p_2(\tau)}{p_2(\infty)}. \tag{9.21}$$

$p_2(\tau)$ can be obtained by solving the system of rate equations (9.10)–(9.13). In the first step, we combine Eq. (9.13) with Eqs. (9.10) and (9.12) and obtain

$$\begin{aligned} \dot{p}_1 &= -(\gamma_{12} + \gamma_{31})p_1 + (\gamma_r + \gamma_{nr} - \gamma_{31})p_2 + \gamma_{31}, \\ \dot{p}_2 &= \gamma_{12}p_1 - (\gamma_r + \gamma_{nr} + \gamma_{23})p_2. \end{aligned} \tag{9.22}$$

This system of coupled differential equations can be solved using Laplace transformation. To this end, we write Eq. (9.22) in matrix form as

$$\dot{\mathbf{p}}(\tau) = \begin{bmatrix} a & b \\ c & d \end{bmatrix} \mathbf{p}(\tau) + \begin{bmatrix} f \\ 0 \end{bmatrix}. \tag{9.23}$$

[7] Assume that a photon emitted by the quantum emitter had just been detected.

Here $\mathbf{p}(\tau)$ is a vector with components p_1 and p_2 and the abbreviations a, b, c, d, f are obtained by comparison with Eq. (9.22). In Laplace space, Eq. (9.23) reads as

$$s\mathbf{p}(s) - \mathbf{p}(0) = \begin{bmatrix} a & b \\ c & d \end{bmatrix} \mathbf{p}(s) + \frac{1}{s} \begin{bmatrix} f \\ 0 \end{bmatrix}, \tag{9.24}$$

where the rules for Laplace transformation have to be observed (see e.g. the table of transformations in [30], page 915). Equation (9.24) can be easily solved for $\mathbf{p}(s)$.

$$\mathbf{p}(s) = \left[s \begin{pmatrix} 1 & 0 \\ 0 & 1 \end{pmatrix} - \begin{pmatrix} a & b \\ c & d \end{pmatrix} \right]^{-1} \begin{pmatrix} f/s + 1 \\ 0 + 0 \end{pmatrix}, \tag{9.25}$$

where the inital condition $p_1(0) = 1$ has been used. The back transformation using the Heaviside expansion theorem yields $\mathbf{p}(\tau)$. The population of interest is p_2, which has the form

$$p_2(\tau) = A_1 \mathrm{e}^{s_1 \tau} + A_2 \mathrm{e}^{s_2 \tau} + A_3, \tag{9.26}$$

with

$$s_1 = \frac{1}{2} \left(a + d - \sqrt{(a-d)^2 + 4bc} \right),$$

$$s_2 = \frac{1}{2} \left(a + d + \sqrt{(a-d)^2 + 4bc} \right),$$

$$A_1 = +c \frac{1 + f/s_1}{s_1 - s_2}, \qquad A_2 = -c \frac{1 + f/s_2}{s_1 - s_2}, \qquad A_3 = \frac{cf}{s_1 s_2}.$$

Using the fact that $p_2(\infty) = A_3$ and making use of $-A_1/A_3 = (1 + A_2/A_3)$ leads to the important result

$$g^2(\tau) = -\left(1 + \frac{A_2}{A_3}\right) \mathrm{e}^{s_1 \tau} + \frac{A_2}{A_3} \mathrm{e}^{s_2 \tau} + 1. \tag{9.27}$$

This expression can be simplified considerably by exploiting the fact that for a typical molecule

$$\gamma_{21} \geq \gamma_{12} \gg \gamma_{23} \geq \gamma_{31}, \tag{9.28}$$

i.e. the triplet population and relaxation rates are both very small compared with the respective singlet rates. With these relations we can derive the following approximate expressions for the parameters s_1, s_2, and A_2/A_3:

$$s_1 \simeq -(\gamma_{12} + \gamma_{21}),$$

$$s_2 \simeq -\left(\gamma_{31} + \frac{\gamma_{12}\gamma_{23}}{\gamma_{12} + \gamma_{21}} \right),$$

$$\frac{A_2}{A_3} \simeq \frac{\gamma_{12}\gamma_{23}}{\gamma_{31}(\gamma_{12} + \gamma_{21})}. \tag{9.29}$$

Figure 9.14 shows plots of $g^2(\tau)$ according to Eqs. (9.27) and (9.29) for three different excitation powers, i.e. different rates γ_{12}, on a logarithmic timescale. The latter allows us to visualize a broad timescale, ranging from sub-nanosecond to hundreds of microseconds. What is common to all curves is that the intensity correlation function tends to zero for

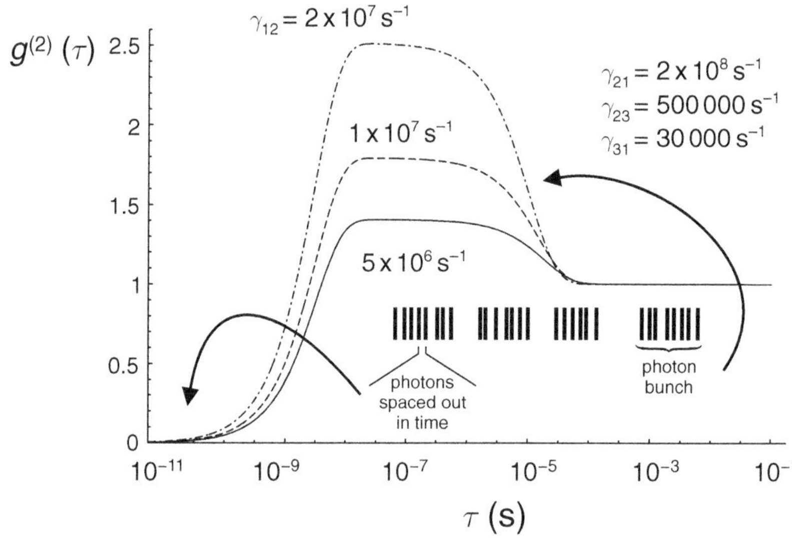

Fig. 9.14 Intensity autocorrelation curves of a three-level system plotted for different excitation rates $\gamma_{12} = 5 \times 10^6\,\text{s}^{-1}$ (solid line), $10 \times 10^6\,\text{s}^{-1}$(dashed line), and $20 \times 10^6\,\text{s}^{-1}$(dash–dotted line) using Eqs. (9.27) and (9.29). Other parameters are $\gamma_{21} = 2 \times 10^8\,\text{s}^{-1}$, $\gamma_{23} = 5 \times 10^6\,\text{s}^{-1}$, and $\gamma_{31} = 3 \times 10^4\,\text{s}^{-1}$. Antibunching is observed for short times, whereas bunching occurs for intermediate times. The inset shows a representation of photon arrivals with bunches of photons separated by dark periods leading to the bunching signature and photons within bunches being spaced out in time leading to the antibunching signature.

short times τ. This antibunching originates from the first term in Eq. (9.27). For small excitation intensities the decay constant s_1 is dominated by the decay rate of the excited state. For longer times, the behavior of $g^2(\tau)$ is characterized by blinking which originates from transitions to the triplet state. Blinking gives rise to photon bunching at intermediate times as illustrated in the inset of Fig. 9.14.

Photon statistics can be experimentally investigated by analyzing time-traces of the emitted intensity. However, to define an intensity it is necessary to bin the detected photons into predefined time intervals. Alternatively, one can use a start–stop configuration that includes two detectors to determine the time differences between consecutively arriving photons (interphoton times) [31]. In the first method, $g^2(\tau)$ is easily calculated from the time-trace. However, only timescales that are larger than the chosen bin-width (typically some microseconds) can be accessed. On the other hand, the start–stop configuration has a time resolution that is limited only by the detector response [29]. A detailed discussion can be found in Ref. [32] and references therein. Figure 9.15 shows an intensity autocorrelation function of a single terylene molecule (see the inset) embedded in a crystalline matrix of p-terphenyl measured using a start–stop configuration. Both antibunching at short times and bunching at longer times can be clearly observed.

The property of a single quantum emitter being able to emit only one photon at a time is of great interest in the field of quantum cryptography, where the polarization state of a single photon defines a qubit. The prominent no-cloning theorem in conjunction with the

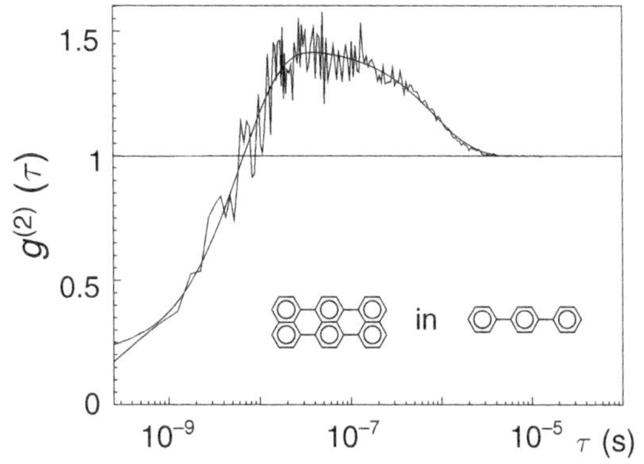

Fig. 9.15 The experimentally obtained second-order autocorrelation function $g^2(\tau)$ of a single molecule (terylene in *p*-terphenyl). Both antibunching and bunching behavior can be observed. The former occurs at short times, whereas the latter is observed for intermediate times at which triplet-state excursions are significant. Adapted from [32].

measurement theorem of quantum mechanics makes it impossible for an eavesdropper to couple a photon out of a stream of single photons without the missing photon being noticed. Single-photon sources can be realized by exciting a two-level system with pulsed laser radiation [33]. It can be shown that the probability of emitting two photons per excitation pulse becomes exceedingly small for pulses that are short compared with the excited-state lifetime of the system (see Problem 9.4).

9.4 Single molecules as probes for localized fields

Besides having interesting statistical properties, a single fluorescent molecule can also serve as a local probe for electric field distributions since they act as point dipoles. For weak excitation intensities ($I \ll I_S$), the fluorescence emission rate (γ_{em}) is nearly independent of the excited-state lifetime and becomes (cf. Eq. (9.3))

$$\gamma_{em} = \frac{1}{2\hbar}\,\mathrm{Im}\{\alpha\}\left|\mathbf{n}_p \cdot \mathbf{E}\right|^2 . \tag{9.30}$$

We assume that the localized excitation field does not bring the dipole approximation into question, i.e. that the field **E** is nearly constant over the size of the quantum system, i.e. on length scales of about 1 nm. For fields that vary more rapidly, higher multipolar transitions must be taken into account.

The absorption dipole of molecules with low symmetry is usually fixed with respect to the molecular framework. Furthermore, if the local environment of the molecule is not changing then the excitation rate is a direct measure of the fluorescence emission rate. Thus, by monitoring γ_{em} as an electric field distribution **E** is raster scanned relative to the rigid molecule, it is possible to map out the projected field strength $|\mathbf{n}_p \cdot \mathbf{E}|^2$. Fluorescent molecules can be fixed in space by embedding them e.g. in a thin transparent polymer film

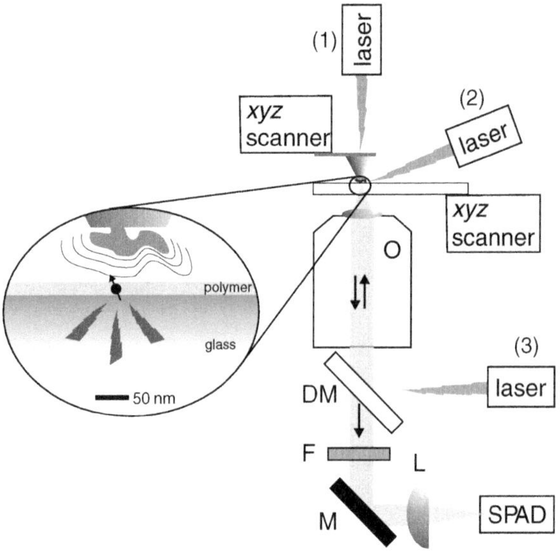

Fig. 9.16 Schematic set-up for metrology of confined fields using single fluorescent molecules with different possible illumination geometries (1), (2), and (3).

on a glass slide. Such films are produced by spin coating of a solution of toluene containing the polymer, e.g. PMMA, and the dye, e.g. DiI, in low concentrations at thicknesses of around 20 nm [34]. The areal density of the dye in the film should be below 1 μm^{-2} in order to avoid clustering of molecules. The molecules will be distributed randomly in the film with respect to depth inside the polymer and orientation of the dipole moment.

Figure 9.16 shows a typical experimental set-up for measuring single-molecule fluorescence with different types of excitation fields, i.e. focused laser radiation and near-field excitation using a local probe. The latter can be self-luminous, as in the case of an aperture probe, or can be externally excited with an irradiating laser beam. The detection path employs a high-NA objective that collects the fluorescence emitted by an excited molecule. Dichroic mirrors and cut-off filters are used to reject the laser excitation line. In essence, the molecule emits as a dipole, and the mapping of fields from object space to image space has been discussed in Chapter 4. However, one needs to take into account that the molecule is emitting not in a homogeneous environment but near an interface. As a consequence (cf. Chapter 10), a randomly oriented molecule emits more than 70% of the emitted photons towards the objective, which increases the collection efficiency. To generate a map of the spatial distribution of $|\mathbf{n}_p \cdot \mathbf{E}|^2$, the single-molecule sample is raster scanned with respect to the fixed excitation field. The emitted fluorescence is continuously recorded with a single-photon detector. The color of each image pixel encodes the respective count-rate. Each molecule in the recorded image is represented by a characteristic pattern that reflects the local field distribution projected along the molecule's dipole axis. Examples of such patterns are shown in Figs. 9.17 and 9.18.

It should be noted that other small particles, such as fluorescent semiconductor quantum dots or small metal particles could also be used to investigate confined fields. However, the

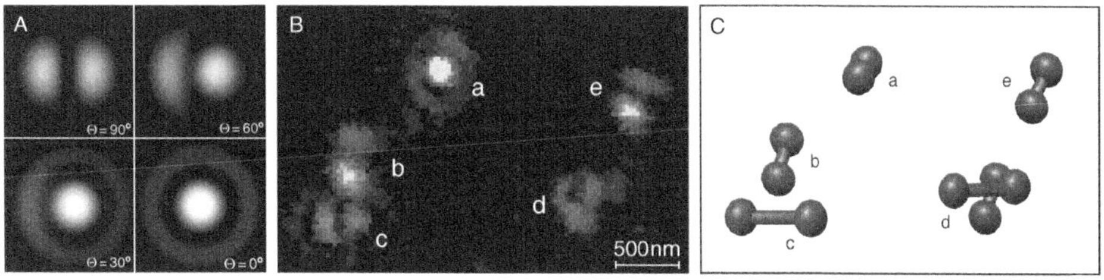

Fig. 9.17 Panel A: calculated fluorescence-rate patterns for a focused, radially polarized excitation beam. The out-of-plane orientation of the probing molecular dipole moment is indicated by the angle Θ ($\Theta = 0$ corresponds to an in-plane molecule, i.e. one oriented perpendicular to the optical axis). Panel B: corresponding experimental patterns obtained with molecules that are randomly oriented in a thin polymer film. Panel C: reconstructed dipole orientations. From [35].

virtue of a well-defined linearly oriented absorption dipole moment is unique to fluorescent molecules. For semiconductor nanocrystals the degenerate dipole moment has to be taken into account.

9.4.1 Field distribution in a laser focus

As an illustration for field mapping we consider the electric field distribution in the focal plane of a strongly focused beam. It represents a confined field that contains field components in all three Cartesian coordinates, i.e. the field in the focus is inhomogeneous as discussed in Chapter 3. For a focused radially polarized beam [35] and an annular (ring-shaped) beam [34] the three field components are of comparable magnitude.

Panel A in Fig. 9.17 shows the calculated fluorescence-rate patterns obtained when a molecule 2 nm below a polymer–air interface is raster scanned through a stationary radially polarized focused beam [35]. The in-plane orientation of the molecular dipole is determined from the orientation of the lobes in the upper-left pattern. The pattern changes as the out-of-plane angle (Θ) of the dipole increases. The lower-right pattern is a map of the longitudinal field component in the focus, which is completely circularly symmetric in the case of a radially polarized beam. In the experiments, the randomly oriented molecules in the polymer film each map a well-defined polarization component in the focus. This results in patterns as displayed in panel B in Fig. 9.17. Knowing the focal field distribution of a radially polarized beam allows us to reconstruct from the experimental patterns in panel B the molecule's dipole orientations (panel C).

A longitudinal field (one with the field vector pointing along the optical axis) can also be generated by a standard fundamental laser beam of which the center of the beam has been blanked out [34]. This type of annular illumination does not alter the general patterns obtained for a strongly focused Gaussian beam (see Chapter 3), but does change the relative intensity of the patterns. Figure 9.18 (left panel) shows calculated fluorescence-rate distributions for molecules close to the surface of the polymer film and as a function of the dipole orientation. Note that the pattern for a molecule oriented in the plane of the film,

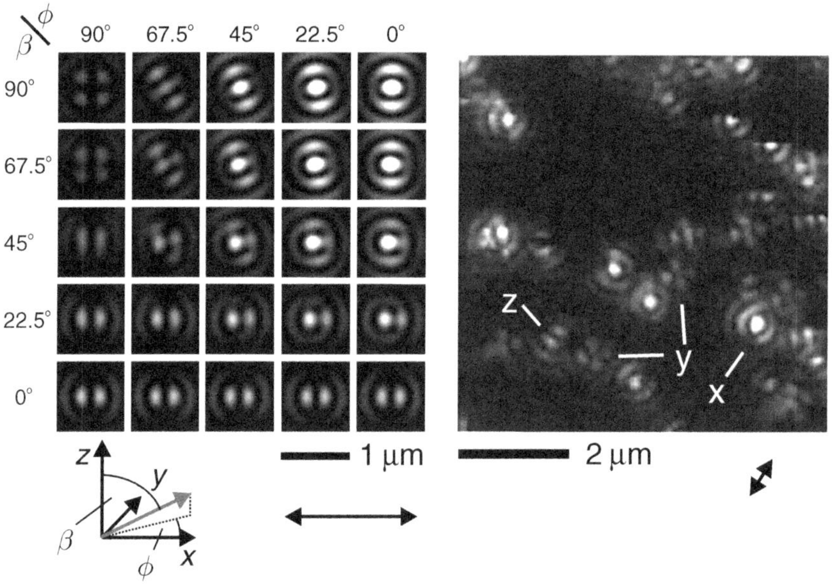

Fig. 9.18 Left panel: calculated fluorescence-rate patterns for a focused annular beam probed by molecules with varying orientations of the dipole moment. Right panel: experimental patterns. Each of these patterns can be assigned to a specific orientation of the molecular absorption dipole. The arrow indicates the polarization direction. Adapted from [34].

perpendicular to the excitation polarization, displays comparable intensity to the pattern of a molecule in the plane of the film with dipole parallel to the excitation polarization. Figure 9.18 (right panel) shows the experimental result. All experimental patterns observed can be assigned to a specific orientation of the molecular absorption dipole.

9.4.2 Probing strongly localized fields

In the previous example, a molecule was used as a probe for the confined field distribution in a laser focus. The same principle can be applied for the characterization of more strongly localized fields. Because of their evanescent nature these fields are bound to material surfaces, which requires that a molecule be brought very close. However, this proximity to the material surface can alter the intrinsic properties of the molecule. For example, the molecule's excited-state lifetime can be altered by the local density of electromagnetic modes, its coupling with other structures can introduce additional relaxation channels (quenching), and strong local fields can even give rise to level shifts similar to the Stark effect. These effects will influence the molecule's fluorescence emission rate. A more detailed discussion of these effects is provided in Chapters 8 and 13. If we assume that the probing molecule is not in direct contact with lossy material surfaces we can, to a first approximation, ignore these perturbing effects. Under this assumption, position-dependent single-molecule fluorescence-rate measurements will qualitatively reflect the vectorial nature of the local field distribution.

Field distribution near subwavelength apertures

The first demonstration of using single molecules to probe localized fields was performed by Betzig and Chichester in 1993 [36]. In their experiments they probed the field distribution near the subwavelength aperture of a near-field probe. Similar experiments were performed by the group of van Hulst [37]. Figure 9.19 shows an electron micrograph of an aperture near-field probe used in such experiments. The end of a tapered metal-coated glass fiber (see also Chapter 6) has been cut with a focused ion beam in order to obtain a flat end-face free of grains and contaminants. As discussed in Chapter 6, the fields near the aperture are excited by coupling laser light into the far end of the fiber. Figure 9.20 shows fluorescence-rate patterns of single $\mathrm{DiIC_{18}}$ molecules that were raster scanned underneath the near-field probe shown in Fig. 9.19. The three images were recorded with different polarizations but represent the same sample area. As predicted by Eq. (9.30), the polarization of the excitation field affects the pattern recorded by a single molecule. The pattern marked by a dashed circle originates from a molecule with dipole pointing along the axis of the near-field probe. It maps the square modulus of the field component along the molecule's dipole axis. The recorded patterns are in qualitative agreement with the predictions of the Bethe–Bouwkamp theory discussed in Chapter 6. According to this theory, the longitudinal field is strongest at the rim of the aperture along the direction of incident polarization. This behavior is nicely supported by the experimental images shown in Fig. 9.20.

Field distribution near tips and particles

Very strong field localization and enhancement can be achieved near sharply pointed metal boundaries. However, because metals are lossy materials at optical frequencies, one can no longer ignore the perturbing influence of the metal on the properties of the molecule. The

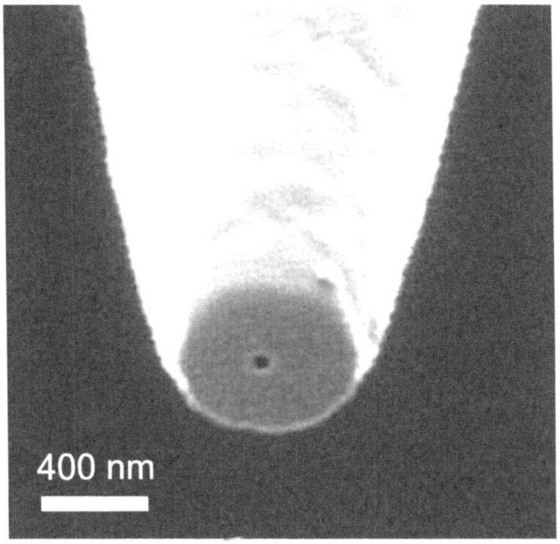

Fig. 9.19 Image of a near-field aperture probe of which the end-face was cut by focused-ion-beam milling. The aperture that shows up in the circular facet has a diameter of 70($\pm$5) nm. The probes have flat end-faces and the apertures have well-defined edges and are circularly symmetric. From [37].

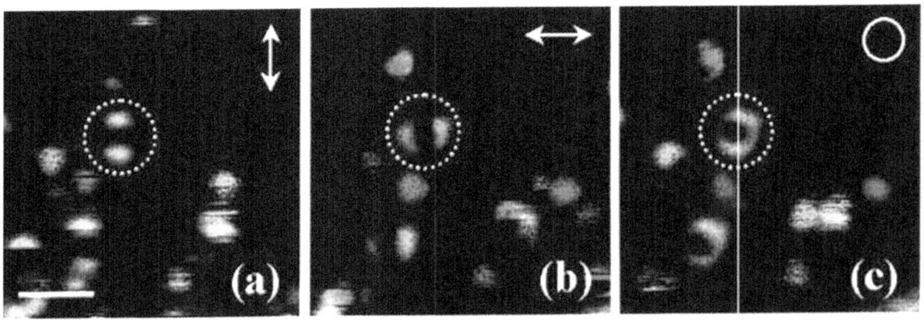

Fig. 9.20 A series of three successive fluorescence maps of the same area (1.2 μm × 1.2 μm) of a sample of single DiIC$_{18}$ molecules embedded in a 10 nm-thin film of PMMA, measured with the 70 nm-aperture probe of Fig. 9.19. The excitation polarization (as measured in the far-field) was changed between linear in the vertical image direction (a), linear in the horizontal image direction (b), and circular (c). The changing polarization affects the molecule's fluorescence-rate pattern. For example, for the molecule in the dashed circle it is oriented perpendicular to the sample plane, i.e. pointing in the direction of the near-field probe. Bar = 300 nm. From [37].

predominant perturbation is fluorescence quenching: an excited molecule can relax to its ground state through non-radiative decay. In this case, the molecule's excitation energy is transferred to the metal, where it is ultimately dissipated to heat. As a consequence, the apparent quantum yield of the molecule is reduced (see also Section 13.3.2).

The example of this section nicely illustrates the competition between enhancement and quenching. We consider a tip-on-aperture near-field probe as discussed in Chapter 6 (cf. Fig. 6.24). In short, a metal tip is grown on the end-face of an aperture-type near-field probe. The light emitted by the aperture illuminates the metal tip and gives rise to a local field enhancement at the tip end. The field distribution that is expected at the tip is that of a vertical dipole at the center of a sphere inscribed into the tip apex as discussed in Chapter 6. The excitation rate of a molecule placed in the vicinity of the tip is, as in previous cases, determined by the projection of the local electric field vector on the absorption dipole axis. Figure 9.21(a) shows the result of an experiment performed by Frey *et al.* [38]. As the illuminated tip is scanned over several molecules attached to the ends of DNA strands deposited on a mica surface, distinct patterns appear, which in most cases consist of two lobes facing each other. A cut through the rotationally symmetric field distribution together with a molecule oriented slightly out of the sample plane is shown in Fig. 9.21(b). The sketch shows that differently oriented molecules are excited in different ways. For example, Fig. 9.21(b) indicates that a molecule with its dipole in the plane of the sample surface will lead to a double-lobed pattern. The direction of the two lobes indicates the orientation of the in-plane component of the absorption dipole moment. If the tip is sitting right above a molecule with its dipole in the sample plane, the excitation is very inefficient and the molecule appears dark. On the other hand, a bright spot is expected for a molecule whose dipole is oriented perpendicular to the sample plane. Experimentally recorded fluorescence patterns from single molecules with various out-of-plane angles are summarized in the upper row of Fig. 9.21(c). Obviously, patterns with a single bright spot are not observed. Instead, vertically oriented molecules appear as a symmetric ring. The reason

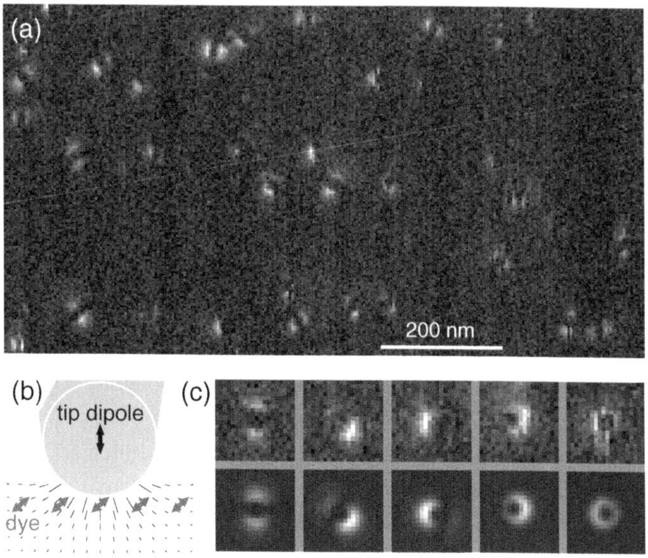

Fig. 9.21 Localized field near a sharp metal tip. (a) Fluorescence-rate patterns of single molecules scanned by a tip-on-aperture probe. (b) Cut through the rotationally symmetric field distribution near a TOA probe sampled at different points by a single fluorescent molecule with an out-of-plane orientation of the absorption dipole moment indicated by the arrows. (c) Comparison of theory with selected experimental patterns (see text). From [38].

for this observation is non-radiative relaxation from the excited state to the ground state. Whenever the molecule is right underneath the tip, fluorescence quenching predominates over the field enhancement, leading to a suppression of the fluorescence. The quenching effect can be included in the calculation of the field patterns by using Eq. (8.141) in Chapter 8, where we analyzed dipole emitters in inhomogeneous environments. In the present consideration, possible changes of the emitter's excited-state lifetime and emission pattern in the presence of the probe have not been treated. Such effects will be covered in the discussion of optical antennas in Chapter 13.

9.5 Conclusion

This chapter discussed the properties of single-quantum emitters such as single molecules, quantum dots, and defect centers in diamond. Because of their small size, these systems are ideal probes for local field distributions and can act as experimental model systems for point dipoles. When a quantum emitter interacts with light, the quantum nature of its internal states gets encoded into the emitted light. Therefore, the quantum emitters discussed in this chapter are promising building blocks for single-photon sources. Depending on the local environment, the intrinsic properties of quantum emitters can change, and hence they can act as local sensors. Single molecules, quantum dots, and defect centers in nano diamond are being used for biophysical studies and also for implementations of quantum logic. Finally, it should be noted that the methods that have been introduced in this chapter can also be applied to other quantum emitters that have not been mentioned here.

Problems

9.1 In a semiconductor, Wannier excitons are the lowest-energy excited states. They form by recombination of an electron–hole pair, e.g. after absorption of a photon with an energy corresponding to the semiconductor's bandgap. The Hamiltonian that describes this excitonic bound state, Eq. (9.2), has the same form as the Hamiltonian of the hydrogen atom. A typical semiconductor used to prepare nanocrystals that emit light in the visible spectral region is CdSe. Its dielectric constant is 10.2, and the effective masses of the electrons and holes are $m_e = 0.12m_0$ and $m_h = 0.5m_0$, respectively, where m_0 is the electron rest mass. Calculate the Bohr radius of the excitons. For nanocrystals smaller than the Bohr radius, quantum confinement effects become important. How does the magnitude of the effective mass influence the Bohr radius?

9.2 The rate of energy dissipation (absorption) by a molecule with dipole moment **p** can be written as $P_{abs}(\omega) = (\omega/2)\mathrm{Im}\big[\mathbf{p}\cdot\mathbf{E}(\omega)\big]$, with **E** being the local exciting field. The dipole moment **p** can be considered to be induced by the same field according to $\mathbf{p} = \overset{\leftrightarrow}{\alpha}\,\mathbf{E}$, where $\overset{\leftrightarrow}{\alpha}$ is the tensorial polarizability of the molecule defined by its dipole orientation. Derive Eqs. (9.3) and (9.4).

9.3 Prove the relations of Eq. (9.16).

9.4 Determine the populations of a two-level system as a function of time for continuous-wave excitation. To simulate the case of pulsed excitation, assume that the excitation pulse has a rectangular shape. Estimate the probability of two photons being emitted due to a single rectangular excitation pulse of a given width. What does the result tell you about the usability of a two-level system as a triggered single-photon source?

9.5 Show that the maximum absorption cross-section of a two-level molecule is $\sigma_{max} = 3\lambda^2/(2\pi)$. Use the expressions for the atomic polarizability α in Appendix A and express α in terms of the absorption cross-section σ. Be careful with orientational factors.

References

[1] L. Mandel and E. Wolf, *Optical Coherence and Quantum Optics*. Cambridge: Cambridge University Press (1995).

[2] C. Cohen-Tannoudji, J. Dupont-Roc, and G. Grynberg, *Atom–Photon Interactions*. New York: Wiley (1998).

[3] A. Yariv, *Quantum Electronics*. New York: Wiley (1975).

[4] R. Loudon, *The Quantum Theory of Light*. Oxford: Oxford University Press (1983).

[5] T. Basché, W. Moerner, M. Orrit, and U. Wild (eds.), *Single-Molecule Optical Detection, Imaging and Spectroscopy*. Weinheim: VCH Verlagsgesellschaft (1997).

[6] R. K. C. Zander and J. Enderlein (eds.), *Single-Molecule Detection in Solution*. Weinheim: Wiley-VCH Verlag (2002).

[7] J. R. Zurita-Sanchez and L. Novotny, "Multipolar interband absorption in a semiconductor quantum dot: I. Electric quadrupole enhancement," *J. Opt. Soc. Am. B* **19**, 1355–1362 (2002).

[8] H. Haken and H. C. Wolf, *Molecular Physics and Elements of Quantum Chemistry*. Hamburg: Springer-Verlag (2004).

[9] Th. Christ, F. Kulzer, P. Bordat, and Th. Basch, "Watching the photooxidation of a single molecule," *Angew. Chem.* **113**, 4323–4326 (2001) and *Angew Chem. Int. Edn. Engl.* **40**, 4192–4195 (2001).

[10] L. E. Brus, "Electron–electron and electron–hole interactions in small semiconductor crystallites: the size dependence of the lowest excited electronic state," *J. Chem. Phys.* **80**, 4403–4409 (1984).

[11] M. Nirmal, D. J. Norris, M. Kuno, *et al.*, "Observation of the 'dark exciton' in CdSe quantum dots," *Phys. Rev. Lett.* **75**, 3728–3731 (1995).

[12] J. Zheng, C. Zhang, and R. M. Dickson, "Highly fluorescent, water-soluble, size-tunable gold quantum dots," *Phys. Rev. Lett.* **93**, 077402-1 (2004).

[13] I. N. Stranski and V. L. Krastanow, *Akad. Wiss. Lit. Mainz Math.-natur. Kl. IIb* **146**, 797–810 (1939).

[14] S. A. Empedocles, R. Neuhauser, and M. G. Bawendi, "Three-dimensional orientation measurements of symmetric single chromophores using polarization microscopy," *Nature* **399**, 126–130 (1999).

[15] F. Koberling, U. Kolb, I. Potapova, *et al.*, "Fluorescence anisotropy and crystal structure of individual semiconductor nanocrystals," *J. Phys. Chem. B* **107**, 7463–7471 (2003).

[16] X. Li, Y. Wu, D. Steel, *et al.*, "An all-optical quantum gate in a semiconductor quantum dot," *Science* **301**, 809–811 (2003).

[17] A. M. Zaitev, "Vibronic spectra of impurity-related centers in diamond," *Phys. Rev. B* **61**, 12909–12922 (2000).

[18] A. M. Zaitev, *Optical Properties of Diamond. A Data Handbook*. Berlin: Springer Verlag (2001).

[19] A. Krüger, *Carbon Materials and Nanotechnology*. Weinheim: Wiley-VCH (2010).

[20] F. Jelezko Image courtesy of Fedor Jelezko and Jörg Wrachtrup.

[21] F. Jelezko and J. Wrachtrup, "Single defect centers in diamond: a review," *Phys. Stat. Sol. (a)* **203**, 3207–3225 (2006).

[22] I. Aharonovich, A. D. Greentree, and S. Prawer, "Diamond photonics," *Nature Photonics* **5**, 397–405 (2011).

[23] C. Bradac, T. Gaebel, N. Naidoo, *et al.*, "Observation and control of blinking nitrogen-vacancy centres in discrete nanodiamonds," *Nature Nanotechnol.* **5**, 345–349 (2010).

[24] J. Wrachtrup and F. Jelezko, "Processing quantum information in diamond," *J. Phys.: Condens. Matter* **18**, S807–S824 (2006).

[25] L. G. Rogers, S. Armstrong, M. J. Sellars, and N. B. Manson, "Infrared emission of the NV centre in diamond: Zeeman and uniaxial stress studies," *New J. Phys.* **10**, 103024 (2008).
[26] E. Rittweger, K. Y. Han, S. E. Irvine, C. Eggeling, and S. W. Hell, "STED microscopy reveals crystal colour centres with nanometric resolution," *Nature Photonics* **3**, 144–147 (2009). Reprinted by permission from Macmillan Publishers Ltd.
[27] C. Bohren and D. Huffman, *Absorption and Scattering of Light by Small Particles*. New York: John Wiley & Sons (1983).
[28] N. Gisin, G. Ribordy, W. Tittel, and H. Zbinden, "Quantum cryptography," *Rev. Mod. Phys.* **74**, 145–195 (2002).
[29] S. Reynaud, *Ann. Phys. (Paris)* **8**, 351 (1983).
[30] G. Arfken and H. Weber, *Mathematical Methods for Physicists*. London: Academic Press (1995).
[31] R. Hanbury-Brown and R. Q. Twiss, "Correlation between photons in two coherent beams of light," *Nature* **177**, 27–29 (1956).
[32] L. Fleury, J. M. Segura, G. Zumofena, B. Hecht, and U. P. Wild, "Nonclassical photon statistics in single-molecule fluorescence at room temperature," *Phys. Rev. Lett.* **84**, 1148–1151 (2000).
[33] B. Lounis and W. E. Moerner, "Single photons on demand from a single molecule at room temperature," *Nature* **407**, 491–493 (2000).
[34] B. Sick, B. Hecht, and L. Novotny, "Orientational imaging of single molecules by annular illumination," *Phys. Rev. Lett.* **85**, 4482–4485 (2000).
[35] L. Novotny, M. Beversluis, K. Youngworth, and T. Brown, "Longitudinal field modes probed by single molecules," *Phys. Rev. Lett.* **86**, 5251–5254 (2001).
[36] E. Betzig and R. Chichester, "Single molecules observed by near-field scanning optical microscopy," *Science* **262**, 1422–1425 (1993).
[37] J. A. Veerman, M. F. García-Parajó, L. Kuipers, and N. F. van Hulst, "Single molecule mapping of the optical field distribution of probes for near-field microscopy," *J. Microsc.* **194**, 477–482 (1999).
[38] H. G. Frey, S. Witt, K. Felderer, and R. Guckenberger, "High-resolution imaging of single fluorescent molecules with the optical near-field of a metal tip," *Phys. Rev. Lett.* **93**, 200801 (2004).

10 Dipole emission near planar interfaces

The problem of dipole radiation in or near planar layered media is of significance to many fields of study. It is encountered in antenna theory, single-molecule spectroscopy, cavity quantum electrodynamics, integrated optics, circuit design (microstrips), and surface-contamination control. The relevant theory was also applied to explain the strongly enhanced Raman effect of adsorbed molecules on noble metal surfaces, and in surface science and electrochemistry for the study of optical properties of molecular systems adsorbed on solid surfaces. Detailed literature on the latter topic is given in Ref. [1]. In the context of nano-optics, dipoles close to a planar interface have been considered by various authors to simulate tiny light sources and small scattering particles [2]. The acoustic analog is also applied to a number of problems such as seismic investigations and ultrasonic detection of defects in materials [3].

In his original paper [4], in 1909, Sommerfeld developed a theory for a radiating dipole oriented vertically above a planar and lossy ground. He found two different asymptotic solutions: space waves (spherical waves) and surface waves. The latter had already been investigated by Zenneck [5]. Sommerfeld concluded that surface waves account for long-distance radio-wave transmission because of their slower radial decay along the Earth's surface compared with that of space waves. Later, when space waves were found to reflect at the ionosphere, the contrary was confirmed. Nevertheless, Sommerfeld's theory formed the basis for all subsequent investigations. In 1911 Hörschelmann [6, 7], a student of Sommerfeld, analyzed the horizontal dipole in his doctoral dissertation and likewise used expansions in cylindrical coordinates. Later, in 1919, Weyl [8] expanded the problem by a superposition of plane and evanescent waves (the angular spectrum representation), and similar approaches were developed by Strutt [9], and Van der Pol and Niessen [10]. Agarwal later used the Weyl representation to extend the theory to quantum electrodynamics [11]. Owing to the overwhelming amount of literature, many aspects of the theory were reinvented over the years, probably owing to the fact that the early literature was written in German. An English version of the early developments is summarized in Sommerfeld's lectures on theoretical physics [12].

At first glance, the calculation of the field of a dipole near planar interfaces seems to be an easy task. The primary dipole field (free-space Green function) possesses a simple mathematical description, and the planar interfaces have reduced dimensionality. Furthermore, the planar interfaces are constant coordinate surfaces for different coordinate systems. It is therefore very astonishing that there is no closed solution for this elementary problem, not even for the vertically oriented dipole which has perfect rotational symmetry. The desired simplicity is obtained only for limiting cases, such as ideally conducting interfaces and the quasi-static limit.

10.1 Allowed and forbidden light

Let us consider the situation shown in Fig. 10.1, where a dipole is located above a layered substrate. We assume that the lower half-space (substrate) is optically denser than the upper half-space (vacuum). If the distance of the dipole from the surface of the topmost layer is less than about one wavelength, *evanescent* field components of the dipole interact with the layered structure and thereby excite other forms of electromagnetic radiation. Their energy can be (1) absorbed by the layer, (2) transformed into propagating waves in the lower half-space, or (3) coupled to modes propagating along the layer. In the second case the plane waves propagate in directions beyond the critical angle of total internal reflection $\alpha_c = \arcsin(n_1/n_3)$, where n_1 and n_3 are the refraction coefficients of the upper and lower half-spaces, respectively. The amplitude of the plane waves depends exponentially on the height of the dipole above the layer. Thus, for dipoles more than a couple of wavelengths from the surface there will be virtually no light coupled into directions beyond the critical angle. This is why the light at supercritical angles is called *forbidden light* [13].

Figure 10.2 illustrates the difference between allowed and forbidden light (cf. Section 2.11.2). Here, we assume that $\varepsilon_3 > \varepsilon_1 > \varepsilon_2$. In configuration (a) a dielectric interface is illuminated by a plane wave incident from the upper medium in such a way that a propagating transmitted wave exists. If a second interface is brought close, the light transmitted into the downmost medium does not depend, apart from interference undulations, on the

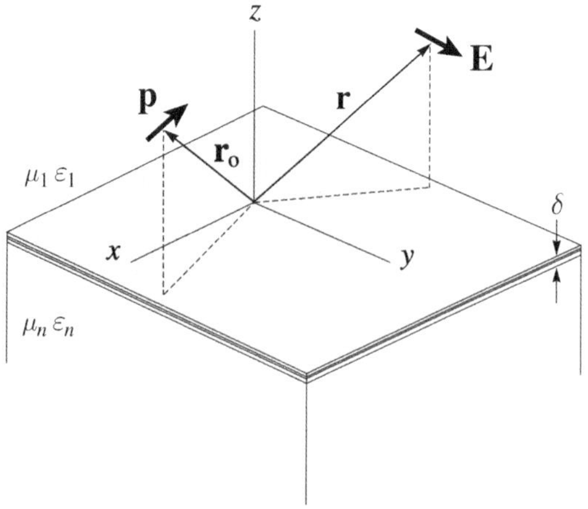

Fig. 10.1 Configuration of the dipole problem. The dipole is located at $\mathbf{r}_0 = (x_0, y_0, z_0)$ and the planar interfaces are characterized by $z = \text{constant}$. The surface of the topmost layer coincides with the coordinate origin. The properties of the upper and lower half-spaces are designated by the indices 1 and n, respectively.

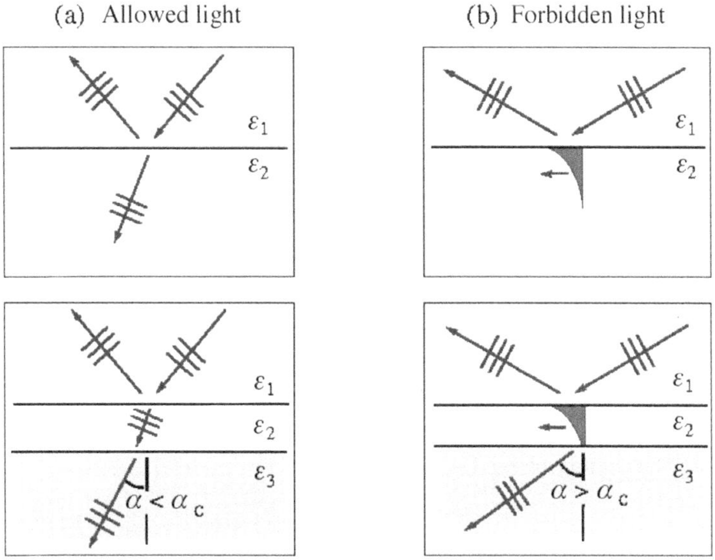

Fig. 10.2 Illustration of allowed and forbidden light. The three media fulfill $\varepsilon_3 > \varepsilon_1 > \varepsilon_2$. The incident wave hits the upper interface in such a way that (a) a transmitted wave exists and (b) the wave is totally reflected.

spacing between the two interfaces, and the transmitted light propagates in a direction that is within the critical angle of total internal reflection. The situation in (b) is quite different from the previous one. Here, the wave hits the upper interface in such a way that no transmitted field exists. Instead, an evanescent wave is formed, decaying exponentially in the normal direction and propagating along the interface. If the second interface is approached, the evanescent wave will be transformed into a propagating wave in the lowest region (optical tunneling). This wave propagates in directions beyond the critical angle of total internal reflection and depends sensitively on the gap between the two interfaces (see also Section 2.14.2).

10.2 Angular spectrum representation of the dyadic Green function

The solution to the problem depicted in Fig. 10.1 has to be expanded by suitable functions satisfying Maxwell's equations. In order to fulfill the boundary conditions analytically, the functions have to be orthogonal on the interfaces. This is true for expansions in both Cartesian and cylindrical coordinates. Both treatments have their advantages and disadvantages, and lead to integrals that cannot be solved analytically. Sommerfeld used expansions in cylindrical waves. This approach is very efficient from a computational point of view since the fields are represented by just a single integral. A detailed account of Sommerfeld's approach can be found in Ref. [2]. Here, we will adopt an expansion in plane and evanescent waves (the angular spectrum representation) because the results are physically more intuitive. Furthermore, with suitable substitutions it is straightforward to transform the results at a later stage from a Cartesian system to a cylindrical system. In order to account for all possible orientations of the dipole we will use the dyadic Green-function formalism outlined earlier in Chapter 2.

Let us first review the dipole fields in a homogeneous, linear, and isotropic medium. In this case, the interfaces in Fig. 10.1 are removed and the entire space is characterized by ε_1 and μ_1. The dyadic Green function $\overleftrightarrow{\mathbf{G}}_0(\mathbf{r},\mathbf{r}_0)$ defines the electric field $\mathbf{E}(\mathbf{r})$ of an electric dipole $\mathbf{p}$ located at $\mathbf{r}_0 = (x_0, y_0, z_0)$ according to

$$\mathbf{E}(\mathbf{r}) = \omega^2\mu_0\mu_1\overleftrightarrow{\mathbf{G}}_0(\mathbf{r},\mathbf{r}_0)\mathbf{p}. \tag{10.1}$$

The material parameters and the oscillation frequency determine the wavenumber k_1 and its longitudinal component k_{z_1}. To represent $\overleftrightarrow{\mathbf{G}}_0$ by an angular spectrum we first consider the vector potential $\mathbf{A}$, which satisfies (cf. Eq. (2.79))

$$\left[\nabla^2 + k_1^2\right]\mathbf{A}(\mathbf{r}) = -\mu_0\mu_1\,\mathbf{j}(\mathbf{r}). \tag{10.2}$$

Here, $\mathbf{j}$ is the current density of the dipole, which reads as

$$\mathbf{j}(\mathbf{r}) = -\mathrm{i}\omega\delta(\mathbf{r}-\mathbf{r}_0)\mathbf{p}. \tag{10.3}$$

Using the definition of the scalar Green function G_0 (cf. Eq. (2.82)) we obtain

$$\mathbf{A}(\mathbf{r}) = \mathbf{p}\frac{k_1^2}{\mathrm{i}\omega\varepsilon_0\varepsilon_1}\frac{\mathrm{e}^{\mathrm{i}k_1|\mathbf{r}-\mathbf{r}_0|}}{4\pi|\mathbf{r}-\mathbf{r}_0|}, \tag{10.4}$$

where we used Eq. (2.84). Notice that the vector potential is polarized in the direction of the dipole moment. We now introduce the Weyl identity defined in Section 2.15.1 and rewrite the vector potential as[1]

$$\mathbf{A}(\mathbf{r}) = \mathbf{p}\frac{k_1^2}{8\pi^2\omega\varepsilon_0\varepsilon_1}\iint\limits_{-\infty}^{\infty}\frac{1}{k_{z_1}}\mathrm{e}^{\mathrm{i}\left[k_x(x-x_0)+k_y(y-y_0)+k_{z_1}|z-z_0|\right]}\,\mathrm{d}k_x\,\mathrm{d}k_y. \tag{10.5}$$

Using $\mathbf{E} = \mathrm{i}\omega[1 + k_1^{-2}\,\nabla\nabla\cdot]\mathbf{A}$ it is straightforward to derive the electric field. Similarly, the magnetic field is calculated using $\mathbf{H} = (\mu_0\mu_1)^{-1}\,\nabla\times\mathbf{A}$. The resulting expression for $\mathbf{E}$ can be compared with Eq. (10.1) which allows us to identify the dyadic Green function as

$$\begin{aligned}\overleftrightarrow{\mathbf{G}}_0(\mathbf{r},\mathbf{r}_0) &= \frac{\mathrm{i}}{8\pi^2}\iint\limits_{-\infty}^{\infty}\overleftrightarrow{\mathbf{M}}\,\mathrm{e}^{\mathrm{i}\left[k_x(x-x_0)+k_y(y-y_0)+k_{z_1}|z-z_0|\right]}\,\mathrm{d}k_x\,\mathrm{d}k_y,\\ \overleftrightarrow{\mathbf{M}} &= \frac{1}{k_1^2k_{z_1}}\begin{bmatrix} k_1^2-k_x^2 & -k_xk_y & \mp k_xk_{z_1}\\ -k_xk_y & k_1^2-k_y^2 & \mp k_yk_{z_1}\\ \mp k_xk_{z_1} & \mp k_yk_{z_1} & k_1^2-k_{z_1}^2\end{bmatrix}.\end{aligned} \tag{10.6}$$

[1] Remember that $k_{z_i} = \sqrt{k_i^2-(k_x^2+k_y^2)}$ with $\mathrm{Im}\{k_{z_i}\} \geq 0$.

Some terms in the matrix $\overleftrightarrow{\mathbf{M}}$ have two different signs. This originates from the absolute value $|z-z_0|$. The upper sign applies for $z>z_0$ and the lower sign for $z<z_0$. Equation (10.6) allows us to express the fields of an arbitrarily oriented dipole in terms of plane waves and evanescent waves.

10.3 Decomposition of the dyadic Green function

In order to apply the Fresnel reflection and transmission coefficients to the dipole fields, it is beneficial to split $\overleftrightarrow{\mathbf{G}}$ into an s-polarized part and a p-polarized part. This decomposition can be accomplished by dividing the matrix $\overleftrightarrow{\mathbf{M}}$ into the two parts

$$\overleftrightarrow{\mathbf{M}}(k_x,k_y) = \overleftrightarrow{\mathbf{M}}^{\mathrm{s}}(k_x,k_y) + \overleftrightarrow{\mathbf{M}}^{\mathrm{p}}(k_x,k_y), \tag{10.7}$$

where we realize that a dipole oriented perpendicular to the planar interfaces in Fig. 10.1 renders a purely p-polarized field. This follows from the fact that the magnetic field of an electric dipole has only an H_ϕ component (cf. Eq. (8.64)), which is parallel to the interfaces for $\mathbf{p}=p\mathbf{n}_z$. Similarly, a magnetic dipole oriented perpendicular to the interfaces leads to a purely s-polarized field. We therefore define the potentials[2]

$$\mathbf{A}^{\mathrm{e}}(\mathbf{r}) = A^{\mathrm{e}}(\mathbf{r})\mathbf{n}_z, \tag{10.8}$$

$$\mathbf{A}^{\mathrm{h}}(\mathbf{r}) = A^{\mathrm{h}}(\mathbf{r})\mathbf{n}_z, \tag{10.9}$$

and relate them to the electric and magnetic fields as

$$\mathbf{E} = \mathrm{i}\omega\left[1+\frac{1}{k_1^2}\nabla\nabla\cdot\right]\mathbf{A}^{\mathrm{e}} - \frac{1}{\varepsilon_0\varepsilon_1}\nabla\times\mathbf{A}^{\mathrm{h}}, \tag{10.10}$$

$$\mathbf{H} = \mathrm{i}\omega\left[1+\frac{1}{k_1^2}\nabla\nabla\cdot\right]\mathbf{A}^{\mathrm{h}} + \frac{1}{\mu_0\mu_1}\nabla\times\mathbf{A}^{\mathrm{e}}. \tag{10.11}$$

Here, $\mathbf{A}^{\mathrm{e}}$ and $\mathbf{A}^{\mathrm{h}}$ render a purely p-polarized field and a purely s-polarized field, respectively. To proceed, we introduce the angular spectrum representation of the potentials A^{e} and A^{h} as

$$A^{\mathrm{e,h}}(x,y,z) = \frac{1}{2\pi}\iint\limits_{-\infty}^{\infty}\hat{A}^{\mathrm{e,h}}(k_x,k_y)\mathrm{e}^{\mathrm{i}\left[k_x(x-x_0)+k_y(y-y_0)+k_{z_1}|z-z_0|\right]}\,\mathrm{d}k_x\,\mathrm{d}k_y, \tag{10.12}$$

and introduce it with Eqs. (10.8) and (10.9) into Eq. (10.10). The resulting expression for the electric field can be compared with the field generated by the dyadic Green function derived in the previous section. This comparison allows us to identify the Fourier spectra $\hat{A}^{\mathrm{e}}$ and $\hat{A}^{\mathrm{h}}$ as

$$\hat{A}^{\mathrm{e}}(k_x,k_y) = \frac{\omega\mu_0\mu_1}{4\pi}\,\frac{\mp\mu_x k_x k_{z_1} \mp \mu_y k_y k_{z_1} + \mu_z(k^2-k_{z_1}^2)}{k_{z_1}(k_x^2+k_y^2)}, \tag{10.13}$$

[2] Note that only $\mathbf{A}^{\mathrm{e}}$ has the units of a vector potential. $\mathbf{A}^{\mathrm{h}}$ is the magnetic analog of the vector potential.

$$\hat{A}^{\mathrm{h}}(k_x, k_y) = \frac{k_1^2}{4\pi} \; \frac{-\mu_x k_y + \mu_y k_x}{k_{z_1}(k_x^2 + k_y^2)}, \tag{10.14}$$

where we used the Cartesian components $\mathbf{p} = (p_x, p_y, p_z)$ for the dipole moment. Finally, by introducing the expressions for $\hat{A}^{\mathrm{e}}$ and $\hat{A}^{\mathrm{h}}$ into Eq. (10.10) and using the definition Eq. (10.1), the s-polarized and p-polarized parts of the dyadic Green function can be determined. The decomposition of the matrix $\overleftrightarrow{\mathbf{M}}$ turns out to be

$$\begin{aligned}
\overleftrightarrow{\mathbf{M}}^{\mathrm{s}} &= \frac{1}{k_{z_1}(k_x^2 + k_y^2)} \begin{bmatrix} k_y^2 & -k_x k_y & 0 \\ -k_x k_y & k_x^2 & 0 \\ 0 & 0 & 0 \end{bmatrix}, \\
\overleftrightarrow{\mathbf{M}}^{\mathrm{p}} &= \frac{1}{k_1^2(k_x^2+k_y^2)} \begin{bmatrix} k_x^2 k_{z_1} & k_x k_y k_{z_1} & \mp k_x(k_x^2 + k_y^2) \\ k_x k_y k_{z_1} & k_y^2 k_{z_1} & \mp k_y(k_x^2 + k_y^2) \\ \mp k_x(k_x^2 + k_y^2) & \mp k_y(k_x^2 + k_y^2) & (k_x^2 + k_y^2)^2/k_{z_1} \end{bmatrix}.
\end{aligned} \tag{10.15}$$

10.4 Dyadic Green functions for the reflected and transmitted fields

Let us assume that the dipole whose primary field is represented by $\overleftrightarrow{\mathbf{G}}_0$ is located above a planar layered interface as shown in Fig. 10.1. We choose a coordinate system with its origin on the topmost interface. Then, the z-coordinate of the dipole (z_0) denotes the height of the dipole above the layered medium. To calculate the dipole's reflected field we simply multiply the individual plane waves in $\overleftrightarrow{\mathbf{G}}$ by the corresponding (generalized) Fresnel reflection coefficients r^{s} and r^{p}. These coefficients are easily expressed as functions of (k_x, k_y) (cf. Eqs. (2.51) and (2.52)). For the reflected field we obtain the new dyadic Green function

$$\begin{aligned}
\overleftrightarrow{\mathbf{G}}_{\mathrm{ref}}(\mathbf{r}, \mathbf{r}_0) &= \frac{\mathrm{i}}{8\pi^2} \iint_{-\infty}^{\infty} \left[\overleftrightarrow{\mathbf{M}}^{\mathrm{s}}_{\mathrm{ref}} + \overleftrightarrow{\mathbf{M}}^{\mathrm{p}}_{\mathrm{ref}} \right] \mathrm{e}^{\mathrm{i}\left[k_x(x-x_0)+k_y(y-y_0)+k_{z_1}(z+z_0)\right]} \, \mathrm{d}k_x \, \mathrm{d}k_y, \\
\overleftrightarrow{\mathbf{M}}^{\mathrm{s}}_{\mathrm{ref}} &= \frac{r^{\mathrm{s}}(k_x, k_y)}{k_{z_1}(k_x^2 + k_y^2)} \begin{bmatrix} k_y^2 & -k_x k_y & 0 \\ -k_x k_y & k_x^2 & 0 \\ 0 & 0 & 0 \end{bmatrix}, \\
\overleftrightarrow{\mathbf{M}}^{\mathrm{p}}_{\mathrm{ref}} &= \frac{-r^{\mathrm{p}}(k_x, k_y)}{k_1^2(k_x^2 + k_y^2)} \begin{bmatrix} k_x^2 k_{z_1} & k_x k_y k_{z_1} & k_x(k_x^2 + k_y^2) \\ k_x k_y k_{z_1} & k_y^2 k_{z_1} & k_y(k_x^2 + k_y^2) \\ -k_x(k_x^2 + k_y^2) & -k_y(k_x^2 + k_y^2) & -(k_x^2 + k_y^2)^2/k_{z_1} \end{bmatrix}.
\end{aligned} \tag{10.16}$$

The electric field in the upper half-space is now calculated by taking the sum of the primary Green function and the reflected Green function as

$$\mathbf{E}(\mathbf{r}) = \omega^2\mu_0\mu_1\left[\overleftrightarrow{\mathbf{G}}_0(\mathbf{r},\mathbf{r}_0) + \overleftrightarrow{\mathbf{G}}_{\mathrm{ref}}(\mathbf{r},\mathbf{r}_0)\right]\mathbf{p}. \tag{10.17}$$

The sum of $\overleftrightarrow{\mathbf{G}}$ and $\overleftrightarrow{\mathbf{G}}_{\mathrm{ref}}$ can be regarded as the new Green function of the upper half-space.

The transmitted field can be expressed in terms of the Fresnel transmission coefficients t^{s} and t^{p} (cf. Eqs. (2.51) and (2.52)). For the lower half-space we obtain

$$\begin{aligned}
\overleftrightarrow{\mathbf{G}}_{\mathrm{tr}}(\mathbf{r},\mathbf{r}_0) &= \frac{\mathrm{i}}{8\pi^2}\iint\limits_{-\infty}^{\infty}\left[\overleftrightarrow{\mathbf{M}}^{\mathrm{s}}_{\mathrm{tr}} + \overleftrightarrow{\mathbf{M}}^{\mathrm{p}}_{\mathrm{tr}}\right]\mathrm{e}^{\mathrm{i}\left[k_x(x-x_0)+k_y(y-y_0)-k_{z_n}(z+\delta)+k_{z_1}z_0\right]}\,\mathrm{d}k_x\,\mathrm{d}k_y,\\
\overleftrightarrow{\mathbf{M}}^{\mathrm{s}}_{\mathrm{tr}} &= \frac{t^{\mathrm{s}}(k_x,k_y)}{k_{z_1}(k_x^2+k_y^2)}\begin{bmatrix} k_y^2 & -k_xk_y & 0\\ -k_xk_y & k_x^2 & 0\\ 0 & 0 & 0\end{bmatrix},\\
\overleftrightarrow{\mathbf{M}}^{\mathrm{p}}_{\mathrm{tr}} &= \frac{t^{\mathrm{p}}(k_x,k_y)}{k_1k_n(k_x^2+k_y^2)}\begin{bmatrix} k_x^2k_{z_n} & k_xk_yk_{z_n} & k_x(k_x^2+k_y^2)k_{z_n}/k_{z_1}\\ k_xk_yk_{z_n} & k_y^2k_{z_n} & k_y(k_x^2+k_y^2)k_{z_n}/k_{z_1}\\ k_x(k_x^2+k_y^2) & k_y(k_x^2+k_y^2) & (k_x^2+k_y^2)^2/k_{z_1}\end{bmatrix}.
\end{aligned} \tag{10.18}$$

The parameter δ denotes the total height of the layered interface. In the case of a single interface, $\delta = 0$. The electric field in the lower half-space is calculated as

$$\mathbf{E}(\mathbf{r}) = \omega^2\mu_0\mu_1\overleftrightarrow{\mathbf{G}}_{\mathrm{tr}}(\mathbf{r},\mathbf{r}_0)\mathbf{p}. \tag{10.19}$$

The function $\overleftrightarrow{\mathbf{G}}_{\mathrm{tr}}$ can be regarded as the new Green function of the lower half-space.

The calculation of the fields inside the layered structure requires the explicit solution of the boundary conditions at the interfaces. This has been done in Ref. [2] for a two-interface structure (a planar layer on top of a planar substrate) and explicit expressions for the field components can be found in Appendix D. The discussion in this chapter does not require knowledge of the fields inside the individual layers. However, to calculate the fields in the upper and lower half-spaces we need to know the generalized Fresnel reflection and transmission coefficients. For a single interface, these coefficients have been stated in Eqs. (2.51) and (2.52) and the generalization to multiple interfaces can be found in Ref. [14]. As an example, the reflection and transmission coefficients of a single layer of thickness d read as

$$r^{(\mathrm{p,s})} = \frac{r_{1,2}^{(\mathrm{p,s})} + r_{2,3}^{(\mathrm{p,s})}\exp(2\mathrm{i}k_{z2}d)}{1 + r_{1,2}^{(\mathrm{p,s})}r_{2,3}^{(\mathrm{p,s})}\exp(2\mathrm{i}k_{z2}d)}, \tag{10.20}$$

$$t^{(\mathrm{p,s})} = \frac{t_{1,2}^{(\mathrm{p,s})}t_{2,3}^{(\mathrm{p,s})}\exp(\mathrm{i}k_{z2}d)}{1 + r_{1,2}^{(\mathrm{p,s})}r_{2,3}^{(\mathrm{p,s})}\exp(2\mathrm{i}k_{z2}d)}, \tag{10.21}$$

where $r_{i,j}^{(\mathrm{p,s})}$ and $t_{i,j}^{(\mathrm{p,s})}$ are the reflection and transmission coefficients for the single interface (i, j).

In order to calculate the fields in the upper and lower half-spaces it is beneficial to transform the expressions for the fields into a cylindrical system. By using the mathematical identities in Eq. (3.57) it is possible to express the fields in terms of a single integral in k_ρ. The magnetic field can be derived by applying Maxwell's equation $\mathrm{i}\omega\mu_0\mu_i\mathbf{H} = \nabla\times\mathbf{E}$, which directly leads to

$$\mathbf{H}(\mathbf{r}) = \begin{cases} -\mathrm{i}\omega\left[\nabla\times(\overleftrightarrow{\mathbf{G}} + \overleftrightarrow{\mathbf{G}}_{\mathrm{ref}})\right]\mathbf{p} & \text{upper half-space,} \\ -\mathrm{i}\omega(\mu_1/\mu_n)\left[\nabla\times\overleftrightarrow{\mathbf{G}}_{\mathrm{tr}}\right]\mathbf{p} & \text{lower half-space.} \end{cases} \tag{10.22}$$

Here, the curl operator acts separately on each column vector of the dyadic Green functions.

As an example, Fig. 10.3 shows the field distribution of a dipole in close proximity to a slab waveguide. The dipole is oriented at $\theta = 60°$ in the (x, z) plane, i.e. $\mathbf{p} = p\,(\sqrt{3}/2, 0, 1/2)$,

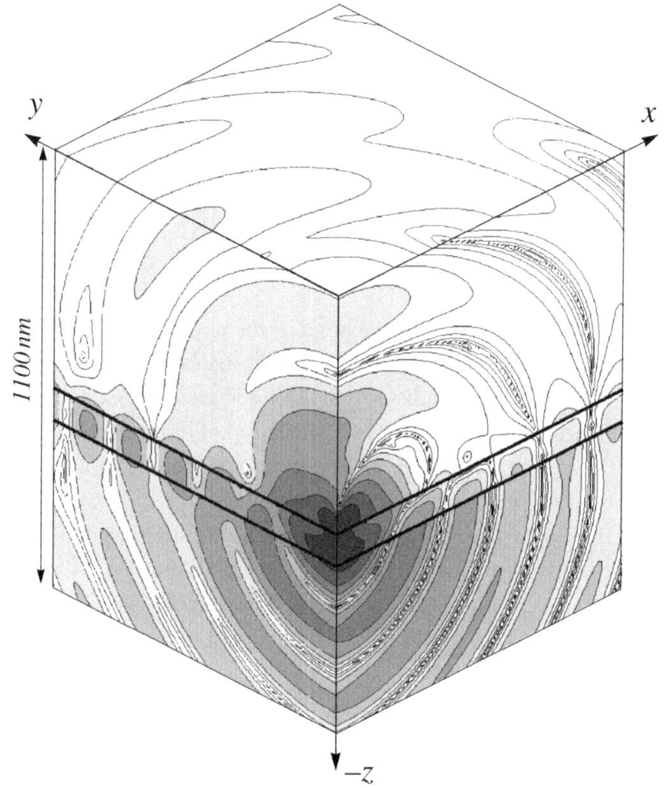

Fig. 10.3 The power density of a dipole above a slab waveguide depicted at a certain time. The dipole is located at $h = 20$ nm and its axis is in the (x, z) plane. $\theta = 60°$, $\lambda = 488$ nm, $d = 80$ nm, $\varepsilon_1 = 1$, $\varepsilon_2 = 5$, $\varepsilon_3 = 2.25$. There is a factor of 2 difference between successive contour lines.

and radiates predominantly into the lower, optically denser medium. The dipole's near-field excites the two lowest modes, TE_0 and TM_0, in the waveguide.

10.5 Spontaneous decay rates near planar interfaces

The normalized rate of energy dissipation P/P_0 of a radiating dipole is defined by Eq. (8.80). Usually, not all of the dipole's energy is transformed into radiation since it can be coupled to other modes supported by the layered structure (phonons, heat, surface modes, waveguide modes, etc.). For an incoherently decaying quantum system with intrinsic quantum yield $q_{\rm i} = 1$, the normalized spontaneous decay rate γ/γ_0 is identical with P/P_0 (cf. Eq. (8.141)) and requires the evaluation of the scattered field $\mathbf{E}_{\rm s}(\mathbf{r}_0)$ at the dipole's origin $\mathbf{r}_0$. In the present situation the scattered field corresponds to the reflected field $\mathbf{E}_{\rm ref}$, which, at its origin, reads as

$$\mathbf{E}_{\rm ref}(\mathbf{r}_0) = \omega^2\mu_0\mu_1\overleftrightarrow{\mathbf{G}}_{\rm ref}(\mathbf{r}_0,\mathbf{r}_0)\mathbf{p}. \tag{10.23}$$

$\overleftrightarrow{\mathbf{G}}_{\rm ref}$ is defined by Eq. (10.16). It is convenient to perform the substitutions

$$k_x = k_\rho\cos\phi, \qquad k_y = k_\rho\sin\phi, \qquad {\rm d}k_x\,{\rm d}k_y = k_\rho\,{\rm d}k_\rho\,{\rm d}\phi, \tag{10.24}$$

which allow us to solve the integral over ϕ analytically.[3] Evaluated at its origin, $\overleftrightarrow{\mathbf{G}}_{\rm ref}$ takes on the diagonal form

$$\overleftrightarrow{\mathbf{G}}_{\rm ref}(\mathbf{r}_0,\mathbf{r}_0) = \frac{\rm i}{8\pi k_1^2}\int_0^\infty \frac{k_\rho}{k_{z_1}}\begin{bmatrix} k_1^2 r^{\rm s} - k_{z_1}^2 r^{\rm p} & 0 & 0 \\ 0 & k_1^2 r^{\rm s} - k_{z_1}^2 r^{\rm p} & 0 \\ 0 & 0 & 2k_\rho^2 r^{\rm p}\end{bmatrix} {\rm e}^{2{\rm i}k_{z_1}z_0}\,{\rm d}k_\rho. \tag{10.25}$$

Using this together with Eq. (10.23) and Eq. (8.80), it is now straightfoward to determine the normalized rate of energy dissipation. For convenience, we perform the substitutions $s = k_\rho/k_1$ and $\sqrt{1-s^2} = k_{z_1}/k_1$. Then, using the abbreviation $s_z = (1-s^2)^{1/2}$, we obtain

$$\frac{P}{P_0} = 1 + \frac{p_x^2+p_y^2}{|\mathbf{p}|^2}\frac{3}{4}\int_0^\infty {\rm Re}\left\{\frac{s}{s_z}\left[r^{\rm s} - s_z^2 r^{\rm p}\right]{\rm e}^{2{\rm i}k_1 z_0 s_z}\right\}{\rm d}s$$
$$+\frac{p_z^2}{|\mathbf{p}|^2}\frac{3}{2}\int_0^\infty {\rm Re}\left\{\frac{s^3}{s_z}r^{\rm p}{\rm e}^{2{\rm i}k_1 z_0 s_z}\right\}{\rm d}s. \tag{10.26}$$

Here, the reflection coefficients are functions of the variable s, i.e. $r^{\rm s}(s)$ and $r^{\rm p}(s)$, and the dipole moment has been written in terms of its Cartesian components as $\mathbf{p} = (p_x, p_y, p_z)$. The integration range $[0\ldots\infty]$ can be divided into the two intervals $[0\ldots 1]$ and $[1\ldots\infty]$. The first interval is associated with the plane waves of the angular spectrum,

[3] Notice the difference from Eq. (3.46), which was arrived at by transforming a planar surface to a spherical surface. Here, the integration is fixed to a planar surface.

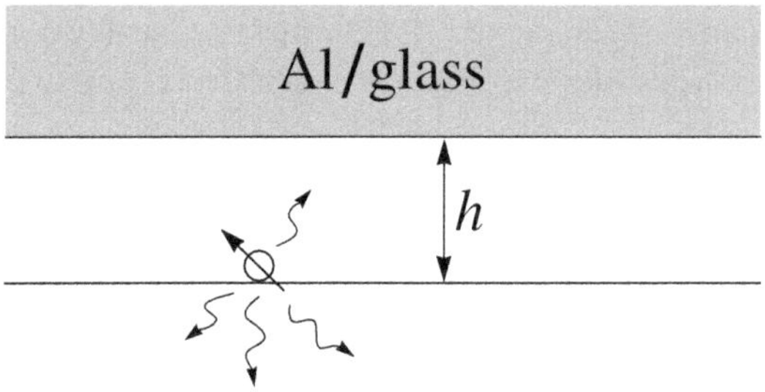

Fig. 10.4 Single-molecule fluorescence near planar interfaces. The molecule is located on the surface of a dielectric substrate and a metal ($\varepsilon = -34.5 + i8.5$) or a glass ($\varepsilon = 2.25$) interface is advanced from above. The applied wavelength is $\lambda = 488$ nm.

i.e. $k_\rho = [0 \ldots k_1]$, whereas the second interval corresponds to the spectrum of evanescent waves $k_\rho = [k_1 \ldots \infty]$. Thus, the dipole interacts with its own reflected plane waves and reflected evanescent waves. The exponential term in the integrands is an exponentially decaying function for evanescent waves, whereas it is oscillatory for plane waves.

According to Eq. (8.116) the normalized rate of energy dissipation is identical with the normalized spontaneous decay rate of a quantum-mechanical two-level system such as a molecule. The normalized lifetime $\tau/\tau_0 = (P/P_0)^{-1}$ of the molecule as a function of the separation h between a substrate and an approaching interface (see Fig. 10.4) is shown in Fig. 10.5. The normalization (τ_0) refers to the situation for which the molecule is located on the glass surface, but the second interface is absent ($h \to \infty$).

The undulations originate from the interference between the propagating fields (plane waves) of the molecule and the reflected fields from the approaching interface. As expected, the undulations are more emphasized for the metal interface and for horizontal dipole orientation. At small h, it can be observed that molecular lifetimes for all configurations decrease. This reduction is caused by the increasing non-radiative decay rate mediated by evanescent field components. Depending on whether the approaching interface is metallic or dielectric, the evanescent field components of the molecule are thermally dissipated or partly converted into fields propagating at supercritical angles in the upper half-space [15]. For the metal interface the lifetime tends to zero [16] as $h \to 0$. In this case, the molecule transfers its excitation energy to the metal and there is no apparent radiation. As a consequence, the fluorescence is quenched.

Figures 10.5(b) and (d) depict the lifetimes for $h < 20$ nm, the distances relevant for near-field optical experiments. For vertically oriented dipoles the lifetimes are always larger in the case of the dielectric interface. This is not so for the horizontal dipole orientation, for which case the two curves intersect. Above $h \approx 8.3$ nm the lifetime of an excited molecule faced by an aluminum interface is higher than in the case of a dielectric interface, but it is lower for separations below $h \approx 8.3$ nm. This lifetime reversal can be transferred to the experimental situation in aperture scanning near-field optical microscopy: a molecule at the center position of the optical probe is faced by the dielectric core, which can be approximated by a planar dielectric interface. For positions below the metal cladding, the situation corresponds to a molecule faced by a planar aluminum interface. Thus, for small probe–sample separations, the lifetime of a molecule with horizontal dipole axis is higher

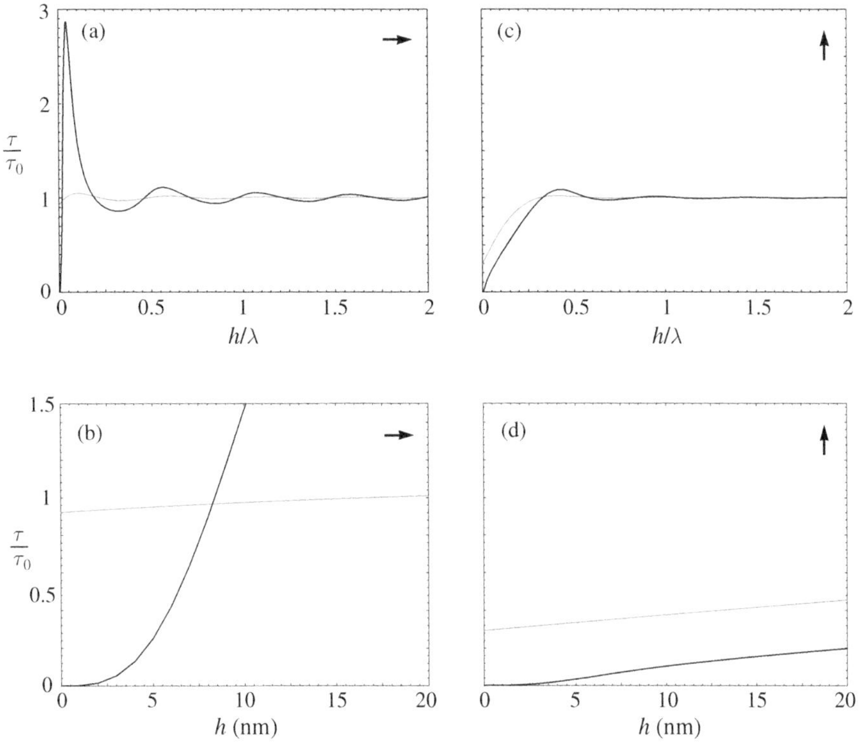

Fig. 10.5 Molecular lifetime as a function of the gap h. The dark curves were obtained for an approaching metal interface, whereas the bright curves refer to an approaching dielectric interface. The arrows indicate the orientation of the dipole axes. The lower figures are close-ups of the upper ones. The normalization with τ_0 corresponds to $h \to \infty$.

at the center position than it is at displaced positions. The contrary is valid for gaps larger than $\approx$8.3 nm. These findings verify experimental observations [17] and reproduce the numerical results reported in Ref. [18].

The point h at which the curves intersect depends on the wavelength of the illuminating light and on the orientation of the dipole axis. For longer wavelengths, aluminum behaves in a more metallic manner, which shifts the intersection point to larger h. At $\lambda = 800$ nm, the dielectric constant of aluminum is $\varepsilon = -63.5 + i47.3$ and the intersection point appears at $h \approx 14.6$ nm.

If a molecule is faced by a finite-sized object, the lateral symmetry is lost and additional effects will occur at the rims of the objects [19, 20].

10.6 Far-fields

In many situations dipoles near a planar layered interface are observed in the far-field zone. To understand how the fields are mapped from the near-field to the far-field we need to derive explicit expressions for the asymptotic fields. The radiation condition usually requires that these fields decay as r^{-1}. However, defining fields at an infinite distance from an infinitely extended object turns out to be a philosophical problem. Furthermore, the existence of closed asymptotic expressions is questionable for reasons of energy conservation: fields that propagate along the layered structure, i.e. guided or surface waves, have to decay as $r^{-1/2}$. There should be a smooth transition between the r^{-1} zone and the $r^{-1/2}$ zone. Thus, it could be concluded that no closed far-field expressions exist for planar media since the decay of the fields depends on the direction of propagation. Nevertheless, closed expressions for the far-field can be derived if the lateral directions, i.e. the regions very close to the layers, are excluded.

One of the advantages of using the angular spectrum representation is the simple and straightforward derivation of the far-field. We learned in Section 3.4 that the far-field $\mathbf{E}_\infty$ observed in the direction of the dimensionless unit vector

$$\mathbf{s} = (s_x, s_y, s_z) = \left(\frac{x}{r}, \frac{y}{r}, \frac{z}{r}\right) \tag{10.27}$$

is determined by the Fourier spectrum $\hat{\mathbf{E}}$ at $z=0$ as

$$\mathbf{E}_\infty(s_x, s_y) = -\mathrm{i}ks_z\hat{\mathbf{E}}(ks_x, ks_y;\, 0)\frac{\mathrm{e}^{ikr}}{r}. \tag{10.28}$$

This equation requires that we express the wavevector $\mathbf{k}$ in terms of the unit vector $\mathbf{s}$. Since we have different optical properties in the upper and lower half-spaces we use the following definitions:

$$\mathbf{s} = \begin{cases} \left(\dfrac{k_x}{k_1}, \dfrac{k_y}{k_1}, \dfrac{k_{z_1}}{k_1}\right) & z > 0, \\[2ex] \left(\dfrac{k_x}{k_n}, \dfrac{k_y}{k_n}, \dfrac{k_{z_n}}{k_n}\right) & z < 0. \end{cases} \tag{10.29}$$

The field $\mathbf{E}$ in the upper and lower half-spaces is determined by the Green functions $\overleftrightarrow{\mathbf{G}}_0$, $\overleftrightarrow{\mathbf{G}}_{\mathrm{ref}}$, and $\overleftrightarrow{\mathbf{G}}_{\mathrm{tr}}$, which are already in the form of an angular spectrum (Eqs. (10.6), (10.16) and (10.18)). We can establish the asymptotic far-field forms of the different Green functions by using the recipe of Eq. (10.28). All that needs to be done is to identify the spatial Fourier spectrum of the Green functions and carry out the algebra. The resulting expressions are given in Appendix D.

In order to have a simple representation of the far-field we choose the origin of the coordinate system on the surface of the uppermost layer such that the dipole is located on the z-axis, i.e.

$$(x_0, y_0) = (0, 0). \tag{10.30}$$

Furthermore, we represent the field in terms of spherical vector coordinates $\mathbf{E} = (E_r, E_\theta, E_\phi)$ by using the spherical angles θ and ϕ (Fig. 10.6). It is important to use the correct signs in the substitutions: in the upper half-space we have $s_z = k_{z_1}/k_1 = \cos\theta$, whereas in the lower half-space the relationship is $s_z = k_{z_n}/k_n = -\cos\theta$. For simpler notation it is convenient to define

$$\tilde{s}_z = \frac{k_{z_1}}{k_n} = \sqrt{(n_1/n_n)^2 - (s_x^2 + s_y^2)} = \sqrt{(n_1/n_n)^2 - \sin^2\theta}, \tag{10.31}$$

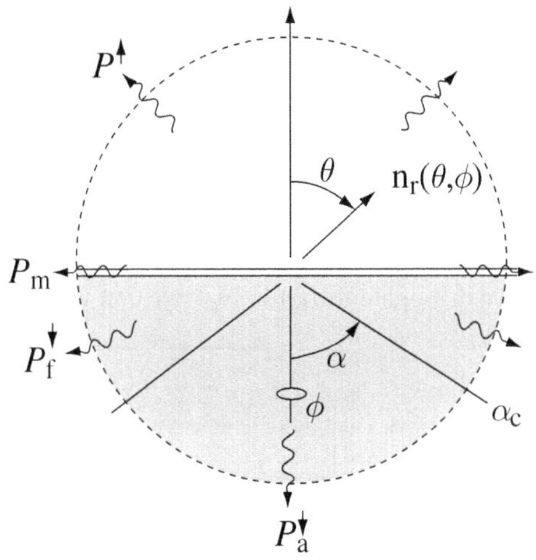

Fig. 10.6 Definition of angles used for the asymptotic far-fields. The radiated power is split into the contributions $P^{\uparrow}$ (radiation into the upper half-space), $P_a^{\downarrow}$ (radiation into the allowed zone), $P_f^{\downarrow}$ (radiation into the forbidden zone), and P_m (radiation dissipated in the layered medium). The total rate of energy dissipation is $P = P^{\uparrow} + P_a^{\downarrow} + P_f^{\downarrow} + P_m + P_i$, with P_i being the intrinsically dissipated power.

where n_1 and n_n are the refractive indices of the upper and lower half-spaces, respectively. Using the index $j \in [1, n]$ to distinguish between upper and lower half-spaces, the far-field can be represented as

$$\mathbf{E} = \begin{bmatrix} E_\theta \\ E_\phi \end{bmatrix} = \frac{k_1^2}{4\pi\varepsilon_0\varepsilon_1} \frac{\exp(ik_j r)}{r} \begin{bmatrix} [\mu_x \cos\phi + \mu_y \sin\phi]\cos\theta\, \Phi_j^{(2)} - \mu_z \sin\theta\, \Phi_j^{(1)} \\ -[\mu_x \sin\phi - \mu_y \cos\phi]\Phi_j^{(3)} \end{bmatrix}, \tag{10.32}$$

with

$$\Phi_1^{(1)} = \left[e^{-ik_1 z_0 \cos\theta} + r^p(\theta) e^{ik_1 z_0 \cos\theta} \right], \tag{10.33}$$

$$\Phi_1^{(2)} = \left[e^{-ik_1 z_0 \cos\theta} - r^p(\theta) e^{ik_1 z_0 \cos\theta} \right], \tag{10.34}$$

$$\Phi_1^{(3)} = \left[e^{-ik_1 z_0 \cos\theta} + r^s(\theta) e^{ik_1 z_0 \cos\theta} \right], \tag{10.35}$$

$$\Phi_n^{(1)} = \frac{n_n \cos\theta}{n_1\, \tilde{s}_z(\theta)} t^p(\theta) e^{ik_n [z_0 \tilde{s}_z(\theta) + \delta\cos\theta]}, \tag{10.36}$$

$$\Phi_n^{(2)} = -\frac{n_n}{n_1} t^p(\theta) e^{ik_n [z_0 \tilde{s}_z(\theta) + \delta\cos\theta]}, \tag{10.37}$$

$$\Phi_n^{(3)} = \frac{\cos\theta}{\tilde{s}_z(\theta)} t^s(\theta) e^{ik_n [z_0 \tilde{s}_z(\theta) + \delta\cos\theta]}. \tag{10.38}$$

A vertically oriented dipole is described by the potential $\Phi_j^{(1)}$, whereas a horizontal dipole is represented by $\Phi_j^{(2)}$ and $\Phi_j^{(3)}$ containing the amounts of p-polarized and s-polarized light, respectively. Let us first discuss the far-fields in the upper half-space. To understand the potentials $\Phi_1^{(1)}$–$\Phi_1^{(3)}$ we analyze the far-field of a dipole in a homogeneous medium. We displace the dipole from the coordinate origin by a distance z_0 along the z-axis. According to Eq. (8.63) the electric field in the far-zone is defined by the term $\exp(ik_1 R)/R$. However, the radial coordinate R is measured from the origin of the dipole, not from the coordinate origin. If we designate the latter by r we can write

$$R = r\sqrt{1 + \frac{z_0^2 - 2z_0 r\cos\theta}{r^2}} \approx r - z_0\cos\theta. \tag{10.39}$$

Only the first two terms in the series expansion of the square root have been retained. It is important to include the second term in the phase of the wave in order to account for diffraction. On the other hand, the second term is meaningless for the amplitude since $r \gg z_0$. Thus, we can write

$$\frac{e^{ik_1 R}}{R} = \frac{e^{ik_1 r}}{r} e^{-ik_1 z_0 \cos\theta}, \tag{10.40}$$

which is known as the Fraunhofer approximation. By comparison we find that the first term in the potentials $\Phi_1^{(1)}$–$\Phi_1^{(3)}$ corresponds to direct dipole radiation. The exponential factor of the second term has a minus sign in the exponent. Therefore, the second term can be identified as radiation from a dipole located a distance z_0 beneath the top surface of the layered medium. The magnitude of this image dipole is weighted by the Fresnel reflection coefficients. This is a remarkable result: in the far-field, a dipole near a layered medium radiates as the superposition of two dipole fields, namely its own field and the field of its image dipole.

The expressions for the transmitted far-field are more complicated. This arises through the term $\tilde{s}_z$ defined in Eq. (10.31). Depending on the optical properties of upper and lower half-spaces, this term can be either real or imaginary. In fact, in many cases the lower half-space (substrate) is optically denser than the upper one. In these situations $\tilde{s}_z$ becomes imaginary for the angular range $\theta = [\pi/2 \ldots \arcsin(n_1/n_n)]$, which exactly corresponds to the forbidden zone discussed before. In the forbidden zone, the exponential factor in the potentials $\Phi_n^{(1)} - \Phi_n^{(3)}$ becomes an exponentially decaying function. Therefore, for separations $z_0 \gg \lambda$ there is no light coupled into the forbidden zone. On the other hand, in the angular range $\theta = [\arcsin(n_1/n_n) \ldots \pi]$ (the allowed zone) the dipole radiation does not depend on the height of the dipole, as we shall see in the next section.

10.7 Radiation patterns

In the far-field, the magnetic field vector is transverse to the electric field vector and the time-averaged Poynting vector is calculated as

$$\langle \mathbf{S} \rangle = \frac{1}{2} \operatorname{Re}\{\mathbf{E} \times \mathbf{H}^*\} = \frac{1}{2} \sqrt{\frac{\varepsilon_0 \varepsilon_j}{\mu_0 \mu_j}} (\mathbf{E} \cdot \mathbf{E}^*) \mathbf{n}_r, \tag{10.41}$$

with $\mathbf{n}_r$ being the unit vector in the radial direction. The radiated power per unit solid angle $d\Omega = \sin\theta \, d\theta \, d\phi$ is

$$P = p(\Omega) d\Omega = r^2 \langle \mathbf{S} \rangle \cdot \mathbf{n}_r, \tag{10.42}$$

where $p(\Omega) = p(\theta, \phi)$ is defined as the radiation pattern. With the far-field in Eq. (10.32) and the corresponding potentials it is straightforward to calculate the normalized radiation patterns as

$$\begin{aligned} \frac{p(\theta,\phi)}{P_0} = \frac{3}{8\pi} \frac{\varepsilon_j}{\varepsilon_1} \frac{n_1}{n_j} \frac{1}{|\mathbf{p}|^2} \Big[& p_z^2 \sin^2\theta \left| \Phi_j^{(1)} \right|^2 \\ & + [p_x \cos\phi + p_y \sin\phi]^2 \cos^2\theta \left| \Phi_j^{(2)} \right|^2 \\ & + [p_x \sin\phi - p_y \cos\phi]^2 \left| \Phi_j^{(3)} \right|^2 \\ & - p_z [p_x \cos\phi + p_y \sin\phi] \cos\theta \, \sin\theta \\ & \times \left[\Phi_j^{*(1)} \Phi_j^{(2)} + \Phi_j^{(1)} \Phi_j^{*(2)} \right] \Big]. \end{aligned} \tag{10.43}$$

Here, P_0 corresponds to the total rate of energy dissipation in a homogeneous (unbounded) medium characterized by ε_1 and μ_1 (cf. Eq. (8.71)). The first term in the brackets of Eq. (10.43) contains the p-polarized contribution of the vertical orientation, whereas the second and third terms contain the p- and s-polarized contributions of the horizontal orientation. Of particular interest is the fourth term, which originates from interferences between the p-polarized terms of the two major orientations. Thus, the p-polarized light of a vertical and a horizontal dipole, which are located at the same point, interfere if the two dipoles radiate coherently. The radiation patterns for arbitrary dipole orientation usually cannot be put together additively. Notice, however, that upon integration over φ the interference term cancels out.

Equation (10.43) allows us to determine the radiation patterns of a dipole near an arbitrarily layered system; in the special case of a single interface it reproduces the formulas obtained by Lukosz and Kunz [15, 21]. As an illustration, Fig. 10.7 shows the radiation patterns of a dipole near a slab waveguide. The radiation in the forbidden zone depends exponentially on the height z_0 of the dipole, whereas the radiation in the allowed zone does not depend on z_0. In the lower half-space the interference term in Eq. (10.43) reads as

$$\left[\Phi_j^{*(1)} \Phi_j^{(2)} + \Phi_j^{(1)} \Phi_j^{*(2)} \right] \propto \left| t^{(p)}(\theta) \right|^2 e^{-2z_0 \operatorname{Im}\{\tilde{s}_z(\theta)\}} \operatorname{Re}\left\{ \frac{\cos\theta}{\tilde{s}_z(\theta)} \right\}. \tag{10.44}$$

In the forbidden zone, $\tilde{s}_z$ is imaginary and the interference term vanishes. Thus, the waves of a vertical and a horizontal dipole at the same position do not interfere in the forbidden zone and the radiation patterns will always be symmetric with respect to ϕ. This rather surprising result was found by Lukosz and Kunz in Ref. [21] for the case of a single interface. Recently, the radiation patterns of Eq. (10.43) have been confirmed for a single molecule

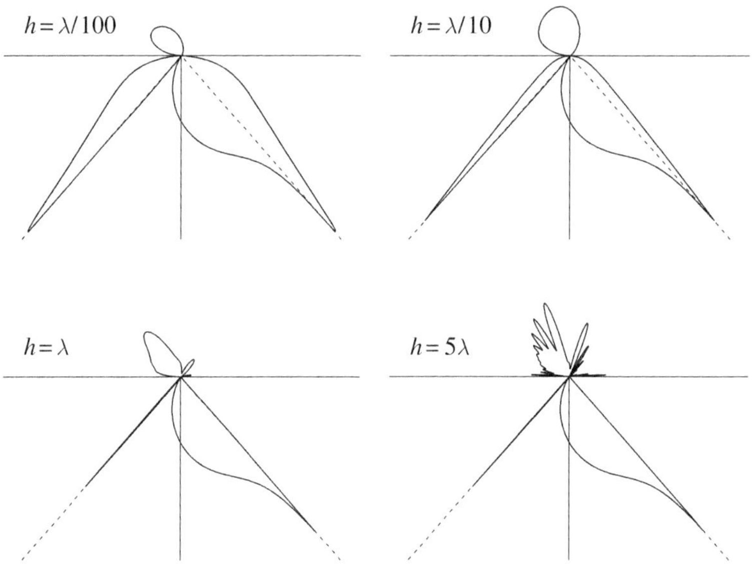

Fig. 10.7 Radiation patterns of a dipole with orientation $\theta=60^\circ$ approaching a planar waveguide; $\lambda=488$ nm, $\delta=80$ nm, $\varepsilon_1=1$, $\varepsilon_2=5$, $\varepsilon_3=2.25$. The different heights $z_0=h$ of the dipole are indicated in the figure. The radiation patterns are shown in the plane defined by the dipole axis and the z-axis. Note that the *allowed light* does not depend on h and that the *forbidden light* is always symmetric with respect to the vertical axis.

near a dielectric interface [22]. It is remarkable that although a single photon is emitted at a time all the interference terms in Eq. (10.43) are retained. Thus, as is well known, the photon travels many paths simultaneously and all the different paths interfere, giving rise to the predicted dipole radiation patterns. Figure 10.8 shows the radiation pattern of a single molecule placed near a glass surface. The pattern has been recorded with a CCD and is compared with the pattern calculated according to Eq. (10.43).

The radiation pattern defined by Eq. (10.43) describes the angular power distribution at an infinite distance from the dipole. However, in practice all distances are finite, and hence the question is when is the "infinite-distance" approximation good enough? The answer depends on the angles (θ, ϕ) of observation. For example, for θ_c, the critical angle of total internal reflection, the fields converge rather slowly towards the analytical far-fields. Light coupled into this angle is generated by dipole fields that propagate parallel to the interface(s). These fields refract at the surface at an infinite lateral distance. Thus, reducing the infinite extent of the layered system will influence the far-field mainly near the critical angle.

The phases of the spherical waves of the upper and lower half-spaces are not identical on the interface. Thus, close to the interface other waveforms must exist in order to compensate for the phase mismatch. In the literature these waves are known as *lateral waves*. Lateral waves decay by radiation into the critical angle θ_c. In the case of a plane interface illuminated under TIR conditions, lateral waves explain the lateral displacement between the incident and the reflected beam (the Goos–Hänchen shift).

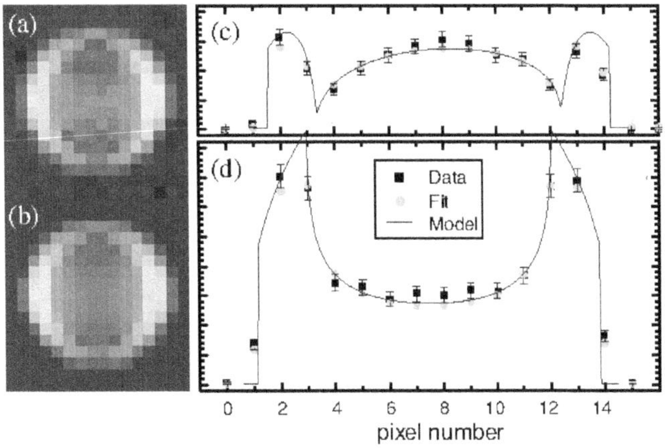

Fig. 10.8 The radiation pattern of a single molecule located near a glass surface. The pattern reflects the photons emitted into the dielectric and imaged onto a CCD with an NA = 1.4 objective lens. (a) Data, (b) fitted pattern using Eq. (10.43), (c) and (d) cross-sections along a horizontal and a vertical line through the center of the pattern, respectively. From [22].

10.8 Where is the radiation going?

Not all of a dipole's dissipated energy is converted into propagating radiation (photons). We have defined the quantum yield Q as the ratio of radiative and total decay rates, i.e. the power released as radiation versus the total dissipated power (cf. Eq. (13.53)). However, in an experiment one cannot detect all of the radiation released and hence one defines the *apparent quantum yield* Q_a as the ratio of *detected* power to totally dissipated power. In this section we analyze how much of a dipole's energy is emitted into the upper half-space, into the lower half-space, and into other modes of radiation (waveguides, surface waves, etc.).

As illustrated in Fig. 10.6, the total rate of energy dissipation is

$$P = P^{\uparrow} + P_{\mathrm{a}}^{\downarrow} + P_{\mathrm{f}}^{\downarrow} + P_{\mathrm{m}} + P_{\mathrm{i}}, \tag{10.45}$$

where $P^{\uparrow}$, $P_{\mathrm{a}}^{\downarrow}$, and $P_{\mathrm{f}}^{\downarrow}$ are the power radiated into the upper half-space, the allowed zone, and the forbidden zone, respectively. P_{m} denotes the power coupled into the layered medium (waveguide modes, surface modes, thermal losses, etc.) and P_{i} is the intrinsically dissipated power. The latter is associated with the intrinsic quantum yield q_{i} defined earlier. In order to derive $P^{\uparrow}$, $P_{\mathrm{a}}^{\downarrow}$, and $P_{\mathrm{f}}^{\downarrow}$, we need to integrate the radiation pattern in Eq. (10.43) over the corresponding angular ranges. It is convenient to use the following substitutions:

$$s = \begin{cases} \sin\theta & z > 0, \\ (n_n/n_1)\sin\theta & z < 0. \end{cases} \tag{10.46}$$

With these substitutions the interval $s = [0 \ldots 1]$ defines the plane-wave components of the dipole field, whereas the interval $s = [1 \ldots \infty]$ is associated with the dipole's evanescent waves. Furthermore, the different angular ranges are mapped as

$$\begin{aligned}
\theta &= [0 \ldots \pi/2] &&\rightarrow \quad s = [0 \ldots 1], \\
\theta &= [\pi/2 \ldots \arcsin(n_1/n_n)] &&\rightarrow \quad s = [(n_n/n_1) \ldots 1], \\
\theta &= [\arcsin(n_1/n_n) \ldots \pi] &&\rightarrow \quad s = [1 \ldots 0].
\end{aligned} \tag{10.47}$$

Hence, we see that the angular range $\theta = [\pi/2 \ldots \arcsin(n_1/n_n)]$, which corresponds to the forbidden zone, is associated with the dipole's evanescent fields. After integration of the radiation pattern in the upper half-space and use of the abbreviation $s_z = (1 - s^2)^{1/2}$ we obtain

$$\begin{aligned}
\frac{P^{\uparrow}}{P_0} = \frac{p_x^2 + p_y^2}{|\mathbf{p}|^2} \Bigg[\frac{1}{2} &+ \frac{3}{8} \int_0^1 \left[s\, s_z \left| r^{\mathrm{p}} \right|^2 + \frac{s}{s_z} \left| r^{\mathrm{s}} \right|^2 \right] \mathrm{d}s \\
&- \frac{3}{4} \int_0^1 \mathrm{Re} \left\{ \left[s\, s_z r^{\mathrm{p}} - \frac{s}{s_z} r^{\mathrm{s}} \right] \mathrm{e}^{2\,\mathrm{i}\, k_1 z_0 s_z} \right\} \mathrm{d}s \Bigg] \\
+ \frac{p_z^2}{|\mathbf{p}|^2} \Bigg[\frac{1}{2} &+ \frac{3}{4} \int_0^1 \frac{s^3}{s_z} \left| r^{\mathrm{p}} \right|^2 \mathrm{d}s + \frac{3}{2} \int_0^1 \mathrm{Re} \left\{ \frac{s^3}{s_z} r^{\mathrm{p}} \mathrm{e}^{2\,\mathrm{i}\, k_1 z_0 s_z} \right\} \mathrm{d}s \Bigg].
\end{aligned} \tag{10.48}$$

Both for the horizontal dipole and for the vertical dipole, there are three different terms. The first one corresponds to direct dipole radiation: half of the dipole's primary field is radiated into the upper half-space. The second term corresponds to the power that is reflected from the interface, and the last term accounts for interferences between the primary dipole field and the reflected dipole field. It is important to notice that the integration runs only over the interval $s = [0 \ldots 1]$. Therefore, only plane-wave components contribute to the radiation into the upper half-space.

To determine radiation into the lower half-space we use the substitution of Eq. (10.46) and integrate over the angular range of the lower half-space. The total radiation in the lower half-space $P^{\downarrow}$ is calculated as

$$\begin{aligned}
\frac{P^{\downarrow}}{P_0} = \frac{3}{8} \frac{p_x^2 + p_y^2}{|\mathbf{p}|^2} \frac{\varepsilon_n}{\varepsilon_1} \frac{n_1}{n_n} &\int_0^{n_n/n_1} s \left[1 - \left(\frac{n_1}{n_n} \right)^2 s^2 \right]^{1/2} \left[|t^{\mathrm{p}}|^2 + \frac{|t^{\mathrm{s}}|^2}{|1 - s^2|} \right] \mathrm{e}^{-2 k_1 z_0 s_z''} \mathrm{d}s \\
+ \frac{3}{4} \frac{p_z^2}{|\mathbf{p}|^2} \frac{\varepsilon_n}{\varepsilon_1} \frac{n_1}{n_n} &\int_0^{n_n/n_1} s^3 \left[1 - \left(\frac{n_1}{n_n} \right)^2 s^2 \right]^{1/2} \frac{|t^{\mathrm{p}}|^2}{|1 - s^2|} \mathrm{e}^{-2 k_1 z_0 s_z''} \mathrm{d}s,
\end{aligned} \tag{10.49}$$

where $s_z'' = \mathrm{Im}\{(1-s^2)^{1/2}\}$. when $n_n > n_1$ it is possible to separate the angular ranges of the allowed zone and the forbidden zone. The allowed light turns out to be

$$\frac{P_{\mathrm{a}}^{\downarrow}}{P_0} = \frac{3}{8}\frac{p_x^2+p_y^2}{|\mathbf{p}|^2}\frac{\varepsilon_n}{\varepsilon_1}\frac{n_1}{n_n}\int_0^1 s\left[1-\left(\frac{n_1}{n_n}\right)^2 s^2\right]^{1/2}\left[|t^{\mathrm{p}}|^2+\frac{|t^{\mathrm{s}}|^2}{1-s^2}\right]\mathrm{d}s$$
$$+\frac{3}{4}\frac{p_z^2}{|\mathbf{p}|^2}\frac{\varepsilon_n n_1}{\varepsilon_1 n_n}\int_0^1 s^3\left[1-\left(\frac{n_1}{n_n}\right)^2 s^2\right]^{1/2}\frac{|t^{\mathrm{p}}|^2}{1-s^2}\,\mathrm{d}s \quad (n_n > n_1). \tag{10.50}$$

Similarily, the forbidden light is determined as

$$\frac{P_{\mathrm{f}}^{\downarrow}}{P_0} = \frac{3}{8}\frac{p_x^2+p_y^2}{|\mathbf{p}|^2}\frac{\varepsilon_n}{\varepsilon_1}\frac{n_1}{n_n}\int_1^{n_n/n_1} s\left[1-\left(\frac{n_1}{n_n}\right)^2 s^2\right]^{1/2}\left[|t^{\mathrm{p}}|^2+\frac{|t^{\mathrm{s}}|^2}{s^2-1}\right]\mathrm{e}^{-2k_1 z_0\sqrt{s^2-1}}\,\mathrm{d}s$$
$$+\frac{3}{4}\frac{p_z^2}{|\mathbf{p}|^2}\frac{\varepsilon_n n_1}{\varepsilon_1 n_n}\int_1^{n_n/n_1} s^3\left[1-\left(\frac{n_1}{n_n}\right)^2 s^2\right]^{1/2}\frac{|t^{\mathrm{p}}|^2}{s^2-1}\mathrm{e}^{-2k_1 z_0\sqrt{s^2-1}}\,\mathrm{d}s$$
$$(n_n > n_1). \tag{10.51}$$

These expressions demonstrate that the allowed light does not depend on the height of the dipole, whereas the forbidden light shows the expected exponential dependence on the dipole's vertical position. Notice that since $s = k_\rho/k_1$ the term with the square root in the integrands corresponds to k_{z_n}/k_n. Assuming that there are no intrinsic losses ($P_{\mathrm{i}} = 0$), the power dissipated by the layered medium (thermal losses, waveguide and surface modes) is calculated as

$$P_{\mathrm{m}} = P - (P^{\uparrow} + P^{\downarrow}), \tag{10.52}$$

where P is determined by Eq. (10.26). For a lossless layered medium that does not support any waveguide modes it can be demonstrated that $P_{\mathrm{m}} = 0$ (see Problem 10.3).

As an illustration of the results developed here, Fig. 10.9 displays the different radiation terms for a dipole located above the dielectric waveguide shown in Fig. 10.3. The dipole is held at a fixed position $z_0 = 20$ nm and the thickness d of the waveguide is varied. While the allowed light is characterized by undulations of periodicity π/k_2, the forbidden light shows an irregular behavior with discontinuities for certain d. The locations of these discontinuities correspond to the cut-off conditions of the waveguide modes. For low d all waveguide modes are beyond cut-off, so that in the time average no energy is coupled into the waveguide ($P_{\mathrm{m}} = 0$). At $d \approx 0.058\lambda$ the fundamental TE_0 mode becomes propagating, and a net energy is coupled to the guide. When d is further increased, other modes can be excited as well. It is remarkable that, as the thickness is increased and one mode after the

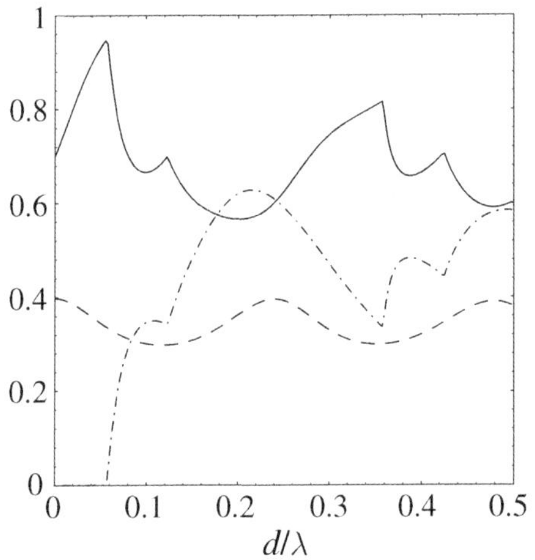

Fig. 10.9 Allowed light ($P_a^{\downarrow}$, dashed curve), forbidden light ($P_f^{\downarrow}$, solid curve), and radiation coupled into the waveguide ($P_m = P - P^{\uparrow} - P^{\downarrow}$, dash–dotted curve) as functions of the thickness d of the slab waveguide characterized in Fig. 10.3. The discontinuities correspond to the cut-offs of the TE_0, TM_0, TE_1, and TM_1 modes. All curves are normalized with the power P_0 emitted in free space.

other with $k_{\parallel} = k_0[n_2 \ldots n_3]$ comes alive, the total power radiated by the dipole shows *no* discontinuities. Instead, the power coupled into the waveguides is roughly compensated for by the forbidden light. Hence, the birth of new modes is indicated in the forbidden light, not by the total power. Finally it should be noted that for lossless media with real dielectric constants, such as the dielectric waveguide considered here, the poles associated with the lateral modes lie on the real s-axis. Numerical integration requires that a small detour around the poles in the integration path be implemented in order to avoid the singularities. Alternatively, a tiny imaginary part can be added to the real dielectric constants.

10.9 Magnetic dipoles

In the microwave regime paramagnetic materials exhibit magnetic transitions (electron spin resonance). In the infrared, small metal particles show magnetic dipole absorption caused by eddy currents of free carriers produced by the magnetic vector of the electromagnetic field. The field of a magnetic dipole in a planar layered medium is therefore important as well. From a theoretical point of view, these fields are dual to the fields of the electric dipole. The field of a magnetic dipole with moment **m** can be derived from the field of an electric dipole moment **p** by simply performing the substitution

$$[\mathbf{E},\ \mathbf{H},\ \mu_0\mu,\ \varepsilon_0\varepsilon,\ \mathbf{p}] \rightarrow [\mathbf{H},\ -\mathbf{E},\ \varepsilon_0\varepsilon,\ \mu_0\mu,\ \mu\mathbf{m}]. \tag{10.53}$$

With these substitutions, the reflection coefficients r^s and r^p are also interchanged. Thus, the field of a vertically oriented magnetic dipole will be purely s-polarized. In this case, no surface waves will be excited. Note that the electric dipole moment has the units $[\mathbf{p}] =$

A m s, whereas the units of the magnetic dipole are $[\mathbf{m}] = \mathrm{A\ m}^2$. The power radiated by an electric dipole with moment $\mathbf{p} = 1$ in a homogeneous medium is $\mu_0 \mu \varepsilon_0 \varepsilon$ times the power radiated by a magnetic dipole with moment $\mathbf{m} = 1$.

10.10 The image dipole approximation

The computational effort can be considerably reduced if retardation is neglected. In this case the fields will still satisfy Maxwell's equations in both half-spaces, but the standard static image theory is applied in order to approximately match the boundary conditions. We will outline the principle of this approximation for a single interface. Since the electromagnetic field is considered in its static limit ($k \to 0$) the electric and magnetic fields are decoupled and can be treated separately. For simplicity, only the electric field is considered.

Figure 10.10 shows an arbitrarily oriented dipole above a planar interface and its induced dipole in the medium below. The distance of the image dipole from the interface is the same as for the primary dipole. However, the magnitude of the image dipole moment is different. The static electric field of the primary dipole in the upper half-space reads as

$$\mathbf{E}_{\text{prim}} = -\nabla\phi\,, \qquad \text{with} \qquad \phi(\mathbf{r}) = \frac{1}{4\pi\,\varepsilon_0\,\varepsilon_1}\,\frac{\mathbf{p}\cdot\mathbf{r}}{r^3}. \tag{10.54}$$

The vector $\mathbf{r}$ denotes the radial vector measured from the position of the primary dipole and r is its magnitude. Similarly, the corresponding radial vector of the image dipole is denoted $\mathbf{r}'$. For simplicity, the dipole moment $\mathbf{p}$ is decomposed into its parallel and vertical parts with respect to the planar interface. Without loss of generality, the parallel component is assumed to point in the x-direction,

$$\mathbf{p} = p_x\mathbf{n}_x + p_z\mathbf{n}_z. \tag{10.55}$$

$\mathbf{n}_x$ and $\mathbf{n}_z$ denote the unit vectors in the x- and z-directions, respectively. In the following, the electric field will be considered for each of the two major orientations separately.

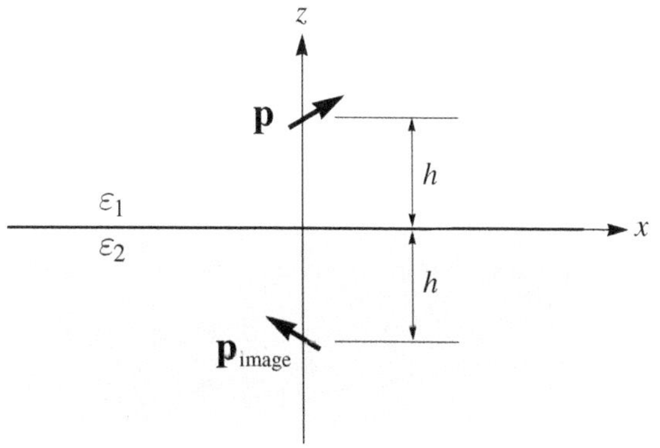

Fig. 10.10 The principle of the image dipole approximation. $\mathbf{p}$ and $\mathbf{p}_{\text{image}}$ denote the primary dipole and the image dipole, respectively. Static image theory is applied to determine the magnitude of $\mathbf{p}_{\text{image}}$.

10.10.1 Vertical dipole

For a dipole $\mathbf{p}=p_z\mathbf{n}_z$, the evaluation of the primary electric field in Eq. (10.54) in Cartesian coordinates leads to

$$\mathbf{E}_{\text{prim}} = \frac{p_z}{4\pi\varepsilon_0\varepsilon_1}\left[\frac{3x(z-h)}{r^5}, \frac{3y(z-h)}{r^5}, \frac{3(z-h)^2}{r^5} - \frac{1}{r^3}\right], \tag{10.56}$$

where h is the height of the dipole above the interface. Assuming an image dipole $\mathbf{p}=p_z\mathbf{n}_z$, a similar expression can be derived for the image field:

$$\mathbf{E}_{\text{image}} = \frac{p_z}{4\pi\varepsilon_0\varepsilon_1}\left[\frac{3x(z+h)}{r'^5}, \frac{3y(z+h)}{r'^5}, \frac{3(z+h)^2}{r'^5} - \frac{1}{r'^3}\right], \tag{10.57}$$

where r' denotes the radial distance measured from the location of the image dipole. A reasonable ansatz for the total field $\mathbf{E}$ in either of the two half-spaces is

$$\mathbf{E} = \begin{cases} \mathbf{E}_{\text{prim}} + A_{\text{v}}\mathbf{E}_{\text{image}} & z > 0 \\ B_{\text{v}}\mathbf{E}_{\text{prim}} & z < 0, \end{cases} \tag{10.58}$$

with the two unknown parameters A_{v} and B_{v}. By requiring the boundary conditions at the interface $z=0$, A_{v} and B_{v} can be determined as

$$A_{\text{v}} = \frac{\varepsilon_2 - \varepsilon_1}{\varepsilon_2 + \varepsilon_1}, \tag{10.59}$$

$$B_{\text{v}} = \frac{\varepsilon_1}{\varepsilon_2}\frac{2\varepsilon_2}{\varepsilon_2 + \varepsilon_1}.$$

A_{v} and B_{v} correspond to the Fresnel reflection and transmission coefficients in the quasi-static limit (cf. Section 2.8.1).

10.10.2 Horizontal dipole

The procedure for a dipole $\mathbf{p}=p_x\mathbf{n}_x$ is similar. The primary and image fields turn out to be

$$\mathbf{E}_{\text{prim}} = \frac{p_x}{4\pi\varepsilon_0\varepsilon_1}\left[\frac{3x^2}{r^5} - \frac{1}{r^3}, \frac{3xy}{r^5}, \frac{3x(z-h)}{r^5}\right], \tag{10.60}$$

$$\mathbf{E}_{\text{image}} = \frac{p_x}{4\pi\varepsilon_0\varepsilon_1}\left[\frac{3x^2}{r'^5} - \frac{1}{r'^3}, \frac{3xy}{r'^5}, \frac{3x(z+h)}{r'^5}\right]. \tag{10.61}$$

The corresponding ansatz for the total field $\mathbf{E}$ in either of the two half-spaces is

$$\mathbf{E} = \begin{cases} \mathbf{E}_{\text{prim}} + A_{\text{h}}\mathbf{E}_{\text{image}} & z > 0, \\ B_{\text{h}}\mathbf{E}_{\text{prim}} & z < 0. \end{cases} \tag{10.62}$$

As before, the unknown parameters A_{h} and B_{h} can be determined by the boundary conditions at $z=0$ as

$$A_{\mathrm{h}} = -\frac{\varepsilon_2 - \varepsilon_1}{\varepsilon_2 + \varepsilon_1}, \tag{10.63}$$

$$B_{\mathrm{h}} = \frac{\varepsilon_1}{\varepsilon_2}\frac{2\,\varepsilon_2}{\varepsilon_2 + \varepsilon_1}.$$

Besides the sign of A_{h}, the two parameters are identical with the parameters A_{v} and B_{v} calculated for the vertical dipole.

10.10.3 Including retardation

Using the parameters A_{v}, B_{v}, A_{h}, and B_{h} the magnitude of the image dipole is

$$|\mathbf{p}_{\mathrm{image}}| = \frac{\varepsilon_2 - \varepsilon_1}{\varepsilon_2 + \varepsilon_1}|\mathbf{p}|. \tag{10.64}$$

As indicated in Fig. 10.10, the horizontal components of $\mathbf{p}_{\mathrm{image}}$ and $\mathbf{p}$ point in different directions if their vertical components have the same direction. To obtain the static field in the upper half-space, the fields of the two dipoles $\mathbf{p}$ and $\mathbf{p}_{\mathrm{image}}$ have to be superposed. The field in the lower half-space simply corresponds to the attenuated primary dipole field. The attenuation is given by the factor $2\varepsilon_2/(\varepsilon_2 + \varepsilon_1)$. Note that the dipoles are considered to be located in the same medium as the point of observation.

So far, the location, orientation, and magnitude of the dipole moments $\mathbf{p}$ and $\mathbf{p}_{\mathrm{image}}$ have been determined. In order to fulfill Maxwell's equations in both half-spaces, the static dipole fields are replaced by their non-retarded forms:

$$\mathbf{E} \sim [\nabla\nabla\cdot]\,\frac{\mathbf{p}}{r} \quad \rightarrow \quad \mathbf{E} \sim \left[k^2 + \nabla\nabla\cdot\right]\frac{\mathbf{p}}{r}\mathrm{e}^{ikr}. \tag{10.65}$$

Although this substitution rescues Maxwell's equations in both half-spaces, it introduces a violation of the boundary conditions. The image dipole approximation therefore has obvious limitations. In order to keep the errors within bounds, the height h of the primary dipole must be small and the fields may be evaluated only in a limited range from the dipole location. In fact, the image dipole approximation leads to reasonable accuracy as long as short-range interactions are considered.

Problems

10.1 Derive Eq. (10.26) and plot the radiative (plane waves), non-radiative (evanescent waves), and total decay rates ($q_{\mathrm{i}} = 1$) as functions of the normalized height z_0/λ for the following situations.
  (1) Horizontal dipole in vacuum above a dielectric substrate ($\varepsilon = 2.25$).
  (2) Vertical dipole in vacuum above a dielectric substrate ($\varepsilon = 2.25$).
  (3) Horizontal dipole in vacuum above an aluminum substrate ($\varepsilon = -34.5 + 8.5\mathrm{i}$, $\lambda = 488$ nm).
  (4) Vertical dipole in vacuum above an aluminum substrate ($\varepsilon = -34.5 + 8.5\mathrm{i}$, $\lambda = 488$ nm).

10.2 Calculate the normalized energy flux $(P_1^{\downarrow}/P_0)$ through a horizontal plane right beneath a dipole that is located above an arbitrary stratified medium. First derive the magnetic field **H** that corresponds to the electric field in Eq. (10.16), and then determine the z-component of the Poynting vector $\langle S_z \rangle$. Use the Bessel-function closure relations (cf. Eq. (3.112)) to integrate $\langle S_z \rangle$ over the horizontal plane. Show that the result is identical with $(P - P_1^{\uparrow} - P_n^{\downarrow})/P_0$ as defined in Section 10.8.

10.3 Demonstrate that for a dipole near a single dielectric interface the total dissipated power P is identical to the total integrated radiation pattern $P^{\uparrow} + P_{\mathrm{a}}^{\downarrow} + P_{\mathrm{f}}^{\downarrow}$. Hint: express the transmission coefficients in terms of the reflection coefficients as

$$t^{\mathrm{s}} = [1 + r^{\mathrm{s}}], \qquad (k_{z_n}/k_{z_1})t^{\mathrm{s}} = (\mu_n/\mu_1)[1 - r^{\mathrm{s}}],$$

$$t^{\mathrm{p}} = (\varepsilon_1/\varepsilon_n)(n_n/n_1)[1 + r^{\mathrm{p}}], \qquad (k_{z_n}/k_{z_1})t^{\mathrm{p}} = (n_n/n_1)[1 - r^{\mathrm{p}}].$$

10.4 Consider a molecule with an emission dipole moment parallel to an aluminum substrate. The emission wavelength is $\lambda = 488$ nm and the dielectric constant of the substrate is $\varepsilon = -34.5 + 8.5\mathrm{i}$. Determine the apparent quantum yield q_{a} defined as the ratio between the energy radiated into the upper half-space and the total dissipated energy. Plot q_{a} as a function of the molecule's vertical position z_0/λ. Use the plot ranges $z_0/\lambda = [0 \ldots 2]$ and $q_a = [0 \ldots 1]$.

10.5 For a dipole sitting on an air/dielectric interface ($n_1 = 1$, $n_2 = 1.5$) calculate the ratio between the energy radiated into the upper half-space and the energy radiated into the lower half-space. Perform the calculations separately for a horizontal and a vertical dipole.

References

[1] H. Metiu, "Surface enhanced spectroscopy," in *Progress in Surface Science*, ed. I. Prigogine and S. A. Rice, vol. 17. New York: Pergamon Press, pp. 153–320 (1984).

[2] See, for example, L. Novotny, "Allowed and forbidden light in near-field optics," *J. Opt. Soc. Am. A* **14**, 91–104 and 105–113 (1997), and references therein.

[3] L. M. Brekhovskikh and O. A. Godin, *Acoustics of Layered Media*. Berlin: Springer-Verlag (1990).

[4] A. Sommerfeld, "Über die Ausbreitung der Wellen in der drahtlosen Telegraphie," *Ann. Phys.* **28**, 665–736 (1909).

[5] J. Zenneck, "Fortpflanzung ebener elektromagnetischer Wellen längs einer ebenen Leiterfläche," *Ann. Phys.* **23**, 846–866 (1907).

[6] H. von Hörschelmann, "Über die Wirkungsweise des geknickten Marconischen Senders in der drahtlosen Telegraphie," *Jahresbuch drahtl. Telegr. Teleph.* **5**, 14–34 and 188–211 (1911).

[7] A. Sommerfeld, "Über die Ausbreitung der Wellen in der drahtlosen Telegraphie," *Ann. Phys.* **81**, 1135–1153 (1926).

[8] H. Weyl, "Ausbreitung elektromagnetischer Wellen über einem ebenen Leiter," *Ann. Phys.* **60**, 481–500 (1919).

[9] M. J. O. Strutt, "Strahlung von Antennen unter dem Einfluβ der Erdbodeneigenschaften," *Ann. Phys.* **1**, 721–772 (1929).

[10] B. Van der Pol and K. F. Niessen, "Über die Ausbreitung elektromagnetischer Wellen über einer ebenen Erde," *Ann. Phys.* **6**, 273–294 (1930).

[11] G. S. Agarwal, "Quantum electrodynamics in the presence of dielectrics and conductors. I. Electrodynamic-field response functions and black-body fluctuations in finite geometries," *Phys. Rev. A* **11**, 230–242 (1975).

[12] A. Sommerfeld, *Partial Differential Equations in Physics*, 5th edn. New York: Academic Press (1967).

[13] B. Hecht, D. W. Pohl, H. Heinzelmann, and L. Novotny, " 'Tunnel' near-field optical microscopy: TNOM-2," in *Photons and Local Probes*, ed. O. Marti and R. Möller. Dordrecht: Kluwer, pp. 93–107 (1995).

[14] W. C. Chew, *Waves and Fields in Inhomogeneous Media*. New York: Van Nostrand Reinhold (1990).

[15] W. Lukosz and R. E. Kunz, "Light emission by magnetic and electric dipoles close to a plane interface. I. Total radiated power," *J. Opt. Soc. Am.* **67**, 1607–1615 (1977).

[16] I. Pockrand, A. Brillante, and D. Möbius, "Nonradiative decay of excited molecules near a metal surface," *Chem. Phys. Lett.* **69**, 499–504 (1994).

[17] J. K. Trautman and J. J. Macklin, "Time-resolved spectroscopy of single molecules using near-field and far-field optics," *Chem. Phys.* **205**, 221–229 (1996).

[18] R. X. Bian, R. C. Dunn, X. S. Xie, and P. T. Leung, "Single molecule emission characteristics in near-field microscopy," *Phys. Rev. Lett.* **75**, 4772–4775 (1995).

[19] L. Novotny, "Single molecule fluorescence in inhomogeneous environments," *Appl. Phys. Lett.* **69**, 3806–3808 (1996).

[20] H. Gersen, M. F. García-Parajó, L. Novotny, *et al.*, "Influencing the angular emission of a single molecule," *Phys. Rev. Lett.* **85**, 5312–5314 (2000).

[21] W. Lukosz and R. E. Kunz, "Light emission by magnetic and electric dipoles close to a plane dielectric interface. II. Radiation patterns of perpendicular oriented dipoles," *J. Opt. Soc. Am.* **67**, 1615–1619 (1977).

[22] M. A. Lieb, J. M. Zavislan, and L. Novotny, "Single molecule orientations determined by direct emission pattern imaging," *J. Opt. Soc. Am. B* **21**, 1210–1215 (2004).

11 Photonic crystals, resonators, and cavity optomechanics

Artificial optical materials and structures have enabled the observation of various new optical effects. For example, photonic crystals are able to inhibit the propagation of certain light frequencies and provide the unique ability to guide light around very tight bends and along narrow channels. With metamaterials, on the other hand, one can achieve negative refraction. The high field strengths in optical microresonators lead to nonlinear optical effects that are important for future integrated optical networks, and the coupling between optical and mechanical degrees of freedom opens up the possibility of cooling macroscopic systems down to the quantum ground state. This chapter explains the basic underlying principles of these novel optical structures.

11.1 Photonic crystals

Photonic crystals are materials with a spatial periodicity in their dielectric constant, a system that was first analyzed by Lord Rayleigh in 1887 [1]. Under certain conditions, photonic crystals can create a photonic bandgap, i.e. a frequency window within which propagation of light through the crystal is inhibited. Light propagation in a photonic crystal is similar to the propagation of electrons and holes in a semiconductor. An electron passing through a semiconductor experiences a periodic potential due to the ordered atomic lattice. The interaction between the electron and the periodic potential results in the formation of energy bandgaps. It is not possible for the electron to pass through the crystal if its energy falls within the range of the bandgap. However, defects in the periodicity of the lattice can locally destroy the bandgap and give rise to interesting electronic properties. If the electron is replaced by a photon and the atomic lattice by a material with a periodic dielectric constant we end up with basically the same effects. However, while atoms arrange themselves naturally to form a periodic structure, photonic crystals need to be fabricated artificially. One exception is gemstone opals, which are formed by spontaneous organization of colloidal silica spheres into a crystalline lattice. In order for a particle to interact with its periodic environment, its wavelength must be comparable to the periodicity of the lattice. Therefore, in photonic crystals the lattice constant must be in the range 100 nm to 1 μm. This size range can be accessed with conventional nanofabrication and self-assembly techniques (see Fig. 11.1).

To calculate the optical modes in a photonic crystal one needs to solve Maxwell's equations in a periodic dielectric medium. Although this task appears quite simple, it is not possible to analytically solve Maxwell's equations for two- or three-dimensional periodic lattices. Instead, numerical techniques have to be invoked. However, many interesting phenomena can be deduced by considering the simpler one-dimensional case, i.e. a periodically layered medium. The understanding and intuition developed here will help us to discuss the properties of the more complex two- and three-dimensional photonic crystals. A more detailed account of photonic crystals can be found in Refs. [3, 4].

11.1.1 The photonic bandgap

Let us consider a material made of an infinite number of planar layers of thickness d oriented perpendicular to the direction z as shown in Fig. 11.2. The dielectric constant of the layers is assumed to alternate between the values ε_1 and ε_2. The optical mode propagating inside the material is characterized by the wavevector $\mathbf{k} = (k_x, k_y, k_z)$. It is further assumed that both materials are non-magnetic, i.e. $\mu_1 = \mu_2 = 1$, and lossless. We can distinguish two kinds of modes, TE modes, for which the electric field vector is always parallel to the boundaries between adjacent layers, and TM modes, for which the magnetic field vector is always parallel to the boundaries. Separation of variables leads to the following ansatz for the complex field amplitudes:

$$\text{TE:} \quad \mathbf{E}(\mathbf{r}) = E(z)\, \mathrm{e}^{\mathrm{i}(k_x x + k_y y)} \mathbf{n}_x, \tag{11.1}$$

$$\text{TM:} \quad \mathbf{H}(\mathbf{r}) = H(z)\, \mathrm{e}^{\mathrm{i}k_x x + k_y y)} \mathbf{n}_x. \tag{11.2}$$

In each layer n, the solution for $E(z)$ and $H(z)$ is a superposition of a forward and a backward propagating wave, i.e.

$$\text{TE:} \quad E_{n,j}(z) = a_{n,j} \mathrm{e}^{\mathrm{i}k_{z_j}(z-nd)} + b_{n,j} \mathrm{e}^{-\mathrm{i}k_{z_j}(z-nd)}, \tag{11.3}$$

$$\text{TM:} \quad H_{n,j}(z) = a_{n,j} \mathrm{e}^{\mathrm{i}k_{z_j}(z-nd)} + b_{n,j} \mathrm{e}^{-\mathrm{i}k_{z_j}(z-nd)}, \tag{11.4}$$

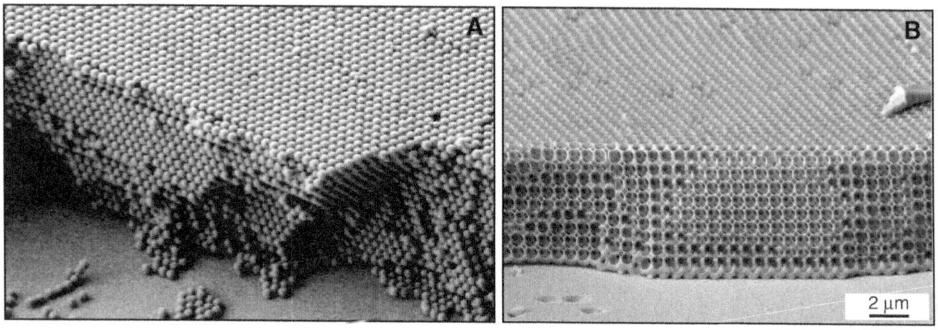

Fig. 11.1 Fabrication of silicon photonic bandgap crystals. (a) A template produced by 855 nm silica spheres deposited on a Si wafer. (b) A photonic crystal obtained after filling the interstitial spaces with high-index Si and removing the template by wet etching. Reprinted with permission from Macmillan Publishers Ltd. [2].

where $a_{n,j}$ and $b_{n,j}$, are constants that depend on the layer number n and the medium ε_j. The longitudinal wavenumber k_{z_j} is defined as

$$k_{z_j} = \sqrt{\frac{\omega^2}{c^2}\,\varepsilon_j - k_{\parallel}^2}\,, \qquad k_{\parallel} = \sqrt{k_x^2 + k_y^2}, \tag{11.5}$$

with $k_{\parallel}$ being the parallel wavenumber. To find the constants $a_{n,j}$ and $b_{n,j}$ we apply the boundary conditions at the interface $z = z_n = nd$ between the nth and the $(n+1)$th layer:

$$\text{TE:} \qquad E_{n,1}(z_n) = E_{n+1,2}(z_n), \tag{11.6}$$

$$\frac{\mathrm{d}}{\mathrm{d}z}E_{n,1}(z_n) = \frac{\mathrm{d}}{\mathrm{d}z}E_{n+1,2}(z_n), \tag{11.7}$$

$$\text{TM:} \qquad H_{n,1}(z_n) = H_{n+1,2}(z_n), \tag{11.8}$$

$$\frac{1}{\varepsilon_1}\frac{\mathrm{d}}{\mathrm{d}z}H_{n,1}(z_n) = \frac{1}{\varepsilon_2}\frac{\mathrm{d}}{\mathrm{d}z}H_{n+1,2}(z_n). \tag{11.9}$$

Equation (11.7) is arrived at by expressing the transverse component of the magnetic field in terms of the electric field by using $\nabla \times \mathbf{E} = \mathrm{i}\omega\mu_0\mathbf{H}$. Similarly, Eq. (11.9) follows from $\nabla \times \mathbf{H} = -\mathrm{i}\omega\varepsilon_0\varepsilon\mathbf{E}$. Inserting Eqs. (11.3) and (11.4) leads to

$$a_{n,1} + b_{n,1} = a_{n+1,2}\mathrm{e}^{-\mathrm{i}k_{z_2}d} + b_{n+1,2}\mathrm{e}^{\mathrm{i}k_{z_2}d}, \tag{11.10}$$

$$a_{n,1} - b_{n,1} = p_m\Big[a_{n+1,2}\mathrm{e}^{-\mathrm{i}k_{z_2}d} - b_{n+1,2}\mathrm{e}^{\mathrm{i}k_{z_2}d}\Big], \tag{11.11}$$

where $p_m \in \{p_{\mathrm{TE}}, p_{\mathrm{TM}}\}$ is a factor that depends on the polarization as

$$p_{\mathrm{TE}} = \frac{k_{z_2}}{k_{z_1}} \quad \text{(TE modes)}, \qquad p_{\mathrm{TM}} = \frac{k_{z_2}}{k_{z_1}}\frac{\varepsilon_1}{\varepsilon_2} \quad \text{(TM modes)}. \tag{11.12}$$

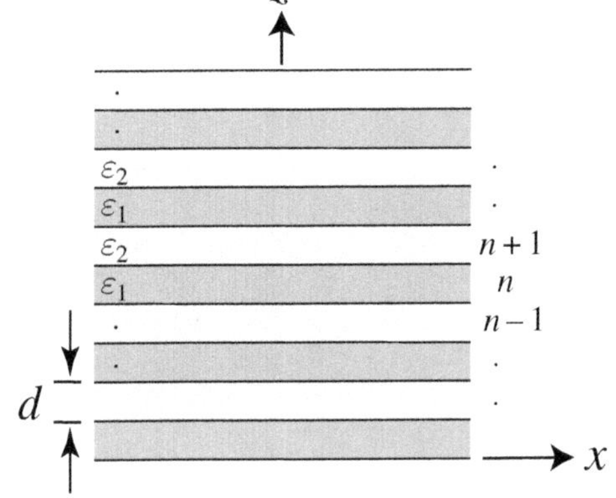

Fig. 11.2 A one-dimensional photonic crystal made of an infinite number of planar layers of thickness d.

For a given mode type we have two equations but four unknowns, i.e. $a_{n,1}$, $b_{n,1}$, $a_{n+1,2}$, and $b_{n+1,2}$. Thus, we need more equations. Evaluating the boundary conditions at the interface $z = z_{n-1} = (n-1)d$ between the $(n-1)$th and nth layers leads to

$$a_{n-1,2} + b_{n-1,2} = a_{n,1}\mathrm{e}^{-\mathrm{i}k_{z_1}d} + b_{n,1}\mathrm{e}^{\mathrm{i}k_{z_1}d}, \tag{11.13}$$

$$a_{n-1,2} - b_{n-1,2} = \frac{1}{p_m}\left[a_{n,1}\mathrm{e}^{-\mathrm{i}k_{z_1}d} - b_{n,1}\mathrm{e}^{\mathrm{i}k_{z_1}d}\right]. \tag{11.14}$$

Although we now have four equations for each mode type, we have also increased the number of unknowns by two, i.e. by including $a_{n-1,2}$ and $b_{n-1,2}$. However, $a_{n-1,2}$ and $b_{n-1,2}$ can be expressed in terms of $a_{n+1,2}$ and $b_{n+1,2}$ with the help of the *Floquet–Bloch theorem* [5, 6]. This theorem states that if E is a field in a periodic medium with periodicity $2d$ then it has to satisfy

$$E(z + 2d) = \mathrm{e}^{\mathrm{i}k_{\mathrm{Bl}}2d}E(z), \tag{11.15}$$

where k_{Bl} is an as-yet-undefined wavevector, called the *Bloch wavevector*. A similar equation holds for the magnetic field $H(z)$. The Floquet–Bloch theorem has to be viewed as an *ansatz*, a trial function for our system of coupled differential equations. Application of the Floquet–Bloch theorem leads to

$$\left[a_{n+1,2} + b_{n+1,2}\mathrm{e}^{-2\mathrm{i}k_{z_2}[z-(n-1)d]}\right] = \mathrm{e}^{\mathrm{i}k_{\mathrm{Bl}}2d}\left[a_{n-1,2} + b_{n-1,2}\mathrm{e}^{-2\mathrm{i}k_{z_2}[z-(n-1)d]}\right]. \tag{11.16}$$

Since this equation has to hold for any position z, we have to require that

$$a_{n+1,2} = a_{n-1,2}\mathrm{e}^{\mathrm{i}k_{\mathrm{Bl}}2d}, \tag{11.17}$$

$$b_{n+1,2} = b_{n-1,2}\mathrm{e}^{\mathrm{i}k_{\mathrm{Bl}}2d}, \tag{11.18}$$

which reduces the number of unknowns from six to four and allows us to solve the *homogeneous* system of equations defined by Eqs. (11.10)–(11.14). The system of equations can be written in matrix form and the determinant must be zero in order to guarantee a solution. The resulting *characteristic equation* turns out to be

$$\cos(2k_{\mathrm{Bl}}d) = \cos(k_{z_1}d)\,\cos(k_{z_2}d) - \frac{1}{2}\left[p_m + \frac{1}{p_m}\right]\sin(k_{z_1}d)\sin(k_{z_2}d). \tag{11.19}$$

Since $\cos(2k_{\mathrm{Bl}}d)$ is always in the range $[-1 \ldots 1]$, solutions cannot exist when the absolute value of the right-hand side is larger than unity. This absence of solutions gives rise to the formation of *bandgaps*. For example, a wave at normal incidence ($k_{z_1} = \sqrt{\varepsilon_1}\omega/c$, $k_{z_2} = \sqrt{\varepsilon_2}\omega/c$) to a photonic crystal with $\varepsilon_1 = 2.25$ and $\varepsilon_2 = 9$ can propagate for $\lambda = 12d$ but not for $\lambda = 9d$.

For each Bloch wavevector k_{Bl} one finds a dispersion relation $\omega(k_\parallel)$. If all possible dispersion relations are plotted on the same graph one obtains a so-called *band diagram*. An example is shown in Fig. 11.3, where the shaded areas correspond to allowed bands for which propagation through the crystal is possible. Notice that propagating modes exist even if one of the longitudinal wavenumbers (k_{z_j}) is imaginary. The Bloch wavevector at the band edges is determined by $k_{\text{Bl}}d = n\pi/2$. For a given direction of propagation characterized by $k_{\|}$ one finds frequency regions for which propagation through the crystal is possible and frequency regions for which propagation is inhibited. However, for a one-dimensional crystal there is no *complete bandgap*, i.e. there are no frequencies for which propagation is inhibited in all directions. If a wave propagating in vacuum is directed onto the photonic crystal, then only modes with $k_{\|}$ smaller than $k = \omega/c$ can be excited. The vacuum light-lines are indicated in Fig. 11.3 and one can find complete frequency bandgaps inside the region $k_{\|} < k$. For these frequencies the photonic crystal is a perfect mirror (omnidirectional reflector), which is technically exploited e.g. for laser high-reflectors.

A complete bandgap is possible in three-dimensional photonic crystals. It is favorable if the dielectric constants of the media differ by a large amount. The volume ratio between the two media is also important. Unfortunately, the solutions for two-dimensional and three-dimensional photonic crystals cannot be found by analytical means, but efficient numerical techniques have been developed over the past few years.

In semiconductors, the valence band corresponds to the topmost filled energy band for which electrons stay bound to the ion cores. If electrons are excited into the next higher band, which is called the conduction band, they become delocalized and conduction through the crystal strongly increases. The situation is similar for photonic crystals: the band below a bandgap is referred to as the *dielectric band* and the band above the bandgap as the *air band*. In the dielectric band, the optical energy is confined inside the material

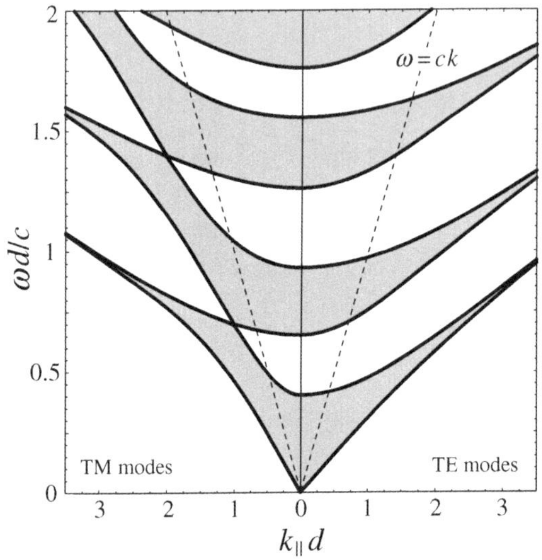

Fig. 11.3 The band diagram for a one-dimensional photonic crystal. The shaded areas are the allowed bands. The diagram represents both TE and TM modes. For a one-dimensional photonic crystal, there are no complete bandgaps, i.e. there are no frequencies for which propagation is inhibited in all directions. Values used: $\varepsilon_1 = 2.33$ (SiO_2) and $\varepsilon_2 = 17.88$ (InSb).

with the higher dielectric constant, whereas in the air band the energy is found to be in the material with lower dielectric constant. Thus, excitation from one band to another promotes the optical energy from the high-dielectric-constant to the low-dielectric-constant material.

A photonic crystal can also strongly affect the spontaneous emission rate of an embedded quantum system such as an atom or a molecule. For example, the excited state of an atom cannot couple to any radiation modes if the transition frequency between the excited state and the ground state lies in the bandgap region of the photonic crystal. In this case, spontaneous emission is severely inhibited and the atom will reside in its excited state (cf. Section 8.4). As discussed later, a localized defect near the atom can have the opposite effect and enhance the emission rate of the atom significantly.

11.1.2 Defects in photonic crystals

Defects in photonic crystals are introduced in order to localize or guide light. While photons with energies within the photonic bandgap cannot propagate through the crystal, they can be confined to defect regions. A line of defects opens up a waveguide: light with a frequency within the bandgap can propagate only along the channel of defects since it is repelled from the bulk crystal. Waveguides in photonic crystals can transport light around tight corners with virtually no loss. Photonic crystal waveguides therefore are of great practical importance for miniaturized optoelectronic circuits and devices. As an example, Fig. 11.4 shows a waveguide T-junction in a photonic crystal. The line defects are created by dislocating certain portions of the crystal and by removing a row of elements [7]. The device functions as a diplexer, i.e. high frequencies are deflected to the left and low frequencies are deflected to the right. To improve the performance, an additional perturbation has been added to the intersection region. Furthermore, photonic crystal waveguides can be composed of air channels, thereby significantly reducing group velocity dispersion. A short pulse of light can travel large distances without being temporally broadened. Technical applications include photonic crystal optical fibers, which may be used for nonlinear

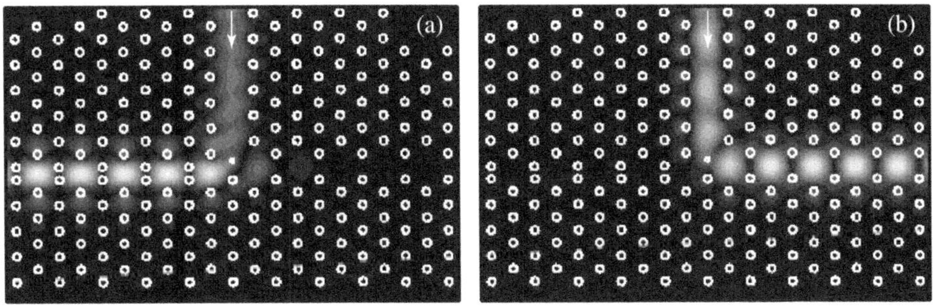

Fig. 11.4 A two-dimensional photonic crystal diplexer. A waveguide T-junction is formed by dislocations and removal of elements. High frequencies are deflected to the left and low frequencies are deflected to the right. The figure shows the computed optical intensity for (a) $\omega = 0.956\pi c/d$ and (b) $\omega = 0.874\pi c/d$, with d being the lattice constant. From [7].

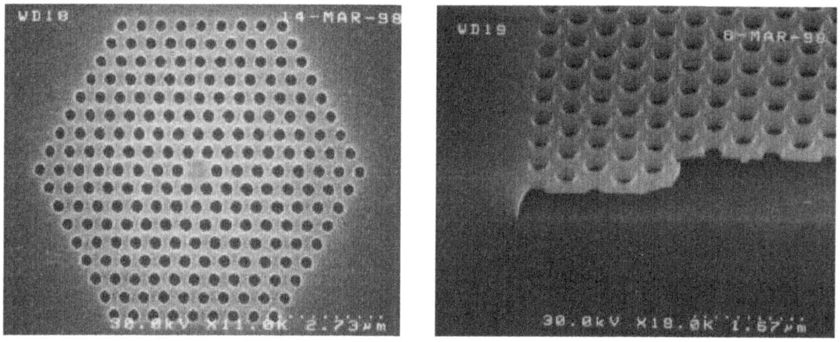

Fig. 11.5 Top view and cross-section of a two-dimensional photonic crystal with a single central defect. The crystal consists of a microfabricated hexagonal array of air holes in InGaAsP and the defect is introduced by a filled central hole. From [8].

white-light continuum generation (using a dielectric band), or dispersion-free propagation of femtosecond-laser pulses (in the air band).

While defect arrays in photonic crystals are introduced primarily for waveguide applications, localized defects are intended to trap light. Optical cavities formed by localized defects can have very high quality factors, a prerequisite for various nonlinear optical effects and laser applications. Figure 11.5 shows a two-dimensional photonic crystal with a single central defect [8]. A laser is formed by embedding the photonic crystal in between two Bragg mirrors acting as the end mirrors of a laser cavity. The lateral confinement is provided by the photonic crystal.

Photonic crystal cavities can also be used to control the spontaneous emission rate of quantum systems located in the defect region. Depending on the physical properties of the cavity, the local density of states (DOS) at the emission wavelength λ_0 of the quantum system can be increased or decreased over the free-space DOS (see Section 8.4). The local DOS at λ_0 depends on the ability of the cavity to store energy at the emission wavelength λ_0. Thus, the higher the quality factor $Q = \omega_0/\Delta\omega$ is, the higher the DOS will be. The density of states in a large cavity can be approximated as

$$\rho = \frac{1}{\omega_0}\frac{D\,Q}{V}, \tag{11.20}$$

where V is the volume of the cavity and D is the mode degeneracy, i.e. the number of cavity modes with the same frequency. The free-space DOS has been derived in Eq. (8.120) as

$$\rho_0 = \frac{1}{\omega_0}\frac{8\pi}{\lambda_0^3}. \tag{11.21}$$

Thus, the spontaneous decay rate is enhanced by a factor of

$$K = \frac{\rho}{\rho_0} = \frac{D}{8\pi}\,Q\,\frac{\lambda_0^3}{V} \tag{11.22}$$

in a photonic crystal cavity. Strong enhancement depends on a small cavity volume and a high Q-factor.

11.2 Metamaterials

The interaction of electromagnetic fields with materials is described by Maxwell's equations and the constitutive relations (2.11)–(2.13), which, in the linear regime, are routinely expressed in terms of the permittivity ε and permeability μ. These parameters account for a material's electromagnetic response averaged over many atoms. This averaging is typically justified as long as the atoms are much smaller than the spatial variations of electromagnetic fields. In free space, these spatial variations are defined by the wavelength of radiation. For example, the refraction of a plane wave entering a dielectric medium with a positive refractive index n can be explained by assuming a speed of light that is reduced by a factor n compared with that in vacuum. The same effect can be explained from an atomistic perspective, namely by vectorially summing the secondary fields propagating at speed c and originating from individual electrons in the material [9]. If the properties of the atoms could be tailored, i.e. if the real atoms could be replaced by artificial scatterers, the optical response of the material could be changed at will within wide boundaries. A metamaterial in this line of thought is therefore considered to be a material that consists of a sufficiently densely packed array of "artificial atoms," i.e. nanoscale optical scatterers, that are much smaller than the wavelength of operation. These artificial atoms scatter light in a predesigned fashion, thus creating novel optical properties such as for example a negative refractive index.

The difference between photonic crystals and metamaterials according to this definition is that in a photonic crystal with a bandgap the scatterers need to be arranged periodically and their lattice constant has to be comparable to the wavelength, because the effect of the bandgap arises from diffraction and destructive interference. In the case of metamaterials the artificial atoms and their distances have to be much smaller than the wavelength, since diffraction should be avoided. The optical response of a metamaterial therefore is that of a (piecewise) homogeneous medium. Note that there is a regime where the concepts of a photonic crystal and a metamaterial do mix [10].

11.2.1 Negative-index materials

Figure 11.6 shows a classification of optical materials according to the signs of their ε and μ. The propagation of light in a homogeneous medium can be characterized by the dispersion relation

$$\mathbf{k} \cdot \mathbf{k} = \varepsilon(\omega)\mu(\omega)\frac{\omega^2}{c^2} = n^2(\omega)\frac{\omega^2}{c^2}. \tag{11.23}$$

For most materials at optical frequencies the magnetic permeability is unity, i.e. $\mu = 1$. Therefore, μ is mostly neglected in the constitutive relations and optical properties and phenomena are solely related to ε. For example, because of $\mu = 1$ the Brewster effect appears only for p-polarization. However, in the case of magneto-dielectrics for which $\mu \neq 1$, a Brewster effect is observed also for s-polarization [10].[1]

Metals typically exhibit a negative ε, while dielectrics, such as glass, exhibit a positive ε. For both materials we typically have $\mu = 1$ at optical frequencies. In 1968 Veselago published a theoretical study [11] in which he considered the properties of a hypothetical material having both a negative ε and a negative μ. He showed that such a material leads to a negative index of refraction $n = \sqrt{\varepsilon\mu}$. He predicted that such negative-index materials would possess a number of intriguing properties, such as anomalous refraction, reversal of both the Doppler shift and Čerenkov radiation, and even reversal of radiation pressure to radiation tension.

It is easy to see that according to (11.23) materials with a negative index of refraction support undamped propagating waves, like a dielectric, since the resulting refractive index is mostly real. Only if ε and μ are both positive or negative can the material support wave propagation. In the other cases the resulting materials are opaque. Such materials fall into the upper-left and lower-right quadrants of Fig. 11.6. An important question is that of why the refractive index is chosen to be negative. Assuming that $\varepsilon = -1$ and $\mu = -1$, one might conclude that $n = \sqrt{\varepsilon\mu} = \sqrt{(-1)\cdot(-1)} = \sqrt{1} = 1$ and thus positive. However, considering that $\varepsilon(\omega)$ and $\mu(\omega)$ are complex functions, the question can be reduced to the decision regarding which branch of the complex square root should be chosen. This ambiguity can be fixed by requiring that for a passive medium without gain $\mathrm{Im}[n(\omega)] \geq 0$ [11, 12]. The resulting root then reads as

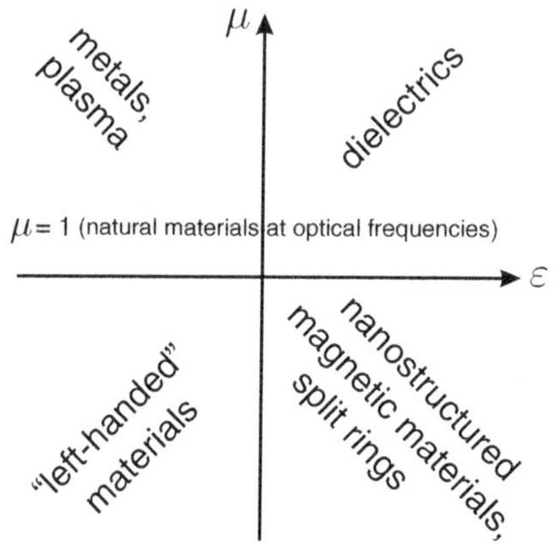

Fig. 11.6 Classification of materials in the (ε, μ) plane (only real parts are considered). Metals and other media with free charge carriers exhibit a negative real part of ε below their plasma frequencies, making them opaque over a broad spectral range. Dielectrics exhibit a positive real part of ε, resulting in good transparency for frequencies below the bandgap. The horizontal gray line indicates the value of $\mu = 1$, which is the value of the magnetic permeability for most materials at optical frequencies. Metamaterials hold promise for creating artificial materials with optical properties that fall into the two lower quadrants, in particular left-handed materials for which both ε and μ are negative.

[1] In Chapter 2 we therefore included μ in all equations.

$$n = \sqrt{|\varepsilon||\mu|}\,\exp\left[\frac{i}{2}\left(\operatorname{arccot}\left(\frac{\mathrm{Re}[\varepsilon]}{\mathrm{Im}[\varepsilon]}\right) + \operatorname{arccot}\left(\frac{\mathrm{Re}[\mu]}{\mathrm{Im}[\mu]}\right)\right)\right], \tag{11.24}$$

which for the case of $\varepsilon, \mu \rightarrow -1$ yields $n = -1$. To give an example, lets assume that both $\varepsilon(\omega)$ and $\mu(\omega)$ exhibit a resonance structure similar to the response of bound electrons in a metal (12.20) according to

$$\begin{aligned} \varepsilon(\omega) &= 1 + \frac{\omega_{\mathrm{p},1}^2}{(\omega_{0,1}^2 - \omega^2) - i\gamma_1\omega}, \\ \mu(\omega) &= 1 + \frac{\omega_{\mathrm{p},2}^2}{(\omega_{0,2}^2 - \omega^2) - i\gamma_2\omega}. \end{aligned} \tag{11.25}$$

It is easy to show by using (11.25) in (11.24) that for sufficiently pronounced resonances a negative real part of $n(\omega)$ can be obtained.[2]

In 2000, it was demonstrated for the first time by Smith *et al.* [13] that negative-index materials operating in the microwave regime *can* be fabricated. Their metamaterial (see Fig. 11.7(a)) is based on a combination of a network of straight wires that mimic a free-electron response to the electric field (see Eq. (12.17)) and millimeter-sized split-ring resonators that are responsible for a magnetic resonance as suggested by Pendry [14]. The concept was soon picked up by other research groups and the wavelength of operation was pushed towards the optical regime. Because of field penetration into the metals, operation at optical frequencies necessitated various modifications of the split-ring resonator design, such as antibonding modes of parallel wire pairs (see Section 13.3.3). A critical issue is the kinetic inductance of metals (see Section 13.3.1), which strongly increases on entering the visible regime. Figure 11.7(b) illustrates the timeline of metamaterial miniaturization [16].

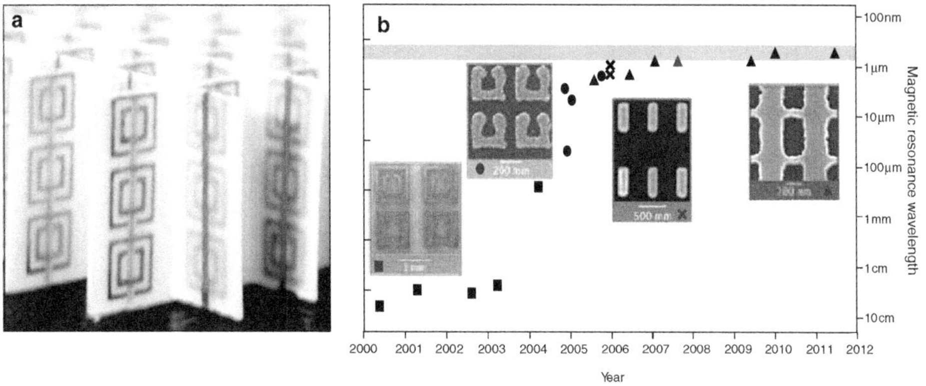

Fig. 11.7 Realization of metamaterials. (a) An artificial material for microwaves based on a network of straight wires and split-ring resonators. From [15]. (b) Timeline of metamaterial miniaturization. Adapted with permission from Macmillan Publishers Ltd from [16].

[2] Note that any negative-index material must be strongly dispersive, i.e. there must exist frequency ranges with a positive refractive index, because otherwise the energy density integrated over all frequencies would be negative.

Note the enormous technological challenge associated with the fabrication of metamaterials at optical frequencies; according to the definition of metamaterials, the "atoms" constituting metamaterials need to be composed of deeply subwavelength structures.

11.2.2 Anomalous refraction and left-handedness

An important demonstration of the surprising effects of negative-index materials is the anomalous refraction of light. Consider the situation depicted in Fig. 11.8, where a plane wave is incident from a medium with index $n = 1$ (left) on a medium with index $n = -1$ (right). According to Snell's law, the angle of refraction turns out to be equal to the angle of incidence but in the negative direction, that is, to the same side of the surface normal as the incident wave. Note that there is no reflected beam in this example since the Fresnel reflection coefficient is zero. Negative refraction can easily be verified by using the boundary conditions for the fields (2.41)–(2.44). While the Poynting vector, **S**, points away from the interface, indicating transport of energy away from the interface, the refracted **k** vector points in the reverse direction! This is a signature of a so-called "backward" propagating wave in which the phase velocity is antiparallel to the group velocity. We further note that in the medium with $n = -1$ the vectors **E**, **H**, and **k** form no longer a right-handed but instead a left-handed tripot, which is the reason why negative-index materials are also called *left-handed materials*. The observed refraction to the "wrong" side is in accordance with Fermat's principle, which states that the light propagates from point to point along the shortest optical path length [9].

The first experimental demonstration of negative refraction was realized using microwaves in 2001 [15]. In 2008, a similar experiment was performed using a three-dimensionally stacked fishnet-type metamaterial using infrared light [17] (see Fig. 11.9).

11.2.3 Imaging with negative-index materials

Negative refraction allows us to devise unconventional optical elements. Veselago had argued that a thin slab of thickness d consisting of a negative-index medium surrounded by air could be used as a focusing lens [11]. A ray-optics beam path of a so-called Veselago

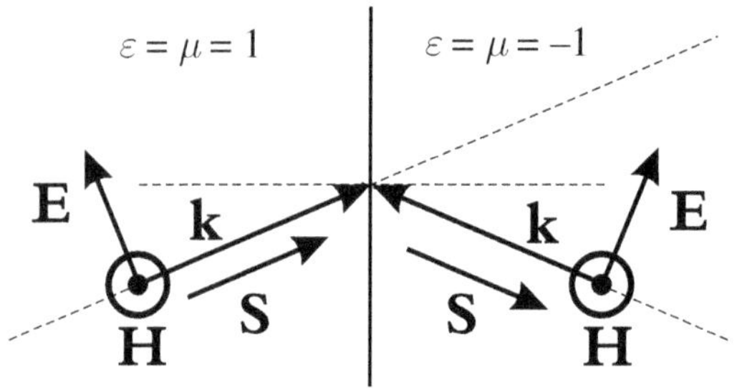

Fig. 11.8 Refraction of a plane wave at the interface between media with positive and negative unity refractive index. Note the opposite directions of **k** and **S** in the medium with the negative refractive index.

lens is sketched in Fig. 11.10. We consider again a slab with $n = -1$ surrounded by a medium with $n = 1$. We choose a point source that is placed to the left of the slab at a distance g from the slab's surface. An image of the point source can be readily constructed using simple ray tracing according to Fig. 11.8. It can be seen that for distances $g < d$ an intermediate image is formed inside the slab at a distance g from the left interface. A second image is formed outside the slab at a distance b from the right interface. By inspection of Fig. 11.10 we find that

$$d = g + b, \tag{11.26}$$

which is the "lens formula" of the system.[3] Evidently, the magnification is 1. In order for the metamaterial to behave as a continuous medium its feature size δ has to fulfill $\delta < \lambda$. On the other hand, operation in the ray-optical regime requires $d > \lambda$. Thus, using (11.26), it follows that $g, b > \lambda > \delta$ [10].

Pendry reconsidered the Veselago lens and pointed out that a negative-index slab provides super-resolution imaging since the evanescent waves of an external source are exponentially amplified inside the negative-index medium, thus making up exactly for the exponential decay outside of the slab [14]. Because such a system would provide unlimited resolution, the idea was termed a "perfect lens" or "super lens" and initiated an enormous research effort. The validity of the "perfect-lens" concept has since been debated and a good summary can be found in [18]. Computer simulations of negative-index materials show that super-resolution can indeed be obtained, but only for thin slabs ($d < \lambda$) and for very small damping [19]. It needs to be emphasized that, in order to amplify evanescent waves, the negative-index material needs to be "loaded" first, which requires time. In the limit in which a point is imaged on a point this time becomes infinitely long. In other words, the perfect lens works "perfectly" for stationary fields only, not for transients. Of course, for finite resolution this restriction is relaxed. Another interesting point is that the perfect lens is based on the requirement that the index of refraction is exactly $n = -1$. What about

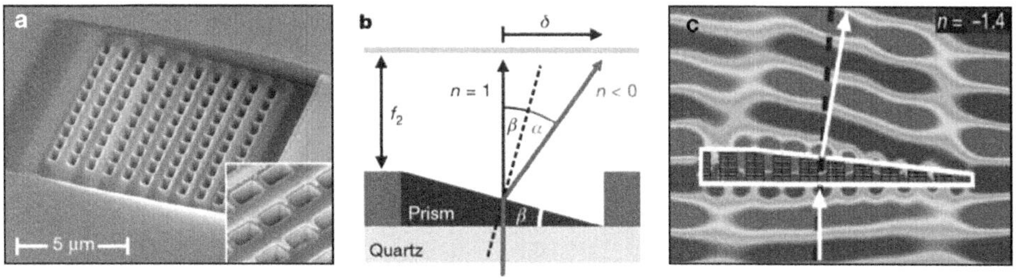

Fig. 11.9 Experimental demonstration of negative refraction of infrared light using a three-dimensionally stacked fishnet metamaterial. (a) A scanning electron microscopy image of the fabricated structure consisting of a nanostructured stack of Ag–MgF_2 layers. (b) A sketch of the experiment, indicating the directions of the refracted wave for different refractive indices. (c) Simulation of the in-plane electric field component for the prism structure at 1763 nm ($n = -1.4$) showing the phase fronts. Reprinted with permission form Macmillan Publishers Ltd [17].

[3] The formula for a standard thin lens is $(1/f) = (1/g) + (1/b)$, with f being the focal length.

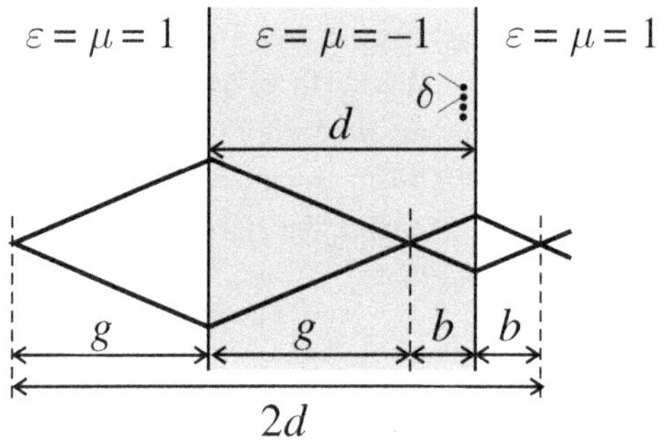

Fig. 11.10 A Veselago lens. Ray-optics picture of point-to-point imaging with a slab of $n = -1$.

a small deviation, i.e. $n = -1 + \Delta n$? It turns out that Δn determines the resolution Δx of the system according to $\Delta x = -2\pi d / \ln(|\Delta n|)$ [20]. Thus, the thicker the slab is, the more susceptible is the lens to deviations from $n = -1$.

In the quasi-static approximation, i.e. the electrostatic limit, electric and magnetic fields are decoupled. For thin slabs with $d \ll \lambda$, the quasi-static approximation holds [14] and a sufficient condition for a perfect lens is that the permittivity ε is negative, irrespective of the value of μ. Such materials no longer need to be metamaterials and can be made of metals or polar dielectrics. Indeed, Rainer Hillenbrand and coworkers have used a SiC film of thickness 440 nm to achieve a negative permittivity at a wavelength of 10 μm, which is close to the surface polariton resonance of SiC. To fulfill the lens formula (11.26), the SiC film was sandwiched between two dielectric layers of equal thickness (220 nm). Near-field optical microscopy was then used to record a super-resolved image of a planar structure that was placed on the other side of the SiC superlens [21].

Although experimental realizations and applications of metamaterials still lag behind the theoretical efforts, there is a constant stream of new ideas regarding how to fabricate and make use of metamaterials. These include anisotropic materials with a negative ε only in certain directions. Such materials give rise to a hyperbolic dispersion relation $\omega(\mathbf{k})$, which in principle allows the propagation of waves with arbitrarily large wavevectors and thus should lead to super-resolution [22, 23]. Hyperbolic metamaterials are also of interest in the field of quantum electrodynamics since they provide a singular density of states over a wide wavelength range. Other ideas involve the use of "transformation optics" [24] to design metamaterials with specific properties, such as invisibility cloaks.

11.3 Optical microcavities

Optical microcavities formed by dielectric spheres have attracted considerable interest in various fields of research. The high quality factors associated with the resonant modes have inspired experiments in cavity quantum electrodynamics and gave rise to sensitive biosensors, and the high energy density in the cavities allowed researchers to observe various nonlinear processes such as switching of coherent light, low-threshold lasing, and stimulated Raman scattering [25].

To understand these processes it is necessary to solve Maxwell's equations for the simple geometry of a sphere. The mathematical basis is identical with the famous Mie theory, and the details can be found in various excellent books such as Ref. [26]. Although the Mie theory is in excellent agreement with experimental measurements, the convergence of the expansions is very slow for spheres with diameters $D \gg \lambda$ [27]. For such spheres it is observed that small variations in the initial conditions (size, dielectric constant) lead to considerable variations of the scattering cross-section. These variations, called ripples, can be associated with sphere resonances. For each ripple peak, light remains trapped for a long time inside the sphere and orbits near the surface by multiple total internal reflections. These resonant modes are called *whispering-gallery modes* or *morphology-dependent resonances*. The Q-factors of the resonant modes are always finite, but can theoretically be as large as 10^{21}. Consequently, the resonant modes are leaky modes and the sphere is a non-conservative system because energy is permanently lost due to radiation. The largest experimentally observed Q-factors are on the order of $Q = 10^{10}$.

Instead of reproducing the full Mie theory, we intend to provide an intuitive picture for the resonances occurring in optical microspheres. This picture was developed by Nussenzveig and Johnson [27, 28], and is called the *effective-potential approach*. It has a direct analogy to the quantum-mechanical theory of a finite spherical well. The finite Q-factors of microspheres can be associated with the phenomenon of tunneling.

Let us consider a homogeneous sphere with dielectric constant ε_1 and radius a surrounded by a homogeneous medium with dielectric constant ε_2. The complex field amplitudes inside and outside of the sphere have to satisfy the vector Helmholtz equation

$$\left[\nabla^2 + \frac{\omega^2}{c^2}\varepsilon_i\right]\mathbf{E}(\mathbf{r}) = \mathbf{0}, \tag{11.27}$$

where $i \in [1, 2]$, depending on whether the field is evaluated inside or outside of the sphere. A similar equation holds for the magnetic field $\mathbf{H}$. Using the mathematical identity

$$\nabla^2 [\mathbf{r} \cdot \mathbf{E}(\mathbf{r})] = \mathbf{r} \cdot \left[\nabla^2\mathbf{E}(\mathbf{r})\right] + 2\nabla \cdot \mathbf{E}(\mathbf{r}), \tag{11.28}$$

setting the last term equal to zero, and inserting the result into Eq. (11.27) leads to the scalar Helmholtz equation

$$\left[\nabla^2 + \frac{\omega^2}{c^2}\varepsilon_i\right] f(\mathbf{r}) = 0, \qquad f(\mathbf{r}) = \mathbf{r} \cdot \mathbf{E}(\mathbf{r}). \tag{11.29}$$

Separation of variables yields

$$f(r, \vartheta, \varphi) = Y_l^m(\vartheta, \varphi) R_l(r), \tag{11.30}$$

with Y_l^m being the spherical harmonics and R_l being a solution of the radial equation

$$\left[\frac{\mathrm{d}}{\mathrm{d}r^2} + \left(\frac{\omega^2}{c^2}\varepsilon_i - \frac{l(l+1)}{r^2}\right)\right] rR_l(r) = 0. \tag{11.31}$$

The solutions of this equation are the spherical Bessel functions (see Section 16.1).

A similar equation is encountered in quantum mechanics. For a spherically symmetric potential $V(\mathbf{r}) = V(r)$ one obtains the radial Schrödinger equation

$$\left[-\frac{\hbar^2}{2m}\frac{\mathrm{d}}{\mathrm{d}r^2} + \left(V(r) + \frac{\hbar^2}{2m}\frac{l(l+1)}{r^2}\right)\right] rR_l(r) = ErR_l(r), \tag{11.32}$$

where $\hbar$ is the reduced Planck constant and m the effective mass. Besides the centrifugal term with $1/r^2$ dependence, the equation is identical in form with the one-dimensional Schrödinger equation. The expression in the round brackets is called the *effective potential*, $V_{\mathrm{eff}}(r)$.

The similarity between the electromagnetic problem and the quantum-mechanical problem allows us to introduce an effective potential V_{eff} and an energy E for the dielectric sphere. From the identity of the two equations in free space ($V = 0$, $\varepsilon_i = 1$) we find

$$E = \frac{\hbar^2}{2m}\frac{\omega^2}{c^2}. \tag{11.33}$$

With this definition, the effective potential of the dielectric sphere turns out to be

$$V_{\mathrm{eff}}(r) = \frac{\hbar^2}{2m}\left[\frac{\omega^2}{c^2}(1 - \varepsilon_i) + \frac{l(l+1)}{r^2}\right]. \tag{11.34}$$

Figure 11.11 shows the effective potential for a dielectric sphere in air. The abrupt change of ε at the boundary of the sphere gives rise to a discontinuity in V_{eff} and thus to a potential well. The horizontal line in Fig. 11.11 indicates the energy E as defined in Eq. (11.33). Notice that, unlike in quantum mechanics, the energy E depends on the shape of the potential well. Thus, a change of V_{eff} will also affect E.

As for quantum-mechanical tunneling, the finite height of the potential barrier gives rise to energy leakage through the barrier. Thus, a resonant mode in the optical microcavity will damp out with a characteristic time defined by the tunneling rate through the barrier. In quantum mechanics, only discrete energy values are possible for the states within the potential well. These values follow from an energy eigenvalue equation defined by the boundary conditions. The situation is similar for the electromagnetic problem, where we can distinguish between two kinds of modes, namely TE modes and TM modes. They are defined as

$$\text{TE modes: } \mathbf{r} \cdot \mathbf{E}(\mathbf{r}) = 0, \tag{11.35}$$

$$\text{TM modes: } \mathbf{r} \cdot \mathbf{H}(\mathbf{r}) = 0. \tag{11.36}$$

For TE modes, the electric field is always transverse to the radial vector, and for TM modes the same holds for the magnetic field.

The boundary conditions at the surface of the sphere ($r = a$) connect the interior fields with the exterior fields. The radial dependence of the interior field is expressed in terms of spherical Bessel functions j_l and the exterior field in terms of spherical Hankel functions of the first kind $h_l^{(1)}$. j_l ensures that the field is regular within the sphere, whereas $h_l^{(1)}$ is required to fulfill the radiation condition at infinity. The boundary conditions lead to a homogeneous system of equations, from which the following *characteristic equations* are derived:

$$\text{TE modes:} \quad \frac{\psi_l'(\tilde{n}x)}{\psi_l(\tilde{n}x)} - \tilde{n}\frac{\zeta_l'(x)}{\zeta_l(x)} = 0, \tag{11.37}$$

$$\text{TM modes:} \quad \frac{\psi_l'(\tilde{n}x)}{\psi_l(\tilde{n}x)} - \frac{1}{\tilde{n}}\frac{\zeta_l'(x)}{\zeta_l(x)} = 0. \tag{11.38}$$

Here, the ratio of interior to exterior refractive indices is denoted by $\tilde{n} = \sqrt{\varepsilon_1/\varepsilon_2}$ and x is the size parameter defined as $x = ka$, with k being the vacuum wavenumber $k = \omega/c = 2\pi/\lambda$. The primes denote differentiations with respect to the argument and ψ_l and ζ_l are Ricatti–Bessel functions defined as

$$\psi_l(z) = z j_l(z), \qquad \zeta_l(z) = z h_l^{(1)}(z). \tag{11.39}$$

For a given angular momentum mode number l, there are many solutions of the characteristic equations. These solutions are labeled with a new index ν, called the radial mode order. As shown in Fig. 11.12, ν indicates the number of peaks in the radial intensity distribution inside the sphere. Among all the possible solutions, only those solutions whose energies according to Eq. (11.33) lie within the range demarcated by the bottom and top of the potential well are considered resonant modes. Notice that the characteristic equations (11.37) and (11.38) cannot be fulfilled for real x, which means that the eigenfrequencies $\omega_{\nu l}$ are *complex*. Consequently, the modes of the microsphere are *leaky modes* and the

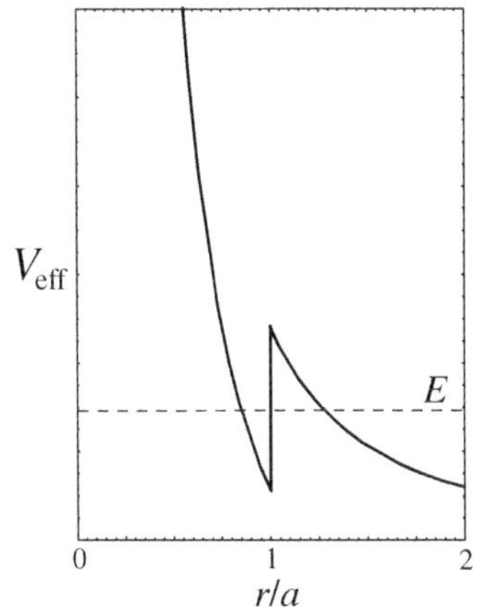

Fig. 11.11 The effective potential V_{eff} for a dielectric sphere according to Eq. (11.34). The radiative decay of a resonant mode can be associated with energy tunneling through the potential barrier. The following parameters were used: $\varepsilon_1 = 2.31$, $\varepsilon_2 = 1$, $\lambda = 800$ nm, $l = 500$, and $a = 50$ μm.

stored energy is continuously dissipated through radiation. The real part of $\omega_{\nu l}$ denotes the center frequency ω_0 of the mode, and the imaginary part indicates half the width $\Delta\omega$ of the resonance. Thus, the Q-factor can be expressed as

$$Q = \frac{\omega_0}{\Delta\omega} = \frac{\mathrm{Re}\{\omega_{\nu l}\}}{2\,|\mathrm{Im}\{\omega_{\nu l}\}|}. \tag{11.40}$$

Because of the dissipative nature of the resonances, the modes are referred to as quasi-normal modes.

To better visualize the classification of modes, we consider the example of a glass sphere ($a = 10\,\mu\text{m}$, $\varepsilon_1 = 2.31$) in air ($\varepsilon_2 = 1$) and we assume an angular momentum mode number of $l = 120$. The wavelength of the mode with the highest Q-factor can be estimated from the geometrical requirement that the circumference of the sphere must be a multiple of the internal wavelength

$$\text{highest-}Q\text{ mode:}\;\; l \approx nka, \tag{11.41}$$

where n is the interior index of refraction. For the present example we find $\lambda \approx 796\,\text{nm}$ or $x \approx 79$ and the spectral separation between adjacent l-modes is $\Delta\lambda \approx \lambda^2/(2\pi an) = 6.6\,\text{nm}$.

Solving Eq. (11.37) for $l = 120$ yields the values (real parts) $\lambda^{\text{TE}}_{1,120} = 743.25\,\text{nm}$, $\lambda^{\text{TE}}_{2,120} = 703.60\,\text{nm}$, $\lambda^{\text{TE}}_{3,120} = 673.35\,\text{nm}$, ... Similarly, the solutions of Eq. (11.38) are $\lambda^{\text{TM}}_{1,120} = 739.01\,\text{nm}$, $\lambda^{\text{TM}}_{2,120} = 699.89\,\text{nm}$, $\lambda^{\text{TM}}_{3,120} = 670.04\,\text{nm}$, ... The $\nu = 1$ modes, with a single energy maximum inside the sphere, have the highest Q-factors. Their wavelengths are in rough agreement with the estimate of $\lambda \approx 796\,\text{nm}$ according to Eq. (11.41). TM modes exhibit shorter wavelengths than do TE modes. Generally, the Q-factor decreases with increasing radial mode number. For the current example, the Q-factor decreases from $\approx 10^{17}$ for the $\nu = 1$ modes to $\approx 10^6$ for the $\nu = 6$ modes. Figure 11.13 shows the spectral positions of the $l = 119$, $l = 120$, and $l = 121$ modes. The spacing between same-l-modes is $\approx 6\,\text{nm}$, in agreement with the previous estimate. Modes are represented as vertical lines, the height of which indicates the Q-factor on a logarithmic scale. Solid lines are TE modes and dashed lines are TM modes. A dense network of modes is formed when all

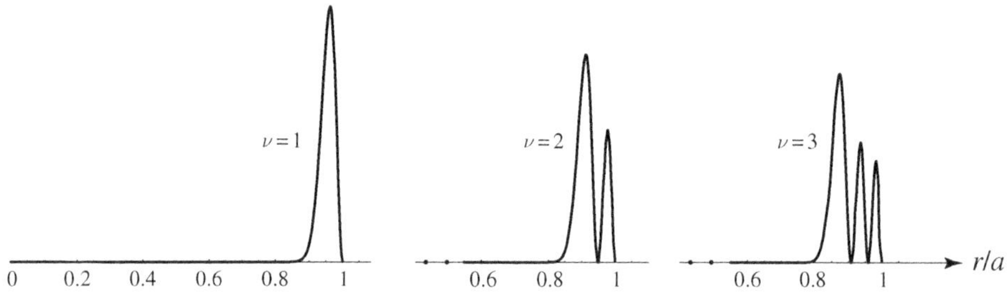

Fig. 11.12 Radial energy distribution of TM modes with angular momentum mode number $l = 120$. The microsphere has a dielectric constant of $\varepsilon = 2.31$. The radial mode number ν indicates the number of energy maxima in the radial direction.

l-modes are plotted on the same axis. Furthermore, since the azimuthal modes (mode number m) are degenerate, each l-mode consists of a multitude of submodes. The degeneracy is lifted by geometrical asymmetries or material imperfections resulting in even more mode frequencies.

The calculated Q-factors account only for radiation losses. For microspheres with $a > 500\,\mu\text{m}$ these Q-factors can be larger than 10^{20}. However, the highest measured Q-factors are on the order of 10^{10}, indicating that other contributions such as surface roughness, shape deformations, absorption, or surface contamination are the limiting factors for a high Q. These factors can be taken into account by defining the total quality factor of a particular microcavity mode as

$$\frac{1}{Q_{\text{tot}}} = \frac{1}{Q} + \frac{1}{Q_{\text{other}}}, \tag{11.42}$$

where Q is the radiation-limited, theoretical quality factor and Q_{other} accounts for all other contributions. Usually, Q can be neglected in comparison with Q_{other}. Near a resonance with angular frequency ω_0, the electric field takes on the form

$$\mathbf{E}(t) = \mathbf{E}_0 \exp\left[\left(\mathrm{i}\omega_0 - \frac{\omega_0}{2Q_{\text{tot}}}\right) t\right], \tag{11.43}$$

and the stored energy density assumes a Lorentzian distribution

$$W_\omega(\omega) = \frac{\omega_0^2}{4Q_{\text{tot}}^2} \frac{W_\omega(\omega_0)}{(\omega - \omega_0)^2 + [\omega_0/(2Q_{\text{tot}})]^2}. \tag{11.44}$$

While spherical microcavities can have nearly atomic-scale surface smoothness and therefore high Q-factors, they are not easily integrated into optoelectronic devices. Toroidal microcavities, as shown in Fig. 11.14, overcome this limitation. They are amenable to wafer-based processing and yield Q-factors in excess of 10^8.

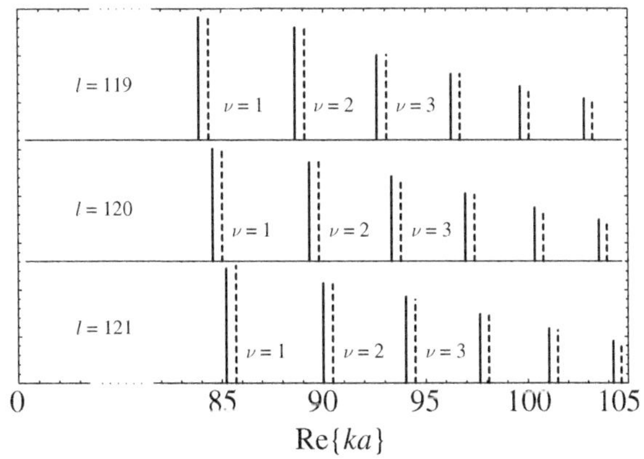

Fig. 11.13 Normalized mode frequencies for a microsphere with $\varepsilon = 2.31$ and angular momentum mode numbers $l = 119$, $l = 120$, and $l = 121$. Solid lines are TE modes and dashed lines are TM modes. The height of the lines indicates the quality factor on a logarithmic scale. The $\nu = 1$ modes have a Q-factor of $\approx 10^{17}$ and the $\nu = 6$ modes have a Q-factor of $\approx 10^6$.

The mode structure of a microsphere gives rise to a discrete photonic density of states ρ, as qualitatively illustrated in Fig. 11.15. ρ depends on the position relative to the microsphere and on the orientation of the transition dipole (see Section 8.4.3). Efficient energy transfer between molecules and other quantum systems can be accomplished only within the narrow frequency windows of individual resonant modes. Also, the excited-state lifetime of a molecule is strongly reduced if its emission frequency coincides with the frequency of a resonant mode. On the other hand, the lifetime can be drastically prolonged if the emission frequency is between two mode frequencies. If the emission bandwidth of a molecule spans several mode frequencies, the fluorescence spectrum will consist of discrete lines. The same is true for the absorption spectrum. Thus, the free-space spectra of emission and absorption are sampled with the discrete mode spectrum of a microcavity. Since energy transfer between molecules depends on the overlap of emission and absorption spectra (see Section 8.6.2), it would be expected, at first glance, that the energy-transfer efficiency is reduced in or near a microcavity because the overlap bandwidth associated with the narrow mode frequencies is drastically reduced compared with the free-space situation. However, for a high-Q cavity this is not the case because the density of states at the frequency of a resonant mode is so high that the overlap integral becomes much larger than that in free space, despite the narrower bandwidth. Arnold and coworkers have shown that energy transfer in a microsphere can be several orders more efficient than it is in free space [25], making microspheres promising candidates for long-range energy transfer. Microspheres have been used in applications such as biosensors, optical switching, and cavity QED. Various other experiments can be thought of, such as two-photon energy transfer, and exciting results can be expected in the near future.

11.3.1 Cavity perturbation

A sharp resonance is a key requirement for ultrasensitive detection in various applications. For example, watches and clocks use high-Q quartz crystals to measure time, some

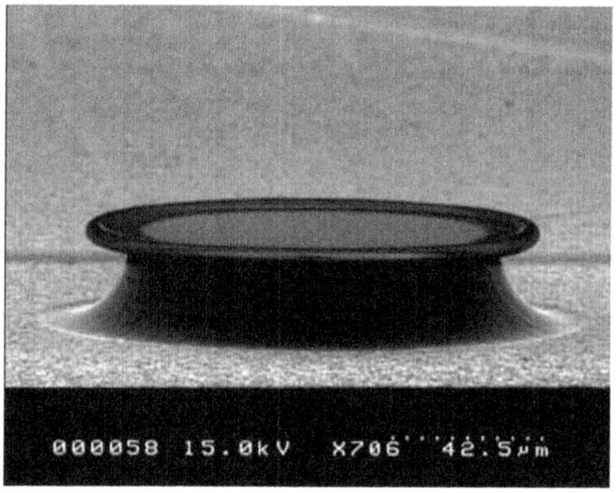

Fig. 11.14 An ultrahigh-Q toroidal microresonator fabricated on-chip by a combination of lithography, dry etching, and a selective reflow process. Adapted from [29].

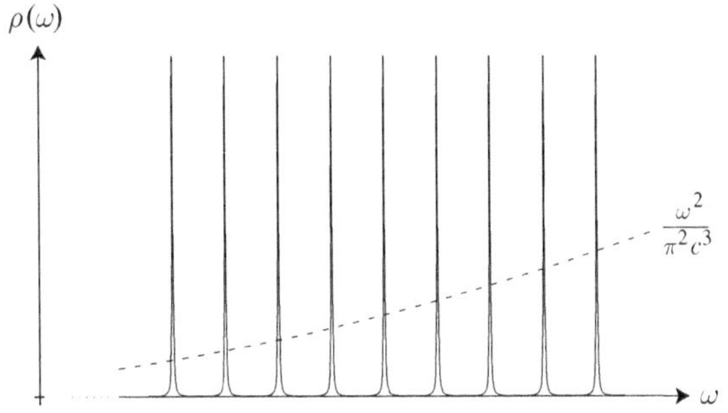

Fig. 11.15 The photonic density of states of a microsphere (solid line) and in free space (dashed line). In a microsphere all energy is concentrated in the narrow frequency windows of individual resonant modes.

biosensing schemes make use of oscillating cantilevers to detect adsorption of molecules, and atomic clocks use atomic resonances as frequency standards. Because of their extremely high Q-factor, optical microcavities, such as microspheres or toroidal resonators, are attractive candidates for various biosensing schemes. A perturbation of the optical microcavity, for example due to particle adsorption or a change of the index of refraction, leads to a shift of the resonance frequency, which can be measured and used as a control signal [30].

To establish an understanding of cavity perturbation we consider the system depicted in Fig. 11.16. A leaky microcavity and its environment are characterized by a spatially varying permittivity $\varepsilon(\mathbf{r})$ and permeability $\mu(\mathbf{r})$. In the absence of any perturbation the system assumes a resonance at frequency ω_0 and the fields are described by

$$\nabla \times \mathbf{E}_0 = \mathrm{i}\omega_0 \mu_0 \mu(\mathbf{r}) \mathbf{H}_0 , \qquad \nabla \times \mathbf{H}_0 = -\mathrm{i}\omega_0 \varepsilon_0 \varepsilon(\mathbf{r}) \mathbf{E}_0, \tag{11.45}$$

with $\mathbf{E}_0(\mathbf{r}, \omega_0)$ and $\mathbf{H}_0(\mathbf{r}, \omega_0)$ denoting the unperturbed complex field amplitudes. A particle with anisotropic material parameters $\boldsymbol{\Delta}\varepsilon(\mathbf{r})$ and $\boldsymbol{\Delta}\mu(\mathbf{r})$ constitutes a perturbation and gives rise to a new resonance frequency ω.[4] Maxwell's curl equations for the perturbed system read as

$$\nabla \times \mathbf{E} = \mathrm{i}\omega\mu_0 \left[\mu(\mathbf{r})\mathbf{H} + \boldsymbol{\Delta}\mu\,(\mathbf{r})\mathbf{H}\right], \tag{11.46}$$

$$\nabla \times \mathbf{H} = -\mathrm{i}\omega\varepsilon_0 \left[\varepsilon(\mathbf{r})\mathbf{E} + \boldsymbol{\Delta}\varepsilon(\mathbf{r})\mathbf{E}\right]. \tag{11.47}$$

Notice that both $\boldsymbol{\Delta}\varepsilon$ and $\boldsymbol{\Delta}\mu$ are zero outside of the volume occupied by the perturbation. Using $\nabla \cdot (\mathbf{A} \times \mathbf{B}) = (\nabla \times \mathbf{A}) \cdot \mathbf{B} - (\nabla \times \mathbf{B}) \cdot \mathbf{A}$ we find

$$\begin{aligned}\nabla \cdot \left[\mathbf{E}_0^* \times \mathbf{H} - \mathbf{H}_0^* \times \mathbf{E}\right] &= \mathrm{i}(\omega - \omega_0)\left[\varepsilon_0\varepsilon(\mathbf{r})\mathbf{E}_0^* \cdot \mathbf{E} + \mu_0\mu(\mathbf{r})\mathbf{H}_0^* \cdot \mathbf{H}\right] \\ &+ \mathrm{i}\omega\left[\mathbf{E}_0^*\varepsilon_0\boldsymbol{\Delta}\varepsilon(\mathbf{r})\mathbf{E} + \mathbf{H}_0^*\mu_0\boldsymbol{\Delta}\mu(\mathbf{r})\mathbf{H}\right].\end{aligned} \tag{11.48}$$

We now consider a fictitious spherical surface ∂V at very large distance from the cavity and integrate Eq. (11.48) over the enclosed volume V (c.f. Fig. 11.16). Using Gauss's theorem, the left-hand side of Eq. (11.48) becomes

[4] $\boldsymbol{\Delta}\varepsilon$ and $\boldsymbol{\Delta}\mu$ are tensors of rank 2.

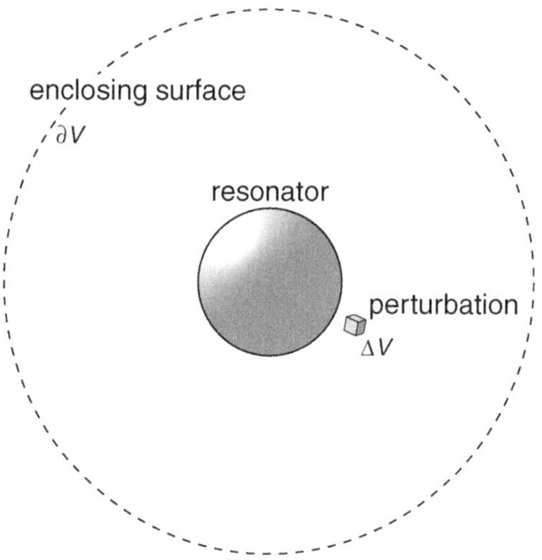

Fig. 11.16 An optical resonator with resonance frequency ω_0 interacts with an external perturbation, giving rise to a new resonance frequency ω. The calculation makes use of a fictitious spherical surface at infinity.

$$\int_{\partial V} \left[\mathbf{H} \cdot \left(\mathbf{n} \times \mathbf{E}_0^*\right) + \mathbf{H}_0^* \cdot (\mathbf{n} \times \mathbf{E})\right] \mathrm{d}a = 0, \tag{11.49}$$

where $\mathbf{n}$ is a unit vector normal to the surface ∂V. The above expression vanishes because of the transversality of the field, i.e. $(\mathbf{n} \times \mathbf{E}_0^*) = (\mathbf{n} \times \mathbf{E}) = 0$ on the surface of the spherical surface. We thus arrive at the equation

$$\frac{\omega - \omega_0}{\omega} = -\frac{\int_V [\mathbf{E}_0^* \varepsilon_0 \boldsymbol{\Delta}\varepsilon(\mathbf{r})\mathbf{E} + \mathbf{H}_0^* \mu_0 \boldsymbol{\Delta}\mu(\mathbf{r})\mathbf{H}] \mathrm{d}V}{\int_V \left[\varepsilon_0 \varepsilon(\mathbf{r}) \mathbf{E}_0^* \cdot \mathbf{E} + \mu_0 \mu(\mathbf{r}) \mathbf{H}_0^* \cdot \mathbf{H}\right] \mathrm{d}V}, \tag{11.50}$$

which is known as the Bethe–Schwinger cavity perturbation formula [31, 32]. Equation (11.50) is an exact formula, but because $\mathbf{E}$ and $\mathbf{H}$ are not known the equation cannot be used in this form. Notice that because $\boldsymbol{\Delta}\varepsilon$ and $\boldsymbol{\Delta}\mu$ are zero outside of the volume occupied by the perturbation the integral in the nominator runs only over the volume of the perturbation ΔV. For situations in which there are no radiation losses and all the energy is contained inside the boundaries of a resonator the surface ∂V can be chosen to coincide with the boundaries.

We assume that the perturbation has a small effect on the cavity. Therefore we write as a first-order approximation $\mathbf{E} = \mathbf{E}_0$ and $\mathbf{H} = \mathbf{H}_0$. After performing these substitutions in Eq. (11.50) we find

$$\frac{\omega - \omega_0}{\omega} \approx -\frac{\int_{\Delta V} [\mathbf{E}_0^* \varepsilon_0 \boldsymbol{\Delta}\varepsilon(\mathbf{r})\mathbf{E}_0 + \mathbf{H}_0^* \mu_0 \boldsymbol{\Delta}\mu(\mathbf{r})\mathbf{H}_0] \mathrm{d}V}{\int_V \left[\varepsilon_0 \varepsilon(\mathbf{r}) \mathbf{E}_0^* \cdot \mathbf{E}_0 + \mu_0 \mu(\mathbf{r}) \mathbf{H}_0^* \cdot \mathbf{H}_0\right] \mathrm{d}V}. \tag{11.51}$$

For a high-Q resonator the radiation losses are small and the integration volume V can be taken over the boundaries of the resonator. To evaluate Eq. (11.51) we first must solve for the fields $\mathbf{E}_0(\mathbf{r})$ and $\mathbf{H}_0(\mathbf{r})$ of the unperturbed cavity. Interestingly, for a weakly dispersive medium the denominator of Eq. (11.51) denotes the total energy of the unperturbed cavity (W_0), whereas the nominator accounts for the energy introduced by the perturbation (ΔW). Hence, $(\omega - \omega_0)/\omega = -\Delta W/W_0$. An increase of the energy by ΔW causes the resonance frequency to *redshift* to $\omega = \omega_0 \left[W_0/(W_0 + \Delta W)\right]$. A *blueshift* can be achieved by perturbing the cavity volume, i.e. by removing ΔW from the cavity.

As an example let us consider a planar cavity with perfectly reflecting end-faces of area A and separated by a distance L. The fundamental mode $\lambda = 2L$ has a resonance frequency $\omega_0 = \pi c/L$, and the electric and magnetic fields inside the cavity are calculated to be $E_0 \sin(\pi z/L)$ and $-\mathrm{i}\sqrt{\varepsilon_0/\mu_0}E_0 \cos(\pi z/L)$, respectively. The coordinate z is perpendicular to the surfaces of the end-faces. The denominator of Eq. (11.51) is easily determined to be $V\varepsilon_0 E_0^2$, where $V = LA$. We place a spherical nanoparticle with dielectric constant $\Delta\varepsilon$ and volume ΔV in the center of the cavity and assume that the field is homogeneous across the dimensions of the particle. The nominator of Eq. (11.51) is calculated to be $\Delta V\, \Delta\varepsilon\, \varepsilon_0 E_0^2$ and the frequency shift is determined to be $(\omega - \omega_0)/\omega = -\Delta\varepsilon\, \Delta V/V$. A better approximation retains the perturbed fields $\mathbf{E}$ and $\mathbf{H}$ in the nominator of Eq. (11.50). Making use of the quasi-static solution for a small spherical particle (c.f. Section 12.3.1), we write $\mathbf{E} = 3\mathbf{E}_0/(2+\Delta\varepsilon)$ and obtain a frequency shift of $(\omega-\omega_0)/\omega = -[3\,\Delta\varepsilon/(2+\Delta\varepsilon)]\Delta V/V$. In both cases the resonance shift scales with the ratio of resonator and perturbation volumes.

11.4 Cavity optomechanics

A mechanical force acting on an optical system can influence the state of the optical field. For example, a force applied to an optical microcavity can be used to control the cavity resonance and the coupling to an external laser field. *Vice versa*, optical radiation acting on a mechanical system can influence the dynamics of the system, for example, through laser heating or radiation pressure. The mutual coupling of mechanical and optical degrees of freedom is being explored in the emerging field of *cavity optomechanics* [33].

The interaction between light and matter sets ultimate limits on the accuracy of optical measurements. Braginsky predicted that the finite response time of light in an optical interferometer can lead to mechanical instabilities [34] and impose limits on the precision of laser-based gravitational interferometers. Later, it was demonstrated that this "dynamic backaction mechanism" can also be used to slow down the motion of a mechanical system and to effectively cool it below the temperature of the environment [35–39].

To conceptually understand optomechanical coupling, let us consider the laser-irradiated mechanical oscillator shown in Fig. 11.17(a). The oscillator consists of a mirror of mass m attached by a spring of stiffness K_0 to a rigid wall. The mirror oscillates with amplitude x_0 at the mechanical resonance frequency $\Omega_0 = \sqrt{K_0/m}$ and periodically modulates the phase of the reflected wave E_r. For mirror velocities much smaller than the speed of light ($x_0\Omega_0 \ll c$) the reflected field can be represented as [40]

$$E_r = -E_0 \, \mathrm{Re}\left\{ e^{-ikx} \sum_{n=-\infty}^{\infty} e^{-i(\omega+n\Omega_0)t} J_{-n}(2kx_0) \right\}, \tag{11.52}$$

where J_{-n} are Bessel functions of order $-n$, and ω and $k = \omega/c$ are the angular frequency and the wavenumber of the incident field, respectively. For oscillation amplitudes much smaller than the wavelength of light ($x_0 \ll \lambda$) we can neglect terms with $|n| > 1$ and end up with three terms of different frequency, namely ω and the two sidebands $\omega \pm \Omega_0$.

Let us now assume that the reflected wave E_r is not emitted into free space but coupled to an optical cavity with a narrow energy spectrum of width $\gamma_0 \ll \Omega_0$. If we choose the cavity spectrum to overlap with one of the reflected sidebands, say $\omega + \Omega_0$, then we effectively suppress the contribution of $\omega - \Omega_0$ (see Fig. 11.17(b)). As a result, the reflected field has more quanta of energy $\hbar(\omega + \Omega_0)$ than quanta of energy $\hbar(\omega - \Omega_0)$, implying that energy is being gained upon reflection. In other words, the laser field extracts energy from the mechanical oscillator, thereby lowering its center-of-mass temperature. This type of laser cooling is referred to as *resolved-sideband cooling*.

To establish a more quantitative understanding of optomechanical coupling we analyze the situation depicted in Fig. 11.18, where the mechanical oscillator forms one of the end mirrors of an optical cavity. The system is characterized by the coupling of the cavity field $E(t)$ and the mechanical displacement $x(t)$ of the mirror. A displacement of the mirror causes a change in the resonance frequency of the cavity and, *vice versa*, a shift in resonance frequency changes the light intensity and hence the force acting on the oscillator. Thus, a displacement of the mirror acts back on itself, a phenomenon referred to as *dynamical back-action*.

We first analyze the field inside the cavity. The field $\mathbf{E}(\mathbf{r}, t)$ satisfies the Helmholtz equation

$$\nabla^2 \mathbf{E}(\mathbf{r}, t) - \frac{1}{c^2} \frac{d^2}{dt^2} \mathbf{E}(\mathbf{r}, t) = 0. \tag{11.53}$$

We express the solution of $\mathbf{E}$ in terms of normalized eigenmodes $\mathbf{u}_n$ using the following ansatz:

$$\mathbf{E}(\mathbf{r}, t) = e^{-i\omega t} \sum_n E_n(t) \, \mathbf{u}_n(\mathbf{r}). \tag{11.54}$$

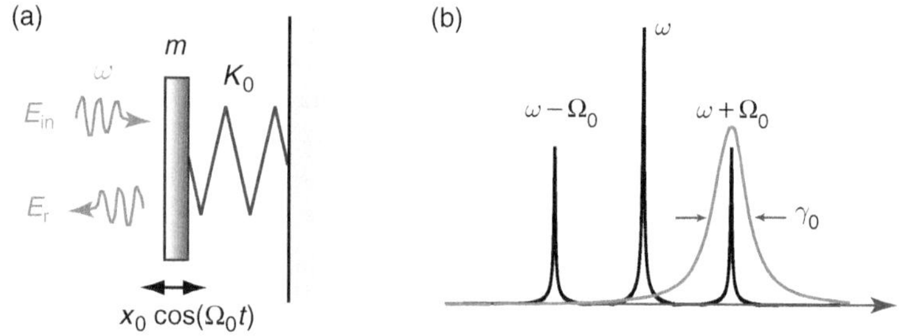

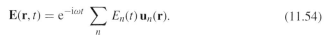

Fig. 11.17 Reflection of light from an oscillating mirror. (a) The oscillating mirror modulates the phase of the incident wave and gives rise to new frequency components ($\omega \pm n\Omega_0$) upon reflection. (b) Resolved sideband cooling. By coupling the optical field to a cavity it is possible to spectrally select one of the frequency components and to suppress the others.

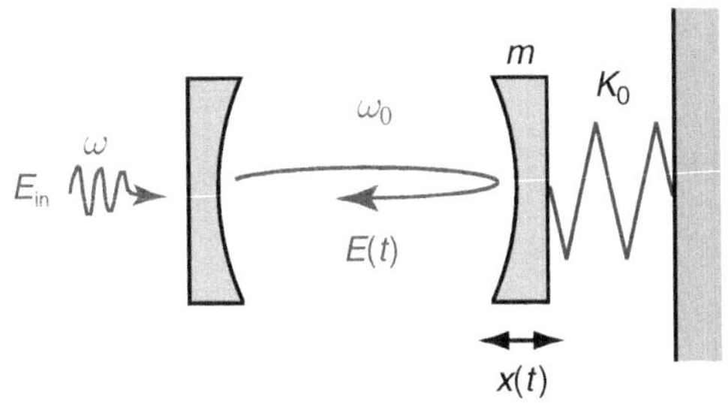

Fig. 11.18 Coupling of an optical cavity and a mechanical oscillator of frequency Ω_0. A displacement of the mirror shifts the resonance frequency of the cavity, which changes the stored energy and hence the force acting on the mirror.

Here, the eigenmodes satisfy the eigenvalue equation $[\nabla^2 + \omega_n^2/c^2]\mathbf{u}_n = 0$, with ω_n being the eigenfrequencies. We assume that the frequency ω is very close to the eigenfrequency ω_0 and retain only the $n = 0$ term in the sum over eigenmodes. Inserting ansatz (11.54) into Eq. (11.53), making use of the eigenvalue equation, multiplying by $\mathbf{u}_0^*$, integrating over the cavity volume, and making use of orthogonality leads to a differential equation for $E_0(t)$. Because $\ddot{E}_0 \ll -\mathrm{i}\omega\dot{E}_0$ for a good cavity, we may ignore second-order derivatives and obtain $\dot{E}_0 = \mathrm{i}(\omega - \omega_0)E_0$, where we used $\omega_0^2 - \omega^2 \approx 2\omega(\omega_0 - \omega)$. To account for cavity losses and for the incoupling and outcoupling of radiation we add two terms and finally obtain

$$\frac{\mathrm{d}}{\mathrm{d}t}E_0(t) = \big[\mathrm{i}(\omega - \omega_0) - \gamma_0\big]E_0(t) + \kappa E_{\mathrm{in}}(t). \tag{11.55}$$

Cavity losses and the outcoupling of radiation are accounted for by the decay rate γ_0, whereas the rate of energy incoupling is described by the coupling constant κ. While ω is the center frequency of the incident radiation and the field in the cavity, ω_0 is the resonance frequency of the cavity. The latter shifts if the cavity length L is changed. Equation (11.55) is the basic equation that governs the dynamics of the field in the cavity. For $\omega = \omega_0$ and $E_{\mathrm{in}} = 0$ we obtain the solution (11.43) with $Q_{\mathrm{tot}} = \omega_0/(2\gamma_0)$. Notice that the cavity decay rate has two contributions, that is $\gamma_0 = \gamma_{\mathrm{ex}} + \gamma_{\mathrm{in}}$, where γ_{ex} is associated with outcoupling of radiation and γ_{in} is due to internal losses. The lifetime of a photon in the cavity is $\tau = 1/(2\gamma_0)$.

We now turn to the equation of motion of the mechanical oscillator, which can be written as

$$\frac{\mathrm{d}^2}{\mathrm{d}t^2}x(t) + \Gamma_0\frac{\mathrm{d}}{\mathrm{d}t}x(t) + \Omega_0^2 x(t) = \frac{1}{m}\big[F_{\mathrm{fluct}}(t) + F_{\mathrm{opt}}(t)\big]. \tag{11.56}$$

We have split the driving force into two terms, a random Langevin force, satisfying $\langle F_{\mathrm{fluct}}(t)F_{\mathrm{fluct}}(t')\rangle = 2m\Gamma_0 k_{\mathrm{B}}T\delta(t - t')$, and a term due to the interaction with light.

Having established the equations for the cavity field and the mechanical oscillator, we now focus on their mutual coupling. First, we note that a small change of x in cavity length L will shift the resonance frequency by $\Delta\omega_0 = -\omega_0\, x/L$. Equation (11.55) must therefore be replaced by

$$\frac{\mathrm{d}}{\mathrm{d}t}E_0(t) = \Big[\mathrm{i}\big(\omega - \omega_0\{1 - x(t)/L\}\big) - \gamma_0\Big]E_0(t) + \kappa E_{\mathrm{in}}(t). \tag{11.57}$$

Second, the optical force F_{opt} depends on the cavity field $E_0(t)$. In the first demonstration of dynamical back-action [36], the optical force was a bolometric force. In later follow-up experiments, the optical force was dominated by radiation pressure [37–39]. In principle, one can consider any relationship between F_{opt} and E_0, such as photophoretic or gradient forces. To keep the discussion within bounds, let us concentrate on the radiation pressure (see Section 14.2), which in terms of the cavity field reads as $F_{\mathrm{opt}}(t) = (\varepsilon_0/2)|E_0|^2(t)nA(1+R)$, where R is the reflectivity of the mirror, n is the index of refraction of the cavity medium, and A is the irradiated mirror area. Thus, the equation of motion (11.56) becomes

$$\frac{\mathrm{d}^2}{\mathrm{d}t^2}x(t) + \Gamma_0\frac{\mathrm{d}}{\mathrm{d}t}x(t) + \Omega_0^2 x(t) = \frac{1}{m}\Big[F_{\mathrm{fluct}}(t) + \frac{\varepsilon_0}{2}n(1+R)A|E_0|^2(t)\Big]. \tag{11.58}$$

The dynamics of the system illustrated in Fig. 11.18 is now entirely described by the coupled equations (11.57) and (11.58). The solution for $E_0(t)$ and $x(t)$ depends on a number of parameters, such as the cavity resonance ω_0, the excitation frequency ω, the excitation intensity $|E_{\mathrm{in}}|^2$, and the quality factors of the cavity ($Q_{\mathrm{c}} = \omega_0/(2\gamma_0)$) and oscillator ($Q_{\mathrm{m}} = \Omega_0/(2\Gamma_0)$).

Instead of solving the two coupled equations, we take a closer look at the equation of motion (11.58). The solution for x depends on $|E_0|^2$, which in turn is a function of x through Eq. (11.57). Thus, a change in x feeds back on itself (dynamical back-action). Now imagine that the energy density in the cavity depends on the rate at which the mirror position changes, that is $|E_0|^2 = C\,\mathrm{d}x/\mathrm{d}t$, with C being a constant. In this case, the last term in Equation (11.58) can be combined with the frictional term $\Gamma_0\,\mathrm{d}x/\mathrm{d}t$ on the left-hand side, thereby increasing or decreasing the damping rate of the oscillator, depending on the sign of C. This is the essence of parametric amplification and cooling in cavity optomechanics and can be exploited for applications, such as switching, storage, and quantum information processing. Figure 11.19 shows an experimental verification of laser-induced cooling of a microlever [36].

Because the optical force acting on the harmonic oscillator leads to a change in the damping constant and oscillator frequency, Eq. (11.58) can be represented in the form

$$\frac{\mathrm{d}^2}{\mathrm{d}t^2}x(t) + (\Gamma_0 + \delta\Gamma)\frac{\mathrm{d}}{\mathrm{d}t}x(t) + (\Omega_0 + \delta\Omega)^2 x(t) = \frac{1}{m}F_{\mathrm{fluct}}(t), \tag{11.59}$$

where $\delta\Gamma$ and $\delta\Omega$ are now functions of the optical force. A formal solution of Eqs. (11.57) and (11.58) yields [39]

$$\delta\Gamma = \frac{\pi^2 R}{(1-R)^2}\,\frac{8n^2\omega_0}{mc^2\Omega_0}\,\frac{\gamma_{\mathrm{ex}}\gamma_0 P_{\mathrm{in}}}{(\omega-\omega_0)^2+\gamma_0^2}\left[\frac{\gamma_0^2}{(\omega-\omega_0+\Omega_0)^2+\gamma_0^2} - \frac{\gamma_0^2}{(\omega-\omega_0-\Omega_0)^2+\gamma_0^2}\right], \tag{11.60}$$

$$\delta\Omega = \frac{\pi^2 R}{(1-R)^2}\frac{4n^2\omega_0}{mc^2\Omega_0}\frac{\gamma_{ex}\gamma_0 P_{in}}{(\omega-\omega_0)^2+\gamma_0^2}\left[\frac{(\omega-\omega_0+\Omega_0)\gamma_0}{(\omega-\omega_0+\Omega_0)^2+\gamma_0^2}+\frac{(\omega-\omega_0-\Omega_0)\gamma_0}{(\omega-\omega_0-\Omega_0)^2+\gamma_0^2}\right], \tag{11.61}$$

where $P_{in} = (1/2)\varepsilon_0 cA|E_{in}|^2$ is the input power. The first term in brackets in Eqs. (11.60) and (11.61) is associated with anti-Stokes scattering, whereas the second term refers to Stokes scattering (see Fig. 11.18(b)). In the *resolved-sideband* regime ($\gamma_0 \ll \Omega_0$) the damping rate Γ can be made negative for red detuned excitation ($\omega < \omega_0$) and positive for blue detuned excitation ($\omega > \omega_0$). Thus, the excitation frequency is a knob for selecting between amplification (red detuning) or cooling (blue detuning).

For $\gamma_0 \gg \Omega_0$ we enter the so-called *weak-retardation regime*. In this regime, the term in brackets in Eq. (11.60) can be approximated as $-4(\omega-\omega_0)\Omega_0\gamma_0^2/[(\omega-\omega_0)^2+\gamma_0^2]^2$. Cooling and amplification can still be accomplished, depending on whether the excitation frequency is red or blue detuned from cavity resonance. However, the cooling or amplification rate is much weaker than that in the resolved-sideband regime.

Let us now look at the equation of motion (11.59) in frequency space Ω. Using the correlation $\langle F_{fluct}(t)F_{fluct}(t')\rangle = 2m\Gamma_0 k_B T\delta(t-t')$ introduced above, together with the Wiener–Khintchine theorem (15.16), we obtain the power spectral density (spectrum)

$$\int_{-\infty}^{\infty}\langle\hat{x}(\Omega)\hat{x}^*(\Omega')\rangle d\Omega' = \frac{k_B T}{\pi m}\frac{\Gamma_0}{([\Omega_0+\delta\Omega]^2-\Omega^2)^2+\Omega^2[\Gamma_0+\delta\Gamma]^2}, \tag{11.62}$$

where $\hat{x}$ is the Fourier transform of x. Integrating both sides over Ω yields the mean-square displacement

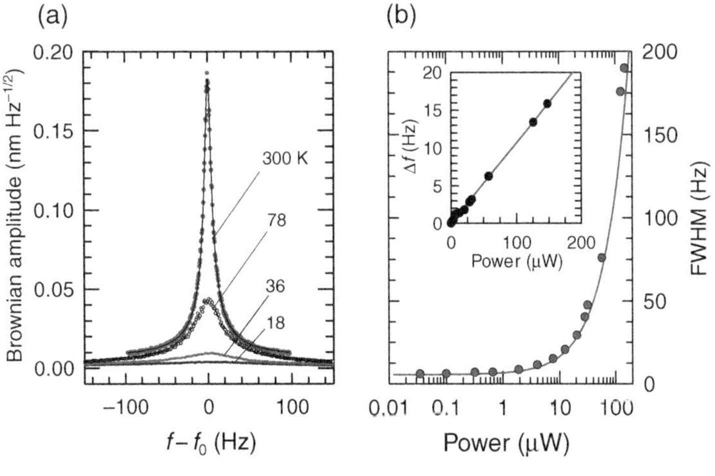

Fig. 11.19 Laser cooling of a microlever. (a) The lineshape of the mechanical resonance for different power levels. The widths of the curves define the effective temperature. (b) The full width at half-maximum (FWHM) of the resonance curves as a function of the laser power. Inset: the frequency shift of the mechanical resonance. Reprinted with permission from Macmillan Publishers Ltd [36].

$$\left\langle x^2 \right\rangle = \left\langle x(0)\, x(0) \right\rangle = \frac{k_B T}{m(\Omega_0 + \delta\Omega)^2} \frac{\Gamma_0}{\Gamma_0 + \delta\Gamma}. \tag{11.63}$$

We now define an effective temperature T_{eff} according to the equipartition theorem, that is $k_B T_{\text{eff}} = m(\Omega_0 + \delta\Omega)^2 \langle x^2 \rangle$. Since $\delta\Omega \ll \Omega_0$ we obtain

$$T_{\text{eff}} = T \frac{\Gamma_0}{\Gamma_0 + \delta\Gamma}, \tag{11.64}$$

where T is the equilibrium temperature in the absence of the optical force ($\delta\Gamma = 0$). Thus, the temperature of the oscillator can be raised or lowered, depending on the sign of $\delta\Gamma$ in Eq. (11.60). However, shot noise associated with the discrete nature of photons imposes a limit to Eq. (11.64). Furthermore, Eq. (11.64) is valid only for $\delta\Gamma \ll \gamma_0$ and $T_{\text{eff}} > T/Q_m$.

In the quantum limit, the mechanical oscillator exhibits discrete states separated in energy by $\hbar(\Omega_0 + \delta\Omega) \approx \hbar\Omega_0$. The mean thermal occupancy $\langle n \rangle$ of an oscillation mode becomes

$$\langle n \rangle = \frac{k_B T_{\text{eff}}}{\hbar\Omega_0}. \tag{11.65}$$

In order to resolve the quantum ground state of the oscillator we require $\langle n \rangle < 1$. For a 1-MHz oscillator, this condition implies $T_{\text{eff}} \sim 50\,\mu\text{K}$.

The Hamiltonian for the combined system "oscillator + cavity" can be derived by considering the energy of the "dressed" cavity field, that is, the cavity that has been acted on by the mechanical oscillator. As discussed before, a displacement of the cavity length L by an amount x leads to a cavity frequency shift of

$$\omega - \omega_0 = -\omega_0 \frac{x}{L}. \tag{11.66}$$

This frequency shift modifies the energy in the cavity. In terms of the single-mode creation and annihilation operators $a^\dagger$ and a we can write the Hamiltonian of the "dressed" cavity as

$$\begin{aligned} \hbar\omega[a^\dagger a + 1/2] &= \hbar\omega_0[a^\dagger a + 1/2] - \hbar\omega_0 \frac{q}{L}[a^\dagger a + 1/2] \\ &= H_{\text{cavity}} + H_{\text{int}}, \end{aligned} \tag{11.67}$$

where we made use of Eq. (11.66), replaced x by the generalized coordinate q, and noted that the first term on the right-hand side corresponds to the unperturbed cavity and the second term accounts for the interaction. After adding the Hamiltonian of the mechanical oscillator, we obtain

$$\begin{aligned} H_{\text{tot}} &= H_{\text{cavity}} + H_{\text{mech}} + H_{\text{int}} \\ &= \hbar\omega_0[a^\dagger a + 1/2] + (1/2)[p^2/m + m\Omega_0^2 q^2] + \hbar g_0 q[a^\dagger a + 1/2], \end{aligned} \tag{11.68}$$

where we introduced the *optomechanical coupling rate* $g_0 = d\omega_0/dq = \omega_0/L$. The splitting of the total Hamiltonian into three terms is similar to the discussion in Section 8.2 (c.f. Eq. (8.35)). The interaction Hamiltonian (the last term in Eq. (11.68)) accounts for the displacement q of the mechanical system by the radiation pressure, which is proportional to the number of photons $n = a^\dagger a$ in the cavity. The force per photon is given by $F_{\text{ph}} = \hbar g_0$. Note that the cavity acts also on the oscillator, thereby shifting its oscillation frequency to

$\Omega = \Omega_0 + \delta\Omega$, but this shift is usually negligibly small ($\delta\Omega \ll \Omega_0$). Since the mechanical motion is harmonic we can represent the operators q and p by corresponding creation and annihilation operators ($b^\dagger$, b) and rewrite the Hamiltonian as

$$H_{\text{tot}} = \hbar\omega_0[a^\dagger a + 1/2] + \hbar\Omega_0[b^\dagger b + 1/2] + \hbar g_0 x_0\,[b^\dagger + b][a^\dagger a + 1/2]. \quad (11.69)$$

Here, $x_0 = \sqrt{\hbar/(2m\Omega_0)}$ is the zero-point oscillation amplitude of the mechanical oscillator. Note that in most calculations the constant (zero-point) terms "1/2" are dropped since they merely account for an offset of the energy eigenstates. The term $g_0 x_0$ is what is referred to as the *single-photon coupling strength*. In order to observe strong coupling between the mechanical oscillator and the optical cavity we require $g \gg \gamma_0, \Gamma_0$. Strong coupling between interacting oscillators is discussed in more detail in Section 8.7. To account for losses, fluctuations, and the coupling of radiation in to and out of the cavity, the Hamiltonian formalism based on Eq. (11.69) needs to be generalized by using so-called quantum Langevin equations.

Finally, we note that optical cooling and heating of a mechanical oscillator implies that the optical force acting on the oscillator is non-conservative, that is

$$W = \oint F_{\text{opt}}(x)dx \neq 0, \quad (11.70)$$

with x being the oscillator position. Condition (11.70) requires that there is a time lag between the action of the oscillator and the response of the cavity. This time-lag is defined by the photon lifetime $\tau = 1/(2\gamma_0)$ and hence by the cavity Q-factor.

Problems

11.1 Consider a one-dimensional photonic crystal made of two alternating dielectric layers with the dielectric constants ε_1 and ε_2 and *different* thicknesses d_1 and d_2. Derive the characteristic equation for TE and TM modes. Plot the dispersion curves $k_x(\omega)$ for $\varepsilon_1 = 17.88$, $\varepsilon_2 = 2.31$, and $d_2/d_1 = 2/3$.

11.2 Estimate the electromagnetic resonance frequency $\omega = \sqrt{1/(LC)}$ of a metallic ring with a single cut by using the capacitance C of a plate capacitor and the inductance L of a solenoid with a single loop. What are the geometrical dimensions required in order to enter the optical regime? How is the resonance shifted by adding a second cut to the ring? Determine why the single-split ring, the double-split ring, and the wire pair (obtained from the double-split ring by bending the wires) act as a magnetic dipoles. Discuss the impact of the kinetic inductance (see Section 13.3.1).

11.3 Apply the boundary conditions for **E** and **H** at the boundary between media with refractive indices $n = 1$ and $n = -1$ and the continuity of the wavevector component parallel to the interface to verify the situation sketched in Fig. 11.8. Also show that there is no reflected wave in the sketched situation.

11.4 Estimate the wavelength of the highest-Q mode of a microsphere with radius $a = 50\,\mu\text{m}$ and dielectric constant $\varepsilon = 2.31$. Determine the spacing $\Delta\lambda$ between modes.

11.5 For a microsphere with $\varepsilon = 2.31$, plot numerically the right-hand sides of Eqs. (11.37) and (11.38) in the complex-ka plane. Assume an angular momentum mode number $l = 10$ and estimate the value ka for modes with radial mode numbers $\nu = 1, 2, 3$.

11.6 The resonance shift $(\omega - \omega_0)$ of a microcavity due to the presence of a small particle with polarizability α has been calculated using the formula [41]

$$\hbar(\omega - \omega_0) = -(\alpha/2)\langle \mathbf{E}(\mathbf{r}_0, t)^2 \rangle, \tag{11.71}$$

where $\mathbf{r}_0$ is the location of the particle and α is the excess polarizability, i.e. that measured against background. Here $\langle .. \rangle$ denotes the time average. This formula is essentially an energy balance for a single photon. Divide on both sides by the "unperturbed" photon energy $\hbar\omega_o = (1/2)\int_V \varepsilon_0\,\varepsilon(\mathbf{r})\,\mathbf{E}_0^* \cdot \mathbf{E}_0\,\mathrm{d}V$ and compare the resulting equation with the Bethe-Schwinger cavity perturbation. Discuss the major differences.

11.7 Consider an optical cavity characterized by resonance frequency ω_0 and decay rate γ_0 coupled by radiation pressure to a mechanical oscillator with resonance frequency Ω_0 and damping Γ_0. Determine the laser frequency ω that provides the highest cooling rate.

11.8 An alternative configuration to Fig. 11.18 is a trapped dielectric particle with polarizability α and volume V_p held by a laser tweezer (frequency ω_T) inside an optical cavity (resonance ω_0) and driven by a field of frequency ω. Determine the mechanical oscillation frequency Ω_0 as a function of the optical-tweezer parameters (power, NA, λ_T, ...) and calculate the back-action on the cavity mode using the cavity-perturbation formula. Assume that the particle can be treated in the dipole limit and that its oscillation is centered in a field minimum of the cavity mode. Determine the cavity-induced damping rate $\delta\Gamma$.

References

[1] Lord Rayleigh, "On the maintenance of vibrations by forces of double frequency, and on the propagation of waves through a medium endowed with a periodic structure" *Phil. Mag. (Series 5)* **24**, 145–159 (1887).

[2] Y. A. Vlasov, X. Z. Bo, J. C. Sturm, and D. J. Norris, "On-chip natural assembly of silicon photonic bandgap crystals," *Nature* **414**, 289–293 (2001).

[3] J. D. Joannopoulos, R. D. Meade, and J. N. Winn, *Photonic Crystals*. Princeton, MA: Princeton University Press (1995).

[4] J. D. Joannopoulos, P. R. Villeneuve, and S. Fan, "Photonic crystals: putting a new twist on light," *Nature* **386**, 143–149 (1997).

[5] G. Floquet, "Sur les équations differentielles linéares à coefficients périodiques," *Ann. Ecole Norm. Supér.* **12**, 47–88 (1883).

[6] F. Bloch, "Über die Quantenmechanik der Elektronen in Kristallgittern," *Z. Phys.* **52**, 555–600 (1929).
[7] E. Moreno, D. Erni, and Ch. Hafner, "Modeling of discontinuities in photonic crystal waveguides with the multiple multipole method," *Phys. Rev. E* **66**, 036618 (2002).
[8] O. J. Painter, A. Husain, A. Scherer, *et al.*, "Two-dimensional photonic crystal defect laser," *J. Lightwave Technol.* **17**, 2082–2089 (1999).
[9] R. P. Feynman, R. B. Leighton, and M. Sands, *The Feynman Lectures on Physics*, vol. 1. Reading, MA: Addison-Wesley (1977).
[10] V. Veselago, L. Braginsky, V. Shklover, and Ch. Hafner, "Negative refractive index materials," *J. Comput. Theor. Nanosci.* **3**, 1–30 (2006).
[11] V. G. Veselago, "The electrodynamics of substances with simultaneously negative values of ϵ and μ,"*Sov. Phys. Usp.* **10**, 509–514 (1968).
[12] J. Kästel and M. Fleischhauer, "Quantum electrodynamics in media with negative refraction," *Laser Phys.* **15**, 135–145 (2005).
[13] D. R. Smith, W. J. Padilla, D. C. Vier, S. C. Nemat-Nasser, and S. Schultz, "Composite medium with simultaneously negative permeability and permittivity," *Phys. Rev. Lett.* **84**, 4184–4187 (2000).
[14] J. P. Pendry, "Negative refraction makes a perfect lens," *Phys. Rev. Lett.* **85**, 3966–3969 (2000).
[15] R. A. Shelby, D. R. Smith, and S. Schultz, "Experimental verification of a negative index of refraction," *Science* **292**, 77–79 (2001). Reprinted with permission from AAAS.
[16] C. M. Soukoulis and M. Wegener, "Past achievements and future challenges in the development of three-dimensional photonic metamaterials," *Nature Photonics* **5**, 523–530 (2011).
[17] J. Valentine, S. Zhang, T. Zentgraf, *et al.*, "Three-dimensional optical metamaterial with a negative refractive index," *Nature* **455**, 376–379 (2008).
[18] R. E. Collin, "Frequency dispersion limits resolution in Veselago lens," *Prog. Electromagn. Res. B* **19**, 233–261 (2010).
[19] C. Hafner, C. Xudong, and R. Vahldieck, "Resolution of negative index slabs," *J. Opt. Soc. Am. A* **23**, 1768–1778 (2006).
[20] R. Merlin, "Analytical solution of the almost-perfect-lens problem," *Appl. Phys. Lett.* **84**, 1290–1292 (2004).
[21] T. Taubner, D. Korobkin, Y. Urzhumov, G. Shvets, and R. Hillenbrand, "Near-field microscopy through a SiC superlens," *Science* **313**, 1595 (2006).
[22] Z. Jacob, L. V. Alekseyev, and E. Narimanov, "Optical hyperlens: far-field imaging beyond the diffraction limit," *Opt. Express* **14**, 8247–8256 (2006).
[23] Z. Liu, H. Lee, Y. Xiong, C. Sun, and X. Zhang, "Far-field optical hyperlens magnifying sub-diffraction-limited objects," *Science* **315**, 1686 (2007).
[24] J. B. Pendry, D. Schurig, and D. R. Smith, "Controlling electromagnetic fields," *Science* **312**, 1780–1782 (2006).

[25] S. Arnold, S. Holler, and S. D. Druger, "The role of MDRs in chemical physics: intermolecular energy transfer in microdroplets," in *Optical Processes in Microcavities*, ed. R. K. Chang and A. J. Campillo. Singapore: World Scientific pp. 285–312 (1996).

[26] C. G. Bohren and D. R. Huffman, *Absorption and Scattering of Light by Small Particles*. New York: John Wiley (1983).

[27] H. M. Nussenzveig, *Diffraction Effects in Semiclassical Scattering*. Cambridge: Cambridge University Press (1992).

[28] B. R. Johnson, "Theory of morphology-dependent resonances: shape resonances and width formulas," *J. Opt. Soc. Am. A* **10**, 343–352 (1993).

[29] S. M. Spillane, T. J. Kippenberg, K. J. Vahala, *et al.*, "Ultrahigh-Q toroidal microresonators for cavity quantum electrodynamics," *Phys. Rev. A* **71**, 013817 (2005).

[30] F. Vollmer and S. Arnold, "Whispering-gallery-mode biosensing: labelfree detection down to single molecules," *Nature Methods* **5**, 591–596 (2008).

[31] J. Schwinger, *The Theory of Obstacles in Resonant Cavities and Waveguides*, MIT Radiation Laboratory Report no. 43-34 (1943).

[32] W. Hauser, *Introduction to the Principles of Electromagnetism*. Reading, MA: Addison-Wesley (1971).

[33] T. J. Kippenberg and K. J. Vahala, "Cavity opto-mechanics," *Opt. Express* **15**, 17172–17205 (2007).

[34] V. B. Braginsky, *Measurement of Weak Forces in Physics Experiments*. Chicago, IL: University of Chicago Press (1977).

[35] P. F. Cohadon, A. Heidmann, and M. Pinard, "Cooling of a mirror by radiation pressure," *Phys. Rev. Lett.* **83**, 3174–3177 (1999).

[36] C. Höhberger Metzger and K. Karrai, "Cavity cooling of a microlever," *Nature* **432**, 1002–1005 (2004).

[37] O. Arcizet, P. F. Cohadon, T. Briant, M. Pinard, and A. Heidmann, "Radiation-pressure cooling and optomechanical instability of a micromirror," *Nature* **444**, 71–74 (2006).

[38] S. Gigan, H. R. Bohm, M. Paternostro, *et al.*, "Self-cooling of a micromirror by radiation pressure," *Nature* **444**, 67–70 (2006).

[39] A. Schliesser, P. Del'Haye, N. Nooshi, K. J. Vahala, and T. J. Kippenberg, "Radiation pressure cooling of a micromechanical oscillator using dynamical backaction," *Phys. Rev. Lett.* **97**, 243905 (2006).

[40] J. Van Bladel and D. De Zutter, "Reflections from linearly vibrating objects: plane mirror at normal incidence," *IEEE Trans. Antennas Propag.* **29**, 629–636 (1981).

[41] S. Arnold, M. Khoshsima, I. Teraoka, S. Holler and F. Vollmer, "Shift of whispering gallery modes in microspheres by protein adsorption," *Opt. Lett.* **28**, 272–274 (2003).

12 Surface plasmons

The interaction of metals with electromagnetic radiation is largely dictated by their free conduction electrons. According to the Drude model, the free electrons oscillate 180° out of phase relative to the driving electric field. As a consequence, most metals possess a negative dielectric constant at optical frequencies, which causes, for example, a very high reflectivity. Furthermore, at optical frequencies the metal's free-electron gas can sustain surface and volume charge-density oscillations, called plasmons, with distinct resonance frequencies. The existence of plasmons is characteristic of the interaction of metal nanostructures with light at optical frequencies. Similar behavior cannot be simply reproduced in other spectral ranges using the scale invariance of Maxwell's equations since the material parameters change considerably with frequency. Specifically, this means that model experiments with, for instance, microwaves and correspondingly larger metal structures cannot replace experiments with metal nanostructures at optical frequencies.

The surface charge-density oscillations associated with surface plasmons at the interface between a metal and a dielectric can give rise to strongly enhanced optical near-fields, which are spatially confined near the metal surface. Similarly, if the electron gas is confined in three dimensions, as in the case of a small particle, the overall displacement of the electrons with respect to the positively charged lattice leads to a restoring force, which in turn gives rise to specific particle–plasmon resonances depending on the geometry of the particle. In particles of suitable (usually pointed) shape, localized charge accumulations that are accompanied by strongly enhanced optical fields can occur.

The study of optical phenomena related to the electromagnetic response of metals has been termed *plasmonics* or *nanoplasmonics*. This field of nanoscience is concerned with the control of light localization and propagation on subwavelength scales. At optical frequencies, a noble metal is characterized by a dielectric function $\varepsilon = -\varepsilon' + i\varepsilon''$ whose real part $|\varepsilon'|$ is typically larger than its imaginary part $|\varepsilon''|$. The opposite is true for metals in the microwave or infrared frequency regime. Hence, we can define plasmonics as the interaction of light with metals under the condition $|\varepsilon'| > |\varepsilon''|$. Many innovative concepts and applications of metal optics have been developed over the past few years, and in this chapter we discuss a few examples. We start out by discussing metals from the perspective of plasma physics, which provides insights into the physics of plasmons. The discussion includes screening and ponderomotive forces, which give rise to a wide range of optical nonlinearities. Since the interaction of light with metal structures is described by the frequency dependence of the metal's complex dielectric function, we continue with a discussion of the fundamental optical properties of metals. We then turn to important solutions of Maxwell's equations for noble-metal structures, i.e. the plane metal–dielectric interface and small metal wires and particles. Where appropriate, applications of surface plasmons in nano-optics are discussed. Finally, it should be noted that optical interactions similar to those discussed here are also encountered for infrared radiation interacting with polar materials. The corresponding excitations are called surface phonon polaritons.

12.1 Noble metals as plasmas

The free conduction electrons in a metal constitute a plasma, a gas of charged particles that responds collectively to electromagnetic fields. Plasmas are the most common form of matter and are found in stellar nebulae, lightning, stars, flames, and the outer atmosphere. At optical frequencies, metals behave as plasmas, with characteristic shape-dependent resonances. To establish an understanding of the collective response of electrons to optical fields we begin this chapter by reviewing the basic properties of a standard plasma.

12.1.1 Plasma oscillations

Consider a charge-neutral material characterized by a rigid ionic lattice and a gas of free electrons. The material is confined by two parallel surfaces with surface normal $\mathbf{n}_z$. A uniform displacement of the gas by a small distance Δz gives rise to a positive surface charge on one side of the material of $\sigma = ne\,\Delta z$, where n is the electron density and e the elementary charge. The surface charge on the opposite side is $-\sigma$. The situation is sketched in Fig. 12.1(a). The surface charges give rise to a uniform homogeneous electric

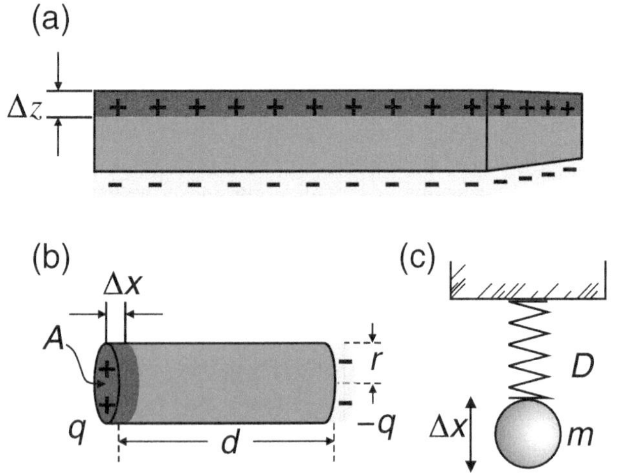

Fig. 12.1 Mass-and-spring model for plasma oscillations. (a) A sketch of a metal slab with a uniform displacement of the electron gas by a small distance Δz causing a homogeneous electric field. (b) A sketch of a metal particle whose electrons have been displaced by an amount Δx. (c) The frequency of the resulting oscillation can be represented by a harmonic oscillator with spring constant D and electron mass m.

field $E = \sigma/\varepsilon_0$ pointing from one surface to the other. This field causes a restoring force, which acts on the electrons and gives rise to the following equation of motion:

$$m\,\Delta\ddot{z} = -eE = -\Delta z\, ne^2/\varepsilon_0, \tag{12.1}$$

with m being the electron mass. The solution of this equation is $\Delta z(t) = \Delta z_0 \cos(\omega_p t)$, where Δz_0 is the displacement at $t = 0$ and ω_p is the plasma frequency defined as

$$\omega_p = \sqrt{\frac{ne^2}{m\varepsilon_0}}. \tag{12.2}$$

Because e and m are constants the only parameter that influences the plasma frequency is the electron density n. Notice that the plasma oscillation is in the direction of the electric field and hence it defines a *longitudinal* mode. In reality, the plasma oscillation is damped due to electron–lattice collisions and also due to radiation losses. In our simple model, we have ignored any spatial variation of the electric field and a more detailed analysis yields $\omega_p(k) \approx \omega_p\left[1 + (3/10)(v_F/\omega_p)^2 k^2 + \cdots\right]$, where v_F is the Fermi velocity, k is the wavevector of the oscillation, and ω_p on the right-hand side is defined by expression (12.2) [1]. It is convenient to express ω_p in terms of an energy $\hbar\omega_p$, which, for most metals, is in the range 2–20 eV. The spectra recorded by electron-energy-loss spectroscopy at metal surfaces exhibit typical peaks at energies that correspond to multiples of $\hbar\omega_p$, referred to as volume plasmons.

Let us now consider plasma oscillations specific to a finite particle, such as a cylindrical rod. When the electron gas in the particle is displaced by Δx, opposite nearly point-like charges $\pm q$ build up at both ends. The magnitude of q depends on n and the cross-sectional area of the cylinder A as $q = neA\,\Delta x$. The Coulomb potential energy of the two charges is then

$$W(\Delta x) = \frac{1}{4\pi\varepsilon_0}\frac{q^2}{d} = \frac{1}{4\pi\varepsilon_0}\frac{(neA)^2}{d}\Delta x^2. \tag{12.3}$$

The restoring force can now be determined as

$$F(\Delta x) = -\frac{\partial W(\Delta x)}{\partial \Delta x} = -\frac{1}{2\pi\varepsilon_0}(ne)^2\frac{A^2}{d}\,\Delta x = -D\,\Delta x, \tag{12.4}$$

from which we deduce a spring constant D. Using a similar equation of motion to that used before we find harmonic plasma oscillations of the system, which can be described by a simple mass-and-spring model. The relevant mass is the mass of all electrons that are involved in the oscillation, that is, $m_{tot} = nmAd$. The approximate resonance frequency ω_{res} of the cylindrical particle turns out to be

$$\omega_{res} = \sqrt{\frac{D}{m_{tot}}} = \frac{\omega_p}{2\sqrt{2}}\frac{1}{R}, \tag{12.5}$$

where we have used Eq. (12.2), $A = \pi r^2$, and $R = d/(2r)$, the *aspect ratio* of the particle. Since we assumed that the charge distributions are localized at the particle ends, we cannot expect the result to quantitatively reproduce the exact resonance frequencies for shorter and thicker particles. However, the experimentally observed trend that the resonance frequency is inversely proportional to the aspect ratio R of a particle and that it occurs close to the visible regime is reproduced. The physical reason for the scaling behavior of the aspect ratio lies in the fact that the electric field of the charge distribution has a dipolar character.

12.1.2 The ponderomotive force

Electrons in a plasma experience a net force if they interact with a non-uniform electric field $\mathbf{E}(\mathbf{r}, t)$. This force can be derived from a potential and is referred to as the ponderomotive force. The easiest derivation starts with the potential energy V_{p} of an electron oscillating at the frequency ω. For a harmonic oscillator, the average potential energy and the average kinetic energy are the same and hence $V_{\mathrm{p}} = \frac{1}{2}m\langle v^2\rangle$. The velocity follows from the equation of motion as $m\dot{\mathbf{v}} = -e\mathbf{E}$, where $\mathbf{E} = \mathbf{E}_0 \cos(\omega t)$. Solving for the velocity yields $\mathbf{v} = -e\mathbf{E}_0 \sin(\omega t)/(m\omega)$ and, after inserting this into the potential V_{p}, we arrive at

$$V_{\mathrm{p}}(\mathbf{r}) = \frac{e^2}{2m\omega^2}\left\langle |\mathbf{E}(\mathbf{r}, t)|^2 \right\rangle, \tag{12.6}$$

which is the ponderomotive potential or the quiver energy of the electron. The force is derived as $\mathbf{F} = -\nabla V_{\mathrm{p}}$. Similarly to the gradient force (c.f. Section 14.4), the ponderomotive force depends on the gradient of the electric field intensity. Comparison with Eq. (14.47) yields that the polarizability of an electron is $\alpha' = -e^2/(m\omega^2)$. The ponderomotive force expels electrons from regions of high field strength, thereby affecting the local electron density. This interaction is a major source of optical nonlinearities in metals.

12.1.3 Screening

Plasmas are charge-neutral but exhibit local fluctuations in their charge densities. Let us denote the local density of positive charges by n_+ and the local density of negative charges by n_-. The electrostatic potential between charges is Φ. Averaged over large distances, the number of positive and negative charges must be the same, but *locally* the charge density can fluctuate. In thermal equilibrium, the *local* charge densities obey

$$\frac{n_-}{n_+} = \exp[e\Phi/(k_{\mathrm{B}}T)], \tag{12.7}$$

where T is the temperature. We now introduce an external charge at $\mathbf{r} = 0$. The corresponding charge density is $\rho_{\mathrm{ext}} = e\delta(\mathbf{r})$. This external charge modifies the local charge distribution and gives rise to an induced charge density $\rho_{\mathrm{ind}} = -e(n_- - n_+)$. According to Gauss's law, the two charge densities give rise to a local field $\nabla \cdot \mathbf{E} = (\rho_{\mathrm{ext}} + \rho_{\mathrm{ind}})/\varepsilon_0$, which can be expressed in terms of the potential Φ as

$$\nabla^2\Phi(\mathbf{r}) = -\frac{e}{\varepsilon_0}\left[\delta(\mathbf{r}) - (n_- - n_+)\right]. \tag{12.8}$$

We used $\mathbf{E} = -\nabla\Phi$ and introduced the expressions for ρ_{ext} and ρ_{ind}. On substituting Eq. (12.7) into this we obtain the so-called Poisson–Boltzmann equation. Typically, $k_B T \gg e\Phi(\mathbf{r})$ and hence we expand the exponential term in Eq. (12.7) as $\exp[e\Phi/(k_B T)] = 1 + [e\Phi/(k_B T)] + \cdots$ and drop all higher-order terms in Φ. We then obtain $\nabla^2\Phi = -(e/\varepsilon_0)\big[\delta(\mathbf{r}) - n_+ e\Phi/(k_B T)\big]$, which can be rearranged as

$$\left[\nabla^2 - \frac{e^2 n_+}{\varepsilon_0 kT}\right]\Phi(\mathbf{r}) = -\frac{e}{\varepsilon_0}\delta(\mathbf{r}). \tag{12.9}$$

The solution of this inhomogeneous differential equation is

$$\Phi(r) = -\frac{e}{4\pi\varepsilon_0}\frac{\exp(-r/\lambda_D)}{r}, \qquad \lambda_D = \sqrt{\frac{\varepsilon_0 k_B T}{e^2 n}}, \tag{12.10}$$

where λ_D is the Debye screening length. Because the plasma is neutral over distances larger than λ_D we have set $n_+ = n_- = n$.

The analysis above is correct for plasmas in which the kinetic energy of electrons is defined by the temperature T. However, this is not the case for metals at room temperature. According to quantum mechanics, the kinetic energy of a free electron obeys the dispersion relation $E(k) = \hbar^2 k^2/(2m)$. The highest energy is the Fermi energy $E_F = E(k_F)$, where $k_F^3 = 3\pi^2 n$. Using the electron density for gold we find $E_F \sim 213 k_B T$ at room temperature, which makes the classical Debye theory clearly invalid for any practical temperatures. We therefore perform the replacement $(3/2)k_B T \to E_F$ in Eq. (12.10) and obtain the screening potential

$$\Phi(r) = -\frac{e}{4\pi\varepsilon_0}\frac{\exp(-r/\lambda_{TF})}{r}, \tag{12.11}$$

where

$$\lambda_{TF} = \sqrt{\frac{\pi^2\hbar^2\varepsilon_0}{me^2 k_F}}. \tag{12.12}$$

λ_{TF} is referred to as the Thomas–Fermi screening length. The screening potential states that the Coulomb potential of a point charge in a plasma is shielded on length scales larger than λ_{TF}. This shielding is established by the exponential decay of the screening potential for distances $r > \lambda_{TF}$. In other words, in a metal charges do feel each other only if their separation is shorter than λ_{TF}. For gold, $\lambda_{TF} \sim 59$ pm, which is smaller than the mean distance $\bar{r} \sim 160$ pm between electrons defined by the electron density as $\bar{r} = (n \cdot 4\pi/3)^{-1/3}$. Therefore, for most practical purposes, electron–electron interactions can be ignored in real metals. The velocity v_F of a Fermi electron follows from $mv_F^2/2 = E_F$ and can be used to calculate the distance traveled in one plasma oscillation period. For gold we find $d = v_F/\omega_p \sim 1$ nm and hence the electron sea travels further during an oscillation period than the mean distance between electrons.

Notice that the screening potential originates from a collective electron response. It is therefore an effective potential, not the potential, that defines the energy of an individual electron, i.e. $\bar{V} = e^2/(4\pi\varepsilon_0\bar{r})$.

12.2 Optical properties of noble metals

The optical properties of metals and noble metals in particular can be described by a complex dielectric function that depends on the frequency of light (see Chapter 2). The properties are determined mainly by the facts that (i) the conduction electrons can move freely within the bulk of material and (ii) interband excitations can take place if the energy of the photons exceeds a threshold energy for the respective metal. In the picture we adopt here, the presence of an electric field leads to a displacement $\mathbf{r}$ of an electron, which is associated with a dipole moment $\mathbf{p}$ according to $\mathbf{p} = e\mathbf{r}$. The cumulative effect of all individual dipole moments of all free electrons results in a macroscopic polarization per unit volume of $\mathbf{P} = n\mathbf{p}$. As discussed in Chapter 2, the macroscopic polarization $\mathbf{P}$ can be expressed as

$$\mathbf{P}(\omega) = \varepsilon_0 \chi_e(\omega) \mathbf{E}(\omega). \tag{12.13}$$

From (2.6) and (2.15) we have

$$\mathbf{D}(\omega) = \varepsilon_0 \varepsilon(\omega) \mathbf{E}(\omega) = \varepsilon_0 \mathbf{E}(\omega) + \mathbf{P}(\omega). \tag{12.14}$$

From this we calculate

$$\varepsilon(\omega) = 1 + \chi_e(\omega), \tag{12.15}$$

the frequency-dependent dielectric function of the metal. The displacement $\mathbf{r}$ and therefore the macroscopic polarization $\mathbf{P}$ and χ_e can be obtained by solving the equation of motion of the electrons under the influence of an external field.

12.2.1 Drude–Sommerfeld theory

As a starting point, we consider only the effects of the free electrons and apply the Drude–Sommerfeld model for the free-electron gas (see e.g. [2]).

$$m_e \frac{\partial^2 \mathbf{r}}{\partial t^2} + m_e \Gamma \frac{\partial \mathbf{r}}{\partial t} = e\mathbf{E}_0 e^{-i\omega t}, \tag{12.16}$$

where e and m_e are the charge and effective mass of the free electrons, and $\mathbf{E}_0$ and ω are the amplitude and frequency of the applied electric field. Note that the equation of motion contains no restoring force, since free electrons are considered. The damping term is proportional to $\Gamma = v_F/l$, where v_F is the Fermi velocity and l is the electron mean free path between scattering events. Solving (12.16) using the ansatz $\mathbf{r}(t) = \mathbf{r}_0 e^{-i\omega t}$ and using the result in (12.15) yields

$$\varepsilon_{\text{Drude}}(\omega) = 1 - \frac{\omega_p^2}{\omega^2 + i\Gamma\omega}. \tag{12.17}$$

Here $\omega_p = \sqrt{ne^2/(m_e\varepsilon_0)}$ is the volume plasma frequency derived in Eq. (12.2). Expression (12.17) can be divided into real and imaginary parts

$$\varepsilon_{\text{Drude}}(\omega) = 1 - \frac{\omega_p^2}{\omega^2 + \Gamma^2} + i\frac{\Gamma\omega_p^2}{\omega(\omega^2 + \Gamma^2)}. \tag{12.18}$$

Using $\hbar\omega_\mathrm{p} = 8.95\,\mathrm{eV}$ and $\hbar\Gamma = 65.8\,\mathrm{meV}$, which are the values for gold, the real and imaginary parts of the dielectric function (12.18) are plotted in Fig. 12.2 as a functions of the wavelength over the extended visible range. We note that the real part of the dielectric constant is negative. One obvious consequence of this behavior is the fact that a plane light wave can penetrate a metal only to a very small extent, since the negative dielectric constant leads to a strong imaginary part of the refractive index $n = \sqrt{\varepsilon}$. Other consequences will be discussed later. The imaginary part of ε describes the dissipation of energy associated with the motion of electrons in the metal (see Problem 12.1).

12.2.2 Interband transitions

Although the Drude–Sommerfeld model gives quite accurate results for the optical properties of metals at low frequencies, for higher frequencies it needs to be supplemented by the response of bound electrons. For example, for gold, at a wavelength shorter than ~550 nm, the measured imaginary part of the dielectric function increases beyond the value predicted by the Drude–Sommerfeld theory. This is because higher-energy photons can promote electrons of lower-lying d bands into the sp conduction band. In a classical picture such transitions may be described as exciting the oscillation of bound electrons. Bound electrons in metals exist e.g. in lower-lying shells of the metal atoms. We apply the same method as used above for the free electrons to describe the response of the bound electrons. The equation of motion for a bound electron reads

$$m\frac{\partial^2 \mathbf{r}}{\partial t^2} + m\gamma\frac{\partial \mathbf{r}}{\partial t} + \alpha\mathbf{r} = e\mathbf{E}_0 \mathrm{e}^{-\mathrm{i}\omega t}. \tag{12.19}$$

Here, m is the *effective* mass of the bound electrons, which is in general different from the effective mass of a free electron in a periodic potential, γ is the damping constant describing mainly radiative damping in the case of bound electrons, and α is the spring

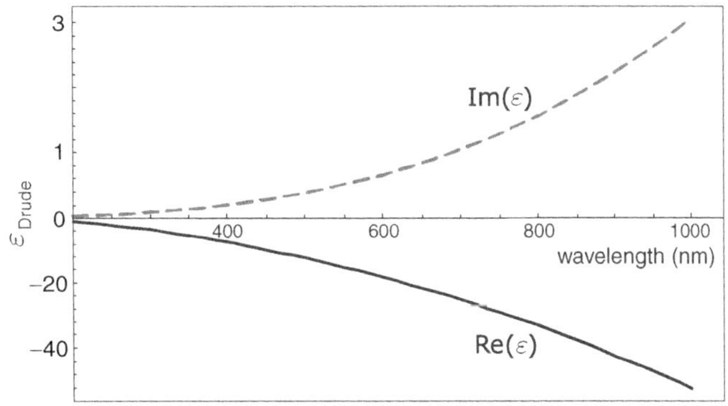

Fig. 12.2 Real and imaginary part of the dielectric constant for gold according to the Drude–Sommerfeld free-electron model ($\hbar\omega_\mathrm{p} = 8.95$ eV, $\hbar\Gamma = 65.8$ meV). The solid line is the real part; the dashed line is the imaginary part. Note the different scales for real and imaginary parts.

constant of the potential that keeps the electron in place. Using the same ansatz as before, we find the contribution of bound electrons to the dielectric function

$$\varepsilon_{\text{Interband}}(\omega) = 1 + \frac{\tilde{\omega}_{\text{p}}^2}{(\omega_0^2 - \omega^2) - \text{i}\gamma\omega}. \tag{12.20}$$

Here $\tilde{\omega}_{\text{p}} = \sqrt{\tilde{n}e^2/(m\varepsilon_0)}$, with $\tilde{n}$ being the density of the bound electrons. $\tilde{\omega}_{\text{p}}$ is introduced in analogy to the plasma frequency in the Drude–Sommerfeld model, albeit, obviously, here with a different physical meaning, and $\omega_0 = \sqrt{\alpha/m}$. Again we can rewrite (12.20) to separate the real and imaginary parts:

$$\varepsilon_{\text{Interband}}(\omega) = 1 + \frac{\tilde{\omega}_{\text{p}}^2(\omega_0^2 - \omega^2)}{(\omega_0^2 - \omega^2)^2 + \gamma^2\omega^2} + \text{i}\frac{\gamma\tilde{\omega}_{\text{p}}^2\omega}{(\omega_0^2 - \omega^2)^2 + \gamma^2\omega^2}. \tag{12.21}$$

Figure 12.3 shows the contribution to the dielectric constant of a metal which derives from bound electrons.[1] Clear resonant behavior is observed for the imaginary part and dispersion-like behavior is observed for the real part. Figure 12.4 is a plot of the dielectric constant (real and imaginary parts) taken from the paper of Johnson and Christy [4] for gold (filled circles). For wavelengths above 650 nm the behavior clearly follows the Drude–Sommerfeld theory. For wavelengths below 650 nm obviously interband transitions become significant. It is possible to model the shape of the curves by appropriately adding up the free-electron (Eq. (12.18)) and interband absorption contributions (Eq. (12.21)) to the complex dielectric function (solid line). Indeed, this much better reproduces the experimental data. One has to introduce a constant offset to (12.21), though, namely $\varepsilon_\infty = 5$, which accounts for the integrated effect of all higher-energy interband transitions not considered in the present model (see e.g. [5]). Also, since only one interband transition is taken into account, the model still fails to reproduce the data below ~500 nm.

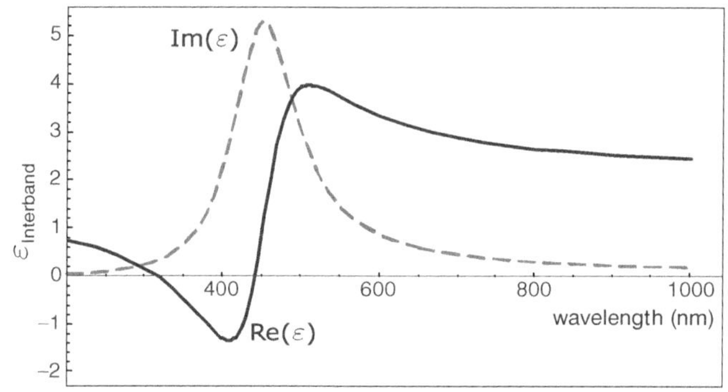

Fig. 12.3 Contribution of bound electrons to the dielectric function of gold. The parameters used are $\hbar\tilde{\omega}_{\text{p}} = 2.96$ eV, $\hbar\gamma = 0.59$ eV, and $\omega_0 = 2\pi c/\lambda$, with $\lambda = 450$ nm. The solid line is the real part and the dashed curve is the imaginary part of the dielectric function associated with bound electrons.

[1] This theory naturally also applies to the behavior of dielectrics, and the dielectric response over a broad frequency range consists of several absorption bands related to different electromagnetically excited resonances [3].

12.3 Surface plasmon polaritons at plane interfaces

By definition surface plasmon polaritons are the quanta of surface-charge-density oscillations. In a classical picture, surface plasmon polaritons are particular solutions of Maxwell's equations (surface modes) that appear for certain boundary conditions. In this section, we consider a plane interface between two media. One medium is characterized by a complex frequency-dependent dielectric function $\varepsilon_1(\omega)$, whereas the dielectric function of the other medium $\varepsilon_2(\omega)$ is assumed to be real. We choose the interface to coincide with the plane $z = 0$ of a Cartesian coordinate system (see Fig. 12.5). We are looking for *homogeneous* solutions of Maxwell's equations that are localized at the interface. A homogeneous solution is an eigenmode of the system, i.e. a solution that exists without external excitation. Mathematically, it is the solution of the wave equation

$$\nabla \times \nabla \times \mathbf{E}(\mathbf{r}, \omega) - \frac{\omega^2}{c^2}\varepsilon(\mathbf{r}, \omega)\mathbf{E}(\mathbf{r}, \omega) = 0, \tag{12.22}$$

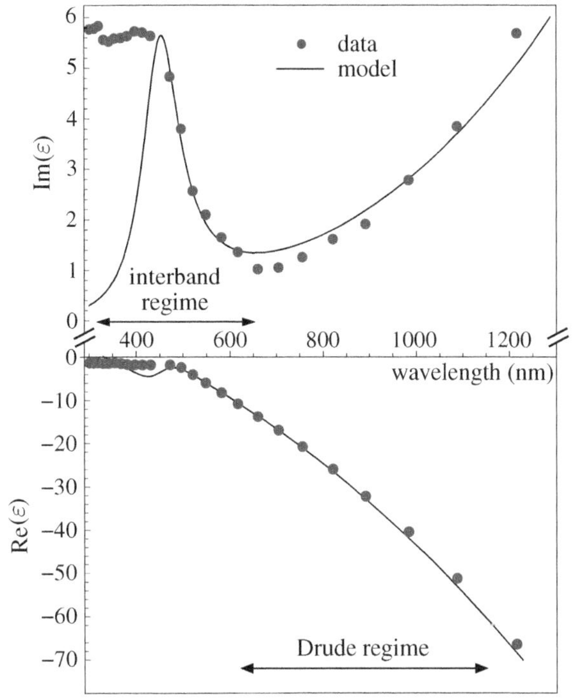

Fig. 12.4 The dielectric function of gold: experimental values and fit. Upper panel: imaginary part. Lower panel: real part. Dots: experimental values taken from [4]. Solid line: model of the dielectric function taking into account the free-electron contribution (Fig. 12.2) and the contribution of a single interband transition (Fig. 12.3). Note the different scales for the abscissae.

with $\varepsilon(\mathbf{r},\omega) = \varepsilon_1(\omega)$ for $z < 0$ and $\varepsilon(\mathbf{r},\omega) = \varepsilon_2(\omega)$ for $z > 0$. A mode that is localized to the interface is characterized by electromagnetic fields that decay exponentially with increasing distance from the interface into both half-spaces but propagates along the interface. It is sufficient to consider only p-polarized waves in both half-spaces because no surface-bound modes exist for the case of s-polarization (see Problem 12.2).

The ansatz for p-polarized plane waves in the half-spaces $j = 1$ and $j = 2$ reads as

$$\mathbf{E}_j = \begin{pmatrix} E_{j,x} \\ 0 \\ E_{j,z} \end{pmatrix} \mathrm{e}^{\mathrm{i}k_x x - \mathrm{i}\omega t} \mathrm{e}^{\mathrm{i}k_{j,z} z}, \quad j = 1, 2. \tag{12.23}$$

The situation is depicted in Fig. 12.5. Since the wavevector parallel to the interface is conserved (see Chapter 2), the following relations hold for the wavevector components:

$$k_x^2 + k_{j,z}^2 = \varepsilon_j k^2, \quad j = 1, 2. \tag{12.24}$$

Here $k = 2\pi/\lambda$, where λ is the vacuum wavelength. Exploiting the fact that the displacement fields in both half-spaces have to be source-free, i.e. $\nabla \cdot \mathbf{D} = 0$, leads to

$$k_x E_{j,x} + k_{j,z} E_{j,z} = 0, \quad j = 1, 2, \tag{12.25}$$

which allows us to rewrite (12.23) as

$$\mathbf{E}_j = E_{j,x} \begin{pmatrix} 1 \\ 0 \\ -k_x/k_{j,z} \end{pmatrix} \mathrm{e}^{\mathrm{i}k_{j,z} z}, \quad j = 1, 2. \tag{12.26}$$

The factor $\mathrm{e}^{\mathrm{i}k_x x - \mathrm{i}\omega t}$ is omitted to simplify the notation. Equation (12.26) is particularly useful when a system of stratified layers is considered (see e.g. [6], p. 40 and Problem 12.4). While (12.24) and (12.25) impose conditions that define the fields in the respective half-spaces, we still have to match the fields at the interface using boundary conditions.

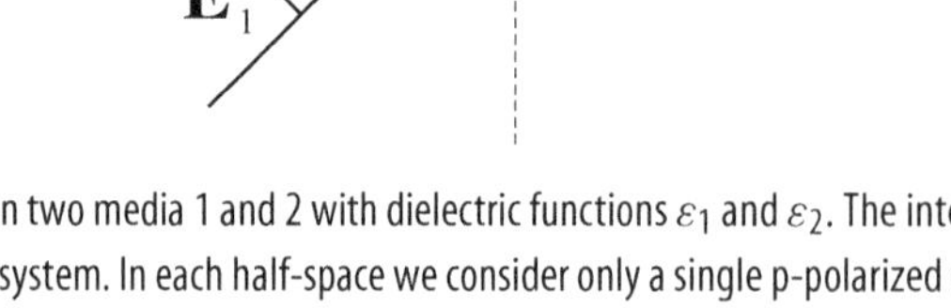

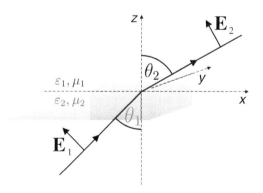

Fig. 12.5 The interface between two media 1 and 2 with dielectric functions ε_1 and ε_2. The interface is defined by $z = 0$ in a Cartesian coordinate system. In each half-space we consider only a single p-polarized wave because we are looking for homogeneous solutions that decay exponentially with distance from the interface.

Requiring continuity of the parallel component of **E** and the perpendicular component of **D** leads to another set of equations, which read as

$$\begin{aligned} E_{1,x} - E_{2,x} &= 0, \\ \varepsilon_1 E_{1,z} - \varepsilon_2 E_{2,z} &= 0. \end{aligned} \tag{12.27}$$

Equations (12.25) and (12.27) form a system of four homogeneous equations for the four unknown field components. The existence of a solution requires that the respective determinant vanishes. This happens either for $k_x = 0$, which does not describe excitations that travel along the interface, or for

$$\varepsilon_1 k_{2,z} - \varepsilon_2 k_{1,z} = 0. \tag{12.28}$$

In combination with (12.24), Eq. (12.28) leads to a dispersion relation, i.e. a relation between the wavevector along the propagation direction and the angular frequency ω,

$$k_x^2 = \frac{\varepsilon_1 \varepsilon_2}{\varepsilon_1 + \varepsilon_2} k^2 = \frac{\varepsilon_1 \varepsilon_2}{\varepsilon_1 + \varepsilon_2} \frac{\omega^2}{c^2}. \tag{12.29}$$

We also obtain an expression for the normal component of the wavevector,

$$k_{j,z}^2 = \frac{\varepsilon_j^2}{\varepsilon_1 + \varepsilon_2} k^2, \qquad j = 1, 2. \tag{12.30}$$

Having derived (12.29) and (12.30), we are in a position to discuss the conditions for an interface mode to exist. For simplicity, we assume that the imaginary part of the complex dielectric function $\varepsilon_1(\omega)$ is small compared with its real part so that it can be neglected for now. A more detailed discussion that justifies this assumption will follow (see also [6]). We are looking for interface waves that propagate along the interface. This requires a real k_x.[2] Looking at (12.29), this can be fulfilled if the sum and the product of the dielectric functions are either both positive or both negative. In order to obtain a "bound" solution, we require that the normal components of the wavevector are purely imaginary in both media, giving rise to exponentially decaying solutions. This can be achieved only if the sum in the denominator of (12.30) is negative. From this we conclude that the conditions for an interface mode to exist are the following:

$$\varepsilon_1(\omega) \cdot \varepsilon_2(\omega) < 0, \tag{12.31}$$

$$\varepsilon_1(\omega) + \varepsilon_2(\omega) < 0, \tag{12.32}$$

which means that the real part of one of the dielectric functions must be negative with an absolute value exceeding that of the other. As we have seen in the previous section, metals, especially noble metals such as gold and silver, have a large negative real part of the dielectric constant together with a small imaginary part. Therefore, localized modes can exist at a metal–dielectric interface. Problem 12.3 discusses a possible solution for positive dielectric constants.

[2] Later we will see that, on taking into account the imaginary part of $\varepsilon_1(\omega)$, k_x becomes complex, which leads to a damped propagation in the x-direction.

12.3.1 Properties of surface plasmon polaritons

Using the results of the previous section, we will now discuss the properties of surface plasmon polaritons (SPPs). To accommodate losses associated with electron scattering (ohmic losses) we have to consider the imaginary part of the metal's dielectric function [7],

$$\varepsilon_1 = \varepsilon_1' + \mathrm{i}\varepsilon_1'', \tag{12.33}$$

with ε_1' and ε_1'' being real. The adjacent medium is assumed to exhibit negligible losses, i.e. ε_2 is a real number. We then naturally obtain a complex parallel wavenumber $k_x = k_x' + \mathrm{i}k_x''$, which defines wave propagation along the metal–dielectric interface.

Plasmon wavelength

The real part k_x' determines the SPP wavelength, while the imaginary part k_x'' accounts for the damping of the SPP as it propagates along the interface. This is easy to see by using the complex k_x in (12.23). The real and imaginary parts of k_x can be determined using (12.29) under the assumption that $|\varepsilon_1''| \ll |\varepsilon_1'|$:

$$k_x' \approx \sqrt{\frac{\varepsilon_1'\varepsilon_2}{\varepsilon_1' + \varepsilon_2}}\,\frac{\omega}{c}, \tag{12.34}$$

$$k_x'' \approx \sqrt{\frac{\varepsilon_1'\varepsilon_2}{\varepsilon_1' + \varepsilon_2}}\,\frac{\varepsilon_1''\varepsilon_2}{2\varepsilon_1'(\varepsilon_1' + \varepsilon_2)}\,\frac{\omega}{c}, \tag{12.35}$$

in formal agreement with Eq. (12.29). For the SPP wavelength we thus obtain

$$\lambda_{\mathrm{SPP}} = \frac{2\pi}{k_x'} \approx \sqrt{\frac{\varepsilon_1' + \varepsilon_2}{\varepsilon_1'\varepsilon_2}}\,\lambda, \tag{12.36}$$

where λ is the wavelength in vacuum. Assuming that $\varepsilon_2 = -\delta\varepsilon_1'$, where $\delta < 1$, then $\lambda_{\mathrm{SPP}} = \sqrt{1-\delta}\cdot\lambda/\sqrt{\varepsilon_2}$, which shows that the plasmon wavelength is always shorter than the wavelength in the transparent medium.

Plasmon propagation length

The propagation length of the SPP along the interface is determined by k_x'', which, according to (12.23), is responsible for an exponential damping of the electric field amplitude. The 1/e decay length of the electric field is $1/k_x''$ ($1/(2k_x'')$ for the intensity). This damping is caused by ohmic losses and results in a heating of the metal. Using $\varepsilon_2 = 1$ and the dielectric functions of silver ($\varepsilon_1 = -18.2+0.5\mathrm{i}$) and gold ($\varepsilon_1 = -11.6+1.2\mathrm{i}$) at a wavelength of 633 nm, we obtain 1/e intensity propagation lengths of ~60 μm and ~10 μm, respectively.

We note that we extracted all losses responsible for plasmon damping from the metal's bulk dielectric function. This is a good approximation as long as the characteristic dimensions of the metal structures are larger than the electron mean free path. For smaller dimensions, there is an increasing chance that electrons scatter from interfaces. In other words, close to interfaces additional loss mechanisms that locally increase the imaginary part of the metal's dielectric function have to be taken into account. It is difficult to correctly account for these non-local losses since the exact parameters are not known. Nevertheless, since the fields associated with surface plasmons penetrate into the metal by more than 10 nm, non-local effects associated with the first few atomic layers can usually be safely ignored.

Plasmon evanescent-field decay length

The electric field of an SPP decays exponentially with distance from the interface. The decay lengths into the dielectric and the metal can be obtained from (12.30) to first order in $|\varepsilon_1''| / |\varepsilon_1'|$ using (12.33) as

$$k_{1,z} = \frac{\omega}{c}\sqrt{\frac{\varepsilon_1'^2}{\varepsilon_1' + \varepsilon_2}}\left[1 + \mathrm{i}\frac{\varepsilon_1''}{2\varepsilon_1'}\right], \tag{12.37}$$

$$k_{2,z} = \frac{\omega}{c}\sqrt{\frac{\varepsilon_2^2}{\varepsilon_1' + \varepsilon_2}}\left[1 - \mathrm{i}\frac{\varepsilon_1''}{2(\varepsilon_1' + \varepsilon_2)}\right]. \tag{12.38}$$

Using the same parameters for silver and gold as before and safely neglecting the very small imaginary parts, we obtain for the 1/e decay lengths ($1/k_{1,z}$ = 23 nm, $1/k_{2,z}$ = 421 nm) and ($1/k_{1,z}$ = 28 nm, $1/k_{2,z}$ = 328 nm), respectively. This shows that the decay into the metal is much shorter than that into the dielectric. It also shows that a sizable amount of the SPP electric field can reach through a thin enough metal film. The decay of the SPP into the air half-space was observed directly in [8] using a scanning tunneling optical microscope.

Intensity enhancement

An important parameter is the intensity enhancement near the interface due to the excitation of surface plasmons. This parameter can be obtained by evaluating the ratio of the incoming intensity and the intensity right above the metal interface. We skip this discussion for the moment and come back to it after the next section (see Problem 12.4).

12.3.2 Thin-film surface plasmon polaritons

So far we have considered surface plasmons propagating along the interface between two infinitely extended half-spaces. In experimental realizations, however, we often encounter the situation that a thin metallic film is deposited on top of a dielectric substrate. In the case of a multilayer system, in principle each metal/dielectric interface can support SPPs. If the films are thin enough, i.e. of thickness comparable to the electric-field decay lengths, see (12.37) and (12.38), surface plasmons of different interfaces can couple with each other, leading to mode hybridization (see Section 8.7.1). Even a single metal film sandwiched between two dielectrics exhibits such surface plasmon interactions. For example, if the two dielectrics are the same one obtains "even" and "odd" modes [9]. For odd modes the propagation length is significantly increased because the field is pushed out of the metal. Similar properties are observed for metal–insulator–metal systems, which are important e.g. for the realization of plasmonic circuitry. For asymmetric systems with different dielectrics, e.g. a metal film sandwiched between glass and air, one can exploit the fact that the speed of light in the higher-index material is less than that in the lower-index material and therefore the light-line has a smaller slope.

Since surface plasmons are solutions of the *homogeneous* wave equation (absence of driving terms), we find conditions for the existence of plasmons from the characteristic equation which results from a set of boundary conditions. The characteristic equation is derived by setting the determinant of the matrix that describes the boundary conditions to zero. For a single interface the characteristic equation corresponds to Eq. (12.28). It is the condition that makes the Fresnel reflection coefficient r^{p} go to infinity or, in other words, it is the condition for a pole of r^{p} (c.f. Eq. (2.51)). For a metal film of finite thickness we proceed in exactly the same way; that is, we derive the modes of the system from the poles of the reflection coefficient. The latter must now account for two interfaces, namely the top surface and the bottom surface of the metal film. The poles are found from the zeros of the denominator of the reflection coefficient, which yields the following characteristic equation (c.f. Eq. (10.20)):

$$1 + r^{\mathrm{p}}_{1,2}(k_x) r^{\mathrm{p}}_{2,3}(k_x) \exp[2\mathrm{i}k_{2z}d] = 0. \tag{12.39}$$

Here, $r^{\mathrm{p}}_{1,2}$ and $r^{\mathrm{p}}_{2,3}$ are the reflection coefficients of the top surface and the bottom metal according to Eq. (2.51), k_{2z} is the normal component of the $\mathbf{k}$ vector in the metal film, and d is the thickness of the film. The solutions of Eq. (12.39) have to be sought in the complex-k_x plane. The real part yields the SPP propagation constant and the imaginary part the propagation length. For a metal film bounded by two different dielectrics one finds four distinct modes [10], of which two are so-called *leaky waves* that are mainly important for transient processes. The other two modes are largely non-radiative and correspond to the odd and even modes discussed above.

It is interesting to follow the plasmon mode that is localized on the metal surface bounded by the medium with lower dielectric constant as a function of the metal thickness d. For large d the two metal surfaces do not interact and we recover the single interface solution derived before. In this regime, the in-plane wavevector k_x has a positive imaginary part, giving rise to surface plasmon attenuation. As we reduce the thickness, the attenuation decreases and reaches zero at a critical thickness d_{crit}, which is typically in the range 50–100 nm for noble-metal films in the visible. For this thickness, k_x is purely real despite the fact that the metal film is lossy. Because $\mathrm{Im}\{k_x\} = 0$ the field in the medium with lower dielectric constant is a pure evanescent wave and the field in the medium with higher dielectric constant is a plane wave. The energy supplied by this plane wave *exactly* compensates for the attenuation by the metal film. Using the Kretschmann configuration (see the next section) it is possible for the system to be in perfect resonance with this situation. Interestingly, as we reduce the thickness d further, the imaginary part of k_x becomes negative, giving rise to plasmon amplification. However, this situation requires excitation by a wave that exponentially increases with distance from the metal surface and therefore has limited validity [10].

Notice that a complex k_x (finite propagation length) implies a complex k_z, which means that the field on the metal surface is neither a propagating wave nor an evanescent wave. To understand this point, we insert $\mathbf{E}\exp(i\mathbf{k}\cdot\mathbf{r})$ into the Helmholtz equation and obtain $\mathbf{k}\cdot\mathbf{k} = \varepsilon\omega/c$. For a medium with real ε (the dielectrics bounding the metal film) the right-hand side is real and hence $\mathrm{Re}\{\mathbf{k}\}\cdot\mathrm{Im}\{\mathbf{k}\} = 0$. Thus, as shown in Fig. 12.6, the real part of $\mathbf{k}$ and its imaginary part are perpendicular, which means that there is a direction along which the field is purely propagating and a direction along which the field is purely evanescent. In general, the propagation direction ($\mathrm{Re}\{\mathbf{k}\}$) is *not* parallel to the metal surface. Only for metals that match the critical thickness d_{crit} do we find a propagation direction parallel to the surface of the metal.

Finally, it has to be emphasized that our discussion is concerned with modes, namely solutions of the wave equation in the absence of excitation. In an experimental situation we always have a driving term, e.g. an incident laser beam. The best coupling to the system under study is obtained when the excitation is resonant with a mode of the system. As discussed in the next section, resonant excitation of SPPs can be established with the Otto or Kretschmann configurations, which are widely used in applications of plasmonics.

12.3.3 Excitation of surface plasmon polaritons

The plasmon dispersion relation

In order to excite SPPs we have to fulfill both energy and momentum conservation. To see how this can be done, we have to analyze the dispersion relation of the surface waves, i.e. the relation between the energy in terms of the angular frequency ω and the momentum in terms of the wavevector in the propagation direction k_x given by Eqs. (12.29) and (12.34). In order to plot this dispersion relation we assume that $\varepsilon_2 = 1$ irrespective of ω,

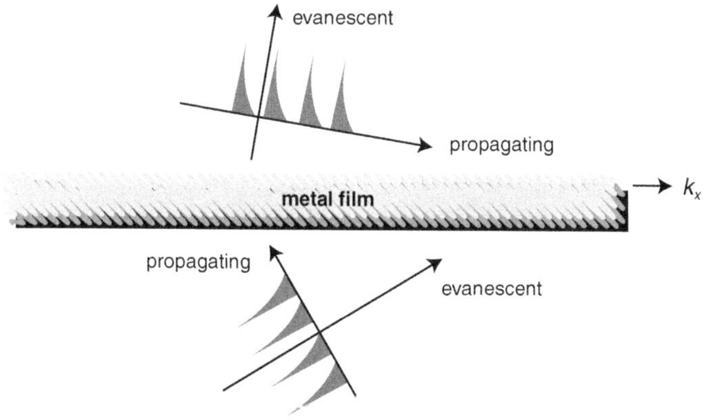

Fig. 12.6 Illustration of the mode structure of a thin metal film. In the dielectrics above and below the metal film the fields have a propagation direction along which $\mathrm{Im}\{\boldsymbol{k}\} = 0$ and an evanescent direction along which $\mathrm{Re}\{\boldsymbol{k}\} = 0$. The two directions are perpendicular.

e.g. air or vacuum. For $\varepsilon_1(\omega)$ we use the measured dielectric functions [3] for gold and silver. Figure 12.7 shows the respective plots. The dashed line represents the dispersion relation of light in the medium with $\varepsilon_2 = 1$, the so-called light-line $\omega = ck$ in air. The surface plasmon dispersion relations show two branches, a high-energy and a low-energy branch, which are clearly disconnected only in the case of silver. For gold, the two branches are connected. The high-energy branch, called the Brewster mode, does not describe true surface waves, since according to (12.30) the z-component of the wavevector in the metal is no longer purely imaginary. It corresponds to wave propagation into the metal, that is $k_x = (\omega/c)\sqrt{\varepsilon_1(\omega)}$. This branch will not be considered further. The low-energy branch corresponds to a true surface wave, the surface plasmon polariton (SPP). "Polariton" refers to an excitation that corresponds to a coupled electromagnetic and matter wave. In the case of surface plasmons, the matter wave is a surface charge oscillation. The surface plasmon wavevector k_x is always larger than the corresponding wavevector of a freely propagating photon. Therefore surface plasmons cannot decay into propagating photons. As the wavelength decreases, the difference in wavevector increases. As the energy approaches a limiting frequency[3], the wavevector assumes a maximum value and the damping of the surface plasmon, i.e. $\text{Im}(k_x)$, strongly increases. For even higher frequencies, the surface plasmon dispersion continuously converts into the higher-energy Brewster mode described above. The effect is also referred to as back-bending of the surface plasmon dispersion relation. The back-bending effect has been experimentally verified (see Ref. [11]) and imposes a limit on the maximum surface plasmon wavenumber k_x that can be achieved in an experiment. Usually, this maximum k_x is smaller than $\sim 3\omega_p/c$.

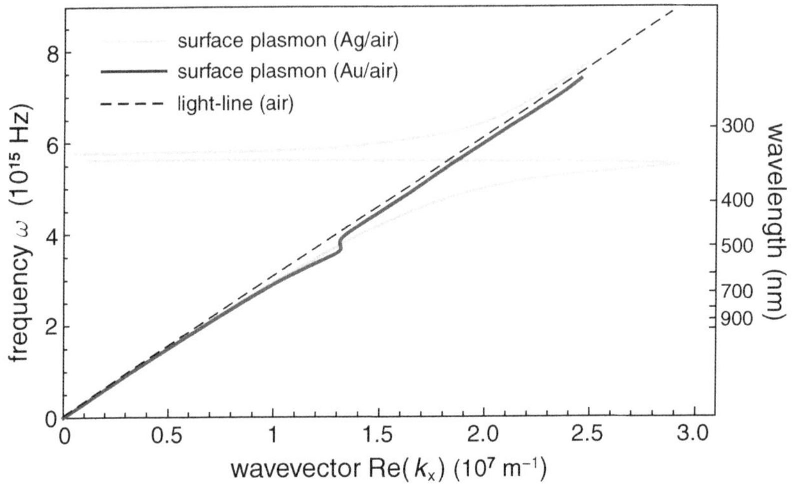

Fig. 12.7 The dispersion relation of surface plasmon polaritons at a gold/air interface (dark line) and at a silver/air interface (gray line) obtained by using the measured $\varepsilon_1(\omega)$ [3] in Eq. (12.29). The dashed straight line is the light-line $\omega = ck_x$ in air.

[3] In the case of a Drude-type free-electron gas this limiting frequency is $\omega_p/\sqrt{1+\varepsilon_2}$.

Excitation configurations

An important feature of surface plasmons is that for a given energy $\hbar\omega$ the wavevector k_x is always larger than the wavevector of light in free space; that is, the plasmon dispersion curve is to the right of the light-line. This is obvious on inspecting (12.29) and also from Figs. 12.7 and 12.8(a), where the light-line ω/c is plotted as a dashed line. The surface plasmon dispersion asymptotically approaches the light-line for small energies. The physical reason for the large surface plasmon momentum is the strong coupling between light and surface charges. The light field has to "drag" the electrons along the metal surface. Consequently, surface plasmons on plane interfaces cannot be excited by light of any frequency incident from free space. Excitation of surface plasmons by light is possible only if a wavevector component of the exciting light can be increased over its free-space value. There are several ways to achieve this increase of the wavevector component. The conceptually simplest solution is to excite surface plasmons by means of evanescent waves created at the interface between a medium with refractive index $n > 1$. The light-line in this case is tilted by a factor of n since $\omega = ck/n$. This situation is shown in Fig. 12.8(a), which sketches the SPP dispersion relation with the light-line in air and the tilted light-line in glass.

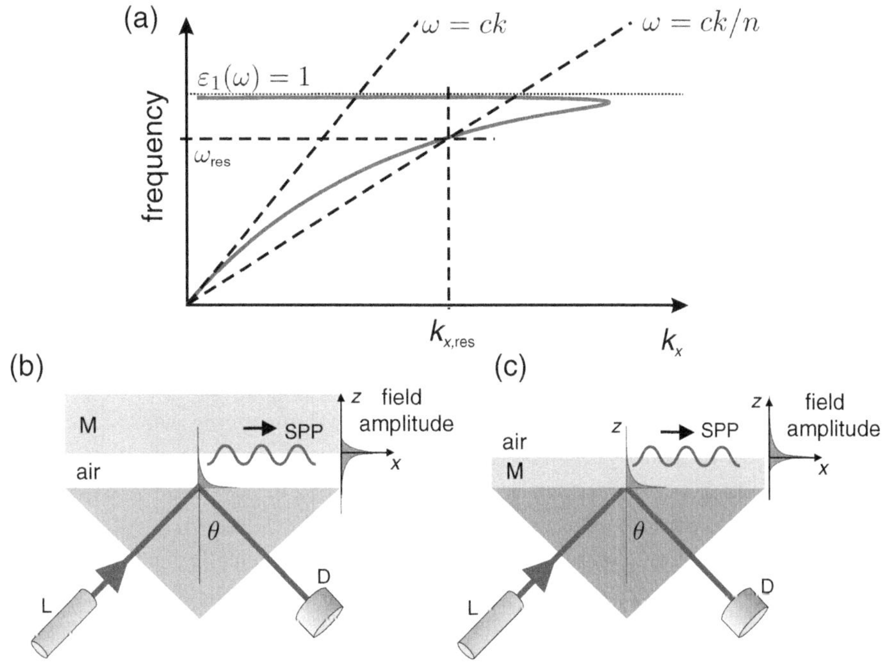

Fig. 12.8 Excitation of surface plasmons. (a) A close-up sketch of the surface plasmon dispersion relation with the light-lines in air and in glass. Experimental arrangements to realize the energy and momentum conservation sketched in (a): (b) Otto configuration and (c) Kretschmann configuration. L, laser; D, detector; M, metal layer.

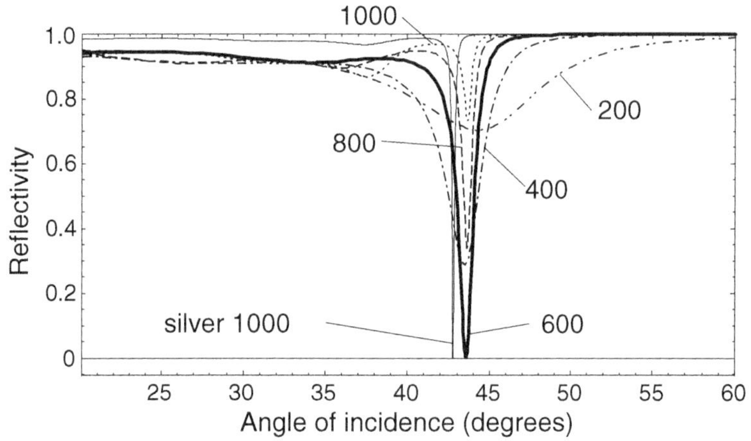

Fig. 12.9 Excitation of surface plasmons in the Otto configuration. The reflectivity of the exciting beam is plotted as a function of the incident angle and for different air gaps (in nanometers). The curves are evaluated for a gold surface. For comparison, a single trace is also plotted for silver, for which the resonance is much sharper because of lower damping. $\lambda = 632.8$ nm.

Figures 12.8(b) and (c) sketch possible experimental arrangements that realize this idea. In the Otto configuration [12] the tail of an evanescent wave at a glass/air interface interacts with a metal/air interface that supports surface plasmons. For a sufficiently large separation between the two interfaces the evanescent wave is only weakly influenced by the presence of the metal. By tuning the angle of incidence of the totally reflected beam inside the prism, the resonance condition for excitation of surface plasmons, i.e. the matching of the parallel wavevector components, can be fulfilled. The excitation of surface plasmons will show up as a minimum in the reflected light. The reflectivity of the system as a function of the angle of incidence and of the separation between interfaces is shown in Fig. 12.9. For the angle of incidence a clear resonance is observed at 43.5°. For small separations the resonance is broadened and shifted due to radiation damping of the surface plasmons, whose evanescent tails can then couple back into the glass, allowing the surface plasmons to rapidly decay radiatively by transforming evanescent fields into propagating waves in the glass. For separations that are too large, surface plasmons can no longer be efficiently excited since the evanescent-field decay length is too short and the resonance vanishes.

The Otto configuration proved to be experimentally inconvenient because of the challenging control of the tiny air gap between the glass and the metal surface. In 1971 Kretschmann came up with an alternative method to excite surface plasmons [13]. In this method, a thin metal film is deposited on top of a glass prism. The geometry is sketched in Fig. 12.8(c). To excite surface plasmons at the metal/air interface an exponentially damped wave created at the glass/metal interface penetrates through the metal layer. Here, arguments similar to those for the Otto configuration apply. If the metal is too thin, the SPP will be strongly damped because of radiation damping into the glass. If the metal film is too thick the SPP can no longer be efficiently excited due to absorption in the metal. Figure 12.10 shows the reflectivity of the excitation beam as a function of the metal film thickness and the angle of incidence. As before, the resonant excitation of surface plasmons is characterized by a dip in the reflectivity curves. For the occurrence of a minimum in the reflectivity curves both in the Otto and in the Kretschmann configuration, the following physical interpretation can be given. The minimum can be thought of as being due to destructive interference between the totally reflected light and the light emitted by the SPP wave due to radiation damping.

Another alternative way to excite surface plasmons is the use of grating couplers [7]. Here, the increase of the wavevector necessary to match the surface plasmon momentum is achieved by adding a reciprocal-lattice vector of the grating to the free-space wavevector. This requires in principle that the metal surface is structured with the right periodicity a over an extended spatial region. The new parallel wavevector then reads as $k_x' = k_x + 2\pi n/a$, with $2\pi n/a$ being a reciprocal-lattice vector of the grating. This principle was used to enhance the interaction of subwavelength holes with surface plasmons in silver films [14].

12.3.4 Surface plasmon sensors

The distinct resonance condition associated with the excitation of surface plasmons has found application in various sensors. For example, the position of the dip in the reflectivity curves can be used as an indicator for environmental changes. With this method, the adsorption onto or removal of target materials from the metal surface can be detected with

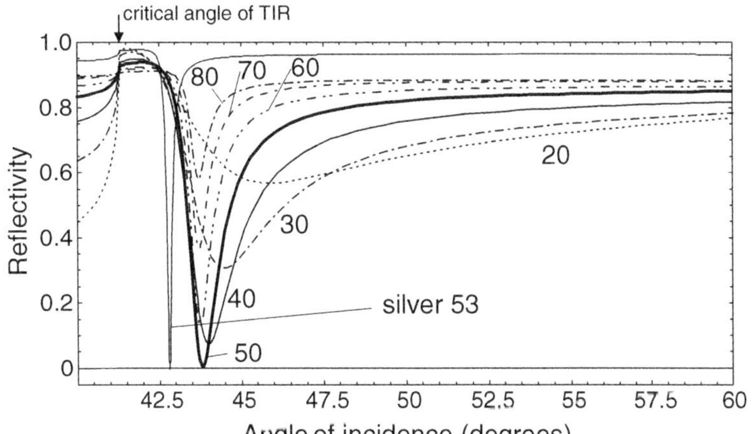

Fig. 12.10 Excitation of surface plasmons in the Kretschmann configuration. The reflectivity of the exciting beam is plotted as a function of the angle of incidence and for different thicknesses of a gold film on glass (in nanometers). For comparison a single trace for a silver film is plotted. Note the much sharper resonance due to the smaller damping of silver. The critical angle of total internal reflection shows up as a discontinuity marked by an arrow. $\lambda = 632.8$ nm.

submonolayer accuracy. Figure 12.11 illustrates this capability by a simulation. It shows the effect of a 3-nm layer of water on top of a 53-nm-thick silver film on glass. A strongly shifted plasmon resonance curve can be observed. Assuming that the angle of incidence of the excitation beam has been adjusted to the dip in the reflectivity curve, the deposition of a minute amount of material increases the signal (reflectivity) drastically. This means that the full dynamic range of a low-noise intensity measurement can be used to measure a coverage ranging between 0 and 3 nm. Consequently, SPP sensors are very attractive for applications ranging from biological binding assays to environmental sensing. For a review see e.g. [15].

The reason for the high sensitivity lies in the fact that the excitation of SPPs is associated with enhanced fields near the metal surface. In the Kretschmann configuration, this enhancement factor can be determined by evaluating the ratio of the intensity above the metal and the incoming intensity. In Fig. 12.11(b) this enhancement factor is calculated and plotted as a function of the angle of incidence both for gold and for silver for a 50-nm thin film.

12.4 Surface plasmons in nano-optics

Near-field optical probes as well as single-quantum emitters lead to additional possibilities for exciting surface plasmons [16–18]. The parallel components of the wavevector (k_x) necessary for SPP excitation are also present in confined optical near-fields in the vicinity of subwavelength apertures, metallic particles, or even fluorescent molecules. If such confined fields are brought close enough to a metal surface, coupling to surface plasmons can be accomplished *locally*. Figure 12.12 shows the principal arrangements. A metal film resides on a (hemispherical) glass prism to allow leakage radiation to escape and to be

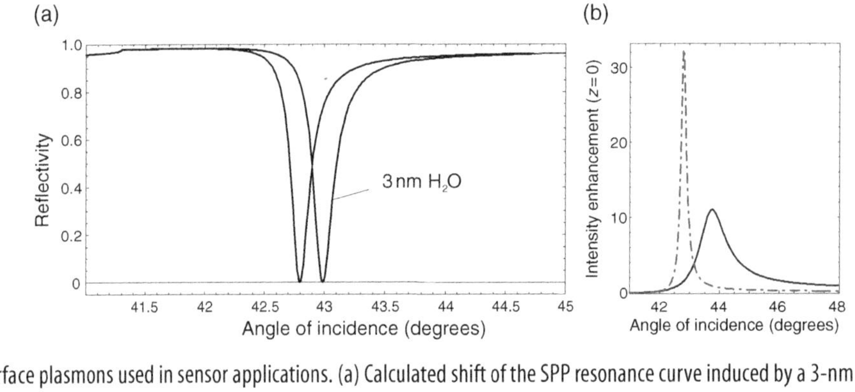

Fig. 12.11 Surface plasmons used in sensor applications. (a) Calculated shift of the SPP resonance curve induced by a 3-nm layer of water ($n = 1.33$) adsorbed on a 53-nm-thick silver film. (b) Intensity enhancement near the metal surface as a function of the angle of incidence in the Kretschmann configuration. For silver ($\varepsilon_1 = -18.2 + 0.5i$, dash–dotted line) and gold ($\varepsilon_1 = -11.6 + 1.2i$, solid line) at a wavelength of 633 nm we observe maximum intensity enhancements of $\sim$32 and $\sim$10, respectively.

recorded. In order to efficiently launch surface plasmons, the exciting light field needs to be confined such that its angular spectrum contains a sizeable amplitude of evanescent-field components that match the parallel wavevector k_x of the surface plasmon.

As an illustration, Fig. 12.13(a) shows the excitation of surface plasmons with an oscillating dipole placed near the surface of a thin silver film deposited on a glass surface. The figure depicts contour lines of constant power density evaluated at a certain instant of time and displayed on a logarithmic scale. The surface plasmons propagating on the top surface decay radiatively, as manifested by the wavefronts in the lower medium. The situation is reciprocal to the situation of the Kretschmann configuration discussed earlier. Also shown in Fig. 12.13(a) is the excitation of surface plasmons at the metal/glass interface. However, at the wavelength of 370 nm, these plasmons are strongly damped and therefore do not propagate over long distances. Figure 12.13(b) shows the radiation pattern evaluated in the lower medium (glass). It corresponds to the radiation collected with a high-NA lens, collimated and projected onto a photographic plate. The circle in the center indicates the critical angle of total internal reflection of an air/glass interface $\theta_c = \arcsin(1/n)$, with n being the index of refraction of glass. Obviously, the plasmon radiates into an angle beyond θ_c. In fact, the emission angle corresponds to the Kretschmann angle discussed previously (cf. Fig. 12.10). Surface plasmons can be excited only with p-polarized field components since there needs to be a driving force on the free charges towards the interface. This is the reason why the radiation pattern exhibits two lobes. The coupling of fluorophores to surface plasmons (see Fig. 12.12(c)) can drastically improve the sensitivity of fluorescence-based assays in medical diagnostics, biotechnology, and gene expression. For finite distances between metal and fluorophores (<200 nm) the coupling to surface plasmons leads to fluorescence-signal enhancement and high directionality of the emission. For example, an immunoassay for the detection of the cardiac marker myoglobin has been developed in Ref. [20].

The dipole is an ideal excitation source in the sense that its angular spectrum is very broad. More realistic sources have finite dimensions. The size of the source and its distance to the metal surface determine the spatial spectrum that is available for the excitation

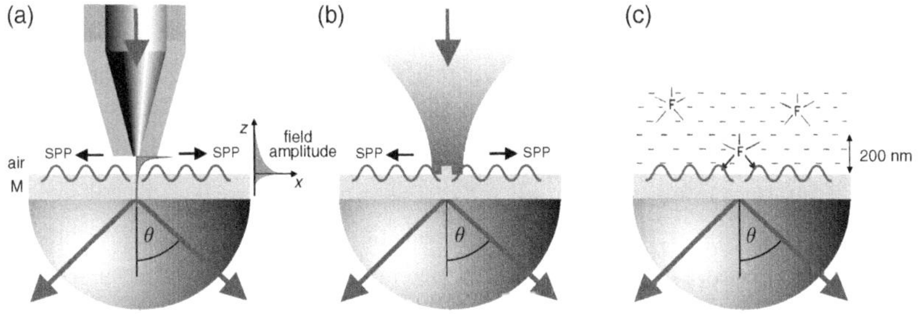

Fig. 12.12 Local excitation of surface plasmons on a metal film using different sources of confined light fields: (a) a subwavelength light source such as an aperture probe [16], (b) an irradiated nanoparticle [19], and (c) fluorescent molecules [17]. In all cases, surface plasmons are excited by evanescent components of the angular spectrum that match the parallel wavevector k_x of the surface plasmon. The arrows at the metal film indicate the emission of leakage radiation.

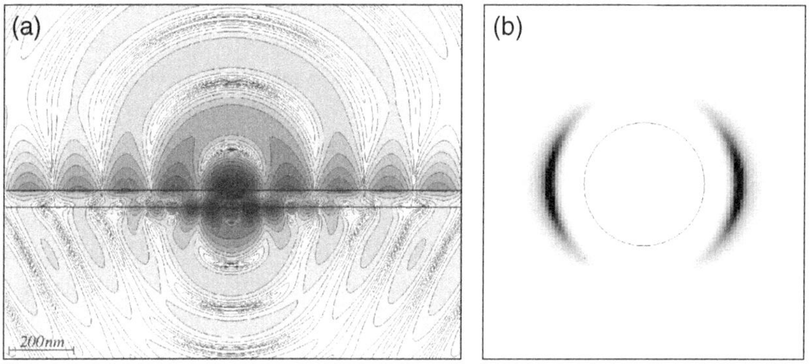

Fig. 12.13 Excitation of surface plasmons with a dipole source placed 5 nm above a 50 nm-thick silver layer supported by a glass substrate. The excitation wavelength is 370 nm and the dipole moment is parallel to the interface. (a) Lines of constant power density (with a factor of 2 between successive contour lines) depicted at a certain instant of time. The figure shows the surface plasmon propagation along the top surface of the silver film and also the radiative decay into the lower half-space. (b) The radiation pattern in the lower medium. The circle indicates the critical angle of total internal reflection at an air/glass interface. The two lobes result from the radiative decay of surface plasmons excited by the dipole source.

of surface plasmons. If the source is too far from the metal surface only plane-wave components of the angular spectrum reach the metal surface and hence coupling to surface plasmons is inhibited. Figure 12.14(a) shows a sketch of the spatial spectrum (spatial Fourier transform) of a confined light source evaluated in planes at different distances from the source (see the inset). The angular spectrum is broad close to the source but narrows with increasing distance from the source. The same figure also shows the spatial spectrum of a surface plasmon supported by a silver film. The excitation of the surface plasmon is possible because of the overlap of the spatial spectrum of the source and the surface plasmon. Owing to the decrease in field confinement for increasing distance from the source, a characteristic distance dependence for the surface plasmon excitation efficiency is expected. As discussed before, in a thin-film configuration, surface plasmon excitation can be monitored by observing the plasmon's leakage radiation into the glass half-space.

Figure 12.14(b) shows, for thin gold and silver films deposited on a glass hemisphere, the integrated intensity of surface plasmon leakage radiation as a function of the distance between the source (aperture) and the metal surface. The curve labeled MMP indicates a numerical simulation. All curves clearly show a dip for very small distances. This dip is due to the perturbation of the surface plasmon resonance condition by the proximity of the probe, i.e. the coupling between probe and sample (see also Fig. 12.9 as an illustration of this effect). Leakage radiation can also be used to visualize the propagation length of surface plasmons. This is done by imaging the metal/glass interface onto a camera using a high-NA microscope objective that can capture the leakage radiation above the critical angle (see Fig. 12.14(c)). The extent of the SPP propagation is in good agreement with Eq. (12.29), although the presence of leakage radiation indicates a reduced propagation length because of this additional loss channel. By varying the separation between the excitation source and the metal surface and by varying the excitation polarization it is possible to control the intensity and the direction in which surface plasmons are launched.

While the excitation of surface plasmons in Fig. 12.14 was accomplished with a near-field aperture probe, the example in Fig. 12.15 shows the same experiment but with a laser-irradiated nanoparticle acting as the excitation source. In this experiment, the surface plasmon propagation is visualized by observing the fluorescence intensity of a thin layer of fluorophores deposited onto the metal surface using a spacer layer. A double-lobed emission pattern is observed due to the fact that surface plasmons can be excited only by p-polarized field components of the near-field. Control over the direction of emission is possible via the choice of the polarization of the excitation beam [18].

An interplay between surface plasmons launched by an aperture probe and surface plasmons excited by particle scattering has been studied in Ref. [16]. Figure 12.16 shows experimentally recorded surface plasmon interference patterns on a smooth silver film with some irregularities. The periodicity of the fringes of 240 ± 5 nm is exactly half the surface plasmon wavelength. The contrast in this image is obtained by recording the intensity of the leakage radiation as the aperture probe is raster scanned over the sample surface. Thus, the fringes are due to surface plasmon standing waves that build up between the probe and the irregularities that act as scattering centers. Strongest leakage radiation is obtained for probe–scatterer distances that are integer multiples of half the surface plasmon wavelength. The observation that surface plasmons originating from different scattering centers on a surface can interfere suggests the possibility of building optical elements for surface plasmon nano-optics by employing suitable arrangements of scatterers [21–23].

12.4.1 Plasmons supported by wires and particles

The electromagnetic field associated with SPPs on plane interfaces is localized in the direction normal to the interface. To establish field confinement in two or three dimensions we

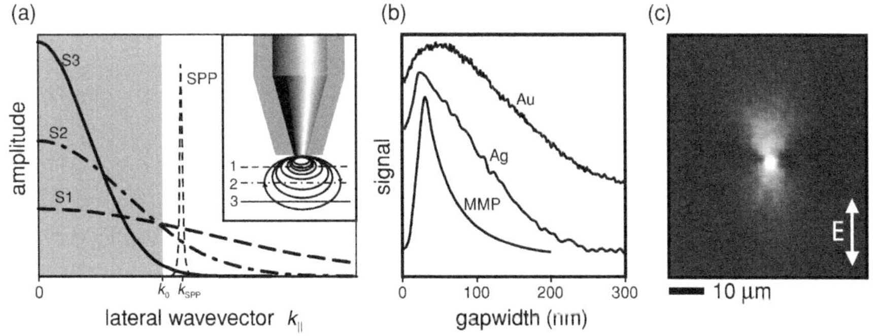

Fig. 12.14 Local excitation of surface plasmons with a near-field aperture probe. (a) A sketch of the overlap of the spatial spectra of the source (evaluated in planes at different distances from the source) and the surface plasmon on a silver film. (b) Distance dependence of the coupling. The dip at short distances is a result of probe–sample coupling, i.e. the presence of the probe locally modifies the plasmon resonance condition. (c) An image of plasmon propagation recorded by focusing the leakage radiation onto an image plane.

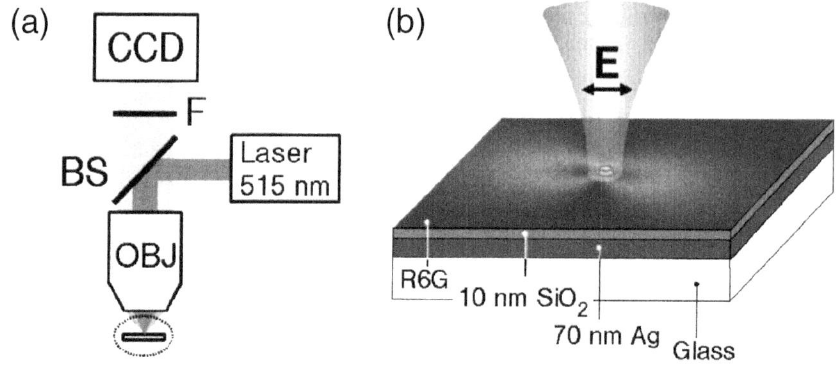

Fig. 12.15 Excitation of surface plasmons by a subwavelength-scale protrusion located on the top surface of a metal film. (a) The set-up. (b) A close-up of the particle–beam interaction area. In this experiment, the surface plasmons are detected by observing the fluorescence intensity of a thin layer of fluorescent molecules deposited on a dielectric spacer layer. From [18].

need to consider metal particles of finite size, such as metal wires or nanoparticles. As we will see, metal wires can support SPP modes that propagate along the wire, much as for the case of a plane interface. Owing to the negative real part of the dielectric constant of metals these modes are usually localized to the wire surface. As for the plane-interface plasmons, these wire modes do not couple to free-space radiation and need appropriate coupling mechanisms in order to be excited or to decay into photons, such as e.g. a grating coupler or a wire discontinuity. In addition to propagating SPPs, a wire also supports transverse plasmons, that is, charge oscillations transverse to the wire axis. Similar modes also appear in small metal particles. These transverse modes are of radiative nature and can couple to propagating radiation. Figure 12.17 illustrates the surface plasmon modes supported by plane interfaces, thin wires, and particles. While Figs. 12.17(a) and (b) display propagating modes, Fig. 12.17(c) corresponds to a transverse dipolar mode.

To determine the surface plasmon eigenmodes of thin wires we need to solve the homogeneous wave equation (12.22) taking into account the respective boundary conditions. For thin wires and small particles we can invoke the quasi-static approximation. This approximation neglects retardation and all points of the particle or the wire circumference oscillate in phase. This is possible only if the relevant characteristic size of the object (wire or particle radius) is much smaller than the metal's skin depth d ($d = \lambda/(4\pi\sqrt{\varepsilon})$). For a small particle this leads to a picture in which the whole free-electron gas is periodically displaced relative to the stationary lattice of the particle.

In the quasi-static approximation the Helmholtz equation reduces to the Laplace equation, which is much easier to solve. A detailed discussion can be found e.g. in [24]. The solutions that are obtained here are quasi-static near-fields. For example, the electric field of an oscillating dipole

$$\mathbf{E}(r\mathbf{n},t) = \frac{1}{4\pi\varepsilon_0}\left[k^2(\mathbf{n}\times\mathbf{p})\times\mathbf{n}\frac{e^{ikr}}{r} + \left[3\mathbf{n}(\mathbf{n}\cdot\mathbf{p}) - \mathbf{p}\right]\left(\frac{1}{r^3} - \frac{ik}{r^2}\right)e^{ikr}\right]e^{-i\omega t}, \quad (12.40)$$

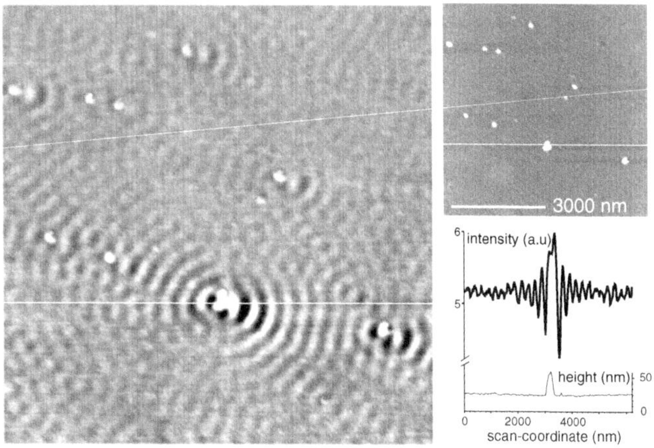

Fig. 12.16 Interference of locally excited surface plasmons. Left panel: integrated leakage radiation from a silver film with some protrusions recorded as an aperture probe is raster scanned over the sample surface. The fringes correspond to surface plasmon standing-wave patterns that build up between the protrusions and the aperture probe. Right panel: shear-force topography of the area shown in the optical image and line cuts along the white line through both the topography and the optical image.

can be approximated in the near-field zone $kr \ll 1$ as

$$\mathbf{E}(r\mathbf{n}, t) = \frac{1}{4\pi\varepsilon_0}\left[3\mathbf{n}(\mathbf{n}\cdot\mathbf{p}) - \mathbf{p}\right]\frac{\mathrm{e}^{-\mathrm{i}\omega t}}{r^3}, \tag{12.41}$$

which results from setting $k = 0$. The resulting field corresponds to that of a static dipole with added harmonic time dependence $\exp(-\mathrm{i}\omega t)$, which is the reason why it is termed *quasi*-static. In the quasi-static limit the electric field can be represented by a potential as $\mathbf{E} = -\nabla\Phi$. The potential has to satisfy the Laplace equation

$$\nabla^2\Phi = 0 \tag{12.42}$$

and the boundary conditions between adjacent materials (see Chapter 2). In the following we will analyze the solutions of (12.42) for a thin metal wire and a spherical nanoparticle.

Transverse plasmon resonances of a thin wire

Let us consider a thin cylindrical metal wire with radius a centered at the origin and extending along the z-axis to infinity. The wire is illuminated by an x-polarized plane wave with its wavevector along the y-direction. The geometry is sketched in Fig. 12.18. This illumination geometry excites the transverse plasmon resonance of a long wire, which does not transport energy along the wire. To tackle this problem we introduce cylindrical coordinates

$$\begin{aligned} x &= \rho\cos\varphi, \\ y &= \rho\sin\varphi, \\ z &= z, \end{aligned} \tag{12.43}$$

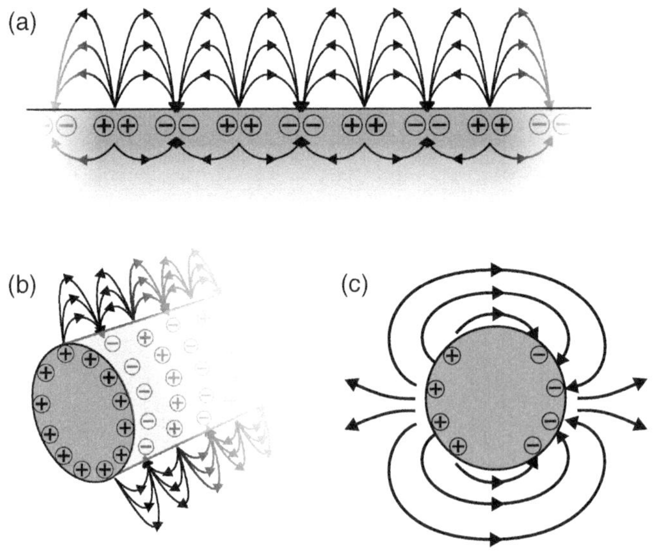

Fig. 12.17 Charge distribution and field lines for SPP modes in different geometries. (a) A propagating plane-interface plasmon. (b) A propagating wire plasmon (TM_0 mode). (c) A transverse wire plasmon and a particle plasmon with a dipolar field pattern.

and express the Laplace equation as

$$\frac{1}{\rho}\frac{\partial}{\partial\rho}\left(\rho\,\frac{\partial\Phi}{\partial\rho}\right)+\frac{1}{\rho^2}\left(\frac{\partial^2\Phi}{\partial\varphi^2}\right)=0. \tag{12.44}$$

Here, we have accounted for the fact that there is no z-dependence. The Laplace equation (12.44) can be separated using the ansatz $\Phi(\rho,\varphi)=R(\rho)\Theta(\varphi)$, yielding

$$\frac{1}{R}\left(\rho\,\frac{\partial}{\partial\rho}\left(\rho\,\frac{\partial R}{\partial\rho}\right)\right)=-\frac{1}{\Theta}\left(\frac{\partial^2\Theta}{\partial\varphi^2}\right)\equiv m^2. \tag{12.45}$$

The angular part has solutions of the form

$$\Theta(\varphi)=c_1\cos(m\varphi)+c_2\sin(m\varphi), \tag{12.46}$$

which implies that m must be an integer in order to ensure the 2π periodicity of the solution. The radial part has solutions of the form

$$R(\rho)=\begin{cases}c_3\rho^m+c_4\rho^{-m}, & m>0,\\ c_5\ln\rho+c_6, & m=0,\end{cases} \tag{12.47}$$

with the same m as introduced in (12.45). Because of the symmetry imposed by the polarization of the exciting electric field (along the x-axis) only $\cos(m\varphi)$ terms need to be

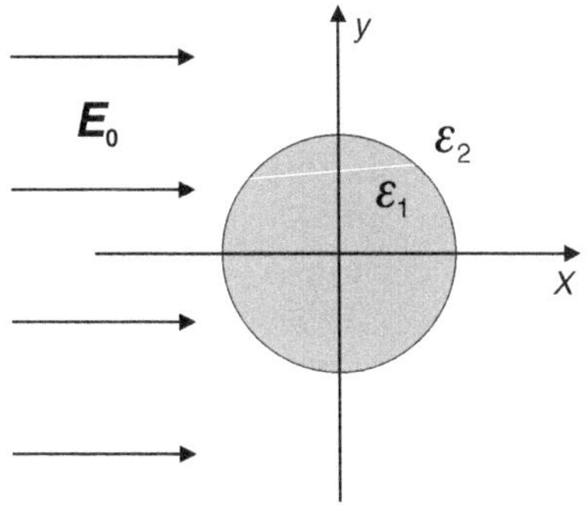

Fig. 12.18 A cut through a thin wire that is illuminated by an x-polarized plane wave.

considered. Furthermore, for $m = 0$ the logarithm in (12.47) diverges at the origin and hence must be rejected. We therefore use the expansion

$$\Phi(\rho < a) = \Phi_1 = \sum_{n=1}^{\infty} \alpha_n \rho^n \cos(n\varphi),$$

$$\Phi(\rho > a) = \Phi_2 = \Phi_{\text{scatter}} + \Phi_0 = \sum_{n=1}^{\infty} \beta_n \rho^{-n} \cos(n\varphi) - E_0 \rho \cos(\varphi), \qquad (12.48)$$

where α_n and β_n are constants to be determined from the boundary conditions on the wire surface $\rho = a$ and ϕ_0 is the potential associated with the exciting field. In terms of the potential Φ the boundary conditions read as

$$\begin{aligned} \left[\frac{\partial \Phi_1}{\partial \varphi}\right]_{\rho=a} &= \left[\frac{\partial \Phi_2}{\partial \varphi}\right]_{\rho=a}, \\ \varepsilon_1 \left[\frac{\partial \Phi_1}{\partial \rho}\right]_{\rho=a} &= \varepsilon_2 \left[\frac{\partial \Phi_2}{\partial \rho}\right]_{\rho=a}, \end{aligned} \qquad (12.49)$$

following from the continuity requirement for the tangential component of the electric field and the normal component of the electric displacement. Here, ε_1 and ε_2 are the complex dielectric constants of the wire and the surroundings, respectively. In order to evaluate (12.49) we use the fact that the functions $\cos(n\varphi)$ are orthogonal. On introducing (12.48) into (12.49) we immediately see that α_n and β_n vanish for $n > 1$. For $n = 1$ we obtain

$$\alpha_1 = -E_0 \frac{2\varepsilon_2}{\varepsilon_1 + \varepsilon_2}, \qquad \beta_1 = a^2 E_0 \frac{\varepsilon_1 - \varepsilon_2}{\varepsilon_1 + \varepsilon_2}. \qquad (12.50)$$

With these coefficients the solutions for the electric field $\mathbf{E} = -\nabla\phi$ turn out to be

$$\mathbf{E}_1 = E_0 \frac{2\varepsilon_2}{\varepsilon_1 + \varepsilon_2} \mathbf{n}_x, \qquad (12.51)$$

$$\mathbf{E}_2 = E_0 \mathbf{n}_x + E_0 \frac{\varepsilon_1 - \varepsilon_2}{\varepsilon_1 + \varepsilon_2} \frac{a^2}{\rho^2} \left(1 - 2\sin^2\varphi\right) \mathbf{n}_x + 2E_0 \frac{\varepsilon_1 - \varepsilon_2}{\varepsilon_1 + \varepsilon_2} \frac{a^2}{\rho^2} \sin\varphi \cos\varphi \, \mathbf{n}_y, \qquad (12.52)$$

where we re-introduced Cartesian coordinates with the unit vectors $\mathbf{n}_x, \mathbf{n}_y, \mathbf{n}_z$. Figure 12.19 shows the electric field and the intensity around the wire as described by Eqs. (12.51) and (12.52). Notice the field maxima along the direction of polarization (see also Chapter 6).

In most applications the dispersion (frequency dependence) of the dielectric medium surrounding the metal can be ignored and one can assume a constant ε_2. However, the metal's dielectric function is strongly wavelength-dependent. The solution for the fields is characterized by the denominator $\varepsilon_1 + \varepsilon_2$. Consequently, the electric field amplitude assumes a maximum when $\mathrm{Re}(\varepsilon_1(\lambda)) = -\varepsilon_2$. This is the resonance condition for a wire excited by a plane wave polarized perpendicular to the wire axis. The shape of the resonance is determined by the dielectric function $\varepsilon_1(\lambda)$. As in the case of the plane interface discussed earlier, changes in the dielectric constant of the surrounding medium (ε_2) lead to shifts of the resonance (see below). Notice that no resonances exist if the electric field is polarized along the wire axis. As in the plane-interface case, the excitation of surface plasmons relies on an accumulation of surface charge at the surface of the wire. In order to drive the charges to the interface, the electric field needs to have a polarization component normal to the metal surface.

Propagating surface plasmon polaritons on thin wires

To obtain surface plasmon propagation *along* a cylindrical wire one needs to solve the full vector wave equation. Such an analysis has been done in Refs. [25, 26]. Propagating solutions are found by solving a set of four homogeneous equations that result from boundary conditions. The characteristic equation of this system of equations yields

$$\frac{\varepsilon_1(\lambda)}{\kappa_1 a}\frac{J_1(\kappa_1 a)}{J_0(\kappa_1 a)} - \frac{\varepsilon_2}{\kappa_2 a}\frac{H_1^{(1)}(\kappa_2 a)}{H_0^{(1)}(\kappa_2 a)} = 0, \tag{12.53}$$

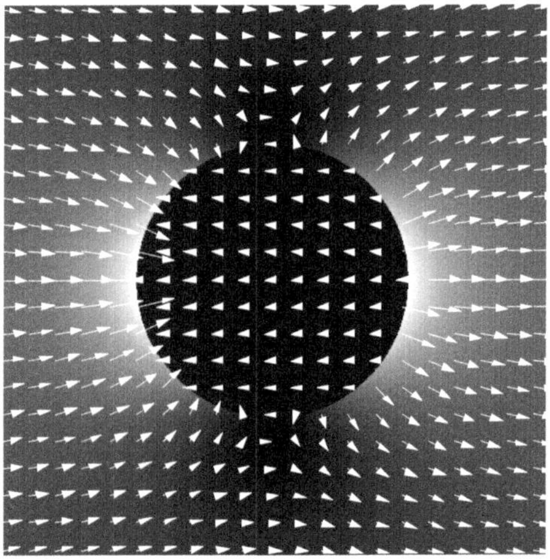

Fig. 12.19 Near-field distribution around a gold wire in the quasi-static limit $\varepsilon_1 = -18, \varepsilon_2 = 2.25$. Grayscale, $|E|^2$; arrows, direction and magnitude of the electric field.

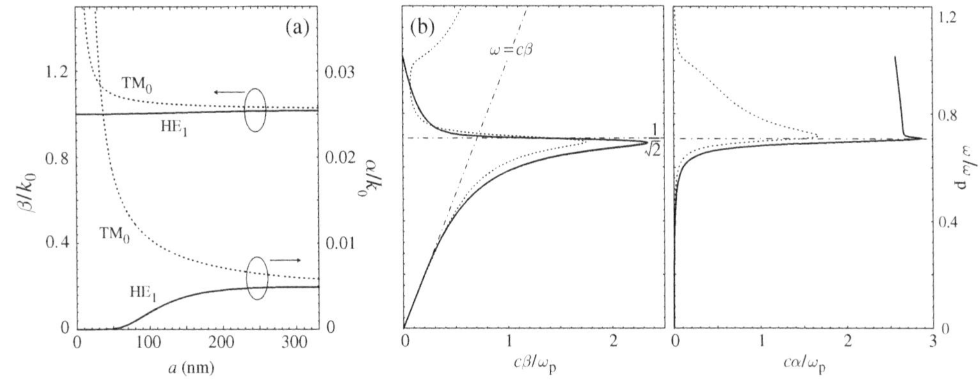

Fig. 12.20 (a) The propagation constant $k_z = \beta + i\alpha$ of the two lowest surface modes supported by an aluminum wire at a wavelength of $\lambda = 488$ nm. a denotes the wire diameter and $k_0 = \omega/c$. (b) Frequency dispersion of the HE_1 surface mode of an $a = 50$ nm aluminum wire. ω_p denotes the plasma frequency of aluminum. The dotted line indicates the corresponding dispersion on a plane interface. Notice the backbending effect discussed earlier.

where J_n and $H_n^{(1)}$ are cylindrical Bessel and Hankel functions of the first kind, respectively, and a is the wire radius. The transverse components of the wavevectors in medium $i \in \{1, 2\}$ are defined as $\kappa_i = k_0 \left[\varepsilon_i - (k_z/k_0)^2\right]^{1/2}$, with $k_0 = 2\pi/\lambda$. The propagation along the wire axis z is determined by the factor

$$\exp[i(k_z z - \omega t)], \tag{12.54}$$

where $k_z = \beta + i\alpha$ is the complex propagation constant. β and α are denoted the phase constant and the attenuation constant, respectively. For the two best propagating surface modes, Fig. 12.20(a) shows the propagation constant of an aluminum cylinder as a function of the cylinder radius a. The TM_0 mode exhibits a radial polarization, i.e. the electric field is axially symmetric. On the other hand, the HE_1 mode has a $\cos\varphi$ angular dependence and, as the radius a tends to zero, it converts to an unattenuated plane wave ($k_z \approx \omega/c$) that is infinitely extended. The situation is different for the TM_0 mode. As the radius a is decreased, its phase constant β becomes larger and the transverse field distribution becomes better localized. However, the attenuation constant α also increases and hence for wires that are too thin the surface plasmon propagation length becomes very small. Figure 12.21 shows an example of SPP propagation on a 120 nm-diameter silver wire. The plasmon is excited by focusing light that is polarized along the wire direction onto the input end (I) of the wire. Emission of far-field photons due to radiative decay of wire plasmons is observed at the distal end (D).

It has been pointed out that both the phase velocity and the group velocity of the TM_0 mode tend to zero as the diameter a is decreased [28]. Therefore, a pulse propagating along a wire whose diameter is adiabatically thinned down never reaches the end of the wire, i.e. its tip. On the other hand, as the wire becomes thinner the lateral field confinement increases, causing a "build-up" of fields towards the wire end (adiabatic focusing). Notice that the modes propagating on the surface of metal wires had already been analyzed in

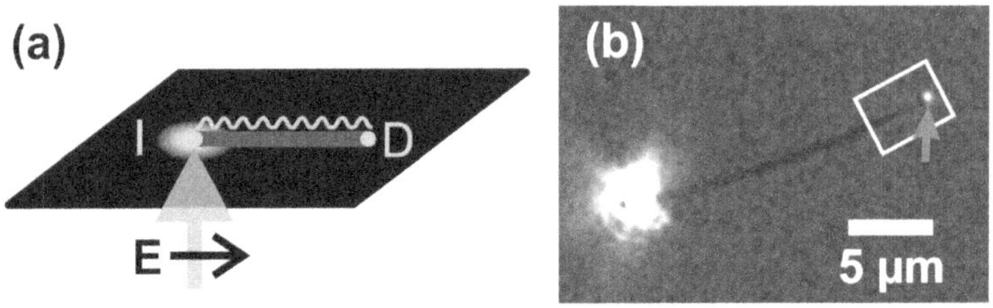

Fig. 12.21 Excitation, propagation, and detection of a wire plasmon on a 120 nm-thick silver nanowire. (a) The excitation geometry. Light with a polarization parallel to the wire is focused to the input end (I) of the wire. The plasmon propagates and is re-radiated at the distal end (D). (b) A wide-field microscopy image of the 18.6 μm-long silver wire. The arrow indicates the weak emission spot at the distal end. Note that a considerable part of the SPP intensity that arrives at D is reflected (~25%). Reprinted with permission from [27].

1909 [29]. It was realized that single wires can transport energy almost free of losses but at the expense of having poor localization, i.e. the fields extend in the surrounding medium over very large distances. Therefore, transmission lines at radio frequencies consist of two or more wires.

Plasmon resonances of a small spherical particle

The fundamental plasmon resonance of a small spherical particle of radius a in the quasi-static limit can be found in much the same way as the transverse plasmon resonance of a thin wire. Here, we have to express the Laplace equation (12.42) in spherical coordinates (r, θ, φ) as

$$\frac{1}{r^2 \sin\theta}\left[\sin\theta \frac{\partial}{\partial r}\left(r^2 \frac{\partial}{\partial r}\right) + \frac{\partial}{\partial \theta}\left(\sin\theta \frac{\partial}{\partial \theta}\right) + \frac{1}{\sin\theta}\frac{\partial^2}{\partial \varphi^2}\right]\Phi(r, \theta, \varphi) = 0. \qquad (12.55)$$

The solutions are of the form

$$\Phi(r, \theta, \varphi) = \sum_{l,m} b_{l,m} \cdot \Phi_{l,m}(r, \theta, \varphi). \qquad (12.56)$$

The $b_{l,m}$ are constant coefficients to be determined from the boundary conditions and the $\Phi_{l,m}$ are of the form

$$\Phi_{l,m} = \left\{ \begin{array}{c} r^l \\ r^{-l-1} \end{array} \right\} \left\{ \begin{array}{c} P_l^m(\cos\theta) \\ Q_l^m(\cos\theta) \end{array} \right\} \left\{ \begin{array}{c} \mathrm{e}^{\mathrm{i}m\varphi} \\ \mathrm{e}^{-\mathrm{i}m\varphi} \end{array} \right\}, \qquad (12.57)$$

where $P_l^m(\cos\theta)$ are the associated Legendre functions and $Q_l^m(\cos\theta)$ are Legendre functions of the second kind [30]. Linear combinations of the functions in the upper and the lower row of (12.57) may have to be chosen according to the particular problem, to avoid having infinities at the origin or at infinite distance. Again, the continuity of the tangential electric fields and the normal components of the electric displacements at the surface of the sphere imply that

$$\left[\frac{\partial \Phi_1}{\partial \theta}\right]_{r=a} = \left[\frac{\partial \Phi_2}{\partial \theta}\right]_{r=a},$$
$$\varepsilon_1\left[\frac{\partial \Phi_1}{\partial r}\right]_{r=a} = \varepsilon_2\left[\frac{\partial \Phi_2}{\partial r}\right]_{r=a}. \qquad (12.58)$$

Here, Φ_1 is the potential inside the sphere and $\Phi_2 = \Phi_{\text{scatter}} + \Phi_0$ is the potential outside the sphere consisting of the potentials of the scattered and the incoming fields. For the incoming electric field we assume, as for the case of the wire, that it is homogeneous and directed along the x-direction. Consequently, $\Phi_0 = -E_0 x = -E_0 r P_1^0(\cos\theta)$. Evaluation of the boundary conditions leads to

$$\Phi_1 = -E_0 \frac{3\varepsilon_2}{\varepsilon_1 + 2\varepsilon_2} r\cos\theta,$$
$$\Phi_2 = -E_0 r\cos\theta + E_0 \frac{\varepsilon_1 - \varepsilon_2}{\varepsilon_1 + 2\varepsilon_2} a^3 \frac{\cos\theta}{r^2} \qquad (12.59)$$

(see Problem 12.7). The most important differences from the solution for the wire are the distance dependence $1/r^2$ rather than $1/r$ and the modified resonance condition with ε_2 multiplied by a factor of 2 in the denominator. It is also important to note that the field is independent of the azimuth angle φ, which is a result of the symmetry implied by the direction of the applied electric field. Finally, the electric field can be calculated from (12.59) using $\mathbf{E} = -\nabla\Phi$ and turns out to be

$$\mathbf{E}_1 = E_0 \frac{3\varepsilon_2}{\varepsilon_1 + 2\varepsilon_2}(\cos\theta\, \mathbf{n}_r - \sin\theta\, \mathbf{n}_\theta) = E_0 \frac{3\varepsilon_2}{\varepsilon_1 + 2\varepsilon_2}\, \mathbf{n}_x, \qquad (12.60)$$

$$\mathbf{E}_2 = E_0(\cos\theta\, \mathbf{n}_r - \sin\theta\, \mathbf{n}_\theta) + \frac{\varepsilon_1 - \varepsilon_2}{\varepsilon_1 + 2\varepsilon_2} \frac{a^3}{r^3} E_0(2\cos\theta\, \mathbf{n}_r + \sin\theta\, \mathbf{n}_\theta). \qquad (12.61)$$

The field distribution near a resonant gold or silver nanoparticle looks qualitatively similar to the plot shown in Fig. 12.19 for the thin wire. However, the field is more strongly localized near the surface of the particle. An interesting feature is that the electric field inside the particle is homogeneous, as expected for the case of a particle of diameter smaller than the skin depth. Another important finding is that the scattered field (the second term in (12.61)) is identical with the electrostatic field of a dipole $\mathbf{p}$ located at the center of the sphere. The dipole is induced by the external field $\mathbf{E}_0$ and has the value $\mathbf{p} = \varepsilon_2\alpha(\omega)\mathbf{E}_0$, with α denoting the polarizability[4]

$$\alpha(\omega) = 4\pi\varepsilon_0 a^3 \frac{\varepsilon_1(\omega) - \varepsilon_2}{\varepsilon_1(\omega) + 2\varepsilon_2}. \qquad (12.62)$$

This relationship can be easily verified by comparison with Eq. (12.41). The scattering cross-section of the sphere is then obtained by dividing the total radiated power of the sphere's dipole (see e.g. Chapter 8) by the intensity of the exciting plane wave. This results in

$$\sigma_{\text{scatt}} = \frac{k^4}{6\pi\varepsilon_0^2}|\alpha(\omega)|^2, \qquad (12.63)$$

[4] Notice that we use dimensionless (relative) dielectric constants, i.e. the vacuum permeability ε_0 is *not* contained in ε_2.

with k being the wavevector in the surrounding medium. Notice that the polarizability (12.62) violates the optical theorem in the dipole limit, i.e. scattering is not accounted for. This inconsistency can be corrected by allowing the particle to interact with itself (radiation reaction). As discussed in Problem 8.5, the inclusion of radiation reaction introduces an additional term into (12.62). See also Problem 16.4.

Figure 12.22 shows plots of the normalized scattering cross-sections of gold and silver particles in different media. Note that the resonance for the silver particles is in the ultraviolet spectral range, while for gold the maximum scattering occurs around 530 nm. A redshift of the resonance is observed if the dielectric constant of the environment is increased.

The power removed from the incident beam due to the presence of a particle is not only due to scattering but also due to absorption. The sum of absorption and scattering is called *extinction*. Therefore, we also need to calculate the power that is dissipated inside the particle. Using Poynting's theorem, we know that the power dissipated by a point dipole is determined as $P_{\text{abs}} = (\omega/2)\text{Im}\left[\mathbf{p} \cdot \mathbf{E}_0^*\right]$. Using $\mathbf{p} = \varepsilon_2 \alpha \mathbf{E}_0$, with ε_2 being real, and the expression for the intensity of the exciting plane wave in the surrounding medium, we find for the absorption cross-section

$$\sigma_{\text{abs}} = \frac{k}{\varepsilon_0} \text{Im}[\alpha(\omega)] \,. \tag{12.64}$$

Again, k is the wavevector in the surrounding medium. It turns out that σ_{abs} scales with a^3, whereas σ_{scatt} scales with a^6. Consequently, for large particles extinction is dominated by scattering, whereas for small particles it is dominated by absorption. This effect can be used to detect extremely small metal particles down to diameter 2.5 nm, which are used as labels in biological samples [31]. The transition between the two size regimes is characterized by a distinct color change. For example, small gold particles absorb green and blue light and

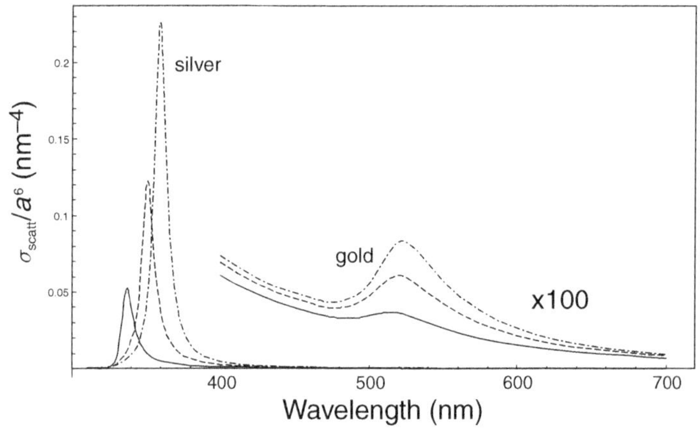

Fig. 12.22 Plots of the scattering cross-section of spherical gold and silver particles in different environments normalized by a^6, with a denoting the particle radius. Solid line, vacuum ($n = 1$). Dashed line, water ($n = 1.33$). Dash–dotted line, glass ($n = 1.5$).

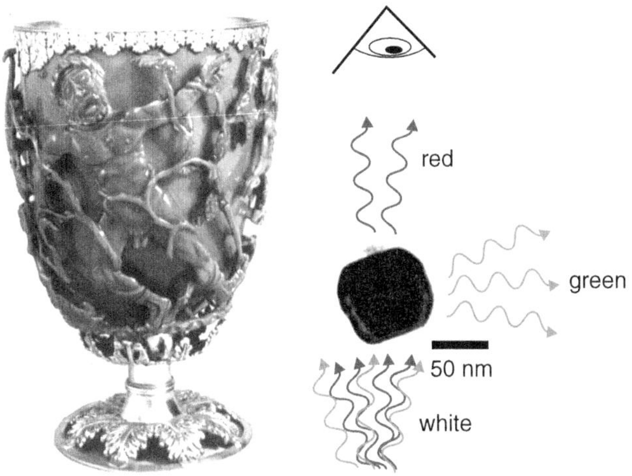

Fig. 12.23 The ancient Roman Lycurgus cup illuminated by a light source from behind. Light absorption by the embedded gold/silver alloy particles (right) leads to a red color of the transmitted light, whereas scattering at the particles yields a greenish color as indicated to the right. From D. J. Borber and I. C. Freestone, *Archeometry* **32**, 1 (1990).

thus render a red color. On the other hand, larger gold particles scatter predominantly in the green and hence render a greenish color. A very nice illustration of these findings is colored glasses. The famous Lycurgus cup shown in Fig. 12.23 was made by ancient Roman artists and is today exhibited at the British Museum, London. When illuminated by a white light source from behind, the cup shows an amazingly rich shading of colors ranging from deep green to bright red. For a long time it was not clear what causes these colors. Today it is known that they are due to nanometer-sized gold/silver particles embedded in the glass. The colors are determined by an interplay of absorption and scattering.

Plasmon resonances of non-spherical particles

For non-spherical particles, due to the broken symmetry, the degeneracy between collective electron oscillations along different directions is lifted. One way to obtain the plasmon resonances of non-spherical particles is to model them as prolate spheroids and to apply the quasi-static approximation [4]. The result is a polarizability that reads as

$$\alpha(\omega) = V\varepsilon_0 \frac{\varepsilon_1(\omega) - \varepsilon_2}{L_i\varepsilon_1(\omega) + (1 - L_i)\varepsilon_2}. \tag{12.65}$$

Here V is the volume of the spheroid and the L_i are geometrical factors that depend on the aspect ratio and describe the longitudinal and transverse plasmon resonances of the spheroid. For aspect ratios ranging from 1 to 3 the resonances cover the visible up to the infrared spectral range, while the resonance frequency decreases linearly with the aspect ratio as discussed at the beginning of this chapter (see Eq. (12.5).

For strongly elongated particles the quasi-static approximation eventually breaks down. In order to provide a qualitative understanding of the longitudinal plasmon resonances of rod-like nanoparticles with a constant cross-section one can take the following point of view. The particle is treated as a finite piece of wire, which supports a propagating plasmon in the TM_0 mode, for which the complex propagation constant $k_z(\omega) = \beta(\omega) + i\alpha(\omega)$ is known; see e.g. Fig. 12.20(b). The ends of the wire act as discontinuities at which the mode is partly reflected. The respective reflection coefficient is a complex number $R(\omega) = |R(\omega)| \exp[i\Phi_R(\omega)]$ and depends on the exact geometry of the termination. As a consequence, the condition for longitudinal resonances can be expressed via the accumulated phase per round trip as

$$\beta(\omega)L_{\text{res}} + \Phi_R(\omega) = n\pi, \tag{12.66}$$

where L_{res} is the rod length for which a resonance occurs at a fixed ω and n is the order of the resonance. For a fixed rod length the respective resonance frequencies can be found. The concept is illustrated in Fig. 12.24. It is important to note that the resonance condition provides only the resonance frequencies, not the width of the resonance or its amplitude. Since wires of finite length are building blocks of optical antennas, we will pick up this topic again in Chapter 13.

Local interactions with particle plasmons: sensing applications

The resonance condition of a particle plasmon depends sensitively on the dielectric constant of the environment. Thus, similarly to the case of a plane interface, a gold or silver particle can be used as a sensing element since its resonance will shift upon local dielectric changes, e.g. due to the specific binding of certain ligands after chemical functionalization of the particle's surface. The advantage of using particle resonances as opposed to resonances of plane interfaces is associated with the much smaller dimensions of the particle and hence the larger surface-to-volume ratio. One can envision anchoring differently functionalized particles onto substrates at extremely high densities and using such arrangements as sensor chips for multiparameter sensing of various chemical compounds, as demonstrated by the detection of single-base-pair mismatches in DNA (see e.g. [32]).

Resonance shifts of small noble-metal particles were also applied in the context of near-field optical microscopy. The observation of the resonance shift of a metal particle as a function of a changing environment was demonstrated by Fischer and Pohl in

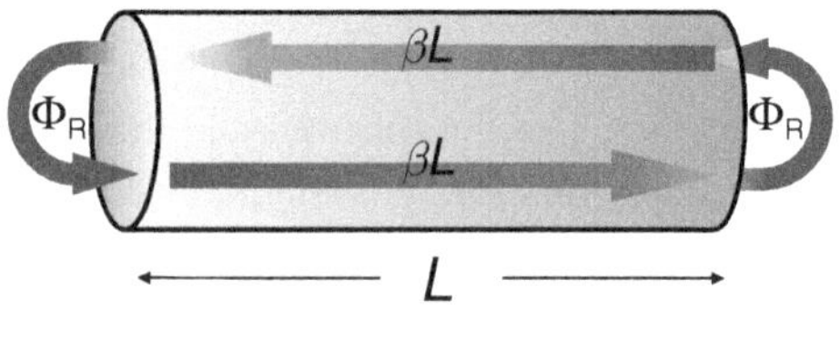

Fig. 12.24 The Fabry–Pérot model of the longitudinal resonances of a nano rod. The accumulated phase per round trip is the sum of the propagation phase $2\beta L$ and the reflection phase $2\Phi_R$, which must add up to 2π in order to obtain a standing wave.

1989 [33]. Similar experiments were performed later using gold particles attached to a tip [34] (c.f. Chapter 6).

12.4.2 Plasmon resonances of more complex structures

Simple, highly symmetric structures, such as isolated spherical nanoparticles or nano rods, exhibit plasmon resonances that can easily be assigned to characteristic surface charge distributions. More complex structures, however, often yield multi-featured resonance spectra that are difficult to interpret at first sight [35]. It has been shown that plasmon resonances of more complex structures can often be viewed as the result of a "hybridization" of elementary plasmons of simpler substructures [36]. To give an example, consider the resonances of a hollow metallic shell as shown in Fig. 12.25(a). The elementary resonances of this particle are found by decomposing it into a solid metal sphere and a spherical cavity in bulk metal. Figure 12.25(b) shows how the elementary modes can be combined to form hybridized modes. A low-energy (redshifted) hybrid mode is obtained for an in-phase oscillation of the elementary plasmons, whereas the anti-phase combination represents a higher-energy mode that is shifted to higher energies. The degree of interaction between the elementary modes and therefore the mode splitting is determined by the interaction strength of the elementary modes, which in the present example is determined by the shell thickness [37]. Plasmon hybridization can be understood using the framework of strong coupling discussed in Section 8.7 and will be discussed in the context of optical antennas (Chapter 13).

12.4.3 Surface-enhanced Raman scattering

The energy spectrum of molecular vibrations can serve as an unambiguous characteristic fingerprint for the chemical composition of a sample. Raman scattering is named after Sir Chandrasekhara V. Raman, who first observed the effect in 1928 [38]. Raman scattering can be viewed as a mixing process similar to the amplitude modulation used in radio signal transmission: the time-harmonic optical field (the carrier) is mixed with

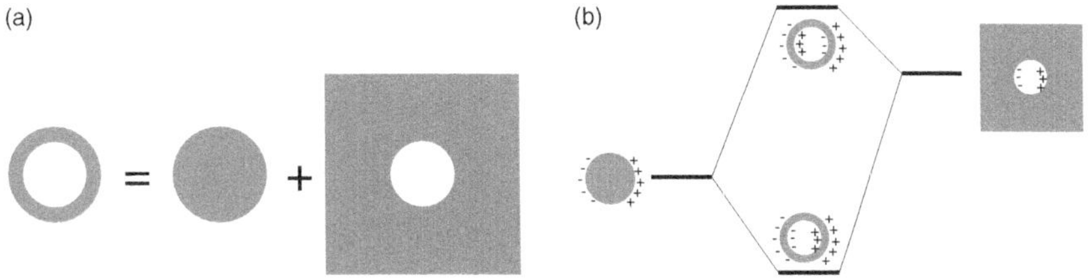

Fig. 12.25 Generation of multi-featured surface plasmon resonances by hybridization of elementary modes for the example of a gold nano-shell [36]. (a) A nano-shell decomposed into elementary structures. (b) Energies of elementary and hybridized modes.

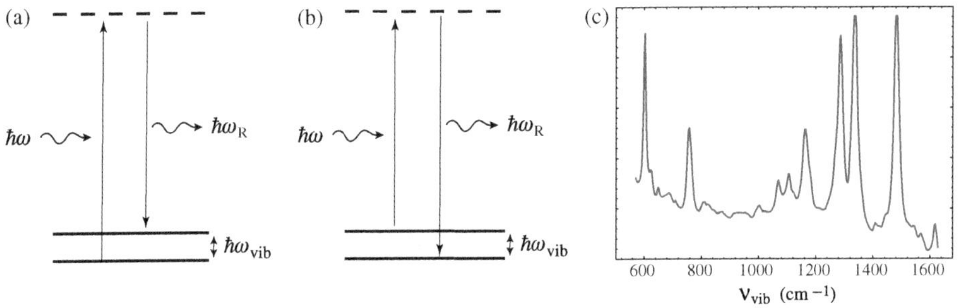

Fig. 12.26 Raman scattering refers to the spectroscopic process in which a molecule absorbs a photon with frequency ω and subsequently emits a photon at a different frequency ω_R, which is offset with respect to ω by a vibrational frequency ω_{vib} of the molecule, i.e. $\omega_R = \omega \pm \omega_{vib}$. Absorption and emission are mediated by a virtual state, i.e. a vacuum state that does not match any molecular energy level. (a) If $\omega > \omega_R$, one speaks of Stokes Raman scattering. (b) If $\omega < \omega_R$, the process is designated anti-Stokes Raman scattering. (c) The Raman-scattering spectrum representing the vibrational frequencies of Rhodamine 6G. The spectrum is expressed in wavenumbers $\nu_{vib}\,(\text{cm}^{-1}) = [1/\lambda\,(\text{cm})] - [1/\lambda_R\,(\text{cm})]$, with λ and λ_R being the wavelengths of the incident and scattered light, respectively.

the molecular vibrations (the signal). This mixing process gives rise to scattered radiation that is frequency-shifted from the incident radiation by an amount that corresponds to the vibrational frequencies of the molecules (ω_{vib}). The vibrational frequencies originate from oscillations between the constituent atoms of the molecules and, according to quantum mechanics, these oscillations persist even at ultralow temperatures. Because the vibrations depend on the particular molecular structure, the vibrational spectrum constitutes a characteristic fingerprint of a molecule. A formal description based on quantum electrodynamics can be found in Ref. [39]. Figure 12.26 shows the energy-level diagrams for Stokes and anti-Stokes Raman scattering together with an experimentally measured spectrum for Rhodamine 6G.

It is not the purpose of this section to go into the details of Raman scattering, but it is important to emphasize that Raman scattering is an extremely weak effect. The Raman scattering cross-section is typically 14–15 orders of magnitude smaller than the fluorescence cross-section of efficient dye molecules. The field enhancement associated with surface plasmons, as described above, has hence been extensively investigated as a means for increasing the interaction strength between a molecule and optical radiation. The most prominent example is surface-enhanced Raman scattering (SERS).

In 1974 it was reported that the Raman-scattering cross-section can be considerably increased if the molecules are adsorbed onto roughened metal surfaces [40]. In the following decades SERS became an active research field [41]. Typical enhancement factors for the Raman signal observed from rough metal substrates as compared with bare glass substrates are on the order of 10^6–10^7, and, using resonance enhancement (excitation frequency near an electronic transition frequency), enhancement factors as high as 10^{12} have been reported. The determination of these enhancement factors was based on ensemble measurements. However, later the authors of two independent *single-molecule* studies

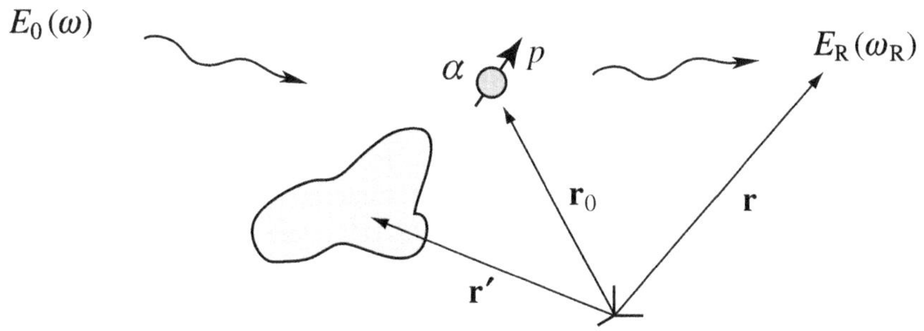

Fig. 12.27 The general configuration encountered in surface-enhanced spectroscopy. The interaction between a molecule with polarizability α and the exciting field E_0 gives rise to a scattered field E_R. Placing metal nanostructures (coordinate $\mathbf{r}'$) near the molecule enhances both the exciting field and the radiated field.

reported giant enhancement factors of 10^{14} [42, 43]. These studies not only shed new light on the nature of SERS but also made Raman scattering as efficient as fluorescence measurements (cross-sections of $\approx 10^{-16}$ cm^2). The interesting outcome of these single-molecule studies is that the average enhancement factor coincides with previous ensemble measurements, but, while most of the molecules remain unaffected by the metal surface, only a few make up the detected signal. These are the molecules with the giant enhancement factors of 10^{14}. These molecules are assumed to be located in a favorable local environment (hot spots) characterized by strongly enhanced electric fields.

It is accepted that the largest contribution to the giant signal enhancement stems from the enhanced electric fields at rough metal surfaces. The highest field enhancements are found at junctions between metal particles or in cracks on surfaces (see e.g. [36, 42]). It is commonly assumed that the Raman-scattering enhancement scales with the fourth power of the electric field enhancement factor. At first glance this seems odd as one would expect that this implies that Raman scattering is a nonlinear effect, scaling with the square of the excitation intensity. However, this is not so. In the following we will provide a qualitative explanation that is based on a scalar phenomenological theory [44]. It is straightforward to rigorously expand this theory, but the mathematical details would obscure the physical picture. Notice that the theory outlined in the following is not specific to Raman scattering but applies also to any other linear interaction, such as Rayleigh scattering and fluorescence.[5]

Let us consider the situation depicted in Fig. 12.27. A molecule located at $\mathbf{r}_0$ is placed in the vicinity of metal nanostructures (particles, tips, ...) that act as a local field-enhancing device. The interaction of the incident field E_0 with the molecule gives rise to a dipole moment associated with Raman scattering according to

$$p(\omega_R) = \alpha(\omega_R, \omega)[E_0(\mathbf{r}_0, \omega) + E_s(\mathbf{r}_0, \omega)], \tag{12.67}$$

[5] In the case of fluorescence, one needs to take into account that the excited-state lifetimes can be drastically reduced near metal surfaces.

where ω is the frequency of the exciting radiation and ω_R is a particular vibrationally shifted frequency ($\omega_R = \omega \pm \omega_{vib}$). The polarizability α is modulated at the vibrational frequency ω_{vib} of the molecule and gives rise to the frequency-mixing process. The molecule is interacting with the local field $E_0 + E_s$, where E_0 is the local field in the absence of the metal nanostructures and E_s is the enhanced field originating from the interaction with the nanostructures (the scattered field). E_s depends linearly on the excitation field E_0 and hence it can be qualitatively represented as $f_1(\omega)E_0$, with f_1 designating the field-enhancement factor.

The electric field radiated by the induced dipole p can be represented by the system's Green function G, which accounts for the presence of the metal nanostructures, as

$$E(\mathbf{r}_\infty, \omega_R) = \frac{\omega_R^2}{\varepsilon_0 c^2} G(\mathbf{r}_\infty, \mathbf{r}_0)\, p(\omega_R) = \frac{\omega_R^2}{\varepsilon_0 c^2} [G_0(\mathbf{r}_\infty, \mathbf{r}_0) + G_s(\mathbf{r}_\infty, \mathbf{r}_0)]\; p(\omega_R). \quad (12.68)$$

As in the case of the exciting local field, we split the Green function into a free-space part G_0 (corresponding to the absence of metal nanostructures) and a scattered part G_s originating from the interaction with the metal nanostructures. We represent G_s qualitatively as $f_2(\omega_R)G_0$, with f_2 being a second field-enhancement factor.

Finally, combining Eqs. (12.67) and (12.68), using the relations $E_s = f_1(\omega)E_0$ and $G_s = f_2(\omega_R)G_0$, and calculating the intensity $I \propto |E|^2$ yields

$$I(\mathbf{r}_\infty, \omega_R) = \frac{\omega_R^4}{\varepsilon_0^2 c^4} \Big|[1 + f_2(\omega_R)]G_0(\mathbf{r}_\infty, \mathbf{r}_0)\alpha(\omega_R, \omega)[1 + f_1(\omega)]\Big|^2 I_0(\mathbf{r}_0, \omega). \quad (12.69)$$

Thus, we find that the Raman-scattered intensity scales linearly with the excitation intensity I_0 and that it depends on the factor

$$\Big|[1 + f_2(\omega_R)][1 + f_1(\omega)]\Big|^2. \quad (12.70)$$

In the absence of metal nanostructures, we obtain the scattered intensity by setting $f_1 = f_2 = 0$. On the other hand, in the presence of the nanostructures we assume that $f_1, f_2 \gg 1$ and hence the overall Raman-scattering enhancement becomes

$$f_{Raman} = \Big|f_2(\omega_R)\Big|^2 \Big|f_1(\omega)\Big|^2. \quad (12.71)$$

Provided that $|\omega_R \pm \omega|$ is smaller than the spectral response of the metal nanostructure, the Raman-scattering enhancement scales roughly with the fourth power of the electric field enhancement. It should be kept in mind that our analysis is qualitative and ignores the vectorial nature of the fields and the tensorial properties of the polarizability. Nevertheless, a rigorous self-consistent formulation along the lines outlined here is possible. Besides the field-enhancement mechanism described, an additional enhancement associated with SERS is a short-range "chemical" enhancement, which results from the direct contact of the molecule with the metal surface. This direct contact results in a modified ground-state electronic charge distribution, which gives rise to a modified polarizability α. Further enhancement can be accomplished through resonant Raman scattering, for which the excitation frequency is near an electronic transition frequency of the molecule, i.e. the virtual levels shown in Fig. 12.26 come close to an electronic state of the molecule.

12.5 Nonlinear plasmonics

In addition to their unique linear optical properties, metals also possess a strongly nonlinear response, which can be exploited for local frequency mixing, switching, and modulation. In nonlinear laser crystals, efficient optical frequency conversion is made possible by phase matching, namely the coherent addition of the nonlinear response upon propagation through the crystal. However, in order for phase matching to occur, the crystal needs to be many wavelengths in size. For material structures smaller than the wavelength, the nonlinear response is defined by the intrinsic material nonlinearities and by the ability to efficiently couple radiation into and out of the material structures. Therefore, traditional nonlinear laser crystals, such as lithium niobate ($LiNbO_3$), are no longer the material of choice for frequency conversion on the nanometer scale. Much stronger nonlinear effects can be achieved with metal nanostructures. For example, the third-order nonlinear susceptibility of gold is $\chi^{(3)} \sim 1\,\text{nm}^2\,\text{V}^{-2}$, which is more than two orders of magnitude larger than that of $LiNbO_3$. As an illustration, Fig. 12.28 shows bursts of photons of frequency $2\omega_1 - \omega_2$ generated by modulating the distance between two gold nanoparticles that are irradiated by laser pulses of frequencies ω_1 and ω_2.

The nonlinear properties of materials are typically expressed in terms of nth-order nonlinear susceptibilities $\chi^{(n)}$, which relate the induced polarization $\mathbf{P}$ to the local excitation fields $\mathbf{E}_i$. For example, the efficiency of sum-frequency generation is defined by $\text{Re}[\chi^{(2)}]$, the efficiency of four-wave mixing by $\text{Re}[\chi^{(3)}]$, and the efficiency of two-photon absorption by $\text{Im}[\chi^{(3)}]$. For a more detailed discussion, the interested reader is referred to textbooks on nonlinear optics [45].

Different physical mechanisms have been discussed as the possible origins of metal nonlinearities [46]. They include higher-order multipole interactions, hot-electron contributions, and interband transitions. Another contribution arises from the ponderomotive potential V_p discussed in Section 12.1.2. V_p modifies the electron density and thereby affects the Drude model that describes the dielectric function of the metal. More specifically, the potential V_p disturbs the Fermi–Dirac distribution and offsets the electron density as [47]

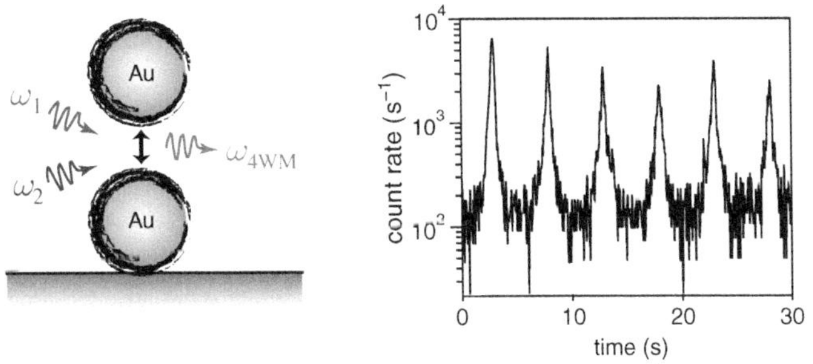

Fig. 12.28 Four-wave mixing from a pair of gold nanoparticles. Narrow-band photon bursts at frequency $\omega_{4WM} = 2\omega_1 - \omega_2$ can be generated by modulating the distance between the gold nanoparticles.

$$n(\mathbf{r}) = \frac{1}{3\pi^2}\left(\frac{2m}{\hbar^2}\right)^{3/2}\left[E_F - V_p(\mathbf{r})\right]^{3/2}. \tag{12.72}$$

Thermal effects can be ignored as long as the Fermi energy E_F is far from the conduction-band edge. E_F can be expressed in terms of the equilibrium plasma frequency ω_p (c.f. Eq. (12.2)) and, because $V_p \ll E_F$, we can approximate the term in brackets as $(E_F - V_p)^{3/2} \approx \left[1 - (3/2)(V_p/E_F)\right] E_F^{3/2}$. If we treat the metal as a free-electron gas with a modified electron density according to Eq. (12.72) we obtain for the dielectric function $\varepsilon(\omega, \mathbf{r}) = 1 - n(\mathbf{r})e^2/(m\varepsilon_0\omega^2)$, where we ignored any damping. Notice that ε turns into a *nonlocal* function; that is, it depends through the ponderomotive potential on the location $\mathbf{r}$. Using the expression for V_p from Eq. (12.6), we arrive at

$$\varepsilon(\mathbf{r}, \omega) = \left[1 - \frac{\omega_p^2}{\omega^2}\right] + \frac{3}{2}\frac{e^4}{\omega^4}\left[\frac{\omega_p}{3\pi^2\varepsilon_0\hbar^3 me}\right]^{2/3}\left\langle|\mathbf{E}(\mathbf{r}, t)|^2\right\rangle. \tag{12.73}$$

The first term in brackets is recognized as the linear dielectric function, while the second term is identified as a nonlinear term that depends on the local intensity. Thus, the ponderomotive force gives rise to a third-order nonlinear susceptibility with magnitude

$$\chi^{(3)} = \frac{3}{2}\frac{e^4}{\omega^4}\left[\frac{\omega_p}{3\pi^2\varepsilon_0\hbar^3 me}\right]^{2/3}. \tag{12.74}$$

Using $\lambda = 2\pi c/\omega = 800\,\text{nm}$ and $\lambda_p = 2\pi c/\omega_p = 138\,\text{nm}$ (gold), we obtain $\chi^{(3)} = 0.15\,\text{nm}^2\,\text{V}^{-2}$, which is in good agreement with experimental values [48]. Equation (12.74) predicts that the nonlinearity of metals increases as we go to lower frequencies. However, this trend has its limits because we ignored the conductive contribution to the dielectric response (the imaginary part of ε) and hence the theory outlined here is valid only in the visible-to-mid-infrared frequency range.

In most cases, metals possess inversion symmetry, for which second-order nonlinearities ($\chi^{(2)}$) are greatly suppressed. However, this argument is based on the dipole approximation and does not consider higher-order multipolar terms in the light–matter interaction. Furthermore, inversion symmetry can be broken by the surface of the material structure. For example, two closely spaced nanoparticles possess inversion symmetry only if they are of equal size. Thus, the second-order nonlinear response of metals can be geometrically engineered. Strong second-harmonic generation and sum-frequency generation have been observed from non-centrosymmetric configurations, such as metal tips and pyramidal nanoparticles.

12.6 Conclusion

In this chapter we have discussed the basic properties of surface plasmons. We have pointed out the nature of these modes as being a hybrid between local optical fields and associated electron-density waves in a metal. As nano-optics in general deals with optical fields in the close vicinity of nanostructures it is obvious that such collective excitations play a major role in the field. There are many applications and prospects of surface plasmons that we could not mention here. The study of plasmons on metal nanostructures has developed into a research field of its own called "plasmonics." For more information, the interested reader is referred to Ref. [9] and references therein.

Problems

12.1 Study the effect of a complex dielectric function on the propagation of a plane wave. What happens if a plane wave is normally incident on the interface between a transparent dielectric (positive, real ε) and a metal?

12.2 Show that for an ansatz similar to Eq. (12.23), but with s-polarized waves, a reflected wave has to be added to fulfill the boundary conditions and Maxwell's equations simultaneously.

12.3 Show that, if we do not demand the solution to be a surface wave, i.e. if the perpendicular wavevector, Eq. (12.30), may be real, then we arrive at the well-known condition for the Brewster effect.

12.4 Write a program that plots the reflectivity of a system of (at least up to four) stratified layers as a function of the angle of incidence using the notation of (12.26). Study a system consisting of glass, gold, and air with a thickness of the gold layer of about 50 nm between the glass and the air half-spaces. Plot the reflectivity for light incident from the glass side and from the air side. What do you observe? Study the influence of thin layers of additional materials on top of or below the gold. A few nanometers of titanium or chromium are often used to enhance the adhesion of gold to glass. What happens if a monolayer of proteins ($\sim$5 nm in diameter, refractive index $\sim$1.33) is adsorbed on top of the gold layer? Hint: consider a stratified layer of thickness d (medium 1) between two homogeneous half-spaces (media 0 and 2). According to (12.26) the fields in each medium for p-polarization read as

$$\mathbf{E}_0 = E_0^+ \begin{pmatrix} 1 \\ 0 \\ -k_x/k_{0,z} \end{pmatrix} \mathrm{e}^{\mathrm{i}k_{0,z}z} + E_0^- \begin{pmatrix} 1 \\ 0 \\ k_x/k_{0,z} \end{pmatrix} \mathrm{e}^{-\mathrm{i}k_{0,z}z}, \tag{12.75}$$

$$\mathbf{E}_1 = E_1^+ \begin{pmatrix} 1 \\ 0 \\ -k_x/k_{1,z} \end{pmatrix} \mathrm{e}^{\mathrm{i}k_{1,z}z} + E_1^- \begin{pmatrix} 1 \\ 0 \\ k_x/k_{1,z} \end{pmatrix} \mathrm{e}^{-\mathrm{i}k_{1,z}(z-d)}, \tag{12.76}$$

$$\mathbf{E}_2 = E_2^+ \begin{pmatrix} 1 \\ 0 \\ k_x/k_{2,z} \end{pmatrix} \mathrm{e}^{\mathrm{i}k_{2,z}(z-d)}. \tag{12.77}$$

Exploiting the continuity of $\mathbf{E}_\parallel$ and $\mathbf{D}_\perp$ yields after some manipulation

$$\begin{pmatrix} E_0^+ \\ E_0^- \end{pmatrix} = \frac{1}{2} \begin{pmatrix} 1+\kappa_1\eta_1 & 1-\kappa_1\eta_1 \\ 1-\kappa_1\eta_1 & 1+\kappa_1\eta_1 \end{pmatrix} \begin{pmatrix} 1 & 0 \\ 0 & \mathrm{e}^{\mathrm{i}k_{1,z}d} \end{pmatrix} \begin{pmatrix} E_1^+ \\ E_1^- \end{pmatrix} \tag{12.78}$$

as well as

$$\begin{pmatrix} E_1^+ \\ E_1^- \end{pmatrix} = \begin{pmatrix} e^{-ik_{1,z}d} & 0 \\ 0 & 1 \end{pmatrix} \frac{1}{2} \begin{pmatrix} 1+\kappa_2\eta_2 & 1-\kappa_2\eta_2 \\ 1-\kappa_2\eta_2 & 1+\kappa_2\eta_2 \end{pmatrix} \begin{pmatrix} E_2^+ \\ 0 \end{pmatrix}, \quad (12.79)$$

where $\kappa_i = k_{i-1,z}/k_{i,z}$ and $\eta_i = \varepsilon_i/\varepsilon_{i-1}$. Equations (12.78) and (12.79) can be combined to give

$$\begin{pmatrix} E_0^+ \\ E_0^- \end{pmatrix} = \mathbf{T}_{0,1} \cdot \Phi_1 \cdot \mathbf{T}_{1,2} \begin{pmatrix} E_2^+ \\ 0 \end{pmatrix}. \quad (12.80)$$

Here

$$\mathbf{T}_{0,1} = \frac{1}{2} \begin{pmatrix} 1+\kappa_1\eta_1 & 1-\kappa_1\eta_1 \\ 1-\kappa_1\eta_1 & 1+\kappa_1\eta_1 \end{pmatrix}, \quad (12.81)$$

$$\mathbf{T}_{1,2} = \frac{1}{2} \begin{pmatrix} 1+\kappa_2\eta_2 & 1-\kappa_2\eta_2 \\ 1-\kappa_2\eta_2 & 1+\kappa_2\eta_2 \end{pmatrix}, \quad (12.82)$$

and

$$\Phi_1 = \begin{pmatrix} e^{-ik_{1,z}d} & 0 \\ 0 & e^{ik_{1,z}d} \end{pmatrix}. \quad (12.83)$$

From this we can infer a general relation connecting the fields outside an arbitrary system of stratified layers which reads as

$$\begin{pmatrix} E_0^+ \\ E_0^- \end{pmatrix} = \mathbf{T}_{0,1} \cdot \Phi_1 \cdot \mathbf{T}_{1,2} \cdot \Phi_2 \cdot \cdots \cdot \mathbf{T}_{n,n+1} \begin{pmatrix} E_{n+1}^+ \\ 0 \end{pmatrix}. \quad (12.84)$$

The reflectivity $R(\omega, k_x)$ can be calculated from (12.84) as

$$R(\omega, k_x) = \frac{|E_0^-|^2}{|E_0^+|^2}, \quad (12.85)$$

from which E_{n+1}^+ cancels out. To test the program plot the reflectivity of a glass/air interface and find the Brewster angle.

12.5 Extend the program you have just written to determine the amount of intensity enhancement obtained right above the metal layer by determining the ratio between the incoming intensity and the intensity just above the metal layer.

12.6 Prove that Eq. (12.41) actually is exactly the electrostatic field of a point dipole, except that it oscillates in time with $e^{i\omega t}$.

12.7 Solve the Laplace equation (12.55) for a spherical particle and verify the results (12.59) and (12.60).

12.8 The nonlinear response of a thin metal film can be described by a surface nonlinearity. Incident fields induce a nonlinear surface polarization $\mathbf{P}$, which acts as a source current for fields at the nonlinear frequency ω. Consider a polarization current $\mathbf{P} = [P_x, 0, P_z]^{\mathrm{T}} \exp(ik_x x)\delta(z)$ confined to the $z = 0$ plane between two dielectric media with dielectric constants ε_1 and ε_2, respectively, and calculate the fields $\mathbf{E}$ emitted into the two half-spaces according to

$$\mathbf{E}(\mathbf{r}) = \frac{\omega^2}{\varepsilon_0 c^2} \int_{\text{surface}} \overleftrightarrow{\mathbf{G}}(\mathbf{r}, \mathbf{r}')\mathbf{P}(\mathbf{r}')\mathrm{d}^2 r'.$$

Use the Weyl identity (c.f. Chapter 10) and assume that $\sqrt{\varepsilon_2} > ck_x/\omega > \sqrt{\varepsilon_1}$.

References

[1] L. D. Landau, E. M. Lifshitz, and L. P. Pitaevskii, *Electrodynamics of Continuous Media*, 2nd edn. Amsterdam: Elsevier (1984).

[2] N. W. Ashcroft and N. D. Mermin, *Solid State Physics*. Philadelphia, PA: Saunders College Publishing (1976).

[3] C. F. Bohren and D. R. Huffman, *Absorption and Scattering of Light by Small Particles*. New York: John Wiley & Sons (1983).

[4] P. B. Johnson and R. W. Christy, "Optical constants of the noble metals," *Phys. Rev. B* **6**, 4370–4379 (1972).

[5] P. G. Etchegoin, E. C. Le Ru, and M. Meyer, "An analytic model for the optical properties of gold," *J. Chem. Phys.* **125**, 164705 (2006).

[6] K. Welford, "The method of attenuated total reflection," in *Surface Plasmon Polaritons*. Bristol: IOP Publishing, pp. 25–78 (1987).

[7] H. Raether, *Surface Plasmons on Smooth and Rough Surfaces and on Gratings*. Berlin: Springer-Verlag (1988).

[8] O. Marti, H. Bielefeldt, B. Hecht, *et al.*, "Near-field optical measurement of the surface plasmon field," *Opt. Commun.* **96**, 225–228 (1993).

[9] S. A. Maier, *Plasmonics: Fundamentals and Applications*. New York: Springer (2007).

[10] J. J. Burke, G. I. Stegeman, and T. Tamir, "Surface-polariton-like waves guided by thin, lossy metal films," *Phys. Rev. B* **33**, 5186–5201 (1986).

[11] E. T. Arakawa, M. W. Williams, R. N. Hamm, and R. H. Ritchie, "Effect of damping on surface plasmon dispersion," *Phys. Rev. Lett.* **31**, 1127–1130 (1973).

[12] A. Otto, "Excitation of nonradiative surface plasma waves in silver by the method of frustrated total reflection," *Z. Phys.* **216**, 398–410 (1968).

[13] E. Kretschmann, "Die Bestimmung optischer Konstanten von Metallen durch Anregung von Oberflachenplasmaschuingungen," *Z. Phys.* **241**, 313–324 (1971).

[14] H. J. Lezec, A. Degiron, E. Devaux, *et al.*, "Beaming light from a subwavelength aperture," *Science* **297**, 820–822 (2002).

[15] J. Homola, S. S. Yee, and G. Gauglitz, "Surface plasmon resonance sensors: review," *Sensors Actuators B* **54**, 3–15 (1999).

[16] B. Hecht, H. Bielefeldt, L. Novotny, Y. Inouye, and D. W. Pohl, "Local excitation, scattering, and interference of surface plasmons," *Phys. Rev. Lett.* **77**, 1889–1893 (1996).

[17] J. R. Lakowicz, "Radiative decay engineering 3. Surface plasmon-coupled directional emission," *Anal. Biochem.* **324**, 153–169 (2004).

[18] H. Ditlbacher, J. R. Krenn, N. Felidj, *et al.*, "Fluorescence imaging of surface plasmon fields," *Appl. Phys. Lett.* **80**, 404–406 (2002).

[19] L. Novotny, B. Hecht, and D. W. Pohl, "Interference of locally excited surface plasmons," *J. Appl. Phys.* **81**, 1798–1806 (1997).

[20] E. Matveeva, Z. Gryczynski, I. Gryczynski, J. Malicka, and J. R. Lakowicz, "Myoglobin immunoassay utilizing directional surface plasmon-coupled emission," *Angew. Chem.* **76**, 6287–6292 (2004).

[21] S. I. Bozhevolnyi and V. Coello, "Elastic scattering of surface plasmon polaritons: modelling and experiment," *Phys. Rev. B* **58**, 10899–10910 (1998).

[22] A. Bouhelier, Th. Huser, H. Tamaru, *et al.*, "Plasmon optics of structured silver films," *Phys. Rev. B* **63**, 155404 (2001).

[23] H. Ditlbacher, J. R. Krenn, G. Schider, A. Leitner, and F. R. Aussenegg, "Two-dimensional optics with surface plasmon polaritons," *Appl. Phys. Lett.* **81**, 1762–1764 (2002).

[24] M. Kerker, *The Scattering of Light and Other Electromagnetic Radiation*. New York: Academic Press, p. 84 (1969).

[25] L. Novotny and C. Hafner, "Light propagation in a cylindrical waveguide with a complex, metallic, dielectric function," *Phys. Rev. E* **50**, 4094–4106 (1994).

[26] L. Novotny, "Effective wavelength scaling for optical antennas," *Phys. Rev. Lett.* **98**, 266802 (2007).

[27] H. Ditlbacher, A. Hohenau, D. Wagner, *et al.*, "Silver nanowires as surface plasmon resonators," *Phys. Rev. Lett.* **95**, 257403 (2005). Copyright 2005 American Physical Society.

[28] M. I. Stockman, "Nanofocusing of optical energy in tapered plasmonic waveguides," *Phys. Rev. Lett.* **93**, 137404 (2004).

[29] D. Hondros, "Über elektromagnetische Drahtwellen," *Ann. Phys.* **30**, 905–950 (1909).

[30] G. B. Arfken and H. J. Weber, *Mathematical Methods for Physicists*. London: Academic Press (1995).

[31] D. Boyer, Ph. Tamarat, A. Maali, B. Lounis, and M. Orrit, "Photothermal imaging of nanometer-sized metal particles among scatterers," *Science* **297**, 1160–1163 (2002).

[32] S. J. Oldenburg, C. C. Genicka, K. A. Clarka, and D. A. Schultz, "Base pair mismatch recognition using plasmon resonant particle labels," *Anal. Biochem.* **309**, 109–116 (2003).

[33] U. Ch. Fischer and D. W. Pohl, "Observation on single-particle plasmons by near-field optical microscopy," *Phys. Rev. Lett.* **62**, 458–461 (1989).

[34] T. Kalkbrenner, M. Ramstein, J. Mlynek, and V. Sandoghdar, "A single gold particle as a probe for apertureless scanning near-field optical microscopy," *J. Microsc.* **202**, 72–76 (2001).

[35] A. M. Michaels, J. Jiang, and L. Brus, "Ag nanocrystal junctions as the site for surface-enhanced Raman scattering of single Rhodamine 6G molecules," *J. Phys. C* **104**, 11965–11971 (2000).

[36] E. Prodan, C. Radloff, N. J. Halas, and P. Nordlander, "A hybridization model for the plasmon response of complex nanostructures," *Science* **302**, 419–422 (2003).

[37] J. B. Jackson, S. L. Westcott, L. R. Hirsch, J. L. West, and N. J. Halas, "Controlling the surface enhanced Raman effect via the nanoshell geometry," *Appl. Phys. Lett.* **82**, 257–259 (2003).

[38] C. V. Raman and K. S. Krishnan, "A new type of secondary radiation," *Nature* **121**, 501–502 (1928).

[39] M. Diem, *Introduction to Modern Vibrational Spectroscopy*. New York: Wiley-Interscience (1993).

[40] M. Fleischmann, P. J. Hendra, and A. J. McQuillan, "Raman spectra of pyridine adsorbed at a silver electrode," *Chem. Phys. Lett.* **26**, 163–166 (1974).

[41] A. Otto, I. Mrozek, H. Grabhorn, and W. Akemann, "Surface enhanced Raman scattering," *J. Phys.: Condens. Matter* **4**, 1143–1212 (1992).

[42] S. Nie and S. R. Emory, "Probing single molecules and single nanoparticles by surface enhanced Raman scattering," *Science* **275**, 1102–1106 (1997).

[43] K. Kneipp, Y. Wang, H. Kneipp, *et al.*, "Single molecule detection using surface enhanced Raman scattering (SERS)," *Phys. Rev. Lett.* **78**, 1667–1670 (1997).

[44] S. Efrima and H. Metiu, "Classical theory of light scattering by an adsorbed molecule. I. Theory," *J. Chem. Phys.* **70**, 1602–1613 (1979).

[45] See, for example, R. W. Boyd, *Nonlinear Optics*, 3rd edn. San Diego, CA: Academic Press (2008).

[46] M. Scalora, M. A. Vincenti, D. de Ceglia, "Second- and third-harmonic generation in metal-based structures," *Phys. Rev. A* **82**, 043828 (2010).

[47] P. Ginzburg, A. Hayat, N, Berkovitch, and M. Orenstein, "Nonlocal ponderomotive nonlinearity in plasmonics," *Opt. Lett.* **35**, 1551–1553 (2010).

[48] J. Renger, R. Quidant, N. van Hulst, and L. Novotny, "Surface-enhanced nonlinear four-wave mixing," *Phys. Rev. Lett.* **104**, 046803 (2010).

13 Optical antennas

An optical antenna is a mesoscopic structure that enhances the local light–matter interaction. Similarly to their radiowave analogs, optical antennas mediate the information and energy transfer between the free radiation field and a localized receiver or transmitter. The degree of localization and the magnitude of transduced energy indicate how good an antenna is. We thus define an optical antenna as *a device designed to efficiently convert free-propagating optical radiation to localized energy, and vice versa* [1]. In this sense, even a standard lens is an antenna, but since the degree of localization is limited by diffraction, the lens is a poor antenna. To characterize the quality and the properties of an antenna, radio engineers have introduced antenna parameters, such as gain and directivity. Optical antennas hold promise for controllably enhancing the performance and efficiency of optoelectronic devices, such as photodetectors, light emitters, and sensors.

Although many of the properties and parameters of *optical* antennas are similar to those of their radiowave and microwave conuterparts, there are important differences resulting from their small size and the plasmon resonances of metal nanostructures. In this chapter we introduce the basic principles of optical antennas, building on the background of both radiowave antenna engineering and plasmonics.

13.1 Significance of optical antennas

The length scale of free radiation is determined by the wavelength λ, which is on the order of 500 nm. However, the characteristic size of the source generating this radiation is significantly smaller, typically sub-nanometer. To illustrate this, let us consider a simple particle-in-a-box model with energy difference $\Delta E_{12} = hc/\lambda$ between the ground state and the first excited state. Using $\lambda = 500$ nm, we readily find that the box size needs to be $\sim$1 nm. Thus, there is a mismatch of almost three orders of magnitude between the wavelength of radiation and the electronic confinement. Since the wavelength is also the relevant scale for diffraction effects, e.g. in the focusing of light, this mismatch prevents photons from being confined to the size of a quantum emitter. This leads, for example, to the inefficient absorption of light by a single quantum emitter under ambient conditions even when it is illuminated with a tightly focused laser beam (see Chapter 9). Similar arguments explain the small cross-section for the generation of excitons in a semiconductor material – a fundamental process for solar energy conversion. A further consequence of the mismatch is the rather long lifetime of the excited state of a quantum emitter in vacuum.

Because the size of a quantum emitter is so much smaller than the wavelength of light, the "birth" of a photon is a highly inefficient process [5]. This is illustrated by considering the total power emitted by a time-harmonic point dipole, Eq. (8.71). Assuming that the dipole **p** has a small, but non-negligible, size Δl and that it is oscillating at the frequency ω, we can express the dipole moment in terms of the current as $|\mathbf{p}| = I\,\Delta l/\omega$, with I being the peak amplitude of the current. We then find that the total radiated power can be written as

$$P_{\rm rad} = \frac{\pi}{3} I^2 Z_{\rm w} \left(\frac{\Delta l}{\lambda}\right)^2 , \qquad (13.1)$$

where $Z_{\rm w} = \sqrt{\mu_0/\varepsilon_0} = 377\ \Omega$ is the wave impedance of free space. The radiated power turns out to be proportional to the square of the length-to-wavelength ratio and is therefore very small. Note that we can represent Eq. (13.1) as $P_{\rm rad} = (1/2) R_{\rm rad} I^2$, where $R_{\rm rad}$ defines the radiation resistance. Evidently, the smaller Δl is, the smaller $R_{\rm rad}$ will be, and the less efficient it is to release energy from the emitter.

The radiation emitted by a quantum emitter is composed of discrete quanta of energy $E = \hbar\omega = (hc/\lambda)$ and hence $P = E\gamma$, with γ being the photon-emission rate. γ specifies how fast the emitter is being cycled between the ground state and the excited state. Evidently, the maximum value of γ is defined by the emitter's excited-state lifetime τ, that is, $\gamma_{\rm max} = 1/\tau$. Typically, τ is on the order of nanoseconds, and hence the maximum number of photons that can be emitted per unit time is relatively small, which limits the use of single quantum emitters as sources of single photons [6] and their detectability in

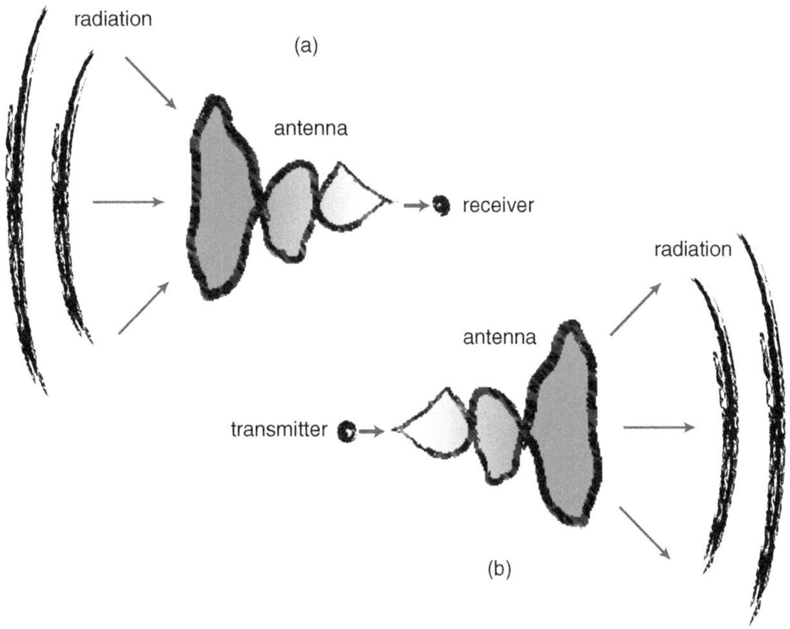

Fig. 13.1 The optical antenna principle. (a) Receiving antenna. (b) Transmitting antenna. Arrows indicate the direction of energy flow. The two configurations are related by the principle of reciprocity (see Section 2.13).

spectroscopic applications as well as in sensor devices. Furthermore, there is plenty of time for transitions to dark states or for the excited-state energy to be dissipated through alternative non-radiative channels, including photochemical processes. By coupling a quantum emitter to an optical antenna we can reduce τ and thereby improve the photon-emission rate and reduce the likelihood of dark-state transitions.

Thus, optical antennas enable (i) the confinement of optical radiation to nanoscale dimensions and (ii) the efficient release of radiation from localized sources, thereby enhancing the light–matter interaction.

13.2 Elements of classical antenna theory

As a basis for our discussion we recall basic elements of classical antenna theory. The classical theory of antennas, which is documented in many textbooks (see e.g. [7, 8]), uses Maxwell's equations to describe the interplay of time-dependent currents and electromagnetic waves. Most of the characteristic features of *classical* antennas are related to the two facts that (i) antenna wires can be treated as perfect conductors (there is no field penetration into the metal and respective boundary conditions) and (ii) critical dimensions, such as the antenna feed-gap and wire thickness, can be made negligibly small compared with the wavelength.

The electromagnetic field emitted by an antenna can be expressed in terms of the current density $\mathbf{j}(\mathbf{r})$ and the charge density $\rho(\mathbf{r})$ of the antenna elements. $\mathbf{j}(\mathbf{r})$ and $\rho(\mathbf{r})$ are related by the continuity relation (2.5). It is most common to express the electromagnetic field in terms of the vector potential $\mathbf{A}(\mathbf{r})$ and the scalar potential $\Phi(\mathbf{r})$, which, using the Lorenz gauge (c.f. Eq. (2.78)), satisfy a set of four scalar Helmholtz equations

$$\left[\nabla^2 + k^2\right] \mathbf{A}(\mathbf{r}) = -\mu_0\mu\mathbf{j}(\mathbf{r}), \tag{13.2}$$

$$\left[\nabla^2 + k^2\right] \Phi(\mathbf{r}) = -\frac{1}{\varepsilon_0\varepsilon}\rho(\mathbf{r}). \tag{13.3}$$

The solutions of these equations can be expressed in terms of the scalar Green function G_0 (c.f. Eq. (2.84)) as

$$\mathbf{A}(\mathbf{r}) = \mu_0\mu \int_V \mathbf{j}(\mathbf{r}')\, G_0(\mathbf{r},\mathbf{r}')\, \mathrm{d}V', \tag{13.4}$$

$$\Phi(\mathbf{r}) = \frac{1}{\varepsilon_0\varepsilon} \int_V \rho(\mathbf{r}')\, G_0(\mathbf{r},\mathbf{r}')\, \mathrm{d}V'. \tag{13.5}$$

These solutions define the field distribution, radiation pattern, and radiated power. The electric and magnetic fields are found by straightforward differentiation according to Eqs. (2.75) and (2.76).

Note that the procedure outlined above depends on a-priori knowledge of the sources $\mathbf{j}(\mathbf{r})$ and $\rho(\mathbf{r})$. It turns out, however, that it is quite difficult to accurately determine the current distribution on the antenna elements. For popular center-fed antennas consisting of

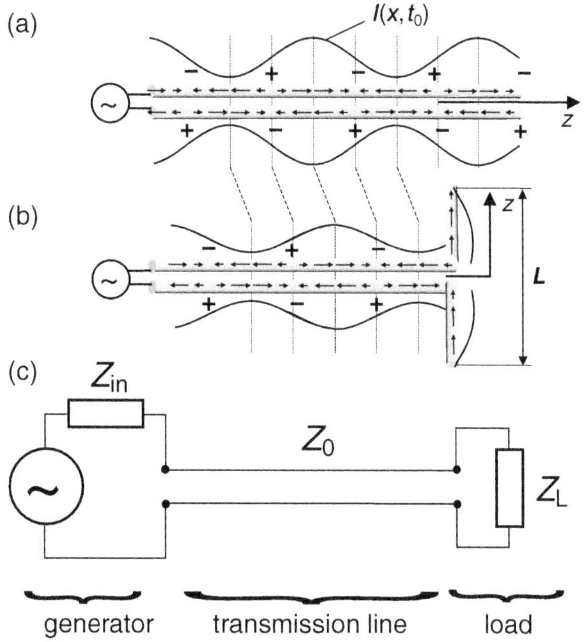

Fig. 13.2 A harmonically driven two-wire transmission line terminated by (a) an open end and (b) a finite-length antenna. For a given instant of time, the arrows indicate the magnitude and direction of the current, plus and minus signs indicate local charge accumulation, and the solid line indicates the standing current wave; (c) equivalent circuit of the system including the internal impedance of the generator, $Z_{\rm in}$, the characteristic impedance of the transmission line, Z_0, and the impedance of the antenna, $Z_{\rm L}$, acting as a load.

thin wires and small feed-gaps an approximate current distribution can be found by solving an integral equation that accounts for radiation reaction (see e.g. [9] for details). Here, for reasons of simplicity, we discuss important antenna parameters under the assumption that the current distribution has a sinusoidal shape. Specifically, as sketched in Fig. 13.2(a), we consider a two-wire transmission line terminated by an open end and driven by a high-frequency voltage source.

The truncated transmission line itself, although it sustains time-harmonic currents with a spatially varying amplitude, does not radiate into the far-field if the gap between the wires is small, since each current element of one wire has its counterpart in the other wire oscillating 180° out of phase. Radiation therefore largely cancels out in the far-field, although there is a strong near-field that is localized between the wires. Since good conductors are being considered, the wavelength of the standing wave is practically the same as the wavelength in free space. For an infinitely long transmission line the local ratio of the voltage between the wires and the current through the wires is a constant called the *characteristic impedance*, $Z_0 = U(z)/I(z)$, which is independent of the position z along the line. It depends solely on the materials used and on the geometry of the transmission line [10]. It has to be emphasized that the characteristic impedance Z_0 is different from the wave impedance Z_W, despite their having the same units. The former is the ratio of voltages and currents, whereas the latter is the ratio of the electric and magnetic fields. For example, a parallel-plate waveguide with a spacing of $d < \lambda/2$ has a wave impedance of 377 Ω, which is similar to that of free space, but its characteristic impedance depends on the particular geometrical parameters.

Let us now consider the situation depicted in Fig. 13.2(b), where the parallel wires are bent by 90° at a distance of $L/2$ from the open end – one upwards and one downwards. We assume that the bending does not significantly affect the sinusoidal current distribution along the bent part of the wires, which is true for thin wires. Considering that the current is zero at the wire ends, we write

$$I(z) = I_{\max} \sin\left[k\left(\frac{1}{2}L - |z|\right)\right]. \tag{13.6}$$

$z = \pm L/2$ corresponds to the wire ends and $z = 0$ to the location of the bend. The maximum current amplitude becomes $I_{\max} = I(0)/\sin(\frac{1}{2}kL)$, which is to be expected from a simple standing-wave model. The actual current amplitude, however, differs from that found in the unbent transmission line. The reason is that the antenna arms can be thought of as a resonant circuit with a complex impedance $Z_L \neq Z_0$, leading in general to a reflection at the bending point and a shift in the standing-wave pattern as sketched in Fig. 13.2(b). It is then natural to define the input impedance of an antenna by the ratio of the voltage measured over the input terminals and the current flowing into each antenna arm, $Z_L = U(0)/I(0) = R_L + iX_L$. As for any frequency-dependent complex impedance, the equivalent circuit of the antenna shows a resonance for that driving frequency for which $\mathrm{Im}(Z_L) = X_L = 0$, which also leads to a maximum in the current amplitude. We will refer to such a resonance as an "antenna resonance."

The power dissipated by the antenna is determined by the real part of the antenna impedance R_L, which includes ohmic losses, R_{nr}, as well as losses due to radiation, R_{rad}, and accordingly

$$R_L = R_{rad} + R_{nr}. \tag{13.7}$$

Once the radiation resistance is known, the radiated power can be calculated as $P_{rad} = (1/2)R_{rad}I(0)^2$. A corresponding relation holds for the nonradiative power dissipated into heat. However, since ohmic losses for radio-frequency antennas are very small, R_{nr} can often be neglected.

The equivalent circuit for the system made of the antenna, the transmission line, and an electrical source is shown in Fig. 13.2(c). The equivalent circuit allows one to describe the relevant parameters of the circuit. In particular, the power delivered to the antenna can be maximized by seeking impedance matching between the transmission line and the antenna. In an unmatched situation it is possible that the antenna is on resonance (i.e. $\mathrm{Im}(Z_L) = X_L = 0$) but very little power is delivered to it via the transmission line because of a large impedance mismatch. This is a situation that occurs, for example, for an antenna with $L = \lambda$ in which the current vanishes in the gap according to Eq. (13.6). Although it has favorable properties, such an antenna cannot be fed by connecting wires at the feed-gap because the antenna impedance diverges. The question of how to feed an antenna in an optimal way is also of importance for nanoantennas at optical frequencies, as we will discuss later on.

For a given source current $I(0)$ the radiation resistance defines the total radiated power. However, for an antenna it is important not only to efficiently radiate but also to direct the radiation towards a target, such as a receiving antenna. To visualize the angular distribution of radiation one plots the radiation pattern $p(\theta, \phi)$. For a thin linear antenna with a sinusoidal current distribution (Eq. (13.6)) we find [7]

$$p(\theta, \phi) \sim \left| \frac{\cos\left(\frac{1}{2}kL\cos\theta\right) - \cos\left(\frac{1}{2}kL\right)}{\sin\theta} \right|^2, \tag{13.8}$$

where the angle θ is measured from the direction of the antenna wires and ϕ is the azimuthal angle. For small antenna arms all current elements are in phase and hence the radiation pattern is very similar to that of a Hertzian dipole ($L \ll \lambda$) except that its angular dependence becomes slightly narrower. Only when the antenna length increases beyond λ are current elements on the same wire that oscillate 180° out of phase introduced, causing strong interference effects and radiation cancellation in some directions. In this case, the radiation pattern features multiple lobes (see Fig. 13.3). The radiation pattern can be further influenced by deviating from the linear shape of the antenna or by adding additional wires as passive elements at well-chosen positions, as is done in the famous Yagi–Uda antenna design.

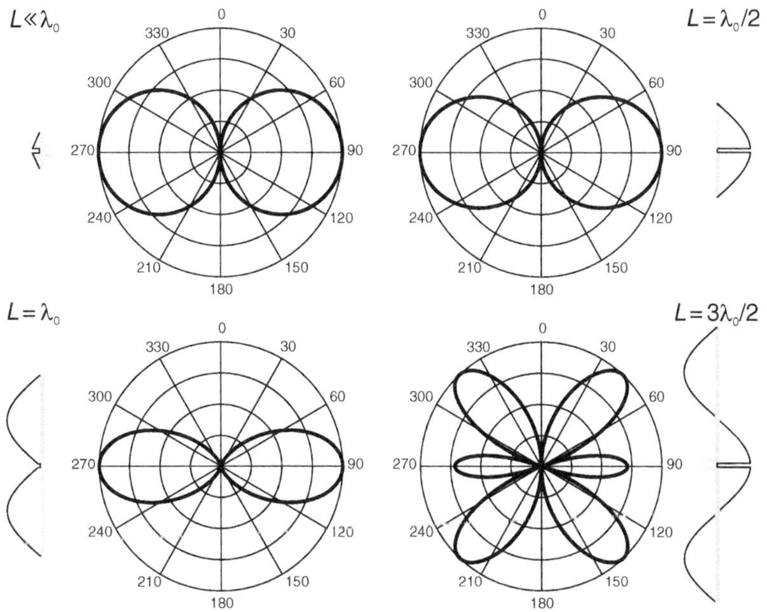

Fig. 13.3 Normalized radiation patterns for a point-like dipole ($L \ll \lambda_0$) and for perfectly conducting thin-wire antennas of length $L = \lambda_0/2$, λ_0, and $3\lambda_0/2$ [7]. The gap antenna attached to an impedance-matched waveguide effectively behaves as a single-wire antenna. A sketch of the current standing wave is provided beside each emission pattern.

13.3 Optical antenna theory

The physics described by Maxwell's equations in vacuum is scale-invariant, i.e. independent of the wavelength. However, this is no longer true in the presence of matter, whose frequency-dependent properties enter Maxwell's equations via the material equations. Metal nanowires, the constituting elements of optical antennas, can no longer be considered to be perfect conductors at optical frequencies. Since typical wire diameters are smaller than the skin depth, electromagnetic fields fully penetrate the wire and induce volume currents as opposed to the pure surface currents that determine the behavior of radiowave antennas. Furthermore, noble-metal nanowires can support wire plasmon modes with wavelengths less than that for free space, which dominate the antenna behavior at optical frequencies (see Chapter 12). While optical antenna design can adapt some design principles known from radiowave antenna technology, important antenna parameters need to be recalculated considering the volume currents and the reduced wavelength.

A major difference between radiowave and optical antennas is the way in which a receiver or transmitter is connected to the antenna. As illustrated in Fig. 13.2, a transmitter or receiver is typically connected to a radiowave antenna by means of an impedance-matched transmission line. However, at optical frequencies, the small size of receivers and transmitters prevents them from being wired to antenna elements in the traditional fashion. Instead, interconnects become part of the antenna design, and, in the extreme limit, receivers and transmitters become discrete quantum objects such as molecules, quantum dots, or tunnel junctions that couple to the antenna by energy or charge transfer.

The objective of optical antenna design is equivalent to that of classical antenna design, namely to optimize the energy transfer between a localized source or receiver and the free radiation field. As illustrated in Fig. 13.4, optical antennas can enhance several distinct

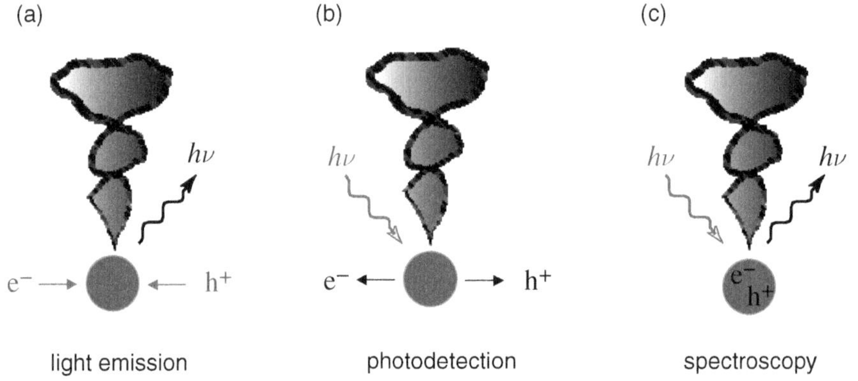

Fig. 13.4 Antenna-coupled optical interactions. The circle indicates a material system in which (a) charges are combined to produce radiation, (b) radiation generates charge separation, or (c) incident radiation generates polarization currents that give rise to secondary radiation.

photophysical processes. In light-emitting devices (LEDs), an electron and a hole combine to emit a photon. The reverse process takes place in photodetection.

Optical antennas can have a variety of different shapes depending on the specific purpose they are intended to fulfill. Figure 13.5 shows a selection of structures that have been investigated so far. In most of the structures single metal wires are used as fundamental building blocks. Structures consisting of multiple wires are used to tailor the polarization response and the emission pattern much as in radiowave antenna engineering.

13.3.1 Antenna parameters

Figure 13.6 illustrates a generic antenna problem. It consists of a transmitter and a receiver, both represented by dipoles **p**. The antenna is introduced to enhance the transmission efficiency from the transmitter to the receiver. This enhancement can be achieved by increasing the total amount of radiation released by the transmitter or by altering the radiation pattern such that more power is directed towards the receiver. To quantify these processes radio engineers have introduced specific parameters, which we shall discuss in the following.

Antenna efficiency

The total power P that is dissipated by an antenna-coupled emitter is the sum of the radiated power P_{rad} and the power dissipated into heat and other channels (P_{loss}). The antenna efficiency is defined as

$$\eta_{\mathrm{rad}} = \frac{P_{\mathrm{rad}}}{P} = \frac{P_{\mathrm{rad}}}{P_{\mathrm{rad}} + P_{\mathrm{loss}}} . \tag{13.9}$$

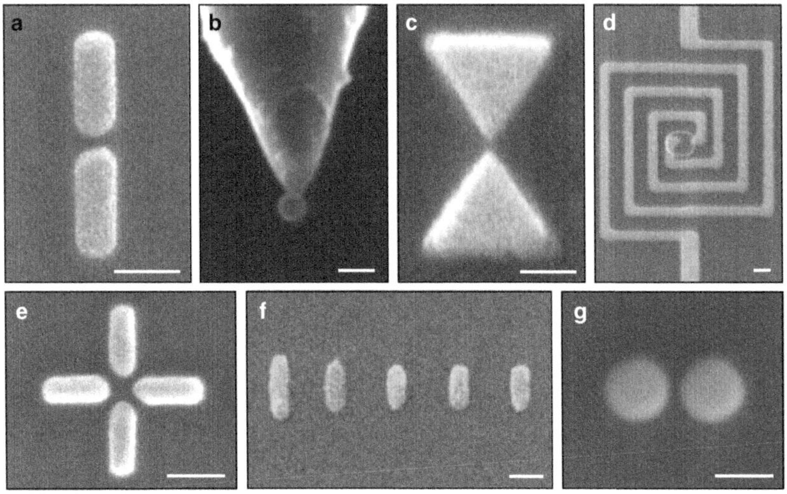

Fig. 13.5 SEM images of optical antennas: (a) coupled-dipole antenna, (b) nanoparticle antenna, (c) bowtie antenna, (d) square-spiral antenna [11], (e) cross antenna, (f) Yagi–Uda antenna [12], and (g) Hertzian dimer antenna. All scale bars 100 nm.

While P is most conveniently determined by calculating the field $\mathbf{E}$ at the dipole's position according to Eq. (8.74), $P_{\rm rad}$ requires the calculation of the energy flux through a surface enclosing both the dipole and the antenna.

Intrinsic efficiency

It is useful to distinguish dissipation in the antenna and the transmitter, which is not accomplished by Eq. (13.9). We therefore define the intrinsic efficiency of the emitter as

$$\eta_{\rm i} = \frac{P^{\rm o}_{\rm rad}}{P^{\rm o}_{\rm rad} + P^{\rm o}_{\rm intrinsic\ loss}}, \tag{13.10}$$

where the superscripts $^{\rm o}$ designate the absence of the antenna. With this definition of $\eta_{\rm i}$ we can rewrite Eq. (13.9) as

$$\eta_{\rm rad} = \frac{P_{\rm rad}/P^{\rm o}_{\rm rad}}{P_{\rm rad}/P^{\rm o}_{\rm rad} + P_{\rm antenna\ loss}/P^{\rm o}_{\rm rad} + (1-\eta_{\rm i})/\eta_{\rm i}}. \tag{13.11}$$

For an emitter with $\eta_{\rm i} = 1$ (no intrinsic loss) the antenna can only reduce the efficiency. Note, however, that even if the efficiency gets reduced, the emission rate of the emitter can be enhanced (see Section 13.4). For emitters with low $\eta_{\rm i}$ the antenna can increase η.

Radiation pattern

The transmission efficiency from emitter to receiver can also be improved by directing the radiation in the direction of the receiver. To account for the angular distribution of the radiated power we define the normalized angular power density $p(\theta, \phi)$, or radiation pattern, as

$$\int_0^{\pi}\int_0^{2\pi} p(\theta,\phi) \sin\theta \, {\rm d}\phi \, {\rm d}\theta = P_{\rm rad}. \tag{13.12}$$

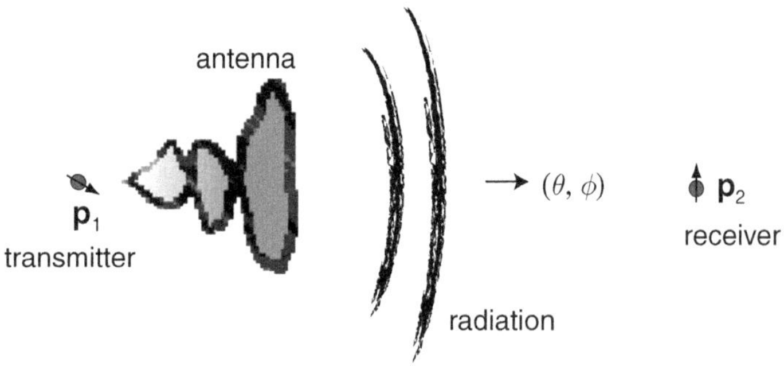

Fig. 13.6 The generic antenna problem. An antenna enhances the transmission efficiency from the transmitter to the receiver.

Directivity

The directivity is a measure of an antenna's ability to concentrate the radiated power into a certain direction. It corresponds to the angular power density relative to a hypothetical isotropic radiator. Formally,

$$D(\theta, \phi) = \frac{4\pi}{P_{\text{rad}}} p(\theta, \phi). \tag{13.13}$$

When the direction (θ, ϕ) is not explicitly stated one usually refers to the direction of maximum directivity, i.e. $D_{\text{max}} = (4\pi/P_{\text{rad}})\text{Max}\left[p(\theta, \phi)\right]$.

Because the fields at a large distance from an antenna are transverse, they can be written in terms of two polarization directions, $\mathbf{n}_\theta$ and $\mathbf{n}_\phi$. The partial directivities are then defined as

$$D_\theta(\theta, \phi) = \frac{4\pi}{P_{\text{rad}}} p_\theta(\theta, \phi) \quad \text{and} \quad D_\phi(\theta, \phi) = \frac{4\pi}{P_{\text{rad}}} p_\phi(\theta, \phi). \tag{13.14}$$

Here, p_θ and p_ϕ are the normalized angular powers measured after polarizers aligned in the directions $\mathbf{n}_\theta$ and $\mathbf{n}_\phi$, respectively. Because $\mathbf{n}_\theta \cdot \mathbf{n}_\phi = 0$ we have

$$D(\theta, \phi) = D_\theta(\theta, \phi) + D_\phi(\theta, \phi). \tag{13.15}$$

The influence of an optical antenna on the radiation pattern of a single molecule was studied by van Hulst *et al.* [12, 13], and it has been shown that the antenna provides a high level of control for the direction and polarization of the emitted photons (see Fig. 13.7).

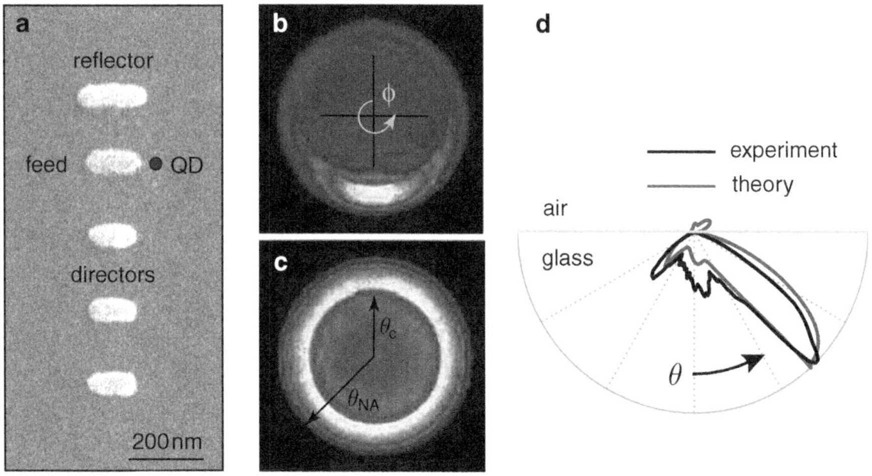

Fig. 13.7 An optical Yagi–Uda antenna directing the photon emission from a single quantum dot (QD). (a) An SEM image of the antenna fabricated on a quartz surface. It consists of a reflector, three directors, and a feed element to which the QD is coupled. (b) The radiation pattern, showing photon emission in the direction of the antenna. (b) The reference radiation pattern of a QD emitting in the absence of the antenna. θ_c indicates the critical angle and θ_{NA} the maximum angle accessible, as determined by the numerical aperture (NA) of the collection lens. (c) Comparison of the experimental and theoretical radiation patterns. Adapted from Ref. [12].

Effective area

$D(\theta,\phi)/(4\pi)$ defines the fraction of the total radiated power that is radiated in the direction of (θ,ϕ). In other words, it is the fraction of power radiated into the infinitesimal unit solid angle $\mathrm{d}\Omega = \sin\theta\,\mathrm{d}\theta\,\mathrm{d}\phi$. At a distance $R \gg \lambda$ from the antenna, this fraction of power spreads over an area of $\mathrm{d}A = R^2\,\mathrm{d}\Omega$, which assumes that we're in free space and that there are no obstructions and other inhomogeneities. Thus, on writing $p(\theta,\phi) = \mathrm{d}P_{\mathrm{rad}}/\mathrm{d}\Omega = [\mathrm{dP}_{\mathrm{rad}}/\mathrm{d}A]\,[\mathrm{d}A/\mathrm{d}\Omega] = I(\theta,\phi)R^2$ we find

$$D(\theta,\phi) = 4\pi R^2\,\frac{I(\theta,\phi)}{P_{\mathrm{rad}}}, \tag{13.16}$$

where I is the intensity.

Let us now express $D(\theta,\phi)$ in terms of an effective aperture A_{eff}. This aperture is placed at the location of the antenna and is irradiated by a plane wave. Its surface normal points in the direction of (θ,ϕ). After the wave has passed through the aperture it spreads out due to diffraction. We want to adjust the size of A_{eff} such that the diffracted beam has a directivity $D(\theta,\phi)$. Assuming that the radius w_0 of the aperture is larger than λ, diffraction theory yields (c.f. Fig. 3.1 and Eq. (3.16))

$$\theta = \lambda/(\pi w_0). \tag{13.17}$$

In terms of the power emitted by the aperture (P_{rad}), the effective aperture area ($A_{\mathrm{eff}} = \pi w_0^2$), and the beam area $\pi(R\theta)^2$ we can now express the beam intensity at a distance R as $I = A_{\mathrm{eff}}P_{\mathrm{rad}}/(\lambda^2R^2)$. Inserting this into Eq. (13.16) yields

$$A_{\mathrm{eff}} = \frac{\lambda^2}{4\pi}D(\theta,\phi). \tag{13.18}$$

Thus, in terms of radiation, an antenna behaves like an aperture that directs radiation in the direction of observation (θ,ϕ). Note that, if we reverse the flow of radiation, then A_{eff} defines the power received by the antenna, that is, $P_{\mathrm{received}} = A_{\mathrm{eff}}I_{\mathrm{A}}$, with I_A being the intensity at the aperture. A_{eff} then assumes the meaning of an absorption cross-section, which will be discussed later on.

Gain

The combination of antenna efficiency and directivity is referred to as the antenna gain,

$$G = \frac{4\pi}{\mathrm{P}}p(\theta,\phi) = \eta_{\mathrm{rad}}D. \tag{13.19}$$

D and G are usually measured in decibels (dB). Since perfectly isotropic radiators do not exist in reality, it is often more practical to refer to an antenna of a known directional pattern. The *relative* gain is then defined as the ratio of the power gain in a given direction to the power gain of a reference antenna in the same direction. A dipole antenna is the standard choice as a reference because of its relatively simple radiation pattern.

The Chu limit

In most antenna problems, G is the quantity of relevance since it defines the transmission efficiency from emitter to receiver. An interesting question is whether there is an upper limit for G. Evidently, the larger we make an antenna the more engineering degrees we have available and the larger G will be. Similarly, G can be optimized by reducing the bandwidth of transmission frequencies and making the antenna more resonant. A limit for G can indeed be derived for a fixed volume V and a fixed relative bandwidth B (the bandwidth divided by the center frequency) [14, 15]

$$GB \leq c_0 V/\lambda^3, \tag{13.20}$$

where c_0 is a constant of order unity that depends on the geometry of V. This limit goes back to L. G. Chu's work in 1948 and is based on ideal conductors and a field expansion in spherical harmonics [14]. Chu's theory was later expanded and generalized by Gustafsson and coworkers [15].

Reciprocity

According to the reciprocity theorem (see Eq. (2.102)) we can interchange the fields and sources. For a pair of dipoles (c.f. Fig. 13.6) we obtain $\mathbf{p}_1 \cdot \mathbf{E}_2 = \mathbf{p}_2 \cdot \mathbf{E}_1$, where $\mathbf{E}_1$ ($\mathbf{E}_2$) is the field of dipole $\mathbf{p}_1$ ($\mathbf{p}_2$) evaluated at the location of $\mathbf{p}_2$ ($\mathbf{p}_1$). The separation between the two dipoles is assumed to be sufficiently large ($kR \gg 1$) to ensure that they interact only via their far-fields. Furthermore, the direction of $\mathbf{p}_2$ is chosen to be transverse to the vector connecting the two dipoles.

In the classical picture, we assume that dipole $\mathbf{p}_1$ has been induced by the field $\mathbf{E}_2$ of dipole $\mathbf{p}_2$ according to $\mathbf{p}_1 = \overleftrightarrow{\alpha}_1 \mathbf{E}_2$, where $\overleftrightarrow{\alpha}_1 = \alpha_1 \mathbf{n}_{\mathrm{p}_1}\mathbf{n}_{\mathrm{p}_1}$ is the polarizability tensor. Here, $\mathbf{n}_{\mathrm{p}_1}$ is the unit vector in the direction of $\mathbf{p}_1$. According to Eq. (8.74), the power absorbed by the dipole at $\mathbf{r}_1$ is

$$P_{\mathrm{abs}} = (\omega/2)\mathrm{Im}\left\{\mathbf{p}_1^* \cdot \mathbf{E}_2(\mathbf{r}_1)\right\} = (\omega/2)\mathrm{Im}\{\alpha_1\}\left|\mathbf{n}_{\mathrm{p}_1} \cdot \mathbf{E}_2(\mathbf{r}_1)\right|^2. \tag{13.21}$$

We now substitute into Eq. (13.21) the reciprocity relation in the form $|\mathbf{p}_1|\mathbf{n}_{\mathrm{p}_1} \cdot \mathbf{E}_2 = |\mathbf{p}_2|\mathbf{n}_{\mathrm{p}_2} \cdot \mathbf{E}_1$, and obtain

$$P_{\mathrm{abs}} = (\omega/2)|\mathbf{p}_2/\mathbf{p}_1|^2\ \mathrm{Im}\{\alpha_1\}\left|\mathbf{n}_{\mathrm{p}_2} \cdot \mathbf{E}_1(\mathbf{r}_2)\right|^2. \tag{13.22}$$

The term $\left|\mathbf{n}_{\mathrm{p}_2} \cdot \mathbf{E}_1(\mathbf{r}_2)\right|^2$ corresponds to the power a photodetector at $\mathbf{r}_2$ would read if it were placed behind a polarizer oriented in the direction $\mathbf{n}_{\mathrm{p}_2}$.

We now invoke the partial directivities defined in Eqs. (13.14). In terms of the field $\mathbf{E}$ evaluated at $\mathbf{r}_2 = (R, \theta, \phi)$ the partial directivity D_θ reads

$$D_\theta(\theta, \phi) = 4\pi \frac{|\mathbf{n}_\theta \cdot \mathbf{E}(R, \theta, \phi)|^2}{\int_{4\pi} |\mathbf{E}(R, \theta, \phi)|^2\, \mathrm{d}\Omega}, \tag{13.23}$$

where Ω is the unit solid angle and $\mathbf{n}_\theta$ the unit polar vector. $D_\phi(\theta, \phi)$, referring to radiation polarized in azimuthal direction $\mathbf{n}_\phi$, is expressed similarly.

To proceed, we choose the dipole $\mathbf{p}_2$ to point in the direction of $\mathbf{n}_\theta$. Equation (13.22) can then be represented as

$$P_{\text{abs},\theta}(\theta,\phi) = (\omega/2)|\mathbf{p}_2/\mathbf{p}_1|^2 \,\text{Im}\{\alpha_1\}\frac{P_{\text{rad}}}{2\pi\,\varepsilon_0 cR^2}\,D_\theta(\theta,\phi). \tag{13.24}$$

Here, $P_{\text{rad}} = (1/2)\varepsilon_0 cR^2 \int_{4\pi} |\mathbf{E}(R,\theta,\phi)|^2\, d\Omega$ is the total radiated power. $P_{\text{abs},\theta}(\theta,\phi)$, specifies the power absorbed by dipole $\mathbf{p}_1$ when excited by the field of dipole $\mathbf{p}_2$ located at (R,θ,ϕ) and oriented in the $\mathbf{n}_\theta$ direction. Because $kR \gg 1$ the field exciting dipole $\mathbf{p}_1$ and the antenna is essentially a plane wave polarized in the direction $\mathbf{n}_\theta$.

We now remove the antenna and write down an equation similar to Eq. (13.24). Dividing one of the two equations by the other yields

$$\frac{P_{\text{abs},\theta}(\theta,\phi)}{P^{\text{o}}_{\text{abs},\theta}(\theta,\phi)} = \frac{P_{\text{rad}}}{P^{\text{o}}_{\text{rad}}}\,\frac{D_\theta(\theta,\phi)}{D^{\text{o}}_\theta(\theta,\phi)}, \tag{13.25}$$

where the superscipts ^{o} carry the same meaning as in Eq. (13.10), namely the absence of an antenna. Using the fact that P_{abs} is proportional to the excitation rate γ_{exc} and P_{rad} is proportional to the radiative decay rate γ_{rad} (see, for example, Eq. (8.116)) we can rewrite Eq. (13.25) as

$$\frac{\gamma_{\text{exc},\theta}(\theta,\phi)}{\gamma^{\text{o}}_{\text{exc},\theta}(\theta,\phi)} = \frac{\gamma_{\text{rad}}}{\gamma^{\text{o}}_{\text{rad}}}\,\frac{D_\theta(\theta,\phi)}{D^{\text{o}}_\theta(\theta,\phi)}\,, \tag{13.26}$$

which states that the enhancement of the excitation rate due to the presence of the antenna is proportional to the enhancement of the radiative rate, a relationship that has been used qualitatively in various studies [16–18]. Note that the same analysis can be repeated with $\mathbf{n}_\phi$ instead of $\mathbf{n}_\theta$, which corresponds to polarization rotated by 90°.

Antenna aperture

The antenna aperture describes the efficiency with which incident radiation is captured. It corresponds to the area of incident radiation that interacts with the antenna, and is defined as

$$\sigma_{\text{A}}(\theta,\phi,\mathbf{n}_{\text{pol}}) = \frac{P_{\text{abs}}}{I}, \tag{13.27}$$

where P_{abs} denotes the power absorbed by the receiver and I is the intensity of radiation incident from (θ,ϕ) and polarized in the direction $\mathbf{n}_{\text{pol}}$. If the direction or polarization is not specified one usually refers to the one that yields the maximum aperture. Formally, the antenna aperture is equivalent to the absorption cross-section.

The antenna increases the optical energy density that falls on a target and thereby increases its efficiency. For a detector that is small compared with the wavelength λ the received power is calculated according to Eq. (8.74) as

$$P_{\text{abs}} = (\omega/2)\text{Im}\{\alpha\}\left|\mathbf{n}_{\mathbf{p}}\cdot\mathbf{E}\right|^2. \tag{13.28}$$

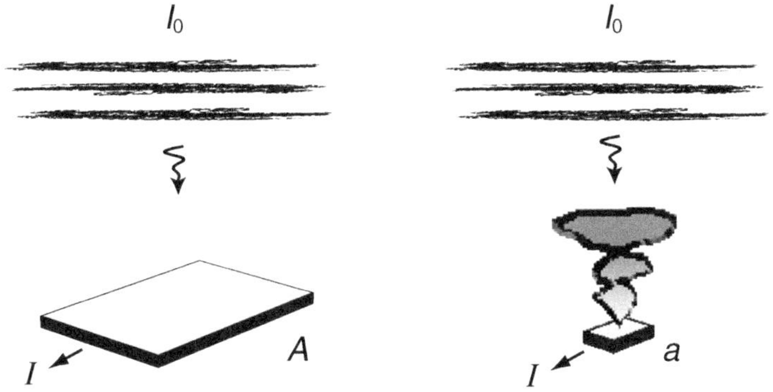

Fig. 13.8 Reduction of the area of the photodetector. Left: light with intensity I_0 is incident on a photodetector with area A to produce a photocurrent I. Right: same situation for an antenna-coupled detector. The detector area can be reduced ($a \ll A$) without affecting the response I/I_0. The reduced detector area improves the signal-to-noise ratio and reduces the response time and energy consumption.

Here, $\mathbf{n_p}$ is the unit vector in the direction of the absorption dipole $\mathbf{p}$, and $\mathbf{E}$ is the field at the location of the detector. If we denote the field at the target in the absence of the antenna as $\mathbf{E}_0$ we can represent the antenna aperture as

$$\sigma_A = \sigma_A^0 |\mathbf{n_p} \cdot \mathbf{E}|^2 / |\mathbf{n_p} \cdot \mathbf{E}_0|^2, \tag{13.29}$$

where σ_A^0 is the aperture in the absence of the antenna and $\mathbf{E}$ is the field at the target in the presence of the antenna. Thus, we find that the absorption enhancement corresponds to the local intensity-enhancement factor.

Theoretical and experimental studies have shown that intensity enhancements of 10^4–10^6 are readily achievable [19] and hence, for typical molecules with free-space cross-sections of $\sigma_A^0 = 1\,\text{nm}^2$, we find that a layer of molecules spaced 0.1–1 μm apart can absorb all of the incident radiation if each molecule is coupled to an optical antenna. Of course, this estimate ignores the coupling between antennas and therefore has limited validity. Note that in general both σ_A and σ_A^0 depend on the direction of incidence (θ, ϕ) and the polarization direction $\mathbf{n}_{\text{pol}}$.

The application of optical antennas in photodetectors is particularly promising (see Fig. 13.8). The main reason is that, according to Eq. (13.29), an optical antenna increases the absorption cross-section and hence the light flux that impinges on a detector. Thus, an antenna-coupled detector requires a much smaller detector area for the same signal output. Because the dark current i_D of a photodetector scales with the detector area we find that the noise current $i_N = (2ei_D\,\Delta f)^{1/2}$ and the noise equivalent power (NEP) defined as

$$\text{NEP} = \frac{i_N}{\eta}\frac{h\nu}{e} \tag{13.30}$$

both scale with the square root of the area. Here, η is the quantum efficiency, ν the frequency, and Δf the bandwidth. The NEP corresponds to the lowest power a detector can detect with a signal-to-noise ratio (SNR) of one. In order to eliminate the dependence on detector area, one defines the detectivity $D^* = (A\,\Delta f)^{1/2}/\text{NEP}$, with units of Jones

$= \text{cm Hz}^{1/2}\ \text{W}^{-1}$. Authors of recent studies have reported values of NEP $= 1.53$ nW and $D^* = 2.15 \times 10^6$ Jones for antenna-coupled infrared detectors based on metal–oxide–metal diodes [20]. Note that a reduced detector area also reduces the power consumption and the response time.

Friis equation

Let us now consider the situation of two antennas, one transmitting (subscript "t") and one receiving (subscript "r"), located in free space and separated by a large distance ($kR \gg 1$). We place the transmitting antenna at the origin of our coordinate system and the receiving antenna at (R, θ, ϕ). Multiplying Eq. (13.16) with the antenna efficiency η_{rad} and making use of the definition of the gain G in Eq. (13.19) renders the intensity I_{t} radiated by the transmitter at the location of the receiver

$$I_{\text{t}}(\theta, \phi) = \frac{P_{\text{t}}}{4\pi R^2} G_{\text{t}}(\theta, \phi), \tag{13.31}$$

where P_{t} is the power with which the transmitting antenna is being driven. The power incident on the receiving antenna is $P_{\text{rad}} = \sigma_{\text{A}} I_{\text{t}}$ and the power that is generated by the receiving antenna is $P_{\text{r}} = \eta_{\text{rad,r}} \text{P}_{\text{rad}}$. We now make use of Eq. (13.18) and the fact that σ_{A} is equivalent to the effective area A_{eff} and obtain

$$P_{\text{r}} = G_{\text{r}}(\theta, \phi) \frac{\lambda^2}{4\pi} I_{\text{t}}(\theta, \phi). \tag{13.32}$$

The combination of Eqs. (13.31) and (13.32) finally yields

$$\frac{P_{\text{r}}}{P_{\text{t}}} = \left[\frac{\lambda}{4\pi R} \right]^2 G_{\text{r}}(\theta, \phi) G_{\text{t}}(\theta, \phi), \tag{13.33}$$

which is referred to as the *Friis equation* [21]. In deriving this equation we have neglected antenna reflections, inhomogeneities in space, and polarization effects. These can be included in a more detailed derivation. The Friis equation states that the transmission efficiency is better for long wavelengths, which is a consequence of diffraction (c.f. Eq. (13.17)). The Friis equation follows from simple dimensional considerations. First, it is evident that the transmission efficiency between two antennas must be proportional to the product of their gains G. Second, on grounds of energy conservation, it has to scale as $1/R^2$. Finally, we multiply by λ^2 to match the units.

Effective wavelength

Radiowave antennas have design rules that relate to the wavelength of incident radiation λ. For example, a half-wave antenna has a length L of $\lambda/2$, and a Yagi–Uda antenna has separations between elements that correspond to certain fractions of λ. Because all elements are proportional to λ, it is straightforward to scale the antenna design from one wavelength to another. However, this scaling fails at optical frequencies because the penetration of radiation into metals can no longer be neglected. Owing to the finite electron density, there is a delay between the driving field and the electronic response, resulting in a skin depth that is typically larger than the diameter of the antenna elements. As a consequence, electrons in metals do not respond to the wavelength λ of the incident radiation but to an effective wavelength $\lambda_{\rm eff}$, which is determined by a simple linear scaling rule [22]

$$\lambda_{\rm eff} = n_1 + n_2 \frac{\lambda}{\lambda_{\rm p}}, \tag{13.34}$$

where n_1 and n_2 are geometric constants and $\lambda_{\rm p}$ is the plasma wavelength. This wavelength scaling rule follows from a Fabry–Pérot model for a metal wire of finite length (see Fig. 12.24). The proportionality between $\lambda_{\rm eff}$ and λ is a consequence of the small-radius approximation ($R \to 0$) for the wire's propagation constant. According to Eq. (13.34), an optical half-wave antenna is not $\lambda/2$ in length but has a shorter length of $\lambda_{\rm eff}/2$. The difference between λ and $\lambda_{\rm eff}$ depends on geometrical factors, but is typically in the range of a factor of 2–5 for most metals that are used for optical antenna fabrication. For example, according to Fig. 13.9, to build a half-wave antenna with a gold wire of radius $R = 5\,$nm that responds to incident light of wavelength $\lambda = 800\,$nm, we need to cut the wire length to $\Delta l = \lambda_{\rm eff}/2 \approx 160\,$nm rather than $\Delta l = \lambda/2 = 400\,$nm.

Because the wavelength scaling rule is linear in λ we can, in principle, downscale established radiowave antenna designs to the optical frequency regime. However, the antenna dimensions will not simply scale with the ratio of wavelengths, but with the ratio $\lambda_{\rm eff}(\lambda_{\rm opt})/\lambda_{\rm rf}$, where $\lambda_{\rm opt}$ and $\lambda_{\rm rf}$ are the design wavelengths in the optical and radiowave frequency regimes, respectively.

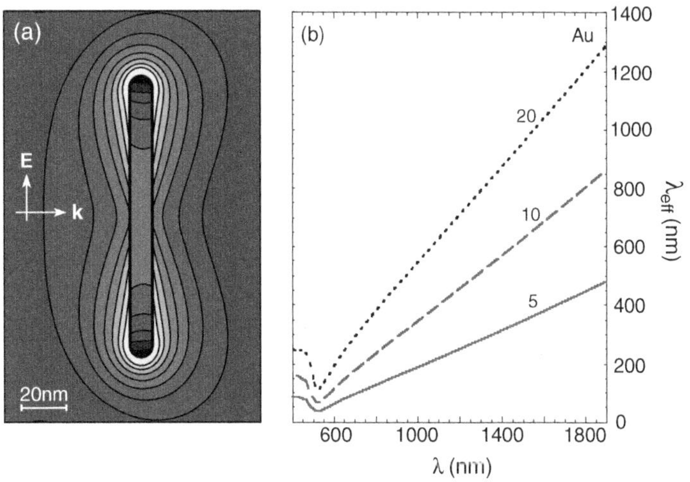

Fig. 13.9 Effective wavelength scaling for linear gold antennas. (a) Intensity distribution (E^2, with a factor of 2 difference between contour lines) for a gold half-wave antenna irradiated with a plane wave ($\lambda = 1150\,$nm). (b) Effective wavelength scaling for gold rods of different radii (5, 10, and 20 nm). The dip at $\lambda = 550\,$nm is a consequence of interband transitions (bound electrons).

Radiation resistance

As discussed in Section 13.2, the radiation resistance R_{rad} is defined by $P_{\text{rad}} = (1/2)I(0)^2 R_{\text{rad}}$, where P_{rad} is the power radiated by the antenna and $I(0)$ is the source current at the feed-point. The larger R_{rad} the better the antenna radiates. For a dipole antenna, Eq. (13.1) yields

$$R_{\text{rad}} = \frac{2\pi}{3} Z_{\text{w}} \left(\frac{\Delta l}{\lambda}\right)^2 , \tag{13.35}$$

where Δl is the antenna length. For a half-wave antenna we would typically have $\Delta l = \lambda/2$.[1] However, according to the wavelength scaling rule discussed above, we have to choose $\Delta l = \lambda_{\text{eff}}/2$. Consequently, the radiation resistance of a half-wave antenna at optical frequencies is smaller by a factor of $(\lambda_{\text{eff}}/\lambda)^2$ than that of its radiowave analogue, which implies that the antenna radiates poorly and that most energy remains stored as reactive power, which is characteristic of a resonator! A significant portion of this reactive power is finally dissipated to heat. Thus, a major challenge for optical antennas is to increase the radiation resistance and thereby improve their radiation efficiency.

Lumped-circuit elements

The concept of lumped circuit elements is a well-established method to simplify the analysis of complex electrical circuits that otherwise would have to be treated as physical systems with distributed parameters. Lumped elements can be defined as long as the characteristic size of each element in the circuit is much smaller than the operating wavelength. The concept of impedance is then introduced to describe the properties of each element and the interactions between different lumped elements in the circuit by using Kirchhoff's laws. The applicability of Kirchhoff's voltage law locally requires the quasi-static approximation to be valid, i.e. $\nabla \times \mathbf{E} \simeq 0$. When the object size is increased and becomes comparable to the operating wavelength this might no longer be valid. Although antennas are clearly not subwavelength elements, their feed-points are usually very close to each other, and in the specific region of the feed-gap the condition $\nabla \times \mathbf{E} \simeq 0$ is fulfilled, thus justifying the introduction of an input impedance. For coupled-dipole antennas one can therefore calculate an input impedance by taking the voltage to (displacement) current ratio across the subwavelength feedgap. In this way, numerical simulations can provide impedance values which compare well with those of standard radiofrequency antennas [23].

For the case in which a transmission line is attached to drive an optical antenna the impedance discontinuity at the connection points between the leads and the antenna that acts as a load will result in a reflection of the forward-traveling voltage wave. The related reflection coefficient is then given by [10]

[1] A half-wave antenna is no longer a dipole antenna. Equation (13.35) yields a value that is too large by a factor of 2.7.

$$\Gamma = \frac{Z_L - Z_0}{Z_L + Z_0}, \tag{13.36}$$

where Z_0 is the characteristic impedance of the transmission line and Z_L is the antenna impedance. Experimentally, Γ can be obtained by imaging the standing-wave pattern that builds up in the transmission line due to reflection at the antenna by means of near-field [24, 25] or photoemission-electron microscopy techniques [26]. The characteristic impedance of the transmission line, Z_0, can be calculated by two-dimensional field simulations. Equation (13.36) therefore provides a practical approach by which to determine antenna impedances [27].

Kinetic inductance

Inductance is a property of an electrical circuit. It generates a 90° phase shift between the applied current and the induced voltage. Commonly, inductance is associated with a coil of wire that generates a magnetic field in the center, a so-called inductor. At optical frequencies, however, another form of inductance comes into play, namely the kinetic inductance. This inductance originates from the inertia of charge carriers, that is, from their inability to immediately respond to the driving field.

Let us recall that the current density $\mathbf{j}$ in a medium can be written as a sum of a conduction current density $\mathbf{j}_c$, a polarization current density $-i\omega\mathbf{P}$, and a magnetization current density $\nabla \times \mathbf{M}$ (see Eq. (2.10)). At optical frequencies we can mostly ignore the last term. We now write $\mathbf{j}_c = \sigma\mathbf{E}$ and $\mathbf{P} = \varepsilon_0(\varepsilon' - 1)\mathbf{E}$, with ε' being the real part of the frequency-dependent dielectric constant. We now express the conductivity in terms of the imaginary part of ε as $\sigma = \omega\varepsilon_0\varepsilon''$ (see Eq. (2.32)) and obtain

$$\mathbf{j} = \omega\varepsilon_0\big[\varepsilon'' - i(\varepsilon' - 1)\big]\mathbf{E}. \tag{13.37}$$

Because ε'' and ε' are real, the first term generates a current that is in phase with the electric field, whereas the second term gives rise to a current that is 90° out of phase with $\mathbf{E}$, as in an inductor. It is this term that is referred to as the kinetic inductance.

To understand the frequency regime for which the kinetic inductance becomes relevant we replace ε by the Drude model discussed in Section 12.2.1. Using the fact that the damping constant Γ is much smaller than the plasma frequency ω_p, we find

$$\mathbf{j} = \varepsilon_0 \frac{\omega_p^2}{\omega}\left(\frac{\Gamma}{\omega} + i\right)\mathbf{E}. \tag{13.38}$$

Thus, for low frequencies ($\omega \ll \Gamma$) we can neglect the kinetic inductance, but at optical frequencies ($\omega > \Gamma$) we can no longer ignore it. The appearance of the kinetic inductance is a further challenge for *optical* antenna design and prevents a straightforward downscaling of established radiowave-antenna concepts.

To estimate the magnitude of the kinetic inductance, we consider a half-wave antenna made of a silver wire of radius $R = 5$ nm and cut to a length of $\Delta l = 150$ nm. The kinetic inductance becomes $L_{kin} = \Delta l/(\pi R^2 \omega_p^2 \varepsilon_0) = 1.1$ pH, which corresponds to the inductance of a single-loop coil of radius 500 nm made of infinitely thin and perfectly conducting wire.

The density of states

Arguably, one of the most important quantities in a discussion of antennas is the impedance, which is defined in circuit theory in terms of the source current I and voltage V as $Z = V/I$. This definition assumes that the source is connected to the antenna via a current-carrying transmission line, as illustrated in Fig. 13.2. But optical antennas are typically fed by localized light emitters, not by real currents. Thus, the definition of the antenna input impedance needs some adjustments. A viable alternative definition involves the local density of electromagnetic states (LDOS) discussed in Section 8.4.3, which can be expressed in terms of the Green-function tensor $\overleftrightarrow{\mathbf{G}}$, and which accounts for the energy dissipation of a dipole in an arbitrary inhomogeneous environment. An optical antenna enhances the LDOS, thereby making it possible for the emitter to dissipate its energy more easily.

We recall from Section 8.4.3 that the partial LDOS can be expressed in terms of the system's dyadic Green function as

$$\rho_{\mathbf{p}}(\mathbf{r}_0, \omega) = \frac{6\omega}{\pi c^2}\left[\mathbf{n_p} \cdot \mathrm{Im}\left\{\overleftrightarrow{\mathbf{G}}(\mathbf{r}_0, \mathbf{r}_0, \omega)\right\} \cdot \mathbf{n_p}\right], \tag{13.39}$$

where $\mathbf{n_p}$ is a unit vector pointing in the direction of dipole $\mathbf{p}$. The Green function used in Eq. (13.39) is indirectly defined by the electric field $\mathbf{E}$ at the observation point $\mathbf{r}$ generated by a dipole $\mathbf{p}$ located at $\mathbf{r}_0$ (c.f. Eq. (2.68)). The *total* LDOS (ρ) is obtained by assuming that the quantum emitter has no preferred dipole axis and averaging Eq. (13.39) over different dipole orientations.

In terms of the LDOS we can express the total power dissipated by an oscillating dipole as

$$P = \frac{\pi\,\omega^2}{12\,\varepsilon_0}|\mathbf{p}|^2 \rho_{\mathbf{p}}(\mathbf{r}_0, \omega). \tag{13.40}$$

Evidently, for free space $\rho_{\mathbf{p}} = \omega^2/(\pi^2 c^3)$ and hence $P^{\mathrm{o}} = |\mathbf{p}|^2\omega^4/(12\pi\varepsilon_0 c^3)$, which is the classical dipole-radiation formula. Thus, we can express the LDOS in terms of the normalized power dissipation from a dipole as

$$\rho_{\mathbf{p}}(\mathbf{r}_0, \omega) = \frac{\omega^2}{\pi^2 c^3} P/P^0. \tag{13.41}$$

An optical antenna is a means to locally "engineer" the LDOS and to enhance the power dissipation from a dipole emitter placed close by.

Greffet and coworkers have established an analogy between the LDOS and the antenna resistance Re$\{Z\}$ [28]. The latter accounts for the total power dissipation, including the radiated and absorbed power. If we express the dipole $\mathbf{p}$ in Eq. (13.40) in terms of the current density $\mathbf{j} \sim \mathrm{i}\omega\mathbf{p}$ and the power P in terms of a resistance according to $P \sim \mathrm{Re}\{Z\}|\mathbf{j}|^2$ we arrive at

$$\mathrm{Re}\{Z\} = \frac{\pi}{12\,\varepsilon_0}\rho_{\mathbf{p}}(\mathbf{r}_0, \omega), \tag{13.42}$$

and hence an equivalence of LDOS and Re{Z}. The units of Re{Z} are ohms per area instead of the usual ohms. Notice that Z depends both on the location $\mathbf{r}_0$ and on the orientation $\mathbf{n_p}$ of the receiving or transmitting dipole. As discussed by Greffet *et al.*, the imaginary part of Z accounts for the energy stored in the near-field (reactive power).

13.3.2 Antenna-coupled light–matter interactions

In free space, the momentum of a photon with energy $E = \hbar\omega$ is $p_{\rm ph} = \hbar\omega/c$. On the other hand, the momentum of an unbound electron with the same energy is $p_{\rm e} = (2m^*\hbar\omega)^{1/2}$, which is a factor of $[2\,m^*c^2/(\hbar\omega)]^{1/2} \approx 10^2$–$10^3$ larger than the photon momentum. Therefore, the photon momentum can be neglected in electronic transitions, i.e. optically excited transitions are vertical in an electronic band diagram. However, near optical antennas the photon momentum is no longer defined by its free-space value. Instead, the localized optical fields are associated with a broad momentum distribution whose bandwidth $p_{\rm ph} = \pi\hbar/\Delta$ is given by the spatial confinement Δ, which can be as small as 1–10 nm. Thus, in the optical near-field the photon momentum can be increased by a factor of $\lambda/\Delta \sim 100$, which brings it into the range of the electron momentum, especially in materials with small effective mass m^*. Hence, localized optical fields can give rise to "diagonal" transitions in an electronic band diagram, thereby increasing the overall absorption strength represented by Im$\{\alpha\}$. The increase of photon momentum in optical near-fields has been discussed in the context of photoelectron emission [29] and photoluminescence [30].

The strong field confinement near optical antennas also has implications on selection rules in atomic or molecular systems. The light–matter interaction involves matrix elements of the form $\langle f|\hat{\mathbf{p}}\cdot\hat{\mathbf{A}}|i\rangle$, with $\hat{\mathbf{p}}$ and $\hat{\mathbf{A}}$ being the momentum and field operators, respectively (c.f. Eq. (8.39)). As long as the quantum wavefunctions of states $|i\rangle$ and $|f\rangle$ are much smaller than the spatial extent over which $\hat{\mathbf{A}}$ varies, it is legitimate to pull $\hat{\mathbf{A}}$ out of the matrix element. The remaining expression $\langle f|\hat{\mathbf{p}}|i\rangle$ is what defines the dipole approximation and leads to standard dipole selection rules. However, the localized fields near optical antennas give rise to spatial variations of $\hat{\mathbf{A}}$ of a few nanometers and hence it might no longer be legitimate to invoke the dipole approximation. This is especially the case in semiconductor nanostructures, where the low effective mass gives rise to quantum orbitals with large spatial extent. In situations where the field confinement becomes comparable to quantum confinement it is possible to expand the light–matter interaction in a multipole series (see Section 8.1). Theoretical studies have shown that higher-order multipoles have different selection rules [31]. Additional "transition channels" are opened up in near-field interactions, which can be exploited for boosting the sensitivity of photodetection. Once the field confinement becomes stronger than the quantum confinement, the multipole series no longer converges and transition rates are solely defined by the local overlap of ground-state and excited-state wavefunctions. In this limit, an optical antenna can be used to spatially map out the quantum wavefunctions, providing direct optical images of atomic orbitals. However, this would require antennas with field confinements of better than 1 nm.

13.3.3 Coupled-dipole antennas

One of the simplest antenna geometries is the optical half-wave antenna consisting of a metal rod whose length Δl is resonant with the incident wavelength λ. As discussed before, the lowest resonance is reached when $\Delta l = \lambda_{\text{eff}}/2$, which is generally smaller than $\lambda/2$ by a factor of 2–5. Additional degrees of freedom arise when two linear antenna elements are coupled end-to-end, as shown in Fig. 13.10. We will call this geometry a *coupled-dipole antenna*. It is also referred to as a gap antenna. The antenna resonance can be adjusted by altering the length of the antenna segments, the gap size, and also the material in the gap. It has been shown that the gap properties sensitively influence the antenna impedance and its radiation efficiency [23]. In the following we will analyze coupled-dipole antennas from a perspective of *mode hybridization* (c.f. Fig. 12.25).

When an external field irradiates the coupled-dipole antenna, it creates oscillating surface charges on the two antenna elements. Each element can be thought of as a spring with a respective effective mass attached to it [32]. The induced charges give rise to Coulomb interactions between the two elements, which can be accounted for by a spring coupling the two elements together. We thus end up with a mechanical analog consisting of two coupled harmonic oscillators, as discussed in Section 8.7.1.

Figure 13.11(a) illustrates the coupled harmonic-oscillator system. The coupling of the two antenna arms results in the appearance of two eigenmodes of different frequencies. One eigenmode exhibits in-phase oscillation of the two springs, while the other eigenmode

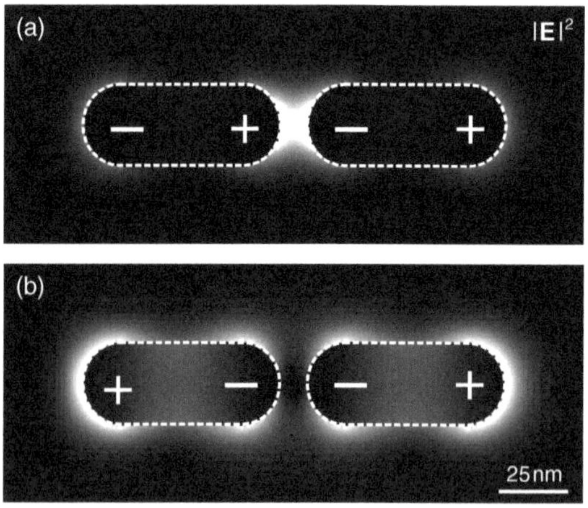

Fig. 13.10 Normalized near-field intensity-enhancement eigenmode patterns of the bonding (upper panel) and the antibonding (lower panel) antenna mode for an antenna consisting of two cylindrical gold nanowires with spherical end caps in vacuum. The intensity in a plane including the long wire axis is shown. The plus and minus signs indicate the momentary surface charge distribution on each nanowire. Note the strong field localization in the feed-gap for the bonding-mode resonance, while for the antibonding-mode resonance there is a node line in the feed-gap and strong enhancement at the rod ends.

is characterized by an anti-phase oscillation. In the former case, the interaction spring shifts the resonance to lower frequencies, whereas in the latter case the interaction shifts the resonance to higher frequencies. This simple and intuitive classical model contains the most characteristic features of strongly coupled systems. However, for the electromagnetically coupled antenna arms the coupling strength is not the same for the two modes and depends on the gapwidth. This difference is due to the different charge distributions of the two modes (see Fig. 13.10). For the *bonding mode*, the presence of opposite charges on either side of the gap generates an attractive force and hence an increase of the overall surface charge density. The opposite is the case for the *antibonding mode*. Therefore the coupling strength increases for the in-phase oscillation while it slightly decreases for the out-of-phase mode. The modification of the coupling strength leads to a characteristic redshift of the in-phase mode and a small blueshift of the anti-phase mode.

Figure 13.11(b) illustrates the dependence of the mode splitting on the interaction strength. The latter increases with decreasing gapwidth. As a consequence, the splitting between the bonding-mode resonance and the antibonding-mode resonance becomes larger when the gap is reduced from 16 nm to 6 nm, as shown in Fig. 13.11(b). The antibonding resonance can be described in terms of two counter-oscillating dipoles, which is characteristic for a *dark mode*. The coupling to the radiation field is strongly suppressed for a dark mode, and the reduced radiation rate results in a narrow linewidth (sharp resonance). On the other hand, the bonding mode defines a *bright mode* with a wider linewidth. Figure 13.10 shows the near-fields of the two eigenmodes. The + and − signs symbolize the momentary charge distribution. For the bonding mode, opposite charges at the gap give

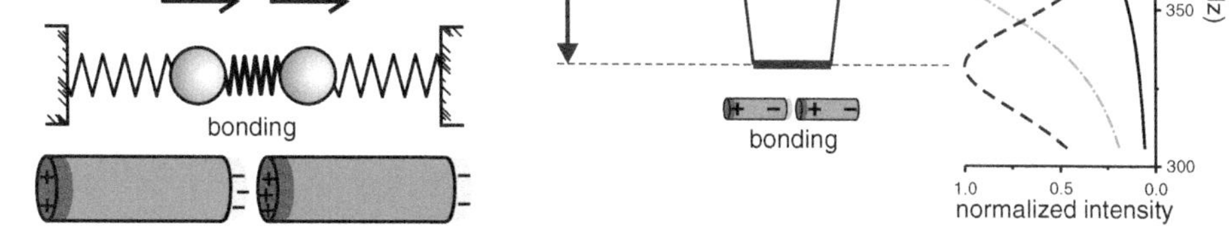

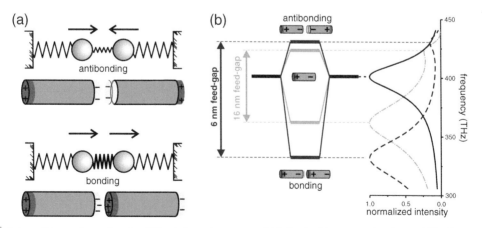

Fig. 13.11 Interparticle coupling and mode splitting in two-wire antennas. (a) A sketch of the mass-and-spring model for the coupling between two plasmonic oscillators. Upper panel, antibonding mode; lower panel, bonding mode. (b) The energy-level diagram and simulated near-field intensity spectra for 30-nm-high, 50-nm-wide, and 110-nm-long (arm length) symmetric two-wire gold antennas with gaps of 6 nm (black dashed line) and 16 nm (gray dash–dotted line), as well as for a single-wire antenna with the same dimensions (black solid line).

rise to strongly enhanced fields, whereas for the antibonding mode the field is suppressed due to the existence of equal charges at the gap.

Alù and Engheta have shown that the gap between the two antenna arms can be modeled as a load impedance $Z_{\rm load}$ [23] and that the load can be tuned by filling the gap with suitable materials. For a gap in the form of a disk of height t, radius R, and dielectric constant $\varepsilon_{\rm gap}$, the load impedance is calculated as

$$Z_{\rm load} = \mathrm{i}\frac{t}{\pi\omega R^2\varepsilon_{\rm gap}}. \tag{13.43}$$

Figure 13.12 shows the effect of such antenna loading. The antenna resonance shifts depending on $\varepsilon_{\rm gap}$ of the gap. Note that if the gap material is the same as that of the antenna arms we obtain a "half-wave" antenna; that is, the gap becomes perfectly impedance-matched. The scenario is similar to a traditional radiowave $\lambda/2$ antenna, which consists of two segments of length $\lambda/4$ separated by a tiny feed-gap connected to an impedance-matched transmission line ($Z = 73 + \mathrm{i}42\,\Omega$) that supplies the antenna with current. The perturbation introduced by the feed-gap is essentially eliminated by impedance matching and discontinuity of electrical current across the gap is largely eliminated.

To conclude this section we note that the loading of a coupled-dipole antenna with a nonlinear material is of interest for applications such as optical switching. Theoretical studies have shown that optical bistability can be achieved with a gap material having a Kerr nonlinear coefficient of $n_2 \approx 10^{-12}\,\mathrm{cm^2\,W}$ and with threshold intensities of $I < 1\,\mathrm{GW\,cm^{-2}}$ [33]. The threshold intensity can be reduced by orders of magnitude by choosing a nonlinear material with gain.

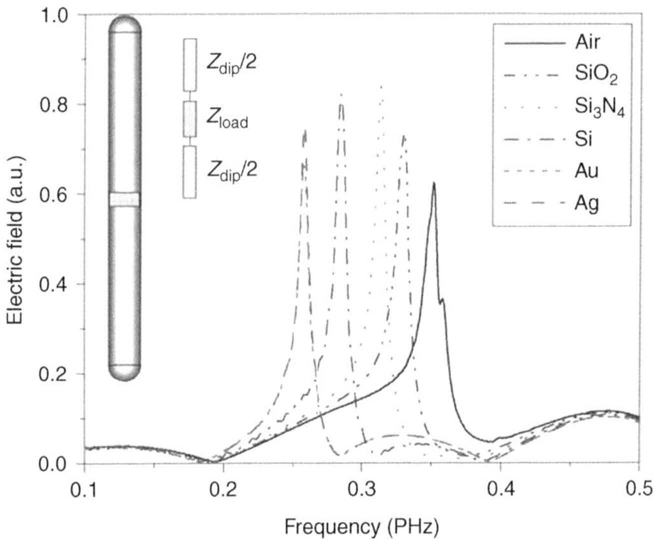

Fig. 13.12 Loading of a coupled-dipole antenna. The antenna response can be tuned by the gap material and the gap can be modeled as a load impedance $Z_{\rm load}$. From [23].

13.4 Quantum emitter coupled to an antenna

Radiofrequency antennas are connected to transmission lines that either drive the antenna or collect the received signal. Signal transduction can be optimized by proper impedance matching. At optical frequencies, however, antennas are typically used as isolated elements and a description in terms of circuit theory is not straightforward. As discussed before, the emission and absorption of light by a discrete quantum system can be described in terms of the local density of states (LDOS), or simply by the Green function, and here we use this concept to understand the interaction of a quantum emitter with an optical antenna.

While the spontaneous decay rate of a quantum emitter is related to Einstein's A coefficient, stimulated emission is described by Einstein's B coefficient. In a homogeneous medium, the two coefficients are proportional and the proportionality constant depends solely on the frequency and on the index of refraction. Various laser parameters depend on the product of the A and B coefficients. Examples are the saturation intensity $I_{\rm sat}$ and the gain coefficient g. By engineering the local environment of quantum emitters we may gain access to controlling different laser parameters.

Let us consider a single quantum emitter interacting with an optical antenna. As shown in Fig. 13.13, we model the quantum emitter as a four-level system, with state $|1\rangle$ being the ground state. External laser radiation controls the transitions between states $|1\rangle$ and $|2\rangle$ and between states $|3\rangle$ and $|4\rangle$. An external "pump" laser excites the system from the ground

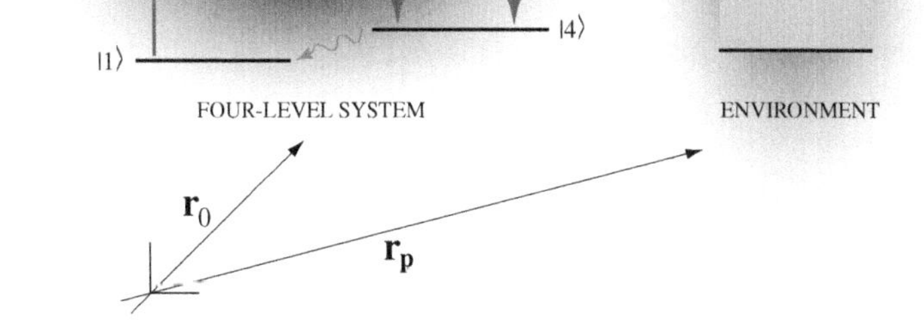

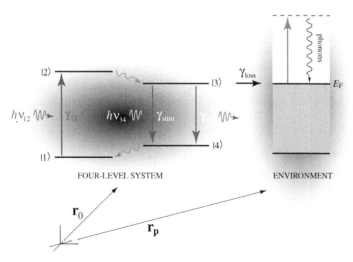

Fig. 13.13 Schematic energy-level diagram of a four-level quantum system coupled to its local environment, such as an optical antenna. The environment enhances the local field, thereby increasing the pump rate γ_{12} and the stimulated-emission rate $\gamma_{\rm stim}$. The local environment also influences the local density of states $\rho(\mathbf{r}_0, \omega_{34})$, thereby affecting the balance between the different rates.

state $|1\rangle$ to the excited state $|2\rangle$ with a rate of γ_{pump}, and a "depletion" laser forces state $|3\rangle$ to relax to state $|4\rangle$ by stimulated emission.

For simplicity, we assume that the relaxation from level $|2\rangle$ to level $|3\rangle$ is fast (internal conversion, vibrational relaxation) and that it isn't influenced by the lasers and the local environment. We also assume that the physical dimensions (orbitals) of the four-level system are much smaller than the length scale of light confinement imposed by the local environment. In this regime, all transitions are treated in the dipole approximation and the system can be considered to be point-like.

Let us first consider the excitation rate γ_{12} from state $|1\rangle$ to state $|2\rangle$. We restrict our analysis to the case of weak excitation. In this limit, the system resides predominantly in the ground state and the dynamics is described by first-order perturbation theory; excited-state saturation and Rabi flopping are outside of this framework. γ_{12} can then be represented in terms of the laser frequency ω_{12} and the local excitation field $\mathbf{E}$ at $\mathbf{r}_0$ as

$$\gamma_{12} = 3\frac{\sigma_{12}}{\hbar\omega_{12}} I_{12} = \frac{3}{2}\frac{\varepsilon_0 c\sigma_{12}}{\hbar\omega_{12}} |\mathbf{n}_{12} \cdot \mathbf{E}(\mathbf{r}_0, \omega_{12})|^2 . \tag{13.44}$$

Here, σ_{12} is the absorption cross-section and $\mathbf{n}_{12}$ is the unit vector defined by the transition dipole moment $\langle 1|\hat{\mathbf{p}}|2\rangle$. The factor 3 appearing in Eq. (13.44) compensates for the orientational averaging of σ_{12}. The local field can be written as $\mathbf{E} = (1 + \overleftrightarrow{\mathbf{f}}_{12})\mathbf{E}_0$, with $\overleftrightarrow{\mathbf{f}}_{12}$ being the local field enhancement factor. For sufficiently strong $\overleftrightarrow{\mathbf{f}}$, the *effective* absorption cross-section, which relates to the incident field, can be written as $\sigma_{\text{eff}} \approx \sigma_{12}|\overleftrightarrow{\mathbf{f}}_{12} \cdot \mathbf{E}_0|^2/|\mathbf{E}_0|^2$. The field-enhancement factor depends on the properties of the local environment and on the frequency. It has to be derived from rigorous field calculations or near-field measurements. As an example consider the field distributions plotted in Fig. 13.10(a) showing that a single emitter placed in the gap between the two nanowires would experience the largest effective absorption cross-section.

After excitation, the system relaxes rapidly from state $|2\rangle$ to state $|3\rangle$. Once in state $|3\rangle$, the system relaxes to state $|4\rangle$ by different decay mechanisms, by stimulated emission, by spontaneous emission, or by non-radiative energy transfer to the antenna material or local environment. The latter pathway is usually associated with non-radiative losses leading to fluorescence quenching. The transition rate γ_{34} between states $|3\rangle$ and $|4\rangle$ can be written as

$$\gamma_{34} = 1/\tau + \gamma_{\text{stim}}, \tag{13.45}$$

where τ is the lifetime of state $|3\rangle$ and γ_{stim} is the stimulated emission rate. Similarly to Eq. (13.44), the stimulated emission rate is related to the local field at frequency ω_{34} as

$$\gamma_{\text{stim}} = 3\frac{\sigma_{34}}{\hbar\omega_{34}} I_{34} = \frac{3}{2}\frac{\varepsilon_0 c\sigma_{34}}{\hbar\omega_{34}} |\mathbf{n}_{34} \cdot \mathbf{E}(\mathbf{r}_0, \omega_{34})|^2 , \tag{13.46}$$

where σ_{34} is the cross-section for stimulated emission.

Within the validity of Fermi's golden rule (see Appendix B), the excited-state lifetime τ in Eq. (13.45) is derived from the partial local density of electromagnetic states $\rho(\mathbf{r}_0, \omega_{34})$ as (see Section 8.4)

$$\frac{1}{\tau} = \gamma_{\text{rad}} + \gamma_{\text{loss}} = \gamma_{\text{rad}}^{\text{o}} \frac{2\pi c^3}{\omega_{34}^2} \rho(\mathbf{r}_0, \omega_{34}), \tag{13.47}$$

where $\gamma_{\rm rad}^{\rm o}$ is the free-space decay rate (c.f. Eq. (8.121)) and $\gamma_{\rm loss}$ is the rate of energy transfer to the local environment. The local density of states is calculated via the Green function according to (c.f. Eq. (8.115))

$$\rho(\mathbf{r}_0, \omega_{34}) = \frac{6\,\omega_{34}}{\pi c^2}\left[\mathbf{n}_{34} \cdot \mathrm{Im}\left\{\overset{\leftrightarrow}{\mathbf{G}}(\mathbf{r}_0, \mathbf{r}_0; \omega_{34})\right\} \cdot \mathbf{n}_{34}\right] \tag{13.48}$$

and can be calculated by simply replacing the molecule by a classical dipole and evaluating the fields at the origin. Evidently, these fields are influenced by the local environment.

We have now established the procedure for calculating the transition rates γ_{12} and γ_{34}. The rates require the calculation of (1) the local excitation field, (2) the local depletion field, and (3) the local density of states. If the objective is to enhance the emission of photons, then the ratio $\gamma_{\rm rad}/\gamma_{\rm loss}$ needs to be maximized. It has been demonstrated that $\gamma_{\rm loss}$ becomes dominant for very short separations between the quantum emitter and material boundaries [17] and hence a minimum distance between emitter and material boundaries of the antenna needs to be ensured.

Because of the short-range energy-transfer dependence, $\gamma_{\rm loss}$ can be estimated by ignoring the curvature of the material boundaries. In this case, the quantum emitter interacts with its mirror image and the energy-transfer rate can be expressed as [17]

$$\gamma_{\rm loss} = \gamma_{\rm rad}^{\rm o}\,\frac{3}{16}\,\mathrm{Im}\left\{\frac{\varepsilon - 1}{\varepsilon + 1}\right\}\frac{1}{(k_{34}\,z)^3}\,[\mathbf{n}_{34} \cdot \mathbf{n}_{\parallel} + 2\mathbf{n}_{34} \cdot \mathbf{n}_{\perp}], \tag{13.49}$$

where z is the distance between the molecule and the antenna material's surface, $k_{34} = \omega_{34}/c$, and ε is the material's dielectric function at frequency ω_{34}. The last term in brackets is an orientational factor, with $\mathbf{n}_{\parallel}$ and $\mathbf{n}_{\perp}$ denoting normal vectors parallel and perpendicular to the material surface, respectively. For emitters in the gap between two interfaces the expression of Eq. (13.49) has to be adapted accordingly.

In the situation studied here, the processes $|1\rangle \rightarrow |2\rangle$ and $|3\rangle \rightarrow |4\rangle$ are decoupled. Therefore, the photoemission rate can be represented as

$$\gamma_{\rm em} = \gamma_{12}(1 - \gamma_{\rm loss}/\gamma_{34}), \tag{13.50}$$

where the term in brackets denotes the probability of transitioning from state $|3\rangle$ to state $|4\rangle$ radiatively, i.e. by emitting a photon. Figure 13.14 shows the enhanced emission of photons by a single quantum emitter placed near a gold nanoparticle antenna. For distances of $z \approx 5$ nm the photon-emission rate can be enhanced sevenfold, but for shorter distances the photon-emission rate drops rapidly because of energy transfer according to Eq. (13.49).

Notice that we assumed a quantum emitter with an intrinsic quantum yield of unity, which means that there is no intrinsic non-radiative decay. Hence, the local environment can only decrease the emitter's quantum yield. The photoluminescence enhancement shown in Fig. 13.14 is therefore a consequence of excitation-rate enhancement according to the first term in Eq. (13.50). If, however, the quantum emitter has a low intrinsic quantum yield, the local environment is also able to enhance the quantum yield and bring out a very strong photoluminescence enhancement. For example, a gold nanoparticle has been used to enhance the photoluminescence from rare-earth ions by two orders of magnitude [35], and a gold bowtie antenna has been employed to enhance the photoemission of a near-infrared dye by three orders of magnitude [19]. To account for a finite intrinsic quantum yield we need to include an additional term γ_{int} in Eq. (13.45).

Laser parameters, such as the saturation intensity I_{sat} and the laser gain coefficient g, depend on the balance of spontaneous- and stimulated-emission rates. Spontaneous emission is characterized by the excited-state lifetime τ, whereas stimulated emission is represented by the effective cross-section

$$\sigma_{\text{eff}} = \sigma_{34} \left| \frac{\mathbf{n}_{34} \cdot [1 + \overleftrightarrow{\mathbf{f}}_{34}]\mathbf{E}_0(\mathbf{r}_0, \omega_{34})}{\mathbf{n}_{34} \cdot \mathbf{E}_0(\mathbf{r}_0, \omega_{34})} \right|^2 . \tag{13.51}$$

Here, $\mathbf{E}_0$ is the incident field and $\overleftrightarrow{\mathbf{f}}_{34}$ is the local field-enhancement factor. In terms of τ and σ_{eff} the saturation intensity and the laser gain coefficient are defined as

$$I_{\text{sat}} = \frac{\hbar\omega_{34}}{\tau\sigma_{\text{eff}}}, \quad g \propto \gamma_{12}\frac{\tau\sigma_{\text{eff}}}{1 + I/I_{\text{sat}}}. \tag{13.52}$$

It is evident that the product $\tau\sigma_{\text{eff}}$ is of central importance. Hence, favorably engineered optical antennas and local environments in general could provide access to novel laser host materials.

13.5 Quantum yield enhancement

The excitation energy of a molecule or any other quantum system can be dissipated either radiatively or non-radiatively. Radiative relaxation is associated with the emission of a photon, whereas non-radiative relaxation can have various pathways, such as coupling to

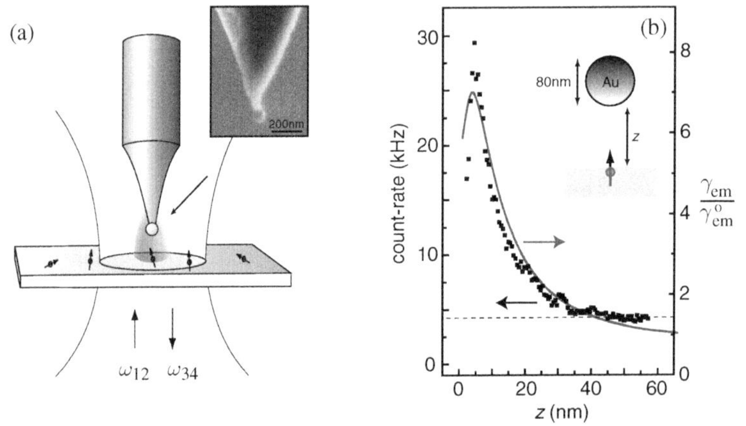

Fig. 13.14 Photoluminescence enhancement by an 80 nm gold particle. (a) Illustration of the experiment. Inset: SEM image of a gold particle attached to an optical fiber. (b) The photoluminescence rate as a function of the particle–emitter distance z (solid curve; theory; dots; experiment). The horizontal dashed line indicates the background level. From Ref. [34].

vibrations, energy transfer to the environment, or quenching by other molecules. Often it is desired to generate conditions that maximize the radiative output of a quantum system. A useful measure for this output is the *quantum yield* defined as

$$Q = \frac{\gamma_{\rm rad}}{\gamma_{\rm rad} + \gamma_{\rm nr}}, \tag{13.53}$$

where $\gamma_{\rm rad}$ and $\gamma_{\rm nr}$ are the radiative and non-radiative decay rates, respectively. Notice that Q is formally equivalent to the antenna efficiency $\eta_{\rm rad}$ (c.f. Eq. (13.9)). We simply have to express "powers" in terms of "rates." In a homogeneous environment, Q is identical to the intrinsic quantum yield $q_{\rm i}$ defined in Section 8.5.1. However, $\gamma_{\rm rad}$ and $\gamma_{\rm nr}$ are functions of the local environment and thus are affected by the presence of an optical antenna.

To determine the quantum yield in a particular environment, it is necessary to divide the total decay rate in Eq. (8.141) into a radiative and a non-radiative part:

$$\gamma = \gamma_{\rm rad} + \gamma_{\rm nr}. \tag{13.54}$$

The two contributions can be determined by calculating the balance between radiation emitted to the far-field, $P_{\rm rad}$, and radiation absorbed in the environment, $P_{\rm abs}$.

Let us denote the decay rate of an excited molecule in free space as $\gamma^{\rm o} = \gamma^{\rm o}_{\rm rad} + \gamma^{\rm o}_{\rm nr}$. Notice that $\gamma^{\rm o}_{\rm nr}$ accounts for non-radiative losses *inside* the molecule only, i.e. it is an intrinsic molecular property. The intrinsic quantum yield of the isolated molecule is defined as $q_{\rm i} = \gamma^{\rm o}_{\rm rad}/(\gamma^{\rm o}_{\rm rad} + \gamma^{\rm o}_{\rm nr})$. The interaction of the molecule with its local environment introduces an additional non-radiative rate $\gamma_{\rm loss}$, thereby modifying the quantum yield to $Q = \gamma_{\rm rad}/(\gamma_{\rm rad} + \gamma^{\rm o}_{\rm nr} + \gamma_{\rm loss})$. Using the definition of $q_{\rm i}$, this can be recast as

$$Q = \frac{\gamma_{\rm rad}/\gamma^{\rm o}_{\rm rad}}{\gamma_{\rm rad}/\gamma^{\rm o}_{\rm rad} + \gamma_{\rm loss}/\gamma^{\rm o}_{\rm rad} + (1 - q_{\rm i})/q_{\rm i}}. \tag{13.55}$$

Here, $\gamma_{\rm rad}$ is the radiative rate in the presence of the optical antenna. We assumed that the antenna does not influence the intrinsic non-radiative rate $\gamma^{\rm o}_{\rm nr}$. Therefore, $\gamma_{\rm nr} = \gamma^{\rm o}_{\rm nr} + \gamma_{\rm loss}$. Note that (13.55) is identical to Eq. (13.11) if we replace all rates γ with corresponding powers P.

To understand the significance of the intrinsic quantum yield $q_{\rm i}$ we consider a quantum emitter placed at a variable distance from an optical antenna. For large separations between emitter and antenna we have $\gamma_{\rm loss} \to 0$ and $\gamma_{\rm rad} \to \gamma^{\rm o}_{\rm rad}$, and hence $Q = q_{\rm i}$. On the other hand, for an emitter with high intrinsic quantum yield ($q_{\rm i} = 1$) we obtain $Q = \gamma_{\rm rad}/(\gamma_{\rm rad} + \gamma_{\rm loss})$ and hence $\gamma_{\rm loss}$ is the only non-radiative decay channel. By bringing the emitter close to the particle we increase $\gamma_{\rm loss}$ and therefore lower the quantum yield Q. A 100%-efficient emitter simply cannot be made more efficient. The situation is different, however, for an emitter with a low $q_{\rm i}$. Here, the local environment, such as a single nanoparticle, can increase a molecule's quantum efficiency, an effect that was observed in 1983 by Wokaun and coworkers [36]. The distance between emitter and particle is very critical. For distances that are too large there is no interaction between emitter and particle, and for distances that are too small all the energy is dissipated into heat.

As an illustration for quantum yield enhancement we consider the situation shown in Fig. 13.14, where a single molecule interacts with a gold particle of radius a and dielectric

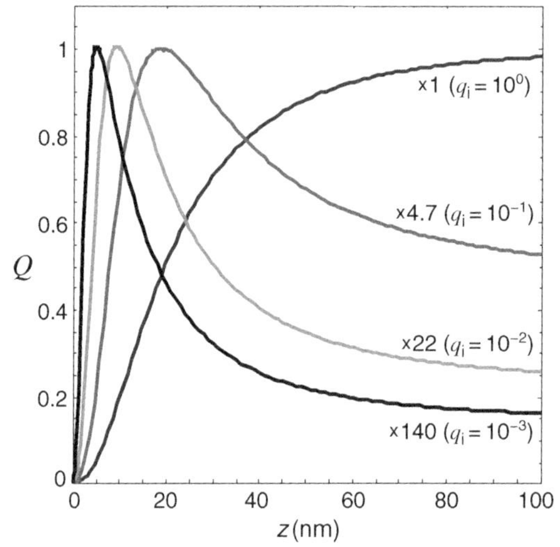

Fig. 13.15 Quantum yield as a function of the separation between a gold nanoparticle and a molecule with different q_i. The lower q_i is, the higher the quantum yield enhancement can be. The curves are scaled to the same maximum value. Here $\lambda = 650$ nm and $a = 40$ nm.

constant ε. If we represent the molecule by a radiating dipole $\mathbf{p}$ pointing in the direction of the particle's origin then the normalized radiative rate can be calculated as [17]

$$\frac{\gamma_{\text{rad}}}{\gamma_r^o} = \frac{|\mathbf{p}+\mathbf{p}_{\text{ind}}|^2}{|\mathbf{p}|^2} = \left|1 + 2\,\frac{\varepsilon-1}{\varepsilon+2}\,\frac{a^3}{(a+z)^3}\right|^2 , \tag{13.56}$$

where $\mathbf{p}_{\text{ind}}$ is the induced dipole in the gold particle and z is the distance between the molecule and the surface of the particle. On the other hand, the energy dissipation in the gold particle can be calculated according to Eq. (13.49). In Fig. 13.15 we plot the quantum yield Q according Eq. (13.55) as a function of the molecule–particle distance z and for different values of the intrinsic quantum yield q_i. The gold particle enhances the quantum yield of a molecule with $q_i \sim 0.001$ by roughly a factor of 10. Much higher enhancements are possible for favorably designed particles, such as two-wire and bowtie optical antennas; see Figs. 13.5(b) and (d).

Notice that, while Eq. (13.56) has a term $\varepsilon+2$ in the denominator, Eq. (13.49) features a term $\varepsilon+1$. The origin of this difference is that the particle scatters radiation as a sphere, but it dissipates power as if it were an infinitely extended plane. For optimum quantum yield enhancement it is therefore favorable to operate near the $\varepsilon = -2$ resonance as opposed to the $\varepsilon = -1$ resonance. For a dielectric function modeled by a Drude-type electron gas (see Chapter 12) the emission wavelength of the molecule therefore needs to be red-detuned from the particle's surface plasmon resonance.

In the case of weak excitation, the number of emitted photons is proportional to the product of the excitation rate and the quantum yield, $\gamma_{12} \cdot Q$. If in the presence of an antenna the reduction of the quantum yield is larger than the increase in the excitation rate then the number of emitted photons per unit time is reduced (see Fig. 13.14 for the case of small separations). If such a coupled system is driven into saturation the number of photons emitted per unit time is determined by the saturation emission rate, which is proportional to the product of the enhanced radiative decay rate and the quantum yield. Consequently, in the low-excitation-rate regime it is more favorable to tune the antenna resonance to the absorption peak of the quantum emitter, whereas in the saturated regime the antenna should be tuned to the wavelength at which the emission is at a maximum.

13.6 Conclusion

In this chapter we summarized the basic principles of *optical antennas*. The chapter defines concepts and terminology, but many aspects and applications have been left out. The field of *optical antennas* is still in its infancy, and new ideas and developments are emerging at a rapid pace. Today, the building blocks for optical antennas are plasmonic nanostructures fabricated from the bottom up by colloidal chemistry or top-down with nanofabrication techniques, such as electron-beam lithography and focused-ion-beam milling. It is conceivable that future optical antenna designs will draw inspiration from biological systems, such as light-harvesting proteins and use molecular systems as antenna building blocks.

Problems

13.1 Consider an antenna in the simple form of a spherical nanoparticle with radius a and dielectric constant ε. This antenna is used to enhance the radiation properties of a dipole quantum emitter as described in Section 13.4. For simplicity we assume that the dipole moment $\mathbf{p}$ is pointing in the direction of the nanoparticle's origin and that the interaction between emitter and nanoparticle can be treated in the quasi-static limit. The distance of the dipole from the particle's surface is z.

(a) The nanoparticle is irradiated by a plane wave with electric field $\mathbf{E}_0$. The electric field vector is parallel to the dipole $\mathbf{p}$. Treat the nanoparticle as a polarizable sphere with polarizability α and determine the field $\mathbf{E}$ at the location of the emitter. From this, determine the normalized excitation rate $\gamma_{\rm exc}/\gamma^{\rm o}_{\rm exc}$, where $\gamma^{\rm o}_{\rm exc}$ refers to the absence of the nanoparticle.

(b) Assume now that the emitter is radiating and that it induces a dipole $\mathbf{p}_{\rm ind}$ in the nanoparticle. The radiated power is then proportional to $|\mathbf{p}+\mathbf{p}_{\rm ind}|^2$. Calculate the normalized radiative rate $\gamma_{\rm rad}/\gamma^{\rm o}_{\rm rad}$, where $\gamma^{\rm o}_{\rm rad}$ refers to the absence of the nanoparticle.

(c) On the basis of the previous two results, discuss to what extent the directivity $D(\theta,\phi)$ of the dipole's emission is influenced by the nanoparticle.

(d) Using Eq. (13.49), determine the normalized non-radiative rate $\gamma_{\rm loss}/\gamma^{\rm o}_{\rm rad}$.

(e) Assume that $q_{\rm i}=1$ and calculate the quantum yield $Q(z)$.

(f) Plot the normalized fluorescence emission rate $\gamma_{\text{em}}/\gamma_{\text{em}}^{\text{o}}$ as a function of z. Assume that the particle is made of gold and that $a = 80\,\text{nm}$. The wavelength of excitation and emission is $\lambda = 650\,\text{nm}$. What is the maximum enhancement?

(g) Determine the density of states $\rho_z(z)$ and discuss the different terms.

13.2 Assume that the equivalent circuit of an optical antenna is an *RLC* series resonant circuit. Assume that in the limit of very small dimensions the capacitance scales as $C(x) = \varepsilon_0 A/d = \varepsilon_0 x$, where x is a characteristic size of the antenna. Similarly, the inductivity is assumed to be given by $L = \mu_0 x$. Show that, assuming small ohmic and radiation damping, the characteristic size x is on the order of $\lambda/(2\pi)$. Furthermore, show that the quality factor of the resonance scales as the wave impedance of free space divided by the sum of ohmic and radiation resistance.

13.3 Consider a side-by-side aligned nano-rod dimer. Using the plasmon hybridization model find the four fundamental eigenmodes of such a system considering that the dimer according to its symmetry can be excited by two independent directions of polarizations. Classify the modes with respect to dark and bright resonances.

13.4 To understand the principle of a multi-element antenna, such as a Yagi–Uda antenna, we consider three dipoles $\mathbf{p}_1$, $\mathbf{p}_2$, and $\mathbf{p}_3$ placed on an axis x. As shown in the figure below, the dipoles are parallel to each other and oriented perpendicular to the x axis. $\mathbf{p}_1$ is a driven dipole (the feed element) and is placed at $x = 0$. $\mathbf{p}_2$ is a parasitic element and is intended to operate as a director, whereas $\mathbf{p}_3$ is also a parasitic element but intended to work as a reflector.

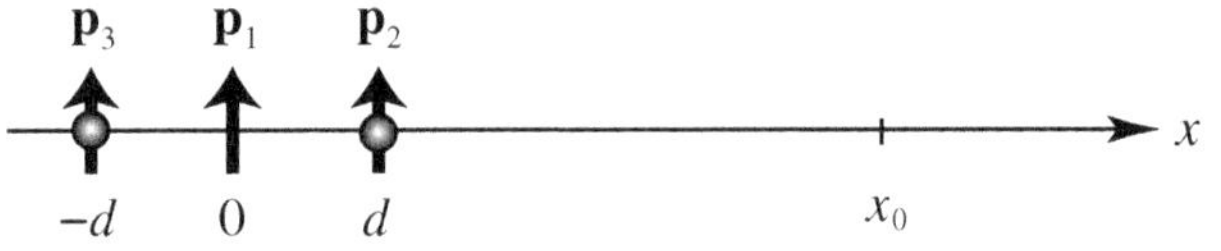

(a) Use the Green function $\overleftrightarrow{\mathbf{G}}$ defined in Section 8.3.1 and evaluate the phase $\phi(x)$ of the field $\mathbf{E}_1$ of dipole $\mathbf{p}_1$. Plot ϕ as a function of x/λ, where λ is the wavelength of radiation. Hint: the phase at $x = 0$ is $\phi(0) = 180°$.

(b) Find the distance of the first extremum of ϕ, i.e. determine the distance $x = d$ for which the phase assumes a first maximum or minimum.

We want that the fields from the different dipoles are in-phase at a distant point $x = x_0 \gg \lambda$, that is, that they constructively interfere in the forward direction. Since $\mathbf{p}_1$ is the only driven dipole, we require that the phase along different paths

$$\phi_{0\to x_0} = \int_0^{x_0} \frac{\mathrm{d}\phi}{\mathrm{d}x}\,\mathrm{d}x \tag{13.57}$$

is the same (within multiples of 2π).

(c) Let us consider the direct path (path 1) from 0 to x_0 and an indirect path (path 2) via the director dipole. The phase along the latter can be written as $\phi_{0\to x_0} = \phi_{0\to d} + \phi_{\text{interaction}} + \phi_{d\to x_0}$, where $\phi_{\text{interaction}}$ is the phase due to the interaction with the director dipole. Assume that $d = \lambda/5$ and calculate the value of $\phi_{\text{interaction}}$ that is required to make the phase along paths 1 and 2 equal.

(d) Repeat the above but for a path (path 3) via the reflector dipole instead of the director dipole. What is the required interaction phase $\phi_{\text{interaction}}$?

(e) $\phi_{\text{interaction}}$ is directly related to the polarizability α of the director or reflector. Assume that α is a Lorentzian function and discuss what is the best choice for the center frequencies both for the director and for the reflector.

References

[1] P. Bharadwaj, B. Deutsch, and L. Novotny, "Optical antennas," *Adv. Opt. Photonics* **1**, 438–483 (2009).

[2] S. Walter, E. Bolmont, and A. Coret, "La correspondance entre Henri Poincaré et les physiciens, chimistes et ingénieurs," in *Publications of the Henri Poincaré Archives*, Chapter 8. Basel: Birkhäuser (2007).

[3] R. Feynman. "There's plenty of room at the bottom," *Caltech Eng. Sci.* **23**(5) 22–36 (1960).

[4] D. W. Pohl, "Near field optics seen as an antenna problem," in *Near-field Optics: Principles and Applications – The Second Asia–Pacific Workshop on Near Field Optics* (1999).

[5] O. Keller, "Near-field optics: the nightmare of the photon," *J. Chem. Phys.* **112**, 7856–7863 (2000).

[6] B. Lounis and M. Orrit, "Single-photon sources," *Rep. Prog. Phys.*, **68**, 1129–1179 (2005).

[7] K. F. Lee, *Principles of Antenna Theory*. New York: John Wiley and Sons (1984).

[8] C. A. Balanis, *Antenna Theory: Analysis and Design*, 2nd edn. New York: John Wiley and Sons (1997).

[9] R. W. P. King and C. Harrison, *Antennas and Waves: A Modern Approach*. Cambridge, MA: MIT Press (1970).

[10] D. K. Cheng, *Field and Wave Electromagnetics*, 2nd edn. New York: Addison Wesley (1989).

[11] J. Alda, J. M. Rico-Garcia, J. M. Lopez-Alonso, and G. Boreman, "Optical antennas for nano-photonic applications," *Nanotechnology* **16**, S230–S234 (2005).

[12] A. G. Curto, G. Volpe, T. H. Taminiau, *et al.*, "Unidirectional emission of a quantum dot coupled to a nanoantenna," *Science* **329**, 930–933 (2010). Reprinted with permission from AAAS.

[13] H. Gersen, M. F. García-Parajó, L. Novotny, J. A. Veerman, L. Kuipers, and N. F. van Hulst, "Influencing the angular emission of a single molecule," *Phys. Rev. Lett.* **85**, 5312–5315 (2000).

[14] L. J. Chu, "Physical limitations of omni-directional antennas," *Appl. Phys.* **19**, 1163–1175 (1948).

[15] M. Gustafsson, C. Sohl and G. Kristensson, "Physical limitations on antennas of arbitrary shape," *Proc. Roy. Soc. A* **463**, 2589–2607 (2007).

[16] K. T. Shimizu, W. K. Woo, B. R. Fisher, H. J. Eisler, and M. G. Bawendi, "Surface-enhanced emission from single semiconductor nanocrystals," *Phys. Rev. Lett.* **89**, 117401 (2002).

[17] P. Bharadwaj and L. Novotny, "Spectral dependence of single molecule fluorescence enhancement," *Opt. Express* **15**, 14266–14274 (2007).

[18] T. H. Taminiau, F. D. Stefani, and N. F. van Hulst, "Enhanced directional excitation and emission of single emitters by a nano-optical Yagi–Uda antenna," *Opt. Express* **16**, 10858-10866 (2008).

[19] A. Kinkhabwala, Z. Yu, S. Fan, *et al.*, "Large single-molecule fluorescence enhancements produced by a bowtie nanoantenna," *Nature Photonics* **3**, 654–657 (2009).

[20] J. A. Bean, B. Tiwari, G. H. Bernstein, P. Fay and W. Porod, "Thermal infrared detection using dipole antenna-coupled metal-oxide-metal diodes," *J. Vac. Sci. Technol. B* **27**, 11–14 (2009).

[21] H. T. Friis, "A note on a simple transmission formula," *Proc. IRE* **34**, 254–256 (1946).

[22] L. Novotny, "Effective wavelength scaling for optical antennas," *Phys. Rev. Lett.* **98**, 266802 (2007).

[23] A. Alù and N. Engheta, "Tuning the scattering response of optical nanoantennas with nanocircuit loads," *Nature Photonics* **2**, 307–310 (2008). Reprinted with permission from Macmillan Publishers Ltd.

[24] J. Dorfmüller, R. Vogelgesang, R. T. Weitz, *et al.*, "Fabry–Pérot resonances in one-dimensional plasmonic nanostructures," *Nano Lett.* **9**, 2372–2377 (2009).

[25] P. M. Krenz, R. L. Olmon, B. A. Lail, M. B. Raschke, and G. D. Boreman, "Near-field measurement of infrared coplanar strip transmission line attenuation and propagation constants," *Opt. Express* **18**, 21678–21686 (2010).

[26] L. Douillard, F. Charra, Z. Korczak, *et al.*, "Short range plasmon resonators probed by photoemission electron microscopy," *Nano Lett.* **8**, 935–940 (2008).

[27] J.-S. Huang, T. Feichtner, P. Biagioni, and B. Hecht, "Impedance matching and emission properties of optical antennas in a nanophotonic circuit," *Nano Lett.* **9**, 1897–1902 (2009).

[28] J.-J. Greffet, M. Laroche, and F. Marquier, "Impedance of a nanoantenna and a single quantum emitter," *Phys. Rev. Lett.* **105**, 117701 (2010).

[29] V. M. Shalaev,"Electromagnetic properties of small-particle composites," *Phys. Rep.* **272**, 61–137 (1996).

[30] M. R. Beversluis, A. Bouhelier, and L. Novotny, "Continum generation from single gold nanostructures through near-field mediated intraband transitions," *Phys. Rev. B* **68**, 115433 (2003).

[31] J. R. Zurita-Sanchez and L. Novotny, "Multipolar interband absorption in a semiconductor quantum dot. I. Electric quadrupole enhancement," *J. Opt. Soc. Am. B* **19**, 1355–1362 (2002); "Multipolar interband absorption in a semiconductor quantum dot. II. Magnetic dipole enhancement," *J. Opt. Soc. Am. B* **19**, 2722–2726 (2002).

[32] W. Rechberger, A. Hohenau, A. Leitner, *et al.*, "Optical properties of two interacting gold nanoparticles," *Opt. Commun.* **220**, 137–141 (2003).

[33] F. Zhou, Y. Liu, Z.-Y. Li, and Y. Xia, "Analytical model for optical bistability in nonlinear metal nano-antennae involving Kerr materials," *Opt. Express* **18**, 13337–13344 (2010).

[34] P. Anger, P. Bharadwaj, and L. Novotny, "Enhancement and quenching of single molecule fluorescence," *Phys. Rev. Lett.* **96**, 113002 (2006).

[35] P. Bharadwaj and L. Novotny, "Plasmon enhanced photoemission from a single $Y_3N@C_{80}$ fullerene," *J. Phys. Chem. C* **14**, 7444–7447 (2010).

[36] A. Wokaun, H.-P. Lutz, A. P. King, U. P. Wild, and R. R. Ernst, "Energy transfer in surface enhanced luminescence," *J. Chem. Phys.* **79**, 509–514 (1983).

14 Optical forces

As early as 1619 Johannes Kepler suggested that the mechanical effect of light might be responsible for the deflection of the tails of comets entering our Solar System. The classical Maxwell theory showed in 1873 that the radiation field carries with it momentum and that "light pressure" is exerted on illuminated objects. In 1905 Einstein introduced the concept of the photon and showed that energy transfer between light and matter occurs in discrete quanta. Momentum and energy conservation was found to be of great importance in microscopic events. Discrete momentum transfer between photons (X-rays) and other particles (electrons) was experimentally demonstrated by Compton in 1925 and the recoil momentum transferred from photons to atoms was observed by Frisch in 1933 [1]. Important studies on the action of photons on neutral atoms were carried out in the 1970s by Letokhov and other researchers in the USSR and by Ashkin's group at the Bell Laboratories in the USA. The latter group proposed bending and focusing of atomic beams and trapping of atoms in focused laser beams. Later work by Ashkin and coworkers led to the development of "optical tweezers." These devices allow optical trapping and manipulation of macroscopic particles and living cells with typical sizes in the range of 0.1–10 μm [2, 3]. Milliwatts of laser power produce piconewtons of force. Owing to the high field gradients of evanescent waves, stronger forces are to be expected in optical near-fields.

The idea that an object might cool through its interaction with the radiation field had been suggested in 1929 by Pringsheim [4]. However, the first proposal to cool atoms in counter-propagating laser beams was made by Hänsch and Schawlow in 1975 [5]. This proposal was the starting point for a series of exciting experiments that led to the 1997 Nobel Prize in physics. The mechanical force in laser trapping and cooling experiments can be understood on a semiclassical basis whereby the electromagnetic field is treated classically and the particle being trapped is treated as a quantized two-level system [6]. However, the quantum theory of photons is used for the correct interpretation of the results [7]. Furthermore, the photon concept asserts that there are quanta of energy and momentum transfer between the radiation field and the atom.

In this chapter we use classical electrodynamics to derive the conservation law for linear momentum in an optical field. The net force exerted on an arbitrary object is entirely determined by Maxwell's stress tensor. In the limiting case of an infinitely extended object, the formalism renders the known formulas for radiation pressure. Similarly, in the small-object limit, we obtain the familiar expressions for gradient and scattering forces. Using the expression for the atomic polarizability derived in Appendix A, it is possible to derive the forces acting on atoms and molecules in optical traps.

14.1 Maxwell's stress tensor

The general law for forces in electromagnetic fields is based on the conservation law for linear momentum. We therefore derive this conservation law in the following. Later we will discuss two different limits, the dipolar limit and the limit of the planar interface. For simplicity, we consider Maxwell's equations in vacuum. In this case we have $\mathbf{D} = \varepsilon_0 \mathbf{E}$ and $\mathbf{B} = \mu_0 \mathbf{H}$. Later we will relax this constraint. The conservation law for linear momentum is entirely a consequence of Maxwell's equations,

$$\nabla \times \mathbf{E}(\mathbf{r}, t) = -\frac{\partial \mathbf{B}(\mathbf{r}, t)}{\partial t}, \tag{14.1}$$

$$\nabla \times \mathbf{B}(\mathbf{r}, t) = \frac{1}{c^2}\frac{\partial \mathbf{E}(\mathbf{r}, t)}{\partial t} + \mu_0 \mathbf{j}(\mathbf{r}, t), \tag{14.2}$$

$$\nabla \cdot \mathbf{E}(\mathbf{r}, t) = \frac{1}{\varepsilon_0}\rho(\mathbf{r}, t), \tag{14.3}$$

$$\nabla \cdot \mathbf{B}(\mathbf{r}, t) = 0, \tag{14.4}$$

and of the force law

$$\begin{aligned} \mathbf{F}(\mathbf{r}, t) &= q\left[\mathbf{E}(\mathbf{r}, t) + \mathbf{v}(\mathbf{r}, t) \times \mathbf{B}(\mathbf{r}, t)\right] \\ &= \int_V \left[\rho(\mathbf{r}, t)\mathbf{E}(\mathbf{r}, t) + \mathbf{j}(\mathbf{r}, t) \times \mathbf{B}(\mathbf{r}, t)\right] \mathrm{d}V. \end{aligned} \tag{14.5}$$

The first expression applies to a single charge q moving with velocity $\mathbf{v}$ and the second expression to a distribution of charges and currents satisfying the continuity equation

$$\nabla \cdot \mathbf{j}(\mathbf{r}, t) + \frac{\partial \rho(\mathbf{r}, t)}{\partial t} = 0, \tag{14.6}$$

which is a direct consequence of Maxwell's equations. The force law connects the electromagnetic world with the mechanical one. The two terms in the first expression are basically definitions of the electric and magnetic fields.

If we operate on Maxwell's first equation by $\times \varepsilon_0 \mathbf{E}$ and on the second equation by $\times \mu_0 \mathbf{H}$, and then add the two resulting equations, we obtain

$$\varepsilon_0(\nabla \times \mathbf{E}) \times \mathbf{E} + \mu_0(\nabla \times \mathbf{H}) \times \mathbf{H} = \mathbf{j} \times \mathbf{B} - \frac{1}{c^2}\left[\frac{\partial \mathbf{H}}{\partial t} \times \mathbf{E}\right] + \frac{1}{c^2}\left[\frac{\partial \mathbf{E}}{\partial t} \times \mathbf{H}\right]. \tag{14.7}$$

We have omitted the arguments $(\mathbf{r}, t)$ for the different fields and we used $\varepsilon_0 \mu_0 = 1/c^2$. The last two expressions in Eq. (14.7) can be combined to give $(1/c^2)\mathrm{d}/\mathrm{d}t\,[\mathbf{E} \times \mathbf{H}]$. For the first expression in Eq. (14.7) we can write

$$\varepsilon_0(\nabla\times\mathbf{E})\times\mathbf{E} = \varepsilon_0\begin{bmatrix} \frac{\partial}{\partial x}(E_x^2-E^2/2)+\frac{\partial}{\partial y}(E_xE_y)+\frac{\partial}{\partial z}(E_xE_z) \\ \frac{\partial}{\partial x}(E_xE_y)+\frac{\partial}{\partial y}(E_y^2-E^2/2)+\frac{\partial}{\partial z}(E_yE_z) \\ \frac{\partial}{\partial x}(E_xE_z)+\frac{\partial}{\partial y}(E_yE_z)+\frac{\partial}{\partial z}(E_z^2-E^2/2) \end{bmatrix} - \varepsilon_0\mathbf{E}\,\nabla\cdot\mathbf{E}$$

$$= \nabla\cdot[\varepsilon_0\mathbf{EE}-(\varepsilon_0/2)E^2\overleftrightarrow{\mathbf{I}}]-\rho\mathbf{E}, \tag{14.8}$$

where Eq. (14.3) has been used in the last step. The notation **EE** denotes the outer product, $E^2 = E_x^2 + E_y^2 + E_z^2$ is the electric field strength, and $\overleftrightarrow{\mathbf{I}}$ denotes the unit tensor. A similar expression can be derived for $\mu_0(\nabla\times\mathbf{H})\times\mathbf{H}$. Using these two expressions in Eq. (14.7), we obtain

$$\nabla\cdot\left[\varepsilon_0\mathbf{EE}+\mu_0\mathbf{HH}-\frac{1}{2}(\varepsilon_0E^2+\mu_0H^2)\overleftrightarrow{\mathbf{I}}\right] = \frac{\mathrm{d}}{\mathrm{d}t}\frac{1}{c^2}(\mathbf{E}\times\mathbf{H})+\rho\,\mathbf{E}+\mathbf{j}\times\mathbf{B}. \tag{14.9}$$

The expression in brackets on the left-hand side is called Maxwell's stress tensor in vacuum, which is usually denoted as $\overleftrightarrow{\mathbf{T}}$. In Cartesian components it reads as

$$\overleftrightarrow{\mathbf{T}} = \left[\varepsilon_0\mathbf{EE}+\mu_0\mathbf{HH}-\frac{1}{2}(\varepsilon_0E^2+\mu_0H^2)\overleftrightarrow{\mathbf{I}}\right]$$

$$= \begin{bmatrix} \varepsilon_0(E_x^2-E^2/2)+\mu_0(H_x^2-H^2/2) & \varepsilon_0E_xE_y+\mu_0H_xH_y & \varepsilon_0E_xE_z+\mu_0H_xH_z \\ \varepsilon_0E_xE_y+\mu_0H_xH_y & \varepsilon_0(E_y^2-E^2/2)+\mu_0(H_y^2-H^2/2) & \varepsilon_0E_yE_z+\mu_0H_yH_z \\ \varepsilon_0E_xE_z+\mu_0H_xH_z & \varepsilon_0E_yE_z+\mu_0H_yH_z & \varepsilon_0(E_z^2-E^2/2)+\mu_0(H_z^2-H^2/2) \end{bmatrix}. \tag{14.10}$$

After integration of Eq. (14.9) over an arbitrary volume V that contains all sources ρ and **j** we obtain

$$\int_V \nabla\cdot\overleftrightarrow{\mathbf{T}}\,\mathrm{d}V = \frac{\mathrm{d}}{\mathrm{d}t}\frac{1}{c^2}\int_V(\mathbf{E}\times\mathbf{H})\mathrm{d}V + \int_V(\rho\,\mathbf{E}+\mathbf{j}\times\mathbf{B})\mathrm{d}V. \tag{14.11}$$

The last term is recognized as the mechanical force (cf. Eq. (14.5)). The volume integral on the left can be transformed into a surface integral using Gauss's integration law

$$\int_V \nabla\cdot\overleftrightarrow{\mathbf{T}}\,\mathrm{d}V = \int_{\partial V}\overleftrightarrow{\mathbf{T}}\cdot\mathbf{n}\,\mathrm{d}a. \tag{14.12}$$

∂V denotes the surface of V, **n** the unit vector perpendicular to the surface, and $\mathrm{d}a$ an infinitesimal surface element. We then finally arrive at the conservation law for linear momentum:

$$\int_{\partial V}\overleftrightarrow{\mathbf{T}}(\mathbf{r},t)\cdot\mathbf{n}(\mathbf{r})\,\mathrm{d}a = \frac{\mathrm{d}}{\mathrm{d}t}\left[\mathbf{G}_{\mathrm{field}}+\mathbf{G}_{\mathrm{mech}}\right] \tag{14.13}$$

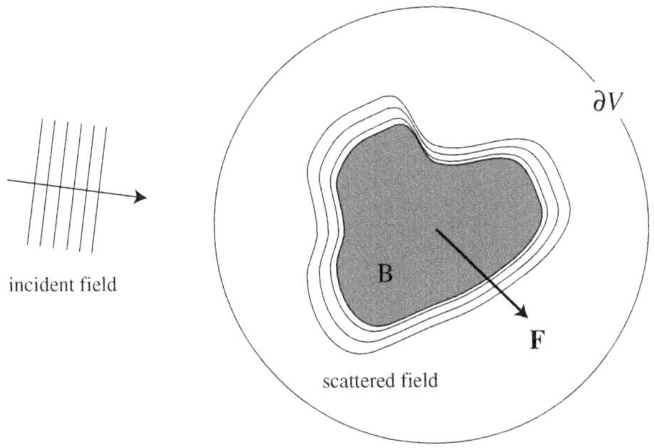

Fig. 14.1 The mechanical force $\mathbf{F}$ acting on the object B is entirely determined by the electric and magnetic fields at an arbitrary surface ∂V enclosing B.

Here, $\mathbf{G}_{\text{mech}}$ and $\mathbf{G}_{\text{field}}$ denote the mechanical momentum and the field momentum, respectively. In Eq. (14.13) we have used Newton's expression of the mechanical force $\mathbf{F} = \mathrm{d}/\mathrm{d}t\, \mathbf{G}_{\text{mech}}$ and the definition of the field momentum (Abraham density)

$$\mathbf{G}_{\text{field}} = \frac{1}{c^2} \int_V (\mathbf{E} \times \mathbf{H}) \mathrm{d}V. \tag{14.14}$$

This is the momentum carried by the electromagnetic field within the volume V. It is created by the dynamic terms in Maxwell's curl equations. The time derivative of the field momentum is zero when it is averaged over one oscillation period and hence the average mechanical force becomes

$$\langle \mathbf{F} \rangle = \int_{\partial V} \langle \overleftrightarrow{\mathbf{T}}(\mathbf{r}, t) \rangle \cdot \mathbf{n}(\mathbf{r}) \mathrm{d}a \quad , \tag{14.15}$$

with $\langle ... \rangle$ denoting the time average. Equation (14.15) is of general validity. It allows the mechanical force acting on an arbitrary body within the closed surface ∂V to be calculated. The force is entirely determined by the electric and magnetic fields on the surface ∂V. It is interesting to note that no material properties enter the expression for the force; the entirety of the information is contained in the electromagnetic field. The only material constraint is that the body is rigid. If the body deforms when it is subject to an electromagnetic field, we have to include electrostrictive and magnetostrictive forces. Since the enclosing surface is arbitrary, the same results are obtained irrespective of whether the fields are evaluated at the surface of the body or in the far-field. It is important to note that the fields used to calculate the force are the self-consistent fields of the problem, which means that they are a superposition of the incident and the scattered fields. Therefore, prior to calculating the force, one has to solve for the electromagnetic fields. If the object B is surrounded by a medium that can be represented accurately enough by a non-dispersive dielectric constant ε and magnetic susceptibility μ (Fig. 14.1), the mechanical force can be calculated in the same way if we replace Maxwell's stress tensor Eq. (14.10) by

$$\overleftrightarrow{\mathbf{T}} = \left[\varepsilon_0\varepsilon\mathbf{EE} + \mu_0\mu\mathbf{HH} - \frac{1}{2}(\varepsilon_0\varepsilon E^2 + \mu_0\mu H^2)\overleftrightarrow{\mathbf{I}}\right] . \tag{14.16}$$

14.2 Radiation pressure

Here, we consider the radiation pressure on a medium with an infinitely extended planar interface as shown in Fig. 14.2. The medium is irradiated by a monochromatic plane wave at normal incidence to the interface. Depending on the material properties of the medium, part of the incident field is reflected at the interface. On introducing the complex reflection coefficient r, the electric field outside the medium can be written as the superposition of two counter-propagating plane waves

$$\mathbf{E}(\mathbf{r},t) = E_0\,\mathrm{Re}\Big\{[\mathrm{e}^{ikz} + r\mathrm{e}^{-ikz}]\mathrm{e}^{-\mathrm{i}\omega t}\Big\}\,\mathbf{n}_x. \tag{14.17}$$

Using Maxwell's curl equation (14.1), we find for the magnetic field

$$\mathbf{H}(\mathbf{r},t) = \sqrt{\varepsilon_0/\mu_0}\;E_0\,\mathrm{Re}\Big\{[\mathrm{e}^{ikz} - r\mathrm{e}^{-ikz}]\mathrm{e}^{-\mathrm{i}\omega t}\Big\}\,\mathbf{n}_y. \tag{14.18}$$

To calculate the radiation pressure P, we integrate Maxwell's stress tensor on an infinite planar surface A parallel to the interface as shown in Fig. 14.2. The radiation pressure can be calculated by using Eq. (14.15) as

$$P\mathbf{n}_z = \frac{1}{A}\int_A \langle\overleftrightarrow{\mathbf{T}}(\mathbf{r},t)\rangle \cdot \mathbf{n}_z\,\mathrm{d}a. \tag{14.19}$$

We do not need to consider a closed surface ∂V since we are interested in the pressure exerted on the interface of the medium rather than in the mechanical force acting on the

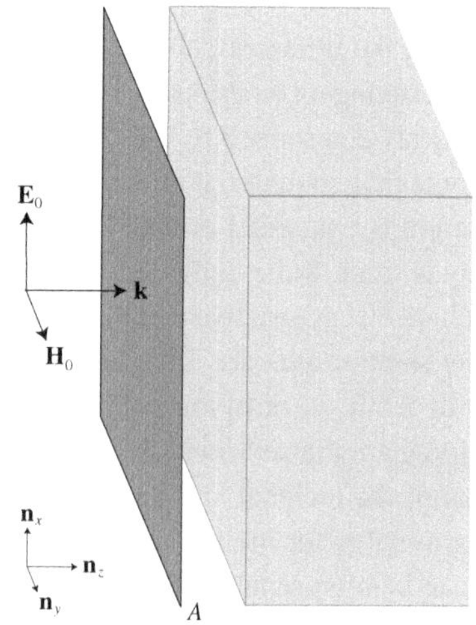

Fig. 14.2 The configuration used to derive the radiation pressure.

medium. Using the fields of Eqs. (14.17) and (14.18), we find that the first two terms in Maxwell's stress tensor Eq. (14.10) give no contribution to the radiation pressure. The third term yields

$$\langle \overleftrightarrow{\mathbf{T}}(\mathbf{r},t)\rangle \cdot \mathbf{n}_z = -\frac{1}{2}\langle \varepsilon_0 E^2 + \mu_0 H^2\rangle \mathbf{n}_z = \frac{\varepsilon_0}{2} E_0^2 [1 + |r|^2]\mathbf{n}_z. \tag{14.20}$$

Using the definition of the intensity of a plane wave $I_0 = (\varepsilon_0/2)cE_0^2$, with c being the vacuum speed of light, we can express the radiation pressure as

$$P = \frac{I_0}{c}(1 + R), \tag{14.21}$$

with $R = |r|^2$ being the reflectivity. For a perfectly absorbing medium we have $R = 0$, whereas for a perfectly reflecting medium $R = 1$. Therefore, the radiation pressure on a perfectly reflecting medium is twice as high as for a perfectly absorbing medium.

14.3 Lorentz force density

Let us return to the force law defined in Eq. (14.5) and assume that all charges and currents are secondary sources associated with the polarization $\mathbf{P}$. Then, according to Eq. (2.10), we can express the current density as $\mathbf{j} = \partial\mathbf{P}/\partial t$, and the charge density follows from the conservation law Eq. (14.6) as $\rho = -\nabla \cdot \mathbf{P}$. Inserting this into the force law gives

$$\mathbf{F}(\mathbf{r},t) = \int_V \mathbf{f}(\mathbf{r},t)\mathrm{d}V, \quad \mathbf{f}(\mathbf{r},t) = -\mathbf{E}\,(\nabla \cdot \mathbf{P}) + \frac{\partial \mathbf{P}}{\partial t} \times \mathbf{B}. \tag{14.22}$$

Here, $\mathbf{f}(\mathbf{r},t)$ denotes the Lorentz force density. Equation (14.22) makes no use of material properties and therefore is generally valid. Using constitutive relations, the polarizability $\mathbf{P}$ can be expressed in terms of the electric field $\mathbf{E}$ in a linear or nonlinear way.

14.4 The dipole approximation

A quantized two-level system such as an atom with transitions restricted to two states is well described by a dipole. The same is true for a macroscopic particle with dimensions much smaller than the wavelength of the illuminating light (a Rayleigh particle). To derive the electromagnetic force acting on a dipole located at $\mathbf{r}_0$ we could introduce the polarizability $\mathbf{P} = \mathbf{p}\delta(\mathbf{r} - \mathbf{r}_0)$ into the Lorentz force density $\mathbf{f}$. Likewise, we could express the field $\mathbf{E}$ in the Maxwell stress tensor by an incident field and the field scattered by a dipole. While both of these procedures are formally correct, it is more intuitive to approach the force acting on a dipole from a microscopic perspective. We will return to the Lorentz force density later and show that it produces the same results.

Consider two oppositely charged particles with masses m_1 and m_2, separated by a tiny distance $|\mathbf{s}|$ and illuminated by an arbitrary electromagnetic field $\mathbf{E}$, $\mathbf{B}$, as shown in

Fig. 14.3. In the non-relativistic limit, the equation of motion for each particle follows from Eq. (14.5) by setting $\mathbf{F}$ equal to $m_1\ddot{\mathbf{r}}_1$ and $m_2\ddot{\mathbf{r}}_2$, respectively. The dots denote differentiation with respect to time. Since the particles are bound to each other we have to consider their binding energy U. Including this contribution, the equation of motion for the two particles reads as

$$m_1\ddot{\mathbf{r}}_1 = q[\mathbf{E}(\mathbf{r}_1,t) + \dot{\mathbf{r}}_1 \times \mathbf{B}(\mathbf{r}_1,t)] - \nabla U(\mathbf{r}_1,t), \tag{14.23}$$

$$m_2\ddot{\mathbf{r}}_2 = -q[\mathbf{E}(\mathbf{r}_2,t) + \dot{\mathbf{r}}_2 \times \mathbf{B}(\mathbf{r}_2,t)] + \nabla U(\mathbf{r}_2,t). \tag{14.24}$$

The two particles constitute a two-body problem, which is most conveniently solved by introducing the center-of-mass coordinate

$$\mathbf{r} = \frac{m_1}{m_1+m_2}\mathbf{r}_1 + \frac{m_2}{m_1+m_2}\mathbf{r}_2. \tag{14.25}$$

Expressing the problem in terms of $\mathbf{r}$ allows us to separate the internal motion of the two particles from the center-of-mass motion. The electric field at the position of the two particles can be represented by a Taylor expansion as

$$\begin{aligned}
\mathbf{E}(\mathbf{r}_1) &= \sum_{n=0}^{\infty} \frac{1}{n!}\Big[(\mathbf{r}_1-\mathbf{r})\cdot\nabla\Big]^n \mathbf{E}(\mathbf{r}) = \mathbf{E}(\mathbf{r}) + [(\mathbf{r}_1-\mathbf{r})\cdot\nabla]\,\mathbf{E}(\mathbf{r}) + \cdots, \\
\mathbf{E}(\mathbf{r}_2) &= \sum_{n=0}^{\infty} \frac{1}{n!}\Big[(\mathbf{r}_2-\mathbf{r})\cdot\nabla\Big]^n \mathbf{E}(\mathbf{r}) = \mathbf{E}(\mathbf{r}) + [(\mathbf{r}_2-\mathbf{r})\cdot\nabla]\,\mathbf{E}(\mathbf{r}) + \cdots.
\end{aligned} \tag{14.26}$$

A similar expansion can be found for $\mathbf{B}(\mathbf{r}_1)$ and $\mathbf{B}(\mathbf{r}_2)$. For $|\mathbf{s}| \ll \lambda$, λ being the wavelength of the radiation field, the expansions can be truncated after the second term (the dipole approximation). A straightforward calculation using Eqs. (14.23)–(14.26) and the definition of the dipole moment

$$\mathbf{p} = q\mathbf{s}, \tag{14.27}$$

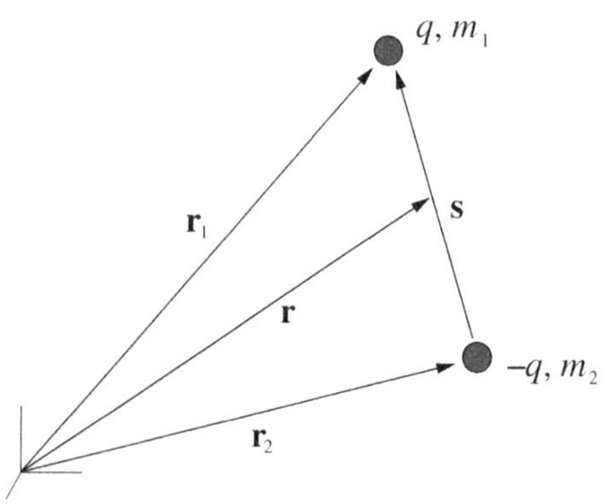

Fig. 14.3 Graphical representation of the symbols used to derive the mechanical force in the dipolar limit. $\boldsymbol{r}$ denotes the center-of-mass coordinate. The two particles are bound to each other by the potential U.

where $\mathbf{s} = \mathbf{r}_1 - \mathbf{r}_2$ leads to the following formula for the total force $\mathbf{F} = (m_1 + m_2)\ddot{\mathbf{r}}$ acting on the system of particles

$$\mathbf{F} = (\mathbf{p} \cdot \nabla)\mathbf{E} + \dot{\mathbf{p}} \times \mathbf{B} + \dot{\mathbf{r}} \times (\mathbf{p} \cdot \nabla)\mathbf{B}. \tag{14.28}$$

Here, we have omitted the arguments $(\mathbf{r}, t)$ for clarity. The brackets in $(\mathbf{p} \cdot \nabla)\mathbf{E}$ indicate that the inner product $\mathbf{p} \cdot \nabla = (p_x, p_y, p_z) \cdot (\partial/\partial x, \partial/\partial y, \partial/\partial z)$ has to be evaluated prior to operating on $\mathbf{E}$. Equation (14.28) is the central equation of this section. It renders the mechanical force exerted by the electromagnetic field on the two particles represented by the dipole moment $\mathbf{p}$. The force consists of three terms: the first originates from the inhomogeneous electric field, the second is the familiar Lorentz force, and the third is due to movement in the inhomogeneous magnetic field. For non-relativistic speeds ($|\dot{\mathbf{r}}| \ll c$), the third term is much smaller than the other two terms, and it will be omitted in the following discussion. It is interesting to note that the fields appearing in Eq. (14.28) correspond to the exciting field. It is assumed that the system represented by the dipole does not change the fields. This is different from the general formalism based on Maxwell's stress tensor, in which the self-consistent fields are considered.

We now return to the Lorentz force density $\mathbf{f}(\mathbf{r}, t)$ and show that it reproduces the terms in Eq. (14.28). We consider the ith Cartesian component of $\mathbf{f}$ and denote it as f_i, where $i \in [x, y, z]$. The first term in the expression of f_i (c.f. Eq. (14.22)) can be expressed as

$$\int_V -E_i(\nabla \cdot \mathbf{P})\mathrm{d}V = \int_V (\mathbf{P} \cdot \nabla)E_i \,\mathrm{d}V - \int_{\partial V} (\mathbf{P}E_i) \cdot \mathbf{n}\, \mathrm{d}a, \tag{14.29}$$

where we made use of Gauss's theorem. We now place a dipole with polarization $\mathbf{P} = \mathbf{p}\delta(\mathbf{r} - \mathbf{r}_0)$ into the volume V. The last term in Eq. (14.29) vanishes if we make sure that the dipole's origin is not on the surface of the volume, i.e. $\mathbf{r}_0 \notin \partial V$. In this case, only the first term on the right-hand side of Eq. (14.29) contributes. After integration this term yields $(\mathbf{p} \cdot \nabla)E_i$, which corresponds to the first term in Eq. (14.28). The second term is readily reproduced if we substitute $\mathbf{P} = \mathbf{p}\delta(\mathbf{r} - \mathbf{r}_0)$ into the expression for $\mathbf{f}$ in Eq. (14.22).

14.4.1 Time-averaged force

The second term in Eq. (14.28) can be represented as

$$\dot{\mathbf{p}} \times \mathbf{B} = -\mathbf{p} \times \frac{\mathrm{d}}{\mathrm{d}t}\mathbf{B} + \frac{\mathrm{d}}{\mathrm{d}t}(\mathbf{p} \times \mathbf{B}) = \mathbf{p} \times (\nabla \times \mathbf{E}) + \frac{\mathrm{d}}{\mathrm{d}t}(\mathbf{p} \times \mathbf{B}). \tag{14.30}$$

We have approximated $\mathrm{d}\mathbf{B}/\mathrm{d}t$ by $\partial\mathbf{B}/\partial t$ because the velocity of the center of mass is assumed to be small compared with c. After dropping the last term in Eq. (14.28) for the same reason, we obtain

$$\mathbf{F} = (\mathbf{p} \cdot \nabla)\mathbf{E} + \mathbf{p} \times (\nabla \times \mathbf{E}) + \frac{\mathrm{d}}{\mathrm{d}t}(\mathbf{p} \times \mathbf{B}), \tag{14.31}$$

which can be rewritten as

$$\mathbf{F} = \sum_i p_i \,\nabla E_i + \frac{\mathrm{d}}{\mathrm{d}t}(\mathbf{p} \times \mathbf{B}), \quad i = x, y, z. \tag{14.32}$$

In the time average, the last term vanishes, and the force can be cast into the concise form

$$\langle \mathbf{F} \rangle = \sum_i \Big\langle p_i(t) \nabla E_i(t) \Big\rangle \; , \tag{14.33}$$

where $\langle \ldots \rangle$ denotes the time average. We have included the arguments of $\mathbf{p}$ and $\mathbf{E}$ in order to distinguish them from their corresponding complex amplitudes introduced below.

Notice that the optical force is additive and that we can sum up the forces acting on individual dipolar subvolumes to obtain the net force acting on a macroscopic body. If we represent Eq. (14.33) by a dipole per unit volume (the polarization $\mathbf{P}$) and add up all the forces acting on the volume we obtain

$$\langle \mathbf{F} \rangle = \int_V \sum_i \langle P_i(\mathbf{r},t) \nabla E_i(\mathbf{r},t) \rangle \mathrm{d}V, \tag{14.34}$$

where P_i denotes the Cartesian components of the polarization $\mathbf{P}$.

14.4.2 Monochromatic fields

Consider a dipolar particle irradiated by an arbitrary monochromatic electromagnetic wave with angular frequency ω. In this case the fields can be represented as[1]

$$\begin{aligned} \mathbf{E}(\mathbf{r},t) &= \ \mathrm{Re}\{\underline{\mathbf{E}}(\mathbf{r})\mathrm{e}^{-\mathrm{i}\omega t}\}, \\ \mathbf{B}(\mathbf{r},t) &= \ \mathrm{Re}\{\underline{\mathbf{B}}(\mathbf{r})\mathrm{e}^{-\mathrm{i}\omega t}\}. \end{aligned} \tag{14.35}$$

If there is a linear relationship between dipole and fields, the dipole assumes the same time dependence and can be written as

$$\mathbf{p}(t) = \mathrm{Re}\{\underline{\mathbf{p}}\mathrm{e}^{-\mathrm{i}\omega t}\}. \tag{14.36}$$

We assume that the particle has no static dipole moment. In this case, to first order, the induced dipole moment is proportional to the electric field at the particle's position $\mathbf{r}$,

$$\underline{\mathbf{p}} = \alpha(\omega)\underline{\mathbf{E}}(\mathbf{r}_0), \tag{14.37}$$

where α denotes the polarizability of the particle. Its form depends on the nature of the particle. For a two-level system, explicit expressions for α are derived in Appendix A. Generally, α is a tensor of rank two, but for atoms and molecules it is legitimate to use a scalar representation since only the projection of $\mathbf{p}$ along the direction of the electric field is important.

The cycle-average of Eq. (14.28) reads as

$$\langle \mathbf{F} \rangle = \frac{1}{2} \mathrm{Re} \Big\{ (\underline{\mathbf{p}}^* \cdot \nabla)\underline{\mathbf{E}} - \mathrm{i}\omega(\underline{\mathbf{p}}^* \times \underline{\mathbf{B}}) \Big\} , \tag{14.38}$$

[1] For clarity, we will designate the complex amplitudes of the fields by an underline.

where we have dropped the third term as discussed before. The two terms on the right-hand side can be combined as done before and we obtain

$$\langle \mathbf{F} \rangle = \sum_i \frac{1}{2} \mathrm{Re}\left\{ \underline{p}_i^* \, \nabla \underline{E}_i \right\}. \tag{14.39}$$

After introducing the linear relationship (14.37) and rearranging the different terms, we find

$$\langle \mathbf{F} \rangle = \frac{\alpha'}{2} \sum_i \mathrm{Re}\{\underline{E}_i^* \, \nabla \underline{E}_i\} + \frac{\alpha''}{2} \sum_i \mathrm{Im}\{\underline{E}_i^* \, \nabla \underline{E}_i\}, \tag{14.40}$$

where we used $\alpha = \alpha' + \mathrm{i}\alpha''$. The first term can be written as $(\alpha'/4)\nabla(\underline{\mathbf{E}}^* \cdot \underline{\mathbf{E}})$, which implies that the force $\langle \mathbf{F}_{\mathrm{grad}} \rangle$ associated with the first term is conservative, i.e. $\nabla \times \langle \mathbf{F}_{\mathrm{grad}} \rangle = 0$. On the other hand, the force associated with the second term $\langle \mathbf{F}_{\mathrm{scatt}} \rangle$ cannot be represented as the gradient of a potential and hence it is not conservative. We thus find that two different terms determine the average mechanical force: the first is denoted the *gradient force* (or dipole force) and the second one is called the *scattering force*. The gradient force originates from field inhomogeneities. It is proportional to the dispersive part (real part) of the complex polarizability. On the other hand, the scattering force is proportional to the dissipative part (imaginary part) of the complex polarizability. The scattering force can be regarded as a consequence of momentum transfer from the radiation field to the particle. The dipole force accelerates polarizable particles towards extrema of the radiation field. Therefore, a tightly focused laser beam can trap a particle in all dimensions at its focus. However, the scattering force pushes the particle in the direction of propagation and, if the focus of the trapping laser is not tight enough, the particle can be pushed out of the trap. Because of radiation reaction, α'' and the scattering force never vanish, even for a lossless particle (c.f. Eq. (8.222) in Problem 8.5). For a small homogeneous sphere with a real quasi-static polarizability α' the imaginary part turns out to be

$$\alpha'' = \frac{k^3}{6\pi\varepsilon_0} \alpha'^2, \tag{14.41}$$

where $k = n(2\pi/\lambda)$, n being the index of refraction of the surrounding medium. The last term in Eq. (14.40) together with α'' determine the scattering force for a small loss-free particle.

Let us take a closer look at the scattering force in Eq. (14.40). Using the identity $\sum_i \underline{E}_i^* \, \nabla \underline{E}_i = (\underline{\mathbf{E}} \cdot \nabla)\underline{\mathbf{E}}^* + \underline{\mathbf{E}} \times (\nabla \times \underline{\mathbf{E}}^*)$ together with Maxwell's equation $\nabla \times \underline{\mathbf{E}} = \mathrm{i}\omega\mu_0 \underline{\mathbf{H}}$ we can represent the scattering force as [8]

$$\langle \mathbf{F}_{\mathrm{scatt}} \rangle = \frac{\alpha''}{2} \sum_i \mathrm{Im}\{\underline{E}_i^* \, \nabla \underline{E}_i\} = \frac{\sigma}{c} \langle \mathbf{S} \rangle + c\sigma \left[\nabla \times \langle \mathbf{L} \rangle \right], \tag{14.42}$$

where $\sigma = \alpha'' k/\varepsilon_0$ is the absorption cross-section, $\langle \mathbf{S} \rangle$ is the time-averaged Poynting vector (c.f. Eq. (2.59)), and $\langle \mathbf{L} \rangle = [\varepsilon_0/(4\mathrm{i}\omega)](\underline{\mathbf{E}} \times \underline{\mathbf{E}}^*)$. While the term with the Poynting vector represents the radiation pressure, the term with $\langle \mathbf{L} \rangle$ is a force associated with the spin density of the light field. This spin curl force has been shown to be relevant in light fields with non-uniform helicities [8].

The physical origin of the gradient force and the scattering force becomes more intuitive if we represent the complex amplitude of the electric field in terms of the *real* amplitude E_0 and phase ϕ as

$$\underline{\mathbf{E}}(\mathbf{r}) = E_0(\mathbf{r})\mathrm{e}^{\mathrm{i}\phi(\mathbf{r})}\mathbf{n}_E, \tag{14.43}$$

with $\mathbf{n}_E$ denoting the unit vector in the direction of the polarization. It is important to emphasize that this representation is approximate and applicable only to fields that vary slowly in space. In general, the different vector components of the fields have different phases. Nevertheless, for most situations Eq. (14.43) is a good approximation and allows us to cast the cycle-averaged force in Eq. (14.39) into the following form:

$$\langle\mathbf{F}\rangle = \frac{\alpha'}{4}\,\nabla E_0^2 + \frac{\alpha''}{2} E_0^2\,\nabla\phi, \tag{14.44}$$

where we used $\nabla E_0^2 = 2E_0\,\nabla E_0$. Notice that ϕ can be written in terms of the local $\mathbf{k}$ vector as $\phi = \mathbf{k}\cdot\mathbf{r}$, which renders $\nabla\phi = \mathbf{k}$.

If we introduce Eq. (14.43) into Eq. (14.35), the time-dependent electric field can be written as

$$\mathbf{E}(\mathbf{r},t) = E_0(\mathbf{r})\cos[\omega t - \phi(\mathbf{r})]\mathbf{n}_E. \tag{14.45}$$

The corresponding magnetic field is determined by $\partial\mathbf{B}/\partial t = -\nabla\times\mathbf{E}$, which, together with $\mathbf{E}$ leads to the relationships

$$E_0^2\,\nabla\phi = 2\omega\langle\mathbf{E}\times\mathbf{B}\rangle, \qquad E_0^2 = 2\langle|\mathbf{E}|^2\rangle, \tag{14.46}$$

with $\langle...\rangle$ denoting the cycle-average. Substituting into Eq. (14.44) gives

$$\langle\mathbf{F}\rangle = \frac{\alpha'}{2}\,\nabla\langle|\mathbf{E}|^2\rangle + \omega\alpha''\langle\mathbf{E}\times\mathbf{B}\rangle, \tag{14.47}$$

where $|\mathbf{E}|$ denotes the time-dependent magnitude of the electric field vector. Equation (14.47) directly proves that the scattering force is proportional to the average field momentum defined in Eq. (14.14).

It has to be emphasized that Eqs. (14.43)–(14.47) are approximate and one needs to work with Eq. (14.40) if the fields are localized to dimensions smaller than $\lambda/2$.

14.4.3 Self-induced back-action

So far we have assumed that the external electric field $\mathbf{E}$ is not affected by the particle being trapped. This assumption is justified in free space but not necessarily in cavities or near material boundaries. In Section 11.3.1 we have already seen that a particle can detune the resonance frequency of a resonator and hence affect the field that is acting on it. This effect has been named *self-induced back-action* [9].

The expressions for the force in Eqs. (14.39) and (14.40) remain valid as long as we account for the modification of the field $\mathbf{E}$ by the particle. Assuming that this modification is weak, we can expand the field in a perturbation series as

$$\mathbf{E}(\mathbf{r}) = \mathbf{E}_0(\mathbf{r}) + \frac{\omega^2}{c^2\varepsilon_0}\overleftrightarrow{\mathbf{G}}_\mathrm{s}(\mathbf{r},\mathbf{r})\alpha(\omega)\mathbf{E}_0(\mathbf{r}) + \frac{\omega^4}{c^4\varepsilon_0^2}\overleftrightarrow{\mathbf{G}}_\mathrm{s}(\mathbf{r},\mathbf{r})\alpha(\omega)\overleftrightarrow{\mathbf{G}}_\mathrm{s}(\mathbf{r},\mathbf{r})\alpha(\omega)\mathbf{E}_0(\mathbf{r}) + \cdots, \tag{14.48}$$

where $\overleftrightarrow{\mathbf{G}}_\mathrm{s}$ is the scattering part of the Green function (c.f. Section 8.3.3), defined as $\overleftrightarrow{\mathbf{G}}_\mathrm{s} = \overleftrightarrow{\mathbf{G}} - \overleftrightarrow{\mathbf{G}}_0$, with $\overleftrightarrow{\mathbf{G}}$ being the total Green function and $\overleftrightarrow{\mathbf{G}}_0$ being its free-space part.

The first term in Eq. (14.48) (zeroth order) ignores the particle's back-action and accounts only for the incident field. The second term (first order) accounts for the scattering of the incident field $\mathbf{E}_0$ by the particle. The scattered field is reflected from the environment and then interacts with the particle again to produce a correction to the field $\mathbf{E}_0$. The interaction order increases by one for every subsequent term. A series expansion for the force acting on the particle can be derived by inserting Eq. (14.48) into Eq. (14.40). Note however, that the convergence of the series needs to be verified for the particular configuration under study. If the series does not converge, one has to proceed with the Maxwell stress-tensor formalism, which often requires computational methods.

14.4.4 Saturation behavior for near-resonance excitation

Saturation is a nonlinear effect that limits the magnitude of the induced dipole moment $\mathbf{p}$. Differently from most nonlinear effects, saturation does not affect the monochromatic time dependence of the induced dipole (see Appendix C). Therefore, the linear relationship in Eq. (14.37) is valid even for saturation. The steady-state polarizability for a two-level atom excited near its resonance is derived in Appendix A. Using the projection of the transition dipole moment along the direction of the electric field ($\mathbf{p}_{12} \cdot \mathbf{n}_E$), the polarizability can be written as

$$\alpha(\omega) = \frac{(\mathbf{p}_{12} \cdot \mathbf{n}_E)^2}{\hbar} \frac{\omega_o - \omega + \mathrm{i}\gamma/2}{(\omega_0 - \omega)^2 + \mathrm{i}\gamma^2/4 + \omega_\mathrm{R}^2/2}. \tag{14.49}$$

Here, ω_0 is the transition frequency, $\omega_\mathrm{R} = (\mathbf{p}_{12} \cdot \mathbf{n}_E)E_0/\hbar$ the Rabi frequency, and γ the spontaneous decay rate. Substituting α into Eq. (14.44) leads to

$$\langle \mathbf{F} \rangle = \hbar \frac{\omega_\mathrm{R}^2/2}{(\omega_0 - \omega)^2 + \gamma^2/4 + \omega_\mathrm{R}^2/2}\left[(\omega - \omega_0)\frac{\nabla E_0}{E_0} + \frac{\gamma}{2}\nabla\phi\right], \tag{14.50}$$

where we used $\gamma \ll \omega_0$. Introducing the so-called *saturation parameter* p as

$$p = \frac{I}{I_\mathrm{sat}} \frac{\gamma^2/4}{(\omega - \omega_0)^2 + \gamma^2/4}, \tag{14.51}$$

with the intensity I and the saturation intensity I_sat defined as

$$I = \frac{\varepsilon_0 c}{2} E_0^2, \qquad I_\mathrm{sat} = 4\pi\varepsilon_0 \frac{\hbar^2 c\gamma^2}{16\pi(\mathbf{p}_{12} \cdot \mathbf{n}_E)^2} = \frac{\gamma^2}{2\omega_\mathrm{R}^2} I, \tag{14.52}$$

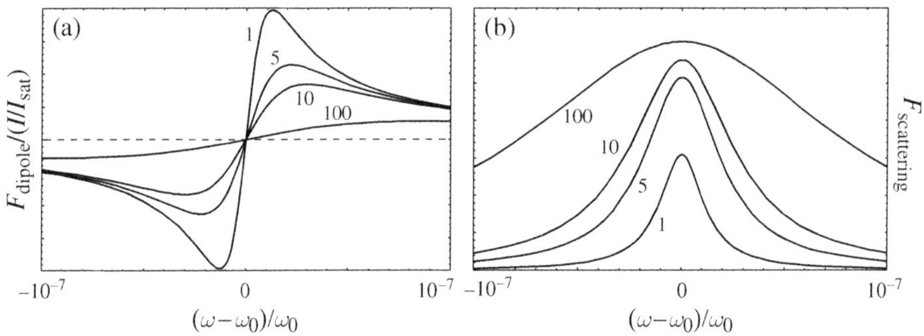

Fig. 14.4 Dipole force and scattering force for sodium atoms ($1/\gamma = 16.1$ ns, $\lambda_0 = 590$ nm) as a function of the excitation frequency ω. The numbers in the figures indicate the value of the ratio I/I_{sat}.

allows us to write the cycle-averaged force in the form

$$\langle \mathbf{F} \rangle = \frac{\hbar p}{1+p}\left[(\omega-\omega_0)\frac{\nabla E_0}{E_0} + \frac{\gamma}{2}\nabla\phi\right] . \tag{14.53}$$

This formula was originally developed by Gordon and Ashkin using a quantum-mechanical derivation [10]. The present derivation uses quantum mechanics only for the calculation of the atomic polarizability (see Appendix A). It follows from quantum theory that the scattering force originates from cycles of absorption and spontaneous emission, whereas the dipole force is due to cycles of absorption and stimulated emission. Notice that the maximum value for the saturation parameter p is obtained for exact resonance, i.e. $\omega = \omega_0$. In this case, the factor $p/(1+p)$ cannot exceed the value of unity, which limits the maximum value of the force (saturation). For an intensity of $I = I_{\text{sat}}$ ($I_{\text{sat}} \approx 1.6$ mW for rubidium atoms) the force amounts to half of the maximum force. For frequencies $\omega < \omega_0$ (red detuning) the dipole force is proportional to $-\nabla E_0$, causing an atom to be attracted towards regions of high intensity. On the other hand, for frequencies $\omega > \omega_0$ (blue detuning) atoms are repelled from regions of high intensity because the dipole force is proportional to ∇E_0. The dipole force vanishes for exact resonance. Figure 14.4 shows qualitatively the frequency behavior of the dipole and scattering force for different excitation intensities. Using $\mathbf{k} = \nabla\phi$ and conditions far from saturation, the scattering force can be written as

$$\langle \mathbf{F}_{\text{scatt}} \rangle = \hbar\mathbf{k}\,\frac{\gamma}{2}\,\frac{I}{I_{\text{sat}}}\,\frac{\gamma^2/4}{(\omega-\omega_0)^2+\gamma^2/4}, \qquad I \ll I_{\text{sat}}, \tag{14.54}$$

which has a maximum for exact resonance. The influence of saturation on the scattering force is illustrated in Fig. 14.5.

In atom-manipulation experiments the scattering force is used to cool atoms down to extremely low temperatures, thereby bringing them almost to rest. Under ambient conditions atoms and molecules move at speeds of about $1000\,\text{m}\,\text{s}^{-1}$ in random directions. Even at temperatures as low as -270 °C the speeds are on the order of $100\,\text{m}\,\text{s}^{-1}$. Only for temperatures close to absolute zero (-273 °C) does the motion of atoms slow down significantly. The initial idea regarding how to slow down the motion of atoms is based on the Doppler effect. It was first proposed by Hänsch and Schawlow in 1975 [5]. Neutral atoms are irradiated by pairs of counter-propagating laser beams. If an atom moves against the propagation direction of one of the laser beams, the frequency as seen from the atom will shift towards higher frequencies (blueshift) according to the Doppler effect. On the other hand, an atom moving in the direction of beam propagation will experience a shift towards lower frequencies (redshift). If the laser frequency is tuned to slightly below a resonance transition, an atom will predominantly absorb a photon when it moves against laser-beam propagation (c.f. Eq. (14.54)). The absorption process slows the atom down according to momentum conservation. Once the atom has been excited, it will eventually reemit its excitation energy by spontaneous emission, which is a random process and does not favor any particular direction. Thus, averaged over many absorption–emission cycles, the atom moving towards the laser will lose velocity and effectively cool. To slow the atom down in all dimensions one requires six laser beams opposed in pairs and arranged in three directions at right angles to each other. In whichever direction the atom tries to move it will be met by photons of the right energy and pushed back into the area where the six laser beams intersect. The movement of the atoms in the intersection region is similar to the movement in a hypothetical viscous medium (optical molasses). It can be calculated that two-level atoms cannot be cooled below a certain temperature, called the Doppler limit [7]. For sodium atoms the limiting temperature is 240 μK, corresponding to speeds of 30 cm s^{-1}. However, it was experimentally found that much lower temperatures could be attained. After surpassing another limit, the so-called recoil limit, which states that the speed of an atom should not be less than that imparted by a single photon recoil, temperatures as low as 0.18 μK have been generated for helium atoms. Under these conditions the helium atoms move at speeds of only 2 cm s^{-1}. Once the atoms are sufficiently cold they would fall out of the optical molasses due to gravity. To prevent this from happening, an initial trapping scheme based on the dipole force allowed one to grip the atoms at the focal point of a tightly focused beam [11]. Unfortunately, the optical dipole trap was not strong enough for most applications and a new three-dimensional trap based on the scattering force has been developed. This kind of trap is now called the magneto-optical trap. Its restoring force comes from a combination of oppositely directed circularly polarized

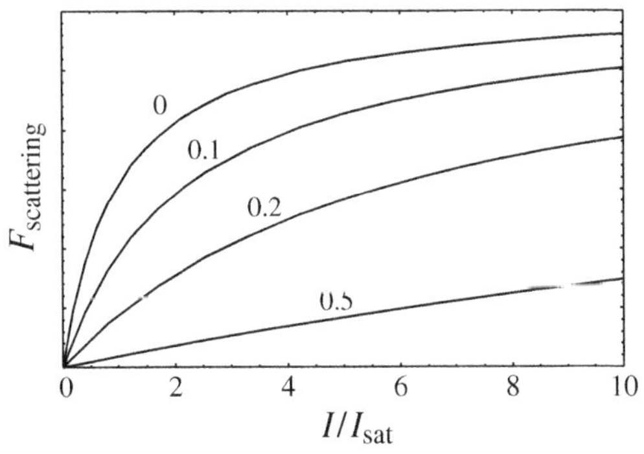

Fig. 14.5 The scattering force for sodium atoms ($1/\gamma = 16.1$ ns, $\lambda_0 = 590$ nm) as a function of I/I_{sat}. The numbers in the figure indicate the frequency detuning in units of $10^7\,(\omega - \omega_0)/\omega_0$.

laser beams and a weak, varying, inhomogeneous magnetic field with a minimum in the intersection region of the laser beams. The magnetic field strength increases with distance from the trap center and gives rise to a force towards the trap center.

14.4.5 Beyond the dipole approximation

In principle, any macroscopic object can be regarded as being composed of individual dipolar subunits. The self-consistent solution for the electric and magnetic fields generated by these dipoles is (see Section 2.12)

$$\begin{aligned} \underline{\mathbf{E}}(\mathbf{r}) &= \underline{\mathbf{E}}_0(\mathbf{r}) + \frac{\omega^2}{\varepsilon_0 c^2} \sum_{n=1}^{N} \overleftrightarrow{\mathbf{G}}(\mathbf{r}, \mathbf{r}_n) \underline{\mathbf{p}}_n, \\ \underline{\mathbf{H}}(\mathbf{r}) &= \underline{\mathbf{H}}_0(\mathbf{r}) - \mathrm{i}\omega \sum_{n=1}^{N} \left[\nabla \times \overleftrightarrow{\mathbf{G}}(\mathbf{r}, \mathbf{r}_n) \right] \underline{\mathbf{p}}_n, \quad \mathbf{r} \neq \mathbf{r}_n, \end{aligned} \tag{14.55}$$

where we used the complex representation of the time-harmonic fields. $\overleftrightarrow{\mathbf{G}}$ denotes the dyadic Green function, $\underline{\mathbf{p}}_n$ the electric dipole moment at $\mathbf{r} = \mathbf{r}_n$, and $\underline{\mathbf{E}}_0$, $\underline{\mathbf{H}}_0$ the exciting field. The system is assumed to consist of N individual dipoles. To first order, the dipole moment $\underline{\mathbf{p}}_n$ is

$$\underline{\mathbf{p}}_n = \alpha(\omega) \underline{\mathbf{E}}(\mathbf{r}_n). \tag{14.56}$$

By combining Eqs. (14.55) and (14.56) we obtain implicit equations for the fields $\underline{\mathbf{E}}$ and $\underline{\mathbf{H}}$, which can be solved by matrix-inversion techniques. In principle, the mechanical force acting on an arbitrary object made of single dipolar subunits can be determined by using Eq. (14.40) in combination with Eqs. (14.55) and (14.56). However, if we require that the object does not deform under the influence of the electromagnetic field, the internal forces must cancel out and the mechanical force is entirely determined by the fields outside of the object. In this case, the mechanical force can be determined by solving for the

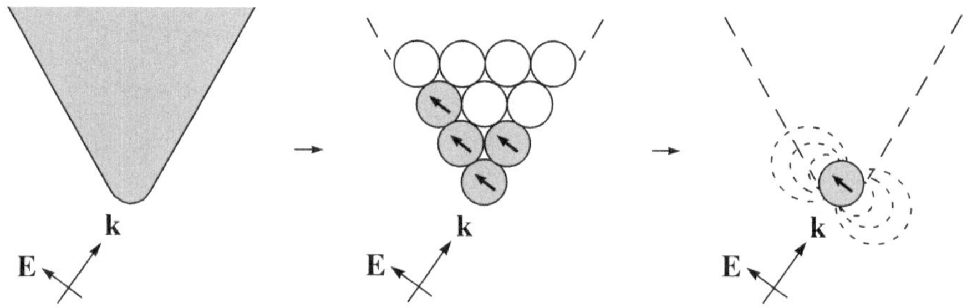

Fig. 14.6 Illustration of the coupled-dipole approach. A macroscopic object is subdivided into individual microscopic dipolar subunits. Each dipole moment can be calculated self-consistently by using the Green-function formalism. In a rough approximation the field in front of a metal tip can be replaced by the field of a single dipole. However, the parameters of the polarizability have to be deduced from a rigorous calculation.

fields outside the object and evaluating Maxwell's stress tensor according to Eqs. (14.10) and (14.15). See Figure 14.6.

14.5 Optical tweezers

In 1986 Ashkin and coworkers showed that a single tightly focused laser beam could be used to hold, in three dimensions, a microscopic particle near the beam focus. This has now become established as a powerful non-invasive technique known as *optical tweezers* [2]. Optical tweezers have found widespread application, especially in biology, and have been used to manipulate dielectric spheres, living cells, DNA, bacteria, and metallic particles. Optical tweezers are routinely applied to measure the elasticity, force, torsion, and position of a trapped object. Forces measured with optical tweezers are typically in the range 1–10 pN. While trapping of small particles (diameters $d \ll \lambda$) is well explained by the dipole force (the first term in Eq. (14.44)), a theory for trapping of larger particles requires an extension of the dipole approximation by including higher multipole orders, similar to Mie scattering. The trapping force can be represented in the form

$$\langle \mathbf{F}(\mathbf{r}) \rangle = \mathbf{Q}(\mathbf{r})\, \frac{\varepsilon_s^2 P}{c}, \tag{14.57}$$

where ε_s is the dielectric constant of the surrounding medium, P is the power of the trapping beam, and c is the vacuum speed of light. The dimensionless vector $\mathbf{Q}$ is called the trapping efficiency. In the dipole limit and in the absence of particle losses, $\mathbf{Q}$ depends on the normalized gradient of the light intensity and the polarizability α given by

$$\alpha(\omega) = 3\varepsilon_0 V_0 \frac{\varepsilon(\omega) - \varepsilon_s(\omega)}{\varepsilon(\omega) + 2\varepsilon_s(\omega)}, \tag{14.58}$$

where V_0 and ε are the particle's volume and dielectric constant, respectively. Figure 14.7 shows the maximum axial trapping efficiency Max$[Q_z(x = 0, y = 0, z)]$ for a polystyrene particle ($\varepsilon = 2.46$) with variable radius r_0 irradiated by a focused Gaussian beam. For small particles ($r_0 < 100$ nm) the trapping efficiency scales as r_0^3 in accordance with the dipole approximation and Eq. (14.58). However, for larger particles, the dipole approximation becomes inaccurate.

As illustrated in Fig. 14.8, a simple ray-optical analysis can be applied to describe trapping of particles larger than the wavelength. In this model, every refraction of a light ray at the particle surface transfers momentum from the trapping laser to the particle. The time rate of change of the momentum is the trapping force. The total force can be calculated by representing the light beam as a collection of rays (see Section 3.5) and summing the forces due to each of the rays. Stable trapping requires that there is a position for which the net force on the particle is zero and any displacement results in a restoring force towards the "zero-force" position. The reader is referred to the work of Ashkin for further details on optical trapping in the ray-optics regime [13].

An important concept in applications of laser tweezers is the trap *stiffness* k. For small displacements x from the equilibrium position, the trapping potential can be approximated by a harmonic function and the restoring force becomes linearly dependent on x,

$$\langle F \rangle = kx. \tag{14.59}$$

In principle, k is a tensor of rank two since the stiffness depends on the direction of displacement. For a single-beam gradient trap it is often sufficient to distinguish between transverse and longitudinal stiffness. The trap stiffness depends on the particle's polarizability, the excitation power and the field gradients. Figure 14.9 illustrates the linear approximation for a paraxial Gaussian beam. The trap stiffness can be measured experimentally by using the viscous drag force F_d acting on a particle inside a medium with relative velocity v. For a spherical particle with radius r_0, F_d is described by Stokes' law,

$$\langle F_d \rangle = 6\pi \eta r_0 v. \tag{14.60}$$

Here, η is the viscosity of the medium ($10^{-3}\,\text{N s m}^{-2}$ for water) and it is assumed that inertial forces are negligible (small Reynolds number). Thus, on moving the surrounding medium with velocity v past a stationary trapped particle of known size, Stokes' law determines the force $\langle F_d \rangle$ exerted on the particle. This force has to be balanced by the trapping force $\langle F \rangle$ in Eq. (14.59), which allows us to determine the stiffness k by measuring the displacement x. There are different ways to establish a relative speed v between a particle and a surrounding medium: (1) the medium is pumped past a stationary particle using a flow chamber, (2) the chamber containing the medium is moved past a stationary particle using piezo-transducers or a motorized stage, and (3) the optical trap is moved using beam-steering methods while the medium remains stationary. No matter what the method is, the calibration of k relies on an accurate measurement of the displacement x. Most commonly, x is determined by refocusing the scattered light from the trapped particle onto a position-sensitive detector, such as a silicon quadrant detector [14].

Brownian motion has to be taken into account if the depth of the trapping potential is not negligible compared with the energy $k_B T$. Stable trapping often requires a trap

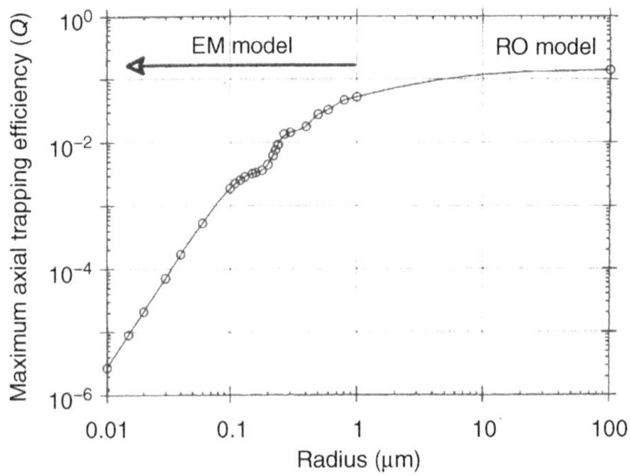

Fig. 14.7 Calculated maximum axial trapping efficiency $\text{Max}[Q_z(x=0, y=0, z)]$ for a polystyrene particle ($\varepsilon = 2.46$) with variable radius r_0 irradiated by a focused Gaussian beam. The surrounding medium is water ($\varepsilon_s = 1.77$) and the numerical aperture is 1.15. Reprinted with permission from [12].

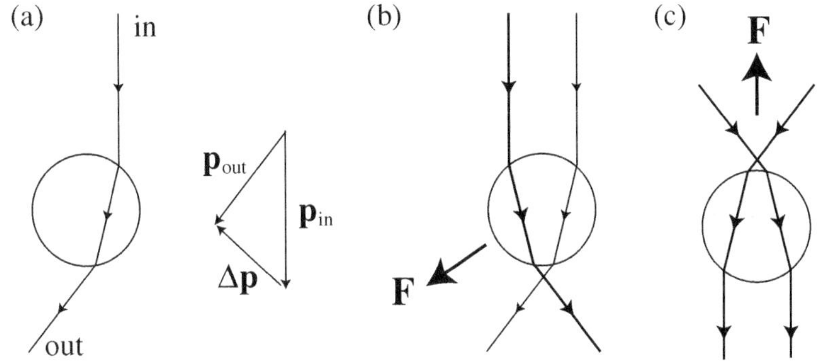

Fig. 14.8 Illustration of the ray-optics picture for optical trapping of particles larger than the wavelength. (a) A single ray is refracted twice at the surface of the particle. The net momentum change $\Delta\boldsymbol{p}$ is calculated by taking the vectorial difference of the momenta of incoming and outgoing rays. Momentum conservation requires that the momentum transferred to the particle is $-\Delta\boldsymbol{p}$. (b) Refraction of two light rays with different intensities. The particle is pulled towards the higher intensity. (c) Axial trapping of a particle in a single-beam trap. A particle initially located beneath the focus is pulled towards the focus.

depth of $\approx 10 k_B T$. Brownian motion leads to noise in force measurements, giving rise to a characteristic power spectrum [3]. Unfortunately, the Langevin equation cannot be solved for a trapping potential with finite depth. Therefore, to answer questions regarding trap stability it is necessary to solve the Fokker–Planck equation [15].

14.6 Angular momentum and torque

Besides energy and momentum, an electromagnetic field can also carry angular momentum, which exerts a mechanical torque on an irradiated structure. This torque can be calculated from a conservation law for angular momentum similar to Eq. (14.13):

$$-\int_{\partial V} \left[\overleftrightarrow{\mathbf{T}}(\mathbf{r},t) \times \mathbf{r}\right] \cdot \mathbf{n}(\mathbf{r}) \mathrm{d}a = \frac{\mathrm{d}}{\mathrm{d}t}\left[\mathbf{J}_{\text{field}} + \mathbf{J}_{\text{mech}}\right]. \tag{14.61}$$

As before, ∂V denotes the surface of a volume enclosing the irradiated structure, $\mathbf{n}$ is the unit vector perpendicular to the surface, and $\mathrm{d}a$ is an infinitesimal surface element. $\mathbf{J}_{\text{field}}$ and $\mathbf{J}_{\text{mech}}$ denote the total mechanical and electromagnetic angular momentum, respectively, and $[\overleftrightarrow{\mathbf{T}} \times \mathbf{r}]$ is the angular-momentum flux-density pseudotensor. The mechanical torque $\mathbf{N}$ acting on the irradiated structure is defined as

$$\mathbf{N} = \frac{\mathrm{d}}{\mathrm{d}t}\mathbf{J}_{\text{mech}}. \tag{14.62}$$

For a monochromatic field the time-averaged torque can be represented as

$$\langle \mathbf{N} \rangle = -\int_{\partial V} \left\langle \overleftrightarrow{\mathbf{T}}(\mathbf{r},t) \times \mathbf{r}\right\rangle \cdot \mathbf{n}(\mathbf{r}) \mathrm{d}a, \tag{14.63}$$

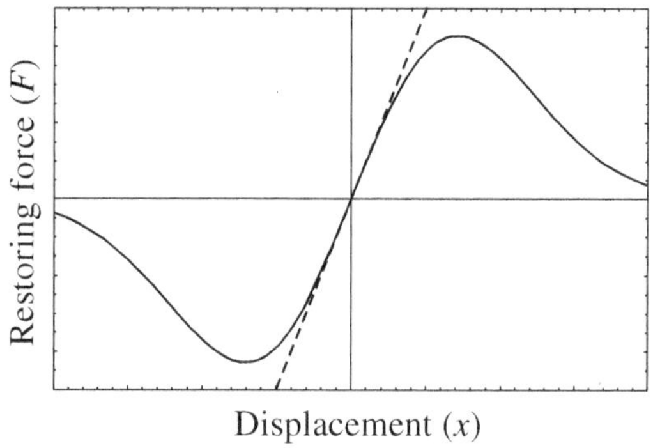

Fig. 14.9 Linear approximation (dashed line) to the restoring force (solid line) for a single-beam gradient trap. The slope of the linear approximation is denoted as the trap stiffness k. It depends on the particle polarizability, laser power, and field gradients.

where we have used the fact that $\langle \mathrm{d}\mathbf{J}_{\mathrm{field}}/\mathrm{d}t \rangle = 0$. Equation (14.63) allows us to calculate the mechanical torque acting on an arbitrary body within the closed surface ∂V. The torque is entirely determined by the electric and magnetic fields on the surface ∂V.

One of the first demonstrations of angular momentum transfer from an optical beam to an irradiated object was performed by Beth in 1936 [16]. He measured the torque on a suspended birefringent half-wave plate as circularly polarized light passed through it. This experiment provided evidence that the angular momentum per photon in a pure circularly polarized state is $\hbar$. Since Beth's experiment, various demonstrations have been performed, demonstrating that optical beams with non-vanishing angular field momentum can indeed be used to promote a trapped particle into a spinning state [17] and applications as optical and biological micromachines have been suggested [18].

14.7 Forces in optical near-fields

Optical near-fields are mainly composed of evanescent-field terms that decay rapidly with distance from the source. This fast decay leads to strong field gradients and thus to strong dipole forces. Evanescent waves created by total internal reflection at a glass/air interface have been used as atomic mirrors. In these experiments, an atomic beam incident on the interface is deflected by the dipole force exerted by the evanescent field if the light frequency is tuned to the blue side of an electronic resonance [19]. Evanescent fields have also been used to accelerate micrometer-sized particles along a plane surface and along planar waveguides by means of the scattering force [20]. Optical near-field traps have been proposed for atom trapping [21] and also for the manipulation of polarizable particles with diameters down to 10 nm [20]. The strongest dipole forces arise from strongly enhanced fields near material edges, corners, gaps, and tips. Therefore, as an application of the theory developed in Section 14.4 we calculate the forces near a sharp metal tip.

The electric field distribution for a laser-illuminated gold tip is strongly polarization dependent [20]. Figure 14.10 shows the electric field distribution (calculated with the

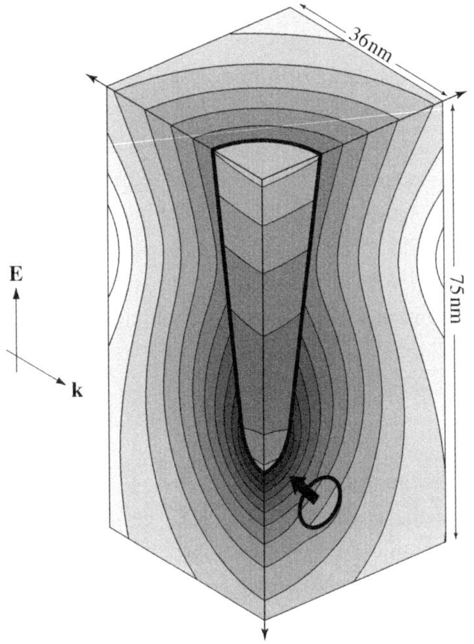

Fig. 14.10 Trapping of a dielectric particle by a laser-irradiated gold tip in water. The figure shows contour lines of $E^2 = \underline{\mathbf{E}} \cdot \underline{\mathbf{E}}^*$ (with a factor of 2 between successive lines) for plane-wave excitation with $\lambda = 810$ nm and polarization along the tip axis. The diameters of tip and particle are 10 nm. The arrow indicates the direction of the trapping force.

MMP method) near a sharply pointed gold tip irradiated with a monochromatic plane wave polarized along the tip axis. The field lines are slightly distorted by a small particle in the vicinity of the tip. The arrow indicates the trapping force acting on the particle. While the intensity at the foremost part of the tip is strongly enhanced over the intensity of the excitation light, no enhancement beneath the tip is observed when the exciting light is polarized perpendicular to the tip axis. Calculations for platinum and tungsten tips show lower enhancements, whereas the field beneath a glass tip is reduced compared with the excitation field.

With the field distribution around the tip determined, the force acting on the particle can be calculated by evaluating Maxwell's stress tensor. However, in order to avoid elaborate computations, we represent both the tip and the particle by point dipoles. The dipole force acting on a Rayleigh particle can be easily calculated as (cf. Eq. (14.47))

$$\langle \mathbf{F} \rangle = (\alpha'/2)\nabla\langle|\mathbf{E}|^2\rangle = (\alpha'/2)\nabla(\underline{\mathbf{E}} \cdot \underline{\mathbf{E}}^*), \tag{14.64}$$

where α' is the real part of the polarizability of the particle and $\mathbf{E}$ is the electric field in the absence of the particle. The particle tends to move to the higher-intensity region where its induced dipole has lower potential energy. We neglect the scattering force (the second term in Eq. (14.47)) because of the small particle size. The assumptions inherent in Eq. (14.64) are that the external field is homogeneous across the particle and that the particle does not alter the field $\mathbf{E}$ in Eq. (14.64). These assumptions, however, do not hold for the particle shown in Fig. 14.10. The intensity contours are distorted around the particle and the field inside is highly inhomogeneous. Nevertheless, it will be shown later by comparison with the exact solutions that the point-dipole approximation leads to reasonable results.

The situation to be analyzed is shown in Fig. 14.11. The metal tip is illuminated by a plane wave at right angles such that the polarization is parallel to the tip axis.

Calculations show that the spatial distribution of the fields close to the metal tip is similar to the field of an on-axis dipole $\mathbf{p}_t$. Without loss of generality, we place this dipole at the origin of the coordinate system. The dipole moment $\mathbf{p}_t$ can be expressed in terms of the computationally determined enhancement factor, f, for the electric field intensity ($|\mathbf{E}|^2$) as

$$\mathbf{E}(x=0, y=0, z=r_t) = \frac{2\mathbf{p}_t}{4\pi\varepsilon_0\varepsilon_s r_t^3} \equiv \sqrt{f}\,\mathbf{E}_0, \tag{14.65}$$

where r_t denotes the tip radius ($z = r_t$ is the foremost end of the tip), ε_s is the dielectric constant of the environment, and $\mathbf{E}_0$ is the electric field amplitude of the exciting plane wave. Equation (14.65) allows us to calculate the dipole moment of the tip as a function of the tip size and the enhancement factor. Since we consider tip–particle distances d for which $kd \ll 1$, we retain only the dipole's near-field, from which we calculate

$$\underline{\mathbf{E}} \cdot \underline{\mathbf{E}}^* = \frac{|\mathbf{p}_t|^2}{(4\pi\varepsilon_0\varepsilon_s)^2}\,\frac{1+3(z/r)^2}{r^6}, \tag{14.66}$$

where $r = \sqrt{x^2+y^2+z^2}$.

We assume that the coupling between tip and particle can be neglected. In this limit, the incident field $\mathbf{E}_0$ excites a dipole moment $\mathbf{p}_t$ in the tip and the fields generated by $\mathbf{p}_t$ induce a dipole moment $\mathbf{p}$ in the particle. Using Eq. (14.66) together with the expression

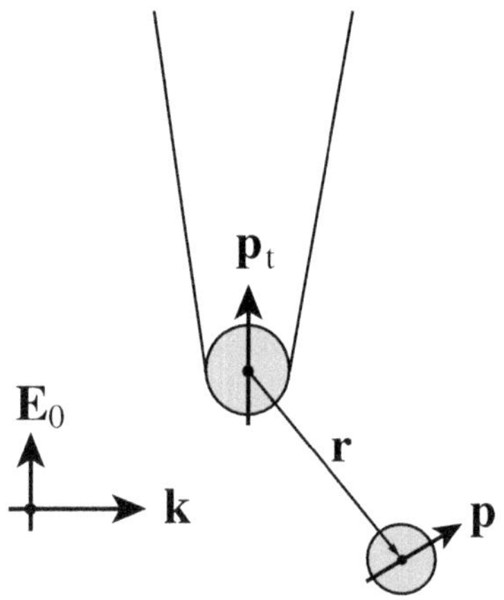

Fig. 14.11 Trapping of a particle by a laser-illuminated metal tip. The tip is illuminated by a plane wave polarized along the tip axis. Both the tip and the particle are represented by dipoles.

for $\alpha(\omega)$ in Eq. (14.58), the force acting on the particle located at (x, y, z) is determined by Eq. (14.64) as

$$\langle \mathbf{F} \rangle = -\frac{3 r_t^6 f E_0^2 \alpha'}{4 r^6} \left[\rho (1 + 4z^2/r^2) \mathbf{n}_\rho + 4(z^3/r^2) \mathbf{n}_z \right]. \tag{14.67}$$

Here, $\mathbf{n}_z$ and $\mathbf{n}_\rho$ denote the unit vectors along the tip axis and in the transverse direction, respectively, and the transverse distance is $\rho = \sqrt{x^2 + y^2}$. The minus sign indicates that the force is directed towards the tip. We find that $\langle \mathbf{F} \rangle$ is proportional to the enhancement factor f, the intensity of the illuminating light $I_0 = (1/2)\sqrt{\varepsilon_0 \varepsilon_s / \mu_0}\, E_0^2$, the real part of the particle polarizability α', and the sixth power of the tip radius a_t. It has to be kept in mind that f and r_t are not independent parameters; their relationship can be determined only by rigorous calculations.

We now calculate the potential energy of the particle in the field of the tip dipole (the trapping potential) as

$$V_{\text{pot}}(\mathbf{r}) = -\int_{\infty}^{\mathbf{r}} \langle \mathbf{F}(\mathbf{r}') \rangle \mathrm{d}\mathbf{r}'. \tag{14.68}$$

The integration path from $\mathbf{r}$ to ∞ is arbitrary because $\mathbf{F}$ is a conservative vector field. After carrying out the integration we find

$$V_{\text{pot}}(\mathbf{r}) = -r_t^6 f E_0^2 \alpha' \, \frac{1 + 3z^2/r^2}{8 r^6}. \tag{14.69}$$

The maximum value of V_{pot} is reached exactly in front of the tip at $z = r_0 + r_t$, r_0 being the particle's radius. Figure 14.12 shows V_{pot} along the tip axis and along a transverse axis immediately in front of the tip. Since in aqueous environments the trapping forces compete with Brownian motion, the potential is normalized with $k_B T$ (k_B is the Boltzmann constant, $T = 300$ K). Additionally, the curves are scaled with the incident intensity I_0.

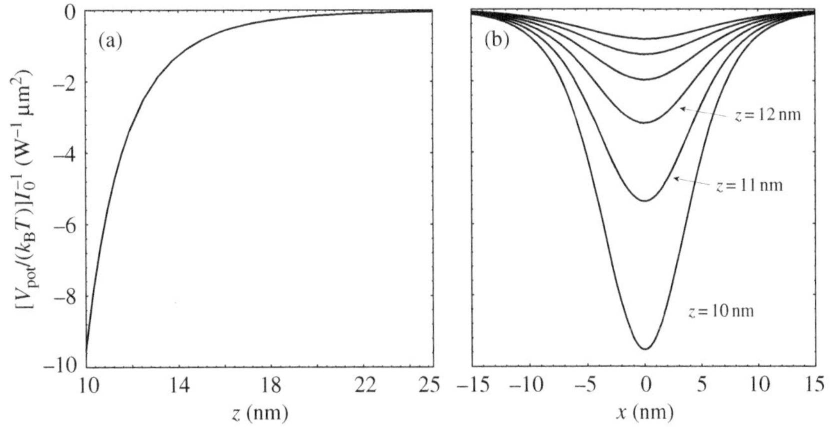

Fig. 14.12 The trapping potential V_{pot} along the tip axis (a) and along a transverse direction at $z = r_t + r_0$ beneath the tip. The radii of tip and particle are $r_t = r_0 = 5$ nm. The forces are normalized with $k_B T$ and the incident intensity I_0.

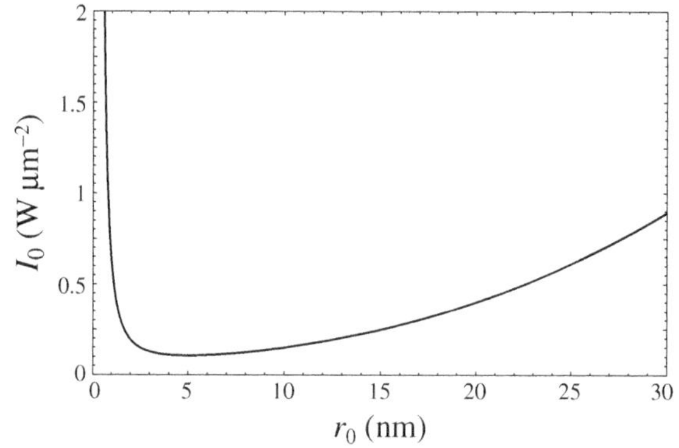

Fig. 14.13 The minimum trapping intensity I_0 as a function of the particle radius r_0. The tip radius is $r_t = 5$ nm.

Let us assume for the following that a sufficient condition for trapping is $V_{pot} > k_B T$. We can then calculate the intensity required in order to trap a particle of a given size. Using the expression for the particle polarizability and evaluating Eq. (14.69) at $\mathbf{r} = (r_t + r_0)\mathbf{n}_z$, we find

$$I_0 > \frac{k_B T c}{4\pi \sqrt{\varepsilon_s}} \operatorname{Re}\left\{ \frac{\varepsilon_p + 2\,\varepsilon_s}{\varepsilon_p - \varepsilon_s} \right\} \frac{(r_t + r_0)^6}{f\, r_t^6\, r_0^3}. \tag{14.70}$$

The curve for which the equality holds is shown in Fig. 14.13. The minimum in the curve indicates that the incident intensity and the tip radius can be adjusted to selectively trap particles with sizes in a limited range. Particles that are too small are not trapped because their polarizability is too small. On the other hand, for particles that are too big the minimum separation between tip and particle $(r_t + r_0)$ becomes too large. As a rule of thumb, the particle size should be in the range of the tip size.

Notice that, instead of calculating first the trapping force, the potential $V_{pot}(\mathbf{r})$ could have been more easily determined by considering the interaction energy of the particle in the dipole approximation. With $\mathbf{E}$ being the field of the tip dipole $\mathbf{p}_t$ it is easy to show that

$$V_{pot}(\mathbf{r}) = -\mathbf{p} \cdot \mathbf{E}(\mathbf{r}) = -(\alpha'/2) E^2(\mathbf{r}) \tag{14.71}$$

leads to the same result as Eq. (14.69).

The simple two-dipole model applied here renders a trapping potential whose general shape is in very good agreement with the results in Ref. [20]. A comparison shows that the forces calculated here are off by a factor ≈2–3. Experiments have shown that the formation of eddy currents in the aqueous environment does have an effect on the trapping scheme. These eddy currents are generated by laser heating of the metal tip.

14.8 Conclusion

We have discussed light-induced forces acting on polarizable objects. These forces are conveniently described by Maxwell's stress-tensor formalism, which yields both gradient forces and radiation pressure for arbitrarily shaped objects. For objects much smaller than the wavelength of light the fields can be represented by a multipole series and the lowest dipole term yields the familiar gradient force and scattering force. The former is the key ingredient in optical tweezers, whereas the latter constitutes the recipe for atomic cooling. In general, the forces are semiclassical in nature, which means that the fields can be treated classically whereas the material properties (polarizabilities) require a quantum treatment. Because of the strong field gradients associated with optical near-fields, gradient forces can potentially be exploited for translating, manipulating, and controlling nanoscale structures. However, near-fields are strongest at material interfaces and hence additional counteracting forces (van der Waals, electrostatic) are needed in order to create stable trapping beyond the material boundaries.

Problems

14.1 A spherical glass particle in water is trapped at the focus of a monochromatic paraxial Gaussian beam with $\lambda = 800\,\text{nm}$ and variable NA (see Section 3.2). The polarizability of the particle is

$$\alpha = 3\varepsilon_0 V_0 \frac{\varepsilon - \varepsilon_\text{w}}{\varepsilon + 2\varepsilon_\text{w}}, \tag{14.72}$$

where V_0 is the volume of the particle, and the dielectric constants of glass and water are $\varepsilon = 2.25$ and $\varepsilon_\text{w} = 1.76$, respectively.

(1) Show that for small transverse displacements (x) from the focus the force is proportional to x. Determine the spring constant as a function of NA, d_0, λ, and P_0, where d_0 is the particle diameter and P_0 the laser power.

(2) Is it possible to derive in the same way a spring constant for longitudinal displacements z? If yes, calculate the corresponding spring constant as a function of NA, d_0, and P_0.

(3) Assume NA $= 1.2$ and $d_0 = 100\,\text{nm}$. What laser power is necessary in order to create a trapping potential $V > 10k_\text{B}T$, where k_B is Boltzmann's constant and $T = 300\,\text{K}$ is the ambient temperature? What is the restoring force for a transverse displacement of $x = 100\,\text{nm}$?

14.2 Consider the total internal reflection of a plane wave with wavelength $\lambda = 800\,\text{nm}$ incident at an angle $\theta = 70°$ from the normal of a glass/air interface ($\varepsilon = 2.25$). The plane wave is incident from the glass side and is s-polarized. The normal of the interface is parallel to the gravitational axis and the air side is pointing to the bottom. A tiny glass particle is trapped on the air side in the evanescent field generated by the totally internally reflected plane wave. Calculate the minimum required intensity I of the plane wave to prevent the glass particle from falling down (α given by Eq. (14.72) with $\varepsilon_\text{w} = 1$). The specific density of glass is $\rho = 2.2 \times 10^3\,\text{kg}\,\text{m}^{-3}$ and the particle diameter is $d_0 = 100\,\text{nm}$. What happens if the particle size is increased?

14.3 A particle is placed into the field of two counter-propagating plane waves of identical amplitudes, phases, and polarizations. The gradient force retains the particle in a transverse plane formed by the constructive interference of the two waves. The intensity of a single plane wave is I and the polarizability of the particle is α. Calculate the energy required to promote the particle from one constructive interference plane to the next as a function of I.

14.4 Calculate the mutual attraction force between two identical dipolar particles that are irradiated by a plane wave polarized along the axis defined by the two particle centers. Plot the force as a function of the interparticle distance and use suitable normalizations for the axes.

14.5 Evaluate Maxwell's stress tensor on a spherical surface enclosing a Rayleigh particle irradiated by a plane wave. What does the result tell you?

References

[1] R. Frisch, "Experimenteller Nachweis des Einsteinschen Strahlungsrückstosses," *Z. Phys.* **86**, 42–45 (1933).

[2] A. Ashkin, "Optical trapping and manipulation of neutral particles using lasers," *Proc. Nat. Acad. Sci.* **94**, 4853–4860 (1987).

[3] K. Svoboda and S. T. Block, "Biological applications of optical forces," *Annu. Rev. Biophys. Biomol. Struct.* **23**, 247–285 (1994).

[4] B. Pringsheim, "Zwei Bemerkungen über den Unterschied von Lumineszenz- und Temperaturstrahlung," *Z. Phys.* **57**, 739–741 (1929).

[5] T. W. Hänsch and A. L. Schawlow, "Cooling of gases by laser radiation," *Opt. Commun.* **13**, 68–69 (1975).

[6] Y. Shimizu and H. Sasada, "Mechanical force in laser cooling and trapping," *Am. J. Phys.* **66**, 960–967 (1998).

[7] S. Stenholm, "The semiclassical theory of laser cooling," *Rev. Mod. Phys.* **58**, 699–739 (1986).

[8] S. Albaladejo, M. I. Marques, M. Laroche, and J. J. Saenz, "Scattering forces from the curl of the spin angular momentum of a light field," *Phys. Rev. Lett.* **102**, 113602 (2009).

[9] M. L. Juan, R. Gordon, Y. Pang, F. Eftekhari, and R. Quidant, "Self-induced back-action optical trapping of dielectric nanoparticles," *Nature Phys.* **5**, 915–919 (2009).

[10] J. P. Gordon and A. Ashkin, "Motions of atoms in a radiation trap," *Phys. Rev. A* **21**, 1606–1617 (1980).

[11] S. Chu, J. E. Bjorkholm, A. Ashkin, and A. Cable, "Experimental observation of optically trapped atoms," *Phys. Rev. Lett.* **57**, 314–317 (1986).

[12] W. H. Wright, G. J. Sonek, and M. W. Berns, "Radiation trapping forces on microspheres with optical tweezers," *Appl. Phys. Lett.* **63**, 715–717 (1993). Copyright 1993 American Institute of Physics.

[13] A. Ashkin, "Forces of a single-beam gradient laser trap on a dielectric sphere in the ray optics regime," *Biophys. J.* **61**, 569–582 (1992).

[14] F. Gittes and C. F. Schmidt, "Interference model for back-focal-plane displacement detection in optics tweezers," *Opt. Lett.* **23**, 7–9 (1998).

[15] R. Zwanzig, *Nonequilibrium Statistical Mechanics*. Oxford: Oxford University Press (2001).

[16] R. A. Beth, "Mechanical detection and measurement of the angular momentum of light," *Phys. Rev.* **50**, 115–125 (1936).

[17] See, for example, T. A. Nieminen, N. R. Heckenberg, and H. Rubinsztein-Dunlop, "Optical measurement of microscopic torques," *J. Mod. Opt.* **48**, 405–413 (2001).

[18] See, for example, L. Paterson, M. P. MacDonald, J. Arlt, *et al.*, "Controlled rotation of optically trapped microscopic particles," *Science* **292**, 912–914 (2001).

[19] For a review see C. S. Adams, M. Sigel, and J. Mlynek, "Atom optics," *Phys. Rep.* **240**, 143–210 (1994).

[20] S. Kawata and T. Tani, "Optically driven Mie particles in an evanescent field along a channeled waveguide," *Opt. Lett.* **21**, 1768–1770 (1996).

[21] S. K. Sekatskii, B. Riedo, and G. Dietler, "Combined evanescent light electrostatic atom trap of subwavelength size," *Opt. Commun.* **195**, 197–204 (2001).

[22] L. Novotny, R. X. Bian, and X. S. Xie, "Theory of nanometric optical tweezers," *Phys. Rev. Lett.* **79**, 645–648 (1997).

15 Fluctuation-induced interactions

The thermal and zero-point motion of electrically charged particles inside materials gives rise to a fluctuating electromagnetic field. Quantum theory tells us that the fluctuating particles can only assume discrete energy states and, as a consequence, the emitted fluctuating radiation takes on the spectral form of blackbody radiation. However, while the familiar blackbody radiation formula is strictly correct at thermal equilibrium, it is only an approximation for non-equilibrium situations. This approximation is reasonable at large distances from the emitting material (far-field) but it can strongly deviate from the true behavior close to material surfaces (near-field).

Because fluctuations of charge and current in materials lead to dissipation via radiation, no object at finite temperature can be in thermal equilibrium in free space. Equilibrium with the radiation field can be achieved only by confining the radiation to a finite space. However, in most cases the object can be considered to be close to equilibrium and the non-equilibrium behavior can be described by linear-response theory. In this regime, the most important theorem is the *fluctuation–dissipation theorem*. It relates the rate of energy dissipation in a non-equilibrium system to the fluctuations that occur spontaneously at different times in equilibrium systems.

The fluctuation–dissipation theorem is of relevance for the understanding of fluctuating fields near nanoscale objects and optical interactions at nanoscale distances (e.g. the van der Waals force). This chapter is intended to provide a detailed derivation of important aspects in fluctuational electrodynamics.

15.1 The fluctuation–dissipation theorem

The fluctuation–dissipation theorem has its roots in Nyquist's relation for voltage fluctuations across a resistor. However, it was Callen and Welton who derived the theorem in its general form [1]. The derivation presented here is purely classical. A substitution at the end of the derivation introduces the Planck constant into the theorem. We consider a nanoscale system with characteristic dimensions much smaller than the wavelength of light (see Fig. 15.1). This allows us to treat the interaction with the system in the electric-dipole approximation. The theory can be easily extended by including higher-order multipolar terms. The nanoscale system consists of a finite number of charged particles with N degrees

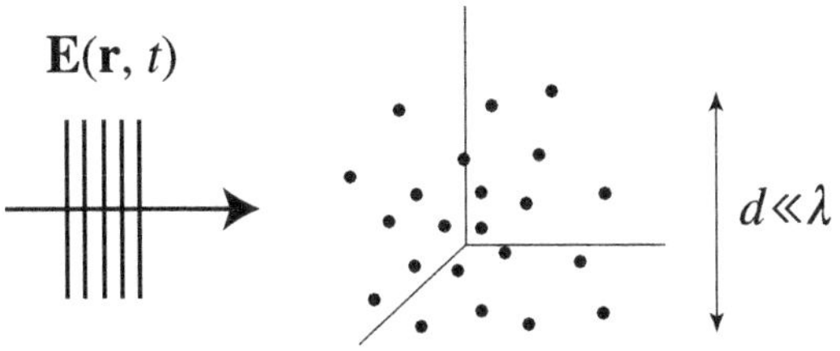

Fig. 15.1 Interaction of an optical field with a system of particles initially at thermal equilibrium. The state of the system is defined by the phase-space coordinate $s=[q_1 \dots q_N;\, p_1 \dots p_N]$, with q_j and p_j being the coordinates and conjugate momenta, respectively. If the characteristic length scale d of the system is small compared with the wavelength λ, the interaction energy between the optical field and the system is given by the electric dipole approximation $\delta H = -\mathbf{p}(s,t)\cdot\mathbf{E}(t)$, where $\mathbf{p}$ is the electric dipole moment.

of freedom. At *thermal equilibrium*, the probability of the system's dipole moment $\mathbf{p}$ being in state $s=[q_1 \dots q_N;\, p_1 \dots p_N]$ is given by the distribution function

$$f_{\text{eq}}(s) = f_0 \mathrm{e}^{-H_0(s)/(k_\text{B}T)}, \tag{15.1}$$

where f_0 is a normalization constant ensuring that $\int f_{\text{eq}}\,\mathrm{d}s = 1$. H_0 is the equilibrium Hamiltonian of the system, k_B the Boltzmann constant, and T the temperature. q_j and p_j denote the generalized coordinates and conjugate momenta, respectively. s is a point in phase-space and can be viewed as a short-hand for all the coordinates and momenta of the system. At thermal equilibrium the ensemble average of $\mathbf{p}$ is defined as

$$\langle \mathbf{p}(s,t)\rangle = \frac{\int f_{\text{eq}}(s)\mathbf{p}(s,t)\mathrm{d}s}{\int f_{\text{eq}}(s)\mathrm{d}s} = \langle \mathbf{p}\rangle, \tag{15.2}$$

where the integration runs over all coordinates $[q_1 \dots q_N;\, p_1 \dots p_N]$. Because of equilibrium the ensemble average is independent of time.

15.1.1 The system response function

Let us consider an external field $\mathbf{E}(\mathbf{r},t)$ that perturbs the equilibrium of the system. Assuming that the characteristic dimension d of the system is much smaller than the wavelength λ, we can apply the dipole approximation and the Hamiltonian of the perturbed system becomes

$$H = H_0 + \delta H = H_0 - \mathbf{p}(s,t)\cdot\mathbf{E}(t) = H_0 - \sum_{k=x,y,z} p_k(s,t)E_k(t). \tag{15.3}$$

Owing to the external perturbation $\mathbf{E}(t)$ the expectation value of $\mathbf{p}$ will deviate from its equilibrium average $\langle\mathbf{p}\rangle$. We will designate the expectation value of $\mathbf{p}$ in the perturbed system by $\bar{\mathbf{p}}$ in order to distinguish it from $\langle\mathbf{p}\rangle$. We assume that the deviation

$$\delta\bar{\mathbf{p}}(t) = \bar{\mathbf{p}}(t) - \langle\mathbf{p}\rangle \tag{15.4}$$

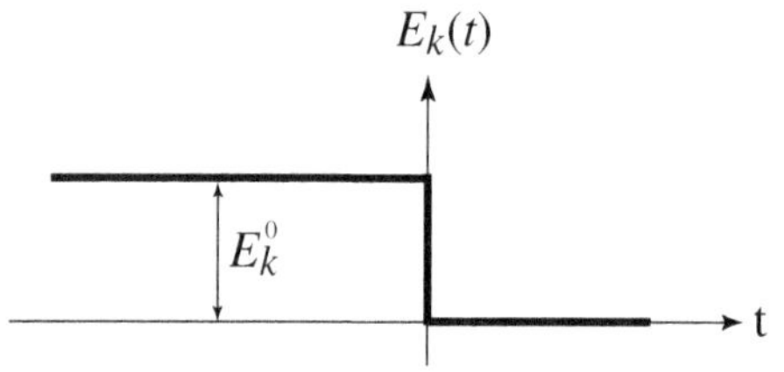

Fig. 15.2 The time dependence of the considered perturbation. The perturbation ensures complete relaxation of the system at times $t = 0$ (immediately before the step) and $t \to \infty$.

is small and that it depends linearly on the external perturbation, i.e.

$$\delta\bar{p}_j(t) = \frac{1}{2\pi}\sum_k \int_{-\infty}^{t} \tilde{\alpha}_{jk}(t-t')\,E_k(t')\mathrm{d}t', \quad j,k = x,y,z. \tag{15.5}$$

Here, $\tilde{\alpha}_{jk}$ is the response function of the system. We have assumed that the system is stationary, $\tilde{\alpha}_{jk}(t,t') = \tilde{\alpha}_{jk}(t-t')$, and causal, $\tilde{\alpha}_{jk}(t-t') = 0$ for $t' > t$. Equation (15.5) states that the response at time t depends not only on the perturbation at time t but also on the perturbations prior to t. The "memory" of the system is contained in $\tilde{\alpha}_{jk}$. Our goal is to determine $\tilde{\alpha}_{jk}$ as a function of the statistical equilibrium properties of the system. It is convenient to consider the perturbation shown in Fig. 15.2, which promotes the system from one completely relaxed (equilibrated) state to another [2]. The relaxation time can be intuitively associated with the memory of the response function. Evaluating Eq. (15.5) for the perturbation shown in Fig. 15.2 gives

$$\delta\bar{p}_j(t) = \frac{E_k^0}{2\pi}\int_{-\infty}^{0} \tilde{\alpha}_{jk}(t-t')\mathrm{d}t' = \frac{E_k^0}{2\pi}\int_{t}^{\infty} \tilde{\alpha}_{jk}(\tau)\,\mathrm{d}\tau, \tag{15.6}$$

which can be solved for $\tilde{\alpha}_{jk}$ as

$$\tilde{\alpha}_{jk}(t) = -\frac{2\pi}{E_k^0}\Theta(t)\,\frac{\mathrm{d}}{\mathrm{d}t}\delta\bar{p}_j(t). \tag{15.7}$$

Here, we assumed that $\tilde{\alpha}_{jk}$ and its time derivative tend to zero for times $t \to \infty$ and we introduced the Heaviside step function $\Theta(t)$ to ensure causality; $\tilde{\alpha}_{jk}(t-t') = 0$ for $t' > t$.[1] According to Eq. (15.7), we find $\tilde{\alpha}_{jk}$ if we calculate the time derivative of $\delta\bar{p}_j$ at time t.

The expectation value of $\mathbf{p}$ at time t is determined by the distribution function $f(s)$ at the initial time $t=0$ according to (see Fig. 15.3)

$$\bar{\mathbf{p}}(t) = \frac{\int f(s)\mathbf{p}(s,t)\mathrm{d}s}{\int f(s)\mathrm{d}s}. \tag{15.8}$$

[1] $\Theta(t) = 0$ for $t < 0$, $\Theta(t) = 1/2$ for $t = 0$, and $\Theta(t) = 1$ for $t > 0$.

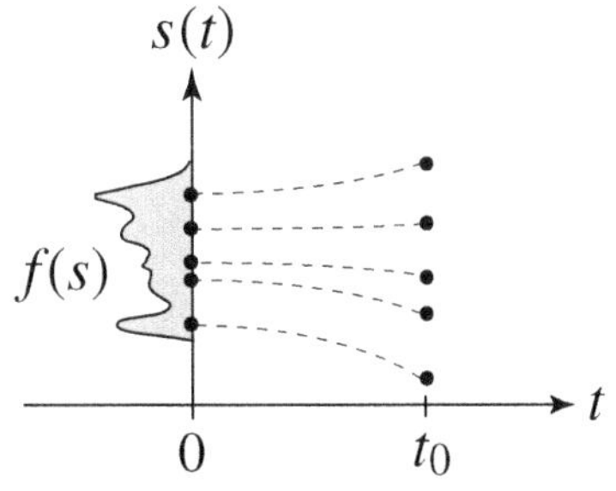

Fig. 15.3 Newton's equations of motion map each phase-space point s at time $t=0$ into a phase-space point at time t_0. The dipole moment at time t_0 can be expressed as $\mathbf{p}[s(t_0)] = \mathbf{p}[s(0), t_0] = \mathbf{p}[s, t_0]$ and its ensemble average at time t_0 is determined by the initial distribution function $f(s)$.

Because of thermal equilibrium at time $t=0$, the distribution function reads as

$$f(s) \propto \mathrm{e}^{-[H_0+\delta H]/(k_\mathrm{B}T)} = f_\mathrm{eq}(s)\,\mathrm{e}^{-\delta H(s)/(k_\mathrm{B}T)} = f_\mathrm{eq}(s)\left(1 - \frac{1}{k_\mathrm{B}T}\,\delta H(s) + \cdots\right), \tag{15.9}$$

where $f_\mathrm{eq}(s)$ is given by Eq. (15.1). The last term in brackets is the series expansion of $\exp[-\delta H/(k_\mathrm{B}T)]$. By insertion into Eq. (15.8) and retaining only terms up to linear order in δH we obtain[2]

$$\bar{\mathbf{p}}(t) = \left\langle \mathbf{p} \right\rangle - \frac{1}{k_\mathrm{B}T}\left[\left\langle \delta H(s)\mathbf{p}(s,t)\right\rangle - \left\langle \mathbf{p}(s,t)\right\rangle\left\langle \delta H(s)\right\rangle\right], \tag{15.10}$$

where $\langle\ldots\rangle$ denotes the expectation value in the absence of the perturbation, i.e. the expectation value calculated by using the distribution function f_eq in Eq. (15.1). Since $\delta H(s)$ is the perturbation at time $t=0$ we have $\delta H(s) = -p_k(s,0)E_k^0$ and Eq. (15.10) can be rewritten as

$$\delta\bar{p}_j(t) = \bar{p}_j(t) - \langle p_j\rangle = -\frac{E_k^0}{k_\mathrm{B}T}\left[\left\langle p_j\right\rangle\left\langle p_k\right\rangle - \left\langle p_k(0)p_j(t)\right\rangle\right]$$
$$= \frac{E_k^0}{k_\mathrm{B}T}\Big\langle [p_k(0) - \langle p_k\rangle][p_j(t) - \langle p_j\rangle]\Big\rangle = \frac{E_k^0}{kT}\left\langle \delta p_k(0)\delta p_j(t)\right\rangle, \tag{15.11}$$

where we used Eq. (15.2) and defined $\delta p_j(t) = [p_j(t) - \langle p_j\rangle]$. On introducing this result into Eq. (15.7) we finally find

$$\tilde{\alpha}_{jk}(t) = -\frac{2\pi}{k_\mathrm{B}T}\Theta(t)\,\frac{\mathrm{d}}{\mathrm{d}t}\left\langle \delta p_k(0)\delta p_j(t)\right\rangle \quad \text{(classical)}. \tag{15.12}$$

This important result is often referred to as the time-domain fluctuation–dissipation theorem. It states that the system's response to a weak external field can be expressed in terms

[2] $[1 - \langle\delta H\rangle/(k_\mathrm{B}T)]^{-1} \approx [1 + \langle\delta H\rangle/(k_\mathrm{B}T) - \cdots]$.

of the system's fluctuations in the absence of the external field! Notice that the correlation function $\langle \delta p_k(0)\delta p_j(t)\rangle$ is a property of the stationary equilibrium system and that the correlation function can be offset by an arbitrary time τ as

$$\left\langle \delta p_k(0)\delta p_j(t)\right\rangle = \left\langle \delta p_k(\tau)\delta p_j(t+\tau)\right\rangle. \tag{15.13}$$

For many problems it is convenient to express Eq. (15.12) in the frequency domain by using the Fourier transforms[3]

$$\alpha_{jk}(\omega) = \frac{1}{2\pi}\int_{-\infty}^{\infty}\tilde{\alpha}_{jk}(t)\mathrm{e}^{\mathrm{i}\omega t}\,\mathrm{d}t, \qquad \delta\hat{p}_j(\omega) = \frac{1}{2\pi}\int_{-\infty}^{\infty}\delta p_j(t)\,\mathrm{e}^{\mathrm{i}\omega t}\,\mathrm{d}t. \tag{15.14}$$

The correlation function in the frequency domain $\langle \delta\hat{p}_j(\omega)\delta\hat{p}_k^*(\omega')\rangle$ can be calculated by substituting the Fourier transforms for $\delta\hat{p}_j(\omega)$ and $\delta\hat{p}_k^*(\omega')$ as

$$\begin{aligned}\left\langle \delta\hat{p}_j(\omega)\,\delta\hat{p}_k^*(\omega')\right\rangle &= \frac{1}{4\pi^2}\iint_{-\infty}^{\infty}\left\langle \delta p_j(\tau')\,\delta p_k(\tau)\right\rangle \mathrm{e}^{\mathrm{i}[\omega\tau'-\omega'\tau]}\,\mathrm{d}\tau'\,\mathrm{d}\tau\\ &= \frac{1}{4\pi^2}\iint_{-\infty}^{\infty}\left\langle \delta p_k(\tau)\,\delta p_j(t+\tau)\right\rangle \mathrm{e}^{\mathrm{i}[\omega-\omega']\tau}\,\mathrm{e}^{\mathrm{i}\omega t}\,\mathrm{d}\tau\,\mathrm{d}t,\end{aligned} \tag{15.15}$$

where we used the substitution $\tau' = \tau + t$. Because of stationarity, the correlation function in the integrand does not depend on τ and the integration over τ reduces to a delta function.[4] The final relation is known as the *Wiener–Khintchine theorem*,

$$\left\langle \delta\hat{p}_j(\omega)\delta\hat{p}_k^*(\omega')\right\rangle = \delta(\omega-\omega')\frac{1}{2\pi}\int_{-\infty}^{\infty}\left\langle \delta p_k(\tau)\delta p_j(t+\tau)\right\rangle \mathrm{e}^{\mathrm{i}\omega t}\,\mathrm{d}t, \tag{15.16}$$

which demonstrates that spectral components that belong to different frequencies are uncorrelated. The integral on the right-hand side is known as the *spectral density*. To obtain a spectral representation of the fluctuation–dissipation theorem, we need to Fourier transform Eq. (15.12). The right-hand side leads to a convolution between the spectrum of the step function, $\hat{\Theta}(\omega)$,[5] and the spectrum of $\mathrm{d}/\mathrm{d}t\,\langle \delta p_k(0)\delta p_j(t)\rangle$. To get rid of the imaginary part of $\hat{\Theta}$ we solve for $\alpha_{jk}(\omega) - \alpha_{kj}^*(\omega)$ instead of $\alpha_{jk}(\omega)$. Making use of stationarity, the Wiener–Khintchine theorem, and the fact that $\langle \delta p_k(\tau)\delta p_j(t+\tau)\rangle$ is real, we obtain

$$\left[\alpha_{jk}(\omega) - \alpha_{kj}^*(\omega)\right]\delta(\omega-\omega') = \frac{2\pi\mathrm{i}\omega}{k_{\mathrm{B}}T}\left\langle \delta\hat{p}_j(\omega)\delta\hat{p}_k^*(\omega')\right\rangle \quad \text{(classical)}. \tag{15.17}$$

This is the analogue of Eq. (15.12) in the frequency domain. The factor $k_{\mathrm{B}}T$ can be identified as the average energy per degree of freedom of a particle in the system (the equipartition principle). This average energy is based on the assumption that the energy distribution of electromagnetic modes is continuous. However, according to quantum

[3] Because the function $\delta p_j(t)$ is a stochastic process it is not square integrable and therefore its Fourier transform is not defined. However, these difficulties can be overcome by the theory of generalized functions, and it can be shown that the Fourier transform can be used in symbolic form [3].

[4] $\int_{-\infty}^{\infty}\exp(\mathrm{i}xy)\mathrm{d}y = 2\pi\,\delta(x)$.

[5] $\hat{\Theta}(\omega) = \frac{1}{2}\delta(\omega) - 1/(2\pi\mathrm{i}\omega)$.

mechanics these modes can only assume discrete energy values separated by $\Delta E = \hbar\omega$ and, as a consequence, the average energy $k_B T$ has to be substituted as

$$k_B T \rightarrow \frac{\hbar\omega}{\exp[\hbar\omega/(k_B T)] - 1} + \hbar\omega, \tag{15.18}$$

which corresponds to the mean energy of the quantum oscillator (first term) plus the zero-point energy $\hbar\omega$ (second term). We choose $\hbar\omega$ instead of $\hbar\omega/2$ in order to be consistent with quantum theory, which requires that $\left\langle \delta\hat{p}_j(\omega)\delta\hat{p}_k^*(\omega')\right\rangle$ is an antinormally ordered quantity for $\omega > 0$ (see Section 15.1.4).

In the limit $\hbar \rightarrow 0$ or $\hbar\omega \ll k_B T$ the substitution (15.18) recovers the classical value of $k_B T$. Rewriting the right-hand side of Eq. (15.18) as $\hbar\omega/\{1 - \exp[-\hbar\omega/(k_B T)]\}$ and substituting into Eq. (15.17) renders the quantum version of the fluctuation–dissipation theorem [4, 5]:

$$\left\langle \delta\hat{p}_j(\omega)\delta\hat{p}_k^*(\omega')\right\rangle = \frac{1}{2\pi \mathrm{i}\omega}\left[\frac{\hbar\omega}{1 - \mathrm{e}^{-\hbar\omega/(k_B T)}}\right]\left[\alpha_{jk}(\omega) - \alpha_{kj}^*(\omega)\right]\delta(\omega - \omega'). \tag{15.19}$$

While *dissipation* is associated with the right-hand side, the left-hand side represents *fluctuations* of the equilibrium system. It is important to notice that quantum mechanics leads to dissipation even for temperatures at absolute zero. The remaining fluctuations affect only *positive* frequencies! This can easily be seen by noting the following limit:

$$\lim_{T\rightarrow 0}\left[\frac{1}{1 - \mathrm{e}^{-\hbar\omega/(k_B T)}}\right] = \Theta(\omega) = \begin{cases} 1 & \omega > 0, \\ 1/2 & \omega = 0, \\ 0 & \omega < 0. \end{cases} \tag{15.20}$$

The fluctuation–dissipation theorem can be generalized to include the spatial dependence of the sources. It turns out that as long as the system's response function is local, i.e. $\tilde{\varepsilon}_{jk}(\mathbf{r}, t) = \tilde{\varepsilon}_{jk}(t)$ or $\varepsilon_{jk}(\mathbf{k}, \omega) = \varepsilon_{jk}(\omega)$, fluctuations at two distinct spatial coordinates are *uncorrelated* [6]. For a fluctuating current density $\delta\mathbf{j}(\mathbf{r}, t)$ in an *isotropic* and *homogeneous* medium with dielectric constant $\varepsilon(\omega)$, Eq. (15.19) can be generalized as [7]

$$\left\langle \delta\hat{j}_j(\mathbf{r}, \omega)\delta\hat{j}_k^*(\mathbf{r}', \omega')\right\rangle = \frac{\omega\varepsilon_0}{\pi}\varepsilon''(\omega)\left[\frac{\hbar\omega}{1 - \mathrm{e}^{-\hbar\omega/(k_B T)}}\right]\delta(\omega - \omega')\delta(\mathbf{r} - \mathbf{r}')\delta_{jk}. \tag{15.21}$$

ε'' is the imaginary part of ε, $\delta\hat{j}$ denotes the Fourier transform of δj, and the Kronecker delta δ_{jk} is a consequence of *isotropy*.

15.1.2 Johnson noise

We finally write the fluctuation–dissipation theorem in the form originally developed by Callen and Welton [1]. We note that the fluctuating dipole moment $\delta\mathbf{p}$ is related to a local stochastic electric field $\delta\mathbf{E}$ according to

$$\delta\hat{p}_j(\omega) = \sum_k \alpha_{jk}(\omega)\delta\hat{E}_k(\omega), \quad j, k = x, y, z, \tag{15.22}$$

which directly follows from the time-domain relationship of Eq. (15.5) on using the definitions of Fourier transforms in Eq. (15.14). Substituting the linear relationship into Eq. (15.19) leads to

$$\left\langle \delta\hat{E}_j(\omega)\delta\hat{E}_k^*(\omega')\right\rangle = \frac{1}{2\pi\mathrm{i}\omega}\left[\frac{\hbar\omega}{1-\mathrm{e}^{-\hbar\omega/(k_\mathrm{B}T)}}\right]\left[\alpha_{kj}^{*\,-1}(\omega)-\alpha_{jk}^{-1}(\omega)\right]\delta(\omega-\omega'). \quad (15.23)$$

This equation renders the local electric field correlation *induced* by the fluctuating dipole. Integrating on both sides over ω' and applying the Wiener–Khintchine theorem leads to

$$\frac{1}{2\pi\mathrm{i}\omega}\left[\frac{\hbar\omega}{1-\mathrm{e}^{-\hbar\omega/(k_\mathrm{B}T)}}\right]\left[\alpha_{kj}^{*\,-1}(\omega)-\alpha_{jk}^{-1}(\omega)\right] = \frac{1}{2\pi}\int\limits_{-\infty}^{\infty}\left\langle \delta E_k(\tau)\,\delta E_j(t+\tau)\right\rangle \mathrm{e}^{\mathrm{i}\omega t}\,\mathrm{d}t. \quad (15.24)$$

Further integration over ω gives rise to a delta function on the right-hand side, which allows the time integral to be evaluated. The final result reads as

$$\left\langle \delta E_k(\tau)\delta E_j(\tau)\right\rangle = \frac{1}{2\pi}\int\limits_{-\infty}^{\infty}\frac{1}{\mathrm{i}\omega}\left[\frac{\hbar\omega}{1-\mathrm{e}^{-\hbar\omega/(k_\mathrm{B}T)}}\right]\left[\alpha_{kj}^{*\,-1}(\omega)-\alpha_{jk}^{-1}(\omega)\right]\mathrm{d}\omega. \quad (15.25)$$

We now apply this formula to charge fluctuations in a resistor. The fluctuating current density can be expressed in terms of the fluctuating dipole moment as $\delta j = \mathrm{d}/\mathrm{d}t\,[\delta p]\delta(\mathbf{r}-\mathbf{r}')$. Assuming an isotropic resistor ($j=k$), the relationship between current and field in the spectral domain becomes $\delta\hat{j}(\omega) = -\mathrm{i}\omega\alpha(\omega)\delta(\mathbf{r}-\mathbf{r}')\delta\hat{E}$, which allows us to identify the term $[-\mathrm{i}\omega\alpha(\omega)\delta(\mathbf{r}-\mathbf{r}')]^{-1}$ with the resistivity $\rho(\omega)$. Assuming that $\rho(\omega)$ is real, we can express Eq. (15.25) as

$$\left\langle \delta E^2\right\rangle = \frac{1}{\pi}\int_{-\infty}^{\infty}\left[\frac{\hbar\omega}{1-\mathrm{e}^{-\hbar\omega/(k_\mathrm{B}T)}}\right]\rho(\omega)\delta(\mathbf{r}-\mathbf{r}')\mathrm{d}\omega, \quad (15.26)$$

which can be rewritten in terms of the voltage V and resistance R as

$$\begin{aligned}\left\langle \delta V^2\right\rangle &= \frac{1}{\pi}\int_{-\infty}^{\infty}\left[\frac{\hbar\omega}{1-\mathrm{e}^{-\hbar\omega/(k_\mathrm{B}T)}}\right]R(\omega)\mathrm{d}\omega\\ &= \frac{1}{\pi}\int_0^{\infty}\left\{\left[\frac{\hbar\omega}{1-\mathrm{e}^{-\hbar\omega/(k_\mathrm{B}T)}}\right]R(\omega) - \left[\frac{\hbar\omega}{1-\mathrm{e}^{-\hbar\omega/(k_\mathrm{B}T)}}-\hbar\omega\right]R(-\omega)\right\}\mathrm{d}\omega\\ &= \frac{2}{\pi}\int_0^{\infty}\left[\frac{\hbar\omega}{\mathrm{e}^{\hbar\omega/(k_\mathrm{B}T)}-1}+\frac{1}{2}\hbar\omega\right]R(\omega)\mathrm{d}\omega. \end{aligned} \quad (15.27)$$

We reduced the integration range to $[0\ldots\infty]$ and made use of $R(\omega)=-R(-\omega)$. The left-hand side can be identified with the mean-square voltage fluctuations. For temperatures $k_\mathrm{B}T \gg \hbar\omega$, which is fulfilled for any practical frequencies at room temperature, we can replace the expression in brackets by its classical limit $k_\mathrm{B}T$. Furthermore, for a system with finite bandwidth $B=(\omega_\mathrm{max}-\omega_\mathrm{min})/(2\pi)$ and a frequency-independent resistance we obtain

$$\left\langle \delta V^2\right\rangle = 4k_\mathrm{B}TBR. \quad (15.28)$$

This is the familiar formula for the *white noise*, also called *Johnson noise*, generated in electrical circuits by resistors. In a bandwidth of 10 kHz and at room temperature, a resistor of 10 MΩ generates a voltage of $\approx 40\ \mu V_{\mathrm{rms}}$.

15.1.3 Dissipation due to fluctuating external fields

We have derived the dissipation of a system as a function of its charge fluctuations. Here we intend to express the dissipation in terms of the fields that the fluctuating charges generate. The current density $\delta\hat{\mathbf{j}}$ in Eq. (15.21) generates an electric field according to

$$\delta\hat{\mathbf{E}}(\mathbf{r},\omega) = \mathrm{i}\omega\mu_0 \int_{V_0} \overleftrightarrow{\mathbf{G}}(\mathbf{r},\mathbf{r}_0;\omega)\delta\hat{\mathbf{j}}(\mathbf{r}_0,\omega)\mathrm{d}^3\mathbf{r}_0, \tag{15.29}$$

where all currents are confined in the source region V_0. Multiplying the above expression by the corresponding expression for the field $\delta\hat{\mathbf{E}}(\mathbf{r}',\omega')$, taking the ensemble average, and applying Eq. (15.21) gives

$$\left\langle \delta\hat{E}_j(\mathbf{r},\omega)\delta\hat{E}_k^*(\mathbf{r}',\omega')\right\rangle = \frac{\omega^3}{\pi c^4\varepsilon_0}\left[\frac{\hbar\omega}{1-\mathrm{e}^{-\hbar\omega/(k_\mathrm{B}T)}}\right]\delta(\omega-\omega') \times \sum_n \int_{V_0} G_{jn}(\mathbf{r},\mathbf{r}_0;\omega)\varepsilon''(\omega)G_{kn}(\mathbf{r}',\mathbf{r}_0;\omega)\mathrm{d}^3\mathbf{r}_0. \tag{15.30}$$

We now note that the dielectric properties of the source region are defined not only by ε'' but also by $\overleftrightarrow{\mathbf{G}}$ because its definition depends on the factor $k^2 = (\omega/c)^2\varepsilon(\omega)$ (cf. Eq. (2.87)). Therefore, it is possible to rewrite the above equation for the electric field correlations by using the identity [8, 9]

$$\frac{\omega^2}{c^2}\sum_n \int_{V_0} G_{jn}(\mathbf{r},\mathbf{r}_0;\omega)\varepsilon''(\omega)G_{kn}(\mathbf{r}',\mathbf{r}_0;\omega)\mathrm{d}^3\mathbf{r}_0 = \mathrm{Im}\left\{G_{jk}(\mathbf{r},\mathbf{r}';\omega)\right\}, \tag{15.31}$$

which can be derived by using $G_{ij}(\mathbf{r}',\mathbf{r};\omega) = G_{ji}(\mathbf{r},\mathbf{r}';\omega)$, requiring that the Green function is zero at infinity, and making use of the definition of $\overleftrightarrow{\mathbf{G}}$ (Eq. (2.87)). In order for $\overleftrightarrow{\mathbf{G}}$ to be zero at infinity, $\overleftrightarrow{\mathbf{G}}$ has to consist of an outgoing and an incoming part, ensuring that there is no net energy transport, i.e. the time-averaged Poynting vector has to be zero for any point in space. This condition ensures that all charges are in equilibrium with the radiation field [10].

The fluctuation–dissipation theorem for the electric field can now be expressed in terms of the Green function alone as

$$\left\langle \delta\hat{E}_j(\mathbf{r},\omega)\delta\hat{E}_k^*(\mathbf{r}',\omega')\right\rangle = \frac{\omega}{\pi c^2\varepsilon_0}\left[\frac{\hbar\omega}{1-\mathrm{e}^{-\hbar\omega/(k_\mathrm{B}T)}}\right]\mathrm{Im}\left\{G_{jk}(\mathbf{r},\mathbf{r}';\omega)\right\}\delta(\omega-\omega'). \tag{15.32}$$

This result establishes the correspondence between field fluctuations (left side) and dissipation (right side), which is expressed in terms of the imaginary part of the Green function.

As before, the result is strictly valid only at equilibrium, i.e. when the field and the sources are at the same temperature.

15.1.4 Normal and antinormal ordering

Let us split the electric field $\mathbf{E}(t)$ at an arbitrary space point $\mathbf{r}$ into two parts as

$$\mathbf{E}(t) = \mathbf{E}^{+}(t) + \mathbf{E}^{-}(t) = \int_{0}^{\infty} \hat{\mathbf{E}}(\omega) e^{-i\omega t}\, d\omega + \int_{-\infty}^{0} \hat{\mathbf{E}}(\omega) e^{-i\omega t}\, d\omega, \tag{15.33}$$

where $\hat{\mathbf{E}}(\omega)$ is the Fourier spectrum of $\mathbf{E}(t)$ (see Section 2.5). The functions $\mathbf{E}^{+}$ and $\mathbf{E}^{-}$ are no longer real functions but are so-called complex analytic signals [3]. $\mathbf{E}^{+}$ is defined by the *positive* frequencies of $\hat{\mathbf{E}}$, whereas $\mathbf{E}^{-}$ is defined by the *negative* frequencies of $\hat{\mathbf{E}}$. Because $\mathbf{E}(t)$ is real we have $\hat{\mathbf{E}}^{*}(\omega) = \hat{\mathbf{E}}(-\omega)$, which implies that $\mathbf{E}^{-} = \left[\mathbf{E}^{+}\right]^{*}$. Let us also define the (inverse) Fourier transforms of $\mathbf{E}^{+}$ and $\mathbf{E}^{-}$:

$$\mathbf{E}^{+}(t) = \int_{-\infty}^{\infty} \hat{\mathbf{E}}^{+}(\omega) e^{-i\omega t}\, d\omega, \qquad \mathbf{E}^{-}(t) = \int_{-\infty}^{\infty} \hat{\mathbf{E}}^{-}(\omega) e^{-i\omega t}\, d\omega. \tag{15.34}$$

Obviously, the spectra are related to the original spectrum $\hat{\mathbf{E}}$ as

$$\hat{\mathbf{E}}^{+}(\omega) = \begin{cases} \hat{\mathbf{E}}(\omega) & \omega > 0, \\ 0 & \omega < 0, \end{cases} \qquad \hat{\mathbf{E}}^{-}(\omega) = \begin{cases} 0 & \omega > 0, \\ \hat{\mathbf{E}}(\omega) & \omega < 0. \end{cases} \tag{15.35}$$

In quantum mechanics, $\hat{\mathbf{E}}^{-}$ is associated with the creation operator $\hat{a}^{\dagger}$ and $\hat{\mathbf{E}}^{+}$ with the annihilation operator $\hat{a}$ (see Section 8.4). The sequence $\hat{\mathbf{E}}^{-}\hat{\mathbf{E}}^{+}$ describes the probability of photon absorption and the sequence $\hat{\mathbf{E}}^{+}\hat{\mathbf{E}}^{-}$ the probability of photon emission [3]. The important thing is that in quantum mechanics the two processes are not the same, i.e. $\hat{\mathbf{E}}^{+}$ and $\hat{\mathbf{E}}^{-}$ do not commute. Therefore, we need to calculate separately the correlations of $\hat{\mathbf{E}}^{-}\hat{\mathbf{E}}^{+}$ (normal ordering) and $\hat{\mathbf{E}}^{+}\hat{\mathbf{E}}^{-}$ (antinormal ordering).

We now turn our attention to the fluctuating field $\delta\hat{\mathbf{E}}(\mathbf{r},t)$ with zero average value and we decompose its Fourier spectrum into positive- and negative-frequency parts. Using the results from Ref. [4] and procedures similar to those used to derive Eq. (15.32), we find

$$\begin{aligned} &\left\langle \delta\hat{E}_j^{-}(\mathbf{r},\omega)\,\delta\hat{E}_k^{+*}(\mathbf{r}',\omega') \right\rangle \\ &\quad = \frac{\omega\Theta(-\omega)}{\pi c^2 \varepsilon_0} \left[\frac{\hbar\omega}{1 - e^{-\hbar\omega/(k_B T)}} \right] \mathrm{Im}\left\{ G_{jk}(\mathbf{r},\mathbf{r}';\omega) \right\} \delta(\omega - \omega'), \end{aligned} \tag{15.36}$$

$$\begin{aligned} &\left\langle \delta\hat{E}_j^{+}(\mathbf{r},\omega)\,\delta\hat{E}_k^{-*}(\mathbf{r}',\omega') \right\rangle \\ &\quad = \frac{\omega\Theta(\omega)}{\pi c^2 \varepsilon_0} \left[\frac{\hbar\omega}{1 - e^{-\hbar\omega/(k_B T)}} \right] \mathrm{Im}\left\{ G_{jk}(\mathbf{r},\mathbf{r}';\omega) \right\} \delta(\omega - \omega'), \end{aligned} \tag{15.37}$$

where $\Theta(\omega)$ is the unit step function. Hence the correlation of the normally ordered operators is zero for positive frequencies. Similarly, the correlation of the antinormally ordered operators is zero for negative frequencies.

It can be shown that $\langle \delta\hat{E}_j^- \, \delta\hat{E}_k^{-*} \rangle = \langle \delta\hat{E}_j^+ \, \delta\hat{E}_k^{+*} \rangle = 0$, and hence the correlations for the total field $\hat{E} = \hat{E}^- + \hat{E}^+$ are simply the sum of the correlations for the normally and antinormally ordered fields given above. This recovers our result Eq. (15.32) and allows us to interpret the correlation $\langle \delta\hat{E}_j \, \delta\hat{E}_k^* \rangle$ as a sequence of absorption and emission events.

For completeness, we also state the fluctuation–dissipation theorem for symmetrized correlation functions. The quantity of interest is

$$\frac{1}{2}\Big\langle \Big[\delta\hat{E}_j(\mathbf{r},\omega)\delta\hat{E}_k^*(\mathbf{r}',\omega') + \delta\hat{E}_k(\mathbf{r},\omega)\delta\hat{E}_j^*(\mathbf{r}',\omega') \Big] \Big\rangle. \tag{15.38}$$

Using Eqs. (15.36) and (15.37) it is straightforward to show that the above expression equals

$$\frac{\omega}{\pi c^2 \varepsilon_0}\,\hbar\omega \left[\frac{1}{2} + \frac{1}{e^{\hbar\omega/(k_B T)} - 1} \right] \mathrm{Im}\big\{ G_{jk}(\mathbf{r},\mathbf{r}';\omega) \big\} \delta(\omega - \omega'). \tag{15.39}$$

Thus, the only difference compared with Eq. (15.32) is the replacement of a factor of 1 by 1/2. Consequently, for $T = 0$ the symmetrized correlations are no longer zero at negative frequencies.

15.2 Emission by fluctuating sources

The energy density of an arbitrary fluctuating electromagnetic field in vacuum is given by (cf. Eq. (2.57))

$$W(\mathbf{r},t) = \frac{\varepsilon_0}{2}\delta\mathbf{E}(\mathbf{r},t)\cdot\delta\mathbf{E}(\mathbf{r},t) + \frac{\mu_0}{2}\delta\mathbf{H}(\mathbf{r},t)\cdot\delta\mathbf{H}(\mathbf{r},t). \tag{15.40}$$

For simplicity, we will skip the position vector $\mathbf{r}$ in the arguments. Assuming stationary fluctuations, the average of W becomes

$$\overline{W} = \int_{-\infty}^{\infty} \overline{W}_\omega(\omega)\mathrm{d}\omega = \frac{\varepsilon_0}{2}\Big\langle \delta\mathbf{E}(t)\cdot\delta\mathbf{E}(t) \Big\rangle + \frac{\mu_0}{2}\Big\langle \delta\mathbf{H}(t)\cdot\delta\mathbf{H}(t) \Big\rangle. \tag{15.41}$$

The mean-square value of $\delta\mathbf{E}$ can be expressed as

$$\Big\langle \delta\mathbf{E}(t)\cdot\delta\mathbf{E}(t) \Big\rangle = \frac{1}{2\pi} \iint_{-\infty}^{\infty} \Big\langle \delta\mathbf{E}(t)\cdot\delta\mathbf{E}(t+\tau) \Big\rangle \, e^{i\omega\tau} \, \mathrm{d}\omega \, \mathrm{d}\tau, \tag{15.42}$$

with a similar expression for $\delta\mathbf{H}$. We can now identify the spectral energy density $\overline{W}_\omega$ in Eq. (15.41) as[6]

$$\overline{W}_\omega(\omega) = \int_{-\infty}^{\infty} \Big[\frac{\varepsilon_0}{4\pi}\Big\langle \delta\mathbf{E}(t)\cdot\delta\mathbf{E}(t+\tau) \Big\rangle + \frac{\mu_0}{4\pi}\Big\langle \delta\mathbf{H}(t)\cdot\delta\mathbf{H}(t+\tau) \Big\rangle \Big] \, e^{i\omega\tau} \, \mathrm{d}\tau. \tag{15.43}$$

[6] Keep in mind that $\overline{W}_\omega$ is defined for positive and negative frequencies.

After multiplication on both sides by $\delta(\omega-\omega')$, making use of the Wiener–Khintchine theorem (cf. Eq. (15.16)), and reintroducing the spatial dependence we obtain

$$\overline{W}_{\omega}(\mathbf{r},\omega)\delta(\omega-\omega') = \frac{\varepsilon_0}{2}\left\langle \delta\hat{\mathbf{E}}^{*}(\mathbf{r},\omega)\cdot\delta\hat{\mathbf{E}}(\mathbf{r},\omega')\right\rangle + \frac{\mu_0}{2}\left\langle \delta\hat{\mathbf{H}}^{*}(\mathbf{r},\omega)\cdot\delta\hat{\mathbf{H}}(\mathbf{r},\omega')\right\rangle, \tag{15.44}$$

where $\delta\hat{\mathbf{E}}$ and $\delta\hat{\mathbf{H}}$ are the Fourier transforms of $\delta\mathbf{E}$ and $\delta\mathbf{H}$, respectively. In the far-field, $|\delta\hat{\mathbf{H}}| = |\delta\hat{\mathbf{E}}|\sqrt{\varepsilon_0/\mu_0}$ and the electric and magnetic energy densities turn out to be equal.

We would like to determine the spectral energy density $\overline{W}_{\omega}$ due to a distribution of fluctuating currents $\delta\mathbf{j}$ in an arbitrary polarizable reference system. We assume that the latter can be accounted for by a dyadic Green function $\overleftrightarrow{\mathbf{G}}(\mathbf{r},\mathbf{r}',\omega)$. Using the volume-integral equations discussed in Section 8.3.1, we obtain

$$\delta\hat{\mathbf{E}}(\mathbf{r},\omega) = \mathrm{i}\omega\mu_0\int_V \overleftrightarrow{\mathbf{G}}(\mathbf{r},\mathbf{r}',\omega)\delta\hat{\mathbf{j}}(\mathbf{r}',\omega)\mathrm{d}V', \tag{15.45}$$

$$\delta\hat{\mathbf{H}}(\mathbf{r},\omega) = \int_V \left[\nabla\times\overleftrightarrow{\mathbf{G}}(\mathbf{r},\mathbf{r}',\omega)\right]\delta\hat{\mathbf{j}}(\mathbf{r}',\omega)\mathrm{d}V'. \tag{15.46}$$

After introducing these equations into the expression for $\overline{W}_{\omega}$, the averages over the fields reduce to averages over the currents.[7] The latter can then be eliminated by using the fluctuation–dissipation theorem given in Eq. (15.21). Integration over ω' leads to

$$\overline{W}_{\omega}(\mathbf{r},\omega) = \frac{\omega}{\pi c^2}\left[\frac{\hbar\omega}{1-\mathrm{e}^{-\hbar\omega/(k_\mathrm{B}T)}}\right] \times \sum_{j,k}\int_V \varepsilon''(\mathbf{r}',\omega)\left[\frac{\omega^2}{c^2}\left|[\overleftrightarrow{\mathbf{G}}(\mathbf{r},\mathbf{r}',\omega)]_{jk}\right|^2 + \left|[\nabla\times\overleftrightarrow{\mathbf{G}}(\mathbf{r},\mathbf{r}',\omega)]_{jk}\right|^2\right]\mathrm{d}V', \tag{15.47}$$

with $[\overleftrightarrow{\mathbf{G}}]_{jk}$ and $[\nabla\times\overleftrightarrow{\mathbf{G}}]_{jk}$ denoting the jkth elements of the tensors $\overleftrightarrow{\mathbf{G}}$ and $(\nabla\times\overleftrightarrow{\mathbf{G}})$, respectively. The first term in the brackets originates from the electric contribution to $\overline{W}_{\omega}$ whereas the second term is due to the magnetic field. In general, the result for $\overline{W}_{\omega}$ can be written in the form

$$\overline{W}_{\omega}(\mathbf{r},\omega) = \overline{w}(\omega,T)N(\mathbf{r},\omega), \tag{15.48}$$

where $\overline{w}(\omega,T)$ is the average energy per mode. $N(\mathbf{r},\omega)$ depends only on the dielectric properties $\varepsilon(\omega)$ and the Green function of the reference system. It has a similar meaning to the *local density of states* defined previously. In fact, as will be shown later, $N(\mathbf{r},\omega)$ is identical with the local density of states if the system considered is an equilibrium system. In a non-equilibrium system, $N(\mathbf{r},\omega)$ comprises only a fraction of the total number of possible modes.

[7] The fields due to a set of discrete fluctuating dipoles can be written in a similar form (see Section 8.3.1). $\overline{W}_{\omega}$ can then be derived by using the fluctuation–dissipation theorem of Eq. (15.23).

15.2.1 Blackbody radiation

Consider a body that is made of fluctuating point sources. Thermal equilibrium with the radiation field implies that the averaged Poynting vector vanishes at all points $\mathbf{r}$ in space (there is no net heat transport). In this case we can use the fluctuation–dissipation theorem Eq. (15.32). In free space, the two terms in Eq. (15.47) turn out to be identical and we obtain [10]

$$\overline{W}_{\omega}(\mathbf{r},\omega) = \left[\frac{\hbar\omega}{1-\mathrm{e}^{-\hbar\omega/(k_{\mathrm{B}}T)}}\right]\frac{\omega}{\pi c^2}\sum_j \mathrm{Im}\left\{[\overleftrightarrow{\mathbf{G}}(\mathbf{r},\mathbf{r},\omega)]_{jj}\right\} \quad \text{(equilibrium).} \qquad (15.49)$$

Remember that the total energy is given by integration over positive and negative frequencies. Let us replace the term in brackets by an antisymmetric part and a symmetric part as

$$\frac{\hbar\omega}{2} + \left[\frac{\hbar\omega}{2} + \frac{\hbar\omega}{\mathrm{e}^{\hbar\omega/(k_{\mathrm{B}}T)}-1}\right]. \qquad (15.50)$$

Considering that $\mathrm{Im}\{\overleftrightarrow{\mathbf{G}}\}$ is an odd function of ω, we can drop the first term in the above expression because its contribution cancels out on integrating over positive and negative frequencies. The remaining integral can be written over positive frequencies only as

$$\overline{W} = \int_0^{\infty} \overline{W}^{+}_{\omega}(\omega)\mathrm{d}\omega = \int_0^{\infty} \overline{w}(\omega,T)N(\mathbf{r},\omega)\mathrm{d}\omega, \qquad (15.51)$$

where

$$\overline{w}(\omega,T) = \left[\frac{\hbar\omega}{2} + \frac{\hbar\omega}{\mathrm{e}^{\hbar\omega/(k_{\mathrm{B}}T)}-1}\right],$$

$$N(\mathbf{r},\omega) = \frac{2\omega}{\pi c^2}\sum_j \mathrm{Im}\left\{[\overleftrightarrow{\mathbf{G}}(\mathbf{r},\mathbf{r},\omega)]_{jj}\right\} = \frac{2\omega}{\pi c^2}\,\mathrm{Im}\left\{\mathrm{Tr}[\overleftrightarrow{\mathbf{G}}(\mathbf{r},\mathbf{r},\omega)]\right\}.$$

$N(\mathbf{r},\omega)$ is identical with the local density of states (cf. Eq. (8.118)) and $\overline{w}(\omega,T)$ corresponds to the average energy of a quantum oscillator. $\overline{W}^{+}_{\omega}(\omega)$ is the spectral energy density defined over positive frequencies only.

By expanding the exponential term $\exp(ikr)$ in $\overleftrightarrow{\mathbf{G}}$ into a series, it has been shown in Section 8.3.3 that $\mathrm{Im}\{\overleftrightarrow{\mathbf{G}}\}$ is *not singular* at its origin. Using the free-space Green function, we obtain $\mathrm{Im}\{[\overleftrightarrow{\mathbf{G}}(\mathbf{r},\mathbf{r},\omega)]_{jj}\} = \omega/(6\pi c)$ and Eq. (15.51) becomes

$$\overline{W}^{+}_{\omega}(\omega) = \left[\frac{\hbar\omega}{2} + \frac{\hbar\omega}{\mathrm{e}^{\hbar\omega/(k_{\mathrm{B}}T)}-1}\right]\frac{\omega^2}{\pi^2 c^3}. \qquad (15.52)$$

This is the celebrated *Planck blackbody radiation* formula which renders the electromagnetic energy per unit volume in the frequency range $[\omega \ldots \omega + \mathrm{d}\omega]$ (Fig. 15.4). It is strictly valid only for an equilibrium system.

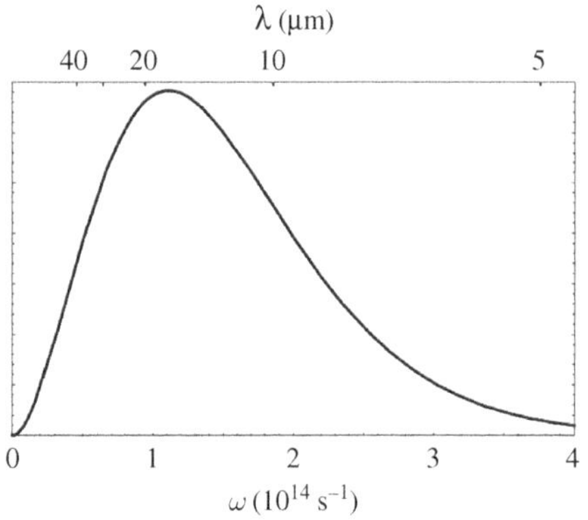

Fig. 15.4 The blackbody radiation spectrum $\overline{W}_\omega^+$ for $T = 300$ K. Equilibrium conditions require that the net Poynting vector vanishes everywhere.

15.2.2 Coherence, spectral shifts, and heat transfer

Thermal equilibrium between matter and the radiation field is practically never encountered. Therefore, the spectral energy density has to be calculated by Eq. (15.47) and the local density of states N becomes position-dependent. Shchegrov *et al.* [7] calculated $N(\mathbf{r}, \omega)$ near a planar material surface and found that it strongly depends on the distance to the surface. Figure 15.5 shows the spectral energy density at $T = 300$ K above an SiC half-space. At large distances from the surface (Fig. 15.5, top), the spectrum looks like a blackbody spectrum multiplied by the SiC emissivity. The latter is responsible for the dip in the spectrum. The emitted radiation is incoherent, with a typical coherence length of $\approx\lambda/2$ (a Lambertian source). At distances considerably smaller than λ, the spectrum is dominated by a single peak (Fig. 15.5, bottom) that originates from a surface mode (surface phonon polariton). The narrow linewidth of the peak leads to increased coherence and thus to almost monochromatic fields. The sequence of figures clearly indicates that the spectrum changes on propagation.

The observed increase of $\overline{W}_\omega$ near material surfaces has implications for *radiative heat transfer*. Radiative heat transfer will occur between two bodies that are kept at different temperatures. However, even a single body in free space will lose its thermal energy by continuous radiation. Mulet *et al.* [11] showed that the radiative heat transfer between two bodies can be increased by several orders of magnitude as the spacing between the bodies is decreased. This increase originates from the interaction of surface waves localized near the interfaces. The interaction gives rise to heat transfer limited to a narrow spectral window.

Near-field heat transfer has been systematically investigated by Greffet and co-workers [12]. They measured the thermal conductance G between a heated surface and a microsphere as a function of the separation d and temperature difference ΔT. As shown in Fig. 15.6, the experimental data show a steep distance dependence that can be accurately described by

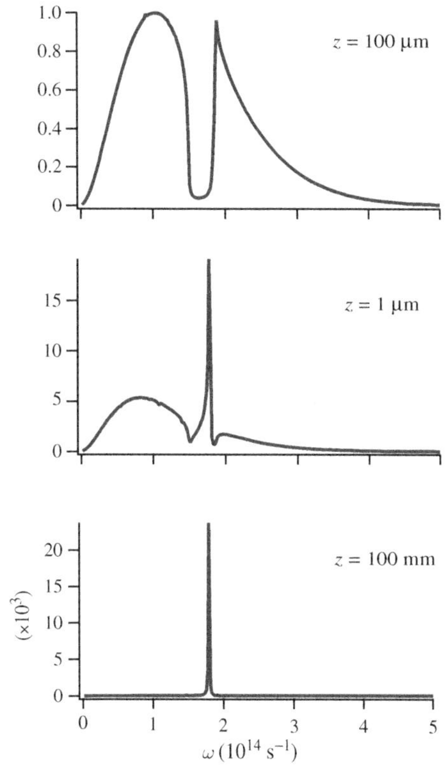

Fig. 15.5 Spectra of thermal emission from a semi-infinite sample of SiC at $T = 300$ K evaluated at three different heights z above the surface. From [7].

$$G(d, \Delta T) = G_{\mathrm{ff}} + \frac{H}{\Delta T}\delta(d), \tag{15.53}$$

where G_{ff} corresponds to far-field heat transfer and the second term accounts for near-field heat transfer. H is a constant and $\delta(d)$ is a short-range distance function. Besides its fundamental interest, near-field heat transfer also finds application in heat-assisted magnetic recording. Areal bit densities as high as 1.5 Pb m^{-2} have been demonstrated by use of a heated optical antenna [13].

Thermal near-fields affect not only the spectral energy density of the emitted radiation but also their *spatial coherence*. A measure for spatial coherence is given by the electric-field cross-spectral density tensor W_{jk} defined as

$$W_{jk}(\mathbf{r}_1, \mathbf{r}_2, \omega)\,\delta(\omega - \omega') = \left\langle \delta\hat{E}_j(\mathbf{r}_1, \omega)\,\delta\hat{E}_k^*(\mathbf{r}_2, \omega')\right\rangle. \tag{15.54}$$

Carminati and Greffet [14] have evaluated W_{jk} near surfaces of different materials. They find that an opaque material not supporting a surface mode (e.g. tungsten) can have a spatial coherence length much smaller than the well-known $\lambda/2$ coherence length of blackbody radiation. The coherence length can be arbitrarily small, being limited only by non-local effects close to the material surface. On the other hand, near material surfaces supporting surface modes (e.g. silver) the correlation length can reach several tenths of λ.

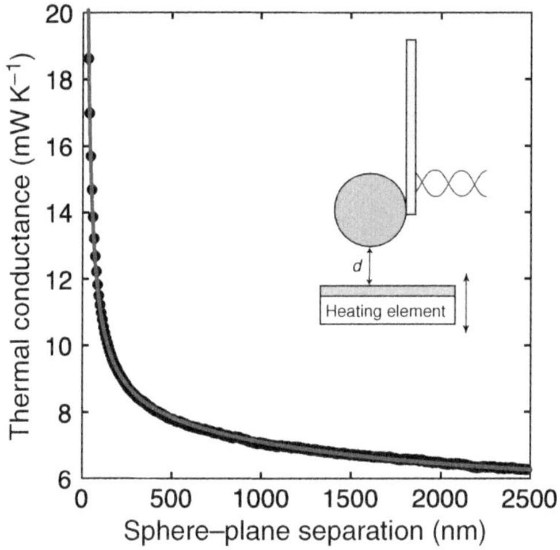

Fig. 15.6 Thermal conductance between a sphere of diameter 40 μm and a heated plate as a function of the gap distance. The temperature difference between the sphere and the plate is $\Delta T = 21$ K. The inset illustrates the experimental arrangement. Reprinted by permission from Macmillan Publishers Ltd [12].

15.3 Fluctuation-induced forces

Fluctuating charges in a neutral body give rise to fluctuating electromagnetic fields that interact with the charges in other bodies. As a consequence, electromagnetic fields mediate between the charge fluctuations in separate bodies. The resulting charge correlations give rise to an electromagnetic force that is referred to as the *dispersion* force. For short distances between two bodies, the force is called the *van der Waals* force, whereas at larger separations it is designated the *Casimir* force. Although these forces are small on macroscopic scales, they cannot be ignored on the scales of nanostructures. For example, two parallel conducting plates each of area 1 μm^2 placed 5 nm apart will experience an attractive force of $\approx$2 nN. This force is sufficient to squash a biomolecule! Dispersion forces are also responsible for weak molecular binding and for adhesion of particles to interfaces. For example, geckos climb even the most slippery surfaces with ease and can hang from glass using a single toe. The secret behind this extraordinary climbing skill lies in the millions of tiny keratin hairs on the surface of each foot. Although the dispersion force associated with each hair is minuscule, the millions of hairs collectively produce a powerful adhesive effect. The "gecko effect" is applied to the design of strongly adhesive tapes.

In this section we derive the force acting on a small polarizable particle in an arbitrary environment following the steps of Ref. [5]. To simplify the notation we assume that all fluctuations have zero average. This allows us to write $\mathbf{p}(t) = \delta\mathbf{p}(t)$ and $\mathbf{E}(t) = \delta\mathbf{E}(t)$. To calculate the force acting on a polarizable particle located at $\mathbf{r} = \mathbf{r}_0$ we use the expression for the gradient force derived in Section 14.4 (cf. Eq. (14.33)). However, we have to consider that both the field **E** and the dipole moment **p** have fluctuating and induced parts. Therefore,

$$\langle \mathbf{F}(\mathbf{r}_0)\rangle = \sum_i \left[\left\langle p_i^{(\mathrm{in})}(t) \nabla E_i^{(\mathrm{fl})}(\mathbf{r}_0, t) \right\rangle + \left\langle p_i^{(\mathrm{fl})}(t) \nabla E_i^{(\mathrm{in})}(\mathbf{r}_0, t) \right\rangle \right], \tag{15.55}$$

where $i = \{x, y, z\}$. The first term describes the field fluctuations (spontaneous and thermal) that correlate to the induced dipole moment according to

$$\hat{\mathbf{p}}^{(\mathrm{in})}(\omega) = \alpha_1(\omega) \hat{\mathbf{E}}^{(\mathrm{fl})}(\mathbf{r}_0, \omega), \tag{15.56}$$

where we assumed an isotropic polarizability. For later purposes, we denote the properties of the particle by an index 1. The second term in Eq. (15.55) originates from the particle's dipole fluctuations and the corresponding induced field according to

$$\hat{\mathbf{E}}^{(\mathrm{in})}(\mathbf{r}, \omega) = \frac{\omega^2}{c^2} \frac{1}{\varepsilon_0} \overleftrightarrow{\mathbf{G}}(\mathbf{r}, \mathbf{r}_0; \omega) \hat{\mathbf{p}}^{(\mathrm{fl})}(\omega). \tag{15.57}$$

Here, $\overleftrightarrow{\mathbf{G}}$ is the Green function of the reference system and **r** denotes an arbitrary field point as visualized in Fig. 15.7. Correlations between the fluctuating field and the fluctuating dipole are zero because they originate from different physical systems. Likewise, there are no correlations between the induced quantities.

After expressing **p** and **E** in Eq. (15.55) by their Fourier transforms and making use of the fact that $\mathbf{E}(t) = \mathbf{E}^*(t)$, we obtain

$$\begin{aligned} \langle \mathbf{F}(\mathbf{r}_0)\rangle = & \sum_i \iint_{-\infty}^{\infty} \left\langle \hat{p}_i^{(\mathrm{in})}(\omega) \nabla \hat{E}_i^{*\,(\mathrm{fl})}(\mathbf{r}_0, \omega') \right\rangle \mathrm{e}^{\mathrm{i}(\omega'-\omega)t} \, \mathrm{d}\omega' \, \mathrm{d}\omega \\ & + \sum_i \iint_{-\infty}^{\infty} \left\langle \hat{p}_i^{(\mathrm{fl})}(\omega) \nabla \hat{E}_i^{*\,(\mathrm{in})}(\mathbf{r}_0, \omega') \right\rangle \mathrm{e}^{\mathrm{i}(\omega'-\omega)t} \, \mathrm{d}\omega' \, \mathrm{d}\omega. \end{aligned} \tag{15.58}$$

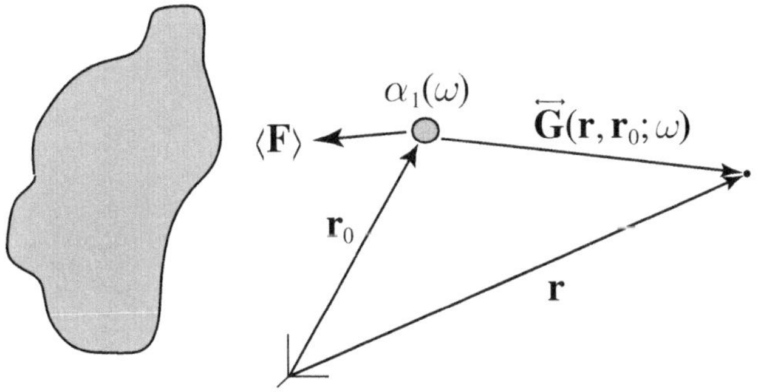

Fig. 15.7 The dispersion force acting on a polarizable particle located at $\mathbf{r} = \mathbf{r}_0$. The force originates from correlated charge fluctuations in the particle and other bodies in the environment. The latter are accounted for by the Green function $\overleftrightarrow{\mathbf{G}}$.

Introducing the linear relationships (15.56) and (15.57) and arranging terms allows us to express the first term as a function of $\hat{\mathbf{E}}^{(\mathrm{fl})}$ and the second term as a function of $\hat{\mathbf{p}}^{(\mathrm{fl})}$ as

$$\begin{aligned}\big\langle \mathbf{F}(\mathbf{r}_0)\big\rangle &= \sum_i \iint_{-\infty}^{\infty} \alpha_1(\omega)\nabla_2\Big\langle \hat{E}_i^{*\,(\mathrm{fl})}(\mathbf{r}_0,\omega)\hat{E}_i^{*\,(\mathrm{fl})}(\mathbf{r}_0,\omega')\Big\rangle \mathrm{e}^{\mathrm{i}(\omega'-\omega)t}\,\mathrm{d}\omega'\,\mathrm{d}\omega \\ &+ \sum_{i,j} \iint_{-\infty}^{\infty} \frac{\omega'^2}{c^2}\frac{1}{\varepsilon_0}\nabla_1 G_{ij}^*(\mathbf{r}_0,\mathbf{r}_0;\omega')\Big\langle \hat{p}_i^{(\mathrm{fl})}(\omega)\hat{p}_j^{*\,(\mathrm{fl})}(\omega')\Big\rangle \mathrm{e}^{\mathrm{i}(\omega'-\omega)t}\,\mathrm{d}\omega'\,\mathrm{d}\omega, \qquad (15.59)\end{aligned}$$

where ∇_n specifies that the gradient has to be taken with respect to the nth spatial variable in the argument. Using the fluctuation–dissipation theorems for dipole and field (Eqs. (15.19) and (15.32)) and the fact that

$$\nabla_1 \overleftrightarrow{\mathbf{G}}(\mathbf{r},\mathbf{r}_0;\omega) = \nabla_2 \overleftrightarrow{\mathbf{G}}(\mathbf{r},\mathbf{r}_0;\omega) \qquad (15.60)$$

allows us to write the force in the compact form

$$\big\langle \mathbf{F}(\mathbf{r}_0)\big\rangle = \sum_i \int_{-\infty}^{\infty} \frac{\omega}{\pi c^2 \varepsilon_0}\left[\frac{\hbar\omega}{1-\mathrm{e}^{-\hbar\omega/(k_B T)}}\right] \mathrm{Im}\Big\{\alpha_1(\omega)\nabla_1 G_{ii}(\mathbf{r}_0,\mathbf{r}_0;\omega)\Big\}\,\mathrm{d}\omega. \qquad (15.61)$$

Notice that the force is determined by the properties of the environment that is encoded in the Green function $\overleftrightarrow{\mathbf{G}}$. The force vanishes in the absence of any objects, i.e. when $\overleftrightarrow{\mathbf{G}}$ equals the free-space Green function. Equation (15.61) allows us to calculate the force acting on a small polarizable particle in an arbitrary environment. The equation is valid for an isotropic particle but it can be generalized to account for anisotropic polarizabilities such as for molecules with fixed transition dipole moments.

15.3.1 The Casimir–Polder potential

In this section we derive the force acting on a particle with polarizability α_1 due to another particle with polarizability α_2. As indicated in Fig. 15.8, the two particles are separated by a distance R. For short distances, the force varies as R^{-7}, whereas for larger distances the force assumes an R^{-8} dependence. The stronger distance dependence at large distances

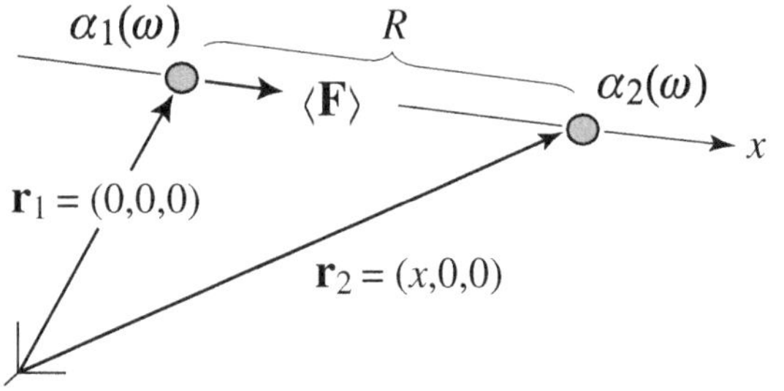

Fig. 15.8 Definition of coordinates for the calculation of the dispersion force between two polarizable particles.

is counter-intuitive since the decay of electromagnetic fields becomes weaker on going from the near-field to the far-field. It will be shown that, for temperature $T = 0$, the force at all distances can be deduced from a single potential $U(R)$, called the *Casimir–Polder potential*. Finite temperatures affect the force only marginally [5], hence we will restrict the analysis to the case $T = 0$.

The force in Eq. (15.61) is defined by the Green function $\overleftrightarrow{\mathbf{G}}$. Therefore, let us derive the Green function accounting for the presence of a polarizable particle with polarizability α_2 centered at $\mathbf{r}_2$. The field $\mathbf{E}$ at $\mathbf{r}$ due to a dipole at $\mathbf{r}_1$ can be expressed as

$$\hat{\mathbf{E}}(\mathbf{r},\omega) = \frac{\omega^2}{c^2}\frac{1}{\varepsilon_0}\overleftrightarrow{\mathbf{G}}^0(\mathbf{r},\mathbf{r}_1;\omega)\hat{\mathbf{p}}_1(\omega) + \hat{\mathbf{E}}_{\mathrm{s}}(\mathbf{r},\omega), \tag{15.62}$$

with $\overleftrightarrow{\mathbf{G}}^0$ denoting the free-space Green dyadic. The scattered field $\hat{\mathbf{E}}_{\mathrm{s}}$ originates from the particle at $\mathbf{r}_2$ and is determined as

$$\begin{aligned}\hat{\mathbf{E}}_{\mathrm{s}}(\mathbf{r},\omega) &= \frac{\omega^2}{c^2}\frac{1}{\varepsilon_0}\overleftrightarrow{\mathbf{G}}^0(\mathbf{r},\mathbf{r}_2;\omega)\hat{\mathbf{p}}_2(\omega)\\ &= \frac{\omega^2}{c^2}\frac{1}{\varepsilon_0}\left[\frac{\omega^2}{c^2}\frac{1}{\varepsilon_0}\overleftrightarrow{\mathbf{G}}^0(\mathbf{r},\mathbf{r}_2;\omega)\alpha_2(\omega)\overleftrightarrow{\mathbf{G}}^0(\mathbf{r}_2,\mathbf{r}_1;\omega)\right]\hat{\mathbf{p}}_1(\omega).\end{aligned} \tag{15.63}$$

Combining Eqs. (15.62) and (15.63) allows us to identify the Green function of the system of "free-space plus particle at $\mathbf{r}_2$" as

$$\overleftrightarrow{\mathbf{G}}(\mathbf{r},\mathbf{r}_1;\omega) = \overleftrightarrow{\mathbf{G}}^0(\mathbf{r},\mathbf{r}_1;\omega) + \frac{\omega^2}{c^2}\frac{1}{\varepsilon_0}\overleftrightarrow{\mathbf{G}}^0(\mathbf{r},\mathbf{r}_2;\omega)\alpha_2(\omega)\overleftrightarrow{\mathbf{G}}^0(\mathbf{r}_2,\mathbf{r}_1;\omega). \tag{15.64}$$

The gradient of $\overleftrightarrow{\mathbf{G}}$ evaluated at its origin $\mathbf{r} = \mathbf{r}_1$ is

$$\nabla_1\overleftrightarrow{\mathbf{G}}(\mathbf{r}_1,\mathbf{r}_1;\omega) = \frac{\omega^2}{c^2}\frac{1}{\varepsilon_0}\alpha_2(\omega)\left[\nabla_1\overleftrightarrow{\mathbf{G}}^0(\mathbf{r}_1,\mathbf{r}_2;\omega)\right]\overleftrightarrow{\mathbf{G}}^0(\mathbf{r}_2,\mathbf{r}_1;\omega). \tag{15.65}$$

Let us choose the coordinates as $\mathbf{r}_1 = 0$ and $\mathbf{r}_2 = (x,0,0) = x\mathbf{n}_x$. We then obtain for the sum of the diagonal elements of $\nabla\overleftrightarrow{\mathbf{G}}$

$$\sum_i \nabla_1 G_{ii}(\mathbf{r}_1,\mathbf{r}_1;\omega) = \frac{\omega^2}{c^2}\frac{1}{\varepsilon_0}\,\alpha_2(\omega)\,\sum_i\left[\frac{\partial}{\partial x}G^0_{ii}(i,0;\omega)\right]G^0_{ii}(i,0;\omega), \tag{15.66}$$

where we made use of the properties of the free-space Green function $\overleftrightarrow{\mathbf{G}}^0$. Using the explicit form of $\overleftrightarrow{\mathbf{G}}^0$ in the above expression (cf. Section 8.3.1) gives

$$\begin{aligned}\sum_i \nabla_1 G_{ii}(\mathbf{r}_1,\mathbf{r}_1;\omega) &= \frac{c^2}{\omega^2}\frac{1}{\varepsilon_0}\frac{\exp(2\mathrm{i}x\omega/c)}{8\pi^2x^7}\alpha_2(\omega)\\ &\quad\times\left[-9+18\mathrm{i}\left(\frac{\omega}{c}x\right)+16\left(\frac{\omega}{c}x\right)^2-8\mathrm{i}\left(\frac{\omega}{c}x\right)^3\right.\\ &\quad\left.-3\left(\frac{\omega}{c}x\right)^4+\mathrm{i}\left(\frac{\omega}{c}x\right)^5\right]\mathbf{n}_x\\ &= \sum_i \nabla_1 G_{ii}(x;\omega).\end{aligned} \tag{15.67}$$

We now introduce this Green function into the force formula (15.61), which, for $T = 0$, reads as

$$\langle \mathbf{F}(x) \rangle = \frac{\hbar}{\pi c^2 \varepsilon_0} \int_0^\infty \omega^2 \, \mathrm{Im} \left\{ \alpha_1(\omega) \sum_i \nabla_1 G_{ii}(x;\omega) \right\} \mathrm{d}\omega. \tag{15.68}$$

Here, we made use of the fact that contributions with negative frequencies vanish (cf. Eq. (15.20)).

It is straightforward to show that $\nabla \times \langle \mathbf{F} \rangle = 0$ and hence the force is conservative. Therefore, we can derive the force from a potential U by integration over the variable x. We obtain

$$U = -\int \langle F(x) \rangle \mathrm{d}x = \frac{\hbar}{16\pi^3 \varepsilon_0^2 x^6} \, \mathrm{Im} \int_0^\infty \alpha_1(\omega)\alpha_2(\omega) \mathrm{e}^{2\mathrm{i}x\omega/c} \times \left[-3 + 6\mathrm{i}\left(\frac{\omega}{c}x\right) + 5\left(\frac{\omega}{c}x\right)^2 - 2\mathrm{i}\left(\frac{\omega}{c}x\right)^3 - \left(\frac{\omega}{c}x\right)^4 \right] \mathrm{d}\omega. \tag{15.69}$$

We now substitute the integration variable as $\tilde{\omega} = \omega c$ and replace the interparticle distance by R. We then realize that the integrand is analytic in the upper half-space of the integration variable and that the integrand goes to zero as $\tilde{\omega} \rightarrow \infty$. Therefore, we can integrate along the imaginary axis using

$$\int_0^\infty f(\tilde{\omega}) \mathrm{d}\tilde{\omega} = \mathrm{i} \int_0^\infty f(\mathrm{i}\eta) \mathrm{d}\eta. \tag{15.70}$$

By combining these mathematical tricks we obtain for the interparticle potential

$$U = -\frac{\hbar c}{16\pi^3 \varepsilon_0^2 R^6} \int_0^\infty \alpha_1(\mathrm{i}c\eta)\alpha_2(\mathrm{i}c\eta) \mathrm{e}^{-2\eta R} [3 + 6\eta R + 5(\eta R)^2 + 2(\eta R)^3 + (\eta R)^4] \mathrm{d}\eta. \tag{15.71}$$

We made use of the fact that $\alpha_i(\Omega)$ is purely real on the imaginary axis $\Omega = \mathrm{i}\eta$. Equation (15.71) is the celebrated Casimir–Polder potential, which is valid for any interparticle separation R. Our result agrees with rigorous calculations based on quantum electrodynamics using fourth-order perturbation theory [15]. The derivation presented here allows us to incorporate higher-order corrections by simply adding additional interaction terms to the Green function $\overleftrightarrow{\mathbf{G}}$ in Eq. (15.64). The force can be retrieved from the potential using $\langle \mathbf{F} \rangle = -\nabla U$.

It is interesting to evaluate the potential for the limiting cases of large and small interparticle distances. For short distances we retain only the first term in the bracket, set $\exp(-2\eta R) = 1$, and obtain

$$U(R \to 0) = -\frac{6\hbar}{32\pi^3\varepsilon_0^2}\frac{1}{R^6}\int_0^\infty \alpha_1(\mathrm{i}\eta)\alpha_2(\mathrm{i}\eta)\mathrm{d}\eta. \tag{15.72}$$

This is the van der Waals potential valid for short interparticle distances R. The potential depends on the dispersive properties of the particle polarizabilities and scales with the inverse sixth power of the particle separation R.

To obtain the limit for large R, we make the substitution $u = \eta R$ in Eq. (15.71), which leads to the following expression for the interparticle potential:

$$U = -\frac{\hbar c}{16\pi^3\varepsilon_0^2 R^7}\int_0^\infty \alpha_1(\mathrm{i}cu/R)\alpha_2(\mathrm{i}cu/R)\mathrm{e}^{-2u}\Big[3 + 6u + 5u^2 + 2u^3 + u^4\Big]\mathrm{d}u. \tag{15.73}$$

Then, in the large-distance limit ($R \to \infty$), one can replace the polarizabilities by their static values $\alpha_i(0)$. After moving the polarizabilities out of the integral one obtains

$$U(R \to \infty) = -\frac{\hbar c}{16\pi^3\varepsilon_0^2}\frac{\alpha_1(0)\alpha_2(0)}{R^7}\int_0^\infty \mathrm{e}^{-2u}\Big[3 + 6u + 5u^2 + 2u^3 + u^4\Big]\,\mathrm{d}u. \tag{15.74}$$

Finally, using the equality

$$\int_0^\infty u^n \mathrm{e}^{-2u}\,\mathrm{d}u = \frac{n!}{2^{n+1}} \quad \forall\, n \geq 0, \tag{15.75}$$

one can analytically perform the integration in Eq. (15.74). We then obtain the Casimir–Polder interparticle potential in the limit of large distances as

$$U(R \to \infty) = -\frac{23\hbar c}{64\pi^3\varepsilon_0^2}\frac{\alpha_1(0)\alpha_2(0)}{R^7}. \tag{15.76}$$

This result is a pure manifestation of vacuum fluctuations and it is referred to as the Casimir potential, which was first derived in 1948 by Hendrik Casimir [16]. It is remarkable that the potential scales with the inverse seventh power of the interparticle distance R. Thus, the force decays more rapidly for large distances than it does for short distances. This behavior is opposite to the distance dependence of the electromagnetic energy density, which shows the fastest decay (R^{-6}) close to the sources. The Casimir potential depends only on the static ($\omega = 0$) polarizabilities of the particles and hence it does not matter what their spectral properties are. Notice that in deriving the Casimir–Polder potential we considered only the gradient force and neglected the influence of the scattering force. The scattering force is non-conservative and it must be zero if the particle(s) remain in equilibrium with the vacuum field.

It has to be emphasized that the Casimir–Polder potential originates solely from zero-point fluctuations and does not account for thermal fluctuations. At room temperature, thermally induced forces are usually more than one order of magnitude weaker than the forces associated with vacuum fluctuations [5].

15.3.2 Electromagnetic friction

Electromagnetic interactions between two charge-neutral objects give rise not only to conservative dispersion forces but also to a *non-conservative* friction force if the two objects are in motion relative to each other. This friction force is associated only with thermal fluctuations and it brings the motion of an object ultimately to rest. Although this force is small, it has direct consequences for the development of nano-electro-mechanical systems (NEMS) and for various proposals in the field of quantum information. Electromagnetic friction leads to increased decoherence in miniaturized particle traps such as ion traps and atom chips, and limits the Q-factor of mechanical resonances.

Let us consider a small, charge-neutral particle such as an atom, molecule, or cluster, or a nanoscale structure that is small compared with all relevant wavelengths λ. In this limit, the particle is represented by the polarizability $\alpha(\omega)$. The particle is placed in an arbitrary environment characterized by the Green function $\overleftrightarrow{\mathbf{G}}$ and we assume that the motion of its center-of-mass coordinate $x(t)$ is governed by the classical Langevin equation

$$m\,\frac{\mathrm{d}^2}{\mathrm{d}t^2}x\,(t) + \int_{-\infty}^{t} \gamma\big(t-t'\big)\frac{\mathrm{d}}{\mathrm{d}t'}x\big(t'\big)\mathrm{d}t' + mw_0^2 x(t) = F_x(t). \tag{15.77}$$

Here, m is the mass of the particle, $\gamma(t)$ is the damping coefficient originating from thermal electromagnetic field fluctuations, w_0 is the natural frequency of the oscillating particle, and $F_x(t)$ is the stochastic force. Note that the restoring force $mw_0^2x(t)$ is added only for generality and does not influence the final result. In thermal equilibrium, $F_x(t)$ is a stationary stochastic process with zero ensemble average. The spectral force spectrum $S_{\mathrm{F}}(\omega)$ is given by the Wiener–Khintchine theorem as (cf. Eq. (15.16))

$$S_{\mathrm{F}}(\omega) = \frac{1}{2\pi}\int_{-\infty}^{\infty} \langle F_x(\tau)\,F_x(0)\rangle \mathrm{e}^{\mathrm{i}\omega\tau}\,\mathrm{d}\tau, \tag{15.78}$$

where ω is the angular frequency. Furthermore, at thermal equilibrium S_{F} is related to the friction coefficient by the fluctuation–dissipation theorem. Because the motion of the macroscopic particle is classical, we consider the classical limit, i.e.

$$k_{\mathrm{B}}T\,\hat{\gamma}(\omega) = \pi S_{\mathrm{F}}(\omega), \tag{15.79}$$

with $\hat{\gamma}(\omega)$ being the Fourier transform of $\gamma(t)$ defined *only* for $t > 0$.

In Eq. (15.77), we assumed a general friction force term whose magnitude at time t depends on the particle's velocity at earlier times. We now consider that the interaction time of the thermal bath with the particle is short compared with the particle's dynamics, thus the change of velocity of the particle during the interaction time is very small. In this Markovian approximation friction has no memory and thus

$$F_{\mathrm{friction}}(t) = -\gamma_0\frac{\mathrm{d}}{\mathrm{d}t}x(t), \quad \gamma_0 = \int_0^{\infty} \gamma(t)\mathrm{d}t. \tag{15.80}$$

On evaluating Eq. (15.79) at $\omega = 0$ and using Eq. (15.80), we find that the damping constant is related to the force spectrum by

$$k_{\mathrm{B}}T\gamma_0 = \pi S_{\mathrm{F}}(\omega = 0). \tag{15.81}$$

This is the final expression that relates the linear-velocity damping coefficient to the force spectrum. To calculate γ_0, we need to solve for the force spectrum, which, in turn, is defined by the electromagnetic fields due to fluctuating currents in the environment and the fluctuating dipole (cf. Eq. (15.55)).

Using the Wiener–Khintchine theorem (15.78), the Fourier transform of the dipole force (15.55), and the stationarity of the fluctuations we obtain

$$\left\langle \hat{F}_x^*(\omega')\hat{F}_x(\omega)\right\rangle = S_{\mathrm{F}}(\omega)\delta(\omega-\omega')$$
$$= \sum_{i,j=1}^{3}\Bigg\langle\left[\left(\hat{p}_j^{*\,(\mathrm{fl})}(\omega')+\hat{p}_j^{*\,(\mathrm{in})}(\omega')\right)\otimes\left(\frac{\partial}{\partial x}\hat{E}_j^{*\,(\mathrm{fl})}(\omega')+\frac{\partial}{\partial x}\hat{E}_j^{*\,(\mathrm{in})}(\omega')\right)\right]$$
$$\times\left[\left(\hat{p}_i^{(\mathrm{fl})}(\omega)+\hat{p}_i^{(\mathrm{in})}(\omega)\right)\otimes\left(\frac{\partial}{\partial x}\hat{E}_i^{(\mathrm{fl})}(\omega)+\frac{\partial}{\partial x}\tilde{E}_i^{(\mathrm{in})}(\omega)\right)\right]\Bigg\rangle, \tag{15.82}$$

where $\otimes$ denotes convolution. Each of the additive terms in $\langle\tilde{F}_x^*(\omega')\tilde{F}_x(\omega)\rangle$ is a fourth-order frequency-domain correlation function. Because the fluctuation–dissipation theorem involves second-order correlations and not fourth-order correlations it is not possible to find a solution using near-equilibrium statistical mechanics. However, there is a way out: thermal fluctuating fields can be thought of as arising from the superposition of a large number of radiating oscillators with a broadband spectrum. Consequently, the central-limit theorem applies. The same is true for the dipole fluctuations because of their broad thermal spectrum. Stochastic processes with Gaussian statistics have the property that a fourth-order correlation function can be expressed by a sum of pair-products of second-order correlation functions. Thus, Eq. (15.82) can be calculated by knowing the second-order correlations of the thermal electromagnetic fields and the electric dipole fluctuations. At *thermal equilibrium*, these correlation functions are given by the fluctuation–dissipation theorems Eqs. (15.19) and (15.32). Thus, we have all the ingredients to calculate the damping coefficient γ_0 in Eq. (15.81). We replace the induced terms in Eq. (15.82) by the fluctuating terms using the linear relationships Eqs. (15.56) and (15.57). Then we introduce the fluctuation–dissipation theorems Eqs. (15.19) and (15.32). Finally, we make use of Eq. (15.81), through which we find the spectrum of the damping constant γ_0. The four additive terms in Eq. (15.82) lead to four additive damping constants, of which two are negligibly small.

It can be shown that friction disappears as $T \to 0$, which indicates that friction is associated only with thermal fluctuations and not with quantum zero-point fluctuations. In fact, this result is also implied by the requirement that zero-point fluctuations are invariant under the Lorentz transformation [17]. Furthermore, another remarkable result is that friction is present even in empty space as long as the temperature is finite. Thus, an object moving in empty space ultimately comes to rest. In the free-space limit we obtain

$$\gamma_0 = \frac{\hbar^2}{18\pi^3 c^8 \varepsilon_0^2 k_{\mathrm{B}} T}\int_0^\infty |\alpha(\omega)|^2\omega^8\eta(\omega,T)\mathrm{d}\omega + \frac{\hbar^2}{3\pi^2 c^5\varepsilon_0 k_{\mathrm{B}} T}\int_0^\infty \mathrm{Im}[\alpha(\omega)]\omega^5\eta(\omega,\mathrm{T})\mathrm{d}\omega, \tag{15.83}$$

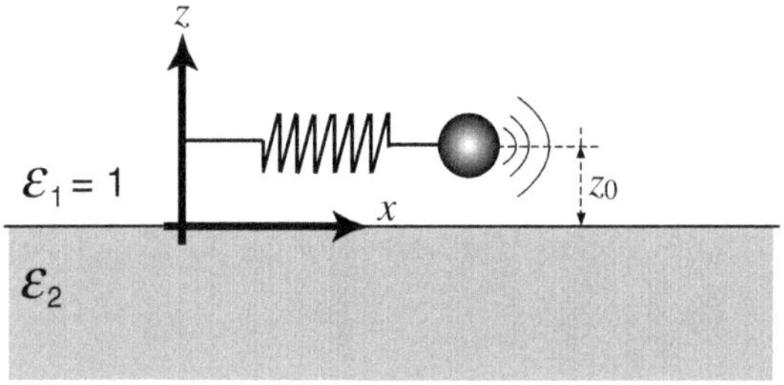

Fig. 15.9 A particle in vacuum moves parallel to a planar substrate with a dielectric function $\varepsilon_2(\omega)$.

where

$$\eta(\omega, T) \equiv \left[1/(e^{\hbar\omega/(k_B T)} - 1)\right]\left[1 + 1/(e^{\hbar\omega/(k_B T)} - 1)\right]. \tag{15.84}$$

The first term in Eq. (15.83) is consistent with the result by Boyer [17], whereas the second term was independently derived in Refs. [18, 19].

In Ref. [19], electromagnetic friction has been analyzed for the special case of a polarizable spherical particle (of radius a) placed near a semi-infinite half-space (substrate) with a complex dielectric constant $\varepsilon_2(\omega)$. Similar studies have been presented in Refs. [20, 21]. It is assumed that the particle is moving parallel to the surface (in the x-direction) at a vertical height of z_0 (see Fig. 15.9). These studies revealed not only a steep distance dependence of the damping constant but also a strong dependence on the material properties of the particle and substrate.

The friction between a particle and a planar substrate can be explained by the following qualitative physical picture [19]. The fluctuating currents in the particle and substrate generate a fluctuating electromagnetic field. This field polarizes the particle and induces an electric dipole with a corresponding image dipole beneath the surface of the substrate. The motion of the particle gives rise to motion of the image dipole. The Joule losses associated with the motion of the image dipole become larger with increasing resistivity of the substrate. As a consequence, the damping coefficient increases too. Physically, more work is needed to move the induced dipole beneath the surface as the resistivity increases and, consequently, the damping coefficient becomes larger. In the limit of a perfect dielectric the induced dipole cannot be displaced and damping becomes infinitely strong. On the one hand, it is surprising to find this result for a perfect (lossless) dielectric since there is no intrinsic dissipation. On the other hand, a lossless dielectric does not exist from the point of view of causality (Kramers–Kronig relations) and the fluctuation–dissipation theorem (fluctuations imply dissipation). Nevertheless, in the limit $T \to 0$, the damping coefficient vanishes even for a perfect dielectric. Notice that, because γ_0 is much weaker for metals, local friction measurements render metals transparent and reveal buried dielectric structures. This property can be used for subsurface imaging in metals and for the localization of defects.

15.4 Conclusion

In this chapter we have derived the fluctuation–dissipation theorem. The theorem is of fundamental importance and finds applications in different fields of science and technology. For example, the theorem explains Brownian motion in fluids and Johnson noise in electrical resistors. Applied to electromagnetic fields and sources, the theorem yields Planck's blackbody radiation spectrum, explains radiative heat transfer, and predicts the electromagnetic spectrum near material surfaces. We have applied the fluctuation–dissipation theorem to derive the dispersion force acting between separate objects and found that for small objects the force can be represented by the Casimir–Polder potential. For objects in relative motion, thermal fluctuations give rise to a dissipative interaction force (friction) even in the absence of mechanical contact between the objects. It is fascinating that the fluctuation–dissipation theorem accounts for so many apparently different physical phenomena. However, it should be kept in mind that the theorem can become insufficient for systems that are strongly out of equilibrium. In these situations, the system's response depends on the particular dynamics of the system's constituents.

Problems

15.1 Derive Eq. (15.10) by using the series expansion of Eq. (15.9) for the distribution function.

15.2 Equation (15.47) describes the spectral energy density W_ω as a function of the dielectric constant $\varepsilon(\mathbf{r}, \omega)$. Derive a similar equation for a system of N particles with coordinates $\mathbf{r}_n$ and polarizabilities $\alpha_n(\omega)$. Hint: use the fluctuation–dissipation theorem in Eq. (15.19).

15.3 Determine the spectral energy density W_ω originating from fluctuating sources in a small particle (diameter $\ll\lambda$) with polarizability α. Show that the electric and magnetic energy densities are identical and that there is *no* near-field contribution. Hint: use the Green functions defined in Eqs. (8.55) and (8.57).

15.4 The polarizability of an aluminum cluster can be approximated by the quasi-static formula

$$\alpha(\omega) = 3\varepsilon_0 V_0 \frac{\varepsilon(\omega) - 1}{\varepsilon(\omega) + 2}, \tag{15.85}$$

where V_0 is the volume of the cluster and ε is the dielectric constant of aluminum. The latter is described by the Drude model as

$$\varepsilon(\omega) = 1 - \frac{\omega_\mathrm{p}^2}{\omega^2 + \mathrm{i}\gamma\omega}, \tag{15.86}$$

with ω_p and γ being the plasma frequency and the damping constant, respectively. A good approximation is obtained using the values $\hbar\omega_\mathrm{p} = 15.565\,\mathrm{eV}$ and $\hbar\gamma = 0.608\,\mathrm{eV}$. Calculate the mean square of the fluctuating dipole moment in the frequency range $[\omega \ldots \omega + \mathrm{d}\omega]$ and plot this quantity as a function of frequency for temperatures $kT \ll \hbar\omega_{\mathrm{p}}$. Determine the total radiated power.

15.5 Derive the force formula (15.61) starting with Eq. (15.55) and following the steps outlined in Section 15.3.

15.6 The polarizability of a helium atom can be approximated by a single Lorentzian function as

$$\alpha(\omega) = \frac{(e^2/m_{\mathrm{e}})f_0}{\omega_0^2 - \omega^2 - \mathrm{i}\omega\gamma_0},$$

where the resonance frequency ω_0 accounts for all ${}^1\mathrm{S} \rightarrow {}^1\mathrm{P}^0$ transitions. The oscillator strength is related to the static polarizability by $f_0 = \alpha(0)\omega_0^2\,(m_{\mathrm{e}}/e^2)$, and γ_0 is the effective linewidth.

(1) Derive $\alpha(\mathrm{i}\eta)$ and show that it is real. Make use of $\gamma_0 \ll \omega_0$.

(2) The van der Waals potential between two helium atoms can be represented as $U_{\mathrm{v}} = -C_6/R^6$. Calculate the coefficient C_6 and express it in terms of $\alpha(0)$ and $\hbar\omega_0$. The resulting expression is known as *London's empirical formula*. Hint:

$$\int_{\infty}^{\infty} \frac{1}{(A^2 + x^2)^2}\,\mathrm{d}x = \frac{\pi}{2A^3}. \tag{15.87}$$

(3) Determine the distance R_0 for which U_{v} is equal to the Casimir potential (U_{c}). Use $\lambda_0 = 2\pi c/\omega_0 \approx 58\,\mathrm{nm}$.

(4) The static polarizability is $\alpha(0) = 2.280 \times 10^{-41}\,\mathrm{C\,m^2\,V^{-1}}$ and ω_0 is given by $\lambda_0 \approx 58\,\mathrm{nm}$. Plot the Casimir–Polder potential (U_{cp}) as a function of R. Include the curves for U_{v} and U_{c} and discuss the validity of these approximations. Provide the value of U_{cp} at the distance R_0.

References

[1] H. B. Callen and T. A. Welton, "Irreversibility and generalized noise," *Phys. Rev.* **83**, 34–40 (1951).

[2] D. Chandler, *Introduction to Modern Statistical Mechanics*. New York: Oxford University Press (1987).

[3] L. Mandel and E. Wolf, *Optical Coherence and Quantum Optics*. New York: Cambridge University Press (1995).

[4] G. S. Agarwal, "Quantum electrodynamics in the presence of dielectrics and conductors. I. Electromagnetic-field response functions and black-body fluctuations in finite geometries," *Phys. Rev. A* **11**, 230–242 (1975).

[5] C. Henkel, K. Joulain, J.-P. Mulet, and J.-J. Greffet, "Radiation forces on small particles in thermal fields," *J. Opt. A: Pure Appl. Opt.* **4**, S109–S114 (2002).

[6] S. M. Rytov, Yu. A. Kravtsov, and V. I. Tatarskii, *Principles of Statistical Radiophysics, Volume 3: Elements of Random Fields*. Berlin: Springer-Verlag (1987).

[7] A. V. Shchegrov, K. Joulain, R. Carminati, and J.-J. Greffet, "Near-field spectral effects due to electromagnetic surface excitations," *Phys. Rev. Lett.* **85**, 1548–1551 (2000).

[8] H. T. Dung, L. Knöll, and D.-G. Welsch, "Three-dimensional quantization of the electromagnetic field in dispersive and absorbing inhomogeneous dielectrics," *Phys. Rev. A* **57**, 3931–3942 (1998).

[9] O. D. Stefano, S. Savasta, and R. Girlanda, "Three-dimensional electromagnetic field quantization in absorbing and dispersive bounded dielectrics," *Phys. Rev. A* **61**, 023803 (2000).

[10] W. Eckhardt, "First and second fluctuation–dissipation theorem in electromagnetic fluctuation theory," *Opt. Commun.* **41**, 305–308 (1982).

[11] J. P. Mulet, K. Joulain, R. Carminati, and J. J. Greffet, "Nanoscale radiative heat transfer between a small particle and a plane surface," *Appl. Phys. Lett.* **78**, 2931–2933 (2001).

[12] E. Rousseau, A. Siria, G. Jourdan, *et al.*, "Radiative heat transfer at the nanoscale," *Nature Photonics* **3**, 514–517 (2009).

[13] B. C. Stipe, T. C. Strand, C. C. Poon, *et al.*, "Magnetic recording at 1.5 Pb m^{-2} using an integrated plasmonic antenna," *Nature Photonics* **4**, 484–488 (2010).

[14] R. Carminati and J.-J. Greffet, "Near-field effects in spatial coherence of thermal sources," *Phys. Rev. Lett.* **82**, 1660–1663 (1999).

[15] D. P. Craig and T. Thirunamachandran, *Molecular Quantum Electrodynamics*. Mineola, NY: Dover Publications (1998).

[16] H. B. G. Casimir, "On the attraction between two perfectly conducting plates," *Proc. Koninkl Ned. Akad. Wetenschap* **51**, 793–795 (1948).

[17] T. H. Boyer, "Derivation of the blackbody radiation spectrum without quantum assumptions," *Phys. Rev.* **182**, 1374–1383 (1969).

[18] V. Mkrtchian, V. A. Parsegian, R. Podgornik, and W. M. Saslow, "Universal thermal radiation drag on neutral objects," *Phys. Rev. Lett.* **91**, 220801 (2003).

[19] J. R. Zurita-Sanchez, J.-J. Greffet, and L. Novotny, "Near-field friction due to fluctuating fields," *Phys. Rev. A* **69**, 022902 (2004).

[20] A. I. Volokitin and B. N. J. Persson, "Dissipative van der Waals interaction between a small particle and a metal surface," *Phys. Rev. B* **65**, 115419 (2002).

[21] M. S. Tomassone and A. Widom, "Electronic friction forces on molecules moving near metals," *Phys. Rev. B* **56**, 4938–4943 (1997).

16 Theoretical methods in nano-optics

A key problem in nano-optics is the determination of electromagnetic field distributions near nanoscale structures and the associated radiation properties. A solid theoretical understanding of field distributions holds promise for new, optimized designs of near-field optical devices, in particular by exploitation of field-enhancement effects and favorable detection schemes. Calculations of field distributions are also necessary for image-reconstruction purposes. Fields near nanoscale structures often have to be reconstructed from experimentally accessible far-field data. However, most commonly the inverse scattering problem cannot be solved in a unique way, and calculations of field distributions are needed in order to provide prior knowledge about source and scattering objects and to restrict the set of possible solutions.

Analytical solutions of Maxwell's equations provide a good theoretical understanding, but can be obtained for simple problems only. Other problems have to be strongly simplified. A pure numerical analysis allows us to handle complex problems by discretization of space and time but computational requirements (usually given by CPU time and memory) limit the size of the problem and the accuracy of results is often unknown. The advantage of pure numerical methods, such as the finite-difference time-domain (FDTD) method and the finite-element (FE) method, is the ease of implementation. We do not review these pure numerical techniques since they are well documented in the literature. Instead we review two commonly used *semi-analytical* methods in nano-optics: the multiple-multipole method (MMP) and the volume-integral method. The latter exists in different implementations such as the coupled-dipole method, the dipole–dipole approximation, and the method of moments. Both the MMP and the volume-integral method are semi-analytical methods since they render an analytical expansion for the electromagnetic field by numerical means.

16.1 The multiple-multipole method

The MMP is a compromise between a pure analytical and a pure numerical approach. It is a well-established technique for solving Maxwell's equations in arbitrarily shaped, isotropic, linear, and piecewise homogeneous media [1]. The method is suited to analyzing extended structures since only the boundaries between homogeneous media need to be discretized and not the media themselves, as in methods such as finite elements and finite differences. The MMP technique provides an analytical expression for the solution of the electromagnetic field and it also provides a reliable validation of the results, since the errors can be calculated explicitly. In the past, the method was used for solving problems in various areas, such as antenna design, electromagnetic compatibility, bioelectromagnetics, and waveguide theory, as well as optics.

With the MMP technique, the electromagnetic field $\mathbf{F} \in \{\mathbf{E}, \mathbf{H}\}$ within individual media (domains) D_i is expanded by analytical solutions of Maxwell's equations:

$$\mathbf{F}^{(i)}(\mathbf{r}) \approx \sum_j A_j^{(i)} \mathbf{F}_j(\mathbf{r}).$$

The basis functions $\mathbf{F}_j$ (partial fields) are any known solutions of the vector Helmholtz equation, such as plane waves, multipole fields, and waveguide modes, among others. The expansions of the different subdomains are numerically matched on the interfaces, i.e., the parameters $A_j^{(i)}$ of the series expansions result from numerical matching of the boundary conditions. Consequently, Maxwell's equations are exactly fulfilled inside the domains but are only approximated on the boundaries. There are various methods similar to the MMP technique that are based on fictitious sources.

In linear, isotropic, homogeneous media, the electric field $\mathbf{E}$ and magnetic field $\mathbf{H}$ must satisfy the vector Helmholtz equation

$$(\nabla^2 + k^2)\mathbf{F} = 0. \tag{16.1}$$

The fields are assumed to be harmonic in time but the factor $\exp(-i\omega t)$ will not be used explicitly. The value of k is given by the dispersion relation $k^2 = (\omega/c)^2\mu\varepsilon$, where ω, c, and ε are the angular frequency, the vacuum velocity of light, and the dielectric constant, respectively.

The general solution to Eq. (16.1) can be constructed from a scalar function f that satisfies the scalar Helmholtz equation

$$(\nabla^2 + k^2)f(\mathbf{r}) = 0. \tag{16.2}$$

A common representation of the solutions of Eq. (16.1) is given by the two independent and mutually perpendicular vector fields [2]

$$\mathbf{M}(\mathbf{r}) = \nabla \times \mathbf{c}f(\mathbf{r}), \tag{16.3}$$

$$\mathbf{N}(\mathbf{r}) = \frac{1}{k}\nabla \times \mathbf{M}(\mathbf{r}), \tag{16.4}$$

which are called vector harmonics. In general, $\mathbf{c}$ is an arbitrary constant vector, but it can be shown that $\mathbf{c}$ may also represent the radial vector $\mathbf{R}$ in spherical coordinates. To prove that $\mathbf{M}$ and $\mathbf{N}$ are indeed solutions of the vector Helmholtz equation, they can be inserted into Eq. (16.1). With the help of vector identities it can then be shown that Eq. (16.1) reduces to Eq. (16.2). Thus, the problem of finding solutions to the field equations reduces to the simpler problem of finding solutions to the scalar Helmholtz equation. The vector harmonics $\mathbf{M}$ and $\mathbf{N}$ represent transverse or solenoidal solutions

$$\nabla \cdot \begin{matrix}\mathbf{M}(\mathbf{r})\\ \mathbf{N}(\mathbf{r})\end{matrix} = 0. \tag{16.5}$$

For the vector wave equation (Eq. (2.31)), for which the fields are not necessarily divergence-free, an additional longitudinal solution can be determined [3],

$$\mathbf{L}(\mathbf{r}) = \nabla f(\mathbf{r}), \tag{16.6}$$

which satisfies

$$\nabla \times \mathbf{L}(\mathbf{r}) = 0. \tag{16.7}$$

In electromagnetic theory, the electric and magnetic fields are always divergence-free in linear, isotropic, homogeneous, and source-free domains as long as the boundaries are not part of the domains (unbounded media). In this case the vector harmonic **L** must be excluded from the expansion of the fields [4] and the electromagnetic field can be entirely expanded in terms of the two vector harmonics **M** and **N**.

In the MMP technique the infinite space is divided into subdomains D_i. The interfaces between the individual subdomains usually follow the physical boundaries given by the material properties, but fictitious boundaries may be defined as well. In every D_i the scalar fields f may be approximated by a series expansion

$$f^{(i)}(\mathbf{r}) \approx \sum_j a_j^{(i)} f_j(\mathbf{r}), \tag{16.8}$$

in which the basis functions f_j cover any of the known analytical solutions of the Helmholtz equation (16.2). In order not to overburden the notation, the domain index (i) will be omitted. Of special importance for MMP are solutions in spherical coordinates.

In spherical coordinates $\mathbf{r} = (R, \vartheta, \varphi)$ the solutions of Eq. (16.2) can be written in the well-known form

$$f_{nm}(\mathbf{r}) = b_n(kR) Y_n^m(\vartheta, \varphi). \tag{16.9}$$

Y_n^m are the spherical harmonics and $b_n \in [j_n, y_n, h_n^{(1)}, h_n^{(2)}]$ are the spherical Bessel functions, of which only two are linearly independent. Solutions that use Bessel functions of the first kind (j_n) for the radial dependence are called *normal expansions*, whereas solutions using one of the other three radial functions are called *multipoles*. Similar relations exist for cylindrical solutions, which are relevant for two-dimensional problems. Multipoles with Hankel functions of the first kind ($h_n^{(1)}$) (radiative multipoles) have particularly favorable characteristics. They represent outgoing waves and fulfill Sommerfeld's radiation condition at infinity. Because they are singular in their origin, they must be located outside the domain in which the field is expanded. Normal expansions, on the other hand, remain finite at the origin but do not fulfill the radiation condition. Therefore they can be used only in finite domains.

To obtain the vector harmonics **M** and **N** in spherical coordinates, it is advantageous to set the vector **c** in Eq. (16.3) equal to the radial vector **R** [3]. It was initially required that **c** be a constant vector; but this would not hold for the choice $\mathbf{c} = \mathbf{R}$. However, for the spherical coordinate system it can in fact be shown that two independent solutions can be obtained from a radial vector [3].

With the choice $\mathbf{c}=\mathbf{R}$, the solution $\mathbf{M}$ is tangential to any spherical surface R = constant and reads

$$\mathbf{M}(\mathbf{r}) = (\nabla \times \mathbf{R})f(\mathbf{r}) = \begin{bmatrix} 0 \\ \sin^{-1}\vartheta\, \partial/\partial\varphi \\ -\partial/\partial\vartheta \end{bmatrix} f(\mathbf{r}), \tag{16.10}$$

where $\mathbf{M} = [M_R, M_\vartheta, M_\varphi]$. Apart from a factor $\mathrm{i}\hbar$ the operator $(\nabla \times \mathbf{R})$ is equal to the quantum-mechanical angular momentum operator. The vector field $\mathbf{N}$ can be derived from Eqs. (16.4) and (16.10).

There are many ways to relate the vector harmonics $\mathbf{M}$ and $\mathbf{N}$ to the electric and magnetic fields. Since the electromagnetic field in sourceless, linear, isotropic, and homogeneous media is entirely defined by two scalar fields (potentials) that satisfy the scalar Helmholtz equation and fulfill the appropriate boundary conditions [5, 6], one usually introduces two potentials from which all other field vectors can be deduced. In Mie scattering theory these potentials are most commonly chosen to correspond to Debye potentials [7]. The MMP technique follows a similar but somewhat easier approach that was first suggested by Bouwkamp and Casimir [5] and is also used in Jackson's book [8]. The two potentials are related to the electric and magnetic fields by

$$f^{\mathrm{e}}(\mathbf{r}) = \frac{A^{\mathrm{e}}}{n(n+1)}\mathbf{R}\cdot\mathbf{E}^{\mathrm{e}}, \tag{16.11}$$

$$f^{\mathrm{m}}(\mathbf{r}) = \frac{A^{\mathrm{m}}}{n(n+1)}\mathbf{R}\cdot\mathbf{H}^{\mathrm{m}}, \tag{16.12}$$

and are both explicitly given by Eq. (16.9). The factor $n(n+1)$ is introduced for later convenience and the amplitudes A^{e} and A^{m} are necessary in order to retain dimensionless potentials. f^{e} and f^{m} define two independent solutions $[\mathbf{E}^{\mathrm{e}}, \mathbf{H}^{\mathrm{e}}]$ and $[\mathbf{E}^{\mathrm{m}}, \mathbf{H}^{\mathrm{m}}]$. With Maxwell's equations and the help of vector identities it can be shown that the field defined by f^{e} corresponds to

$$\mathbf{H}^{\mathrm{e}}(\mathbf{r}) = -\mathrm{i}\omega\varepsilon_0\varepsilon A^{\mathrm{e}}(\nabla\times\mathbf{R})f^{\mathrm{e}}(\mathbf{r}) = -\mathrm{i}\omega\varepsilon_0\varepsilon A^{\mathrm{e}}\,\mathbf{M}(\mathbf{r}), \tag{16.13}$$

$$\mathbf{E}^{\mathrm{e}}(\mathbf{r}) = -\frac{1}{\mathrm{i}\omega\varepsilon_0\varepsilon}\nabla\times\mathbf{H}^{e}(\mathbf{r}) = kA^{\mathrm{e}}\mathbf{N}(\mathbf{r}). \tag{16.14}$$

Since the radial component of the magnetic field vanishes, this solution is called *transverse magnetic* (TM). Similarly, the potential f^{m} defines a *transverse electric* solution (TE) given by

$$\mathbf{E}^{\mathrm{m}}(\mathbf{r}) = \mathrm{i}\omega\mu_0\mu A^{\mathrm{m}}(\nabla\times\mathbf{R})f^{\mathrm{m}}(\mathbf{r}) = i\omega\mu_0\mu A^{\mathrm{m}}\mathbf{M}(\mathbf{r}), \tag{16.15}$$

$$\mathbf{H}^{\mathrm{m}}(\mathbf{r}) = \frac{1}{\mathrm{i}\omega\mu_0\mu}\nabla\times\mathbf{E}^{\mathrm{m}}(\mathbf{r}) = kA^{\mathrm{m}}\mathbf{N}(\mathbf{r}). \tag{16.16}$$

The general solution is obtained by combining the TE and TM solutions. A complete multipole expansion of order N, i.e. an expansion in which both m and n (in Eq. (16.9)) run from 0 to N, contains $N(N+2)$ parameters each for the TE and the TM cases. In the MMP technique these parameters have to be determined from the boundary conditions.

Close to its origin, a multipole function decreases with $\rho^{-(n+1)}$ and therefore affects mainly its immediate neighborhood. This fact led to the idea of using several origins for

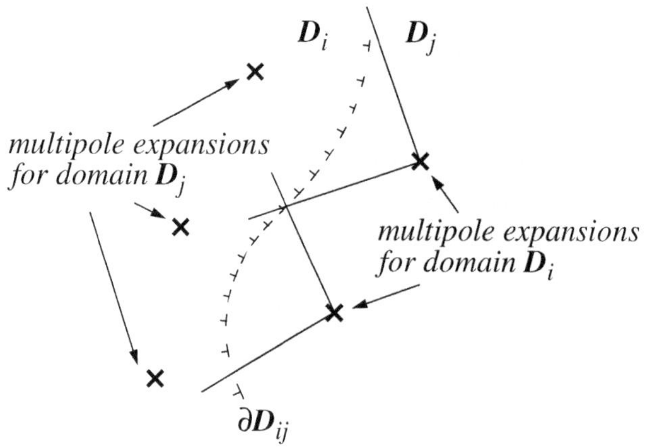

Fig. 16.1 A schematic representation of the principle of the multiple-multipole method. Multipole expansions located in domain D_j approximate the electromagnetic field inside domain D_i and vice versa. On each side of the boundary ∂D_{ij} the field close to a point on the interface is defined mainly by the closest multipole (indicated by the sectors for the field in domain D_i).

multipole expansions. Such a *multiple-multipole* approach achieves a better convergence for boundaries deviating considerably from spherical surfaces. Usually several multipoles are set along the boundary of the domain in which the field is expanded (Fig. 16.1). To avoid numerical dependences, the origins have to be sufficiently separated. The highest possible degree and order of an individual multipole are limited by a spatial sampling criterion that depends on the boundary discretization and on the proximity of the multipole to the boundary [1]. Figure 16.2 shows the MMP modeling for the example of a single scatterer in free space. The interior field of the scatterer is entirely expanded by multipoles, although a normal expansion in the interior could be used to support the multipoles. For the exterior field only multipoles with $b_n = h_n^{(1)}$ can be used in order to fulfill the radiation condition at infinity. Note that for complete expansions $m, n \to \infty$ the fields of the individual multipoles would be linearly dependent. This is not so for the finite expansions used in the numerical implementation. Hence, it is very advantageous to use several origins, since the computational requirements can be considerably reduced.

The unknown parameters $a_j^{(i)}$ in Eq. (16.8) have to be determined from the electric and magnetic boundary conditions. This is done by matching the expansions of adjacent domains D_i and D_j at discrete points $\mathbf{r}_k$ on their interface ∂D_{ij} according to

$$\mathbf{n}(\mathbf{r}_k) \times [\mathbf{E}_i(\mathbf{r}_k) - \mathbf{E}_j(\mathbf{r}_k)] = \mathbf{0}, \tag{16.17}$$

$$\mathbf{n}(\mathbf{r}_k) \times [\mathbf{H}_i(\mathbf{r}_k) - \mathbf{H}_j(\mathbf{r}_k)] = \mathbf{0}, \tag{16.18}$$

$$\mathbf{n}(\mathbf{r}_k) \cdot [\epsilon_i(\mathbf{r}_k)\mathbf{E}_i(\mathbf{r}_k) - \epsilon_j(\mathbf{r}_k)\mathbf{E}_j(\mathbf{r}_k)] = 0, \tag{16.19}$$

$$\mathbf{n}(\mathbf{r}_k) \cdot [\mu_i(\mathbf{r}_k)\mathbf{H}_i(\mathbf{r}_k) - \mu_j(\mathbf{r}_k)\mathbf{H}_j(\mathbf{r}_k)] = 0, \tag{16.20}$$

where $\mathbf{n}(\mathbf{r}_k)$ defines the normal vector to the boundary ∂D_{ij} in point $\mathbf{r}_k$. If conditions (16.17) and (16.18) are exactly fulfilled everywhere on the boundary (analytical solution),

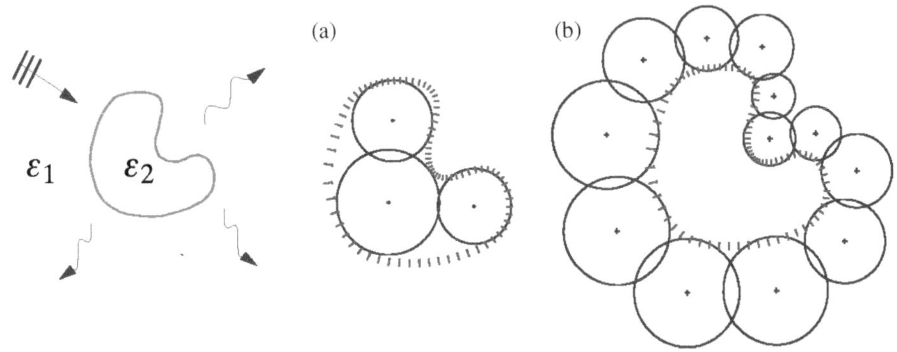

Fig. 16.2 MMP modeling of a single scatterer. (a) Multipoles for the outer domain, (b) multipoles for the inner domain. The circles indicate the area of greatest influence. The boundary is discretized and the normal vector to each matching point is indicated.

then conditions (16.19) and (16.20) are automatically satisfied. To produce more balanced errors, all the boundary conditions are considered in the MMP method. Problems with numerical dependences can be reduced by using an over-determined system of equations (more equations than unknowns) that is solved in the least-squares sense, i.e. by minimizing the squared (optionally weighted) error in the matching points. This procedure leads to a smoother error distribution along the boundaries than is obtained by the usual point matching. In addition, the error in each matching point can be computed and is a measure for the quality of the result. If the result is not accurate enough, additional or more appropriate basis functions have to be included in the series expansion Eq. (16.8). The choice of a suitable set of basis functions is the most difficult task in MMP modeling since no optimum can be determined in a unique way. Therefore, prior knowledge about the solution makes it possible to define favorable basis functions. A cylindrical structure, for example, would be expanded in cylindrical waveguide modes rather than in multipoles. Usually the solution of a given problem is improved by an iterative and interactive procedure. Automatic algorithms based on simple rules have been developed for the placement of the multipole origins and for the determination of the allowed maximum orders and degrees.

Once the system of equations has been solved and the parameters determined, the electromagnetic field can be readily computed at any point, because the solution is given in analytical form (Eq. (16.8)). Note that Maxwell's equations are exactly fulfilled within the individual domains, whereas they are approximated on the boundaries. The quality of this approximation depends on the choice of the expansion functions and on the numerical algorithm that is used for solving the unknown parameters.

The system of equations leads to a dense $M \times N$ matrix, which is commonly solved by means of the Givens procedure [9]. The computational time is proportional to MN^2, where M is the number of equations and N the number of parameters. Symmetries allow considerable reduction of the computational effort.

As an example, Fig. 16.3 shows the MMP model of a two-dimensional aperture near-field optical microscope in the probe–sample region [10]. The structure consists of five domains characterized by their distinct dielectric constants. For each of the domains the

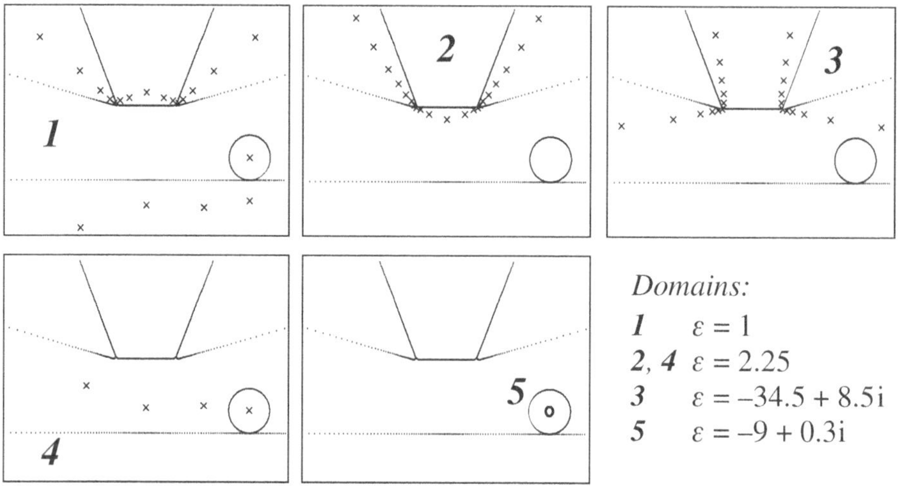

Fig. 16.3 The MMP model of a two-dimensional aperture-type near-field optical microscope. The locations of the multipole origins for the respective shaded domains are indicated by small crosses. Domain 5 is expanded by a normal expansion. The structure consists of a vacuum gap (1), a truncated glass wedge (2) embedded in an aluminum screen (3), and a planar transparent glass substrate (4) carrying a cylindrical silver particle (5).

corresponding multiple-multipole expansion is indicated by the crosses. The entire field is excited by a plane wave at $\lambda = 488\,\text{nm}$, which hits the structure from above at normal incidence. The interior of the cylindrical silver particle is expanded by a normal expansion. All multipole expansions have maximum orders of less than $N = 5$, leading to $2N + 1 = 11$ unknowns per origin. The resulting fields ($|\mathbf{E}|^2$) of the model are shown in Fig. 16.4 for the two principal polarizations. In s-polarization the electric field is always parallel to the boundaries. The contour lines are continuous across the boundary because of the continuity of the tangential electric field components. For p-polarization the field is characterized by the formation of maxima at the edges of the slit (field enhancement). In p-polarization a dipole moment is induced in the particle. Therefore the field close to the particle resembles a dipole field. Although the near-field interaction is stronger for p-polarization, the influence on the propagation of the field is more pronounced for s-polarization [10].

16.2 Volume-integral methods

Small particles can often be approximated by dipolar cells as in the case of Rayleigh scattering. The induced dipole moment in such a particle is proportional to the local field at the dipole's position. As long as a single particle is considered, the local field corresponds to the illuminating incident field. However, if an ensemble of particles is considered, the local field is a superposition of the incident radiation and all the partial fields scattered by the surrounding particles. It thus turns out that each particle is dependent on all the other

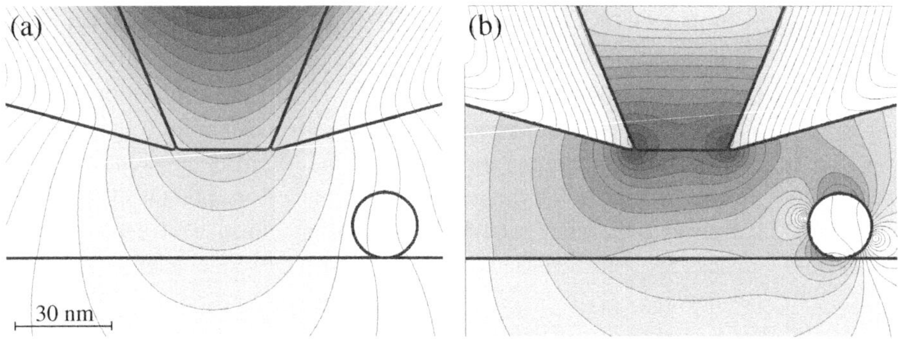

Fig. 16.4 Contours of constant $|\boldsymbol{E}|^2$ for the model of Fig. 16.3 (log scale, with a factor of $\sqrt{2}$ difference between successive lines): (a) s-polarization and (b) p-polarization.

particles. To solve this problem, a formalism for solving self-consistently the fields of an arbitrary number of coherently interacting particles is needed.

The particles are not required to be spatially separated from each other. They can be joined together to form a macroscopic object. Indeed, the response of matter to incident radiation can be formulated as a collective response of individual dipoles, each occupying a volume element. This superposition of elementary dipole fields (Green's functions) has to be done in a self-consistent way, i.e. the magnitude and the orientation of each individual dipole are functions of the local field defined by the excitation and other surrounding dipoles.

Methods based on this concept usually involve summations over all dipolar centers. In the limit, as the size of the dipolar centers goes to zero, the summations become volume integrals. Therefore, these formalisms are called *volume-integral methods*.

Both a *microscopic* and a *macroscopic* point of view exist for basically the same formalism. While, in the former, microscopic dipolar particles are joined together to form a macroscopic ensemble, the latter considers a macroscopic object that is divided into small homogeneous subunits. It will be shown in Section 16.2.4 that the two formalisms are physically and mathematically equivalent. The method following the microscopic point of view will be denoted the *coupled-dipole method* (CDM) and the method following the macroscopic point of view the *method of moments* (MOM).

Both the CDM and the MOM are well-established methods for solving Maxwell's equations in various fields of study. The CDM is widely used in astrophysics [11] for the investigation of interstellar grains, but it also finds applications in other fields such as meteorological optics and surface-contamination control [12]. The MOM has its origins in electromagnetic practice [13], with special focus on antenna theory. However, the MOM also finds applications in biological investigations, in optical scattering, and in near-field optics [14, 15]. In the literature, the two methods very often bear different names. As an example, the CDM is also called the discrete-dipole approximation (DDA) and the MOM is designated the digitized Green-function method [16] or simply the volume-integral equation method [17]. Furthermore, because of the analogy to quantum mechanics, the volume-integral equation is denoted by some authors as the Lippmann–Schwinger equation [15].

Both the CDM and the MOM can be derived from the same volume-integral equation. In the past, some authors compared inadequate forms of the two methods and stated that one method is superior to the other [18]. However, as shown by Lakhtakia [19] for bianisotropic scatterers in free space, the two methods are fully equivalent to each other. The main difference between the CDM and the MOM is the point of view: while the MOM involves the fields that are *actually present* at a given point **r**, the CDM considers the fields that arrive at the point **r** and thus *excite* the small region ΔV centered at **r**. Lakhtakia distinguishes between *weak* and *strong* forms of the two methods, and we shall adopt the same terminology.

16.2.1 The volume-integral equation

Consider an arbitrary reference system, such as a planar layered substrate, whose dielectric properties are sufficiently well represented by a spatially inhomogeneous dielectric constant $\varepsilon_{\text{ref}}(\mathbf{r})$, **r** being the position vector. For simplicity, the reference frame is assumed to be non-magnetic ($\mu_{\text{ref}} = 1$) and isotropic. All the fields are further assumed to be time-harmonic. The dielectric constant of all space will be denoted as $\varepsilon(\mathbf{r})$. Then, as long as the reference system is unperturbed (no other objects are present), ε is identical to ε_{ref}. In the presence of perturbing objects embedded in the reference system, $\varepsilon(\mathbf{r}) - \varepsilon_{\text{ref}}(\mathbf{r})$ defines the dielectric response of the objects relative to the reference system.

In the absence of any source currents and charges, Maxwell's curl equations read as

$$\nabla \times \mathbf{E}(\mathbf{r}) = \mathrm{i}\omega\mu_0\mathbf{H}(\mathbf{r}), \tag{16.21}$$

$$\nabla \times \mathbf{H}(\mathbf{r}) = -\mathrm{i}\omega\varepsilon_0\varepsilon_{\text{ref}}(\mathbf{r})\mathbf{E}(\mathbf{r}) + \mathbf{j}_{\text{e}}(\mathbf{r}), \tag{16.22}$$

where $\mathbf{j}_{\text{e}}$ is the volume distribution of the induced electric current density,

$$\mathbf{j}_{\text{e}}(\mathbf{r}) = -\mathrm{i}\omega\varepsilon_0[\varepsilon(\mathbf{r}) - \varepsilon_{\text{ref}}(\mathbf{r})]\mathbf{E}(\mathbf{r}). \tag{16.23}$$

From Eqs. (16.21) and (16.22) it follows that **E** has to fulfill the inhomogeneous wave equation

$$\nabla \times \nabla \times \mathbf{E}(\mathbf{r}) - k_0^2\varepsilon_{\text{ref}}(\mathbf{r})\mathbf{E}(\mathbf{r}) = \mathrm{i}\omega\mu_0\,\mathbf{j}_{\text{e}}(\mathbf{r}), \tag{16.24}$$

where the free-space wavenumber k_0 is equal to ω/c. Using the definition of the dyadic Green function (Fig. 16.5) (see Section 2.12)

$$\nabla \times \nabla \times \overleftrightarrow{\mathbf{G}}(\mathbf{r},\mathbf{r}') - k_0^2\varepsilon_{\text{ref}}(\mathbf{r})\overleftrightarrow{\mathbf{G}}(\mathbf{r},\mathbf{r}') = \overleftrightarrow{\mathbf{I}}\delta(\mathbf{r}-\mathbf{r}'), \tag{16.25}$$

the electric field can be represented as

$$\mathbf{E}(\mathbf{r}) = \mathbf{E}_0(\mathbf{r}) + \frac{\mathrm{i}\omega}{\varepsilon_0 c^2}\int_V \overleftrightarrow{\mathbf{G}}(\mathbf{r},\mathbf{r}')\mathbf{j}_{\text{e}}(\mathbf{r}')\mathrm{d}V', \quad \mathbf{r} \notin V, \tag{16.26}$$

where the prime in V' indicates that the integration refers to $\mathbf{r}'$. While $\mathbf{E}_0$ denotes the homogeneous solution ($\mathbf{j}_{\text{e}} = 0$ everywhere), the term on the right-hand side represents the

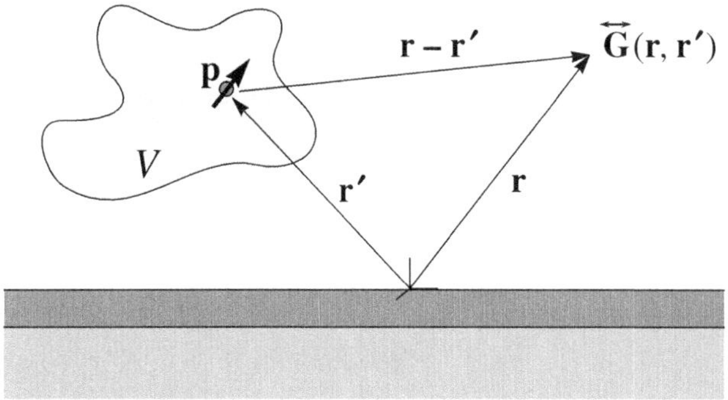

Fig. 16.5 Illustration of the dyadic Green function $\overleftrightarrow{\mathbf{G}}(\mathbf{r},\mathbf{r}')$. $\mathbf{r}'$ and $\mathbf{r}$ represent the dipolar source and the point of observation, respectively. The field at $\mathbf{r}$ depends on the orientation of the dipole at $\mathbf{r}'$. The three columns of $\overleftrightarrow{\mathbf{G}}$ denote the electric fields for the three major orientations of the dipole.

particular solution. A similar procedure can be applied to obtain the magnetic field. One finds

$$\mathbf{H}(\mathbf{r}) = \mathbf{H}_0(\mathbf{r}) + \int_V \left[\nabla\times\, \overleftrightarrow{\mathbf{G}}(\mathbf{r},\mathbf{r}')\right]\mathbf{j}_\mathrm{e}(\mathbf{r}')\mathrm{d}V', \quad \mathbf{r}\notin V. \tag{16.27}$$

Upon substitution of $\mathbf{j}_\mathrm{e}$ into Eqs. (16.26) and (16.27) implicit integral equations (Fredholm equations of the second kind) are obtained for the fields $\mathbf{E}$ and $\mathbf{H}$. These equations are denoted as the *volume-integral equations* for the electric and magnetic fields and form the basis for the MOM.

The current density of an electric dipole with moment $\mathbf{p}_0$ located at $\mathbf{r}=\mathbf{r}_0$ is

$$\mathbf{j}_\mathrm{e}(\mathbf{r}) = -\mathrm{i}\omega\mathbf{p}_0\delta(\mathbf{r}-\mathbf{r}_0), \tag{16.28}$$

where the delta function has the units m^{-3}. On inserting this current into Eqs. (16.26) and (16.27) and assuming the homogeneous solution is zero (no external excitation), the electromagnetic fields can be expressed in terms of $\overleftrightarrow{\mathbf{G}}(\mathbf{r},\mathbf{r}')$ as

$$\mathbf{E}(\mathbf{r}) = \frac{\omega^2}{\varepsilon_0\, c^2}\overleftrightarrow{\mathbf{G}}(\mathbf{r},\mathbf{r}_0)\mathbf{p}_0, \tag{16.29}$$

$$\mathbf{H}(\mathbf{r}) = -\mathrm{i}\omega\left[\nabla\times\, \overleftrightarrow{\mathbf{G}}(\mathbf{r},\mathbf{r}_0)\right]\mathbf{p}_0. \tag{16.30}$$

Thus, the $\mathbf{E}$ field of a dipole with orientation $\mathbf{p}_0 = |\mathbf{p}|\,\mathbf{n}_x$ located at $\mathbf{r}=\mathbf{r}_0$ corresponds to the first column of $\overleftrightarrow{\mathbf{G}}(\mathbf{r},\mathbf{r}_0)$. Similarly, the $\mathbf{E}$ field of a y-oriented (z-oriented) dipole corresponds to the second (third) column of $\overleftrightarrow{\mathbf{G}}(\mathbf{r},\mathbf{r}_0)$. In other words, the columns of $\overleftrightarrow{\mathbf{G}}(\mathbf{r},\mathbf{r}_0)$ render the $\mathbf{E}$ vectors for the three major orientations of the dipole. The same relationship holds for the $\mathbf{H}$ field and $[\nabla\times\, \overleftrightarrow{\mathbf{G}}(\mathbf{r},\mathbf{r}_0)]$. The electromagnetic field of an arbitrarily oriented dipole can therefore simply be represented in terms of $\overleftrightarrow{\mathbf{G}}$ and $[\nabla\times\, \overleftrightarrow{\mathbf{G}}]$.

For later purposes it will be convenient to split $\overleftrightarrow{\mathbf{G}}$ into two separate contributions,

$$\overleftrightarrow{\mathbf{G}}(\mathbf{r},\mathbf{r}_0) = \overleftrightarrow{\mathbf{G}}_0(\mathbf{r},\mathbf{r}_0) + \overleftrightarrow{\mathbf{G}}_\mathrm{s}(\mathbf{r},\mathbf{r}_0). \tag{16.31}$$

$\overleftrightarrow{\mathbf{G}}_\mathrm{s}$ is the scattering part of the Green function and accounts for the secondary electromagnetic field, i.e. the field that is reflected from or transmitted through inhomogeneities in the environment. Similarly, $\overleftrightarrow{\mathbf{G}}_0$ is the primary part of the Green function (Eq. (16.32)) and determines the direct dipole field. While $\overleftrightarrow{\mathbf{G}}_0$ is singular at its origin $\mathbf{r}=\mathbf{r}'$, the scattering part $\overleftrightarrow{\mathbf{G}}_\mathrm{s}$ behaves regularly. $\overleftrightarrow{\mathbf{G}}_0$ will contribute only to the field in the (sub)domain in which it is located. $\overleftrightarrow{\mathbf{G}}_0$ corresponds to the Green dyadic in free space and can be determined in closed analytical form from the scalar Green function $G_0(\mathbf{r},\mathbf{r}')$ according to (cf. Section 2.10)

$$\overleftrightarrow{\mathbf{G}}_0(\mathbf{r},\mathbf{r}') = \left[\overleftrightarrow{\mathbf{I}} + \frac{1}{k^2}\nabla\nabla\right] G_0(\mathbf{r},\mathbf{r}'), \tag{16.32}$$

where $G_0(\mathbf{r},\mathbf{r}')$ is a solution of

$$\nabla^2 G_0(\mathbf{r},\mathbf{r}') + k^2 G_0(\mathbf{r},\mathbf{r}') = -\delta(\mathbf{r}-\mathbf{r}'). \tag{16.33}$$

The solution of this equation is

$$G_0(\mathbf{r},\mathbf{r}') = \frac{1}{4\pi}\frac{\mathrm{e}^{\pm ik|\mathbf{r}-\mathbf{r}'|}}{|\mathbf{r}-\mathbf{r}'|}, \tag{16.34}$$

where the plus sign refers to an outgoing wave and the minus sign to an incoming wave.

So far, the field $\mathbf{E}$ has been derived for points outside the scattering objects ($\mathbf{r} \notin V$). However, if the fields within the source volume ($\mathbf{r} \in V$) are to be evaluated, a *principal volume* V_δ must be introduced in order to exclude the singularity of $\overleftrightarrow{\mathbf{G}}_0$ at $\mathbf{r}=\mathbf{r}'$. In this case, the solution of Eq. (16.24) reads

$$\begin{aligned}\mathbf{E}(\mathbf{r}) = \mathbf{E}_0(\mathbf{r}) &+ \frac{i\omega}{\varepsilon_0 c^2}\int_V \overleftrightarrow{\mathbf{G}}_\mathrm{s}(\mathbf{r},\mathbf{r}')\,\mathbf{j}_\mathrm{e}(\mathbf{r}')\mathrm{d}V' \\ &+ \frac{i\omega}{\varepsilon_0 c^2}\lim_{\delta\to 0}\int_{V-V_\delta} \overleftrightarrow{\mathbf{G}}_0(\mathbf{r},\mathbf{r}')\,\mathbf{j}_\mathrm{e}(\mathbf{r}')\mathrm{d}V' + \frac{\overleftrightarrow{\mathbf{L}}\,\mathbf{j}_\mathrm{e}(\mathbf{r})}{i\omega\varepsilon_0\varepsilon_\mathrm{ref}(\mathbf{r})}, \qquad \mathbf{r}\in V.\end{aligned} \tag{16.35}$$

A similar expression can be found for the magnetic field $\mathbf{H}$. In the limit as the maximum chord length δ approaches zero, the exclusion volume V_δ becomes infinitely small. The source dyadic $\overleftrightarrow{\mathbf{L}}$ accounts for the depolarization of the excluded volume V_δ and turns out to depend entirely on the geometry of the principal volume [20],

$$\overleftrightarrow{\mathbf{L}} = \frac{1}{4\pi}\int_{S_\delta} \frac{\mathbf{n}(\mathbf{r}')(\mathbf{r}'-\mathbf{r})}{|\mathbf{r}-\mathbf{r}'|^3}\,\mathrm{d}S'. \tag{16.36}$$

The limit as $\delta \to 0$ is omitted in the expression for $\overleftrightarrow{\mathbf{L}}$ because the surface integral depends only on the geometry of V_δ. For cubic or spherical principal volumes the source dyadic turns out to be $\overleftrightarrow{\mathbf{L}} = (1/3)\overleftrightarrow{\mathbf{I}}$. As Yaghjian points out, the value of the volume integral also varies with the geometry of the principal volume and in just the right way to keep the sum of the volume and surface integrals independent of the geometry of the principal volume [20].

Equation (16.35) is known as the (electric) volume-integral equation. It can be represented by the simpler equation (16.26) if the Green function is written in the symbolic form

$$\overleftrightarrow{\mathbf{G}}(\mathbf{r},\mathbf{r}') = \text{P.V.}\left[\overleftrightarrow{\mathbf{G}}(\mathbf{r},\mathbf{r}')\right] - \frac{\overleftrightarrow{\mathbf{L}}\delta(\mathbf{r}-\mathbf{r}')}{k_0^2\varepsilon_{\text{ref}}(\mathbf{r}')}. \tag{16.37}$$

The symbol P.V. denotes the principal value and was introduced by van Bladel [21]. A volume integral over P.V.$[\overleftrightarrow{\mathbf{G}}(\mathbf{r},\mathbf{r}')]$ acting on a current $\mathbf{j}(\mathbf{r}')$ implies that an infinitesimal exclusion volume at $\mathbf{r}=\mathbf{r}'$ has to be invoked and that depolarization of this volume must be taken into account. In other words

$$\int_V \text{P.V.}\left[\overleftrightarrow{\mathbf{G}}(\mathbf{r},\mathbf{r}')\right]\mathbf{j}_{\text{e}}(\mathbf{r}')\text{d}V' = \lim_{\delta\to 0}\int_{V-V_\delta}\overleftrightarrow{\mathbf{G}}(\mathbf{r},\mathbf{r}')\,\mathbf{j}_{\text{e}}(\mathbf{r}')\text{d}V' + \frac{\overleftrightarrow{\mathbf{L}}\,\mathbf{j}_{\text{e}}(\mathbf{r})}{k_0^2\varepsilon_{\text{ref}}(\mathbf{r})}. \tag{16.38}$$

In the usual notation, the symbol P.V. is taken out of the integral. The principal-volume notation is stated here for completeness only and will not be used in the following.

The source volume V can be split into N volume elements ΔV_n such that

$$V = \sum_{n=1}^{N}\Delta V_n. \tag{16.39}$$

It is assumed that the individual volume elements are sufficiently small, such that the current density $\mathbf{j}_{\text{e}}$ can be regarded as constant over the dimensions of ΔV_n

$$\mathbf{j}_{\text{e}}(\mathbf{r}) = \mathbf{j}_{\text{e}}(\mathbf{r}_n), \quad \mathbf{r}\in\Delta V_n, \tag{16.40}$$

where $\mathbf{r}_n$ is an arbitrary point inside ΔV_n. In this case, the solution for the field $\mathbf{E}$ can be written as

$$\mathbf{E}(\mathbf{r}) = \mathbf{E}_0(\mathbf{r}) + \sum_{n=1}^{N}\Delta\mathbf{E}_n^0(\mathbf{r}) + \sum_{n=1}^{N}\Delta\mathbf{E}_n^{\text{s}}(\mathbf{r}), \tag{16.41}$$

where $\Delta\mathbf{E}_n^0$ is the primary field generated by the current in the subvolume ΔV_n, and $\Delta\mathbf{E}_n^{\text{s}}$ is the corresponding scattered field. $\Delta\mathbf{E}_n^0$ and $\Delta\mathbf{E}_n^{\text{s}}$ are determined by

$$\Delta\mathbf{E}_n^0(\mathbf{r}) = \begin{cases} \dfrac{\text{i}\omega}{\varepsilon_0 c^2}\left[\displaystyle\int_{\Delta V_n}\overleftrightarrow{\mathbf{G}}_0(\mathbf{r},\mathbf{r}')\text{d}V'\right]\mathbf{j}_{\text{e}}(\mathbf{r}_n), & \mathbf{r}\notin\Delta V_n, \\[2ex] \dfrac{\text{i}\omega}{\varepsilon_0 c^2}\left[\displaystyle\lim_{\delta\to 0}\int_{\Delta V_n - V_\delta}\overleftrightarrow{\mathbf{G}}_0(\mathbf{r},\mathbf{r}')\text{d}V' - \frac{\overleftrightarrow{\mathbf{L}}}{k_0^2\varepsilon_{\text{ref}}(\mathbf{r})}\right]\mathbf{j}_{\text{e}}(\mathbf{r}_n), & \mathbf{r}\in\Delta V_n, \end{cases} \tag{16.42}$$

$$\Delta\mathbf{E}_n^{\text{s}}(\mathbf{r}) = \frac{\text{i}\omega}{\varepsilon_0 c^2}\left[\int_{\Delta V_n}\overleftrightarrow{\mathbf{G}}_{\text{s}}(\mathbf{r},\mathbf{r}')\text{d}V'\right]\mathbf{j}_{\text{e}}(\mathbf{r}_n). \tag{16.43}$$

Owing to the smooth behavior of $\overleftrightarrow{\mathbf{G}}_0$ at $\mathbf{r} \neq \mathbf{r}'$, the integral in the expression for $\mathbf{r} \notin \Delta V_n$ can be approximated by $\Delta V_n \overleftrightarrow{\mathbf{G}}_0(\mathbf{r}, \mathbf{r}_n)$. This approximation cannot be applied for $\mathbf{r} \in \Delta V_n$ because of the strong variation of $\overleftrightarrow{\mathbf{G}}_0$ near $\mathbf{r} = \mathbf{r}'$. Instead, the volume integral has to be carried out explicitly for a given geometry of the principal volume V_δ. Since $\overleftrightarrow{\mathbf{G}}_\mathrm{s}$ is well behaved for all $\mathbf{r}$, the integration of $\overleftrightarrow{\mathbf{G}}_\mathrm{s}$ can be replaced by $\Delta V_n \overleftrightarrow{\mathbf{G}}_\mathrm{s}(\mathbf{r}, \mathbf{r}_n)$ everywhere. For later convenience, the remaining volume integral will be denoted as

$$\overleftrightarrow{\mathbf{M}} = \lim_{\delta \to 0} \int\limits_{\Delta V_n - V_\delta} \overleftrightarrow{\mathbf{G}}_0(\mathbf{r}, \mathbf{r}')\mathrm{d}V'. \tag{16.44}$$

After inserting Eqs. (16.42) and (16.43) into Eq. (16.41) and evaluating the field $\mathbf{E}$ at the positions $\mathbf{r}_k = \mathbf{r}_n$, the following N vector equations are obtained:

$$\begin{aligned}\mathbf{E}(\mathbf{r}_k) = \mathbf{E}_0(\mathbf{r}_k) &+ \frac{\mathrm{i}\omega}{\varepsilon_0 c^2}\left[\overleftrightarrow{\mathbf{M}}(\mathbf{r}_k) - \frac{\overleftrightarrow{\mathbf{L}}(\mathbf{r}_k)}{k_0^2\,\varepsilon_\mathrm{ref}(\mathbf{r}_k)} + \Delta V_k\,\overleftrightarrow{\mathbf{G}}_\mathrm{s}(\mathbf{r}_k, \mathbf{r}_k)\right]\mathbf{j}_\mathrm{e}(\mathbf{r}_k) \\ &+ \frac{\mathrm{i}\omega}{\varepsilon_0 c^2}\sum_{\substack{n=1\\ n\neq k}}^{N} \overleftrightarrow{\mathbf{G}}(\mathbf{r}_k, \mathbf{r}_n)\mathbf{j}_\mathrm{e}(\mathbf{r}_n)\Delta V_n, \quad k = 1, \ldots, N.\end{aligned} \tag{16.45}$$

These N equations are the basis for both the MOM and the CDM. The dyadics $\overleftrightarrow{\mathbf{L}}$ and $\overleftrightarrow{\mathbf{M}}$ are given by Eqs. (16.36) and (16.44), respectively. $\overleftrightarrow{\mathbf{G}}$ is the Green function and $\overleftrightarrow{\mathbf{G}}_\mathrm{s}$ denotes its scattering part. Note that the term in brackets, containing $\overleftrightarrow{\mathbf{M}}$, $\overleftrightarrow{\mathbf{L}}$, and $\overleftrightarrow{\mathbf{G}}_\mathrm{s}$, defines the interaction of the volume element ΔV_k with itself, whereas the sum in the second row accounts for interactions with other dipolar subunits. The various contributions to $\mathbf{E}(\mathbf{r}_k)$ are illustrated in Fig. 16.6 for the case of two single, spatially isolated volume elements.

The dyadics $\overleftrightarrow{\mathbf{M}}$ and $\overleftrightarrow{\mathbf{L}}$ can be evaluated for a specific geometry of V_δ, but their symbolic representation will be maintained for general validity. It can be shown that $\overleftrightarrow{\mathbf{M}}(\mathbf{r}_n)$ approaches zero as the subvolume ΔV_n is reduced arbitrarily. Therefore, in the limit $\Delta V_n \to 0$ the contribution of $\overleftrightarrow{\mathbf{M}}(\mathbf{r}_n)$ can be ignored. The dyadic $\overleftrightarrow{\mathbf{L}}(\mathbf{r}_n)$, on the other hand, does not vanish in the limit $\Delta V_n \to 0$. This dyadic accounts for self-depolarization and its incorporation is absolutely necessary in a self-consistent formalism.

Since Eq. (16.45) considers both $\overleftrightarrow{\mathbf{M}}$ and $\overleftrightarrow{\mathbf{L}}$, the equation represents a so-called *strong* form. The *weak* form is obtained if $\overleftrightarrow{\mathbf{M}}$ is ignored and only $\overleftrightarrow{\mathbf{L}}$ is considered. According to Lakhtakia [19], only comparisons between strong forms *or* weak forms are appropriate. A comparison between the strong form of the MOM and the weak form of the CDM will show the same inconsistency as a comparison between the strong and weak forms of the same method.

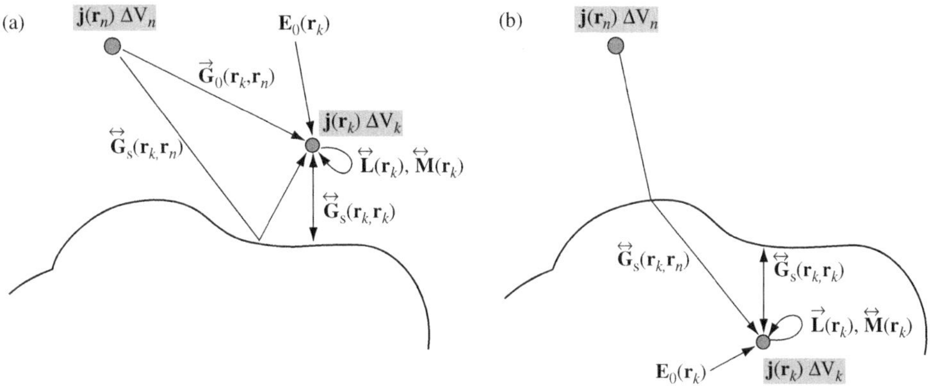

Fig. 16.6 Interaction of a volume element ΔV_k with its environment and with itself. For clarity, only a single interacting volume element ΔV_n is shown. (a) Volume elements ΔV_k and ΔV_n in the same material domain. (b) Volume elements ΔV_k and ΔV_n in different domains. The arrows indicate the "path" of interaction and the symbols denote the quantities involved. The Green function can be split into a primary and a scattering part, $\overleftrightarrow{G} = \overleftrightarrow{G}_0 + \overleftrightarrow{G}_s$, and it fulfills the boundary conditions imposed by the reference system.

16.2.2 The method of moments (MOM)

The method of moments considers the fields that are actually present at a given point $\mathbf{r}$. These fields are directly represented by Eq. (16.45). In order to arrive at a solvable system of equations, the electric current density

$$\mathbf{j}_e(\mathbf{r}) = -i\omega\varepsilon_0[\varepsilon(\mathbf{r}) - \varepsilon_{\text{ref}}(\mathbf{r})]\mathbf{E}(\mathbf{r}) = -i\omega\varepsilon_0\, \Delta\varepsilon(\mathbf{r})\mathbf{E}(\mathbf{r}) \tag{16.46}$$

is introduced into Eq. (16.45). This leads to the following system of equations:

$$\mathbf{E}_0(\mathbf{r}_k) = \sum_{n=1}^{N} \overleftrightarrow{\mathbf{A}}_{kn}\mathbf{E}(\mathbf{r}_n), \quad k = 1, ..., N, \tag{16.47}$$

where the submatrices $\overleftrightarrow{\mathbf{A}}_{kn}$ are given by

$$\overleftrightarrow{\mathbf{A}}_{kn} = \left[\overleftrightarrow{\mathbf{I}} - \left[k_0^2\overleftrightarrow{\mathbf{M}}(\mathbf{r}_k) - \frac{\overleftrightarrow{\mathbf{L}}(\mathbf{r}_k)}{\varepsilon_{\text{ref}}(\mathbf{r}_k)} + \Delta V_k\, k_0^2\overleftrightarrow{\mathbf{G}}_s(\mathbf{r}_k, \mathbf{r}_k)\right]\Delta\varepsilon(\mathbf{r}_k)\right]\delta_{kn}$$

$$- \left[\Delta V_n\, k_0^2\overleftrightarrow{\mathbf{G}}(\mathbf{r}_k, \mathbf{r}_n)\Delta\varepsilon(\mathbf{r}_k)\right](1 - \delta_{kn}). \tag{16.48}$$

Since Eq. (16.47) is a vector matrix equation, $\overleftrightarrow{\mathbf{A}}_{kn}$ is a $3N \times 3N$ submatrix. Different computational schemes, such as the conjugate-gradient method, serve to solve the system of equations. Probably the most difficult task associated with the MOM is that of finding an efficient and reliable algorithm for solving Eq. (16.47). Because the resulting matrices usually have low conditions, a direct solution can become numerically unstable for large systems. To overcome this problem, Martin *et al.* [15] introduced an iterative procedure that is based on Dyson's equation.

The current $\mathbf{j}_e$ given by Eq. (16.46) can also be inserted into Eq. (16.26) or Eq. (16.35) in order to obtain an integral formulation of Eq. (16.47). Furthermore, it should be noted that the formalism is not restricted to isotropic scatterers. Equation (16.47) remains unaffected if $\varepsilon(\mathbf{r})$ is assumed to be a tensor. The extension to bianisotropic scatterers can be found in Ref. [19].

16.2.3 The coupled-dipole method (CDM)

In contrast to the MOM, the CDM considers the field $\mathbf{E}_{\text{exc}}$ that excites a given volume-element. This field is different from the field $\mathbf{E}$ in Eq. (16.45). In order to obtain the field $\mathbf{E}_{\text{exc}}$, the "self-fields" associated with $\overleftrightarrow{\mathbf{M}}$ and $\overleftrightarrow{\mathbf{L}}$ have to be subtracted from the actual field $\mathbf{E}$ to yield

$$\begin{aligned}\mathbf{E}_{\text{exc}}(\mathbf{r}_k) = {} & \mathbf{E}_0(\mathbf{r}_k) + \frac{\mathrm{i}\omega}{\varepsilon_0 c^2} \overleftrightarrow{\mathbf{G}}_{\text{s}}(\mathbf{r}_k, \mathbf{r}_k)\, \mathbf{j}_{\text{e}}(\mathbf{r}_k)\Delta V_k \\ & + \frac{\mathrm{i}\omega}{\varepsilon_0 c^2} \sum_{\substack{n=1 \\ n\neq k}}^{N} \overleftrightarrow{\mathbf{G}}(\mathbf{r}_k, \mathbf{r}_n)\, \mathbf{j}_{\text{e}}(\mathbf{r}_n)\Delta V_n, \quad k = 1, \ldots, N.\end{aligned} \tag{16.49}$$

While the dyadics $\overleftrightarrow{\mathbf{M}}$ and $\overleftrightarrow{\mathbf{L}}$ define the *direct* interaction, the term containing $\overleftrightarrow{\mathbf{G}}_{\text{s}}$ accounts for the *indirect* interaction. The field associated with $\overleftrightarrow{\mathbf{G}}_{\text{s}}(\mathbf{r}_k, \mathbf{r}_k)$ is the field that was emitted at $\mathbf{r} = \mathbf{r}_k$ at former times and now arrives back at $\mathbf{r} = \mathbf{r}_k$ after having been scattered in the environment (Fig. 16.6). Therefore, this field also contributes to the exterior excitation of the volume element at ΔV_k and hence must be included in Eq. (16.49).

Using the microscopic polarizability $\overleftrightarrow{\alpha}_k$, the dipole moment $\mathbf{p}_k$ induced in the volume element ΔV_k can be related to the field $\mathbf{E}_{\text{exc}}(\mathbf{r}_k)$ by

$$\mathbf{p}_k = \overleftrightarrow{\alpha}_k \, \mathbf{E}_{\text{exc}}(\mathbf{r}_k). \tag{16.50}$$

This relationship can be introduced in Eq. (16.49) after expressing the current density in terms of the dipole moment

$$\mathbf{j}_{\text{e}}(\mathbf{r}_k) = -\frac{\mathrm{i}\omega}{\Delta V_k} \mathbf{p}_k. \tag{16.51}$$

The resulting system of equations in matrix form reads as

$$\mathbf{E}_0(\mathbf{r}_k) = \sum_{n=1}^{N} \overleftrightarrow{\mathbf{B}}_{kn} \, \mathbf{E}_{\text{exc}}(\mathbf{r}_n), \quad k = 1, \ldots, N, \tag{16.52}$$

where the submatrices $\overleftrightarrow{\mathbf{B}}_{kn}$ are given by

$$\overleftrightarrow{\mathbf{B}}_{kn} = \left[\overleftrightarrow{\mathbf{I}} - \frac{\omega^2}{\varepsilon_0 c^2} \overleftrightarrow{\mathbf{G}}_{\text{s}}(\mathbf{r}_k, \mathbf{r}_k) \overleftrightarrow{\alpha}_k \right] \delta_{kn} - \left[\frac{\omega^2}{\varepsilon_0 c^2} \overleftrightarrow{\mathbf{G}}(\mathbf{r}_k, \mathbf{r}_n) \overleftrightarrow{\alpha}_n \right] (1 - \delta_{kn}). \tag{16.53}$$

If Eq. (16.52) is multiplied by $\overleftrightarrow{\alpha}_k$ on both sides, a system of equations is obtained for the dipole moments,

$$\overleftrightarrow{\alpha}_k\,\mathbf{E}_0(\mathbf{r}_k) = \sum_{n=1}^{N} \overleftrightarrow{\mathbf{C}}_{kn}\,\mathbf{p}_n, \quad k = 1, ..., N, \tag{16.54}$$

where the submatrices $\overleftrightarrow{\mathbf{C}}_{kn}$ are given by

$$\overleftrightarrow{\mathbf{C}}_{kn} = \left[\overleftrightarrow{\mathbf{I}} - \frac{\omega^2}{\varepsilon_0 c^2}\overleftrightarrow{\alpha}_k \overleftrightarrow{\mathbf{G}}_{\mathrm{s}}(\mathbf{r}_k, \mathbf{r}_k)\right]\delta_{kn} - \left[\frac{\omega^2}{\varepsilon_0 c^2}\overleftrightarrow{\alpha}_k \overleftrightarrow{\mathbf{G}}(\mathbf{r}_k, \mathbf{r}_n)\right](1 - \delta_{kn}). \tag{16.55}$$

Once the dipole moments have been determined, the field can easily be calculated everywhere in space. It has to be emphasized once more that $\mathbf{E}_{\mathrm{exc}}$ is identical with $\mathbf{E}$ only outside the scatterer occupied by the volume V. Inside V the two fields are different. In order to obtain the actual field inside V from $\mathbf{E}_{\mathrm{exc}}$, the "self-field" associated with $\overleftrightarrow{\mathbf{L}}$ and $\overleftrightarrow{\mathbf{M}}$ has to be added to every interior point. However, outside the volume V the field due to the N induced dipoles reads as

$$\mathbf{E}(\mathbf{r}) = \mathbf{E}_0(\mathbf{r}) + \frac{\omega^2}{\varepsilon_0 c^2}\sum_{n=1}^{N} \overleftrightarrow{\mathbf{G}}(\mathbf{r}, \mathbf{r}_n)\mathbf{p}_n, \quad \mathbf{r} \notin V. \tag{16.56}$$

In order to compare the CDM and the MOM, the polarizabilities $\overleftrightarrow{\alpha}_k$ have to be expressed in terms of $\overleftrightarrow{\mathbf{L}}(\mathbf{r}_k)$, $\overleftrightarrow{\mathbf{M}}(\mathbf{r}_k)$, and $\varepsilon(\mathbf{r}_k)$. The requirement for the CDM to be identical with the MOM leads to

$$\overleftrightarrow{\alpha}_k = \Delta V_k\,\varepsilon_0\,\Delta\varepsilon(\mathbf{r}_k)\left[\overleftrightarrow{\mathbf{I}} - \left[k_0^2\overleftrightarrow{\mathbf{M}}(\mathbf{r}_k) - \frac{\overleftrightarrow{\mathbf{L}}(\mathbf{r}_k)}{\varepsilon_{\mathrm{ref}}(\mathbf{r}_k)}\right]\Delta\varepsilon(\mathbf{r}_k)\right]^{-1}. \tag{16.57}$$

This relation follows from the equality of the current density Eq. (16.46) in the MOM and the current density Eqs. (16.50) and (16.51) in the CDM. The exciting field $\mathbf{E}_{\mathrm{exc}}$ has further to be expressed in terms of the actual field $\mathbf{E}$ according to

$$\begin{aligned}\mathbf{E}_{\mathrm{exc}}(\mathbf{r}_k) &= \mathbf{E}(\mathbf{r}_k) - \frac{\mathrm{i}\,\omega}{\varepsilon_0 c^2}\left[\overleftrightarrow{\mathbf{M}}(\mathbf{r}_k) - \frac{\overleftrightarrow{\mathbf{L}}(\mathbf{r}_k)}{k_0^2\,\varepsilon_{\mathrm{ref}}(\mathbf{r}_k)}\right]\mathbf{j}_{\mathrm{e}}(\mathbf{r}_k) \\ &= \left[\overleftrightarrow{\mathbf{I}} - \left[k_0^2\overleftrightarrow{\mathbf{M}}(\mathbf{r}_k) - \frac{\overleftrightarrow{\mathbf{L}}(\mathbf{r}_k)}{\varepsilon_{\mathrm{ref}}(\mathbf{r}_k)}\right]\Delta\varepsilon(\mathbf{r}_k)\right]\mathbf{E}(\mathbf{r}_k),\end{aligned} \tag{16.58}$$

which follows from Eqs. (16.45), (16.46), and (16.49). It remains to be shown that Eq. (16.57) reduces to the known forms for the polarizability.

16.2.4 Equivalence of the MOM and the CDM

In the weak forms of the MOM and the CDM, the contribution of the dyadic $\overleftrightarrow{\mathbf{M}}$ is ignored. In this case Eq. (16.57) can be expressed as (cf. Section 12.4)

$$\overleftrightarrow{\alpha}_k = 3\varepsilon_0\varepsilon_{\mathrm{ref}}(\mathbf{r}_k)\Delta V_k\,\frac{\varepsilon(\mathbf{r}_k) - \varepsilon_{\mathrm{ref}}(\mathbf{r}_k)}{\varepsilon(\mathbf{r}_k) + 2\varepsilon_{\mathrm{ref}}(\mathbf{r}_k)}\,\overleftrightarrow{\mathbf{I}}, \tag{16.59}$$

where we used the explicit value of $\overleftrightarrow{\mathbf{L}} = (1/3)\,\overleftrightarrow{\mathbf{I}}$. Equation (16.59) is recognized as the quasi-static polarizability of an electrically small sphere. Thus, the MOM and the CDM turn out to be identical in their weak forms! Furthermore, since Eq. (16.58) relates the exciting field $\mathbf{E}_{\text{exc}}$ to the actual field $\mathbf{E}$, it can be shown that the field inside the subvolume ΔV_k is

$$\mathbf{E}(\mathbf{r}_k) = \frac{3\varepsilon_{\text{ref}}(\mathbf{r}_k)}{\varepsilon(\mathbf{r}_k) + 2\varepsilon_{\text{ref}}(\mathbf{r}_k)} \mathbf{E}_{\text{ext}}(\mathbf{r}_k). \tag{16.60}$$

This relationship is consistent with the corresponding expression obtained for a small sphere in a homogeneous external field.

In order to compare the strong forms of the MOM and the CDM, an explicit value for the dyadic $\overleftrightarrow{\mathbf{M}}$ has to be determined. The calculation is most easily performed for a spherical principal volume V_δ. In this case, the integral in Eq. (16.44) can be determined and the expression for $\overleftrightarrow{\mathbf{M}}$ reads as [21]

$$\overleftrightarrow{\mathbf{M}}(\mathbf{r}_k) = \frac{2}{3}\frac{1}{k_{\text{ref}}^2(\mathbf{r}_k)}\left[\left[1 - \mathrm{i}k_{\text{ref}}(\mathbf{r}_k)a_k\right]\mathrm{e}^{\mathrm{i}k_{\text{ref}}(\mathbf{r}_k)a_k} - 1\right]\overleftrightarrow{\mathbf{I}}\,. \tag{16.61}$$

Here a_k is the radius of the spherical subvolume $\Delta V_k = (4\pi/3)a_k^3$ and k_{ref} is given by $k_{\text{ref}}^2 = k_0^2\varepsilon_{\text{ref}}$. As expected, $\overleftrightarrow{\mathbf{M}}$ equals zero for $a_k \to 0$. On inserting Eq. (16.61) into the expression for the polarizability (16.57) and using $\overleftrightarrow{\mathbf{L}} = (1/3)\overleftrightarrow{\mathbf{I}}$ one obtains

$$\overleftrightarrow{\alpha}_k = \left[3\varepsilon_0\varepsilon_{\text{ref}}(\mathbf{r}_k)\frac{\Delta\varepsilon(\mathbf{r}_k)\Delta V_k}{\varepsilon(\mathbf{r}_k) + 2\varepsilon_{\text{ref}}(\mathbf{r}_k)}\right]\left[\overleftrightarrow{\mathbf{I}} - \frac{3k_{\text{ref}}^2(\mathbf{r}_k)\Delta\varepsilon(\mathbf{r}_k)}{\varepsilon(\mathbf{r}_k) + 2\varepsilon_{\text{ref}}(\mathbf{r}_k)}\overleftrightarrow{\mathbf{M}}\right]^{-1}. \tag{16.62}$$

The first factor is recognized as the weak form of the polarizability, whereas the second expression defines a correction term that accounts for the finite subvolume V_δ. For $\Delta V_k \to 0$ this term equals $\overleftrightarrow{\mathbf{I}}$. The polarizability α_k in Eq. (16.62) was first explicitly determined by Lakhtakia [19], and it is *this form* which must be considered in a comparison with the strong form of the MOM. It therefore turns out that the strong forms of the MOM and the CDM are also equivalent. Since the strong forms account for the finite size of the subvolumes, they generally lead to faster convergence than the weak forms.

It has been shown by several authors that the electrostatic polarizability given by Eq. (16.59) satisfies neither energy conservation nor the optical theorem for particles modeled by a single dipole [22, 23]. According to the equivalence of the MOM and the CDM, neither does the weak form of the MOM provide physical solutions for single dipolar scatterers! Therefore, the dyadic $\overleftrightarrow{\mathbf{M}}$ is significant even for very small particles. In order to achieve physical solutions, various other forms of the CDM were proposed, but all of them modify the weak form of the CDM to include higher-order terms in $k_{\text{ref}}(\mathbf{r}_k a_k)$.

It is repeated here that the main difference between the MOM and the CDM is the point of view. While the CDM considers the field incident on a subvolume (the exciting field), the MOM deals with the field that is actually present in the subvolume. Therefore, the field of the CDM represents only the solution for the fields outside the scatterer, whereas inside the scatterer the relation between the exciting field $\mathbf{E}_{\text{exc}}$ and the actual field $\mathbf{E}$ is given by Eq. (16.58).

Note that the same formalism can be applied to particles having magnetic polarizabilities. In this case the dual substitutions of Section 10.9 have to be carried out and Green's dyadic $\overleftrightarrow{\mathbf{G}}$ has to be replaced by $(\nabla\times\overleftrightarrow{\mathbf{G}})$.

16.3 Effective polarizability

The effective polarizability $\overleftrightarrow{\alpha}_{\text{eff}}$ is often introduced in order to account for the interaction of a single dipolar particle with its environment. The interaction originates from the fact that part of the field emitted by the dipole at previous times is reflected back and influences the dipole's properties. The dipole moment $\mathbf{p}$ of a polarizable particle located at $\mathbf{r}=\mathbf{r}_0$ with polarizability $\overleftrightarrow{\alpha}(\omega)$ is related to the local exciting field $\mathbf{E}_{\text{local}}=\mathbf{E}_{\text{exc}}(\mathbf{r}=\mathbf{r}_0)$ by

$$\mathbf{p}=\overleftrightarrow{\alpha}(\omega)\mathbf{E}_{\text{local}}, \tag{16.63}$$

where the polarizability is given by Eq. (16.57). Note that $\mathbf{E}_{\text{local}}$ is the field that excites the particle and is thus not equal to the actual field at $\mathbf{r}=\mathbf{r}_0$. $\mathbf{E}_{\text{local}}$ can be split into two contributions,

$$\mathbf{E}_{\text{local}}=\mathbf{E}_0(\mathbf{r}_0)+\mathbf{E}_{\text{s}}(\mathbf{r}_0), \tag{16.64}$$

where $\mathbf{E}_0$ is the exciting field and $\mathbf{E}_{\text{s}}$ is the dipole's field that is reflected back to its position (the scattered field). The latter can be written as

$$\mathbf{E}_{\text{s}}(\mathbf{r}_0)=\frac{\omega^2}{\varepsilon_0 c^2}\overleftrightarrow{\mathbf{G}}_{\text{s}}(\mathbf{r}_0,\mathbf{r}_0)\mathbf{p}. \tag{16.65}$$

Note that $\overleftrightarrow{\mathbf{G}}_{\text{s}}$, unlike $\overleftrightarrow{\mathbf{G}}_0$, has no singularity and hence may be evaluated at its origin. With Eqs. (16.63)–(16.65) it follows that

$$\mathbf{p}-\frac{\omega^2}{\varepsilon_0 c^2}\overleftrightarrow{\alpha}(\omega)\overleftrightarrow{\mathbf{G}}_{\text{s}}(\mathbf{r}_0,\mathbf{r}_0)\mathbf{p}=\overleftrightarrow{\alpha}(\omega)\mathbf{E}_0(\mathbf{r}_0). \tag{16.66}$$

This equation is identical to Eq. (16.54) of the CDM if a single particle is considered. The right-hand side is simply the primary dipole moment $\mathbf{p}_0$, i.e. the dipole moment induced by the exciting field $\mathbf{E}_0$. Thus, Eq. (16.66) can be rewritten as

$$\mathbf{p}-\frac{\omega^2}{\varepsilon_0 c^2}\overleftrightarrow{\alpha}(\omega)\overleftrightarrow{\mathbf{G}}_{\text{s}}(\mathbf{r}_0,\mathbf{r}_0)\mathbf{p}=\mathbf{p}_0, \tag{16.67}$$

which can be solved for $\mathbf{p}$. The self-consistent dipole moment $\mathbf{p}$ is determined by $\overleftrightarrow{\mathbf{G}}_{\text{s}}$, containing the information relating to the optical properties of the environment, and by $\overleftrightarrow{\alpha}$, describing the properties of the particle itself. From Eq. (16.66) an effective polarizability, $\overleftrightarrow{\alpha}_{\text{eff}}$, can be calculated as

$$\mathbf{p}=\overleftrightarrow{\alpha}_{\text{eff}}(\omega)\mathbf{E}_0(\mathbf{r}_0). \tag{16.68}$$

In free space $\overleftrightarrow{\alpha}_{\text{eff}}$ is equal to $\overleftrightarrow{\alpha}$. Inhomogeneities in the environment change the polarizability from $\overleftrightarrow{\alpha}$ to $\overleftrightarrow{\alpha}_{\text{eff}}$. These changes are due to the dipole's interaction with the environment.

If $\overleftrightarrow{\alpha}$ represents a molecule or atom with well-defined transition energies, then the interaction with the environment leads to resonance shifts and alterations of the decay rates (cf. Section 8.5).

16.4 The total Green function

The entire electromagnetic information of a system consisting of an arbitrary number of particles can be represented by a single dyadic function. This function is denoted $\overleftrightarrow{\mathbf{G}}_{\mathrm{t}}$, where the index t stands for "total." The term "particle" represents any dipolar center, whether spatially isolated from others or merged together with others to form a macroscopic medium. Consider an arbitrary number of particles embedded in an inhomogeneous reference system, such as a planarly layered structure. It is assumed that the Green function $\overleftrightarrow{\mathbf{G}}$ accounts for the inhomogeneous reference system. According to Eq. (16.58), the actual field $\mathbf{E}(\mathbf{r})$ outside of all the particles in the system is equal to the exciting field $\mathbf{E}_{\mathrm{exc}}$ because $\mathbf{r}$ is an exterior point. For $\mu = 1$ this field is given by Eqs. (16.52) and (16.53) and reads

$$\mathbf{E}(\mathbf{r}) = \mathbf{E}_0(\mathbf{r}) + \frac{\omega^2}{\varepsilon_0 c^2} \sum_{n=1}^{N} \overleftrightarrow{\mathbf{G}}(\mathbf{r}, \mathbf{r}_n) \overleftrightarrow{\alpha}_n \mathbf{E}(\mathbf{r}_n), \tag{16.69}$$

where $\mathbf{E}_0$ is the field in the absence of the particles. For the present purpose $\mathbf{E}_0$ is the field of an exciting dipole at $\mathbf{r} = \mathbf{r}_k$ with the dipole moment $\mathbf{p}_k$. According to Eq. (16.29) the dipole field can be expressed in terms of the Green function as

$$\mathbf{E}_0(\mathbf{r}) = \frac{\omega^2}{\varepsilon_0 c^2} \overleftrightarrow{\mathbf{G}}(\mathbf{r}, \mathbf{r}_k) \mathbf{p}_k. \tag{16.70}$$

The combination of these two equations leads to

$$\mathbf{E}(\mathbf{r}) = \frac{\omega^2}{\varepsilon_0 c^2} \overleftrightarrow{\mathbf{G}}(\mathbf{r}, \mathbf{r}_k) \mathbf{p}_k + \frac{\omega^2}{\varepsilon_0 c^2} \sum_{n=1}^{N} \overleftrightarrow{\mathbf{G}}(\mathbf{r}, \mathbf{r}_n) \overleftrightarrow{\alpha}_n \mathbf{E}(\mathbf{r}_n). \tag{16.71}$$

If the Green function of the entire system were known, the field at $\mathbf{r}$ could simply be calculated from

$$\mathbf{E}(\mathbf{r}) = \frac{\omega^2}{\varepsilon_0 c^2} \overleftrightarrow{\mathbf{G}}_{\mathrm{t}}(\mathbf{r}, \mathbf{r}_k) \mathbf{p}_k. \tag{16.72}$$

Here, $\overleftrightarrow{\mathbf{G}}_{\mathrm{t}}$ accounts not only for the inhomogeneous reference system, but also for the particles. The field $\mathbf{E}$ in Eq. (16.72) can be substituted for the fields $\mathbf{E}(\mathbf{r})$ and $\mathbf{E}(\mathbf{r}_n)$ in Eq. (16.71) to obtain

$$\overleftrightarrow{\mathbf{G}}_{\mathrm{t}}(\mathbf{r}, \mathbf{r}_k) \mathbf{p}_k = \overleftrightarrow{\mathbf{G}}(\mathbf{r}, \mathbf{r}_k) \mathbf{p}_k + \frac{\omega^2}{\varepsilon_0 c^2} \sum_{n=1}^{N} \overleftrightarrow{\mathbf{G}}(\mathbf{r}, \mathbf{r}_n) \overleftrightarrow{\alpha}_n \overleftrightarrow{\mathbf{G}}_{\mathrm{t}}(\mathbf{r}_n, \mathbf{r}_k) \mathbf{p}_k. \tag{16.73}$$

This equation can be postmultiplied by $\mathbf{p}_k/|\mathbf{p}_k|^2$ to give

$$\overleftrightarrow{\mathbf{G}}_{\mathrm{t}}(\mathbf{r},\mathbf{r}_k) = \overleftrightarrow{\mathbf{G}}(\mathbf{r},\mathbf{r}_k) + \omega^2\mu_0 \sum_{n=1}^{N} \overleftrightarrow{\mathbf{G}}(\mathbf{r},\mathbf{r}_n)\overleftrightarrow{\alpha}_n(\omega)\overleftrightarrow{\mathbf{G}}_{\mathrm{t}}(\mathbf{r}_n,\mathbf{r}_k). \tag{16.74}$$

This is the discrete form of Dyson's equation [15, 24] and was first derived in quantum mechanics. In the present approach it was assumed that only exterior points are considered. The derivation for interior points follows the same steps but with a slightly more complicated expression for Eq. (16.69) including the scattering part of $\overleftrightarrow{\mathbf{G}}$. The beauty of $\overleftrightarrow{\mathbf{G}}_{\mathrm{t}}$ is that it incorporates all information about the environment. Once $\overleftrightarrow{\mathbf{G}}_{\mathrm{t}}$ is known, the field of a dipole placed at an arbitrary location $\mathbf{r}_k$ is readily calculated using Eq. (16.72). The formalism of Dyson's equation is elaborated in more detail by Martin *et al.* [25] for van der Waals interactions and for electromagnetic scattering. In the former, $\overleftrightarrow{\mathbf{G}}_{\mathrm{t}}$ is called the "field susceptibility," whereas for the latter it bears the name "generalized field propagator."

16.5 Conclusion

In this chapter we discussed theoretical methods that are commonly encountered in the field of nano-optics. We primarily concentrated on methods that use numerical means for finding analytical expressions for the fields under study. We left out purely numerical methods because they are well documented elsewhere. The reader should also notice that we did not cover many fascinating theoretical concepts that have been put forward recently. Among those are primarily inverse methods aimed at reconstructing the object space (geometry and material properties of the sample) using the information accessible through detectors in the near-field or far-field [26]. It has been shown that concepts in inverse scattering can be applied to optical near-fields and that, by sampling information under different angles of excitation or detection, *near-field optical tomography* can be established, i.e. the three-dimensional reconstruction of the object space [27].

Problems

16.1 Consider a single dipolar particle with radius a and dielectric constant ε_{p} located above a dielectric half-space with dielectric constant ε. Derive the effective polarizability $\overleftrightarrow{\alpha}_{\mathrm{eff}}$ as a function of the distance d between surface and center of the particle.

16.2 Calculate the effective polarizability of two dielectric spheres with radius a and dielectric constant ε, separated by a distance d.

16.3 Two metal particles with radii $a = 20\,\mathrm{nm}$ are separated by the distance $d = 3a$ (center to center). Calculate the scattering cross-section for an incident plane wave polarized in the direction of the two particles (along the x-axis). The wavelength is $\lambda = 488\,\mathrm{nm}$ and the dielectric constant $\varepsilon = -34.5 + \mathrm{i}8.5$ (aluminum). Compare the result with the first Born approximation (no interaction between the two particles). What separation d is necessary in order to achieve an accuracy of 10% for the first Born approximation?

16.4 The quasi-static polarizability α does not fulfill the optical theorem. To prove this inconsistency, consider the scattering from a single dipolar particle with polarizability α. The optical theorem states that the extinction cross-section of a plane wave propagating in the direction of z and scattering off an arbitrary object reads as

$$\sigma_{\text{ext}} = \frac{4\pi}{k^2}\text{Re}\{\mathbf{X} \cdot \mathbf{n}_E\}\big|_{x,y=0}, \tag{16.75}$$

where k is the wavenumber, $\mathbf{n}_E$ the unit vector in the direction of the incident polarization, and $\mathbf{X}$ the scattering amplitude evaluated for $z \to \infty$. $x, y = 0$ indicates that the value is to be evaluated along the propagation direction. $\mathbf{X}$ is related to the scattered far-field $\mathbf{E}_s$ by

$$\mathbf{E}_s = -\frac{e^{ikR}}{ikR}\mathbf{X}E_0, \tag{16.76}$$

with R being the radial distance from the scattering object and E_0 the amplitude of the incident plane wave. Notice that $\mathbf{X}$ is dimensionless. In the case of a small particle, the field $\mathbf{E}_s$ originates from the dipole $\mathbf{p}$, which in turn is induced by the incident field $\mathbf{E}_0$.

(a) Derive the result $\sigma_{\text{ext}} = (k/\varepsilon_0)\text{Im}\{\alpha\}$.

(b) Calculate the scattering cross-section σ_{scatt} by using the relationship $\mathbf{p} = \alpha\mathbf{E}_0$, the formula for dipole radiation $P_{\text{scatt}} = |\mathbf{p}|^2\omega^4/(12\pi\varepsilon_0 c^3)$, and the incident intensity $I_0 = \varepsilon_0 c\,|\mathbf{E}_0|^2/2$.

(c) Derive the absorption cross-section σ_{abs} using the relationship $P_{\text{abs}} = (\omega/2)\text{Im}\{\mathbf{p} \cdot \mathbf{E}_0^*\}$ and show that it is identical with σ_{ext}.

It follows that the extinction cross-section calculated through the optical theorem accounts only for absorption and not for scattering. Thus, the quasi-static polarizability has to be handled with care. A solution of this dilemma is considered in Problem 8.5, where radiation reaction leads to an additional contribution to the polarizability (cf. Eq. (8.222)).

References

[1] C. Hafner, *The Generalized Multiple Multipole Technique for Computational Electromagnetics*. Boston, MA: Artech (1990).

[2] C. F. Bohren and D. R. Huffmann (eds.), *Absorption and Scattering of Light by Small Particles*. New York: Wiley (1983).

[3] J. A. Stratton, *Electromagnetic Theory*. New York: McGraw-Hill (1941).

[4] W. C. Chew, *Waves and Fields in Inhomogeneous Media*. New York: Van Nostrand Reinhold (1990).

[5] C. J. Bouwkamp and H. B. G. Casimir, "On multipole expansions in the theory of electromagnetic radiation," *Physica* **20**, 539–554 (1954).
[6] H. S. Green and E. Wolf, "A scalar representation of electromagnetic fields," *Proc. Phys. Soc. A* **66**, 1129–1137 (1953).
[7] M. Born and E. Wolf, *Principles of Optics*, 6th edn. Oxford: Pergamon (1970).
[8] J. D. Jackson, *Classical Electrodynamics*, 2nd edn. New York: Wiley (1975).
[9] G. H. Golub and C. F. van Loan, *Matrix Computations*. Baltimore, MA: Johns Hopkins University Press (1989).
[10] L. Novotny, D. W. Pohl, and P. Regli, "Light propagation through nanometer-sized structures: the two-dimensional-aperture scanning near-field optical microscope," *J. Opt. Soc. Am. A* **11**, 1768–1779 (1994).
[11] B. T. Draine and P. J. Flatau, "Discrete-dipole approximation for scattering calculations," *J. Opt. Soc. Am. A* **11**, 1491–1499 (1994).
[12] M. A. Taubenblatt and T. K. Tran, "Calculation of light scattering from particles and structures on a surface by the coupled dipole method," *J. Opt. Soc. Am. A* **10**, 912–919 (1993).
[13] R. F. Harrington, *Field Computation by Moment Methods*. Piscataway, NJ: IEEE Press (1992).
[14] R. Carminati and J. J. Greffet, "Influence of dielectric contrast and topography on the near field scattered by an inhomogeneous surface," *J. Opt. Soc. Am. A* **12**, 2716–2725 (1995).
[15] O. J. F. Martin, A. Dereux, and C. Girard, "Iterative scheme for computing exactly the total field propagating in dielectric structures of arbitrary shape," *J. Opt. Soc. Am. A* **11**, 1073–1080 (1994).
[16] G. H. Goedecke and S. G. O'Brien, "Scattering by irregular inhomogeneous particles via the digitized Green's function algorithm," *Appl. Opt.* **27**, 2431–2438 (1989).
[17] M. F. Iskander, H. Y. Chen, and J. E. Penner, "Optical scattering and absorption by branched chains of aerosols," *Appl. Opt.* **28**, 3083–3091 (1989).
[18] J. I. Hage and M. Greenberg, "A model for the optical properties of porous grains," *Astrophys. J.* **361**, 251–259 (1990).
[19] A. Lakhtakia, "Macroscopic theory of the coupled dipole approximation method," *J. Mod. Phys. C* **3**, 583–603 (1992).
[20] A. D. Yaghjian, "Electric dyadic Green's functions in the source region," *Proc. IEEE* **68**, 248–263 (1980).
[21] J. van Bladel, "Some remarks on Green's dyadic for infinite space," *IRE Trans. Antennas Propag.* **9**, 563–566 (1961).
[22] B. T. Draine, "The discrete-dipole approximation and its application to interstellar graphite grains," *Astrophys. J.* **333**, 848–872 (1988).
[23] A. Lakhtakia, "Macroscopic theory of the coupled dipole approximation method," *Opt. Commun.* **79**, 1–5 (1990).
[24] E. N. Economou, *Green's Functions in Quantum Physics*, 2nd edn. Berlin: Springer-Verlag (1990).

[25] O. J. F. Martin, C. Girard, and A. Dereux, "Generalized field propagator for electromagnetic scattering and light confinement," *Phys. Rev. Lett.* **74**, 526–529 (1995).

[26] P. S. Carney and J. C. Schotland, "Inverse scattering for near-field optics," *Appl. Phys. Lett.* **77**, 2798–2800 (2000).

[27] P. S. Carney and J. C. Schotland, "Near-field tomography," in *Inside Out: Inverse Problems*, ed. G. Uhlman. Cambridge: Cambridge University Press, pp. 133–168 (2003).

Appendix A Semi-analytical derivation of the atomic polarizability

The purpose of this section is to derive the linear polarizability of a two-level quantum system in the dipole approximation. The quantum system might be an atom, a molecule, or a quantum dot. For simplicity, we denote the system as an atom. Once the atomic polarizability is known, the interaction between atom and radiation field can be treated classically in many applications. A generally valid analytical expression for the polarizability cannot be derived. Instead, one has to distinguish among several approximate expressions that depend on the relative spectral properties of the atom and the field. The two most important regimes are *off-resonance* and *near-resonance* excitation. In the former case, the atom resides mostly in its ground state, whereas in the latter case saturation of the excited level becomes significant.

According to quantum mechanics, the behavior of a system of N particles is described by the wavefunction

$$\Psi(\mathbf{r}, t) = \Psi(\mathbf{r}_1, \ldots, \mathbf{r}_N, t), \tag{A.1}$$

where $\mathbf{r}_i$ denotes the spatial coordinate of particle i and t represents the time variable. To make the notation simpler, the entire set of particle coordinates is represented by the single coordinate $\mathbf{r}$, which also includes spin. However, it should be kept in mind that operations on $\mathbf{r}$ are operations on all particle coordinates $\mathbf{r}_1, \ldots, \mathbf{r}_N$. The wavefunction Ψ is a solution of the Schrödinger equation

$$\hat{H}\Psi(\mathbf{r}, t) = \mathrm{i}\hbar \frac{\mathrm{d}}{\mathrm{d}t}\Psi(\mathbf{r}, t). \tag{A.2}$$

$\hat{H}$ denotes the Hamilton operator, also called the Hamiltonian. Its form depends on the details of the system considered.

For an isolated atom with no external perturbation the Hamiltonian is time-independent and it has the general form

$$\hat{H}_0 = \sum_{i,j} \left[-\frac{\hbar^2}{2m_i} \nabla_i^2 + V(\mathbf{r}_i, \mathbf{r}_j) \right]. \tag{A.3}$$

The sum runs over all particles involved in the system. The index in ∇_i specifies operation on the coordinate $\mathbf{r}_i$. $V(\mathbf{r}_i, \mathbf{r}_j)$ is the potential interaction energy of the ith and jth particles. In general, V has contributions from all four fundamental interactions so far known, namely strong, electromagnetic, weak, and gravitational interactions. For the behavior of electrons only the electromagnetic contribution is of importance, and within the electromagnetic interaction the electrostatic potential is dominant. Since the masses of nuclei are much greater than the mass of an electron, the nuclei move much slower than the electrons. This allows the electrons to practically instantaneously follow the nuclear motion. For an electron, the nucleus appears to be at rest. This is the essence of the Born–Oppenheimer approximation which allows us to separate the nuclear wavefunction from the electronic one. We therefore consider a nucleus of total charge qZ, Z being the atomic number. We assume that the nucleus is located at the origin of coordinates ($\mathbf{r} = 0$) surrounded by Z electrons each of charge $-q$. We can restrict the index i in Eq. (A.3) to run only over electron coordinates. In the case of a time-independent Hamiltonian we can separate the t and $\mathbf{r}$ dependence as

$$\Psi(\mathbf{r}, t) = \sum_{n=1}^{\infty} \mathrm{e}^{-(i/\hbar)E_n t} \varphi_n(\mathbf{r}). \tag{A.4}$$

By inserting this wavefunction into Eq. (A.2) and using $\hat{H} = \hat{H}_0$ we obtain the energy eigenvalue equation (time-independent Schrödinger equation)

$$\hat{H}_0 \varphi_n(\mathbf{r}) = E_n \varphi_n(\mathbf{r}), \tag{A.5}$$

where E_n are the energy eigenvalues of the stationary states $|n\rangle$. In the following we restrict ourselves to the case of a two-level atom ($n = [1, 2]$) with the two stationary wavefunctions

$$\begin{aligned} \Psi_1(\mathbf{r}, t) &= \mathrm{e}^{-(i/\hbar)E_1 t} \varphi_1(\mathbf{r}), \\ \Psi_2(\mathbf{r}, t) &= \mathrm{e}^{-(i/\hbar)E_2 t} \varphi_2(\mathbf{r}). \end{aligned} \tag{A.6}$$

In the next step, we expose the atomic system to the radiation field. The system then experiences an external, time-dependent perturbation represented by the interaction Hamiltonian $\hat{H}'(t)$. We obtain for the total Hamiltonian

$$\hat{H} = \hat{H}_0 + \hat{H}'(t), \tag{A.7}$$

where $\hat{H}_0$ represents the unperturbed system according to Eq. (A.5). The size of an atom is on the order of a couple of Bohr radii, $a_\mathrm{B} \approx 0.05\,\mathrm{nm}$. Since $a_\mathrm{B} \ll \lambda$, λ being the wavelength of the radiation field, we can assume that the electric field $\mathbf{E}$ is constant across the dimensions of the atomic system. Assuming time-harmonic fields, we can write

$$\mathbf{E}(\mathbf{r}, t) = \mathrm{Re}\left\{\mathbf{E}(\mathbf{r})\mathrm{e}^{-\mathrm{i}\omega t}\right\} \approx \mathbf{E}_0 \cos(\omega t), \tag{A.8}$$

where we have set the phase of the field equal to zero or, equivalently, we have chosen the complex field amplitude to be real. Each electron in the system experiences the same field strength $\mathbf{E}_0$ and the same time dependence $\cos(\omega t)$. Using the total electric dipole moment of the atom

$$\hat{\mathbf{p}}_\mathrm{a}(\mathbf{r}) = \hat{\mathbf{p}}_\mathrm{a}(\mathbf{r}_1, \ldots, \mathbf{r}_Z) = q \sum_{i=1}^{Z} \hat{\mathbf{r}}_i, \tag{A.9}$$

we find for the interaction Hamiltonian in the dipole approximation

$$\hat{H}' = -\hat{\mathbf{p}}_\mathrm{a}(\mathbf{r}) \cdot \mathbf{E}_0 \cos(\omega t). \tag{A.10}$$

The dipolar interaction Hamiltonian is real and has odd parity, i.e. $\hat{H}'$ changes sign if the inversion operation $\hat{\mathbf{r}}_i \rightarrow -\hat{\mathbf{r}}_i$ is applied to all $\hat{\mathbf{r}}_i$.

To solve the Schrödinger equation (A.2) for the perturbed system we make a time-dependent superposition of the stationary atomic wavefunctions in Eq. (A.6) as

$$\Psi(\mathbf{r},t) = c_1(t)\Psi_1(\mathbf{r},t) + c_2(t)\Psi_2(\mathbf{r},t). \tag{A.11}$$

We choose the time-dependent coefficients c_1 and c_2 such that the normalization condition $\langle\Psi|\Psi\rangle = \int \Psi^*\Psi\, \mathrm{d}V = |c_1|^2 + |c_2|^2 = 1$ is fulfilled. For clarity, we will drop the arguments in the wavefunctions. After inserting this wavefunction into Eq. (A.2), rearranging terms, and making use of Eqs. (A.3) and (A.6) we obtain

$$\hat{H}'(c_1\Psi_1 + c_2\Psi_2) = \mathrm{i}\hbar\,[\Psi_1\dot{c}_1 + \Psi_1\dot{c}_2], \tag{A.12}$$

where the dots denote differentiation with respect to time. It should be kept in mind that the arguments of Ψ and φ are $(\mathbf{r},t)$ and $(\mathbf{r})$, respectively. To eliminate the spatial dependence we multiply Eq. (A.12) from the left by Ψ_1^* on both sides, introduce the expressions in (A.6) for the wavefunctions, and integrate over all space. After repeating the procedure with Ψ_2^* instead of Ψ_1^* we obtain a set of two time-dependent coupled differential equations:

$$\dot{c}_1(t) = c_2(t)(i/\hbar)\mathbf{p}_{12}\cdot\mathbf{E}_0\cos(\omega t)\mathrm{e}^{-(i/\hbar)(E_2-E_1)t}, \tag{A.13}$$

$$\dot{c}_2(t) = c_1(t)(i/\hbar)\mathbf{p}_{21}\cdot\mathbf{E}_0\cos(\omega t)\mathrm{e}^{+(i/\hbar)(E_2-E_1)t}. \tag{A.14}$$

We have introduced the definition of the dipole matrix element between the states $|i\rangle$ and $|j\rangle$ as

$$\mathbf{p}_{ij} = \langle i|\hat{\mathbf{p}}_\mathrm{a}|j\rangle = \int \varphi_i^*(\mathbf{r})\hat{\mathbf{p}}_\mathrm{a}(\mathbf{r})\varphi_j(\mathbf{r})\mathrm{d}V. \tag{A.15}$$

It has to be emphasized again that the integration runs over all electron coordinates $\mathbf{r} = \mathbf{r}_1,\ldots,\mathbf{r}_Z$. In Eqs. (A.13) and (A.14) we have used the fact that $\mathbf{p}_{ii} = 0$. This follows from the odd parity of $\hat{H}'$, which makes the integrands of $\mathbf{p}_{ii}$ odd functions of $\mathbf{r}$. Integration over $\mathbf{r} = [-\infty\ldots 0]$ leads to a result that is the negative of the result associated with integration over $\mathbf{r} = [0\ldots\infty]$. Upon integration over all space the two contributions cancel out. The dipole matrix elements satisfy $\mathbf{p}_{12} = \mathbf{p}_{21}^*$ because $\hat{\mathbf{p}}_\mathrm{a}$ is a Hermitian operator. However, it is convenient to choose the phases of the eigenfunctions φ_1 and φ_2 such that the dipole matrix elements are real, i.e.

$$\mathbf{p}_{12} = \mathbf{p}_{21}. \tag{A.16}$$

In the following, we will assume that $\Delta E = E_2 - E_1 > 0$, and we introduce the *transition frequency*

$$\omega_0 = \Delta E/h, \tag{A.17}$$

for the sake of simpler notation. The state $|1\rangle$ is the ground state and the state $|2\rangle$ the excited state.

Semiclassical theory does not account for spontaneous emission. The spontaneous emission process can only be found by use of a quantized radiation field. To be in accordance with quantum electrodynamics we have to include the effects of spontaneous emission by introducing a phenomenological damping term in Eq. (A.14). The coupled differential equations then have the form

$$\begin{aligned} \dot{c}_1(t) &= c_2(t)(i/\hbar)\mathbf{p}_{12}\cdot\mathbf{E}_0\cos(\omega t)\mathrm{e}^{-\mathrm{i}\omega_0 t}, \\ \dot{c}_2(t)+\gamma/2\,c_2(t) &= c_1(t)(i/\hbar)\mathbf{p}_{21}\cdot\mathbf{E}_0\cos(\omega t)\mathrm{e}^{+\mathrm{i}\omega_0 t}. \end{aligned} \tag{A.18}$$

The introduction of the damping term asserts that an excited atom must ultimately decay to its ground state by spontaneous emission. In the absence of the radiation field, $\mathbf{E}_0 = 0$, Eq. (A.18) can be integrated at once and we obtain

$$c_2(t) = c_2(0)\mathrm{e}^{-(\gamma/2)t}. \tag{A.19}$$

The average lifetime τ of the excited state is $\tau = 1/\gamma$, γ being the spontaneous decay rate. Since there is no direct analytical solution of Eqs. (A.18) we have to find approximate solutions for different types of excitations.

A.1 Steady-state polarizability for weak excitation fields

We assume that the interaction between the atom and the radiation field is weak. The solution for $c_1(t)$ and $c_2(t)$ can then be represented as a power series in $\mathbf{p}_{21}\cdot\mathbf{E}_0$. To derive the first-order term in this series we set $c_1(t) = 1$ and $c_2(t) = 0$ on the right-hand side of Eqs. (A.18). Once we have found the first-order solution we can insert it again into the right-hand side to find the second-order solution and so on. However, we will restrict ourselves to the first-order term. The solution for c_1 is $c_1(t) = 1$, indicating that the atom resides always in its ground state. This solution is the zeroth-order solution, i.e. there is no first-order solution for c_1. The next higher term would be of second order. The first-order solution for c_2 is obtained by a superposition of the homogeneous solution in Eq. (A.19) and a particular solution. The latter is easily found by writing the cosine term as a sum of two exponentials. We then obtain for the first-order solution of c_2

$$c_2(t) = \frac{\mathbf{p}_{21}\cdot\mathbf{E}_0}{2\hbar}\left[\frac{\mathrm{e}^{\mathrm{i}(\omega_0+\omega-\mathrm{i}\gamma/2)t}-1}{\omega_0+\omega-\mathrm{i}\gamma/2}+\frac{\mathrm{e}^{\mathrm{i}(\omega_0-\omega-\mathrm{i}\gamma/2)t}-1}{\omega_0-\omega-\mathrm{i}\gamma/2}\right]\mathrm{e}^{-(\gamma/2)\,t}. \tag{A.20}$$

We are interested in calculating the *steady-state* behavior for which the atom has been subjected to the electric field $\mathbf{E}_0\cos(\omega t)$ for an infinitely long period of time. In this situation the inhomogeneous term disappears and the solution is given by the homogeneous solution alone.

The expectation value of the dipole moment is defined as

$$\mathbf{p}(t) = \langle\Psi|\hat{\mathbf{p}}_{\mathrm{a}}|\Psi\rangle = \int \Psi^*(\mathbf{r})\hat{\mathbf{p}}_{\mathrm{a}}(\mathbf{r})\Psi(\mathbf{r})\mathrm{d}V, \tag{A.21}$$

The integration again runs over all coordinates $\mathbf{r}_i$. Using the wavefunction Ψ of Eq. (A.11) the expression for $\mathbf{p}$ becomes

$$\mathbf{p}(t) = c_1^* c_2 \, \mathbf{p}_{12} \mathrm{e}^{-\mathrm{i}\omega_0 t} + c_1 c_2^* \, \mathbf{p}_{21} \mathrm{e}^{\mathrm{i}\omega_0 t}, \tag{A.22}$$

where we used the definition of the dipole matrix elements of Eq. (A.15) and the property $\mathbf{p}_{ii} = 0$. Using the first-order solutions for c_1 and c_2 we obtain

$$\mathbf{p}(t) = \frac{\mathbf{p}_{12}[\mathbf{p}_{21} \cdot \mathbf{E}_0]}{2\hbar} \times \left[\frac{\mathrm{e}^{\mathrm{i}\omega t}}{\omega_0 + \omega - \mathrm{i}\gamma/2} + \frac{\mathrm{e}^{-\mathrm{i}\omega t}}{\omega_0 - \omega - \mathrm{i}\gamma/2} + \frac{\mathrm{e}^{-\mathrm{i}\omega t}}{\omega_0 + \omega + \mathrm{i}\gamma/2} + \frac{\mathrm{e}^{\mathrm{i}\omega t}}{\omega_0 - \omega + \mathrm{i}\gamma/2} \right]. \tag{A.23}$$

Since the exciting electric field is expressed as $\mathbf{E} = (1/2)\mathbf{E}_0[\exp(\mathrm{i}\omega t) + \exp(-\mathrm{i}\omega t)]$ we rewrite the dipole moment above as

$$\mathbf{p}(t) = \frac{1}{2}\left[\overset{\leftrightarrow}{\alpha}{}^*(\omega) \mathrm{e}^{\mathrm{i}\omega t} + \overset{\leftrightarrow}{\alpha}(\omega) \mathrm{e}^{-\mathrm{i}\omega t} \right] \mathbf{E}_0 = \mathrm{Re}\left\{ \overset{\leftrightarrow}{\alpha}(\omega) \mathrm{e}^{-\mathrm{i}\omega t} \right\} \mathbf{E}_0, \tag{A.24}$$

where $\overset{\leftrightarrow}{\alpha}$ is the *atomic polarizability* tensor

$$\overset{\leftrightarrow}{\alpha}(\omega) = \frac{\mathbf{p}_{12}\mathbf{p}_{21}}{\hbar} \left[\frac{1}{\omega_0 - \omega - \mathrm{i}\gamma/2} + \frac{1}{\omega_0 + \omega + \mathrm{i}\gamma/2} \right]. \tag{A.25}$$

$\mathbf{p}_{12}\mathbf{p}_{21}$ denotes the matrix formed by the outer product between the (real) transition dipole moments. It is convenient to write the polarizability in terms of a single denominator. Furthermore, we realize that the damping term γ is much smaller than ω_0, which allows us to drop terms in γ^2. Finally, we have to generalize the result to a system with more than two states. Besides the different matrix elements, each state differing from the ground state behaves in a similar way to our previous state $|2\rangle$. Thus, each new level is characterized by its natural frequency ω_n, its damping term γ_n, and the transition dipole moments $\mathbf{p}_{1n}$ and $\mathbf{p}_{n1}$. Then the polarizability takes on the form

$$\overset{\leftrightarrow}{\alpha}(\omega) = \sum_n \overset{\leftrightarrow}{\mathbf{f}}_n \left[\frac{e^2/m}{\omega_n^2 - \omega^2 - \mathrm{i}\omega\gamma_n} \right], \quad \overset{\leftrightarrow}{\mathbf{f}}_n = \frac{2m\omega_n}{e^2\hbar} \mathbf{p}_{1n}\mathbf{p}_{n1}, \tag{A.26}$$

where $\overset{\leftrightarrow}{\mathbf{f}}_n$ is the so-called *oscillator strength*,[1] and e and m denote the electron charge and mass, respectively. It is for historical reasons that we have cast the polarizability in the form of Eq. (A.26). Before the advent of quantum mechanics, H. A. Lorentz developed a classical model for the atomic polarizability that, apart from the expression for $\overset{\leftrightarrow}{\mathbf{f}}_n$, is identical with our result. The model considered by Lorentz consists of a collection of harmonic oscillators for the electrons of an atom. Each electron responds to the driving incident field according to the equation of motion

$$\ddot{\mathbf{p}} + \gamma\,\dot{\mathbf{p}} + \omega_0^2 \mathbf{p} = (q^2/m)\,\overset{\leftrightarrow}{\mathbf{f}}\,\mathbf{E}(t). \tag{A.27}$$

In this theory, the oscillator strength is a fitting parameter since there is no direct way to know how much an electron contributes to a particular atomic mode. On the other hand, the

[1] The average over all polarizations reduces the oscillator strength to a scalar quantity with an extra factor of $1/3$.

semiclassical theory directly relates the oscillator strength to the transition dipole matrix elements and thus to the atomic wavefunctions. Furthermore, the f-sum rule tells us that the sum of all oscillator strengths is equal to unity.

If the energy $\hbar\omega$ of the exciting field is close to the energy difference ΔE between two atomic states, the first term in Eq. (A.25) is much larger than the second one. In this case we can discard the second term (the rotating-wave approximation) and the imaginary part of the polarizability becomes a perfect Lorentzian function.

It is important to notice that there is a linear relationship between the exciting electric field **E** and the induced dipole moment **p**. Therefore, a monochromatic field with angular frequency ω produces a harmonically oscillating dipole with the same frequency. This allows us to use the complex notation for **p** and **E** and write

$$\mathbf{p} = \overleftrightarrow{\alpha}\,\mathbf{E}, \tag{A.28}$$

from which we obtain the time dependence of **E** and **p** by simply multiplying by $\exp(-\mathrm{i}\omega t)$ and taking the real part.

A.2 Near-resonance excitation in the absence of damping

In the previous section we required that the interaction between the excitation beam and the atom is weak and that the atom resides mostly in its ground state. This condition can be relaxed if we consider an exciting field whose energy $\hbar\omega$ is close to the energy difference ΔE between two atomic states. As mentioned before, there is no direct analytical solution to the coupled differential equations in Eqs. (A.18). However, a quite accurate solution can be found if we drop the damping term γ and if the energy of the radiation field is close to the energy difference between excited and ground states, i.e.

$$|\hbar\omega - \Delta E| \ll \hbar\omega + \Delta E. \tag{A.29}$$

In this case, we can apply the so-called rotating-wave approximation. After rewriting the cosines in Eqs. (A.18) in terms of exponentials we find exponents with $(\hbar\omega \pm \Delta E)$. In the rotating-wave approximation we retain only terms with $(\hbar\omega - \Delta E)$ because of their dominating contributions. Equations (A.18) then become[2]

$$\frac{\mathrm{i}}{2}\,\omega_{\mathrm{R}} \mathrm{e}^{-\mathrm{i}(\omega_0 - \omega)t} c_2(t) = \dot{c}_1(t), \tag{A.30}$$

$$\frac{\mathrm{i}}{2}\,\omega_{\mathrm{R}} \mathrm{e}^{\mathrm{i}(\omega_0 - \omega)t} c_1(t) = \dot{c}_2(t), \tag{A.31}$$

where we introduced the *Rabi frequency* ω_{R} defined as

$$\omega_{\mathrm{R}} = \frac{|\mathbf{p}_{12} \cdot \mathbf{E}_0|}{\hbar} = \frac{|\mathbf{p}_{21} \cdot \mathbf{E}_0|}{\hbar}. \tag{A.32}$$

[2] We again choose the phases of the atomic wavefunctions such that the transition dipole matrix elements are real.

ω_R is a measure for the strength of the time-varying external field. On inserting the trial solution $c_1(t) = \exp(i\kappa t)$ into the first equation, Eq. (A.30), we find $c_2(t) = (2\kappa/\omega_R)\exp(i[\omega_0 - \omega + \kappa]t)$. On substituting both c_1 and c_2 into the second equation, Eq. (A.31), we find a quadratic equation for the unknown parameter κ, leading to the two solutions κ_1 and κ_2. The general solutions for the amplitudes c_1 and c_2 can then be written as

$$c_1(t) = A e^{i\kappa_1 t} + B e^{i\kappa_2 t}, \tag{A.33}$$

$$c_2(t) = (2/\omega_R) e^{i(\omega_0 - \omega)t} \left[A\kappa_1 e^{i\kappa_1 t} + B\kappa_2 e^{i\kappa_2 t} \right]. \tag{A.34}$$

To determine the constants A and B we require boundary conditions. The probability for finding the atomic system in the excited state $|2\rangle$ is $|c_2|^2$. Similarily, the probability for finding the atom in its ground state $|1\rangle$ is $|c_1|^2$. By using the boundary conditions for the atom initially in its ground state

$$\begin{aligned} |c_1(t=0)|^2 &= 1, \\ |c_2(t=0)|^2 &= 0, \end{aligned} \tag{A.35}$$

the unknown constants A and B can be determined. Using the expressions for κ_1, κ_2, A, and B, we finally find the solution

$$c_1(t) = e^{-(i/2)(\omega_0 - \omega)t} \left[\cos(\Omega t/2) - \frac{i(\omega - \omega_0)}{\Omega} \sin(\Omega t/2) \right], \tag{A.36}$$

$$c_2(t) = \frac{i\omega_R}{\Omega} e^{(i/2)(\omega_0 - \omega)t} \sin(\Omega t/2), \tag{A.37}$$

where Ω denotes the *Rabi flopping frequency* defined as

$$\Omega = \sqrt{(\omega_0 - \omega)^2 + \omega_R^2}. \tag{A.38}$$

It can be easily shown that $|c_1|^2 + |c_2|^2 = 1$. The probability of finding the atom in its excited state becomes

$$|c_2(t)|^2 = \omega_R^2 \frac{\sin^2(\Omega t/2)}{\Omega^2}. \tag{A.39}$$

The transition probability is a periodic function of time. The system oscillates between the levels E_1 and E_2 at the frequency $\Omega/2$, which depends on the detuning $\omega_0 - \omega$ and the field strength represented by ω_R. If ω_R is small we have $\Omega \approx \omega_0 - \omega$ and, in the absence of damping, the results become identical with the results of the previous section.

The expectation value of the dipole moment is defined by Eqs. (A.21) and (A.22). On inserting the solutions for c_1 and c_2 and using Eq. (A.16) we obtain

$$\mathbf{p}(t) = \mathbf{p}_{12} \frac{\omega_R}{\Omega} \left[\frac{\omega - \omega_0}{\Omega} [1 - \cos(\Omega t)] \cos(\omega t) + \sin(\Omega t) \sin(\omega t) \right]. \tag{A.40}$$

We see that the induced dipole moment oscillates at the frequency of the radiation field. However, it does not instantaneously follow the driving field: it has in-phase and quadrature components. Let us write $\mathbf{p}$ in the complex representation as

$$\mathbf{p}(t) = \operatorname{Re}\left\{ \mathbf{p} e^{-i\omega t} \right\}. \tag{A.41}$$

We then find for the complex dipole moment

$$\mathbf{p} = \mathbf{p}_{12}\frac{\omega_R}{\Omega}\left[\frac{\omega-\omega_0}{\Omega}[1-\cos(\Omega t)] + \mathrm{i}\,\sin(\Omega t)\right]. \tag{A.42}$$

To determine the atomic polarizability, defined as

$$\mathbf{p} = \overleftrightarrow{\alpha}\mathbf{E}, \tag{A.43}$$

we have to express the Rabi frequency ω_R by its definition Eq. (A.32) and obtain

$$\overleftrightarrow{\alpha}(\omega) = \frac{\mathbf{p}_{12}\mathbf{p}_{21}}{\hbar}\left[\frac{\omega-\omega_0}{\Omega^2}[1-\cos(\Omega t)] + \mathrm{i}\,\sin(\Omega t)\right]. \tag{A.44}$$

The most remarkable property of the polarizability is its dependence on field strength (through ω_R) and its time dependence. This is different from the polarizability derived in the previous section. In the present case, the time behavior is determined by the Rabi flopping frequency Ω. In practical situations the time dependence disappears within tens of nanoseconds because of the damping term γ, which has been neglected in the present derivation. For the case of exact resonance ($\omega = \omega_0$) the polarizability reduces to a sinusoidal function of $\omega_R t$. This oscillation is much slower than the oscillation of the optical field. For weak interactions ω_R is small and the polarizability becomes a linear function of t.

A.3 Near-resonance excitation with damping

The damping term γ attenuates the purely oscillatory solution derived in the previous section. After a sufficiently long time, the system will relax into the ground state. To calculate the steady-state behavior it is sufficient to solve for the term $c_1c_2^*$, which, together with its complex conjugate, defines the expectation value of the dipole moment (see Eq. (A.22)). In the steady state, the probability of finding the atom in its excited state will be time-independent, i.e.

$$\frac{\mathrm{d}}{\mathrm{d}t}\left[c_2c_2^*\right] = 0 \qquad \text{(steady state)}. \tag{A.45}$$

Furthermore, in the rotating-wave approximation, it can be expected that the time dependence of the off-diagonal matrix element $c_1c_2^*$ will be solely defined by the factor $\exp[-\mathrm{i}(\omega_0-\omega)t]$. Thus,

$$\frac{\mathrm{d}}{\mathrm{d}t}\left[c_1c_2^*\right] = -\mathrm{i}(\omega_0-\omega)\left[c_1c_2^*\right] \qquad \text{(steady state)}, \tag{A.46}$$

with a similar equation for $c_2c_1^*$. Using

$$\frac{\mathrm{d}}{\mathrm{d}t}\left[c_ic_j^*\right] = c_i\dot{c}_j^* + c_j^*\dot{c}_i, \tag{A.47}$$

inserting Eqs. (A.18), applying the rotating-wave approximation, and making use of the steady-state conditions above, we obtain

$$\omega_R \exp[-i(\omega_0 - \omega)t]\left[c_2 c_1^*\right] - \omega_R^* \exp[i(\omega_0 - \omega)t]\left[c_1 c_2^*\right] - 2i\gamma\left[c_2 c_2^*\right] = 0, \quad (A.48)$$

$$\omega_R\left(\left[c_1 c_1^*\right] - \left[c_2 c_2^*\right]\right) - \left[2(\omega_0 - \omega) + i\gamma\right]\exp[i(\omega_0 - \omega)t]\left[c_1 c_2^*\right] = 0, \quad (A.49)$$

$$\omega_R\left(\left[c_1 c_1^*\right] - \left[c_2 c_2^*\right]\right) - \left[2(\omega_0 - \omega) - i\gamma\right]\exp[i(\omega_0 - \omega)t]\left[c_2 c_1^*\right] = 0. \quad (A.50)$$

This set of equations can be solved for $[c_1 c_2^*]$ and gives

$$\left[c_1 c_2^*\right] = e^{-i(\omega_0 - \omega)t}\,\frac{\frac{1}{2}\omega_R(\omega_0 - \omega - i\gamma/2)}{(\omega_0 - \omega)^2 + \gamma^2/4 + \frac{1}{2}\omega_R^2}, \quad (A.51)$$

with the complex-conjugate solution for $[c_2 c_1^*]$. The expectation value of the dipole moment can now be calculated by using Eq. (A.22) and the steady-state solution for the atomic polarizability for near-resonance excitation ($\omega \approx \omega_0$) can be determined as

$$\overleftrightarrow{\alpha}(\omega) = \frac{\mathbf{p}_{12}\mathbf{p}_{21}}{\hbar}\,\frac{\omega_0 - \omega + i\gamma/2}{(\omega_0 - \omega)^2 + \gamma^2/4 + \frac{1}{2}\omega_R^2}. \quad (A.52)$$

The most remarkable difference from the off-resonance case is the appearance of the term ω_R^2 in the denominator. This term accounts for saturation of the excited state, thereby reducing the absorption rate and increasing the linewidth from γ to $(\gamma + 2\omega_R^2)^{1/2}$, which is denoted *saturation broadening*. Thus, the damping constant becomes dependent on the acting electric field strength. For $\omega_R \to 0$, the polarizability reduces to

$$\overleftrightarrow{\alpha}(\omega) = \frac{\mathbf{p}_{12}\mathbf{p}_{21}}{\hbar}\,\frac{1}{\omega_0 - \omega - i\gamma/2}, \quad (A.53)$$

which is identical with the rotating-wave term of Eq. (A.25).

The polarizability can be calculated once the energy levels E_1 and E_2 and the dipole matrix element $\mathbf{p}_{12}$ are known. The latter is defined by Eq. (A.15) through the wavefunctions φ_1 and φ_2. It is thus necessary to solve the energy eigenvalue equation (A.5) for the quantum system being considered, in order to accurately determine the energy levels and the dipole matrix element. However, Eq. (A.5) can be solved analytically only for simple systems, which are often restricted to two interacting particles. Systems with more than two interacting particles have to be treated with approximate methods such as the Hartree–Fock method or numerically.

Appendix B Spontaneous emission in the weak-coupling regime

In this appendix we derive the normalized spontaneous decay rate of an atomic system using quantum electrodynamics. The analysis is based in part on Ref. [1]. In what follows, we concentrate exclusively on the weak-coupling regime. Section B.1 presents the derivation of the decay constant in free space using QED and the Weisskopf–Wigner approximation [2, 3]. Section B.2 is devoted to calculating the spontaneous-emission decay constant in a linear and inhomogeneous medium using the Heisenberg picture [1], which renders a clear connection between classical theory and QED.

B.1 Weisskopf–Wigner theory

We consider a two-level atom interacting with an infinite number of field modes. Each mode is characterized by its polarization and wavevector $\mathbf{k}$. This atom–field system is described by the Jaynes–Cummings Hamiltonian [4]

$$\hat{H} = \hbar\omega_0 |e\rangle\langle e| + \sum_{\mathbf{k}} \hbar\omega_{\mathbf{k}} \hat{a}_{\mathbf{k}}^{\dagger}\hat{a}_{\mathbf{k}} - \sum_{\mathbf{k}} \hbar g_{\mathbf{k}} \left[\hat{a}_{\mathbf{k}}|e\rangle\langle g| + \hat{a}_{\mathbf{k}}^{\dagger}|g\rangle\langle e|\right]. \tag{B.1}$$

Here $|e\rangle$ ($|g\rangle$) is the excited (ground) state of the atom, $\hat{a}_{\mathbf{k}}$ and $\hat{a}_{\mathbf{k}}^{\dagger}$ are the annihilation and creation operators for the mode $\mathbf{k}$,[1] and $g_{\mathbf{k}}$ is the atom–field coupling strength defined as

$$g_{\mathbf{k}} = \sqrt{\frac{\omega_{\mathbf{k}}}{2\varepsilon_0 \hbar V}}\, \hat{\boldsymbol{\epsilon}}_{\mathbf{k}} \cdot \langle g|\hat{\mathbf{p}}|e\rangle, \tag{B.2}$$

where V is the volume, $\hat{\boldsymbol{\epsilon}}$ is the unit vector in the direction of the electric field mode $\mathbf{E}_{\mathbf{k}}$, and $\hat{\mathbf{p}}$ is the dipole-moment operator.

We assume that at $t = 0$ the atom is in the excited state and no photons are present. The initial state is therefore $|e, 0\rangle$, with e and 0 designating the excited atomic state and the initial photon number, respectively. At any later time t the state $|\psi(t)\rangle$ of the system can be expanded as

$$|\psi(t)\rangle = C_0^{\mathrm{e}}(t)\mathrm{e}^{-\mathrm{i}\omega_0 t}|e, 0\rangle + \sum_{\mathbf{k}} C_{1\mathbf{k}}^{\mathrm{g}}(t)\, \mathrm{e}^{-\mathrm{i}\omega_{\mathbf{k}} t}|g, 1_{\mathbf{k}}\rangle, \tag{B.3}$$

[1] We use the compressed notation for which $\mathbf{k}$ designates simultaneously the $\mathbf{k}$-vector and the polarization state. Each $\mathbf{k}$-vector possesses two linearly independent polarization states.

where the Cs are time-dependent expansion coefficients. In the state $|g, 1_{\mathbf{k}}\rangle$ the atom is in the ground state and one photon of mode $\mathbf{k}$ is released. By inserting Eq. (B.3) into the Schrödinger equation, we obtain

$$\frac{\mathrm{d}C_0^{\mathrm{e}}}{\mathrm{d}t} = -\sum_{\mathbf{k}} |g_{\mathbf{k}}|^2 \int_0^t C_0^{\mathrm{e}}(t_1)\, \mathrm{e}^{-\mathrm{i}(\omega_{\mathbf{k}} - \omega_0)(t - t_1)}\, \mathrm{d}t_1. \tag{B.4}$$

In the large-volume limit, i.e. $V \to \infty$, the sum in Eq. (B.4) can be substituted as

$$\sum_{\mathbf{k}} \longrightarrow 2\frac{V}{(2\pi)^3} \int_0^{2\pi} \mathrm{d}\phi \int_0^{\pi} \mathrm{d}\theta \sin\theta \int_0^{\infty} \mathrm{d}k\, k^2, \tag{B.5}$$

where the factor of 2 arises from summing over the two polarization states associated with each $\mathbf{k}$-vector. Assuming that the dipole is oriented along the z-axis, i.e. $\mathbf{p} = \langle g|\hat{\mathbf{p}}|e\rangle = p\,\hat{\mathbf{n}}_z$, the field–atom coupling strength becomes

$$|g_{\mathbf{k}}|^2 = \frac{\omega_{\mathbf{k}}}{2\varepsilon_0 \hbar V}\, p^2 \cos^2\theta. \tag{B.6}$$

After solving the angular integrals, Eq. (B.4) reduces to

$$\frac{\mathrm{d}C_0^{\mathrm{e}}}{\mathrm{d}t} = -\frac{p^2}{6\pi^2 \varepsilon_0 \hbar c^3} \int_0^{\infty} \omega_{\mathbf{k}}^3 \int_0^t C_0^{\mathrm{e}}(t_1)\, \mathrm{e}^{-\mathrm{i}(\omega_{\mathbf{k}} - \omega_0)(t - t_1)}\, \mathrm{d}t_1\, \mathrm{d}\omega_{\mathbf{k}}. \tag{B.7}$$

So far, the derivation has been exact. We now introduce the Weisskopf–Wigner approximation to solve Eq. (B.7). This approximation involves the following two assumptions: (1) the spectrum of the field modes is very broad, and (2) the coefficient C_0^{e} changes slowly in time. Therefore, for times $t_1 \ll t$ the integrand oscillates very rapidly and there is no significant contribution to the value of the integral. The most dominant contribution originates from times $t_1 \approx t$. We therefore evaluate $C_0^{\mathrm{e}}(t_1)$ at the actual time t and move it out of the integrand. In this limit, the atomic decay becomes a memoryless process (Markov process). To evaluate the remaining integral we extend the upper integration limit to infinity since there is no significant contribution for $t_1 \gg t$. Equation (B.7) now reduces to

$$\frac{\mathrm{d}C_0^{\mathrm{e}}}{\mathrm{d}t} = -\frac{p^2}{6\pi^2 \varepsilon_0 \hbar c^3} C_0^{\mathrm{e}}(t) \int_0^{\infty} \omega_k^3 \int_0^{\infty} \mathrm{e}^{-\mathrm{i}(\omega_{\mathbf{k}} - \omega_0)(t - t_1)}\, \mathrm{d}t_1\, \mathrm{d}\omega_{\mathbf{k}}. \tag{B.8}$$

The integration can now be carried out analytically and we obtain

$$\frac{\mathrm{d}C_0^{\mathrm{e}}}{\mathrm{d}t} = -\left(\frac{\gamma_0}{2} + \mathrm{i}\,\Delta\omega\right) C_0^{\mathrm{e}}(t). \tag{B.9}$$

Here, γ_0 is the free-space decay constant

$$\gamma_0 = \frac{\omega_0^3 p^2}{3\pi\varepsilon_0 \hbar c^3} = \frac{\pi\,\omega_0 p^2}{3\varepsilon_0 \hbar} \rho(\omega_0), \tag{B.10}$$

with $\rho(\omega_0)$ being the electromagnetic density of modes. The second term in Eq. (B.9) is the Lamb shift and reads as

$$\Delta\omega = \frac{1}{4\pi\varepsilon_0} \frac{p^2}{3\pi\hbar c^3}\, \mathrm{P}\left\{ \int \frac{\omega_{\mathbf{k}}^3}{\omega_{\mathbf{k}} - \omega_0}\, \mathrm{d}\omega_{\mathbf{k}} \right\}, \tag{B.11}$$

where P denotes the principal value of the integral. Since the integral diverges, it is necessary to introduce a cut-off frequency ω_{f} according to $\hbar\omega_{\mathrm{f}} = 2m_{\mathrm{e}}c^2$ (the energy for "pair" creation). With this correction, the Lamb shift $\Delta\omega$ turns out to be in the range of a few GHz, which is very small compared with the optical transition frequency.

B.2 Inhomogeneous environments

We apply QED to derive the spontaneous decay rate of an atomic system in an inhomogeneous medium characterized by the lossless dielectric constant $\varepsilon(\mathbf{r})$.

Let us consider the vector potential operator $\hat{\mathbf{A}}(\mathbf{r},t)$ which satisfies the generalized Coulomb gauge $\nabla \cdot [\varepsilon(\mathbf{r})\hat{\mathbf{A}}] = 0$. The transverse vector potential can be expanded in a complete set of orthogonal modes $\mathbf{a}_{\mathbf{k}}$ as [5]

$$\hat{\mathbf{A}}(\mathbf{r},t) = \hat{\mathbf{A}}^{+}(\mathbf{r},t) + \hat{\mathbf{A}}^{-}(\mathbf{r},t), \tag{B.12}$$

$$\hat{\mathbf{A}}^{-}(\mathbf{r},t) = \sum_{\mathbf{k}} \sqrt{\hbar/(2\varepsilon_0\omega_{\mathbf{k}}V)}\,\hat{a}_{\mathbf{k}}(t)\mathbf{a}_{\mathbf{k}}(\mathbf{r}), \tag{B.13}$$

$$\hat{\mathbf{A}}^{+}(\mathbf{r},t) = \sum_{\mathbf{k}} \sqrt{\hbar/(2\varepsilon_0\omega_{\mathbf{k}}V)}\,\hat{a}^{\dagger}_{\mathbf{k}}(t)\mathbf{a}^{*}_{\mathbf{k}}(\mathbf{r}). \tag{B.14}$$

Here, $\hat{\mathbf{A}}^{-}$ and $\hat{\mathbf{A}}^{+}$ contain only negative- and positive-frequency components, respectively. The normal modes satisfy the Helmholtz equation

$$\nabla \times \nabla \times \mathbf{a}_{\mathbf{k}}(\mathbf{r}) + \varepsilon_0\varepsilon(\mathbf{r})\frac{\omega_{\mathbf{k}}^2}{c^2}\mathbf{a}_{\mathbf{k}}(\mathbf{r}) = \mathbf{0}, \tag{B.15}$$

and they form an orthonormal and complete set, namely

$$\int \varepsilon(\mathbf{r})\mathbf{a}_{\mathbf{k}'}(\mathbf{r}) \cdot \mathbf{a}^{*}_{\mathbf{k}}(\mathbf{r})\mathrm{d}^3\mathbf{r} = \delta_{\mathbf{k}\mathbf{k}'}, \tag{B.16}$$

$$\int \mathbf{a}^{*}_{\mathbf{k}}(\mathbf{r}')\mathbf{a}_{\mathbf{k}}(\mathbf{r})\mathrm{d}^3\mathbf{k} = \overset{\leftrightarrow}{\delta}_{\perp}(\mathbf{r}'-\mathbf{r}). \tag{B.17}$$

We now express the interaction term in the Hamiltonian (cf. Eq. (B.1)) in terms of the electron momentum operator $\hat{\mathbf{p}}_{\mathrm{m}}$ and the vector potential operator $\hat{\mathbf{A}}$ and obtain

$$\hat{H}_{\mathrm{int}} = -\hat{\mathbf{p}}_{\mathrm{m}} \cdot \hat{\mathbf{A}} = \sum_{\mathbf{k}} \hbar\Big[\kappa^{*}_{\mathbf{k}}\hat{a}^{\dagger}_{\mathbf{k}}|g\rangle\langle e| + \kappa_{\mathbf{k}}\hat{a}_{\mathbf{k}}|e\rangle\langle g|\Big], \tag{B.18}$$

where $\kappa_{\mathbf{k}}$ denotes the coupling constant defined as

$$\kappa_{\mathbf{k}} = -\frac{e}{\hbar m}\sqrt{\hbar/(2\varepsilon_0\omega_{\mathbf{k}}V)}\;\mathbf{p}_{12} \cdot \mathbf{a}_{\mathbf{k}}(\mathbf{r}_0), \tag{B.19}$$

and $\mathbf{p}_{12}$ is the matrix element $\langle g|\hat{\mathbf{p}}_{\mathrm{m}}|e\rangle$.

In QED, spontaneous decay is generated by vacuum fluctuations of the field. These fluctuations give rise to a source current density whose operator is denoted as $\hat{\mathbf{J}}$. The frequency correlation of $\hat{\mathbf{J}}$ can be calculated as

$$\langle \hat{\mathbf{J}}^{+}_{\omega'}(\mathbf{r}')\,\hat{\mathbf{J}}^{-}_{\omega}(\mathbf{r})\rangle = \frac{e^2}{m^2}\mathbf{p}_{12}\,\mathbf{p}_{12}\delta(\omega-\omega')\delta(\omega-\omega_0)\delta(\mathbf{r}-\mathbf{r}')\delta(\mathbf{r}-\mathbf{r}_0)\langle \hat{N}_{\mathrm{e}}\rangle, \tag{B.20}$$

where $\mathbf{r}_0$ is the center of mass of the atom, ω_0 the center frequency of the distribution, and $\hat{N}_{\mathrm{e}} = |e\rangle\langle e|$ the number operator of the excited state. The number operator satisfies the equation

$$\frac{\mathrm{d}\hat{N}_{\mathrm{e}}}{\mathrm{d}t} = \frac{\mathrm{i}}{\hbar}\int \left[\hat{\mathbf{J}}^{+}(\mathbf{r},t)\cdot\hat{\mathbf{A}}^{-}(\mathbf{r},t) - \hat{\mathbf{A}}^{+}(\mathbf{r},t)\cdot\hat{\mathbf{J}}^{-}(\mathbf{r},t)\right]\mathrm{d}^3\mathbf{r}, \tag{B.21}$$

which can be derived by using Heisenberg's equation of motion for the different operators.

Let us denote the Fourier transforms of $\hat{\mathbf{A}}(\mathbf{r},t)$ and $\hat{\mathbf{J}}(\mathbf{r},t)$ by $\hat{\mathbf{A}}_{\omega}(\mathbf{r})$ and $\hat{\mathbf{J}}_{\omega}(\mathbf{r})$, respectively. Then, as a consequence of Heisenberg's equations of motion and the restriction to the *weak-coupling* regime, we can derive the following *quantum* wave equation [1]:

$$\nabla\times\nabla\times\hat{\mathbf{A}}^{-}_{\omega}(\mathbf{r}) - \varepsilon(\mathbf{r})\frac{\omega^2}{c^2}\hat{\mathbf{A}}^{-}_{\omega}(\mathbf{r}) = \frac{1}{\varepsilon_0 c^2}\hat{\mathbf{J}}^{-}_{\omega}(\mathbf{r}). \tag{B.22}$$

Using the definition of the dyadic Green function from Eq. (2.87) (see Section 2.12) the solution for $\hat{\mathbf{A}}^{-}_{\omega}$ can be represented as

$$\hat{\mathbf{A}}^{-}_{\omega}(\mathbf{r}) = \frac{1}{\varepsilon_0 c^2}\int \overleftrightarrow{\mathbf{G}}(\mathbf{r},\mathbf{r}';\omega)\hat{\mathbf{J}}^{-}_{\omega}(\mathbf{r}')\mathrm{d}^3\mathbf{r}', \tag{B.23}$$

where we included ω in the argument of $\overleftrightarrow{\mathbf{G}}$. By applying the inverse Fourier transform we can derive the corresponding solution $\hat{\mathbf{A}}^{-}(\mathbf{r},t)$ in the time domain. Finally, by combining this solution with Eq. (B.20) and Eq. (B.21) we obtain the simple equation

$$\frac{\mathrm{d}\langle\hat{N}_{\mathrm{e}}\rangle}{\mathrm{d}t} = -\gamma\langle\hat{N}_{\mathrm{e}}\rangle, \tag{B.24}$$

with γ being the spontaneous decay rate

$$\gamma = -\frac{2e^2}{\varepsilon_0\hbar c^2 m^2}\mathbf{p}_{12}\cdot\mathrm{Im}\left\{\overleftrightarrow{\mathbf{G}}(\mathbf{r}_0,\mathbf{r}_0;\omega_0)\right\}\cdot\mathbf{p}_{12}. \tag{B.25}$$

In the (generalized) Coulomb gauge the momentum-matrix elements $\mathbf{p}_{12}$ are related to the dipole-matrix elements $\mathbf{p}$ as

$$\mathbf{p}_{12} = (\mathrm{i}m\omega_0/e)\mathbf{p}, \tag{B.26}$$

which allows us to write Eq. (B.25) in terms of $\hat{\mathbf{p}}$. Furthermore, in an inhomogeneous medium, the Green function can be split into a primary (free-space) part $\overleftrightarrow{\mathbf{G}}_0$ and a scattering part $\overleftrightarrow{\mathbf{G}}_{\mathrm{s}}$. Using the fact that the contribution of $\overleftrightarrow{\mathbf{G}}_0$ leads to the free-space decay rate γ_0 (see Eq. (B.10)) we can write the ratio γ/γ_0 as

$$\frac{\gamma}{\gamma_0} = 1 + \frac{6\pi c}{\omega_0\mu^2}\mathbf{p}\cdot\mathrm{Im}\left\{\overleftrightarrow{\mathbf{G}}_{\mathrm{s}}(\mathbf{r}_0,\mathbf{r}_0;\omega_0)\right\}\cdot\mathbf{p}, \tag{B.27}$$

consistent with the classical derivation (Eq. (8.141)) in Section 8.5.

References

[1] Y. Xu, R. K. Lee, and A. Yariv, "Quantum analysis and the classical analysis of spontaneous emission in a microcavity," *Phys. Rev. A* **61**, 033807 (2000).

[2] V. Weisskopf and E. Wigner, "Berechnung der natürlichen Linienbreite auf Grund der Diracschen Lichttheorie," *Z. Phys.* **63**, 54–73 (1930).

[3] Y. Yamamoto and A. Imamoglu, *Mesoscopic Quantum Optics*. New York: John Wiley & Sons (1999).

[4] E. T. Jaynes and F. W. Cummings, "Comparison of quantum and semiclassical radiation theories with application to the beam maser," *Proc. IEEE* **51**, 89–103 (1963).

[5] R. J. Glauber and M. Lewenstein, "Quantum optics of dielectric media," *Phys. Rev. A* **43**, 467–491 (1991).

Appendix C Fields of a dipole near a layered substrate

C.1 Vertical electric dipole

The cylindrical field components of a vertically oriented dipole (Fig. C.1) $\mathbf{p} = (0, 0, p_z)$ read as

$$E_{1\rho} = \rho(z-z_0)\frac{p_z}{4\pi\,\varepsilon_0\,\varepsilon_1}\frac{e^{ik_1R_0}}{R_0^3}\left[\frac{3}{R_0^2} - \frac{3ik_1}{R_0} - k_1^2\right] - \frac{ip_z}{4\pi\,\varepsilon_0\,\varepsilon_1}\int_0^\infty dk_\rho\,J_1(k_\rho\rho)A_1k_\rho k_{1z}e^{ik_{1z}(z+z_0)}, \tag{C.1}$$

$$E_{2\rho} = \frac{ip_z}{4\pi\,\varepsilon_0\,\varepsilon_1}\int_0^\infty dk_\rho\,J_1(k_\rho\rho)\left[A_2e^{-ik_{2z}z} - A_3e^{ik_{2z}z}\right]k_\rho k_{2z}e^{ik_{1z}z_0}, \tag{C.2}$$

$$E_{3\rho} = \frac{ip_z}{4\pi\,\varepsilon_0\,\varepsilon_1}\int_0^\infty dk_\rho\,J_1(k_\rho\rho)A_4k_\rho k_{3z}e^{i(k_{1z}z_0 - k_{3z}z)}, \tag{C.3}$$

$$E_{1\varphi} = E_{2\varphi} = E_{3\varphi} = 0, \tag{C.4}$$

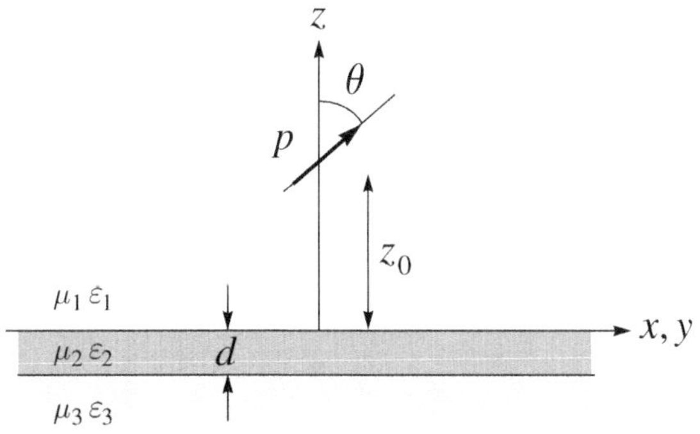

Fig. C.1 An electric dipole with moment $\boldsymbol{p}$ is located at $\boldsymbol{r}_0 = (0, 0, z_0)$ near a layered substrate. The fields in each medium are expressed in cylindrical coordinates $\boldsymbol{r} = (\rho, \varphi, z)$.

$$E_{1z} = \frac{p_z}{4\pi\,\varepsilon_0\,\varepsilon_1}\frac{e^{ik_1R_0}}{R_0}\left[\frac{3(z-z_0)^2}{R_0^4} - \frac{3ik_1(z-z_0)^2}{R_0^3} - \frac{1+k_1^2(z-z_0)^2}{R_0^2} + \frac{ik_1}{R_0} + k_1^2\right]$$

$$+ \frac{p_z}{4\pi\,\varepsilon_0\,\varepsilon_1}\int_0^\infty dk_\rho\, J_0(k_\rho\rho)A_1k_\rho^2 e^{ik_{1z}(z+z_0)}, \tag{C.5}$$

$$E_{2z} = \frac{p_z}{4\pi\,\varepsilon_0\,\varepsilon_1}\int_0^\infty dk_\rho\, J_0(k_\rho\rho)\left[A_2e^{-ik_{2z}z} + A_3e^{ik_{2z}z}\right]k_\rho^2 e^{ik_{1z}z_0}, \tag{C.6}$$

$$E_{3z} = \frac{p_z}{4\pi\,\varepsilon_0\,\varepsilon_1}\int_0^\infty dk_\rho\, J_0(k_\rho\rho)A_4k_\rho^2 e^{i(k_{1z}z_0 - k_{3z}z)}, \tag{C.7}$$

$$H_{1\rho} = H_{2\rho} = H_{3\rho} = 0, \tag{C.8}$$

$$H_{1\varphi} = -\frac{i\,\omega\,p_z}{4\pi}\rho\frac{e^{ik_1R_0}}{R_0^2}\left[\frac{1}{R_0} - ik_1\right]$$

$$- \frac{i\,\omega\,p_z}{4\pi}\int_0^\infty dk_\rho\, J_1(k_\rho\rho)A_1k_\rho e^{ik_{1z}(z+z_0)}, \tag{C.9}$$

$$H_{2\varphi} = -\frac{i\,\omega\,\varepsilon_2\,p_z}{4\pi\,\varepsilon_1}\int_0^\infty dk_\rho\, J_1(k_\rho\rho)\left[A_2e^{-ik_{2z}z} + A_3e^{ik_{2z}z}\right]k_\rho e^{ik_{1z}z_0}, \tag{C.10}$$

$$H_{3\varphi} = -\frac{i\,\omega\,\varepsilon_3\,p_z}{4\pi\,\varepsilon_1}\int_0^\infty dk_\rho\, J_1(k_\rho\rho)A_4k_\rho e^{i(k_{1z}z_0 - k_{3z}z)}, \tag{C.11}$$

$$H_{1z} = H_{2z} = H_{3z} = 0. \tag{C.12}$$

C.2 Horizontal electric dipole

The cylindrical field components of a horizontally oriented dipole $\mathbf{p} = (p_x, 0, 0)$ read as

$$E_{1\rho} = \cos\varphi\,\frac{p_x}{4\pi\,\varepsilon_0\,\varepsilon_1}\frac{e^{ik_1R_0}}{R_0}\left\{\left[k_1^2 + \frac{ik_1}{R_0} - \frac{1}{R_0^2}\right] + \frac{\rho^2}{R_0^2}\left[\frac{3}{R_0^2} - \frac{3ik_1}{R_0} - k_1^2\right]\right\}$$

$$+ \cos\varphi\,\frac{p_x}{4\pi\,\varepsilon_0\,\varepsilon_1}\int_0^\infty dk_\rho\, e^{ik_{1z}(z+z_0)}\left\{\frac{1}{\rho}J_1(k_\rho\rho)\left[k_\rho B_1 - ik_{1z}C_1\right]\right.$$

$$\left. - ik_{1z}J_0(k_\rho\rho)\left[ik_{1z}B_1 - k_\rho C_1\right]\right\}, \tag{C.13}$$

$$E_{2\rho} = \cos\varphi \,\frac{p_x}{4\pi\,\varepsilon_0\,\varepsilon_1} \int\limits_0^\infty \mathrm{d}k_\rho\; \mathrm{e}^{\mathrm{i}k_{1z}z_0} \left\{ \frac{1}{\rho} J_1(k_\rho\rho)\Big[\big[k_\rho B_2 + \mathrm{i}k_{2z}C_2\big]\mathrm{e}^{-\mathrm{i}k_{2z}z} + \big[k_\rho B_3 - \mathrm{i}k_{2z}C_3\big]\mathrm{e}^{\mathrm{i}k_{2z}z}\Big] - \mathrm{i}k_{2z}J_0(k_\rho\rho)\Big[\big[\mathrm{i}k_{2z}B_2 + k_\rho C_2\big]\mathrm{e}^{-\mathrm{i}k_{2z}z} + \big[\mathrm{i}k_{2z}B_3 - k_\rho C_3\big]\mathrm{e}^{\mathrm{i}k_{2z}z}\Big]\right\}, \tag{C.14}$$

$$E_{3\rho} = \cos\varphi \,\frac{p_x}{4\pi\,\varepsilon_0\,\varepsilon_1} \int\limits_0^\infty \mathrm{d}k_\rho\; \mathrm{e}^{\mathrm{i}(k_{1z}z_0 - k_{3z}z)} \left\{ \frac{1}{\rho} J_1(k_\rho\rho)\big[k_\rho B_4 + \mathrm{i}k_{3z}C_4\big] - \mathrm{i}k_{3z}J_0(k_\rho\rho)\big[\mathrm{i}k_{3z}B_4 + k_\rho C_4\big]\right\}, \tag{C.15}$$

$$E_{1\varphi} = \sin\varphi \,\frac{p_x}{4\pi\,\varepsilon_0\,\varepsilon_1} \frac{\mathrm{e}^{\mathrm{i}k_1R_0}}{R_0}\left[\frac{1}{R_0^2} - \frac{\mathrm{i}k_1}{R_0} - k_1^2\right] + \sin\varphi \,\frac{p_x}{4\pi\,\varepsilon_0\,\varepsilon_1} \int\limits_0^\infty \mathrm{d}k_\rho\; \mathrm{e}^{\mathrm{i}k_{1z}(z+z_0)} \left\{ \frac{1}{\rho} J_1(k_\rho\rho)\big[k_\rho B_1 - \mathrm{i}k_{1z}C_1\big] - k_1^2 J_0(k_\rho\rho)B_1\right\}, \tag{C.16}$$

$$E_{2\varphi} = \sin\varphi \,\frac{p_x}{4\pi\,\varepsilon_0\,\varepsilon_1} \int\limits_0^\infty \mathrm{d}k_\rho\; \mathrm{e}^{\mathrm{i}k_{1z}z_0} \left\{ \frac{1}{\rho} J_1(k_\rho\rho)\Big[\big[k_\rho B_2 + \mathrm{i}k_{2z}C_2\big]\mathrm{e}^{-\mathrm{i}k_{2z}z} + \big[k_\rho B_3 - \mathrm{i}k_{2z}C_3\big]\mathrm{e}^{\mathrm{i}k_{2z}z}\Big] - k_2^2 J_0(k_\rho\rho)\big[B_2\mathrm{e}^{-\mathrm{i}k_{2z}z} + B_3\mathrm{e}^{+\mathrm{i}k_{2z}z}\big]\right\}, \tag{C.17}$$

$$E_{3\varphi} = \sin\varphi \,\frac{p_x}{4\pi\,\varepsilon_0\,\varepsilon_1} \int\limits_0^\infty \mathrm{d}k_\rho\; \mathrm{e}^{\mathrm{i}(k_{1z}z_0 - k_{3z}z)} \left\{ \frac{1}{\rho} J_1(k_\rho\rho)\big[k_\rho B_4 + \mathrm{i}k_{3z}C_4\big] - k_3^2 J_0(k_\rho\rho)B_4\right\}, \tag{C.18}$$

$$E_{1z} = \cos\varphi \,\frac{p_x}{4\pi\,\varepsilon_0\,\varepsilon_1}\rho(z-z_0) \frac{\mathrm{e}^{\mathrm{i}k_1R_0}}{R_0^3}\left[\frac{3}{R_0^2} - \frac{3\mathrm{i}k_1}{R_0} - k_1^2\right] - \cos\varphi \,\frac{p_x}{4\pi\,\varepsilon_0\,\varepsilon_1} \int\limits_0^\infty \mathrm{d}k_\rho\; \mathrm{e}^{\mathrm{i}k_{1z}(z+z_0)} k_\rho J_1(k_\rho\rho)\big[\mathrm{i}k_{1z}B_1 - k_\rho C_1\big], \tag{C.19}$$

$$E_{2z} = \cos\varphi \, \frac{p_x}{4\pi\,\varepsilon_0\,\varepsilon_1} \int_0^\infty \mathrm{d}k_\rho \; \mathrm{e}^{\mathrm{i}k_{1z}z_0} \Big\{ k_\rho J_1(k_\rho\rho) \Big[\big[\mathrm{i}k_{2z}B_2 + k_\rho C_2\big]\mathrm{e}^{-\mathrm{i}k_{2z}z} - \big[\mathrm{i}k_{2z}B_3 - k_\rho C_3\big]\mathrm{e}^{\mathrm{i}k_{2z}z} \Big] \Big\}, \tag{C.20}$$

$$E_{3z} = \cos\varphi \, \frac{p_x}{4\pi\,\varepsilon_0\,\varepsilon_1} \int_0^\infty \mathrm{d}k_\rho \; \mathrm{e}^{\mathrm{i}(k_{1z}z_0 - k_{3z}z)} k_\rho J_1(k_\rho\rho)\big[\mathrm{i}k_{3z}B_4 + k_\rho C_4\big], \tag{C.21}$$

$$H_{1\rho} = \sin\varphi \, \frac{\mathrm{i}\,\omega\, p_x}{4\pi}(z - z_0)\frac{\mathrm{e}^{\mathrm{i}k_1R_0}}{R_0^2}\left[\frac{1}{R_0} - \mathrm{i}k_1\right] + \sin\varphi \, \frac{\mathrm{i}\,\omega\, p_x}{4\pi} \int_0^\infty \mathrm{d}k_\rho \; \mathrm{e}^{\mathrm{i}k_{1z}(z+z_0)} \left\{\frac{1}{\rho} J_1(k_\rho\rho)C_1 - \mathrm{i}k_{1z}J_0(k_\rho\rho)B_1\right\}, \tag{C.22}$$

$$H_{2\rho} = \sin\varphi \, \frac{\mathrm{i}\,\omega\,\varepsilon_2\, p_x}{4\pi\,\varepsilon_1} \int_0^\infty \mathrm{d}k_\rho \; \mathrm{e}^{\mathrm{i}k_{1z}z_0} \Big\{\frac{1}{\rho} J_1(k_\rho\rho)\big[C_2\mathrm{e}^{-\mathrm{i}k_{2z}z} + C_3\mathrm{e}^{\mathrm{i}k_{2z}z}\big] - \mathrm{i}k_{2z}J_0(k_\rho\rho)\big[B_2\mathrm{e}^{-\mathrm{i}k_{2z}z} - B_3\mathrm{e}^{\mathrm{i}k_{2z}z}\big]\Big\}, \tag{C.23}$$

$$H_{3\rho} = \sin\varphi \, \frac{\mathrm{i}\,\omega\,\varepsilon_3\, p_x}{4\pi\,\varepsilon_1} \int_0^\infty \mathrm{d}k_\rho \; \mathrm{e}^{\mathrm{i}(k_{1z}z_0 - k_{3z}z)} \left\{\frac{1}{\rho} J_1(k_\rho\rho)C_4 + \mathrm{i}k_{3z}J_0(k_\rho\rho)B_4\right\}, \tag{C.24}$$

$$H_{1\varphi} = \cos\varphi \, \frac{\mathrm{i}\,\omega\, p_x}{4\pi}(z - z_0)\frac{\mathrm{e}^{\mathrm{i}k_1R_0}}{R_0^2}\left[\frac{1}{R_0} - \mathrm{i}k_1\right] - \cos\varphi \, \frac{\mathrm{i}\,\omega\, p_x}{4\pi} \int_0^\infty \mathrm{d}k_\rho \; \mathrm{e}^{\mathrm{i}k_{1z}(z+z_0)} \left\{\frac{1}{\rho} J_1(k_\rho\rho)C_1 + J_0(k_\rho\rho)\big[\mathrm{i}k_{1z}B_1 - k_\rho C_1\big]\right\}, \tag{C.25}$$

$$H_{2\varphi} = \cos\varphi \, \frac{\mathrm{i}\,\omega\,\varepsilon_2\, p_x}{4\pi\,\varepsilon_1} \int_0^\infty \mathrm{d}k_\rho \; \mathrm{e}^{\mathrm{i}k_{1z}z_0} \Big\{\frac{1}{\rho} J_1(k_\rho\rho)\big[C_2\mathrm{e}^{-\mathrm{i}k_{2z}z} + C_3\mathrm{e}^{\mathrm{i}k_{2z}z}\big] - J_0(k_\rho\rho)\Big[\big[\mathrm{i}k_{2z}B_2 + k_\rho C_2\big]\mathrm{e}^{-\mathrm{i}k_{2z}z} - \big[\mathrm{i}k_{2z}B_3 - k_\rho C_3\big]\mathrm{e}^{\mathrm{i}k_{2z}z}\Big]\Big\} \tag{C.26}$$

$$H_{3\varphi} = \cos\varphi \,\frac{\mathrm{i}\,\omega\,\varepsilon_3\,p_x}{4\pi\,\varepsilon_1} \int_0^\infty \mathrm{d}k_\rho\; \mathrm{e}^{\mathrm{i}(k_{1z}z_0 - k_{3z}z)} \left\{ \frac{1}{\rho} J_1(k_\rho\rho)\,C_4 - J_0(k_\rho\rho)\big[\mathrm{i}k_{3z}B_4 + k_\rho C_4\big] \right\}, \tag{C.27}$$

$$H_{1z} = -\sin\varphi\,\frac{\mathrm{i}\,\omega\,p_x}{4\pi}\rho\frac{\mathrm{e}^{\mathrm{i}k_1R_0}}{R_0^2}\left[\frac{1}{R_0} - \mathrm{i}k_1\right] - \sin\varphi\,\frac{\mathrm{i}\,\omega\,p_x}{4\pi}\int_0^\infty \mathrm{d}k_\rho\; \mathrm{e}^{\mathrm{i}k_{1z}(z+z_0)} k_\rho J_1(k_\rho\rho)B_1, \tag{C.28}$$

$$H_{2z} = -\sin\varphi\,\frac{\mathrm{i}\,\omega\,\varepsilon_2\,p_x}{4\pi\,\varepsilon_1}\int_0^\infty \mathrm{d}k_\rho\; \mathrm{e}^{\mathrm{i}k_{1z}z_0} k_\rho J_1(k_\rho\rho)\big[B_2\mathrm{e}^{-\mathrm{i}k_{2z}z} + B_3\mathrm{e}^{\mathrm{i}k_{2z}z}\big], \tag{C.29}$$

$$H_{3z} = -\sin\varphi\,\frac{\mathrm{i}\,\omega\,\varepsilon_3\,p_x}{4\pi\,\varepsilon_1}\int_0^\infty \mathrm{d}k_\rho\; \mathrm{e}^{\mathrm{i}(k_{1z}z_0 - k_{3z}z)} k_\rho J_1(k_\rho\,\rho)B_4. \tag{C.30}$$

C.3 Definition of the coefficients A_j, B_j, and C_j

The coefficients A_j, B_j, and C_j are determined by the boundary conditions on the interfaces. Using the abbreviations

$$\begin{aligned} f_1 &= \varepsilon_2 k_{1z} - \varepsilon_1 k_{2z}, & g_1 &= \mu_2 k_{1z} - \mu_1 k_{2z},\\ f_2 &= \varepsilon_2 k_{1z} + \varepsilon_1 k_{2z}, & g_2 &= \mu_2 k_{1z} + \mu_1 k_{2z},\\ f_3 &= \varepsilon_3 k_{2z} - \varepsilon_2 k_{3z}, & g_3 &= \mu_3 k_{2z} - \mu_2 k_{3z},\\ f_4 &= \varepsilon_3 k_{2z} + \varepsilon_2 k_{3z}, & g_4 &= \mu_3 k_{2z} + \mu_2 k_{3z}, \end{aligned} \tag{C.31}$$

the coefficients read as

$$A_1(k_\rho) = \mathrm{i}\frac{k_\rho(f_1f_4 + f_2f_3\mathrm{e}^{2\mathrm{i}k_{2z}d})}{k_{1z}\,(f_2f_4 + f_1f_3\mathrm{e}^{2\mathrm{i}k_{2z}d})}, \tag{C.32}$$

$$A_2(k_\rho) = \mathrm{i}\frac{2\varepsilon_1 k_\rho f_4}{f_2f_4 + f_1f_3\mathrm{e}^{2\mathrm{i}k_{2z}d}}, \tag{C.33}$$

$$A_3(k_\rho) = \mathrm{i}\frac{2\varepsilon_1 k_\rho f_3\mathrm{e}^{2\mathrm{i}k_{2z}d}}{f_2f_4 + f_1f_3\mathrm{e}^{2\mathrm{i}k_{2z}d}}, \tag{C.34}$$

$$A_4(k_\rho) = \mathrm{i}\frac{4\varepsilon_1\varepsilon_2 k_\rho k_{2z}\mathrm{e}^{\mathrm{i}(k_{2z}-k_{3z})d}}{f_2f_4 + f_1f_3\mathrm{e}^{2\mathrm{i}k_{2z}d}}, \tag{C.35}$$

$$B_1(k_\rho) = \mathrm{i}\frac{k_\rho(g_1g_4 + g_2g_3\mathrm{e}^{2\mathrm{i}k_{2z}d})}{k_{1z}(g_2g_4 + g_1g_3\mathrm{e}^{2\mathrm{i}k_{2z}d})}, \tag{C.36}$$

$$B_2(k_\rho) = \mathrm{i}\frac{\varepsilon_1}{\varepsilon_2}\frac{2\mu_1k_\rho g_4}{g_2g_4 + g_1g_3\mathrm{e}^{2\mathrm{i}k_{2z}d}}, \tag{C.37}$$

$$B_3(k_\rho) = \mathrm{i}\frac{\varepsilon_1}{\varepsilon_2}\frac{2\mu_1k_\rho g_3\mathrm{e}^{2\mathrm{i}k_{2z}d}}{g_2g_4 + g_1g_3\mathrm{e}^{2\mathrm{i}k_{2z}d}}, \tag{C.38}$$

$$B_4(k_\rho) = \mathrm{i}\frac{\varepsilon_1}{\varepsilon_3}\frac{4\mu_1\mu_2k_\rho k_{2z}\mathrm{e}^{\mathrm{i}(k_{2z}-k_{3z})d}}{g_2g_4 + g_1g_3\mathrm{e}^{2\mathrm{i}k_{2z}d}}, \tag{C.39}$$

$$C_1(k_\rho) = \mathrm{i}\frac{k_{1z}}{k_\rho}A_1(k_\rho) + \mathrm{i}\frac{k_{1z}}{k_\rho}B_1(k_\rho), \tag{C.40}$$

$$C_2(k_\rho) = \mathrm{i}\frac{k_{1z}}{k_\rho}A_2(k_\rho) - \mathrm{i}\frac{k_{2z}}{k_\rho}B_2(k_\rho), \tag{C.41}$$

$$C_3(k_\rho) = \mathrm{i}\frac{k_{1z}}{k_\rho}A_3(k_\rho) + \mathrm{i}\frac{k_{2z}}{k_\rho}B_3(k_\rho), \tag{C.42}$$

$$C_4(k_\rho) = \mathrm{i}\frac{k_{1z}}{k_\rho}A_4(k_\rho) - \mathrm{i}\frac{k_{3z}}{k_\rho}B_4(k_\rho). \tag{C.43}$$

In order to stay on the proper Riemann sheet, all square roots

$$k_{jz} = \sqrt{k_j^2 - k_\rho^2}, \quad j \in \{1, 2, 3\} \tag{C.44}$$

have to be chosen such that $\mathrm{Im}\{k_{jz}\} > 0$.

The integrals have to be evaluated numerically. The integration routine has to account for both oscillatory behavior and singularities. It is recommended that the integration range is split into subintervals and that the integration path is extended into the complex-k_ρ plane. For some applications it is advantageous to express the Bessel functions J_n in terms of Hankel functions since they converge rapidly for arguments with an imaginary part. An integration routine that proved very reliable is the so-called Gauss–Kronrod routine.

D

Appendix D Far-field Green functions

In this appendix we state the asymptotic far-field Green functions for a planarly layered medium. It is assumed that the source point $\mathbf{r}_0 = (x_0, y_0, z_0)$ is in the upper half-space ($z > 0$). The field is evaluated at a point $\mathbf{r} = (x, y, z)$ in the far-zone, i.e. $r \gg \lambda$. The optical properties of the upper half-space and the lower half-space are characterized by ε_1, μ_1 and ε_n, μ_n, respectively. The planarly layered medium in between the two half-spaces is characterized by the generalized Fresnel reflection and transmission coefficients. We choose a coordinate system with origin on the topmost surface of the layered medium with the z-axis perpendicular to the interfaces. In this case, z_0 denotes the height of the point source relative to the topmost layer. In the upper half-space, the asymptotic dyadic Green function is defined as

$$\mathbf{E}(\mathbf{r}) = \frac{\omega^2}{\varepsilon_0 c^2}\mu_1 \left[\overleftrightarrow{\mathbf{G}}_0(\mathbf{r}, \mathbf{r}_0) + \overleftrightarrow{\mathbf{G}}_{\text{ref}}(\mathbf{r}, \mathbf{r}_0)\right] \mathbf{p}, \tag{D.1}$$

where $\mathbf{p}$ is the dipole moment of a dipole located at $\mathbf{r}_0$ and $\overleftrightarrow{\mathbf{G}}_0$ and $\overleftrightarrow{\mathbf{G}}_{\text{ref}}$ are the primary and reflected parts of the Green function. In the lower half-space we define

$$\mathbf{E}(\mathbf{r}) = \frac{\omega^2}{\varepsilon_0 c^2}\mu_1 \overleftrightarrow{\mathbf{G}}_{\text{tr}}(\mathbf{r}, \mathbf{r}_0)\mathbf{p}, \tag{D.2}$$

with $\overleftrightarrow{\mathbf{G}}_{\text{tr}}$ being the transmitted part of the Green function. The asymptotic Green functions can be derived by using the far-field forms of the angular spectrum representation.

The primary Green function in the far-zone is found to be

$$\overleftrightarrow{\mathbf{G}}_0(\mathbf{r}, \mathbf{r}_0) = \frac{\exp(\mathrm{i}k_1 r)}{4\pi r} \exp\left[-\mathrm{i}k_1(x_0 x/r + y_0 y/r + z_0 z/r)\right] \times \begin{bmatrix} 1 - x^2/r^2 & -xy/r^2 & -xz/r^2 \\ -xy/r^2 & 1 - y^2/r^2 & -yz/r^2 \\ -xz/r^2 & -yz/r^2 & 1 - z^2/r^2 \end{bmatrix}. \tag{D.3}$$

The reflected part of the Green function in the far-zone is

$$\overleftrightarrow{\mathbf{G}}_{\text{ref}}(\mathbf{r}, \mathbf{r}_0) = \frac{\exp(\mathrm{i}k_1 r)}{4\pi r} \exp\left[-\mathrm{i}k_1\left(x_0\frac{x}{r} + y_0\frac{y}{r} - z_0\frac{z}{r}\right)\right] \times \begin{bmatrix} \frac{x^2}{\rho^2}\frac{z^2}{r^2}\Phi_1^{(2)} + \frac{y^2}{\rho^2}\Phi_1^{(3)} & \frac{xy}{\rho^2}\frac{z^2}{r^2}\Phi_1^{(2)} - \frac{xy}{\rho^2}\Phi_1^{(3)} & -\frac{xz}{r^2}\Phi_1^{(1)} \\ \frac{xy}{\rho^2}\frac{z^2}{r^2}\Phi^{(2)} - \frac{xy}{\rho^2}\Phi_1^{(3)} & \frac{y^2}{\rho^2}\frac{z^2}{r^2}\Phi^{(2)} + \frac{x^2}{\rho^2}\Phi_1^{(3)} & -\frac{yz}{r^2}\Phi_1^{(1)} \\ -\frac{xz}{r^2}\Phi_1^{(2)} & -\frac{yz}{r^2}\Phi_1^{(2)} & \left(1 - \frac{z^2}{r^2}\right)\Phi_1^{(1)} \end{bmatrix}, \tag{D.4}$$

where the potentials are determined in terms of the generalized reflection coefficients of the layered structure as

$$\left.\begin{aligned}\Phi_1^{(1)} &= r^p(k_\rho)\\ \Phi_1^{(2)} &= -r^p(k_\rho)\\ \Phi_1^{(3)} &= r^s(k_\rho)\end{aligned}\right\} \quad k_\rho = k_1\rho/r. \tag{D.5}$$

The transmitted part of the Green function in the far-zone is

$$\begin{aligned}\overleftrightarrow{\mathbf{G}}_{\mathrm{tr}}(\mathbf{r},\mathbf{r}_0) = {} & \frac{\exp[\mathrm{i}k_n(r+\delta z/r)]}{4\pi r}\\ & \times \exp\left[-\mathrm{i}k_1\left(x_0\frac{x}{r} + y_0\frac{y}{r} - z_0\sqrt{1-\left(\frac{n_n^2}{n_1^2}\right)\frac{\rho^2}{r^2}}\right)\right]\\ & \times \begin{bmatrix} \frac{x^2}{\rho^2}\frac{z^2}{r^2}\Phi_n^{(2)} + \frac{y^2}{\rho^2}\Phi_n^{(3)} & \frac{xy}{\rho^2}\frac{z^2}{r^2}\Phi_n^{(2)} - \frac{xy}{\rho^2}\Phi_n^{(3)} & -\frac{xz}{r^2}\Phi_n^{(1)}\\ \frac{xy}{\rho^2}\frac{z^2}{r^2}\Phi_n^{(2)} - \frac{xy}{\rho^2}\Phi_n^{(3)} & \frac{y^2}{\rho^2}\frac{z^2}{r^2}\Phi_n^{(2)} + \frac{x^2}{\rho^2}\Phi_n^{(3)} & -\frac{yz}{r^2}\Phi_n^{(1)}\\ -\frac{xz}{r^2}\Phi_n^{(2)} & -\frac{yz}{r^2}\Phi_n^{(2)} & \left(1-\frac{z^2}{r^2}\right)\Phi_n^{(1)} \end{bmatrix},\end{aligned} \tag{D.6}$$

where δ denotes the overall thickness of the layered structure.

Here, the potentials are determined in terms of the generalized transmission coefficients of the layered structure as

$$\left.\begin{aligned}\Phi_n^{(1)} &= t^{\mathrm{p}}(k_\rho)\,\frac{n_n}{n_1}\,\frac{k_n z/r}{\sqrt{k_1^2-k_\rho^2}}\\ \Phi_n^{(2)} &= -t^{\mathrm{p}}(k_\rho)\,\frac{n_n}{n_1}\\ \Phi_n^{(3)} &= t^{\mathrm{s}}(k_\rho)\,\frac{k_n z/r}{\sqrt{k_1^2-k_\rho^2}}\end{aligned}\right\} \quad k_\rho = k_n\rho/r. \tag{D.7}$$

A vertical dipole is described by the potential $\Phi^{(1)}$ alone and gives rise to purely p-polarized fields. On the other hand, a horizontal dipole is represented by $\Phi^{(2)}$ and $\Phi^{(3)}$ and its field contains both s- and p-polarized components. The coordinates (x, y, z) can be substituted by the spherical angles θ and ϕ. For angles $\alpha = \pi - \theta$ beyond the critical angle $\alpha_c = \arcsin(n_1/n_n)$ the field depends exponentially on the height z_0.

Index

Bold page numbers indicate major discussions